T0189524

Communications in Computer and Information Science   1168

Commenced Publication in 2007
Founding and Former Series Editors:
Phoebe Chen, Alfredo Cuzzocrea, Xiaoyong Du, Orhun Kara, Ting Liu,
Krishna M. Sivalingam, Dominik Ślęzak, Takashi Washio, Xiaokang Yang,
and Junsong Yuan

More information about this series at http://www.springer.com/series/7899

Peggy Cellier · Kurt Driessens (Eds.)

Machine Learning and Knowledge Discovery in Databases

International Workshops of ECML PKDD 2019
Würzburg, Germany, September 16–20, 2019
Proceedings, Part II

 Springer

Editors
Peggy Cellier
Institut National des Sciences Appliquées
Rennes, France

Kurt Driessens 🆔
Maastricht University
Maastricht, The Netherlands

ISSN 1865-0929   ISSN 1865-0937 (electronic)
Communications in Computer and Information Science
ISBN 978-3-030-43886-9   ISBN 978-3-030-43887-6 (eBook)
https://doi.org/10.1007/978-3-030-43887-6

This Springer imprint is published by the registered company Springer Nature Switzerland AG
The registered company address is: Gewerbestrasse 11, 6330 Cham, Switzerland

Preface

The European Conference on Machine Learning and Principles and Practice of Knowledge Discovery in Databases (ECML PKDD) is the premier European machine learning and data mining conference. In 2019, ECML PKDD was held in Würzburg, Germany, during September 16–20.

During the first and last day of the conference, the workshop program allowed a number of specialized and/or new topics to take the fore-front.

A record 46 workshop and tutorial topics were submitted to the 2019 conference. The selection and merging process resulted in 25 workshops taking place over the two days, of which 3 were combined with a tutorial.

The workshop program included the following workshops:

1. The 12th International Workshop on Machine Learning and Music (MML 2019)
2. Workshop on Multiple-aspect analysis of semantic trajectories (MASTER 2019)
3. The 4th Workshop on MIning DAta for financial applicationS (MIDAS 2019)
4. The Second International Workshop on Knowledge Discovery and User Modelling for Smart Cities (UMCit 2019)
5. New Frontiers in Mining Complex Patterns (NFMCP 2019)
6. New Trends in Representation Learning with Knowledge Graphs
7. The Second International Workshop on Energy Efficient Scalable Data Mining and Machine Learning (Green Data Mining)
8. Workshop on Deep Continuous-Discrete Machine Learning (DeCoDeML 2019)
9. Decentralised Machine Learning at the Edge (DMLE 2019)
10. Applications of Topological Data Analysis (ATDA 2019)
11. GEM: Graph Embedding and Mining
12. Interactive Adaptive Learning (AIL 2019)
13. IoT Stream for Data Driven Predictive Maintenance (IoT Steam 2019)
14. Machine Learning for Cybersecurity (MLCS 2019)
15. BioASQ: Large-scale biomedical semantic indexing and question answering
16. The 6th Workshop on Sports Analytics: Machine Learning and Data Mining for Sports Analytics (MLSA 2019)
17. The 4th Workshop on Advanced Analytics and Learning on Temporal Data (AALTD 2019)
18. MACLEAN: MAChine Learning for EArth ObservatioN
19. Automating Data Science
20. The 4th Workshop on Data Science for Social Good (DSSG 2019)
21. The Third Workshop on Advances in managing and mining Large Evolving Graphs (LEG 2019)
22. Data and Machine Learning Advances with Multiple Views (DAMVL 2019)
23. Workshop on Data Integration and Applications (DINA 2019)
24. XKDD Tutorial and XKDD-AIMLAI Workshop
25. The First Workshop SocIaL Media And Harassment (SIMAH 2019)

Of these 25 workshops, 17 workshops decided to select and publish their best papers with Springer. Two workshops were large enough to publish their own proceedings: (i) MIDAS – the 4th Workshop on MIning DAta for financial applicationS and (ii) AALTD – the 4th workshop on Advanced Analytics and Learning on Temporal Data. The 15 other workshops received a total of 200 submitted papers, out of which 70 long and 46 short papers were selected for publication after the conference. These papers are spread over two proceedings volumes.

This two-volume set contains the papers from the following workshops:

1. Automating Data Science
2. XKDD Tutorial and XKDD-AIMLAI Workshop
3. Decentralised Machine Learning at the Edge (DMLE 2019)
4. The Third Workshop on Advances in managing and mining Large Evolving Graphs (LEG 2019)
5. Data and Machine Learning Advances with Multiple Views (DAMVL 2019)
6. New Trends in Representation Learning with Knowledge Graphs
7. The 4th Workshop on Data Science for Social Good (DSSG 2019)
8. The Second International Workshop on Knowledge Discovery and User Modelling for Smart Cities (UMCit 2019)
9. Workshop on Data Integration and Applications (DINA 2019)
10. Machine Learning for Cybersecurity (MLCS 2019)
11. The 6th Workshop on Sports Analytics: Machine Learning and Data Mining for Sports Analytics (MLSA 2019)
12. The First Workshop on SocIaL Media And Harassment (SIMAH 2019)
13. IoT Stream for Data Driven Predictive Maintenance (IoT Stream 2019)
14. The 12th International Workshop on Machine Learning and Music (MML 2019)
15. BioASQ: Large-scale biomedical semantic indexing and question answering

We would like to thank all participants and invited speakers, the workshop organizers and the reviewers, as well as the local organizers for making the workshop program of ECML PKDD 2019 the success that it was. Sincere thanks also goes to Springer for their help in publishing the proceedings.

January 2020   Peggy Cellier
                                                                              Kurt Driessens

Organization

ECML Workshop Chairs/Editors

Peggy Celier   INSA Rennes, France
Kurt Driessens   Maastricht University, The Netherlands

Individual Workshop Chairs/Editors

Tijl De Bie   UGent, Belgium
Luc De Raedt   KU Leuven, Belgium
Jose Hernandez-Orallo   Universitat Politecnica de Valencia, Spain
Adrien Bibal   University of Namur, Belgium
Tassadit Bouadi   University of Rennes/IRISA, France
Benoît Frénay   University of Namur, Belgium
Luis Galárraga   Inria/IRISA, France
Stefan Kramer   Universität Mainz, Germany
Ruggero G. Pensa   University of Turin, Italy
Michael Kamp   University of Bonn, Germany
Yamuna Krishnamurthy   Rolay Holloway University of London, England
Daniel Paurat   Fraunhofer IAIS, Germany
Sabeur Aridhi   University of Lorraine, France
José Antonio de Macedo   Universidade Federal do Ceará, Brazil
Engelbert Mephu Nguifo   University Clermont Auvergne, France
Karine Zeitouni   Université de Versailles Saint-Quentin, France
Stéphane Ayache   Aix-Marseille University, France
Cécile Capponi   Aix-Marseille University, France
Rémi Emonet   Jean-Monnet University, France
Usabelle Guyon   Orsay University, France
Volker Tresp   Ludwig-Maximilians University and Siemens,
                            Germany
Jens Lehmann   Bonn University and Fraunhofer IAIS, Germany
Aditya Mogadala   Saarland University, Germany
Achim Rettinger   Trier University, Germany
Afshin Sadeghi   Fraunhofer IAIS, Germany
Mehdi Ali   Bonn University and Fraunhofer IAIS, Germany
Ricard Gavalda   UPC BarcelonaTech, Spain
Irena Koprinska   University of Sydney, Australia
Joao Gama   University of Porto, Portugal
Rabeah Alzaidy   King Abdullah University of Science and Technology,
                            Saudi Arabia
Marcelo G. Armentano   ISISTAN, CONICET-UNICEN, Argentina
Antonela Tommasel   ISISTAN, CONICET-UNICEN, Argentina

Contents – Part II

Machine Learning for Cybersecurity (MLCS)

**6th Workshop on Sports Analytics: Machine Learning
and Data Mining for Sports Analytics (MLSA)**

**First Workshop on Categorizing Different Types of Online
Harassment Languages in Social Media**

IoT Stream for Data Driven Predictive Maintenance

12th International Workshop on Machine Learning and Music (MML 2019)

**Large-Scale Biomedical Semantic Indexing and Question
Answering (BioASQ)**

Contents – Part I

**Advances in Interpretable Machine Learning and Artificial
Intelligence & eXplainable Knowledge Discovery
in Data Mining (AIMLAI-XKDD)**

Second International Workshop on Knowledge Discovery and User Modeling for Smart Cities (UMCit)

District Heating Substation Behaviour Modelling for Annotating the Performance

Shahrooz Abghari[1(✉)], Veselka Boeva[1], Jens Brage[2], and Christian Johansson[2]

[1] Blekinge Institute of Technology, 371 79 Karlskrona, Sweden
shahrooz.abghari@bth.se
[2] NODA Intelligent Systems AB, 374 35 Karlshamn, Sweden

Abstract. In this ongoing study, we propose a higher order data mining approach for modelling district heating (DH) substations' behaviour and linking operational behaviour representative profiles with different performance indicators. We initially create substation's operational behaviour models by extracting weekly patterns and clustering them into groups of similar patterns. The built models are further analyzed and integrated into an overall substation model by applying consensus clustering. The different operational behaviour profiles represented by the exemplars of the consensus clustering model are then linked to performance indicators. The labelled behaviour profiles are deployed over the whole heating season to derive diverse insights about the substation's performance. The results show that the proposed method can be used for modelling, analyzing and understanding the deviating and suboptimal DH substation's behaviours.

Keywords: Clustering analysis · District heating · Higher order mining · Outlier detection

1 Introduction

A district heating (DH) system provides an entire town, or part of it, with heat. The heat is generated in a central boiler and delivered via a distribution pipe network. The provided heat transfers through DH substations from the distribution network into consumers' buildings. This includes providing both space heating for heating seasons and domestic hot water (DHW) for a whole year. The DH system consists of two sides: *primary* and *secondary*. The primary side includes a central boiler, a distribution network (pre-insulated pipes) and consumers' buildings. The secondary side consists of a heat exchanger, a main piping system of the building, and radiators, convectors, or floor heating for the rooms.

This work is part of the research project *"Scalable resource-efficient systems for big data analytics"* funded by the Knowledge Foundation (grant: 20140032) in Sweden.

P. Cellier and K. Driessens (Eds.): ECML PKDD 2019 Workshops, CCIS 1168, pp. 3–11, 2020.
https://doi.org/10.1007/978-3-030-43887-6_1

The DH substations are made up of different components and each can be a potential source of faults. Faults in substations and the secondary side can be divided into three categories (1) faults resulting in comfort problems such as lack of enough heat, (2) unsolved faults with known cause since their identification are time demanding and costly, and (3) faults that require advanced fault detection systems [1]. Faults in substations do not necessarily result in comfort problems for the consumers, instead in most cases cause sub-optimal behaviour for a long time before they are noticed. Therefore, early detection of faults and deviations can reduce the maintenance cost and help avoid abnormal event progression. Fault detection in DH substations can be performed by monitoring both primary and secondary sides or only primary side.

Gadd and Werner [1] showed that hourly meter readings can be used for detecting faults at DH substations. The authors identified three fault groups: (1) low average annual temperature difference, (2) poor substation control, and (3) unsuitable heat load patterns. The results of the study showed that addressing low average annual temperature differences are the most important issue that can improve efficiency of the DH systems. Nevertheless, unsuitable heat load patterns are probably the easiest and the most cost-effective problem to consider first. In a recent study [2], the authors applied clustering analysis and association rule mining to detect faults in DH substations. In another study, the authors [3] proposed a method based on gradient boosting regression to predict hourly mass flow of a well performing substation. Their built model was tested by manipulating well performed substation data to simulate two different scenarios. Calikus et al. [4] proposed an approach to automatically discover heat load patterns in DH systems. Heat load profiles reflected yearly heat usage in an individual building. Moreover, their discovery is crucial for ensuring effective DH operations and managements.

We propose a higher order mining (HOM)[1] approach for modelling a DH substation's operational behaviour and linking it with two performance indicators. At the modelling step, we use primary side features to build the substation behaviour model by extracting the substation's behaviour patterns on a weekly basis. Heat demand is strongly influenced by social factors, e.g., the need during weekdays versus weekends. However, the social patterns tend to repeat on a weekly basis. Therefore, by considering the time window of a week rather than a day, we can mitigate the social patterns and avoid discovering, e.g., the demand transition between weekdays and weekends. The extracted patterns are used to create weekly behaviour models by clustering them into groups of similar patterns. The built models are further analyzed and integrated into an overall substation model by applying consensus clustering. We consider the exemplars of the consensus clustering model as the substation representative operational behaviour profiles. Further, at the annotating step the exemplars are linked with the two performance indicators. These indicators are calculated by using features from both primary and secondary side data. The annotated behaviour profiles can be deployed over the whole heating season to derive diverse insights about

---

[1] HOM is a sub-field of knowledge discovery that applies to non-primary, derived data or patterns to provide human-consumable results [5].

the substation's performance. They can also be used to quantify the performance of incoming heating weeks.

2 Methods and Techniques

2.1 Sequential Pattern Mining

Sequential pattern mining is the process of finding frequently occurring patterns in a sequence dataset. The records of the sequence dataset contain sequences of events whose orders are important. We use the PrefixSpan algorithm [6] to extract frequent sequential patterns. PrefixSpan applies a prefix-projection method recursively to find sequential patterns. The prefix-based projection enables PrefixSpan to focus only on prefix sub-sequences and project on their corresponding postfix sub-sequences. This yields less projections which in turn reduces both the length and the number of sequences in the projected datasets.

2.2 Clustering Analysis

Affinity Propagation: We use the affinity propagation (AP) algorithm [7] for clustering the extracted patterns. AP is based on the concept of *message passing* between data points. Unlike clustering algorithms, such as k-means [8] which requires the number of clusters as an input, AP estimates the optimal number of clusters from the data. In addition, the chosen exemplars are real data points and representative of the clusters.

Consensus Clustering: Gionis et al. [9] proposed an approach for clustering based on the concept of aggregation, where a number of different clustering solutions are given on some datasets of elements. The objective is to produce a single clustering solution from those elements that agrees as much as possible with the given clustering solutions. Consensus clustering algorithms deal with similar problems to those treated by clustering aggregation techniques. Such algorithms aim to synthesize clustering information about the same phenomenon coming from different sources [10] or from different runs of the same algorithm [11]. In this study, we use the consensus clustering schema proposed in [10] in order to integrate the clustering solutions produced on the datasets collected on a weekly basis for the heating season. The exemplars of the produced clustering solutions are considered and divided into k clusters according to the degree of their similarity by applying the AP algorithm. Subsequently, clusters whose exemplars belong to the same partition are merged in order to obtain the final consensus clustering.

2.3 Distance Measure

The similarity between the extracted patterns are assessed with a dynamic programming version of Levenshtein distance (LD) metric [12]. The LD, also known as edit distance, is a string similarity metric that measures the minimum number of editing operations required to transform one string into the other.

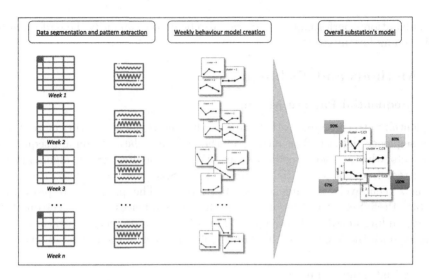

Fig. 1. Schematic illustration of the proposed approach

3   Proposed Method

Our approach has a preprocessing step and two main steps: (1) *Modelling substation's operational behaviour*; (2) *Linking the substation's representative behaviour profiles with performance indicators*. The modelling step consists of three distinctive sub-steps: (i) data segmentation and pattern extraction, (ii) weekly behaviour model creation, and (iii) overall substation's model. The approach is schematically illustrated in Fig. 1.

Data Preprocessing: In order to prepare data for the modelling step all duplicates are removed and missing values are imputed by averaging the neighbouring values. The first and the last missing values are replaced with the next and the previous available values, respectively.

In addition, extreme values that are often a result of faults in measurement tools are smoothed out by a Hampel filter [13], which is a median absolute deviation (MAD) based estimation. The filter computes the median, MAD, and the standard deviation (SD) over the data in a local window. We apply the filter with the default parameters; the size of the window is seven and the threshold for extreme value detection is three, i.e., 3-neighbours on either side of a sample. The threshold for extreme value detection is three. Therefore, in each window a sample with the distance three times the SD from its local median is considered as an extreme value and is replaced by the local median.

We monitor the operational behaviour of substations based on outdoor temperature and the primary side features of the DH system. Our motivation for this choice relates to the fact that the primary side data is always available while the secondary side data requires specific hardware that might not be available at the consumers' building. After discussions with domain experts, we chose five features that have a strong negative correlation with outdoor temperature. The selected features are: (1) primary return temperature, $T_{r,1^{st}}$, (2) primary temperature difference, $\Delta T_{1^{st}}$, (3) primary energy, $Q_{1^{st}}$, (4) primary mass flow rate, $G_{1^{st}}$, and (5) the substation performance indicator based on the hourly consumed energy divided into the hourly mass flow rate, E_s^{E-F}. The fifth feature represents how many units of energy one substation can provide from the consumed volume flow rate.

Z-score normalization is applied on each feature and for every week's period. The normalization is performed to make it possible to assess and compare a substation's operational behaviours in different weeks.

In order to build the DH substation's operational behaviour model using the HOM paradigm, continuous features are converted into categorical features. All five features together build patterns (sequences of events) that represent the operational behaviour of the substation. In this study, we are interested in contextual outlier detection. The context here is referred to as modelling the DH substation's behaviour, during only the heating season. For this purpose we have applied k-means-based discretization method by setting the size of k to four, similar to the number of seasons in Sweden.

1. Modelling DH substation's operational behaviour:

(i) **Data segmentation and pattern extraction:** We extract the substation's behaviour patterns on a weekly basis. The PrefixSpan algorithm is used to find frequent sequential patterns with the length of five in each week. Those sequential patterns that satisfy the user-specified support are considered as frequent ones. The user-specified support threshold is set to be *one* to capture daily patterns, i.e., any patterns that appear at least once will be considered.

(ii) **Weekly behaviour model creation:** The extracted patterns from each week are clustered into groups based on their similarities. Since the aim is to build a DH substation behaviour model for the heating season, all exemplars of the clustering models related to the weeks with the average outdoor temperature above $10\,^{\circ}\mathrm{C}$ are filtered out.

(iii) **Overall substation's model:** The weekly behaviour models built at the previous step are further integrated into an overall substation's behaviour model by applying a consensus clustering technique. The exemplars of the consensus clustering solution are considered as representative profiles for the substation's behaviour, i.e., they can be used to further analyze the substation's behaviour and performance for the whole heating season.

2. Linking behaviour profiles with performance indicators: At this step the derived substation's behaviour profiles are linked to performance indicators. In the current study, we annotate behaviour profiles with two performance indicators: *substation effectiveness* and *grädigkeit*. The two indicators are computed by considering features from both the primary and secondary sides.

Substation effectiveness is computed as $E_s^T = \frac{\Delta T_{1^{st}}}{T_{s,1^{st}} - T_{r,2^{nd}}}$ where, $\Delta T_{1^{st}}$ is the difference between primary supply and return temperatures, $T_{s,1^{st}}$ is the primary supply temperature, and $T_{r,2^{nd}}$ is the return temperature at the secondary side. The efficiency of a well-performed substation should be close to 1 in a normal setting. However, due to the affect of DHW generation on the primary return temperature, the E_s^T can be above 1.

Grädigkeit indicator, also known as the least temperature difference[2], represents the difference between primary and secondary return temperatures and it is computed as $\Delta T_{r,(1^{st},2^{nd})} = T_{r,1^{st}} - T_{r,2^{nd}}$. The grädigkeit of a substation can be greater than or equal to zero, though it can go below zero due to usage of DHW. A lower value of grädigkeit implies better performance.

For each considered performance indicator, we partition the substation's representative behaviour profiles into three categories with respect to the associated performance indicator scores: *low, medium* and *high*. In that way, we have a group of behaviour profiles that represents the substation's sub-optimal performance and two groups of profiles that are linked with satisfactory and optimal substation's performance, respectively. The labelled behaviour profiles can be deployed over the whole heating season in order to further analyze and understand the substation's operational behavior and performance. For example, the profiles from the three different categories can be used to interpret the substation's operational behaviours for particular time intervals. In addition, it is possible to backtrack from these higher order representative profiles to the weekly behaviour models and to the hourly patterns.

4 Results and Discussion

We studied substations' operational behaviour for ten buildings in 2017. We first modeled each substation's weekly operational behaviours. This was performed by grouping the extracted frequent patterns into clusters of similar patterns. We then stored the exemplars of the built clustering model if the average outdoor temperature of the week was less than or equal to 10 °C. This step is motivated by the fact that we want to model the substation's overall operational performance for the whole heating season. The collected exemplars were integrated into a consensus clustering. At last, the obtained consensus clustering model was linked (annotated) with the selected performance indicators. The extracted profiles with respect to each indicator were used to assess behaviour of the substation on a weekly basis.

For the rest of this section we focus on one specific building, B-21. We identified 13 profiles that model the operational behaviour of the substation for the heating season. The extracted profiles were linked with the two performance indicators, *substation effectiveness* and *grädigkeit*. In order to facilitate further analysis, the profiles were sorted from the highest to the lowest performance separately for each indicator. For example, in case of the substation effectiveness the profiles are within a range from 103% to 90%. Regarding the grädigkeit, the profiles are within a range from −2.15 °C to 5.37 °C.

[2] Frederiksen, S., Werner, S.: District heating and cooling, Studentlitteratur Lund (2013).

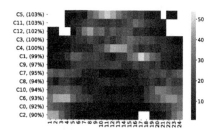

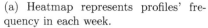

(a) Heatmap represents profiles' frequency in each week.

(b) Heatmap represents profiles' frequency in 24-hour period.

Fig. 2. The deployment of the annotated profiles according to *substation effectiveness* for building B-21 over 2017 heating season. (Color figure online)

Figure 2a shows the substation's effectiveness according to the built profiles for each week. As one can notice the heatmap is sparse and only few weeks, e.g., weeks 3, 4, 7, 10, 14, 15, 17, and 18 represent a high number of frequency for some of the profiles. The heatmap is not easy to interpret and it does not provide interesting information about the substation's weekly behaviour. Figure 2b, on the other hand, provides more information by showing the effectiveness of the same substation at a 24-h period for the whole heating season. For example, one can recognize a yellowish bell shape. Evidently, the substation performed on average 92% at early morning (0:00–5:00) and late evening (20:00–23:00). However, for the rest of the day the performance of the substation is closer to and above 100%. The low performance of the substation might be due to social behaviour, which demonstrates low heat demand in the early morning and late evening.

As mentioned before, we categorize the extracted profiles with respect to their performance indicator labels (substation effectiveness or grädigkeit) into three categories: low, medium, and high. In the case of substation effectiveness *low* represents efficiency below 90%, *medium* indicates efficiency between 90% to 100%, and high stands for efficiency above 100%. Figure 3a shows the overall effectiveness of the substation over the weeks that space heating was required, based on these three categories. As one can see the orange circles, which represent the medium efficiency of the substation, closely follow the curve showing overall substation's effectiveness. This is also valid for the profiles from the other two categories. For example, all profiles linked to optimal performance (blue circles in Fig. 3a) are above the overall substation's effectiveness curve. In Fig. 3a, we can also notice that weeks 19 and 40 represent the end and beginning of the heating season, respectively. The low efficiency of the substation in week 40 might be related to the fact the system required sometime to adjust.

Regarding grädigkeit indicator, *low* represents temperature differences above $3\,^{\circ}C$, *medium* denotes temperatures between 0 to $3\,^{\circ}C$, and high shows temperature differences equal or below to $0\,^{\circ}C$. Figure 3b shows the overall grädigkeit for the studied substation. Similar to Fig. 3a, the medium category is closely

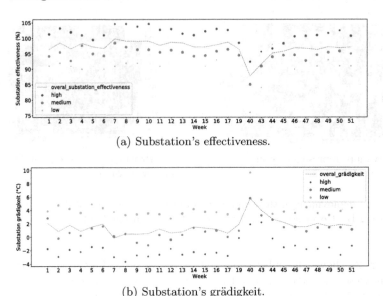

(a) Substation's effectiveness.

(b) Substation's grädigkeit.

Fig. 3. The deployment of the annotated profiles according to performance indicators for building B-21 over 2017 heating season. (Color figure online)

following the curve that represents the overall substation's grädigkeit. Notice that for grädigkeit indicator the temperature differences close to and below zero show a high efficiency.

5   Conclusion and Future Work

We proposed a higher order mining approach for modelling a district heating substation's operational behaviour. The method summarized the substation's behaviour with a series of representative profiles that were linked with two performance indicators. The labelled profiles were deployed over the whole heating season to assess an overall substation's behavior and performance. We applied and studied our method on ten buildings. The initial results showed that the proposed method can be used to analyze and evaluate the operational behaviour of DH substations.

For future work we are interested in studying whether the derived representative behaviour profiles can be used to quantify the performance of incoming heating weeks. In addition we plan to evaluate our approach with other performance indicators.

References

1. Gadd, H., Werner, S.: Fault detection in district heating substations. Appl. Energy **157**, 51–59 (2015)
2. Xue, P., et al.: Fault detection and operation optimization in district heating substations based on data mining techniques. Appl. Energy **205**, 926–940 (2017)
3. Månsson, S., Kallioniemi, P.O.J., Sernhed, K., Thern, M.: A machine learning approach to fault detection in district heating substations. Energy Procedia **149**, 226–235 (2018)
4. Calikus, E., Nowaczyk, S., Sant'Anna, A., Gadd, H., Werner, S.: A data-driven approach for discovery of heat load patterns in district heating. arXiv preprint arXiv:1901.04863 (2019)
5. Roddick, J.F., Spiliopoulou, M., Lister, D., Ceglar, A.: Higher order mining. ACM SIGKDD Explor. Newsl. **10**(1), 5–17 (2008)
6. Pei, J., et al.: PrefixSpan: mining sequential patterns efficiently by prefix-projected pattern growth. In: Proceedings 17th International Conference on Data Engineering, pp. 215–224 (2001)
7. Frey, B.J., Dueck, D.: Clustering by passing messages between data points. Science **315**(5814), 972–976 (2007)
8. MacQueen, J., et al.: Some methods for classification and analysis of multivariate observations. In: Proceedings of the Fifth Berkeley Symposium on Mathematical Statistics and Probability, Oakland, CA, USA, Vol. 1, pp. 281–297 (1967)
9. Gionis, A., Mannila, H., Tsaparas, P.: Clustering aggregation. ACM Trans. Knowl. Disc. Data **1**(1), 4-es (2007). https://doi.org/10.1145/1217299.1217303
10. Boeva, V., Tsiporkova, E., Kostadinova, E.: Analysis of multiple DNA microarray datasets. In: Kasabov, N. (ed.) Springer Handbook of Bio-/Neuroinformatics, pp. 223–234. Springer, Heidelberg (2014). https://doi.org/10.1007/978-3-642-30574-0_14
11. Goder, A., Filkov, V.: Consensus clustering algorithms: comparison and refinement. In: ALENEX, pp. 109–234 (2008)
12. Levenshtein, V.I.: Binary codes capable of correcting deletions, insertions, and reversals. Sov. Phys. Dokl. **10**, 707–710 (1966)
13. Hampel, F.R.: A general qualitative definition of robustness. Ann. Math. Stat. **42**, 1887–1896 (1971)

Modeling Evolving User Behavior
via Sequential Clustering

Veselka Boeva$^{(\boxtimes)}$ and Christian Nordahl

Blekinge Institute of Technology, 371 79 Karlskrona, Sweden
{veselka.boeva,christian.nordahl}@bth.se

Abstract. In this paper we address the problem of modeling the evolution of clusters over time by applying sequential clustering. We propose a sequential partitioning algorithm that can be applied for grouping distinct snapshots of streaming data so that a clustering model is built on each data snapshot. The algorithm is initialized by a clustering solution built on available historical data. Then a new clustering solution is generated on each data snapshot by applying a partitioning algorithm seeded with the centroids of the clustering model obtained at the previous time interval. At each step the algorithm also conducts model adapting operations in order to reflect the evolution in the clustering structure. In that way, it enables to deal with both incremental and dynamic aspects of modeling evolving behavior problems. In addition, the proposed approach is able to trace back evolution through the detection of clusters' transitions, such as splits and merges. We have illustrated and initially evaluated our ideas on household electricity consumption data. The results have shown that the proposed sequential clustering algorithm is robust to modeling evolving behavior by being enable to mine changes and update the model, respectively.

Keywords: Behavior modeling · Clustering evolution · Data mining · Sequential clustering · Household electricity consumption data

1 Introduction

The need for describing and understanding the behavior of a given phenomenon over time led to the emergence of new techniques and methods focused in temporal evolution of data and models [2,6,13]. Data mining techniques and methods that enable to monitor models and patterns over time, compare them, detect and describe changes, and quantify them on their interestingness are encompassed by the paradigm of change mining [5]. The two main challenges of this paradigm are to be able to adapt models to changes in data distribution but also to analyze and understand changes themselves.

Evolving clustering models are referred to incremental or dynamic clustering methods, because they can process data step-wise and update and evolve cluster

This work is part of the research project "Scalable resource efficient systems for big data analytics" funded by the Knowledge Foundation (grant: 20140032) in Sweden.

© Springer Nature Switzerland AG 2020
P. Cellier and K. Driessens (Eds.): ECML PKDD 2019 Workshops, CCIS 1168, pp. 12–20, 2020.
https://doi.org/10.1007/978-3-030-43887-6_2

partitions in incremental learning steps [6,10]. Incremental (sequential) clustering methods process one data element at a time and maintain a good solution by either adding each new element to an existing cluster or placing it in a new singleton cluster while two existing clusters are merged into one [1,7,17]. Dynamic clustering is also a form of online/incremental unsupervised learning. However, it considers not only incrementality of the methods to build the clustering model, but also self-adaptation of the built model. In that way, incrementality deals with the problem of model re-training over time and memory constrains, while dynamic aspects (e.g., data behavior, clustering structure) of the model to be learned can be captured via adaptation of the current model. Lughofer proposes an interesting dynamic clustering algorithm which is also dedicated to incremental clustering of data streams and in addition, it is equipped with dynamic split-and-merge operations [10]. A similar approach defining a set of splitting and merging action conditions is introduced in [8]. Wang et al. also propose a split-merge-evolve algorithm for clustering data into k number of clusters [16]. However, a k cluster output is always provided by the algorithm, i.e. it is not sensitive to the evolution of the data. A split-merge evolutionary clustering algorithm which is robust to evolving scenarios is introduced in [4]. The algorithm is designed to update the existing clustering solutions based on the data characteristics of newly arriving data by either splitting or merging existing clusters. Notice that all these algorithms have the ability to optimize the clustering result in scenarios where new data samples may be added in to existing clusters.

In this paper, we propose a sequential (dynamic) partitioning algorithm that is robust to modeling the evolution of clusters over time. In comparison with the above discussed dynamic clustering algorithms it does not update existing clustering, but groups distinct portions (snapshot) of streaming data so that a clustering model is generated on each data portion. The algorithm initially produces a clustering solution on available historical data. A clustering model is generated on each new data snapshot by applying a partitioning algorithm initialized with the centroids of the clustering solution built on the previous data snapshot. In addition, model adapting operations are performed at each step of the algorithm in order to capture the clusters' evolution. Hence, it tackles both incremental and dynamic aspects of modeling evolving behavior problems. The algorithm also enables to trace back evolution through the identification of clusters' transitions such as splits and merges. We have studied and initially evaluated our algorithm on household electricity consumption data. The results have shown that it is robust to modeling evolving data behavior.

2   Modeling Evolving User Behavior via Sequential Clustering

2.1   Sequential Partitioning Algorithm

In this section, we formally describe the proposed sequential partitioning algorithm. The algorithm idea is schematically illustrated in Fig. 1.

Assume that data sets $D_0, D_1, \ldots, D_n$ are distinct snapshots of a data stream. Further let $C = \{C_i | i = 0, 1, \ldots, n\}$ be a set of clustering solutions (models), such that C_i has been built on a data set D_i. In addition, each clustering solution C_i, for $i = 1, 2, \ldots, n$, is generated by applying a partitioning algorithm (see Sect. 2.2) on data set D_i initialized (seeded) with the centroids of the clustering model built on data set D_{i-1}. The algorithm is initialized by clustering C_0 which is extracted from data set D_0 (available historical data or the initial snapshot).

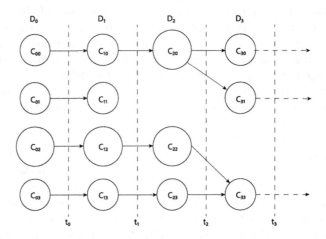

Fig. 1. Schematic illustration of the proposed sequential clustering approach

The basic operations conducted by our algorithm at each time window (on each data snapshot) i are explained below:

1. **Input:** Cluster centroids of partition C_{i-1} $(i = 1, 2, \ldots, n)$.
2. **Clustering step:** Cluster data set D_i by seeding the partitioning algorithm with the centroids of C_{i-1}.
   (a) Initial clustering of D_i.
   (b) Check for empty initial clusters and adapt the partitioning respectively.
   (c) Remove the seeded centroids and finalize the clustering by producing C_i.
3. **Adapting step:** For each cluster $C_{ij} \in C_i$ do the following steps
   (a) Calculate the split condition for C_{ij}.
   (b) If the split condition is satisfied then split C_{ij} into two clusters by applying 2-medoids clustering algorithm and update the list of centroids, respectively.
4. **Output:** Updated clustering partition and list of centroids used to initialize the clustering action that will be conducted on data set D_{i+1}.

Note that at step 2(b) above we check whether there are empty clusters after the initial clustering. If so, this means that the clustering structure is evolving, i.e. some clusters may stop existing while others are merged together. Evidently, the death and merge transitions are part of the clustering step. Therefore the split condition is only checked at step 3^1. We can apply different split conditions. For example, the homogeneity of each cluster C_{ij} may be evaluated and if it is below a given threshold we will perform splitting. Another possibility is to apply the idea implemented by Lughofer [10] in his dynamic split-and-merge algorithm.

In order to trace back the clusters' evolution we can compare the sets of cluster centroids of each pair of partitioning solutions extracted from the corresponding neighboring time intervals (e.g., see Fig. 6). This comparison can be performed by applying some alignment technique, e.g., such as Dynamic Time Warping (DTW) algorithm explained in Sect. 2.3. For example, if we consider two consecutive clustering solutions C_{i-1} and C_i ($i = 1, 2, \ldots, n$), we can easily recognize two scenarios: (i) a centroid of C_{i-1} is aligned to two or more centroids of C_i then the corresponding cluster from C_{i-1} *splits* among the aligned ones from C_i; (ii) a few centroids of C_{i-1} is aligned to a centroid of C_i then the corresponding clusters from C_{i-1} *merge* into the aligned cluster from C_i.

2.2   Partitioning Algorithms

Three partitioning algorithms are commonly used for data analysis to divide the data objects into k disjoint clusters [11]: k-means, k-medians, and k-medoids clustering. The three partitioning methods differ in how the cluster center is defined. In k-means clustering, the cluster center is defined as the mean data vector averaged over all objects in the cluster. In k-medians, the median is calculated for each dimension in the data vector to create the centroid. Finally, in k-medoids clustering, which is a robust version of the k-means, the cluster center is defined as the object with the smallest sum of distances to all other objects in the cluster, i.e., the most centrally located point in a given cluster.

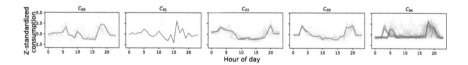

Fig. 2. Initial clustering model produced on the historical data.

2.3   Dynamic Time Warping Algorithm

The DTW alignment algorithm aims at aligning two sequences of feature vectors by warping the time axis iteratively until an optimal match (according to a suitable metrics) between the two sequences is found [15]. Let us consider two matrices $A = [a_1, \ldots, a_n]$ and $B = [b_1, \ldots, b_m]$ with a_i ($i = 1, \ldots, n$) and b_j ($j = 1, \ldots, m$) column vectors of the same dimension. The two vector sequences $[a_1, \ldots, a_n]$ and

---

[1] Step 3 is not implemented into the current version of our sequential clustering algorithm.

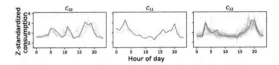

Fig. 3. Clustering model produced on the first new data snapshot. Clusters C_{01} and C_{03} have been empty after the clustering step. Clusters C_{02} and C_{04} are transformed in clusters C_{11} and C_{12}, respectively.

$[b_1, \ldots, b_m]$ can be aligned against each other by arranging them on the sides of a grid, e.g., one on the top and the other on the left hand side. A distance measure, comparing the corresponding elements of the two sequences, can then be placed inside each cell. To find the best match or alignment between these two sequences one needs to find a path through the grid $P = (1,1), \ldots, (i_s, j_s), \ldots, (n, m)$, $(1 \leq i_s \leq n$ and $1 \leq j_s \leq m)$, which minimizes the total distance between A and B.

3   Case Study: Modeling Household Electricity Consumption Behavior

3.1   Case Description

Suppose that a monitoring system for tracking changes in electricity consumption behavior at a household level is developed to be used for some healthcare application. For example, such a system can be used to monitor and look for alterations in the daily routines (sleep-wake cycle) of elderly individuals who have been diagnosed with a neurodegenerative disease. The system is supposed to build and maintain an electricity consumption behavior model for each monitored household. Initially, a model of normal electricity consumption behavior is created for each particular household by using historical data [12]. In order to monitor such a model over time it is necessary to build a new model on each new portion of electricity consumption data and then compare the current model with the new household electricity consumption behavior model. If changes are identified a further analysis of the current electricity consumption behavior model is performed in order to investigate whether these are associated with alterations in the resident's daily routines.

3.2   Data and Experiments

We use electricity consumption data collected from a single randomly selected anonymous household that has been collected with a 1-min interval for a period of 14 months. During those 14 months, there were roughly 2 months worth of data that had not been collected, i.e. zero values which have been removed. We then aggregate the electricity consumption data into a one hour resolution from the one minute resolution.

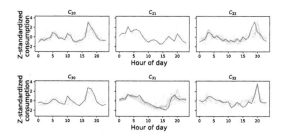

Fig. 4. Clustering models produced on the second (above) and third (below) new data snapshots, respectively.

We divide the data into four parts. The first 50% of the total data represent the historical data (D_0), and the remaining data is evenly distributed into the other three data sets (D_1, D_2 and D_3). In addition, D_2 and D_3 have their contents shifted to simulate a change in their behavior over time. 8% of the contents in D_2 is randomly shifted 1 to 6 h ahead, and 2% of the contents 12 h ahead. Similarly, D_3 has 16% of the data shifted 1 to 6 h ahead and 4% 12 h ahead. We choose these two scenarios to simulate both minor and drastic changes in the sleeping pattern of the resident.

In order to cluster the historical data, we run k-medoids 100 times using randomly initialized cluster medoids, for each k between 2 and 20. DTW is used as the dissimilarity measure and it is restricted to only allow for a maximum warp of two hours. This restriction is in place to allow for some minor alterations in the daily behavior while keeping major alterations in check. The produced clustering solutions are then evaluated using Silhouette Index [14], Connectivity [9], and Average Intra-Cluster distance [3]. The medoids from the best scoring clustering solution are then used as the initial seeds for the next snapshot of data, as explained in Sect. 2.1.

Table 1. Distances between the clustering models generated on the first and second new data snapshots (left), and on the second and third new data snapshots (right), respectively.

	C_{20}	C_{21}	C_{22}		C_{30}	C_{31}	C_{32}
C_{10}	0.142	0.428	0.222	C_{20}	0.123	0.384	0.290
C_{11}	0.373	0.194	0.306	C_{21}	0.468	0.246	0.257
C_{12}	0.235	0.326	0.163	C_{22}	0.212	0.358	0.209

3.3   Results and Discussion

Figure 2 shows the initial clustering model generated on the historical data. As one can see the household electricity consumption behavior is modeled by five different behavior profiles (clusters). C_{03} and C_{04} are the biggest clusters and represent the electricity consumption behavior more typical for working days with clearly recognized morning and evening consumption peaks. Clusters C_{00} and C_{01} are smaller and have an additional consumption peak in the middle of the day, i.e. they model behavior more typical for the weekends. Cluster C_{02} is comparatively big and represents electricity consumption behavior typical for working days with a slightly later start.

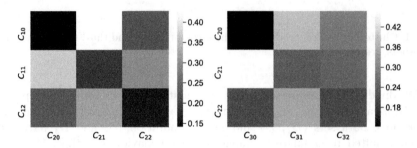

Fig. 5. Heatmaps for distances between the clustering models generated on the first and second new data snapshots (left), and on the second and third new data snapshots (right), respectively.

Figure 3 depicts the clustering model derived from the first new data snapshot. As we can notice the electricity consumption behavior is modeled only by three clusters. C_{01} and C_{03} have been empty after the initial clustering step and their medoids are removed from the list of medoids. Clusters C_{02} and C_{04} are transformed into clusters C_{11} and C_{12}, respectively. It is interesting to observe that C_{12} is also very similar to cluster profile C_{03}. The latter observation is also supported by the DTW alignment between the medoids of the two clustering models given in Fig. 6, where C_{03} and C_{04} are aligned to C_{12}, i.e. they are merged into one cluster. This is also the case for C_{00} and C_{01}, which are replaced by cluster C_{10} at the first time interval.

As it can be seen in Fig. 4 the number of clusters is not changed at the second and third time windows. However, one can easily observe that behavior profile C_{11} evolves its shape over these two time intervals. For example, it moves far from C_{21} and gets closer to C_{20} at the third time window (see Table 1). These observations are also supported by the heatmaps plotted in Fig. 5. One can observe that the respective cells in the heatmap plotted in Fig. 5 (right) have changed their color in comparison with the heatmap in Fig. 5 (left).

We can trace the evolution of the clusters at each step of our algorithm by comparing the sets of cluster centroids of each pair of clusterings extracted from the corresponding consecutive time intervals. This is demonstrated in Fig. 6 which plots the DTW alignment path between the clustering models generated on the historical and first new data sets, respectively. This comparison can be performed on any pair of clustering models generated on the studied data sets. It is also possible to trace back the evolution of a given final cluster down to the initial clustering model.

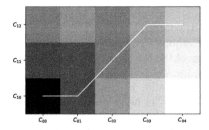

Fig. 6. DTW alignment path between the clustering models generated on the historical and first new snapshot data, respectively.

4 Conclusions and Future Work

In this paper, we have proposed a sequential partitioning algorithm that groups distinct snapshots of streaming data so that a clustering model is generated on each data snapshot. It enables to deal with both incremental and dynamic aspects of modeling evolving behavior problems. In addition, the proposed approach is able to trace back evolution through the detection of clusters' transitions. We have initially evaluated our algorithm on household electricity consumption data. The obtained results have shown that it is robust to modeling evolving data behavior by being enable to mine changes and adapt the model, respectively.

For future work, we aim to further study and evaluate the proposed clustering algorithm on evolving data phenomena in different application domains.

References

1. Ackerman, M., Dasgupta, S.: Incremental clustering: the case for extra clusters. In: Proceedings of NIPS 2014, pp. 307–315 (2014)
2. Aggarwal, C.: On change diagnosis in evolving data streams. IEEE Trans. Knowl. Data Eng. **17**, 587–600 (2005)
3. Baya, A.E., Granitto, P.M.: How many clusters: a validation index for arbitrary-shaped clusters. IEEE/ACM Trans. Comput. Biol. Bioinform. **10**(2), 401–414 (2013)
4. Boeva, V., Angelova, M., Tsiporkova, E.: A split-merge evolutionary clustering algorithm. In: Proceedings of ICAART 2019, pp. 337–346 (2019)
5. Bottcher, M., Hoppner, F., Spiliopoulou, M.: On exploiting the power of time in data mining. In: Proceedings of SIGKDD Explorations, pp. 3–11 (2008)
6. Bouchachia, A.: Evolving clustering: an asset for evolving systems. IEEE SMC Newsl. **36**, 1–6 (2011)
7. Charikar, M., Chekuri, C., Feder, T., Motwani, R.: Incremental clustering and dynamic information retrieval. In: Proceedings STOC 1997, pp. 626–635 (1997)
8. Fa, R., Nandi, A.K.: Smart: Novel self splitting-merging clustering algorithm. In: European Signal Processing Conference. IEEE (2012)
9. Handl, J., Knowles, J., Kell, D.B.: Computational cluster validation in post-genomic data analysis. Bioinformatics **21**(15), 3201–3212 (2005)

10. Lughofer, E.: A dynamic split-and-merge approach for evolving cluster models. Evolving Syst. **3**, 135–151 (2012)
11. MacQueen, J.: Some methods for classification and analysis of multivariate observations. In: Proceedings of the 5th Berkeley symposium on mathematical statistics and probability, vol. 1, pp. 281–297 (1967)
12. Nordahl, C., Boeva, V., Grahn, H., Netz, M.: Profiling of household residents' electricity consumption behavior using clustering analysis. In: Proceedings of ICCS 2019, pp. 779–786 (2019)
13. Oliveira, M., Gama, J.: A framework to monitor clusters evolution applied to economy and finance problems. Intell. Data Anal. **16**, 93–111 (2012)
14. Rousseeuw, P.J.: Silhouettes: a graphical aid to the interpretation and validation of cluster analysis. J. Comput. Appl. Math. **20**, 53–65 (1987)
15. Sakoe, H., Chiba, S.: Dynamic programming algorithm optimization for spoken word recognition. IEEE Trans. Acoust. Speech Signal Proces. **26**(1), 43–49 (1978)
16. Wang, M., Huang, V., Bosneag, A.C.: A novel split-merge-evolve k clustering algorithm. In: IEEE 4th International Conference on Big Data Computing Service and Applications, pp. 229–236 (2018)
17. Zopf, M., et al.: Sequential clustering and contextual importance measures for incremental update summarization. In: Proceedings of COLING 2016, pp. 1071–1082 (2016)

Recognizing User's Activity and Transport Mode Detection: Maintaining Low-Power Consumption

Fitore Muharemi[1]($\boxtimes$), Egzon Syka[2], and Doina Logofatu[1]

[1] Frankfurt University of Applied Sciences, Frankfurt Am Main, Germany
`fitoremuharemi@gmail.com`
[2] University of Bern, Bern, Switzerland

Abstract. Mobile phones are being used for more than just communication because of their wide range of capabilities in aspects of computation and sensing. In this paper, we propose an approach based on supervised learning to detect the user's mode of transport based on the smartphone's built-in accelerometer sensor and the location data. We create a convenient hierarchical classification system, proceeding from a coarse-grained to a fine-grained classification and no requirements of specific position and orientation setting is needed. This study explores how coarse-grained location data from smartphones can be used in combination with accelerometer data to recognize high-level properties of user mobility. Our approach can achieve over 95% accuracy for inferring various transportation modes including tram, bus, train, walking, and stationary. The results suggest that our approach of adding coarse-grained location data improves the accuracy of detection by 10% in comparison with the accelerometer only approach. We present a review of existing approaches for transport mode detection and compare them regarding the type of devices used as sensing unit, the sensors used, the considered transport modes, energy efficiency, and the algorithms used for the classification task.

Keywords: Transport mode · Smartphone-based · Accelerometer · Location data · Supervised learning · AdaBoost · SVM · Random Forest

1 Introduction

Human activity recognition is an important but yet a challenging research area. Efforts to understand human behavior have been subject to many studies through centuries but the explosive spread of smartphones in recent years has provided many fields with a new potential. Modern smartphones are much more than just telecommunication devices, and their use is not anymore limited to the traditional telecommunication field only. They have a wide range of capabilities in aspects of computation and sensing, they are equipped with various sensors that can sense motion, changes in orientation or environment conditions like ambient light, temperature etc., making possible to capture valuable information

© Springer Nature Switzerland AG 2020
P. Cellier and K. Driessens (Eds.): ECML PKDD 2019 Workshops, CCIS 1168, pp. 21–37, 2020.
https://doi.org/10.1007/978-3-030-43887-6_3

for users and about users themselves. These have given rise to mobile-centric context recognition systems, which are able to recognize the context of the carrier—context-awareness. Transportation technologies and strategies are emerging that can help to meet the climate challenge, these include intelligent transportation systems (ITS) and mobility management strategies that can reduce the demand for private vehicles shifting towards less environmentally damaging transportation modes. Studies in this fields require knowledge of transportation activity information and surveys are typically used to collect information for such needs, but surveys conducted through conventional questionnaires to investigate when, where and how people travel have many drawbacks.

Mobile Transport Mode Detection has the potential to solve most of the shortcomings associated with the conventional travel survey methods, including biased response, no response or erroneous time reporting. The objective of our work is to determine the added value of location data when combined together with a set of different features from the [1] paper for accelerometer data and try to overcome shortcomings of [1] and other works [2] that have already tried to fuse this two sensors together but have not used the total potential of accelerometer features.

We implement a fine-grained classification, distinguishing not only between pedestrian and motorized transport but also different modes of pedestrian and motorized transport. To achieve this we base our implementation on the work of [1] paper and then build on it by adding the location data, change filtering algorithms and gravity reduction technique (using gravity sensor). We also discuss the reasons behind using different features and different choices like the sensors that can be used related to what is our primary concern (e.g. accuracy or energy cost), and choices about architecture set-up for instance mobile-based (real-time) prediction versus server-based prediction. Our algorithm uses data from the device's accelerometer, while location data is used only sparsely for obvious reason and later we further discuss some different solution on how to further decrease the power consumption.

2   Related Work

The human activity recognition field has been widely studied, from the studies about the general human activity recognition to more specific like the transport mode detection (TMD). Thus, numerous studies attempt to tackle different aspects of transport mode detection with different approaches.

Related work can be grouped based on main characteristics of approaches like the hardware used, sensors used, the device position and orientation, algorithm complexity and energy awareness, granularity.

In terms of hardware used the related work can be divided into two groups: studies that use dedicated custom devices and those that use smartphones to sense events. [18] is one of the first contributions to the activity recognition problem, in which five biaxial accelerometers worn simultaneously on different parts of the user's body were used to collect the acceleration data recognising everyday activities with an overall accuracy of 84%. However, using dedicated devices has the following disadvantages: (1) cost of devices that we need to purchase is high, (2) in a real-world applications people tend not to like carrying additional devices just for a specific task, (3) the devices need to be distributed to all participants and this is very limiting for almost any kind of application, (4) requires creating a new device infrastructure instead of re-using an existing one. Conversely, using smartphones is very cost effective because we do not need to purchase any devices because most of the people already use them and users will not need to carry additional devices and integration to existing applications is easy. In other words, the infrastructure is already established. For these reasons, most of the studies especially those of TMD field prefer smartphones instead of dedicated custom hardware, and it is also what we are going to use in this paper as the sensing device. Thus, most of the studies that we will discuss here will be of the second group: TMD using smartphones as the sensing device.

Related work based on the sensors used to determine the travel mode can be divided into four groups: GPS only, accelerometer only, GPS with accelerometer and GSM/WiFi sensors group.

GPS-Only Approaches: [8] builds an application Activity Compass which helps guide a cognitively impaired person safely through the community. They use GPS data to classify between three modes of transportation: bus, foot, or car using an unsupervised method of learning a Bayesian model, building a user personalized model based on historical GPS data of the user which means that it is not very robust for the new users. Furthermore, it needs external GIS information about the roads network and bus stops which despite increasing the accuracy it has its limitations. On the other hand [7] do not use external information and mines the knowledge only from the raw GPS data collected in a frequency of one record every two seconds using custom sensing devices. They identify a set of novel features including heading change rate, velocity change rate and stop rate, for a final accuracy of 76.1% with four modes of transport including car, walking, bus and bike. [3] achieves the highest accuracy in this group and can detect various transportation modes including car, bus, train, walking, biking and stationary. It is the first approach that performs fine-grained classification between motorised transportation modes with high accuracy, but in order to do that, it needs knowledge about the underlying transportation network that includes: real-time bus locations, spatial rail and spatial bus stop information which is not always available to us. Among the five classification models that they consider, Random Forest model achieves the highest accuracy.

Accelerometer-Only Approaches: It is worth noting that the works in this group are the most energy efficient because as we will see later in Sect. 3, the accelerometer sensor is less power-hungry compared to other sensors like GPS or GSM/WiFi sensors. [9] is an offline classification architecture performed by training a Support Vector Machine (SVM) with 253 features (250 FFT components plus three statistical features) which detects four modes of transport including walking, running, biking and driving with an accuracy of over 90%. However, the authors point out in the paper that their testing is not as extensive as the other works in this field. It could be said that the most significant work in this group is [1] paper, which consists on extracting a novel set of features capable of capturing time, frequency and statistical characteristics of the acceleration signal. On the contrary to the other TMD algorithms, which seem to fail in recognizing the real characteristics of different modes of transport, [1] algorithm is capable of capturing characteristics of acceleration and breaking patterns for different transportation modes. The main contributions of the approach proposed by [1] are an improved algorithm for estimation of the gravity components, the variety of features used in the classification process and the decomposition of transport mode detection hierarchically into subtasks, proceeding from a coarse-grained classification towards a fine-grained distinction of transport modality. On the contrary, most of the other accelerometer-based solutions estimate gravity components as the mean over a window of fixed duration which does not work during sustained acceleration or when the sensor orientation changes suddenly. In [19], a decision tree classifies a set of features including 32 FFT coefficients and the signal variance. Despite the high number of transportation modes identified in [19], the classification accuracy is only slightly lower than that obtained by [1].

GPS with Accelerometer Approaches: Combining GPS data with accelerometer data is a relatively new approach, but few studies that utilised this combination of sensors have explored the potential that this combination has and have given some encouraging results. For instance, [2] developed a classification system which consists of a decision tree followed by a first-order discrete Hidden Markov Model and achieves an accuracy level of over 93%. The algorithm proposed by [2] analyses GPS speed every second, together with variance and frequency components of the accelerometer signal in order to identify transportation modes, including stationary, walking, running, biking and motorised transport. A study by [10] examines the merits of employing accelerometer data in combination with GPS data in TMD, using Bayesian Belief Network to infer transport modality. Results outline that the approach which combines GPS and accelerometer data yields the best performance.

GSM/WiFi Approaches: The group of sensors that are widely used for TMD except GPS and accelerometer include also GSM and WiFi sensors, with approaches [4,12,13] that make use of them like *the dominant sensor* - relying exclusively on them to identify transport modality, or in a *supporting role* - using them in combination with other sensors usually for specific purposes. [2] proposes to use GSM sensor in order to detect when to start collecting data from the dominant sensors GPS and accelerometer by detecting if the user has gone outdoor and to that end saving energy by not using the power hungry GPS sensor all the time; and [11] uses GSM sensor to derive speed statistics estimates in case of GPS unavailability.

[13] and [12] use changes in the GSM signal environment for coarse-grained detection of transportation modalities including stationary, walking and driving. [12] extracts a set of seven different features to use in a two-stage classification scheme consisting in classifying an instance as stationary or not and if classified as not stationary then determining if the instance was walking or driving. Similarly, [4] combines GSM and WiFi for detecting between dwelling, walking

and driving. Although these solutions obtain accuracies in range 80–90% and being energy efficient compared to GPS solutions, these techniques are susceptible to varying cell densities and cell sizes between different locations, struggling to generalise outside urban areas. As can be seen, all the techniques in this group perform only a coarse-grained detection of transportation modalities since for a fine-grained TMD algorithm including a more sensitive sensor like for instance accelerometer is required.

Comparing all the different solutions available is a challenging task because of the many parameters that have to be considered including identified classes, implementation architecture, sensors used, quality of testing, accuracy and energy cost. A study by [6] compares most of the works mentioned here and concludes that the most interesting and promising works are those by [19] and [1] since:

1. They detect the highest number of relevant transportation modes.
2. They rely only on accelerometer data, no need for external information like GIS.
3. They do not require dedicated devices.
4. State of Art benchmark accuracy: the average accuracy of these two works is 83.5%.

However, the authors point out that caution is recommended with the [19] solution because the dataset is said to include "several hours" but is not actually quantified which leaves suspicion about the quality of their dataset and testing. On the other hand, [1] is thoroughly tested in different scenarios that include different smartphone models, users and countries, with one of the largest dataset found in the literature. So, they conclude that the work by [1] may be considered as the most valuable among the works found in the literature. Our work consists in building a system with the methodology proposed in [1] with some significant changes such as the way we perform gravity components estimation, filter high-frequency signals and then integrate location data from the available location providers (Global Positioning System (GPS), WiFi and mobile cell data). We integrate the location data in the final classifier where the classification of motorized transportation modes is performed and is indeed a key challenge according to the methodology we use.

3   Data Collection

The data used on this experiment are collected over 5 h of transportation data including accelerometer and location data using an Android smartphone (LG G4). We collected accelerometer data at the 60 Hz sample rate, whereas the location data we collected with a frequency of at most every 15 s, using Android's standard options. No measure has been taken to ensure maintaining the location data sample rate constant such as initially gaining a GPS lock or keeping the phone screen on to maintain the connection like in [2]. To ensure that the results of the experiments are not sensitive to the placement of the sensor the data was collected from different smartphone placements. We considered the most common placements for a mobile phone in an urban space: trouser pockets and jacket pockets including any other user choice placement. Furthermore, in contrast to the work of [1] who have used an additional smartphone for the ground truth annotations to avoid disturbing the sensing unit, we have used the same device not only for the annotations but also for other daily activities. Because the ground truth annotation is an inconvenient and sometimes even a dangerous process, we simplified the process by allowing the users to not report instances of being still when annotating modes that were not stationary (i.e., being still at a red light while in motorized transport, or at the bus or train stops etc). These ambiguous instances are labeled with the annotation of the primary mode. Figure 1 shows the distribution of training Fig. 1(a) and testing Fig. 1(b) data based on transport mode.

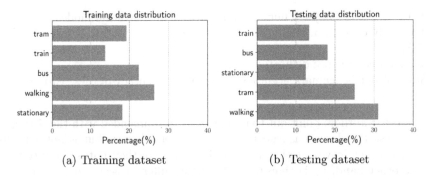

(a) Training dataset   (b) Testing dataset

Fig. 1. Distribution of the collected data based on transport mode

4   Data Preprocessing

Our approach relies on a relatively new practice which is *sensor fusion. Sensor Fusion*[1] is the combining of sensory data or data derived from sensory data such that the resulting information is in some sense better than would be possible when these sources were used individually. The idea here is that each sensor has its strengths and weaknesses, by combining them in a specific way they can compensate each other. In this case, if we knew something about the device orientation we can measure the gravity component in a better way, so what they do in sensor fusion is extract the device orientation from gyroscope which measures the rate of rotation related to the device itself and use it to gravity compensate the accelerometer; the result of using this technique is illustrated in Fig. 2.

The raw measurements generated by the accelerometer sensors are contaminated with environmental noise and jitter that needs to be filtered out. A common practice, in this case, is to preprocess the raw measurements by applying a low-pass filter. There are many algorithms and methods to accomplish this

---

[1] http://www.inforfusion.org/mission.htm.

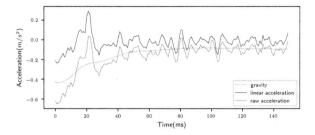

Fig. 2. Separating gravity from accelerometer sensor using sensor fusion.

but moving average filters, and Savitzky-Golay filters are often used to clean up signals, remove noise, perform data averaging and discover meaningful patterns. The subsequent analysis that we will perform later during the feature extraction which includes finding peaks, stationary periods and zero-crossings rate, may be degraded by the presence of too much noise in the signal. Also, as we can see in Fig. 3 it helps with removing what is called as the *zero g bias level* or *offset* that causes non-zero readings even in cases with no acceleration (rest) and in our case that is so high that it can even be measured as a peak. Our approach was to apply a low-pass filter and try to increase signal-to-noise ratio without distorting the signal and maintaining the patterns as much as possible which are essential for us. For this reason, the Savitzky-Golay has the advantage over moving-average because as we can see in Fig. 3, it keeps most of the signal characteristics while reducing noise and offset.

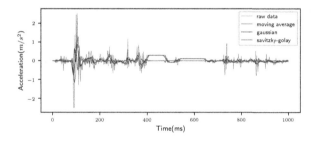

Fig. 3. Separating gravity from accelerometer sensor using sensor fusion.

5   Feature Extraction

We need to transform the collected raw data into more informative and relevant features. Given the significance of the problem, a large number of techniques have been developed that can be used to generate a variety of features sets from different domains for the task of TMD. The most representative domains are the *time* domain, the *statistical* features domain and the *frequency* domain. From the related works that we discussed, only [5] and [1] have used techniques to extract features from this three domains, and a vast majority of works have used only a small subset of statistical features or frequency domain features. Furthermore, in accordance with the approach of [1], we extract features on different levels of granularity. We extracted three sets of features: the frame-based, peak-based and segment-based features that differ on the length of the window, domains and sensors used. There are two important parameters when designing a TMD system that we need to consider: the window size and the sampling frequency, as they both affect the computation and power consumption of the algorithm as well as accuracy. The window size parameter refer to the windows that we use to aggregate the raw sensor data during the feature extraction process; we use different window sizes based on different levels of granularity with the intention to capture various aspects of movement patterns which we will discuss in more detail when we speak about the acceleration based features. Regarding the second parameter of importance the sensor sampling frequency, investigations [14] [17] outline that influences a trade-off between two essential objectives: **increase classification accuracy** – increasing sampling frequency results in improved classification accuracy; **reduce power consumption** – reducing the sampling frequency helps to lower the energy overhead. [17] study asserts that performing classification learning at the same sampling rate as the actual test sampling rate has an only minimal effect on the classification accuracy. Hence, it is possible to train an algorithm at the highest sampling frequency of smartphone sensor as training phase is a one-time activity performed on the server side (offline learning) and then use a much lower sampling frequency in the actual TMD task to reduce the energy overheads that come with high sampling frequency as shown in Fig. 4. Furthermore, [17] report, as expected, that higher sampling frequency usually results in better classification accuracy but that is not always the case as it also depends on the features used, illustrated in Fig. 5 taken from [17] study.

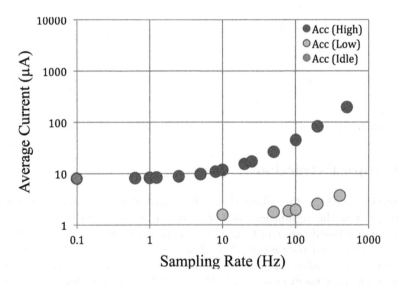

Fig. 4. Power consumption vs sampling rate (from [15])

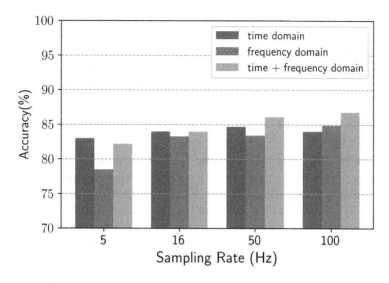

Fig. 5. Sample rate effect on power consumption and classification accuracy

We performed a similar experiment with 1 Hz and 60 Hz sampling frequencies, and we got very similar results for the classification accuracy, but our test was not as extensive as the one of this study for obvious reasons as it was not the intention of our work.

6   Classification Models

Our classification framework is a three-stage hierarchical structure proceeding from a coarse-grained classification towards a fine-grained distinction of transport modality similar to the one of [1]. It consists of three classifiers that infer transport modality on different levels of granularity: at the root of hierarchy the *kinematic motion classifier* which performs a coarse-grained detection between walking and other modalities. When the kinematic classifier detects non-walking activity, the process continues to the classifier on the next level – the *stationary classifier* which distincts between the stationary (pedestrian) and motorised modality. When motorised activity is detected, we proceed to the *motorised classifier* responsible for fine-grained detection between (currently): bus, tram and train modality. The types of classification algorithms used for the task of TMD mostly depends on the system architecture: if the mobile device is only used as a sensing device and the classification is performed on the server which we called the *server-based approach*, then generative models are suitable because of better classification accuracy that they may have. On the other hand, if the detection is intended to run on mobile devices directly, generative models are less popular due to their computational costs and discriminative models are a better fit. Our approach is smartphone-based prediction; so to determine the classifier that most accurately predicts transport mode, we compared: (a) Adaptive Boosting,

(b) Random Forests and (c) Support Vector Machines. We selected these classifiers based on reports and their frequent use in existing literature [1–3, 11].

6.1 Accelerometer Features

We extract a large set of accelerometer features on different levels of granularity and domains, resulting in three sets of features: frame-based, peak-based and segment-based features.

Frame-Based Features. We use this features to capture characteristics of high frequency motion caused such as walking during pedestrian activity or motion from vehicles engine and contact between its wheels and surface. From each frame, we extract features of statistical domain (e.g., mean, variance, min, max and kurtosis), time-domain features (e.g., integral, zero crossings, auto-correlation and mean-crossing rate), and frequency-domain features (e.g., FFT components, energy and entropy).

Peak-Based Features. Acceleration and breaking periods are key periods that characterise the vehicle motion and peak-based features try to capture these patterns to distinguish between different motorised transportation modalities. For that reason we use these features only during the stationary and motorised periods when the classifier which uses frame-based features has failed to detect kinematic movement (pedestrian mode). To extract these features, we use a peak-detection algorithm to find the *peak areas* that correspond to the acceleration or breaking events. The algorithm that we have used is the **smoothed robust z-score algorithm**[2] which is a very robust algorithm based on the principle of dispersion: if a new datapoint is a given x number of standard deviations away from some moving mean (also called standard score or z-score), the algorithm signals which means it identifies a peak area.

Segment-Based Features. These features characterize patterns of acceleration and deceleration periods over an observed segment. We extract these features for the stationary and motorised periods, when the kinematic classifier has failed to recognize pedestrian activity. The features that we consider here are the frequency and duration of stationary periods, the frequency of peak areas, and the variance of peak-based features over the segment.

7 Evaluation and Results

In this section we evaluate performance of the classifiers. We present the mode detection accuracy when we employ a filtering algorithm, compared to the detection accuracy on raw accelerometer data. We also discuss the effect of the window size on the classification performance. Additionally, we use classification feature selection to rank our initial set of features. Given this ranking, we then select the highest rank features to build the final model. We analyse the performance

---

[2] https://stackoverflow.com/a/22640362/930640.

of the transport mode classifier using three quality metrics: accuracy, precision and recall. To test the classifiers, we use the repeated stratified 10-fold cross-validation which is quite frequently used in the literature. [16] concludes that in contrast to the single cross-validation high variance of results, stratification is generally a better scheme, both in terms of bias and variance, when compared to regular cross-validation. Stratified K-Fold is a variation of KFold that returns stratified folds. The folds are made by preserving the percentage of samples for each class. Stratification seeks to ensure that each class is (approximately) equally represented across each fold, which is important in our case because our hierarchical classifiers which usually consist of *one vs all others modes* approach which means that class imbalance is high. Both AdaBoost and RandomForests classifier provide the Out-Of-Bag (OOB) error estimate which is almost identical to that obtained by k-fold cross-validation. Usually, it gives a more pessimistic estimation of error because it trains by a smaller number of samples compared to 10-fold cross-validation. Additionally, we use a separate unseen testing dataset to further validate the obtained results and which is a good estimation about how well the model will perform for new and unseen cases in the future.

The procedure we followed is:

1. Perform repeated stratified k-fold (RSKF) cross validation which gives us a relatively good estimation error(the one we rely at the end)
2. Rebuild the model with the full dataset (the one that was split into folds in the previous step)
3. Use a separate test set to try the final model obtaining a similar(almost surely a higher) error than the one obtained by CV.

When it comes to model evaluation other works apply similar approaches for model evaluation, for example [12] uses 5-fold simple CV and reports those values, but it confirms them in a *controlled* dataset. [2] uses a stratified version of 10-fold cross-validation and validates the results on a separate 3.5 h smaller dataset. However, we consider the approach of [1] as one of the best approaches of how model evaluation should be performed. They use the data from 16 users and perform leave-one-user-out cross-validation which has the following benefits: it guarantees the independence of testing data from the training data which is especially hard to ensure when working with time-series. They take advantage of the independence between different participants data and this allows them to break the strict temporal ordering, at least between individuals data. Unfortunately, such a comprehensive dataset containing many participants currently is not available to us, so we cannot apply this approach of *population-informed* cross-validation. Table 1 shows the effect of window size in the classification performance; the reason why we compare all classifiers with different window size is that using only one window size we might ignore the fact that one algorithm could perform really well with one window length and not that good with others. It is clear that the performance of the classifiers increases with the increase of window length for all the three classifiers, but we also notice an interesting pattern that after windows length of 15 s the performance of the classifiers stabilizes such that the classifiers performance with window length of 15 and 20 s is very similar (sometimes equal).

Table 1. The effect of window size in classifiers detection accuracy using Repeated Stratified 10-fold CV

(%)	Classifier	5 s	10 s	15 s	20 s
Kinematic	AdaB	94.9	95.4	96.0	96.1
	RF	97.9	98.8	99.3	99.6
	LSVM	95.4	95.9	96.4	96.8
Stationary	AdaB	82.9	86.6	89.2	91.2
	RF	92.8	96.6	97.9	98.7
	LSVM	77.0	78.4	81.1	84.2
Motorised	AdaB	99.8	99.9	99.9	99.9
	RF	94.6	98.3	98.6	98.6
	LSVM	90.0	93.2	93.6	93.8

Table 2. The effect of window size on classifiers detection accuracy on the testing set

(%)	Classifier	5 s	10 s	15 s	20 s
Kinematic	AdaB	94.3	97.4	97.8	98.1
	RF	96.2	96.6	97.9	98.5
	LSVM	95.6	96.7	97.3	98.4
Stationary	AdaB	80.6	78.8	81.3	80.1
	RF	81.5	81.5	81.6	80.8
	LSVM	81.8	81.8	82.0	82.1
Motorised	AdaB	64.8	92.3	98.4	99.9
	RF	83.7	86.9	90.4	95.4
	LSVM	73.2	66.5	63.1	62.2

The results, presented on this table, indicate that for the kinematic and stationary classifiers RandomForest classifier using 20 s windows achieve the best results, whereas the motorised classifier achieves near perfect results (99.99% detection accuracy) using AdaBoost at the same window length. Thus, our final classification system, is made up of these two classifier algorithms combined together.

As an external way to corroborate our classification model, we tested the models using the testing dataset, which consists of transportation data (accelerometer + location data) collected in one day. These data are independent of those used to build our model; they are collected in different itinerary and different period of time (approx. 3 months after the collection of the training dataset). The results presented in Table 2 confirm the results of the Table 1 with the only exception being that in stationary classifier we have a drop in the reported detection accuracy. We attribute it to a known human error made during the labeling process; the user forgot to write down the time when he got on the bus after previously was in the stationary mode. Because the dataset contains small amounts of each modality, approximately 20 min of each travel mode, the contribution of that error is noticed much more. However, for the other two classifiers: kinematic and motorised classifier, we observe the reported effect of window size in the performance of the classifiers and we also reconfirm the very good performance of our models, near perfect accuracy for the case of motorised transport mode detection 99.9% accuracy.

For hyperparameters optimization, in our case *the maximum tree depth D* and *the number of estimators(trees) N*, we use traditional techniques like *grid search* using CV for AdaBoost or using OOB error rate for the random forests. We select a suitable values, which balance between the classifier accuracy and classifier complexity. In order to retain classifier simplicity, we opted for the minimal values, after which further increasing the value resulted only in marginal gain in accuracy.

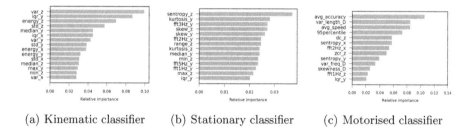

(a) Kinematic classifier   (b) Stationary classifier   (c) Motorised classifier

Fig. 6. Final models: the selected features and their relative importance

Using *sqrt* (square root of the number of features) as the maximum number of features was the best performer. Whereas, regarding the number of estimators(trees), $N = 150$ is optimal and further increasing the number of estimators results only in marginal gain in accuracy. So, we decided to use RFs with 150 and 75 trees for the kinematic and stationary classifier, respectively. Whereas for the motorised classifier with AdaBoost, we found that the parameter that is crucial is the tree depth and regarding the number of trees any classifier with more than 75 trees achieves good results. Both the Random Forest and AdaBoost provide us with a 'tool' that makes very easy to measure the relative importance of each features. In Figs. 6(a), (b), (c) we have presented features with the highest importance for each classifier; in the final classifiers we do not need to use the complete set of features that we extracted but only a subset of features can achieve a very similar performance.

7.1 Effect of Filtering Raw Accelerometer Data on Classification Accuracy

The problem with smoothing is that it is often less beneficial than we might think that is why we decided to test how it affects the algorithm accuracy in our classifiers. The results presented in Table 3 indicate that filtering raw accelerometer values helps with overfitting and is especially noticeable in the case of testing on the new unseen dataset with a lot of variability from the training dataset. Whereas the detection performance reported in Repeated Stratified 10-Fold CV and testing set is almost identical with the exception that filtering helps RandomForest to achieve a better performance in the motorised classifier. So, regarding the filtering of the raw sensor values we find it necessary and beneficial, but further improvements are possible using different filtering techniques or parameters.

Table 3. Detection accuracy comparison for the unseen dataset with/without filtering

Classifier	With filtering	Without filtering
Kinematic	98.3	98.3
Stationary	84.5	74.3
Motorised	70.0	52.2

7.2 Effect of Location Data on Classification Accuracy

Figure 6(c), illustrates the relative importance of features included in the final motorised classifier. We observe that three of five most important features are location based features including *the average accuracy, 95th percentile of speed and the average speed*; giving a strong indication about the predictive power of location based features. Furthermore, the results shown in Fig. 7 quantify the potential that coarse-grained location data have in the aspect of TMD. We observe in Fig. 7 how the inclusion of location data significantly improves F1 score of our approach for all

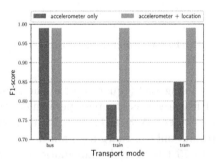

Fig. 7. The effect on F1-score from using location data for motorised transport mode detection.

of the motorised transportation modalities. Our tests show 10% increase in overall detection accuracy when using accelerometer data combined with location data compared to using only accelerometer data.

8   Conclusion

In this paper, we detailed the design, implementation, and evaluation of the transport mode detection system that runs and uses smart-phone sensors: accelerometer and location data. We compared different design choices that need to be made when building such a system: the choice of sensing unit, the sensors selection, the sensor sampling frequency, etc. We gave details about the advantages/disadvantages of different approaches encountered in the related work regarding the system architecture, the achieved accuracy and power consumption ratio, quality of testing, generalization capabilities to different environments and the need for external information, and we motivated our choices. We considered different aspects in our evaluation: the accuracy of transport mode detection, generalization performance, and the robustness of the classifiers. In this study, we employed the proposed solution in [1] as a base classifier and we built on it by addressing its weaknesses and using new approaches on different parts of the system. We extracted a large set of accelerometer features on different levels of granularity and from different domains, that capture different characteristics of user motion, and we also extracted a comprehensive set of location features and we analyzed their discrimination capability. Using a three-stage hierarchical classification system we inferred transport modality on different levels of granularity going from coarse-grained classification towards a fine-grained distinction of transport modality between different motorized transportation modes. The choice of classification algorithms is constrained by the system architecture: for the online mobile-based approaches like ours, the computational cost is important. In order to determine the classifier that most accurately predicts transport mode, a comparison was made between (a) Adaptive Boosting (AdaBoost); (b) Random Forests (RF); (c) Support Vector Machines (SVM). Apart from a good classification algorithm, the sliding window size and appropriate preprocessing are also vital for achieving better results. We compared in detail the importance of smoothing the raw accelerometer values and we also analyzed the importance of the sliding window size in the performance of all three classifiers using windows of different sizes. Using the user location data, we showed that it is possible to address the weakness of previously proposed solution in [1]; that is, to distinguish between motorized modes such as trains, buses, and cars with high accuracy. The results showed that employing location data in the motorized classifier improves the general classification accuracy by 10%, achieving an overall accuracy of over 95%. Future works might comprise energy efficiency strategies that switch off the classification system during extended periods of stationary(indicating that the user is indoors) until notable kinematic movement (e.g. walking for a relatively long period of time) when the user is outside using changes in GSM cell towers to determine the start of outdoor trips, presented in [2] which could provide significant reductions in power consumption. Furthermore, infusing a sensor with low-energy footprint capable of stationary detection (i.e. GSM or WiFi), would also help us in detection between pedestrian and vehicle movement and clear up any ambiguity that often happens between these two modes (caused by overlapping). In any case, the use of changes in GSM/WiFi cell towers has to be exploited in order to determine what is the impact in the classifier performance compared to using the location data and if we could use them instead as they (alone) have a lower power consumption. We also believe that collecting additional data and analyzing features, especially from the spectral domain (higher FFT components can be used to potentially capture some periodical motions that characterize motorized transport), more closely could help further improve the accuracy of our model.

References

1. Hemminki, S., Nurmi, P., Tarkoma, S.: Accelerometer-based transportation mode detection on smartphones. In: Proceedings of the 11th ACM Conference on Embedded Networked Sensor Systems, pp. 13. ACM (2013)
2. Reddy, S., Mun, M., Burke, J., Estrin, D., Hansen, M., Srivastava, M.: Using mobile phones to determine transportation modes. ACM Trans. Sens. Networks (TOSN) (2010). Article 13, ACM
3. Stenneth, L., Wolfson, O., Yu, P.S., Xu, B.: Transportation mode detection using mobile phones and GIS information. In: Proceedings of the 19th ACM SIGSPATIAL International Conference on Advances in Geographic Information Systems, pp. 54–63. ACM (2011)
4. Mun, M., Estrin, D., Burke, J., Hansen, M.: Parsimonious mobility classification using GSM and WiFi traces. In: Proceedings of the Fifth Workshop on Embedded Networked Sensors (HotEmNets) (2008)
5. Figo, D., Diniz, P.C., Ferreira, D.R., Cardoso, J.M.P.: Preprocessing techniques for context recognition from accelerometer data. Pers. Ubiquit. Comput. **14**(7), 645–662 (2010). https://doi.org/10.1007/s00779-010-0293-9
6. Biancat, J., Chiara, B., Brighenti, A.: Review of transportation mode detection techniques. EAI Endorsed Trans. Ambient Syst. **1**(4), 1–10 (2014). Directory of Open Access Journals
7. Zheng, Y., Li, Q., Chen, Y., Xie, X., Ma, W.-Y.: Understanding mobility based on GPS data. In: Proceedings of the 10th International Conference on Ubiquitous Computing, pp. 312–321. ACM (2011)
8. Patterson, D.J., Liao, L., Fox, D., Kautz, H.: Inferring high-level behavior from low-level sensors. In: Dey, A.K., Schmidt, A., McCarthy, J.F. (eds.) UbiComp 2003. LNCS, vol. 2864, pp. 73–89. Springer, Heidelberg (2003). https://doi.org/10.1007/978-3-540-39653-6_6
9. Nham, B., Siangliulue, K., Yeung, S.: Predicting mode of transport from iPhone accelerometer data. In: Machine Learning Final Projects, Stanford University. Citeseer (2008)
10. Feng, T., Timmermans, H.J.P.: Transportation mode recognition using GPS and accelerometer data. Transp. Res. Part C Emerg. Technol. **37**, 118–130 (2013)
11. Xiao, Y., et al.: Transportation activity analysis using smartphones. In: 2012 IEEE Consumer Communications and Networking Conference (CCNC), pp. 60–61. IEEE (2012)
12. Sohn, T., et al.: Mobility detection using everyday GSM traces. In: Dourish, P., Friday, A. (eds.) UbiComp 2006. LNCS, vol. 4206, pp. 212–224. Springer, Heidelberg (2006). https://doi.org/10.1007/11853565_13

13. Muller, I.A.H.: Practical activity recognition using GSM data. In: Proceedings of the 5th International Semantic Web Conference (ISWC), Athens, vol. 1, no. 8. Citeseer (2006)
14. Junker, H., Lukowicz, P., Troster, G.: Sampling frequency, signal resolution and the accuracy of wearable context recognition systems. In: Eighth International Symposium on Wearable Computers. ISWC 2004, vol. 1, pp. 176–177. IEEE (2004)
15. Liu, Q., et al.: Energy-efficient wearable analysis for running. IEEE Trans. Mob. Comput. **16**(9), 2531–2544 (2017)
16. Kohavi, R., et al.: A study of cross-validation and bootstrap for accuracy estimation and model selection. In: IJCAI, Montreal, Canada, vol. 14, no. 2, pp. 1137–1145 (1995)
17. Yan, Z., Subbaraju, V., Chakraborty, D., Misra, A., Aberer, K.: Energy-efficient continuous activity recognition on mobile phones: an activity-adaptive approach. In: 2012 16th International Symposium on Wearable Computers (ISWC), pp. 17–24. IEEE (2012)
18. Bao, L., Intille, S.S.: Activity recognition from user-annotated acceleration data. In: Ferscha, A., Mattern, F. (eds.) Pervasive 2004. LNCS, vol. 3001, pp. 1–17. Springer, Heidelberg (2004). https://doi.org/10.1007/978-3-540-24646-6_1
19. Manzoni, V., Maniloff, D., Kloeckl, K., Ratti, C.: Transportation mode identification and real-time CO_2 emission estimation using smartphones. In: SENSEable City Lab, Massachusetts Institute of Technology. Citeseer (2010)

Can Twitter Help to Predict Outcome of 2019 Indian General Election: A Deep Learning Based Study

Amit Agarwal$^{(\boxtimes)}$, Durga Toshniwal, and Jatin Bedi

Department of Computer Science and Engineering, Indian Institute of Technology
Roorkee, Roorkee, India
aagarwal3@cs.iitr.ac.in, durgatoshniwal@gmail.com, jatinbedi278@gmail.com

Abstract. Traditionally, elections polls have been widely used to analyse trend and to predict likely election results. However, these methods
are very expensive and labour intensive. With the widespread development of several social media platforms, a large amount of unstructured
data become easily available, which in turn could be processed and analysed to extract meaningful information about several topics and events
such as election, sports, natural hazards etc. Hence, in this study, we
utilise twitter data to analyse the 2019 Indian general election and to
predict possible outcomes. We have collected 41 million election related
tweets during the months of April and May 2019. The proposed approach works by initially performing hashtags based tweets segregation to
generate data for training the classification model. Subsequently, we augment different word embeddings with deep learning classification model
to support improved classification accuracy. The model achieves 87.30%
classification accuracy in amalgamation with fasttext word embedding.
Finally, the opinion analysis is performed on the classified tweets to determine possible election outcomes. To evaluate our proposed model, the
output results are validated with the available ground truth and have
shown a close correlation to the actual results. Furthermore, we have also
performed an in-depth analysis of the selected political parties tweets,
to provide an insight into top trending hashtags and mentions used by
people during election campaigns.

Keywords: Sentiment analysis · Deep learning · LSTM · Indian
general election 2019 · Lok Sabha election · Bhartiya Janta Party
(BJP) · Indian National Congress (INC)

1 Introduction

In the year 2019, the tenure of the 16th Lok Sabha got completed and India, the
world's largest democracy witnessed 17th Lok Sabha election in the year 2019.
These elections were held in the seven phases for the 546 parliament constituencies during the month of April and May; the first one held on April 19, 2019 and

© Springer Nature Switzerland AG 2020
P. Cellier and K. Driessens (Eds.): ECML PKDD 2019 Workshops, CCIS 1168, pp. 38–53, 2020.
https://doi.org/10.1007/978-3-030-43887-6_4

the last phase held on May 19, 2019. For any party, these general elections are the biggest exercise to form government at the center (by winning 272 seats or by proving majority with the help of alliances) and thus elect the Prime Minister of India. The main responsibility for conducting elections in a fair manner lies in the hands of the Election Commission of India (ECI). Even though the number of parties registered for the general elections with the ECI has increased to 2354 as compared to 1616 in the year 2014, the main battle is between some major parties namely Bhartiya Janta Party (BJP), Indian National Congress (INC) and some other national parties Aam Aadmi Party, Mahagathbandhan (alliance of several parties like Samajwadi party, Bahujan Samaj party and many more).

In the year 2019, the general elections were different from the previous elections in various ways. Firstly, the people of India were highly inclined towards the positive governance of the current PM Narendra Modi (BJP). Secondly, the various alliances were formed to oppose the current governing party BJP. Thirdly, in the last few years, it has been seen that political parties are increasingly engaging in dialogues with people over social media platforms (specifically Twitter) and people are also increasingly using social platforms to express their views on different political topics. Majority of the political parties leaders/candidates have verified accounts and pages on Twitter, which they are using for their party campaigns.

With the increasing popularity of social media platforms for expressing views or opinions, information dissemination and social interactions, a large amount of real-time data becomes easily available. Over the election duration, this real-time data can be analysed to determine people semantics or opinions on the different political parties or political orientations of the people, which in turn could provide beneficial information on the election results. Hence, there is a need for an automated system that could efficiently utilise such real-time tweet data to predict election results. In the current work, we propose a system to serve the desired purpose. The main research contribution of this work can be summarised as follows:

- Over the decades, standard methods such as polls, surveys have been widely used for predicting election results. Despite being accurate and reliable, these methods require more efforts, costly and are time consuming. Hence, in current work, we propose Twitter as a reliable platform to predict election results from people tweets sentiments/opinions.
- In the current work, we proposed a semi-automated hashtag based approach to initially segregate tweets for building sentiment classification model (to support supervised learning).
- The current work combines deep learning based classification model (Long Short Term Memory Network) with different word embeddings to provide support for improved prediction accuracy.
- In the current work, we perform the State-wise analysis of people opinions on two main national parties of India (BJP and Congress).
- To evaluate the reliability and accuracy of the proposed work, a comparative analysis has been performed between the prediction results and available ground truth.

2   Literature Review

Election results prediction has emerged as a significant area of research in the last few years. Traditional studies in the field have implemented econometric methods to predict election results through demographic information. However, with the emergence of social media platform, real-time people opinions data has become the enrich source of information [9]. This real-time information can be efficiently analysed for the task of decision making and results prediction [1,3,13,20]. In literature, a number of research studies have been implemented on the election related twitter data and some of them are as follows:

Srivastva et al. [20] analysed social network content for analyzing Delhi assembly elections 2015 and trained classification model on the manually annotated dataset for predicting election outcomes. The prediction results are evaluated on the basis of the root-mean squared performance measure. Jose et al. [13] proposed a lexicon analysis or word sense disambiguation based approach for election results prediction. Amit et al. [4] they have performed geo-spatial sentiment analysis for UK-EU referendum by utilizing the Twitter data. Mehndiratta et al. [16] performed twitter based sentiment analysis over 0.25 million public tweets in reference to different politicians or political parties for the 2014 general elections. Ahmed et al. [6] investigate the use of Twitter as a tool for campaigns during the 2014 elections. The study aims to find answers to several research queries including: identifying the party that most frequently uses twitter for campaigns, to identify the dominant issues or topics and many other related queries. Khatua et al. [15] performed sentiment analysis on the 2014 election related data. The author investigated whether Twitter data predict the outcomes of the elections. From the research results of this paper following inferences were drawn: (a) Twitter can serve as an efficient tool for predicting election results (b) Out of the tweet volume and sentiment score, sentiment score can be used as an effective predictor of vote swing.

Ibrahim et al. [12] integrated buzzer (bot) detection tool with twitter analysis to predict results for Indonesia Presidential elections. On the basis of experimental results, the authors have stated that Twitter can serve as an important tool for any political activities. Tsakalidis et al. [22] proposed a novel approach that combined twitter data with opinion polls for Germany, Netherlands and Greece 2014 election results prediction. The model treated twitter based features as time series and implemented three different forecasting models to predict the target results. The output results have proven the effectiveness of combining twitter based features with polls by generating minimum prediction error. Burnap et al. [8] combined prior party support data with current twitter results to predict UK 2015 election results. The approach work by initially calculating tweets and Leader based sentiment score, which in turn are combined to generate party overall positive sentiment score. Subsequently, the integration of the UK 2010 election results with current sentiment scores is utilised to make constituency based election results prediction.

Ebrahimi et al. [10] listed several challenges that occurs while performing sentiment analysis on dynamic events such as elections such as candidate

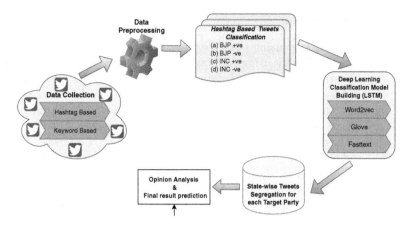

Fig. 1. Methodology of the proposed approach for election results prediction.

dependency (name of both candidates in the same tweet), identifying users' political preferences, availability of limited training dataset and interpretation related challenges. Paul et al. [18] performed spatio-temporal analysis of US elections. The approach combined machine learning models with twitter sentiment analysis for results prediction. The approach works by initially segregating tweets on the basis of their relevance to the target event. Subsequently, sentiment analysis is applied to calculate sentiment polarity related to each tweet. Finally, after being processed by burst event and spatio-temporal analytic modules, machine learning models are trained to the learn abstract representation of the data, which in turn were used for results prediction. The approach provided a prediction accuracy of 84.4%.

As listed above, a number of research studies have been performed on election results prediction and sentiment analysis. However, the majority of these research studies have targeted at United States (US) elections and very limited research work has been carried out on India's general election. Furthermore, to the best of our knowledge, no previous studies have augmented deep learning models with twitter data for election results prediction or target labeled data generation on India's election. Hence, in the current work, we proposed a twitter data based hybrid approach for election results prediction while resolving all major issues associated with analysis on dynamic events (as mentioned in). In addition to this, the current approach combines deep learning model with different word embeddings to achieve 87.30% classification accuracy.

3   Methodology

In this section, we explicate the methodology of the proposed election results prediction approach. Figure 1 shows the broad level diagram of the methodology.

3.1   Dataset Collection

We have collected data by using the Twitter streaming API. Data crawling can be done in two ways, i.e. by using hashtags and the other one by using bounding box. In this study, we collected dataset of 17th Lok Sabha election which held in seven phases for 546 parliament constituencies. We have crawled the dataset by using the hashtags corresponding to BJP and INC. Hashtags used for BJP are ('narendramodi', 'narendramodi_in', 'BJP4India', 'modi', 'NDA', 'BJP') whereas hashtags used for INC are ('RahulGandhi', 'Congress', 'INCIndia', 'INC') and collected around 41 million tweets during April 5-2019 to May-20-2019. Since, collected tweets contains a lot of meta information itself which is off no use. So while downloading the tweets we store only useful information such as userid, tweetid, tweet, location, date and time, RT count, Fav count, hashtags, mentions. After that, we have performed data pre-possessing over the collected tweets.

Data Preprocessing: Twitter post are generally informal, brief, unstructured and often contain grammatical mistakes, misspelling and a lot of noise. It might be due to the 140 character limit imposed on tweets previously, but on November 7, 2017, the restriction is doubled for most of the languages except Chinese, Japanese, and Korean. Authors from the previous studies claim that users knowingly use the abbreviation, shortened, slang words and also uses an amalgamation of prefix and suffix of the word. So it becomes very cumbersome to understand some of the tweets like "Hvy trafic at strt of andheri brdge going 2wrds aiprt & further". Previously Agarwal et al. [2,5] and Subramainum et al. [21] uses different text mining techniques like edit Distance, Longest Common Subsequence and Prefix_Suffix match to handle noisy text in SMS. We have used the same approach so that it could be converted to a readable form like "Heavy traffic at start of andheri bridge going towards airport & further". The various steps involved in data pre-processing are as follows:

– "@" is used in Tweet by user to tag or refer other twitter users so that they may follow on the tweet. So we remove "@" from the tweets.
– #hashtags were used by users before any relevant keyword or phrase to categorize their tweets and show their tweets more easily in twitter search. So hashtags symbol # were removed as they carry no relevance.
– "URL" it is used to shared other web resource. But they carry no relevance information. So we remove it.
– Remove all non-alphabetic terms and stop words (except only Userid, Tweetid, date and time).
– Remove all the repeated characters like ("Trafficcccc" for "traf- fic"), ("stuck-kkk" for "stuck").
– Replacement of Abbreviation and Slang words
– Convert all the characters in to lower case.

3.2   Hashtags Based Tweets Segregation

Hashtags play a predominant role in the identification and characterisation of the tweets. They are basically synonyms that are used to represent the ongoing twitter trends at any particular point of time. In the recent past, various knowledge based or polarity lexicon based research techniques have been proposed to generate sentiments polarity from the tweet hashtags i.e. to distinguish tweets into different target sentiment classes [23]. With the emergence of a topic, new hashtags go on quickly adding to the follow-up tweets of the same topic. So, the process of defining sentiment polarity on the basis of tweets hashtags become very cumbersome. In this context, we proposed a hybrid classification framework to semi-automatically identify and categorize tweets on the basis of initially available trending hashtags. The step by step working of this framework is as follows:

1. Extract trending hashtags in relation to our current topic of interest.
2. Manually analyse and classify hashtags to target sentiment analysis.
3. Build initial tagged tweets dataset on the basis of resultant hashtags to train classification model.
4. Train and Build sentiment polarity classification model.
5. utilise trained model to classify follow up/real-time tweets of a topic.

The set of initially extracted top 20 hashtags along with their target classes are shown in Table 1. Furthermore, the statistics regarding the number of tweets extracted (in relation to each target class) on the basis of initially available hashtags are listed in Table 2.

3.3   Twitter Sentiment Classification

In the proposed work, initially, hashtags-based segmentation (as described in Sect. 3.2) is applied to segment crawled dataset into two different target datasets (BJP, Congress). Each of these data-sets are then individually used to train independent classification models. In the present work, we have implemented the Long-short term memory network (LSTM) [11] model for the classification purpose. LSTM network models [11] have achieved state of the art performance in several computer vision and text mining tasks [12]. These neural networks have the capability of persisting the information or context present in the text. Furthermore, in the present work, these classification models are augmented with different vector representations i.e. word embeddings such as Word2Vec, Glove and fasttext of the tweets to achieve better classification accuracy. The brief introduction to these embeddings are given as follows:

– *Word2Vec* [17]: are basically two layer neural network models used to generate vector representations for the input text corpus. The vector representations are generated from a skip-gram model. The task of the model is to learn the weights for the 300 (embedding dim) neurons in the hidden layer. These weights themselves represent the distributed representation for a particular word. The output of the output layer represents the probability distribution for the remaining words.

Table 1. Sentiment-wise top 20 trending hashtags corresponding to INC and BJP

INC +ve	INC −ve	BJP +ve	BJP −ve
VoteForChange	CongressTerror	PhirEkBaarModiSarkar	AbHogaNyay
AmethiKeDilMeinRahul	CongressMuktBharat	ModiHaiToVikasHai	NyayForEmpowerment
JanSankalpRally	CongressGundiHai	AayegaToModiHi	BJPJumlaManifesto
VaravayiRahulGandhi	CongAdmitsJhoot	DeshKiPasandModi	AbHogaNYAY
VoteNyayVoteCongress	RajivGandhiChorHai	IndiaBoleNaMoPhirSe	NyayYatra
MyVoteForCongress	corruptcongress	NamoAgain	NyayforRajasthan
HogiCongressKiJeet	IndiaRejectsCongress	ModiOnceMore	NYAYforKarnataka
SoniaGandhiRaeBareli	CongressMuktBharat	ModiAgain	ShamelessChowkidars
PriyankaGandhiInAssam	CongAdmitsJhoot	HarVoteModiKo	BJPGameOfThieves
CongressHaiNa	RahulApologizes	BharatKaGarvModi	NYAYforIndia
VoteForCongress	JumlaReturns	IndiaVotesForNaMo	IndiaMaangeNyay
JanaNayakanRaGa	DynastMuktBharat	NaMoForNewIndia	FekuModi
MeraVoteCongressKo	CongressChorHai	IndiaBoleModiDobara	ChowkidarChorHai
CongressForDelhi	RahulControversy	DeshModiKeSaath	AbHogyaNyay
BengalWithRahulGandhi	RahulGandhiChorHai	EveryVoteForModi	JaayegaTohModiHi
SilcharWithCongress	Pappu	IsBaarNaMoPhirSe	ShameOnChowkidars
AmethiKaRahulGandhi	NeechPoliticsOnPM	ModiAaneWalaHai	ModiKaFakeGDP
RahulForBehtarBharat	DynastyMuktBharat	MainBhiChowkidar	ShameOnPMModi
BengalWithCongress	CongInsultsPoor	IndiaWantsModiAgain	RafaleChorChowkidar
MeriAwazMeriCongress	PappuDiwas	ApnaModiAayega	BJPInsultsMartyr

Table 2. Dataset statistic

Political party		INC		BJP	
Sentiment class		+ve	−ve	+ve	−ve
Number of tweets	Total	229645	292650	1271658	504976
	Unique	8377	55888	79658	24068

- *Glove* [19]: The model combines the count based word embedding generation model with the skip-gram model for word generation to improve the training speed. This involves the generation of a co-occurrence matrix(X) for each pair of words.
- *Fasttext* [14]: Fasttext interprets a single word as a collection of n-gram characters obtained from the individual word. Thus, a word like '*trendy*' is represented by the collection of chunks of the original word, that is, the chunks ['*trendy*', '*trend*', '*tren*', '*tre*', '*tr*', '*t*'] represent the original word. This is useful for generating embedding for rare words and the words appearing from out of the vocabulary of the training dataset.

The working of classification model in conjunction with different word embeddings is summarized in Algorithm 1 where $\mathcal{G}$, $\mathcal{W}$ and $\mathcal{F}$ represents glove, word2vec and fasttext embeddings respectively. Furthermore, the corresponding sentiment class prediction results in terms of average prediction accuracy are listed in

Table 3. The term average prediction accuracy denotes the average of the classification model prediction performance on two initially segregated data-sets i.e. Congress and BJP datasets.

3.4 Opinion Analysis Corresponding to Different States

In this section, we make use of the above classification results to perform the BJP (Bharatiya Janata Party) and INC (Indian National Congress) related tweets based opinion analysis. Previous studies [15,16,22] in the field have performed the sentiment analysis (SA) with respect to different political parties as well as some of the work also performed *SA* over the top leaders of different parties. However, to the best of our knowledge, no previous studies have performed opinion analysis with respect to different political parties at state-level. This type of analysis could provide more insight about the users' views corresponding to different political parties at state-level as well as it can play a pivot role in predicting the *Exit polls* results by analyzing the real-time state-wise segregated tweets. So, in the preceding section, we find out users opinion w.r.t different parties and try to find the correlation between the actual election results and our findings [7].

Algorithm 1. *Tweets Sentiments Classification*

Input: *Tweets, Glove:* $\mathcal{G}$, *W2V:* $\mathcal{W}$, *Fasttext:* $\mathcal{F}$
Output: *Classified Tweets, Accuracy*
 1: *Divide Input dataset into X_train, X_test, Y_train, Y_test*
 2: **for** each tweet in *X_train* **do**
 3:     *fit tokenizer on text: tokenizer.fit_on_texts(X_train)*
 4:     *generate d-dimensional encoded representation*
 5:     *Use different word embeddings to generate embedding_matrix for weight_inputs to classification model*
 6: **end for**
 7:     *Train and build Stacked LSTM model.*
 8: **for** each tweet in *X_test* **do**
 9:     *fit tokenizer on text: tokenizer.fit_on_texts(X_test)*
10:     *generate d-dimensional encoded representation*
11: **end for**
12: *Evaluate model performance on test dataset.*

Opinion Analysis Corresponding to BJP (Bharatiya Janata Party). While performing the Opinion analysis, we first segregate the tweets with respect to location information. The location information can be extracted into three ways i.e. Profile Location, GPS enabled tweets (Lat and Log), and Content based location information. In the current study, we segregate the tweets on the basis of location information given in the users' profile and also from GPS enabled tweets

Table 3. Average accuracy prediction by using different word embedding representation method.

Model name	Embedding size	Average prediction accuracy	
		Without stop words	*Without stop words & with Stemming*
LSTM	300	80.50	81.65
Bi-LSTM	300	80.02	80.75
LSTM + word2vec embedding	300	81.03	82.05
LSTM + Glove embedding	300	83.60	85.50
LSTM + fasttext embedding	300	85.78	87.30

(latitude and Longitude basis). Content based location information is not useful in these type of studies because tweets only contain users views corresponding to different political parties as well as their leaders, however it can play a vital role during some emergency situation, crises etc. that contains location information in the tweets itself.

By using profile location information, we are able to collect 12,71,658 positive tweets and 5,04,976 negative tweets with respect to BJP (Bhartiya Janta Party) as well as 2,29,645 positive tweets and 2,92,650 negative tweets corresponding to INC (Indian National Congress) and the corresponding whole statistics are listed in Table 2. While on the basis of latitude and longitude information, we extracted only 657 tweets and 438 tweets, which constitute only 0.036% of total BJP tweets and 0.083% of total INC tweets. After performing state-wise (location based) tweets segregation, we calculated the ratio of positive and negative tweets corresponding to each political party (BJP and INC) as per given Eqs. 1 and 2.

$$pos\_twt = \frac{no.\ of\ positive\ tweets\ corresponding\ to\ each\ party}{number\ of\ total\ tweets} \tag{1}$$

$$neg\_twt = \frac{no.\ of\ negative\ tweets\ corresponding\ to\ each\ party}{number\ of\ total\ tweets} \tag{2}$$

Figure 2(a) shows the state wise individual views in the form of positive and negative with respect to BJP. It clearly shows that in 26 states BJP headed with the higher positive response from total 29 states and 7 Union Territories. A similar state-wise analysis is performed on tweets related to the INC and the corresponding results are demonstrated in Fig. 2(b). From Fig. 2(b), it can be seen that INC received a positive response from 21 states and Union Territories. As the objective of our current research work is to provide an overall estimate of people opinions on the two political parties (BJP and Congress), we further performed a comparative analysis of people opinions on the tweets related to two political parties and the corresponding results are demonstrated in Fig. 3. From

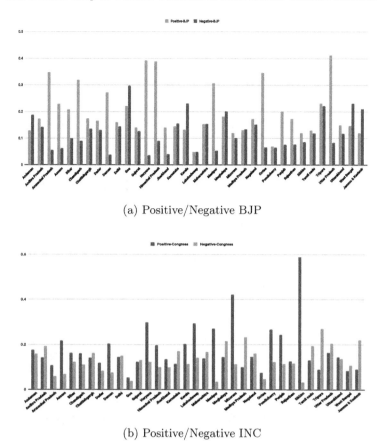

(a) Positive/Negative BJP

(b) Positive/Negative INC

Fig. 2. State-wise opinion analysis with respect to Bharatiya Janata Party (BJP) and Indian National Congress (INC)

the comparative experimental analysis, we came to the conclusion that 28 states and Union territory are in favour of BJP out of 29 states and 7 union territories, whereas 7 states are in favour of INC. In order to validate our research outcomes or election predictions, we have compared our results with the actual Lok Sabha election results[1]. From the comparative analysis, we have found that our results align with the actual outcomes with an absolute error of 0.12%.

For an instance our prediction results shows high positive opinions rate of BJP at several states and the corresponding actual outcomes are (Uttar Pradesh in which BJP won 62 seats out of 80 (62/80), Haryana (10/10), Himachal Pradesh (4/4), Arunachal Pradesh (2/2), Rajasthan (24/25), Chandigarh (1/1), Bihar (33/40), Delhi (7/7), Manipur (1/2), Gujarat (26/26), Madhya Pradesh (28/29), Maharastra (41/48), and Daman & Diu (1/1)) also evident/validates of our findings.

---

[1] http://www.elections.in/results/.

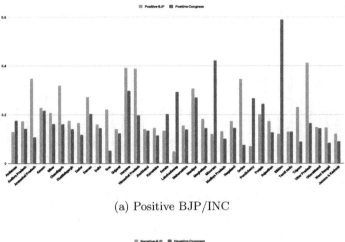

(a) Positive BJP/INC

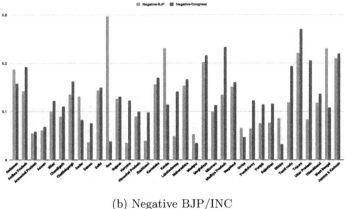

(b) Negative BJP/INC

Fig. 3. State-wise comparative analysis of opinions for Bharatiya Janata Party (BJP) and Indian National Congress (INC)

Whereas in case of INC also our results align with the final results by exhibiting positive response to INC for those states where INC acquire highest seat as compared to BJP i.e Andaman & Nicobar Islands (1/1), Kerala (15/20), Lakshadweep (1/1), Puducherry (1/1), Punjab (8/13), Tamilnadu (8/38). Furthermore, Mizoram and Sikkim are the two states in which INC received highest positive response as compared to the BJP but the seats are won by regional parties. The state-wise comparative analysis of negative response for both the political parties are demonstrated in Fig. 3.

Trending Hashtags w.r.t BJP and INC. People use different hashtags to represent ongoing trends at any particular point of time. So it can play a vital role over the different micro-blogs such as Twitter, Facebook, and Instagram etc. for identification and characterisation of the tweets. In the present work, we have identi-

fied the top trending hashtags corresponding to BJP (Bhartiya Janta Party) and
INC (Indian National Congress) as it can play a significant role during information diffusion. In our collected dataset, there were around 5532 different hashtags
used by BJP and INC Supporters out of which we selected only those hashtags
that were used by at least 10 ($\mathcal{T} = 10$) different peoples. There are 1561 hashtags which satisfied this threshold condition. After that, we have manually identified the top 20 positive and negative hashtags corresponding to two parties as
shown in Table 1. For example, the positive hashtags used for BJP are *('PhirEk-
BaarModiSarkar', 'ModiHaiToVikasHai', 'AayegaToModiHi', 'DeshKiPasand-
Modi', 'NamoAgain')* and negative are *('AbHogaNyay', 'NyayForEmpowerment',
'BJPJumlaManifesto', 'NyayYatra', 'ShamelessChowkidars')* whereas the positive hashtags used for INC are *('VoteForChang', 'AmethiKeDilMeinRahul',*

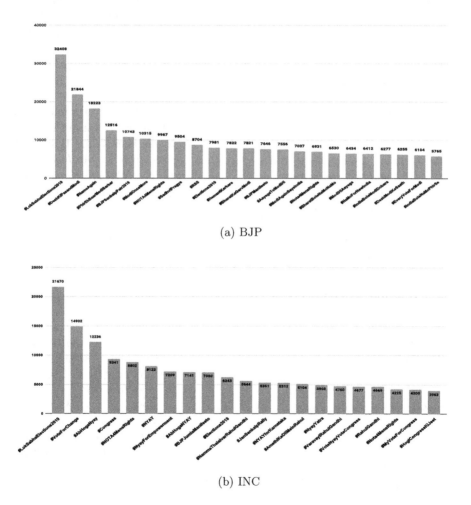

(a) BJP

(b) INC

Fig. 4. Top trending hashtags

'JanSankalpRally', *'VoteNyayVoteCongress'*, *'MyVoteForCongress'*) and negative are *('CongressTerror'*, *'CongressMuktBharat'*, *'CongressGundiHai'*, *'CongressGundiHai'*, *'corruptcongress'*).

Above hashtags were some examples of the positive and negative trending hashtags that were used during the election period. Furthermore, we have analysed a broad picture of both the parties separately by identifying the top 20 hashtags used by the individuals. To extract the top hashtags, we set a threshold value ($T = 500$) over the extracted 1561 selected hashtags and we left with 281 trending hashtags that have frequency more than 500. For the visualization purpose, we have selected 20 hashtags out of these extracted 281 hashtags and the selected 20 hashtags along with their frequency count value are depicted in Fig. 4(a). From the Fig. 4(a), it can be clearly seen that 14 out of the 20 BJP trending hashtags are related to Prime Minister of India i.e (*'DeshKiPasandModi'*, *'NamoAgain'*, *'PhirEkBaarModiSarkar'*, *'ModiONceMOre'*, *'BharatKaGarvModi'*, *'AayegaToModiHi'*, *'ModiAgainsaysIndia'*, *'BharatBoleNaMoNaMo'*, *'ModiHiAayega'*, *'NaMoForNewIndia'*, *'IndiaBoleModiDobara'*, *'DeshModiKSaath'*, *'EveryVoteForModi'*, *'IndiaBoleNamoPhirSe'*) and rest are related to Lok Sabha Election 2019. This clearly shows the influence of PMO Narendra Modi in all states.

The similar kind of analysis has been performed on the hashtags related to the INC and the corresponding top trending 20 hashtags are shown in Fig. 4(b). From the analysis of INC top trending hashtags, it has been found that most of the trending hashtags are related to other parties such as *'AbHogaNyay'*, *'NYAY'*, *'NyayForEmpowement'*, *'BJPJumlaManifesto'*, *'NYAYforKarnataka'*, *'VoteNyayVoteCongress'* and there were no trending hashtags (in top 10) related to any leader like Rahul Gandhi. So it may be one of the reasons to loose in Lok Sabha election as they didn't promote any leaders as other parties like BJP had done. Furthermore, it has also been found that there were no negative hashtags related to INC in BJP trending hashtags.

Top Trending @Mentions w.r.t BJP and INC. In this section, we have find out the top mentions which were tagged by BJP and INC supporters. Figure 5 (a) shows the trending mention related to BJP such as *'@BJP4India'*, *'@narendramodi'*, *'@AmitShah'*, *'@INCIndia'*, *'@RahulGandhi'*, *'@PMOIndia'*, *'@rsprasad'*, *'@ECISVEEP'*, *'@nsitharaman'*, *'@AamAadmiParty'* whereas in Fig. 5 (b) shows the trending mention related to INC such as *'@INCIndia'*, *'@RahulGandhi'*, *'@BJP4India'*, *'@narendramodi'*, *'@priyankagandhi'*, *'@AamAadmiParty'*, *'@ECISVEEP'*, *'@priyankac19'*, *'@PMOIndia'*, *'@AmitShah'*. We have also segregated the trending leaders in both of the parties such as @narendramodi, @AmitShah, @rsprasad, @nsitharaman, @arunjaitley, @PiyushGoyal, @nitin_gadkari, @smritiirani were from BJP. whereas from INC @RahulGandhi, @priyankagandhi, @JhaSanjay, @capt_amarinder, @divyaspandana, @digvijaya_28, @rssurjewala, @ShashiTharoor.

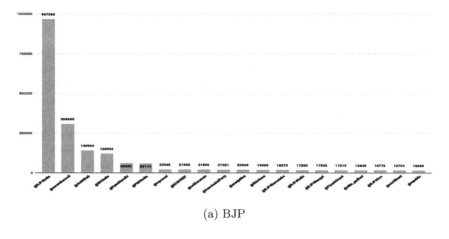

(a) BJP

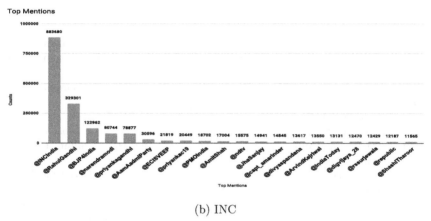

(b) INC

Fig. 5. Top trending @Mentions

4   Conclusion

Twitter has emerged as a successful platform for users' to share their opinions and views on various topics and events, information diffusion, and political activities, thus becoming a rich source of information. So, in the current study, we utilise this abundant source of information to predict the 2019 Indian Lok Sabha election results. The study initially performed hashtag based tweet segregation to generate training data for deep learning model. Subsequently, the trained model is then utilised to categorize tweets into their target political parties sentiment classes. The proposed model succeeds in achieving 87.30% classification accuracy. Finally, we have performed the spatial analysis of classified tweets to predict the state-wise election outcome. From the prediction results of the proposed approach, the following inferences are drawn:

- Twitter can be utilised as a reliable media for election results prediction in the Indian context.
- The amalgamation of deep learning model with fasttext word-embedding provides support for improved classification accuracy.
- The outcome of the comparative analysis of prediction results with actual results has shown that the proposed approach does accurate results prediction and has lower (0.12) absolute error.
- Lastly, from the analysis of top trending hashtags and mentions w.r.t each party, we found some interesting patterns which might be the conclusive factors to the 2019 Lok Sabha election outcomes. For example, most of the trending hashtags used by INC supporters correspond to the other parties, whereas BJP supporters have used hashtags corresponding to their party only.

References

1. Agarwal, A., Toshniwal, D.: Application of lexicon based approach in sentiment analysis for short tweets. In: 2018 International Conference on Advances in Computing and Communication Engineering (ICACCE), pp. 189–193, June 2018. https://doi.org/10.1109/ICACCE.2018.8441696
2. Agarwal, A., Gupta, B., Bhatt, G., Mittal, A.: Construction of a semi-automated model for FAQ retrieval via short message service. In: Proceedings of the 7th Forum for Information Retrieval Evaluation, pp. 35–38. ACM (2015)
3. Agarwal, A., Singh, B., Bedi, J., Toshniwal, D.: A datamining approach for emotions extraction and discovering cricketers performance from stadium to sensex. arXiv preprint arXiv:1809.00310 (2018)
4. Agarwal, A., Singh, R., Toshniwal, D.: Geospatial sentiment analysis using Twitter data for UK-EU referendum. J. Inf. Optim. Sci. **39**(1), 303–317 (2018)
5. Agarwal, A., Toshniwal, D.: Face off: travel habits, road conditions and traffic city characteristics bared using Twitter. IEEE Access **7**, 66536–66552 (2019)
6. Ahmed, S., Jaidka, K., Cho, J.: The 2014 Indian elections on Twitter: a comparison of campaign strategies of political parties. Telematics Inform. **33**(4), 1071–1087 (2016)
7. Bedi, J., Toshniwal, D.: Attribute selection based on correlation analysis. In: Rajsingh, E.B., Veerasamy, J., Alavi, A.H., Peter, J.D. (eds.) Advances in Big Data and Cloud Computing. AISC, vol. 645, pp. 51–61. Springer, Singapore (2018). https://doi.org/10.1007/978-981-10-7200-0_5
8. Burnap, P., Gibson, R., Sloan, L., Southern, R., Williams, M.: 140 characters to victory?: using Twitter to predict the UK 2015 general election. Electoral. Stud. **41**, 230–233 (2016)
9. Dwi Prasetyo, N., Hauff, C.: Twitter-based election prediction in the developing world. In: Proceedings of the 26th ACM Conference on Hypertext & Social Media, pp. 149–158. ACM (2015)
10. Ebrahimi, M., Yazdavar, A.H., Sheth, A.: Challenges of sentiment analysis for dynamic events. IEEE Intell. Syst. **32**(5), 70–75 (2017)
11. Goodfellow, I., Bengio, Y., Courville, A.: Deep Learning. MIT Press, Cambridge (2016)

12. Ibrahim, M., Abdillah, O., Wicaksono, A.F., Adriani, M.: Buzzer detection and sentiment analysis for predicting presidential election results in a Twitter nation. In: 2015 IEEE International Conference on Data Mining Workshop (ICDMW), pp. 1348–1353. IEEE (2015)
13. Jose, R., Chooralil, V.S.: Prediction of election result by enhanced sentiment analysis on Twitter data using word sense disambiguation. In: 2015 International Conference on Control Communication & Computing India (ICCC), pp. 638–641. IEEE (2015)
14. Joulin, A., Grave, E., Bojanowski, P., Mikolov, T.: Bag of tricks for efficient text classification. arXiv preprint arXiv:1607.01759 (2016)
15. Khatua, A., Khatua, A., Ghosh, K., Chaki, N.: Can# twitter_trends predict election results? evidence from 2014 Indian general election. In: 2015 48th Hawaii International Conference on System Sciences, pp. 1676–1685. IEEE (2015)
16. Mehndiratta, P., Sachdeva, S., Sachdeva, P., Sehgal, Y.: Elections again, Twitter may help!!! a large scale study for predicting election results using Twitter. In: Srinivasa, S., Mehta, S. (eds.) BDA 2014. LNCS, vol. 8883, pp. 133–144. Springer, Cham (2014). https://doi.org/10.1007/978-3-319-13820-6_11
17. Mikolov, T., Sutskever, I., Chen, K., Corrado, G.S., Dean, J.: Distributed representations of words and phrases and their compositionality. In: Advances in Neural Information Processing Systems, pp. 3111–3119 (2013)
18. Paul, D., Li, F., Teja, M.K., Yu, X., Frost, R.: Compass: spatio temporal sentiment analysis of us election what Twitter says! In: Proceedings of the 23rd ACM SIGKDD International Conference on Knowledge Discovery and Data Mining, pp. 1585–1594. ACM (2017)
19. Pennington, J., Socher, R., Manning, C.: Glove: global vectors for word representation. In: Proceedings of the 2014 Conference on Empirical Methods in Natural Language Processing (EMNLP), pp. 1532–1543 (2014)
20. Srivastava, R., Kumar, H., Bhatia, M., Jain, S.: Analyzing Delhi assembly election 2015 using textual content of social network. In: Proceedings of the Sixth International Conference on Computer and Communication Technology 2015, pp. 78–85. ACM (2015)
21. Subramaniam, L.V., Roy, S., Faruquie, T.A., Negi, S.: A survey of types of text noise and techniques to handle noisy text. In: Proceedings of The Third Workshop on Analytics for Noisy Unstructured Text Data, pp. 115–122. ACM (2009)
22. Tsakalidis, A., Papadopoulos, S., Cristea, A.I., Kompatsiaris, Y.: Predicting elections for multiple countries using Twitter and polls. IEEE Intell. Syst. 30(2), 10–17 (2015)
23. Wang, X., Wei, F., Liu, X., Zhou, M., Zhang, M.: Topic sentiment analysis in twitter: a graph-based hashtag sentiment classification approach. In: CIKM (2011)

Towards Sensing and Sharing Auditory Context Information Using Wearable Device

Akio Sashima$^{(\boxtimes)}$ and Mitsuru Kawamoto

Human Augmentation Research Center, National Institute of Advanced Industrial
Science and Technology (AIST), Kashiwa, Japan
sashima-akio@aist.go.jp
https://staff.aist.go.jp/sashima-akio/

Abstract. Data-driven information services using wearable devices have
attracted attention in the areas of healthcare, medical care, and educational services. In the services, the users' daily behaviors are modeled
with the sensing data of the physical statuses of individual users, e.g.,
body movements, heart rates, etc. However, to understand human behaviors more deeply, it is also important to know the context information of
the users, such as the surrounding environment and participating activities. In this paper, we describe extracting auditory context information
from ambient sound data sensed by smart watches. First, we describe a
prototype of our wearable ambient sound sensing system by using smart
watches. Then, we describe an analysis of the sound data sensed by the
system. We formalize the context extraction process as unsupervised segmentation of multi-dimensional time-series data and apply non-negative
matrix factorization (NMF) and k-means clustering to the segmentation
at the first step of the study. We confirm that the periods segmented by
the analysis roughly correspond to actual contexts.

Keywords: Smart watch · Auditory context · Non negative matrix
factorization

1 Introduction

Recently, wearable devices are important tools for recent data-driven information
services: healthcare, medical care, and educational services. For example, smart
watches (e.g., Fitbit [1], Apple Watch [2], etc.) can continuously sense users' activities, model the users' daily behaviors based on the sensing data to provide personalized information services. Currently, most of wearable devices are used for sensing
physical statuses of users, e.g., body movements, heart rates, etc. and have often
ignored to know background information of the behaviors, namely context information. However, to understand human behaviors more deeply, it is important to
know context information, such as the surrounding environment and participating activities. If context information is available by using wearable devices, it is
possible to create a new service based on deep understanding of the behaviors.

P. Cellier and K. Driessens (Eds.): ECML PKDD 2019 Workshops, CCIS 1168, pp. 54–59, 2020.
https://doi.org/10.1007/978-3-030-43887-6_5

For example, we have developed a wearable telecare system by using smart watches and wireless biosensors [3]. The service enables family members to share physiological information of cared persons in a peer-to-peer manner. If the family members can get the context information of their cared persons, it can be useful to understand the meaning of physiological data.

In this paper, we describe how to extract auditory context information from ambient sound data sensed by smart watches. Although researches of mobile audio sensing using smartphones have been proposed [4,5], researches using wearable devices to extract context information are few. As researches using wearable devices are still immature compared to smartphones, it is necessary to investigate the potential use of wearable sensing.

This paper is organized as follows: first, we show the prototype of our wearable ambient sound sensing system by using smart watches. Then, we describe an analysis of the sound data sensed by the system. As the first step of the study, we have applied non-negative matrix factorization (NMF) [6] and k-means clustering to unsupervised segmentation of time-series data. At last, based on the implementation and analysis, we discuss issues and future works of sensing and sharing context information by using wearable devices.

2   Wearable Ambient Sound Sensing System

We describe the prototype of our wearable ambient sound sensing system by using smart watches.

2.1   Wearable Device

The wearable ambient sound sensing system uses a commercially available smart watch: Polar M600 [7]. We have implemented the system as an application program on the Wear OS by Google^{TM}smartwatch OS.

The system has sensing and analyzing functions to extract auditory features from ambient sound data. A graphical user interface (GUI) to start/stop the sensing process is also implemented. The GUI using MPAndroidChart [8] shows the recording status of smart watch, such as an auditory feature vector. Currently, communication facilities to share the sensing data with other users are not implemented.

2.2   Auditory Sensing Data

Analyzing processes of the prototype system consist of three processes: sensing ambient sound, converting it to auditory feature vectors, and recording them.

In the sensing process, a microphone of the device senses ambient sound as 16kHz, 16bit, monaural audio format. Then in the converting process, the device immediately converts the data to audio feature vectors, Mel-Frequency Cepstrum Coefficients (MFCC), pitch (F0), and sound pressure (SPL). In the process, the system uses fast Fourier transform (FFT) to obtain the power spectrum of the signal, and calculates F0 and SPL. Then, to derive MFCC, it executes some steps transforming the spectrum so as to reflect human auditory characteristics (e.g., Mel-scale, vocal tract, etc.). We have implemented the sensing and converting processes by using an open source software: TarsosDSP [9]. In the recording process, it records the auditory data in its internal memory. We can retrieve the auditory data through a USB connection.

3   Extracting Context Information

Using our prototype system, we experimentally obtained auditory data of a user who come home from his office by bicycle. After obtaining the data, we have analyzed the data to investigate the feasibility of extracting context information. The auditory data has been analyzed offline.

In this study, we assume that the auditory contexts are latent, continuous periods characterized by auditory features. For example, if a user stayed at home until noon and then moved to a hospital by car, the home, car, and hospital are contexts of the user because they can be characterized by auditory features. As auditory data recorded with the system records is a multi-dimensional time-series data, it means that each context corresponds to a segment of the time-series classified by auditory features.

3.1   Segmentation of Multi-dimensional Time-Series Data

At the first step of the study, we have applied NMF and k-means clustering to segmentation of multi-dimensional time-series data.

The obtained data is 32 dimension time-series data. It represents MFCC, SPL, F0. The length is 48271. At first, we converted the data to the averaged data every 30 s, and normalized it so that it did not contain negative values. The normalized data is shown as a heat map in Fig. 1. In the heat map, the x-axis represents each feature of the data, the y-axis represents the time when the data was recorded. Each cell corresponds to a feature value at each time. Darker cells represent larger values and lighter cells represent smaller values.

As the data still contains artifacts which was caused by body movements. We applied NMF to factorize the data so as to extract important features and reduced the artifacts. In this experiment, the dimension of factorized data is 6. Figure 2 shows a heat map of the factorized data representing a change of the auditory features. The intensity of color indicates the degree of correlation with each factor.

NMF factorizes a matrix according to non-negative constraints. The bases tend to represent local features of the data, namely "parts." As the parts can better correspond to intuitive notions by humans, it is useful to visualize and understand auditory contexts. The bases of other methods, such as principal component analysis and vector quantization, do not always correspond to the intuitive notions [10].

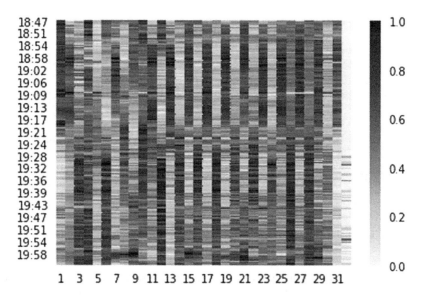

Fig. 1. A heat map of averaged data every 30 s of feature vectors obtained by wearable sensing system.

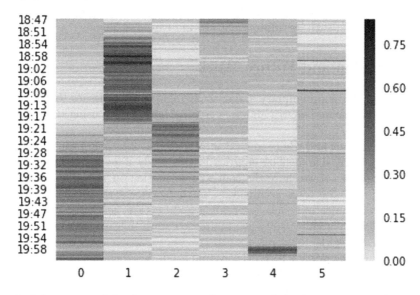

Fig. 2. A heat map of the feature vectors heat map of the feature vectors derived by using non-negative matrix factorization.

Then we classified each vector in the factorized data using k-means clustering $(k = 4)$. The clustering was based on Euclidean distance of the 6 features of each period.

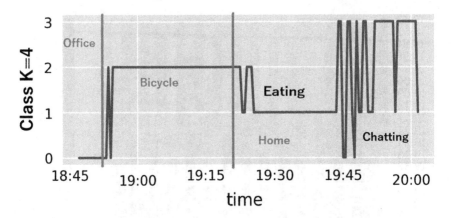

Fig. 3. Transitions of the auditory contexts classified by k-means clustering.

Figure 3 is a line graph that plotting the class labels, 0–3, assigned to each period along a time axis. We can see that consecutive periods tend to be classified into the same cluster. This means that we have succeeded in extracting temporally continuous acoustic features and segmenting the time-series data into four contextual periods. Additionally, the extracted features roughly correspond to actual contexts, such as riding a bicycle and eating at home.

4   Discussion and Future Work

This study is at the preliminary stage, so there are a lot of issues to realize the system that senses and shares auditory context information using the wearable devices.

An issue is when and where context information is extracted from the sensing data and how it is shared. The analysis in this study is performed offline on a personal computer. If we share the context information on wearable devices, we require to design the total architecture of the service system, which includes designing communication, data processing architecture, and user interactions. Designing efficient architecture with limited computation resources of wearable devices is future work.

Considering the privacy issue, the system is not designed to record raw sound data. So it is difficult to validate the correctness of extracted context information based on the raw data. To scale up the prototype to practical service, more sophisticated unsupervised (or sem-unsupervised) approaches should be required.

5   Conclusion

We have shown the prototype of our wearable ambient sound sensing system by using smart watches. We experimentally obtained ambient sound data sensed by the system and analyzed the data to extract context information. In the analysis, we formalized the context extraction process as an unsupervised segmentation of multi-dimensional time-series data and applied NMF and k-means clustering to the segmentation. We confirm that the periods segmented by the analysis roughly corresponded to actual contexts.

Acknowledgement. This study was partly funded by Japan Society for the Promotion of Science (grant number 17K00145) and JST CREST, JPMJCR18A4.

References

1. Fitbit: https://www.fitbit.com/home. Accessed 28 June 2019
2. Apple Watch: https://www.apple.com/watch/. Accessed 28 June 2019
3. Sashima, A., Kawamoto, M., Kurumatani, K.: A peer-to-peer telecare system using smart watches and wireless biosensors. Health Technol. **8**(5), 317–328 (2018). https://doi.org/10.1007/s12553-018-0240-8
4. Lu, H., Pan, W., Lane, N.D., Choudhury, T., Campbell, A.T. : SoundSense: scalable sound sensing for people-centric applications on mobile phones. In: Proceedings of the 7th International Conference on Mobile Systems, Applications, and Services, pp. 165–178, ACM, New York (2009)
5. Lane, N.D., Georgiev, P., Qendro, L.: DeepEar: robust smartphone audio sensing in unconstrained acoustic environments using deep learning. In: Proceedings of the 2015 ACM International Joint Conference on Pervasive and Ubiquitous Computing, pp. 283–294, ACM (2015)
6. Lee, D.D., Seung, H.S.: Algorithms for non-negative matrix factorization. In: Proceedings of the 13th International Conference on Neural Information Processing Systems, pp. 535–541 (2000)
7. Polar M600: https://www.polar.com/us-en/products/sport/M600-GPS-smart watch. Accessed 28 June 2019
8. MPAndroidChart: https://github.com/PhilJay/MPAndroidChart. Accessed 28 June 2019
9. TarsosDSP: https://github.com/JorenSix/TarsosDSP. Accessed 28 June 2019
10. Lee, D.D., Seung, H.S.: Learning the parts of objects by non-negative matrix factorization. Nature **401**, 788–791 (1999)

Workshop on Data Integration and Applications (DINA)

Noise Reduction in Distant Supervision for Relation Extraction Using Probabilistic Soft Logic

Birgit Kirsch[1]([✉]), Zamira Niyazova[1], Michael Mock[1], and Stefan Rüping[1,2]

[1] Fraunhofer IAIS, Sankt Augustin, Germany
{birgit.kirsch,zamira.niyazova,michael.mock,
stefan.ruping}@iais.fraunhofer.de
[2] Fraunhofer Center for Machine Learning, Sankt Augustin, Germany

Abstract. The performance of modern relation extraction systems is to a great degree dependent on the size and quality of the underlying training corpus and in particular on the labels. Since generating these labels by human annotators is expensive, *Distant Supervision* has been proposed to automatically align entities in a knowledge base with a text corpus to generate annotations. However, this approach suffers from introducing noise, which negatively affects the performance of relation extraction systems. To tackle this problem, we propose a probabilistic graphical model which simultaneously incorporates different sources of knowledge such as domain experts knowledge about the context and linguistic knowledge about the sentence structure in a principled way. The model is defined using the declarative language provided by *Probabilistic Soft Logic*. Experimental results show that the proposed approach, compared to the original distantly supervised set, not only improves the quality of such generated training data sets, but also the performance of the final relation extraction model.

Keywords: Probabilistic Soft Logic · Statistical Relational Learning · Distant Supervision · Relation Extraction · Natural Language Processing

1 Introduction

Relation extraction (RE) from text is a learning task with many real-world applications. While raw texts tend to be available in abundance, the bottleneck in practice usually is the availability of annotated (labeled) data, since manual annotation is costly. As a means to alleviate this problem, *Distant Supervision* (Mintz et al. 2009) has been proposed, which automatically labels training data, although at the cost of introducing noise by assigning wrong labels. In practice, again, manual intervention may be required to clean up the training data.

This paper proposes a novel probabilistic approach to reduce noise from distantly supervised training corpora. It incorporates several sources of domain experts and linguistic knowledge in a structured and principled way and models

P. Cellier and K. Driessens (Eds.): ECML PKDD 2019 Workshops, CCIS 1168, pp. 63–78, 2020.
https://doi.org/10.1007/978-3-030-43887-6_6

complex dependencies between training instances using *Probabilistic Soft Logic (PSL)* introduced by Bach et al. (2015). It aims at making better use of the domain experts time by letting them easily integrate valuable knowledge instead of working on the level of single examples.

Most of the existing approaches discussed in Sect. 2, directly integrate denoising-functionality in the training workflow and do not incorporate additional sources of knowledge. In practice it is therefore often necessary to manually denoise training data based on heuristics and hard constraints which are usually applied ad-hoc. To tackle these problems, we suggest a probabilistic approach to integrate heterogeneous sources of knowledge to denoise distantly supervised training samples. Our contribution can be summarized as follows:

1. We propose to integrate domain and linguistic knowledge to combine different heuristics and take into account inconsistencies of different *Named Entity Recognition (NER)*-models that are used to identify entity mentions with the objective to reduce noise that results from error-propagation. We model relational dependencies between possible relation mentions in the corpus based on semantic similarities in order to infer globally consistent training labels.
2. To integrate the knowledge in a structured and principled way, we derive a probabilistic graphical model, a *hinge-loss Markov random field (HL-MRF)*, using a declarative language provided by *PSL* (Bach et al. 2015). This language enables to define *HL-MRFs* in a compact way using comprehensible and easy to extend *First Order Logic Rules*.
3. We measure the influence of the proposed model on the quality of the training set and on downstream-task performance. Experiments are conducted on three publicly available data sets and results are compared with three baselines: *Distant Supervision*, a rule-based approach designed to mimic manual post-processing, and a *SRL* approach similar to *PSL*, a so called *Markov Logic Network (MLN)*. The results show that our probabilistic model outperforms the baselines and not only improves the quality of the training data set but also the performance of the final *RE*-model trained on the data.

The paper is structured as follows: Section 2 discusses related work. Section 3 gives an overview over *HL-MRFs* and the declarative language *PSL*. Section 4 describes the proposed model structure, Sect. 5 presents experimental results and in Sect. 6 we conclude with a summary and discuss future work.

2   Related Work

State-of-the-art information extraction approaches not only focus on identifying independent phrases in text that describe real entities, but also on predicting the semantic relationship between entities, which is referred to as *Relation Extraction (RE)*. The sentences below, for example, represent two mentions of the same relation r. The entity tuple (e_1, e_2) is bound by the relation $r(e_1, e_2) = founder\_of(LarryPage, Google)$.

> *"**Larry Page**, co-founder of **Google**, is now the CEO of Alphabet"*
> *"**Larry Page** and Sergey Brin founded **Google** in 1998"*

Most state-of-the-art relation classification models, such as (Christopoulou et al. 2019; Miwa and Bansal 2016), are based on deep learning architectures which require a large training corpus to generalize. With such data-intense model architectures the acquisition of annotated training data in sufficient size becomes the bottleneck of the overall training workflow. Craven et al. (1999) proposed an alternative paradigm, *Distant Supervision*, which enables to annotate a large training corpus without manual supervision by aligning relations present in an external knowledge base with a text corpus. It later was applied by Mintz et al. (2009) who used Freebase as a knowledge base to generate training data. Given, for example, a knowledge base that contains the relation triple $founder\_of(LarryPage, Google)$, *Distant Supervision* first identifies entity mentions in a text and then assumes that each sentence in which both entities are present is one mention of the relation $founder\_of$.

Following this assumption, the labels assigned to a sentence may often be incorrect, either because of the incompleteness of a knowledge base (Min et al. 2013) or due to a high number of false positive matches (Riedel et al. 2010). Facing this problem, Riedel et al. (2010) proposed *Multi-Instance learning*, which assumes that only at least one sentence containing the entities $e1$ and $e2$ is a representation of a relation present in the knowledge base. Hoffmann et al. (2011) extended the multi-instance model and suggests *Multi-Instance-Multi-Label (MIML)* learning, which assumes that one entity pair (e_1, e_2) can be a representation of multiple relations. Surdeanu et al. (2012) introduced *MIML-RE*, a graphical model to jointly model all entity-pair-instances and labels.

Recent approaches also integrate denoising functionality directly in the training process of neural models. Qin et al. (2018) apply *Deep Reinforcement learning* to automatically detect false positives and integrate these false positives as negative examples in the training set. Lin et al. (2016) employed *Convolutional Neural Networks* (CNN) and used the soft attention weights for all sentences to reduce the weights of noisy relations and Ye and Ling (2019) proposed to not only integrate intra-bag attention to assign weights to a group of sentences sharing the same entity pair but also inter-bag attention to assign weights to groups of sentences that are assigned the same relation label. Wu et al. (2018) use a neural architecture with a neural noise converter to capture the noise and a conditional optimal selector to predict relations based on a bag of sentences with the same entity pair. One alternative approach introduced by Ratner et al. (2016) enables users to define multiple heuristic functions, so called *labeling functions*, that each independently infer noisy labels.

3   Probabilistic Soft Logic

PSL was introduced by (Bach et al. 2015) and is a *statistical relational learning (SRL)* framework which enables to model a probability distribution over a set of random variables. Similar to most *SRL*-approaches, it can be divided in two major building blocks, a *Probabilistic Graphical Model (PGM)* which encodes the probability distribution and a declarative language to compactly define the PGM structure. A *Hinge-Loss - Markov Random Field (HL-MRF)* constitutes the first building block of *PSL* and represents an undirected graphical model which encodes distributions for continuous valued random variables. Given a set of observed variables $X = (X_1, ..., X_n)$, a set of random variables $Y = (Y_1, ..., Y_{n'})$, a set of potential functions $\phi = (\phi_1, ..., \phi_m)$ and a set of weights $\omega = (\omega_1, ..., \omega_m)$, a *HL-MRF* represents the following probability density function over Y conditioned on X:

$$P(Y|X) = \frac{1}{Z(\omega, X)} \exp[-\sum_{j=1}^{m} \omega_j \phi_j(X, Y)] \qquad (1)$$

with Z as a normalization factor. This distribution can be translated in a graph structure with each node representing one random variable. Each node is assumed to be conditionally independent of all nodes in the graph it is not directly connected to via an edge given the value of the nodes it is directly connected to. A subset of fully connected nodes is called a clique. Each potential ϕ_j takes the form of a hinge-loss function:

$$\phi_j(X, Y) = (\max\{l_j(X, Y), 0\})^{p_j} \qquad (2)$$

with l_j representing a linear function and $p_j \in \{1, 2\}$.

Potential functions ϕ_j are real-valued non-negative functions defined per clique and assign a probability mass to each clique state, meaning one assignment of values to all random variables participating in that clique. When a potential function assigns a higher value to one clique state, this state can be interpreted as being more probable than a clique state assigned a lower value.

PSL is a declarative language, which enables to generate templates for potential functions in form of First Order Logic rules. A *PSL*-model is composed of a set of these weighted rules that compactly describe an underlying *HL-MRF* structure. Formula 3 displays an example:

$$w : Friends(A, B) \land PlaysGolf(A) \implies PlaysGolf(B) \qquad (3)$$

with w as the weight assigned to the rule, which indicates its importance. *Friends* and *PlaysGolf* are called *predicates*. A and B are *variables* that serve as placeholders and can be substituted by concrete instances, referred to as *constants*. Given a set of rules and a set constants, the process in which all variables are replaced with these constants is called grounding. Each grounded rule forms a clique in the underlying graph structure representing the HL-MRF, while each grounded predicate participating in that rule represents a random variable mapped to a node in the clique. The weight ω_j of a potential ϕ_j assigned to a grounded rule is determined by the weight of the rule template. Each potential expresses a distance to satisfactions for the rule, which either takes the value 0 if the rule is fully satisfied or a distance value measuring the degree of satisfaction. Squaring a rule potential presented in Formula 2 and setting $p_j = 2$ results in a smoother trade off when trying to satisfy conflicting rules.

Intuitively spoken, assignments of Y are more probable the fewer rules they violate. The weight assigned to each rule template allows to control the influence of one violated rule on the overall probability.

4 HL-MRF Model for Noise Reduction

The following section is dedicated to the *PSL*-model and its structure to denoise a distantly supervised training data set. The *PSL*-model is designed to counter to three major sources of noise which are present in *Distant Supervision*.

Incompleteness of the Knowledge Base (Min et al. 2013)**:** *False negative* relations can be extracted when a given entity mention pair (m_1, m_2) does not have any corresponding relation triple in a knowledge base.

False Positive Relations (Riedel et al. 2010)**:** A sentence containing a relation mention (m_1, m_2) mapped to the relation triple $r(e_1, e_2)$ does not express this relation.

Error Propagation from *Named-Entity-Recognition (NER)-Models:* *NER*-systems applied to identify entity mentions do not deliver 100% accurate results which can lead to additional noise in the training set.

The model consists of several components, each of them addressing one or multiple of these noise sources which will be described in the following.

4.1 Prior Model

The prior model defines which values to assign to all unobserved ground atoms by default. Table 1 displays the rules, variables and predicates.

Rule (1) in Table 1 forces all ground atoms of $HasRel(z, r)$ for each relation mention candidate z and over all relation types r to sum to one. This allows to have only one label for each relation mention candidate. Rules (2) and (3) reflect that all ground atoms for $Ent1Type$ and $Ent2Type$ initially are assigned the truth value 0, meaning that the types for the named entity are not defined. Rule (4) demonstrates the prior belief that, without any further information, the relation type induced by *Distant Supervision* is assumed to be correct.

4.2 Consistency Between Predictions of NER Systems

Distant Supervision usually relies on entities extracted by *NER*-systems. These systems not only identify named entities but also tag them with an entity type according to a fixed set of types such as *PERSON, LOCATION* and *ORGANI-ZATION*. As stated in the problem definition, *NER*-systems applied to identify entity mentions and their entity types do not deliver 100% accurate results and falsely tag entity mentions that do not represent entities or tag them with a wrong type. This can lead to error propagation and the predicted relation types can be inconsistent with the extracted entity types.

To address this problem, we suggest to incorporate an unobserved hidden variable representing the true hidden entity type which is dependent on the

Table 1. Variables, predicates and rules used in prior *PSL*-model

Rules	
$HasRel(z, +r) = 1$	(1)
$!Ent1Type(z, t)$	(2)
$!Ent2Type(z, t)$	(3)
$DSCandRel(z, r) \rightarrow HasRel(z, r)^2$	(4)

Variables	
z - relation mention candidates	variable aligned with a set of constants representing each entity mention pair extracted from a sentence
r - relation type	variable with a set of constants for each relation class including the class *"None"*
t - entity type	t is a variable being substituted with a set of constants representing each entity type

Observed Predicates	
$DSCandRel(z, r)$	takes the value 0 if z is mapped to r according to Freebase, 0 otherwise.

Unobserved Predicates	
$Ent1Type(z, t), Ent2Type(z, t)$	hidden variables reflecting the unknown true entity type t of the entity mention in z
$HasRel(z, r)$	represents target variable that indicates if relation mention candidate z is assigned relation type r

predictions of multiple *NER-systems*. This can be seen as a form of Multiview Learning (Blum and Mitchell 1998). In this case we used the pretrained *NER taggers* provided by *Stanford Corenlp Tool*[1] and *SpaCy*[2]. Additionally we integrated a simple heuristic pattern in order to raise the probability of an entity mention being a true mention by checking if the first letter of an entity is capitalized. Table 2 displays the corresponding predicates and rules. These rules raise the probability of an entity in a relation mention z to be of type t, if both *NER*-taggers induce the type t and the first letter of the entity mention is capitalized. This leads to lower truth values assigned to ground atoms of *Ent1Type* and *Ent2Type*, when predictions of the taggers are inconsistent.

4.3 Sentence Structure Analysis

A variety of *RE* methods incorporate syntactic information to classify relations (for example (Bunescu and Mooney 2005)). One common technique in *NLP* is to extract the grammatical structure and relationships between words of a sentence, incorporate these information into a dependency tree and extract features based on that tree. Figure 1 displays such a dependency tree extracted with *SpaCy*. Words in the sentence are associated with graph nodes, dependencies are represented as directed edges with assigned dependency labels. A dependency path is considered as the concatenation of dependency edges and nodes along a path

---

[1] http://nlp.stanford.edu/software/corenlp.shtml.

[2] An open-source library for Natural Language Processing, https://spacy.io/.

Table 2. Entity rules and predicates

Rules

entity	$Ent1Capital(z) \wedge Ent1TypeSt(z,t) \rightarrow Ent1Type(z,t)^2$ (1)
	$Ent2Capital(z) \wedge Ent2TypeSt(z,t) \rightarrow Ent2Type(z,t)^2$ (2)
	$Ent1Capital(z) \wedge Ent1TypeSp(z,t) \rightarrow Ent1Type(z,t)^2$ (3)
	$Ent2Capital(z) \wedge Ent2TypeSp(z,t) \rightarrow Ent2Type(z,t)^2$ (4)

Observed Predicates

$Ent1TypeSt(z,t)$, Ent2TypeSt(z,t)	is assigned 1 when first (second) entity mention contained in z is predicted to be of entity type t by Stanford NER tagger
$Ent1TypeSp(z,t)$, Ent2TypeSp(z,t)	is assigned 1 when first (second) entity mention contained in z is predicted to be of entity type t by Spacy NER tagger
$Ent1CapitalLetter(z)$, Ent2CapitalLetter(z)	is assigned 1 when first (second) named entity present in a relation mention candidate z contains a capital letter

in the dependency graph. The words in the shortest path between the entities "Rana" and "Pakistan" are marked in bold and indicate a *place of birth* relation between the two entities.

Rana was **born in Pakistan** but eventually became a Canadian citizen.

Fig. 1. Dependency graph extracted by spaCy

To integrate these syntactic information in the model, we suggest the rule set displayed in Table 3. Rules (1) and (2) build upon the work of Bunescu and Mooney (2005), who presented a kernel method based on the shortest dependency path between two entities in the dependency graph assuming that the shortest path can capture all necessary information for relation extraction. Rule (1) increases the probability that a relation mention candidate z labeled as r by *Distant Supervision* really is of relation type r (excluding the "None" relation type) when the length of the shortest dependency is small. Rule (2) increases the probability that a relation mention candidate z labeled as r by *Distant Supervision* is of relation type "None" when the dependency path is long or there is no direct path.

Rules (3) and (4) are inspired by a hypothesis stated in (Chklovski and Pantel 2004), that *"verbs are the primary vehicle for describing events and expressing relations between entities"*. Rule (3) increases the probability of a relation mention z being of relation type r when labeled as r by *Distant Supervision* and when both entities e_1 and e_2 are connected through a predicate in the shortest dependency path. Rule (4), in contrast lowers the probability of a relation mention z being of relation type r when labeled as r by *Distant Supervision* when the verb that is closest in dependency path to e_1 and e_2 is not the same.

Table 3. Syntactic rules for the proposed PSL model.

Observed Predicates

$HasDirectDependency(z)$	assigned 1 when entity mentions present in z are tagged with object- and subject- dependency labels
$DependencyPathLength(z)$	dependency path length normalized using the *minmax normalization*
$HasSimilarEntityVerbs(z)$	assigned 1 when verb closest to first entity mention equals verb closest to second entity mention in the dependency path of z

Rules

synPath	$DSCandRel(z,r) \wedge r \neq None \wedge DependencyPathLength(z) \rightarrow HasRel(z,r)^2$	(1)
	$DSCandRel(z,r) \wedge !DependencyPathLength(z) \rightarrow HasRel(z,'None')^2$	(2)
synVerb	$DSCandRel(z,r) \wedge r \neq None \wedge HasSimilarEntityVerbs(z) \rightarrow HasRel(z,r)^2$	(3)
	$DSCandRel(z,r) \wedge !HasSimilarEntityVerbs(z) \rightarrow HasRel(z,'None')^2$	(4)
synStruct	$DSCandRel(z,r) \wedge r \neq None \wedge HasDirectDependency(z) \rightarrow HasRel(z,r)^2$	(5)
	$DSCandRel(z,r) \wedge !HasDirectDependency(z) \rightarrow HasRel(z,'None')^2$	(6)

Rules (5) and (6) assume the likelihood that two entities participate in a relation is higher when they are tagged with object- or subject-dependency labels.

Integrating the *DSCandRel*-predicate in the body of each rule forces it to be automatically satisfied when a relation mention candidate is not assigned a relation type r by *Distant Supervision*. With that, all rules only contribute to the objective of lowering the noise caused by false positive training samples.

4.4   Context-Based Constraints

Distant Supervision is context-independent and only maps entity mentions identified in a sentence to relation tuples in a knowledge base. Thus, the induced relation types often either do not correspond to the semantic meaning of a sentence or the relation type is not consistent with the entity types participating in the relation. Consider, for example, the following sentence:

"**Gray** moved to the nearby suburb with **Johnston**"

The sentence might be aligned with a relation *place_of_birth*("Gray", "Johnston") present in a knowledge base. When taking into account that the entity type of Johnston is *PERSON* and looking at the context containing the word "moved" it is obvious, that this sentence should not be labeled as a *born_in* relation. We suggest to incorporate this knowledge as displayed in Table 4.

Rules (1) and (2) enable domain experts to specify a set of words that indicate the presence of a specific relation type. Rule (1) raises the probability of z being assigned the "None" relation type, if the shortest dependency path does not contain one of the specified words. Rule (2) raises the probability of z being assigned the *place_of_birth* relation type when one of the words "born, birth or native" appear in the shortest dependency path between entity mentions present in z and the entity types are consistent with the relation type.

Table 4. Context-based rules, predicates and variables

Variables

s - shortest dependency path	variable representing words assigned to the shortest dependency path
w - set of words	variable representing a set of words

Observed Predicates

$HasWord(s, w)$	$=1$ when s contains one word present in w
$ShortDependPath(z, s)$	$=1$ when words assigned to shortest dependency path between entity mentions in z are equal to s

Rules

conFP	$DSCandRel(z, place\_of\_birth) \wedge ShortDependPath(z, s) \wedge$ $\wedge !HasWord(s,'born, birth, native') \rightarrow HasRel(z, None)^2$	(1)
conPred	$Ent1Type(z, PERSON) \wedge Ent2Type(z, LOCATION) \wedge ShortDependPath(z, s) \wedge$ $\wedge HasWord(s,'born, birth, native') \rightarrow HasRel(z, place\_of\_birth)^2$	(2)
conSem	$DSCandRel(z, place\_of\_birth) \wedge !Ent1Type(z, PERSON) \rightarrow !HasRel(z, r)^2$	(3)

Rule (3) lowers the probability of z being assigned the relation type r when the entity types of the entity mentions in r are not consistent with the relation type. In the concrete example, when the first entity mention is not of type *PERSON* it is less likely that z is assigned the *place_of_birth_relation*.

Both Rules (1) and (3) remove false positive examples from the training set, while Rule (2) compensates incompleteness of a knowledge base.

4.5 Semantic Similarity in Noise Reduction

Recently, multiple research works focused on using semantic similarity to perform *RE* (Park et al. 2016; Ru et al. 2018; Grycner et al. 2014). Common approaches for example define metrics to measure similarities between relations and then use them to create clusters for each relation type. This follows the intuition that, if two relation mentions have the same semantic meaning, they are likely to have the same relation type. Thus, we introduce two similarity predicates, explained in Table 5 that calculate a similarity between two relation mention candidates, either based on their dependency path or based on the verb assigned to the relation according to the dependency path.

The verb similarity in this case is determined by calculating the cosine similarity of pre-trained word embeddings provided by (Mikolov et al. 2013). The dependency path similarity is calculated using *WordNet* (Mihalcea et al. 2006).

We suggest to integrate these similarities according to the rule set displayed in Table 5. Rule (1) raises the probability that two relation mention candidates have the same relation type when both are assigned a similar verb according to the dependency path. Rules (2) and (3) force the verb similarity to be transitive and symmetrical. Rules (4), (5) and (6) equivalently model this dependency based on the similarity of the shortest dependency path.

While all rules introduced in the previous subsections only model dependencies between a relation mention candidate, the according entity types and

Table 5. Similarity rules, predicates and variables

Variables

z_1, z_2	relation mention candidates

Similarity Predicates

$SimilarVerb(z_1, z_2)$	assigned value indicates similarity of the verbs present in dependency path of $z1$ and $z2$
$SimilarDependPath(z_1, z_2)$	assigned value indicates similarity of shortest dependency paths extracted from z_1 and z_2

Rules

simVerb	$HasRel(z_1, r) \wedge r \neq None \wedge SimilarVerb(z_1, z_2) \wedge (z_1 \neq z_2) \rightarrow HasRel(z_2, r)^2$	(1)
	$SimilarVerb(z_1, z_2) \wedge SimilarVerb(z_2, z_3) \wedge (z_1 \neq z_3) \rightarrow SimilarVerb(z_1, z_3)^2$	(2)
	$SimilarVerb(z_1, z_2) = SimilarVerb(z_2, z_1)$	(3)
	$HasRel(z_1, r) \wedge r \neq None \wedge SimilarDepenPath(z_1, z_2) \wedge (z_1 \neq z_2) \rightarrow HasRel(z_2, r)^2$	(4)
simPath	$SimilarDepenPath(z_1, z_2) \wedge SimilarDepenPath(z_2, z_3) \wedge (z_1 \neq z_3) \rightarrow SimilarDepenPath(z_1, z_3)^2$	(5)
	$SimilarDepenPath(z_1, z_2) = SimilarDepenPath(z_2, z_1)$	(6)

additional local features, these rules directly incorporate relational dependencies between the target predicates. When performing inference, this forces the model to jointly infer relation types and entity types in order to achieve global consistency. At the same time, this compensates knowledge base incompleteness and reduces the amount of false positive training examples.

5   Experimental Evaluation

This section describes the experiments performed to evaluate the performance of our *PSL* model and presents the final results available on Github[3]. We conduct two types of experiments. The first experiment measures the influence of the proposed model on the quality of the training set. In order to analyze the importance of each rule set (as presented in the different subsections of Sect. 4) individually, we perform the same experiment with different rule subsets. The second experiment measures the influence of the proposed model on the downstream-task performance by using the denoised training corpus to train a RE model using a state-of-the RE framework *CoType* (Ren et al. 2017).

5.1   Experimental Setup: Data and Models

The proposed method is evaluated using the following three different public[4] data sets. In each data set 9 relation types including *"None"* are considered.

KBP Data Set: This data set, also used by Ren et al. (2017), contains a manually annotated set with sentences from the 2013 KBP corpus (Ellis et al. 2012) and the Wiki-KBP corpus, (Ling and Weld 2012), which was constructed via distant supervision by aligning Freebase relations with sentences from English

---

[3] The code is available under: https://github.com/DSDenoisingPSL/DSDenoisePSL.

[4] The data set can be downloaded from https://github.com/shanzhenren/CoType
https://code.google.com/archive/p/relation-extraction-corpus/downloads.

Wikipedia articles. The 2013 KBP corpus is used as a test set and the WIKI-KBP is used as a training corpus for the final relation extraction task.

New York Times News Corpus (NYT): This data set, provided by Riedel et al. (2010), was generated by aligning Freebase relations with sentences from the New York Times news corpus. It includes a test set with manually annotated sentences Hoffmann et al. (2011).

Google Corpus: This data set was released by Google (Sun et al. 2013) and consists of sentences sampled from Wikipedia, aligned by Freebase and judged by humans. Thus, it does not suffer from incompleteness but may contain false positive relations. In order to reduce recall, which was 100 percent, noise was added by increasing the number of false-negative relations and randomly setting labels obtained from Freebase to the *None* type.

In the experiments, the cosine similarity of word vectors was calculated using the Vector Embedding Utility Package provided by Patel et al. (2018) and a pre-trained word2vec-model[5].

5.2   Experimental Setup: Benchmark Methods

The *PSL* model is compared to three baseline models.

Brute-Force model (BF): This model represents a baseline with a set of hard constraints applied to filter the training set. Therefore, each *PSL*-rule was transformed into a constraint using if-else-conditions. This mimics a manual approach, where an expert adds different post-processing after performing *Distant Supervision*, but without any structured way to integrate the different evidences, as it is possible with *PSL*.

Markov Logic Net (MLN): With the objective to evaluate, that *PSL* is a sufficient approach to generate a relational model with the proposed structure, we compared it to a *Markov Logic Network*. Similar to *PSL*, it provides a logic based declarative language to describe an underlying probabilistic graphical model. Thus, the rules described in Sect. 4 can be used to define the model structure. The main difference to *PSL* is that a *MLN* model grounds out to a *Markov Random Field*, in which all random variables take boolean values. Therefore, all observed continuous values, such as the similarities, are converted to boolean values by rounding them. For inference in the *MRF* model the open-source package Tuffy[6] (Niu et al. 2011) was used.

Distant Supervision: Represents data generated by *Distant Supervision*.

5.3   Experimental Results

Training Set Quality: In the first experiment, the performance of the proposed noise reduction method is estimated by measuring the quality of the denoised

---

[5] Can be downloaded under https://code.google.com/archive/p/word2vec/.

[6] http://i.stanford.edu/hazy/tuffy/.

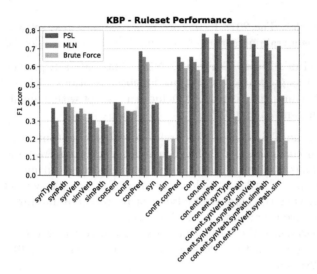

Fig. 2. F1-Score calculated for different rule combinations.

data sets. Therefore, precision, recall and F1-score are calculated on each test data set after the three models, *PSL*, *MLN* and *BF*, have been applied. Figure 2 prototypically displays the F-scores for the *PSL*, *MLN* and *BF* models obtained when applying the different rule combinations to the *KBP* corpus. The plot illustrates F-scores for 19 different rule sets. From left to right, the first 8 rule sets, labeled according to the descriptions in Sect. 4, represent distinct knowledge types, while the last represent the combination of rule sets. For convenience, rule set combinations {*conPred, conFP, conSem*}, {*simVerbs, simPath*} and {*synVerbs, synPath, synType*} denote the *con, sim* and *syn* rule sets.

The results suggest that *PSL* performs at least as good as the baselines and turns out to obtain significantly higher F-scores compared to the brute-force approach when the combined rule sets are incorporated. The *MLN* approach achieves a good performance when multiple rule sets are combined and only the similarity-rule set reduces its F-score significantly. This effect can be explained when looking at the differences between *MLN* and *PSL*: *MLN* uses boolean values for the predicates and this approach does not allow to incorporate a degree of similarity between instances such as *PSL*.

The *KBP* and *NYT* data sets suffer from low recall and each method increases recall significantly. However, the *BF* approach produces lower recall when compared with *PSL* and *MLN*.

For the Google data set the results are different, since the data set is not influenced by low recall values, but it may contain false positive relations. There is no significant improvement when incorporating the *context* rule set. Moreover, the syntax rules *synVerb* significantly decrease the performance of all three approaches. One reason for the degradation is that the Google data set contains mostly a set of sentences where two entities do not belong to the same sentence. Therefore, the short dependency path cannot be captured by a parser.

The results for the first experiment are summarized in Table 6, where those rule sets were chosen which produce the best F-scores for the three approaches and the three data sets. While for two of the three data sets (*KBP* and *NYT*) the *PSL* approach obtains remarkably higher F-scores, for the *Google* corpus *PSL* only performs slightly better than the other models. However, the precision and recall obtained with the *PSL* model are always higher.

Table 6. Performance of the *PSL*, *MLN* and the *BF* model.

Data	Method	Precision	Recall	F1	TP	FP
KBP	Freebase	**0.840**	0.265	0.403	79.0	**15.0**
	$PSL_{con,ent}$	0.806	**0.758**	**0.781**	**220**	53
	$BruteForce_{conPred}$	0.708	0.559	0.625	151	62
	$MLN_{con,ent,synVerb,synPath}$	0.834	0.713	0.769	207	41
NYT	Freebase	0.246	0.251	0.248	62	190
	$PSL_{con,synVerb,synPath,simPath,ent}$	**0.801**	**0.553**	**0.654**	**141**	**35**
	$BruteForce_{conFP,conPred}$	0.746	0.475	0.581	118	40
	$MLN_{con,synPath,ent}$	0.713	0.539	0.614	137	55
Google	Freebase	**0.948**	0.485	0.642	1286	**70**
	$PSL_{con,ent}$	0.931	0.72	**0.8122**	1890	139
	$BruteForce_{conPred}$	0.784	**0.814**	0.799	**2030**	557
	$MLN_{con,ent,synPath}$	0.854	0.548	0.668	1371	234

The results in Figure 2 show that the *PSL* model outperforms the *BF* model over a wide variety of rule sets. The *MLN* model performs worse when the similarity rules are integrated, while the performance of the *BF* model decreases when more rules are added.

Results for Relation Extraction: This subsection describes the results obtained by the second experiments. In a first step, *MLN*, *PSL* and *BT* models are applied to generate a training set. Each model is generated using (1) a context rule sets, (2) the combination of context and entity rule sets and (3) a combination of context, entity and syntax rule sets. In a second step, this training set is used to train a *RE* model with *CoType*. The performance of the trained *RE* model is evaluated based on the different model settings.

Figure 3 shows the F-curves generated using different thresholds during training of the relation extraction model with *CoType*. Figure 3 depicts that for both data sets, *KBP* and *NYT*, the model obtained from the data set processed by *PSL* and *MLN* performs better than the model trained on the original data set, which is produced by aligning the sentences with Freebase. Moreover, the *BF* approach performs undoubtedly poor and decreases the performance of the relation extraction model when trained on the *KBP* data set. Reason could be that

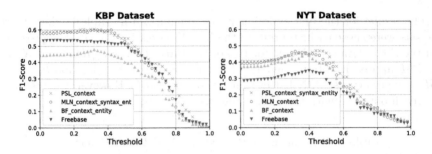

Fig. 3. F1-score achieved with CoType applying different thresholds.

the brute force mechanism predicts numerous false positive relations and this effect influences the final result. The results for the Google data set is different. There is only a slight improvement due to the fact that this data set does not suffer from incompleteness. Therefore, the results are not improved considerably, but the model trained on the data set preprocessed using *PSL* performs better with all thresholds. The results from the second experiment demonstrate that the proposed approach improves the overall performance of the RE model.

6   Conclusions and Outlook

This paper proposes a novel probabilistic approach to denoise training corpora generated for Relation Extraction with *Distant Supervision*. We derive a probabilistic graphical model which incorporates additional knowledge, models relational dependencies between training instances and takes into account consistency between different *NER*-systems. The model structure of a *HL-MRF* is described using a declarative first-order logic based language provided by *PSL*.

The effectiveness of this model was evaluated in two experiments, one to measure the quality of the training set after denoising with the proposed *PSL*-model and one measuring the end performance of a *RE* model trained on the denoised data set. Results are compared with the performance of a *RE* model on the original distantly supervised set and with the performance of two baseline models, a brute-force approach which mimics a manual post-processing by an expert and a Markov Logic Network. The experimental results show that the *PSL*-model outperforms the two baseline models in both experiments and suggest that it not only improves the quality of the training data set generated by *Distant Supervision*, but also the performance of the final *RE* model.

Although the experiments show promising results, further research has to be conducted to compare the performance to end-to-end relation extraction models that incorporate denoising directly in the training process (Ye and Ling 2019; Wu et al. 2018). Additionally, we will focus on validating that the probabilistic model can improve results when applied as a plug-in component combined with an arbitrary end-to-end *RE* model.

Acknowledgements. Funded by the German Federal Ministry of Education and Research under the Competence Center Machine Learning Rhein/Ruhr ML2R (FKZ 01S18038B).

References

Bach, S.H.: Hinge-loss Markov random fields and probabilistic soft logic: a scalable approach to structured prediction. J. Mach. Learn. Res. **18**, 109:1–109:67 (2015)

Blum, A., Mitchell, T.: Combining labeled and unlabeled data with co-training. In: Proceedings of COLT 1998, pp. 92–100. ACM, New York (1998)

Bunescu, R.C., Mooney, R.J.: A shortest path dependency kernel for relation extraction. In: Proceedings of the Conference on Human Language Technology and Empirical Methods in Natural Language Processing, pp. 724–731. Association for Computational Linguistics (2005)

Chklovski, T., Pantel, P.: Verbocean: mining the web for fine-grained semantic verb relations. In: Proceedings of the 2004 Conference on Empirical Methods in Natural Language Processing (2004)

Christopoulou, F., Miwa, M., Ananiadou, S.: A walk-based model on entity graphs for relation extraction (2019)

Craven, M., Kumlien, J., et al.: Constructing biological knowledge bases by extracting information from text sources. In: ISMB, pp. 77–86 (1999)

Ellis, J., Li, X., Griffitt, K., Strassel, S., Wright, J.: Linguistic resources for 2013 knowledge base population evaluations. In: TAC (2012)

Grycner, A., Weikum, G., Pujara, J., Foulds, J., Getoor, L.: A unified probabilistic approach for semantic clustering of relational phrases. In: AKBC 2014 (2014)

Hoffmann, R., Zhang, C., Ling, X., Zettlemoyer, L., Weld, D.S.: Knowledge-based weak supervision for information extraction of overlapping relations. In: Proceedings of 49th Annual Meeting of the Association for Computational Linguistics: Human Language Technologies, pp. 541–550. Association for Computational Linguistics (2011)

Lin, Y., Shen, S., Liu, Z., Luan, H., Sun, M.: Neural relation extraction with selective attention over instances. In: Proceedings of the 54th Annual Meeting of the Association for Computational Linguistics (Volume 1: Long Papers), vol. 1, pp. 2124–2133 (2016)

Ling, X., Weld, D.S.: Fine-grained entity recognition. In: AAAI, vol. 12, pp. 94–100 (2012)

Mihalcea, R., Corley, C., Strapparava, C., et al.: Corpus-based and knowledge-based measures of text semantic similarity. In: AAAI, vol. 6, pp. 775–780 (2006)

Mikolov, T., Chen, K., Corrado, G., Dean, J.: Googlenews-vectors-negative300.bin.gz - efficient estimation of word representations in vector space. arXiv preprint arXiv:1301.3781. https://code.google.com/archive/p/word2vec/ (2013)

Min, B., Grishman, R., Wan, L., Wang, C., Gondek, D.: Distant supervision for relation extraction with an incomplete knowledge base. In: Proceedings of 2013 Conference of the North American Chapter of the Association for Computational Linguistics: Human Language Technologies, pp. 777–782 (2013)

Mintz, M., Bills, S., Snow, R., Jurafsky, D.: Distant supervision for relation extraction without labeled data. In: Proceedings of Joint Conference of the 47th Annual Meeting of the ACL and the 4th International Joint Conference on Natural Language Processing of the AFNLP: Volume 2, pp. 1003–1011 (2009)

Miwa, M., Bansal, M.: End-to-end relation extraction using LSTMs on sequences and tree structures. In: Proceedings of the 54th Annual Meeting of the Association for Computational Linguistics (Volume 1: Long Papers) (2016)

Niu, F., Ré, C., Doan, A., Shavlik, J.: Tuffy: scaling up statistical inference in Markov logic networks using an RDBMS. Proc. VLDB Endow. 4(6), 373–384 (2011)

Park, Y., Kang, S., Seo, J.: Information extraction using distant supervision and semantic similarities. Adv. Electr. Comput. Eng. 16(1), 11–18 (2016)

Patel, A., Sands, A., Callison-Burch, C., Apidianaki, M.: Magnitude: a fast, efficient universal vector embedding utility package. In: Proceedings of the 2018 Conference on EMNLP: System Demonstrations, pp. 120–126 (2018)

Qin, P., Xu, W., Wang, W.Y.: Robust distant supervision relation extraction via deep reinforcement learning. In: Proceedings of the 56th Annual Meeting of the Association for Computational Linguistics (Volume 1: Long Papers), Melbourne, Australia, pp. 2137–2147. ACL, July 2018

Ratner, A.J., De Sa, C.M., Wu, S., Selsam, D., Ré, C.: Data programming: creating large training sets, quickly. In: Lee, D.D., Sugiyama, M., Luxburg, U.V., Guyon, I., Garnett, R. (eds.) Advances in Neural Information Processing Systems, vol. 29, pp. 3567–3575. Curran Associates, Inc. (2016)

Ren, X., Wu, Z., He, W., Qu, M., Voss, C.R., Ji, H., Abdelzaher, T.F., Han, J.: CoType: joint extraction of typed entities and relations with knowledge bases. In: Proceedings of 26th International Conference on World Wide Web, pp. 1015–1024. International World Wide Web Conferences Steering Committee (2017)

Riedel, S., Yao, L., McCallum, A.: Modeling relations and their mentions without labeled text. In: Balcázar, J.L., Bonchi, F., Gionis, A., Sebag, M. (eds.) ECML PKDD 2010. LNCS (LNAI), vol. 6323, pp. 148–163. Springer, Heidelberg (2010). https://doi.org/10.1007/978-3-642-15939-8_10

Ru, C., Tang, J., Li, S., Xie, S., Wang, T.: Using semantic similarity to reduce wrong labels in distant supervision for relation extraction. Inf. Process. Manage. 54(4), 593–608 (2018)

Sun, S., Lao, N., Gupta, R., Orr, D.: 50,000 lessons on how to read: a relation extraction corpus (2013). https://research.googleblog.com/2013/04/50000-lessons-on-how-to-read-relation.html. Accessed 15 Jan 2019

Surdeanu, M., Tibshirani, J., Nallapati, R., Manning, C.D.: Multi-instance multi-label learning for relation extraction. In: Proceedings of the 2012 Joint Conference on Empirical Methods in Natural Language Processing and Computational Natural Language Learning, pp. 455–465. Association for Computational Linguistics (2012)

Wu, S., Fan, K., Zhang, Q.: Improving distantly supervised relation extraction with neural noise converter and conditional optimal selector (2018)

Ye, Z.X., Ling, Z.H.: Distant supervision relation extraction with intra-bag and inter-bag attentions. In: NAACL-HLT (2019)

Privacy-Preserving Record Linkage to Identify Fragmented Electronic Medical Records in the All of Us Research Program

Abel N. Kho[1]([✉]), Jingzhi Yu[1], Molly Scannell Bryan[2,3], Charon Gladfelter[1], Howard S. Gordon[2,3], Shaun Grannis[4], Margaret Madden[1], Eneida Mendonca[4], Vesna Mitrovic[1], Raj Shah[5], Umberto Tachinardi[4], and Bradley Taylor[6]

[1] Northwestern University, Evanston, IL 60611, USA
akho@nm.org
[2] University of Illinois at Chicago, Chicago, IL 60612, USA
[3] Veterans Affairs Medical Center, Chicago, IL 60612, USA
[4] Regenstrief Institute, Indianapolis, IN 46202, USA
[5] Rush University, Chicago, IL 60612, USA
[6] Medical College of Wisconsin, Milwaukee, WI 53226, USA

Abstract. As part of a national study in the United States to recruit one million Americans (All of Us Research Program) and their Electronic Health Record data, we set out to determine the degree to which care is fragmented across a sample of participating health provider organizations (HPOs). We distributed a previously validated Privacy-Preserving Record Linkage (PPRL) tool to participating sites to generate a unique set of keyed encrypted hashes for seven participating institutions across three States in the Upper Midwest of the U.S. An honest broker received the resulting encrypted hashes to identify patients with the same encrypted hashes shared across any combination of more than one institution as a proxy for patients receiving care across institutions. Out of 5,831,238 individuals, we identified 458,680 patients with data at more than one institution. Care fragmentation varied significantly by State and by Institution ranging from 6.1% up to 32.7%. Patients with fragmented care were more likely to be black (11.8% vs 10.8%), and slightly older (Median birth year 1968 vs 1969) compared with patients receiving care at only one participating institution. In contrast, patients who maintained an address in a warmer state ("snowbirds") were the least likely to be black (7.5%) of all study groups. We identified conflicting or inconsistent demographic information in 49.1% of patients with care fragmentation compared with 5.6% of patients without care fragmentation. Privacy-preserving record linkage can be an effective means to identify populations with care fragmentation and poor data quality for focused clinical and data improvement efforts.

Keywords: Record linkage · Privacy preservation · Ecology of care

© Springer Nature Switzerland AG 2020
P. Cellier and K. Driessens (Eds.): ECML PKDD 2019 Workshops, CCIS 1168, pp. 79–87, 2020.
https://doi.org/10.1007/978-3-030-43887-6_7

1 Introduction

1.1 The All of Us Research Program

In 2016, the United States Congress launched the Precision Medicine Initiative (PMI) with $200M in funding in order to advance the development and application of individualized care based on a person's unique lifestyle, environment, and biology. A core foundation of the Precision Medicine Initiative, the All of Us Research Program (AoURP) was initially allocated $130M to create a national cohort of over one million Americans broadly representing the rich diversity of the U.S. population. Widespread adoption of Electronic Health Records (EHRs) across the U.S. was identified early in the design of the AoURP as a potentially rich source of data on patient health conditions and treatments.

The AoURP designated and funded over 40 Health Care Provider Organizations (HPOs) nationally to serve as recruitment centers. As part of the enrollment process, HPOs are required to send EHR data for consented participants to the AoURP Data and Research Center after verifying the identity of the participant and standardizing the EHR data into the Observational Medical Outcomes Partnership (OMOP) data model [1].

1.2 Data Fragmentation Across Institutions

However, healthcare in the United States is delivered across a wide variety of care settings and lacks the availability of a universal patient identifier. As a result, patient records may be fragmented across each location where a patient receives care, and unavailable both for patient care, but also for aggregation for research purposes such as those envisioned by the AoURP. Health Information Exchanges (HIEs) emerged as a means to address data and care fragmentation, and use a master patient index to consistently track the same patient across different care settings but are not available in many regions in the United States, or have struggled to remain financially viable [2]. Some EHR systems can link health records across institutions which use the same EHR system for routine clinical care, but do not currently integrate these data together for research purposes [3]. Because the AoURP aims to aggregate as much information about a participant as possible, investigators at participating HPOs questioned how often participants might receive care at a different care site than the HPO at which they might be enrolled. But without cross-institutional data sharing agreements in place to allow for patient identifiers to be shared across sites, and with many HPOs not part of HIEs, an alternate mechanism to link the same patient record across sites was needed.

1.3 Prior Use of Privacy-Preserving Record Linkage

We previously developed software to generate keyed hashes of patient identifiers that is fully compliant with HIPAA de-identification methods and could enable privacy preserving record linkage across AoURP HPOs [4]. A key finding of the initial linkage across seven healthcare institutions was the significant degree of data fragmentation across care sites ranging from 11 to 28% over a several year span. We subsequently demonstrated similar care fragmentation for specific populations including patients with diabetic ketoacidosis [5] and systemic lupus erythematosus [6]. Notably, we identified worse clinical outcomes for patients with fragmented care vs those without care fragmentation, a finding consistent across each condition we studied. Relevant to a cohort study such as the AoURP, we linked individual data between a longitudinal cohort study (the Multi-Ethnic Study of Atherosclerosis or MESA) and EHR data in our region, and identified gaps in data coverage in both sources of data even for conditions as seemingly obvious as a myocardial infarction [7]. The combination of both multi-institutional EHR data and prospectively collected data for a cohort study created a more complete set of data for a given research study participant than any one source alone.

With this background and with the endorsement of the AoURP Steering Committee, we set out to use our previously validated privacy preserving record linkage method to determine how often patients receive care across participating AoURP institutions within a geographically proximate region of three adjoining States in the Upper Midwest of the United States. Our goal was to identify the degree of data fragmentation across AoURP sites in order to determine whether to pursue additional data sources to fully characterize research cohort participants.

2   Methods

We submitted and received approval for this study of de-identified patient level data from the Northwestern Institutional Review Board. We defined the study population as patients seen at participating institutions from January 1, 2011 through May 1, 2018. We excluded patients aged 90 or over as of April 30, 2018 to comply with HIPAA Safe Harbor restrictions on age. Seven institutions participated in the study, three based in the State of Wisconsin, three in Illinois, and one in Indiana which had access to data from the statewide Health Information Exchange.

At a kickoff meeting hosted in Wisconsin and through subsequent discussion, all participating institutions agreed upon a common data dictionary to define key demographic and clinical fields to extract along with keyed hashes to uniquely identify a patient (Table 1).

Table 1. Key data fields extracted by institutions to characterize the demographics and diagnoses of the study population.

Demographics	Diagnoses
Birth year	Year
Gender	Encounter type (e.g. Inpatient, Emergency Department)
Race	Terminology (ICD9, ICD10, SNOMED)
Ethnicity	Primary diagnosis (yes or no)
Insurance status (most recent)	
3 digit ZIP code	

We distributed an executable software program with known matching performance characteristics as described in our prior publication. Participating institutions installed the software locally, and collectively identified a key to be used to hash the patient identifiers that was kept separate from the group aggregating the data on behalf of the study. Using a combination of last name, first name, date of birth, and social security number (where available), sites encrypted multiple concatenated combinations of these features in order to generate up to 17 secret key encrypted hashes. The central site (Northwestern University) team, acting as an honest broker, received the keyed hashes, along with attached demographic and clinical data as defined by the study data dictionary.

We matched the data across the participating institutions to evaluate the degree of care fragmentation within each State, across States, and across all institutions. Because we included three digit ZIP codes in our data set (which is a broad enough level of geography to still be considered de-identified by HIPAA), we could identify the subpopulation of patients who also have a home address in a considerably warmer region of the United States (the States of Alabama, Arizona, Arkansas, California, Florida, Georgia, Louisiana, Mississippi, New Mexico, and Texas) during the winter months (colloquially referred to as "snowbirds"). We analyzed the differences in demographics between those patients who have fragmented and non-fragmented care, as well as between "snowbirds" and those less capable of escaping the cold winter weather in the Upper Midwest.

Several data fields required additional translation between data terminologies in order to be consistent for further analyses. Diagnoses in EHRs arrived as ICD9, ICD10, and SNOMED codes and required significant re-mapping to a consistent and common terminology, in this case MS-DRG-CM. We identified data quality issues including missing data and data which conflicted across sites.

Due to of the large size of the total number of records, we conducted analyses using Python 3.7 with *pandas* and *numpy* packages.

3   Results

In total, we received records on 5,831,238 individuals across the three states. We identified 458,680 patients with data at more than one institution. Table 2 describes the demographics for our total study population, and the populations of patients with non-fragmented care, fragmented care, and "snowbirds". Demographics information that was declined or missing at the point of recording, as well as patients that had conflicting demographics information from multiple patient records were given the same category. Considerable patient race information were found to be conflicted or missing, and as high as 44.8% in fragmented patients.

Table 2. Demographics of the total study population, patients with non-fragmented care, fragmented care, and "snowbirds".

		Total n = 5,831,238	Non-fragmented n = 5,372,558	Fragmented n = 458,680	Snowbirds n = 79,701
Age	Median birth year	1969	1969	1968	1964
Gender	Female	46.8%	46.2%	54.3%	48.3%
	Male	43.6%	44.0%	45.7%	44.2%
	Other	8.9%	9.7%	0.0%	7.4%
	Conflicted or missing	0.7%	0.0%	8.3%	0.1%
Race	White	58.1%	60.0%	35.4%	62.1%
	Other	15.5%	16.2%	6.9%	14.6%
	Black or African American	10.9%	10.8%	11.8%	7.5%
	Declined or missing or conflicted	12.1%	9.4%	44.8%	11.9%
	Asian or other Pacific Islander	2.5%	2.6%	1.0%	3.4%
	Hispanic or Latino	0.5%	0.6%	0.0%	0.2%
	American Indian/ Alaskan Native	0.4%	0.4%	0.1%	0.3%
Ethnicity	Not Hispanic or Latino	89.5%	89.9%	84.3%	91.7%
	Hispanic or Latino	6.6%	6.8%	4.4%	4.9%
	Conflicted or Missing	3.9%	3.3%	11.4%	3.4%

3.1 Patient with Care Fragmentation

The distribution of patients with care fragmentation was unevenly distributed by State and Institutions. The percent of patients with care fragmentation differed by state ranging from 4.9% to 11.7% (Table 3).

Table 3. Care fragmentation by State.

State	Counts	Total	% of fragmented patients within state
Illinois	328,544	2,811,941	11.7%
Wisconsin	108,996	2,240,339	4.9%
Indiana	88,423	846,241	10.4%

The percent of patients with care fragmentation varied by site ranging from 6.1% to 32.7% (Table 4).

Table 4. Fragmentation by care site.

Site	Counts	Total	% of fragmented patients within site
Northwestern University	253,543	1,931,853	13.1%
Rush University Medical Center	213,946	653,358	32.7%
University of Illinois at Chicago	150,918	516,593	29.2%
University of Wisconsin Madison	72,561	636,585	11.4%
Medical College of Wisconsin	63,252	1,031,119	6.1%
Marshfield Clinic	46,952	646,404	7.3%
Regenstrief Institute	88,423	846,241	10.4%

3.2 Data Quality Issues

We identified a significant percentage of records with conflicting demographic information, with the majority of discrepancies for race (Table 5 and Fig. 1).

Table 5. Number of records with conflicting demographic information by feature.

Race	Ethnicity	Gender	Birth year
466,302	59,888	39,547	3,373

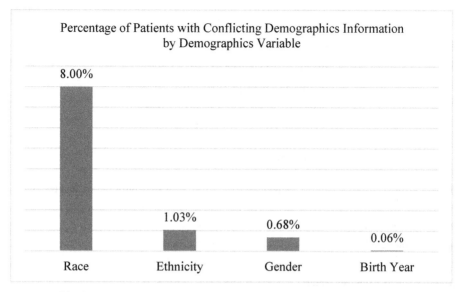

Fig. 1. Most common demographic features with conflicting information.

Patients with care fragmentation had conflicting information at a much higher rate than those without care fragmentation (49.1% vs 5.6%, Table 6)

Table 6. Counts and percentage of patients with conflicting information by fragmentation status.

	# of patients w/conflicted information	# of patients w/o conflicted information	Percentage of patients with conflicted information
Patients that are fragmented within state	225,313	233,367	49.1%
Patients that are not fragmented within state	301,700	5,070,858	5.6%

3.3 Geographic Analysis to Characterize "Snowbirds"

Patients with home addresses (by 3 digit ZIP code) varied by State (Table 7) and by Institution (Table 8).

Table 7. Snowbirds by State

State	Counts	Total	%
Illinois	49,996	2,811,941	1.78%
Wisconsin	26,882	2,240,339	1.20%
Indiana	3,025	846,241	0.36%

Table 8. Snowbirds by Institution

Site	Counts	Total	% of snowbirds out of total patient population
Northwestern University	38,846	1,931,853	2.01%
Medical College of Wisconsin	10,825	1,031,119	1.05%
University of Wisconsin Madison	8,748	636,585	1.37%
Rush University Medical Center	7,763	653,358	1.19%
Marshfield Clinic	7,449	646,404	1.15%
University of Illinois at Chicago	4,333	516,593	0.84%
Regenstrief Institute	3,025	846,241	0.36%

4   Discussion

We used a previously validated privacy preserving record linkage method based on generating keyed hashes of patient identifiers to identify the degree of data fragmentation across a sample of HPOs within the AoURP. Data fragmentation varied from 3.6% to 32.7% with the greatest percentage at sites within IL and the more population-dense Chicago-based institutions. Consistent with prior studies, patients with care fragmentation were more likely to be black and younger. In contrast, patients with the ability to "snowbird" to warmer climes were least likely to be black.

A common problem with linking data across sites is the issue of conflicting data, e.g. one site lists race as "Caucasian" and another site may list race as "unknown". We identified conflicting demographic information for 49.1% of those patients receiving care at more than one institution. Even in patients who receive care at the same institution, demographic information captured over time had conflicting information 5.6% of the time. Race was the most common demographic feature with conflicting information.

There are several limitations to our study. Our study only included a small number of institutions within each State (those that participate in the AoURP), e.g. in the Chicagoland area alone there are over 40 distinct healthcare institutions. Thus our estimates of data fragmentation are likely significant underestimates. Because we focused on sharing only demographic features compliant with HIPAA de-identification criteria, we could not evaluate more specific geographic features beyond 3 digit ZIP code. Geographic features such as home address are likely to change over time for patients as they move, or to be collected in non-standardized fashions, and could be a common feature at risk of conflicting across care sites. We defined "snowbirds" as having a listed address in the EMR from one of several warm winter month states. However, many "snowbirds" may only list their local address so our estimates likely significantly underestimate the population size.

Our study demonstrated the utility of a privacy-preserving record linkage tool to characterize care fragmentation across institutions spanning three contiguous States. Our findings are consistent with prior findings that care fragmentation is associated with at-risk populations but also demonstrates a novel association with significantly higher proportion of conflicting data. We have ongoing work to analyze the differences in insurance status and diagnoses across the study population and to use study results to guide strategies to capture more comprehensive clinical data for patients enrolled in the All of Us Research Program.

Acknowledgements. This study was funded under a supplement to NIH award 1 U2C OD023196-01 (All of Us Research Program Data and Research Center). During the study period, authors EM and UT were faculty at the University of Wisconsin, Madison.

Statement of Conflicts. ANK is an advisor to Datavant, Inc., which supports Privacy-Preserving Record Linkage software. Datavant acquired Health Data Link, Inc. which ANK co-founded based on this earlier version of the software.

References

1. The OMOP data model. https://www.ohdsi.org/data-standardization/the-common-data-model/. Accessed 14 June 2019
2. Holmgren, A.J., Adler-Milstein, J.: Health information exchange in US hospitals: the current landscape and a path to improved information sharing. J. Hosp. Med. **12**(3), 193–198 (2017)
3. Epic Care Everywhere. https://www.epic.com/careeverywhere/. Accessed 14 June 2019
4. Kho, A.N., Cashy, J.P., Jackson, K.L., et al.: Design and implementation of a privacy preserving electronic health record linkage tool in Chicago. J. Am. Med. Inform. Assoc. **22**(5), 1072–1080 (2015)
5. Mays, J.A., Jackson, K.L., Derby, T.A., et al.: An evaluation of recurrent diabetic ketoacidosis, fragmentation of care, and mortality across Chicago, Illinois. Diabetes Care **39**(10), 1671–1676 (2016)
6. Walunas, T.L., Jackson, K.L., Chung, A.H., et al.: Disease outcomes and care fragmentation among patients with systemic lupus erythematosus. Arthritis Care Res (Hoboken) **69**(9), 1369–1376 (2017)
7. Ahmad, F.S., Chan, C., Rosenman, M.B., et al.: Validity of cardiovascular data from electronic sources: The Multi-Ethnic Study of Atherosclerosis and HealthLNK. Circulation **136**(13), 1207–1216 (2017)

Data Integration for the Development
of a Seismic Loss Prediction Model
for Residential Buildings in New Zealand

Samuel Roeslin$^{(\boxtimes)}$ [ID], Quincy Ma [ID], Joerg Wicker [ID], and Liam Wotherspoon [ID]

The University of Auckland, Auckland 1010, New Zealand
s.roeslin@auckland.ac.nz

Abstract. In 2010–2011, New Zealand experienced the most damaging earthquakes in its history. It led to extensive damage to Christchurch buildings, infrastructure and its surroundings; affecting commercial and residential buildings. The direct economic losses represented 20% of New Zealand's GDP in 2011. Owing to New Zealand's particular insurance structure, the insurance sector contributed to over 80% of losses for a total of more than NZ$31 billion. Amongst this, over NZ$11 billion of the losses arose from residential building claims and were covered either partially or entirely from the NZ government backed Earthquake Commission (EQC) cover insurance scheme. In the process of resolving the claims, EQC collected detailed financial loss data, post-event observations and building characteristics for each of the approximately 434,000 claims lodged following the Canterbury Earthquake sequence (CES). Added to this, the active NZ earthquake engineering community treated the event as a large scale outdoor experiment and collected extensive data on the ground shaking levels, soil conditions, and liquefaction occurrence throughout wider Christchurch. This paper discusses the necessary data preparation process preceding the development of a machine learning seismic loss model. The process draws heavily upon using Geographic Information System (GIS) techniques to aggregate relevant information from multiple databases interpolating data between categories and converting data between continuous and categorical forms. Subsequently, the database is processed, and a residential seismic loss prediction model is developed using machine learning. The aim is to develop a 'grey-box' model enabling human interpretability of the decision steps.

Keywords: Seismic loss · Christchurch earthquake sequence · Data aggregation using GIS

1 Background

1.1 The Christchurch Earthquake Sequence

In 2010–2011 New Zealand suffered the costliest natural disaster of its history with a series of earthquakes known as the Canterbury Earthquake sequence

© Springer Nature Switzerland AG 2020
P. Cellier and K. Driessens (Eds.): ECML PKDD 2019 Workshops, CCIS 1168, pp. 88–100, 2020.
https://doi.org/10.1007/978-3-030-43887-6_8

(CES). The CES led to 182 fatalities and extensive building damage across the region, with over NZ$50 billion of economic losses accounting for 20% of New Zealand's GDP [1,24]. The CES began on 4 September 2010 with the Mw 7.1 Darfield earthquake. The Darfield earthquake was centered approximately 40 km west of Christchurch Central Business District (CBD) [12]. It affected mainly unreinforced masonry buildings, induced liquefaction in wider Christchurch and luckily, no lives were lost. In the next 15 months, the Canterbury region experienced numerous aftershocks with around 60 earthquakes above Mw 5 and hundreds over Mw 4, some of these such as the Mw 4.7 aftershock on 26 December 2010 resulted in further damage. Then on 22 February 2011 12.51 pm local time, a Mw 6.2 shallow aftershock occurred directly under Christchurch CBD at a depth of 5 km [13]. This was the most significant event in the CES. It happened near lunch time when office and street pedestrian occupancies were at their peaks. It caused collapses of unreinforced masonry buildings that were not already removed from earlier aftershocks, irrecoverable damaged to many mid-rise and high-rise buildings, and collapse of two notable concrete buildings that led to 135 of the total 182 human casualties in the event [18]. It also prompted liquefaction in Christchurch CBD and eastern residential areas which exacerbated building damage due to foundation displacement. Following this, there were a number of other aftershocks that led to further building damage. In total there were 11,200 aftershocks in the CES.

The CES highlighted a number of civil and earthquake engineering challenges, importance of liquefaction, short-term heightened seismicity, rock slope stability but also impacted the reconstruction and recovery [10]. An estimate of 70% of the Christchurch CBD was demolished or partly reconstructed. Significant parts of the CBD were cordoned off from public access for over 2 years from February 2011 until June 2013 [19]. The CES, being the fourth most costliest insurance event in history globally at the time, also extensively affected the local and global insurance sector regarding seismic building damage [20].

1.2 Seismic Insurance Following the Canterbury Earthquake Sequence

Many countries located near tectonic plate boundaries are exposed to frequent earthquakes. However, insurance uptake for geophysical events remains low (2% in Italy, 5% in Turkey, 9% to 11% in Japan, 10% in Mexico, 26% in Chile, 38% in US, and 80% in New Zealand [1]). New Zealand is an exception with an insurance penetration of 80% [1,20]. Over the two years of the CES, major earthquake events and multiple aftershocks led to 77 events for which more than 650,000 insurance claims have been lodged [17]. Apportionment of the losses by sector is as follow: 59% account for the residential sector and 41% for the commercial sector [2]. Most of the claims for residential buildings were lodged for the main events of the 4 September 2010 and 22 February 2011. However, it was difficult to assess the exact impact of each earthquake and aftershocks on buildings. As the time between the event was too short to permit detailed building assessments following each event, especially for such a large number of affected buildings. This also led to significant legal challenges between claimants, insurers and reinsurers about the damage apportionment between events. Reports shows that 61% of the residential insurance claims were settled by the Earthquake Commission (EQC) and 39% by private insurers [2]. This distribution points the significant participation of EQC.

1.3   The Earthquake Commission

The Earthquake Commission (EQC) is a Crown entity which has for its mission to provide natural disaster insurance for residential property. EQC also manages the Natural Disaster Fund (NDF) and promotes research and education on solutions for reducing the impact of natural disasters. EQC involvement is particularly visible with the EQC insurance EQCover [5]. EQCover provides home and land insurance for natural disaster for every home that is covered by private fire insurance. At the time of the CES, EQC provided coverage for the first NZ$100,000 + 15% Goods and Service Tax (GST) of the building damage, NZ$20,000 + GST for contents and land damage up to the value of the damaged land (since 1 July 2019 the cap for residential building cover was increased to NZ$150,000 but do not include the cover for contents anymore). EQC accessed the NDF and its reinsurance cover to settle the claims. Before the CES, the NDF had a value of NZ$6.1 billion (more than US$4 billion) though this has now been significantly depleted to less than NZ$180 million following the CES and a smaller Kaikoura earthquake in 2016 [8,11].

The CES brought major changes for New Zealand, especially for the insurance industry [16]. EQC increased the annual levy in order to replenish the NDF [4]. Owing to the largely unexpected losses for the private insurers since the CES, there had been a trend of increased scrutiny of the risk profile of any insurance cover. Private insurers are now currently applying risk-based premium pricing for earthquake covers. This had led to increased premiums and at times unavailability of earthquake insurance for some regions in New Zealand.

1.4   EQC's Catastrophe Loss Models

Loss models are important for the insurance and reinsurance sector for quantifying probable losses to ensure adequate provisions in case of a catastrophe. EQC similarly relies on hazard and loss models for adjusting base cover, investment and reinsurance strategies and general planning for response to natural catastrophe [23].

In early attempts to quantify the risk for New Zealand, EQC actuaries estimated possible annual claims from historical data, and probable earthquake intensities. With the evolution of individual computers in the 1980s, new modelling opportunities arose. EQC first employed a computer-based modeling software for loss simulation in 1993. In the past, EQC relied on two models that work in tandem: a system dynamics model (SDM) called 'Logjam' for the management of the claims and a hazard and financial risk management system called 'Minerva' [23]. EQC employed Minerva for estimating claims numbers and losses following a major disaster, as well as for the predicting earthquake loss risk over 10 years in the future to design EQC levy structures and deductibles and to maintain the reserves in the NDF. Minerva relied on an internal database as well as external sources such as the EQC Building Costs or Aon Soils database (Fig. 1a). An earthquake loss subsystem which entails an attenuation and a vulnerability model combined to simulate the losses for any one earthquake event (Fig. 1b). Additionally, it has source models for New Zealand as well as 10-year portfolio models that enable to predict the loss frequency data. Outputs from these possible scenarios are stored in the Minerva database which can then be accessed by the financial management sub-system [27]. Nowadays, EQC works closely with reinsurance companies to ensure that New Zealand retains the necessary international support in case of a disaster [7]. EQC still uses Minerva as an impact estimation tool to predict likely losses for single events and one-year probabilistic analyses.

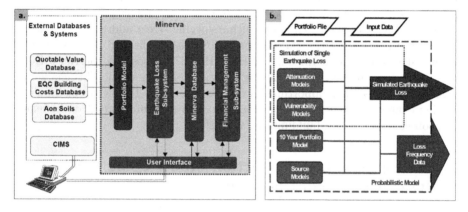

Fig. 1. (a) Overall Minerva system architecture, (b) Schematic diagram of the Earthquake Loss sub system used in Minerva [27]

Without minimizing the great improvement that these tools offered to the New Zealand insurance sector, limitations are still present. Since EQC offers natural disaster insurance for residential building on top of existing private insurance, EQC does not retain a database of its policyholders. It thus uses New Zealand records of real estate property as a base of its calculation [23]. This led to limitations regarding the accuracy of the exact loss prediction per asset. Moreover, the CES highlighted that the existing loss models did not accurately capture liquefaction. Additionally, the models usually took the building stock as undamaged at the time of the earthquake. But in the CES, the time between the events was too short such that the structures could not have been repaired or rebuilt. Cumulative damage occurred in reality but was not taken into account by the loss models [3].

1.5 Earthquake Commission Amendment Bill

On the 18 February 2019, the Earthquake Commission Amendment Bill 2018
(37-2) obtained royal assent [26]. The EQC Amendment Bill introduced changes
including an increase in the time limit to lodge a claim following an earthquake
event from three months to two years, the removal of the insurance cover for
content, but an increase in the cap for the building cover from NZ$100,000
to NZ$150,000. At the same time, the bill brought revisions to the information
sharing provision. EQC is now allowed to share information about the residential
property claims, which have been lodged with EQC. Homeowners and prospec-
tive buyers can now ask EQC to provide them with information on residential
property damage due to a natural disater [6]. The bill also enables EQC to share
information for public good purposes [26] which is favorable to the here presented project. While access to EQC's property and claim database was granted
since November 2017, difficulties arose due to anonymized building coordinates.
Before March 2019, the latitude and longitude of each building in EQC's property database were rounded to approximately 70 m to protect privacy. This lead
to the difficulty to relate each claim with a specific street address thus making
impossible to merge EQC's claim information with additional databases. The
Earthquake Commission Amendment Bill 2018 (37-2) loosened the rules. EQC
is now able to share the exact building location for each claim. This change in
legislation enabled new opportunities for this research. The accurate building
location enabled spatial joining and merging with new information on liquefaction, soil conditions, and building characteristics.

2 Developing a Loss Prediction Model Using EQC's Residential Claim Database

2.1 Exploration of the Database

Following the changes brought by the 2019 Earthquake Commission Amendment
bill, EQC provided access to the claim database for research purposes only. The
exploration made in this paper uses the March 2019 version of the EQC claim
database. Over 95% of the insurance claims for the CES have been settled by that
time. However, revision of the event apportionment is still subjected to review
meaning that the division of the cost between EQC and the private insurers can
still change in future.

The EQC claim database is a wide dataset with 62 variables. It contains
the relevant information related to the claims such as the date of the event, the
opening and closing date of a claim, a unique property number, and the amount
of the claim for the building, content and land. At the time of the CES in 2010–
2011, EQC's liability was capped to the first NZ$100,000 (+GST) of building
damage. Costs above this cap are borne by private insurers if building owner
previously subscribed to adequate insurance coverage. Private insurance could
not disclose information on private claim settlement, leaving the claim database

for this study soft-capped at NZ$100,000 for properties with over NZ$100,000 damage.

CES insurance claims are organized according to the event date when the damage is purported to have stemmed from. For the CES, the EQC database entails 77 different earthquake events. Figure 2 shows the number of claims against the 13 most significant events with more than 1,000 claims lodged. The two most significant events are the 4 September 2010 earthquake (145,000 claims) and 22 February 2011 aftershock (144,300 claims). Among the 62 variables, the database also includes building features. However, not all meta-data were collected in every instance and this led to incomplete data as highlighted in Fig. 3. The original EQC database has 85% of the values missing for critical features regarding the building characteristics (e.g. construction year, primary construction material, number of stories). Furthermore, the building characteristics may be subjective to individual assessor's visual observation.

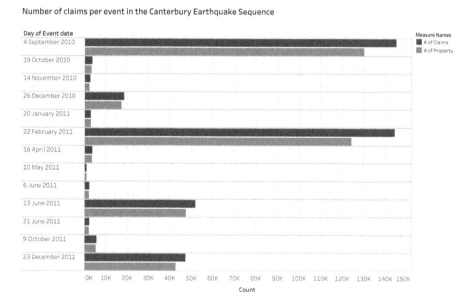

Fig. 2. Number of claims per event in the Canterbury Earthquake Sequence (Source: EQC database for claims on residential buildings)

2.2   Merging of Multiple Databases

To develop a loss prediction model using machine learning, it is necessary to overcome the limitations of missing data for key variables. This is addressed by combining information available in other sources. Figure 4 shows a schematic overview of the databases that are combined with the EQC database.

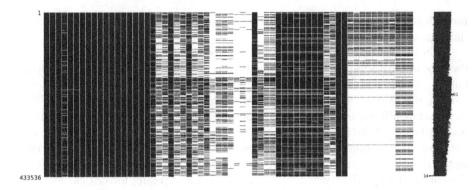

Fig. 3. Graphical overview of the data in the EQC claim database for the Canterbury Earthquake sequence. Each column represent a variable and each claim is a row. White areas represent missing values.

The RiskScape database [15] delivered critical information on buildings characteristics. It contains detailed information on the construction type, use category, building year, floor area, and deprivation index for every building in New Zealand. The Canterbury maps [25] and the New Zealand Geotechnical Database (NZGD) [9] provided records of the location and severity of liquefaction occurrence during CES based on interpretation of observations and LIDAR surveys. Land Information New Zealand (LINZ) [21] and Land Resource Information Systems (LRIS) [22] databases provided further topographical and soil conditions for the buildings of interest. Finally, the GeoNet [14] database provided strong motion seismograph recordings of all events in the CES as recorded at 14 recording stations located throughout Christchurch. This study focused on summary data such as peak ground acceleration (PGA), peak ground velocity (PGV) and peak ground displacement (PGD). This data enabled interpolation layers for all Christchurch to be created through the use of GIS software. Figure 5 presents an example of such an interpolated PGA map.

2.3   Challenges and Lessons Learned

During the process of merging the databases together, several challenges were encountered. These challenges occurred primarily due to the non-exact matching of the coordinates between the databases. Figure 6 shows the location of the EQC claims compared to the actual location of the buildings taken from the RiskScape database. From the map it is to see that the points from the two databases are not close to each other. Additionally, for some property, it can be observed that the EQC database entails two points meaning that multiple claims have been lodged throughout the CES.

As shown on Fig. 4, it was first attempted to join the EQC claim data with RiskScape information using a spatial join function implemented in GIS software. However, due to the distance between the points from EQC and RiskScape the

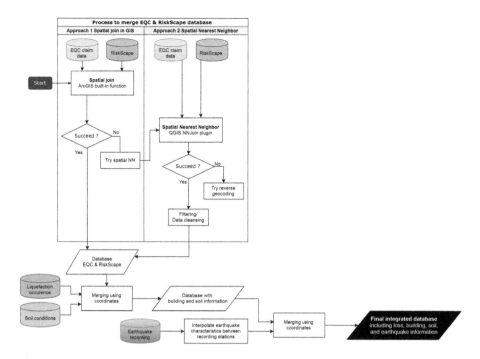

Fig. 4. Overview of the available databases and steps to the final integrated database

software was not able to successfully merge both databases together. It was thus decided to use a spatial nearest neighbor join (NNJoin) [28]. Nevertheless, the RiskScape database entails information for houses as well as secondary buildings such as garages and garden sheds. As shown on Fig. 7, multiple points might be present within the limits of one property tile. Thus, in certain cases the NNJoin led to the join of multiple buildings on one EQC claim. To reduce the number of buildings to the principal property it was not sufficient to filter the merged data by distance. Fortunately the RiskScape database includes information on the building footprint and floor area. It was then possible to select the principal house by filtering the data for each property title on the footprint area. However, it still left the possibility of neighbouring property being incorrectly joined up. To overcome this shortcoming, another approach applying reverse geocoding will be explored in future studies.

In its raw version, EQC's claim database is claim centric. This means one row of data corresponds to one claim, and the total damage to a property can consists of multiple claims or multiple rows of data filed at different dates, particularly due to the nature of multiple events in the CES. The combination of information with additional databases did not change the structure of the original EQC claim database. The final aggregated database retained a claim centric structure. The aim however, is to develop a machine learning model for the loss prediction on a building by building basis. It is thus necessary to have training data that contains

22 Feb 2011 Christchurch earthquake event

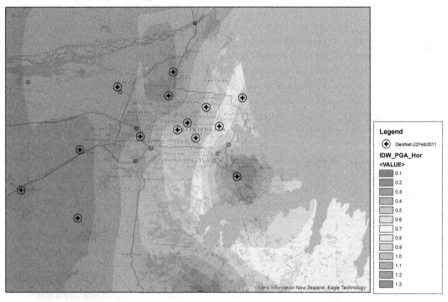

Fig. 5. Location of the GeoNet recording stations in Christchurch and interpolation of the PGA for the 22 February 2011 earthquake

22 Feb 2011 Christchurch earthquake event

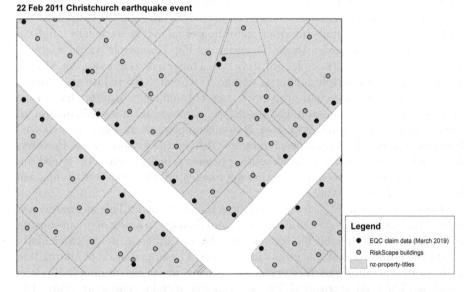

Fig. 6. Comparison of the spatial location of the EQC claim data (blue dots) and the building location from RiskScape database (yellow dots) (Color figure online)

22 Feb 2011 Christchurch earthquake event

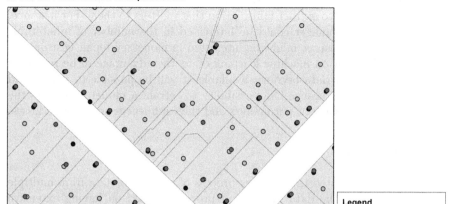

Fig. 7. Comparison of the spatial location of the EQC claim data (blue dots) and the location of NZ street address (pink dots) (Color figure online)

only one unique ID per property. This was achieved by pivoting the database to make it property centric.

3   Future Model Development Using Machine Learning

The combined database will be used as an input for the development of a seismic loss prediction model for residential building in New Zealand. The additional variables obtained through data integration enrich EQC's claim database. Machine learning is applied to process many variables and 'learn' from a large number of instances. Both the 4 September 2010 and 22 February 2011 events led to more than 140,000 claims each. This combined database constitutes the input of a machine learning model for seismic loss prediction.

In the development of the machine learning model, several algorithms such as linear regression, decision tree, support vector machine (SVM), and random forest will be applied. Their prediction accuracy will be compared and the algorithm leading to the most accurate prediction will be retained. The machine learning will be able to extract patterns from the integrated database and evaluate the relative importance of each variables. Nevertheless, particular attention will also be paid to human interpretability of the model. Whenever possible, intrinsically interpretable algorithms are preferred. More complex algorithms are always applied in combination with post hoc methods to allow for human interpretation. The aim is to develop a 'grey-box' model that would produce intermediate output, which allow modelers to look through and validate the predictions at various key intermediate steps. A 'grey-box model' would allow different stakeholders to extract information that matters to them. For instance, a Civil Emergency Manager could be interested in the number of inhabitable dwellings, whilst an insurer might be interested in monetary repair cost only.

A loss model built on machine learning offers the advantage to be retrained easily. Whenever new data becomes available, it will be possible to iterate and improve the model accuracy. The possibility to retrain a model also offers the opportunity to test different parameters and their influences on the final losses.

4   Conclusion

This paper demonstrated the complex process of combining data from multiple sources using GIS. The data integration process focused on having extensive information for each property damaged during the CES. It merged information about the building characteristics, soil type, liquefaction occurrence and seismic demand on top of EQC's claim database. It resulted in a aggregated database that can later be used to develop a seismic loss prediction model for New Zealand using machine learning. It allows for a future analysis of the relationship between variables that are usually not directly considered in a building loss analysis.

Acknowledgments. We acknowledge EQC for generously providing the claim data to realize this study. Many thanks to Geoffrey Spurr for his interpretation, assistance and review of the paper. We gratefully acknowledge the New Zealand Society for Earthquake Engineering (NZSEE) for the financial support. Thanks also goes to Dr. Sjoerd Van Ballegooy for his insightful advices.

References

1. Bevere, L., Balz, G.: Lessons from recent major earthquakes. Technical report, Swiss Reinsurance Company Ltd., Zurich, Switzerland (2012). https://www.swissre.com/institute/library/Expertise-Publication-lessons-from-recent-major-earthquakes.html
2. Deloitte Access Economics: Four years on: Insurance and the Canterbury Earthquakes. Technical report, February, Deloitte Access Economics (2015). https://www.vero.co.nz/documents/newsroom/deloitte-vero-four-years-on-insurance-canterbury-earthquakes-report-february-2015.pdf
3. Drayton, M.J., Verdon, C.L.: Consequences of the Canterbury earthquake sequence for insurance loss modelling. In: 2013 New Zealand Society for Earthquake Engineering Conference, pp. 1–7 (2013). http://db.nzsee.org.nz/2013/Paper_44.pdf
4. Earthquake Commission (EQC): Budget Announcement: EQC levy to increase (2017). https://www.eqc.govt.nz/news/budget-announcement-eqc-levy-to-increase
5. Earthquake Commission (EQC): EQC Insurance (2019). https://www.eqc.govt.nz/what-we-do/eqc-insurance
6. Earthquake Commission (EQC): EQC welcomes Act changes and gets ready to respond (2019). https://www.eqc.govt.nz/news/eqc-welcomes-act-changes-and-gets-ready-to-respond

7. Earthquake Commission (EQC): International reinsurers continue to provide cover to EQC (2019). https://www.eqc.govt.nz/news/international-reinsurers-continue-to-provide-cover-to-eqc

8. Earthquake Commission (EQC): The Natural Disaster Fund (2019). https://www.eqc.govt.nz/about-eqc/our-role/ndf

9. Earthquake Commission (EQC), Ministry of Business Innovation and Employment (MBIE), New Zealand Government: New Zealand Geotechnical Database (NZGD) (2012). https://www.nzgd.org.nz/Default.aspx

10. Elwood, K.J., et al.: Preface earthquake spectra. Earthq. Spectra **30**(1), fmi-fmix (2014). https://doi.org/10.1193/8755-2930-30.1.fmi

11. Feltham, C.: Insurance and reinsurance issues after the Canterbury earthquakes (2011). https://www.parliament.nz/en/pb/research-papers/document/00PlibCIP161/insurance-and-reinsurance-after-canterbury-earthquakes

12. GeoNet: M 7.2 Darfield (Canterbury), 4 September 2010 (2010). https://www.geonet.org.nz/earthquake/story/3366146

13. GeoNet: M 6.2 Christchurch, 22 February 2011 (2011). https://www.geonet.org.nz/earthquake/3468575

14. GeoNet: GeoNet strong-motion FTP site (2012). ftp://ftp.geonet.org.nz/strong/processed/

15. GNS (Geological and Nuclear Sciences Ltd.): RiskScape User Technical Documentation Wiki (2017). https://wiki.riskscape.org.nz/index.php/Overview

16. Greater Christchurch Group - Department of the Prime Minister and Cabinet: Whole of Government Report: Lessons from the Canterbury Earthquake Sequence. Tech. rep., Greater Christchurch Group, Department of the Prime Minister and Cabinet, Christchurch, New Zealand (2017). https://www.dpmc.govt.nz/sites/default/files/2017-07/whole-of-government-report-lessons-from-the-canterbury-earthquake-sequence.pdf

17. Insurance Council of New Zealand (ICNZ): Cantebury Earthquakes (2019). https://www.icnz.org.nz/natural-disasters/canterbury-earthquakes/

18. Kam, W.Y., Pampanin, S., Elwood, K.: Seismic performance of reinforced concrete buildings in the 22 February Christchurch (Lyttelton) earthquake. Bull. N. Z. Soc. Earthq. Eng. **44**(4), 239–278 (2011). http://hdl.handle.net/10092/9006

19. Kim, J.J., Elwood, K.J., Marquis, F., Chang, S.E.: Factors influencing post-earthquake decisions on buildings in Christchurch, New Zealand. Earthq. Spectra **33**(2), 623–640 (2017). https://doi.org/10.1193/072516EQS120M

20. King, A., Middleton, D., Brown, C., Johnston, D., Johal, S.: Insurance: its role in recovery from the 2010–2011 Canterbury earthquake sequence. Earthq. Spectra **30**(1), 475–491 (2014). https://doi.org/10.1193/022813EQS058M

21. Land Information New Zealand (LINZ): LINZ Data Services (2019). https://data.linz.govt.nz/

22. Land Resource Information Systems (LRIS): Christchurch City Soil Map (2010). https://lris.scinfo.org.nz/layer/48148-christchurch-city-soil-map/data/

23. Middleton, D.A.: EQC's use of computer modelling in a catastrophe response. In: Proceedings of 2002 Annual Conference of the New Zealand Society for Earthquake Engineering (2002). https://www.nzsee.org.nz/db/2002/Paper31.PDF

24. Munich RE: NatCatSERVICE (2019). http://natcatservice.munichre.com/

25. New Zealand Government: Canterbury maps - Liquefaction Susceptibility - Final (2017). http://opendata.canterburymaps.govt.nz/datasets/mapped-liquefaction-feb-2011

26. New Zealand Parliament: Earthquake Commission Amendment Bill (2019). https://www.parliament.nz/en/pb/bills-and-laws/bills-proposed-laws/document/BILL_77657/earthquake-commission-amendment-bill

27. Shephard, R.B., Spurr, D.D., Walker, G.R.: The Earthquake Commission's earthquake insurance loss model. In: Proceedings of 2002 Annual Conference of the New Zealand Society for Earthquake Engineering (2002). https://www.nzsee.org.nz/db/2002/Paper32.PDF

28. Tveite, H.: QGIS NNJoin plugin (2019). https://github.com/havatv/qgisnnjoinplugin

Linking IT Product Records

Katsiaryna Mirylenka[1], Paolo Scotton[1(✉)], Christoph Miksovic[1],
and Salah-Eddine Bariol Alaoui[1,2]

[1] IBM Research, Zurich, Switzerland
{kmi,psc,cmi,bar}@zurich.ibm.com
[2] Polytech' Nice, Sophia, France
salaheeddine.bariol-alaoui@etu.univ-cotedazur.fr

Abstract. Today's enterprise decision making relies heavily on insights derived from vast amounts of data from different sources. To acquire these insights, the available data must be cleaned, integrated and linked. In this work, we focus on the problem of linking records that contain textual descriptions of IT products.

Following the insights of domain experts about the importance of alphanumeric substrings for IT product descriptions, we propose a trainable similarity measure that assigns higher weight to alpha-numeric tokens, is invariant to token order and handles typographical errors. The measure is based on Levenshtein distance with trainable parameters that assign more weight to the most discriminative tokens. Not being frequency-based, the parameters capture the semantic specificities of IT product descriptions.

For our task we assess the performance of the most promising lightweight similarity measures, such as (a) edit measure (Levenshtein), (b) frequency-weighted token-based (WHIRL) similarity measure, and (c) the measure based on BERT embeddings after unsupervised retraining. We compare them with the proposed spelling-error-tolerant and order-indifferent hybrid similarity measure that we call the Levenshtein tokenized measure. Using a real-world dataset, we show experimentally that the Levenshtein tokenized measure achieves the best performance for our task.

Keywords: Record linkage · Similarity measure · IT products

1 Introduction

Data have become a precious resource for enterprise decision making. In the IT industry, a company's strategic marketing decisions are often made by considering information about products installed at a customer's sites and products that were already sold by the company to that particular customer. Such information is available through internal and commercial datasets which have heterogeneous representations of items. A fundamental and necessary step to gain insights from such datasets is the ability to link items in the various sets.

P. Cellier and K. Driessens (Eds.): ECML PKDD 2019 Workshops, CCIS 1168, pp. 101–111, 2020.
https://doi.org/10.1007/978-3-030-43887-6_9

In this paper, we focus on the task of linking IT product records, which is crucial for company modeling and future product recommendations [17,18,20]. Linking product records is also a building block of similarity searches [5,6,15,16] and streaming data analysis [13,14,19] for the data series of IT products. A given product can be represented in various more or less "similar" ways in different data sources. The differences across these representations may include formats, synonyms, abbreviations, acronyms and even typographical errors. An example of records to be linked is shown in Fig. 1. The challenge is to detect whether all these representations correspond to the same unique product entity.

Fig. 1. Example of linking records.

We consider a complete dataset of all products of interest, which we call the *master* dataset[1]. The task we wish to accomplish is to match records of a given *query* dataset against the master dataset. The objective is to find the "best" matching catalog entry for each of the items from the query dataset. Both query and master datasets are results of human input. Their vocabulary is not standardized, meaning that product descriptions may contain typos, omissions, and spelling varieties. To find the best matches, we need a quantitative similarity measure to deal with such inconsistencies.

As only a limited amount of ground-truth data is available, it is not feasible to apply supervised machine learning and probabilistic record-linking techniques. In this work, we consider two types of techniques:

1. Record linkage based on advanced contextual word embeddings called BERT (bidirectional encoder representations from transformers), where all the tokens in the product descriptions are correlated. In this case, a BERT network is retrained in an unsupervised manner with the product descriptions available in our master dataset. Then, a similarity search for a query product is performed in the space of retrained BERT product embeddings.
2. The second type of technique are the usual rule-based approaches, where product descriptions are regarded as a string or an arbitrary set of words. The main benefit of these approaches is their limited number of parameters (sometimes no parameters at all) that can be successfully trained, given small ground-truth datasets.

---

[1] Sometimes also called a reference dataset.

The main contributions of this work are the following:

- We analyze and compare lightweight similarity matching techniques from different families to address the record linkage problem for IT products.
- We assess the applicability of context-based word embeddings (BERT) for the task of linking IT product records.
- We propose a hybrid similarity measure called the *Levenshtein tokenized* measure that features trainable weights for alphanumeric[2] tokens. For convenience, we refer to this as the LT measure.
- We demonstrate experimentally that the LT measure outperforms both the Levenshtein measure, a frequency-weighted, token-based measure and the BERT-based measure using real data in our deployment.

2   Related Work

State-of-the-art methods of record linkage include fuzzy or probabilistic record linkage based on supervised machine learning and deep learning models [10,27]. In the context of our application, the amount of training data is limited, making these models infeasible. Thus, we consider either unsupervised machine learning methods or lightweight[3] supervised methods that nevertheless allow for certain statistical inference and parameter tuning.

In 2018, BERT improved the state-of-the-art performance of various NLP tasks such as sentiment analysis or question answering [7]. The principle of BERT is to apply a deep bidirectional transformer architecture to encode long sentences. Essentially, BERT leverages two previously proposed models. The first one is ELMo [22], an LSTM cell architecture that allows contextual word representation. The second is the generative pre-trained transformer (OpenAI GPT) model [23], which uses a left-to-right architecture, where every token can be expected only in the self-attention layers of the transformer. These two models do not allow a word to have context both to its left and its right, thus limiting their performance in some tasks where bi-directional context is important. The major problem when considering a bi-directional context is that a word would itself be taken into account by a bi-directional encoder. BERT uses the "masked language modeling" training objective to predict the missing words, given their bidirectional context.

We use the BERT model to place IT products into the space of BERT embeddings and, then, to make a similarity search within that space. Although we still suspect that the amount of training data might be too small ito retrain such a big network, it is worth comparing this model with much easier hybrid rule-based and machine-learning methods.

Rule-based methods for record linkage are mainly focused on optimal similarity measure searches. There are numerous different algorithms that measure the distance between strings for approximate matching. They implement a similarity function that maps two input strings to a number (a similarity score) such

---

[2] Alphanumeric tokens must contain digits and may contain letters.
[3] Lightweight methods are those with a small number of trainable parameters.

that higher numeric values indicate higher similarity. According to [3], string similarity metrics can be largely classified into edit-distance-based metrics and token-based metrics.

Edit-based measures express similarity by counting the number of primitive operations (insertion, deletion, substitution and transposition) required to convert one string into another. Techniques belonging to this class consider different subsets of these operations. Here are some examples of edit-based measures:

- The *Jaro similarity measure* [11] is designed for short strings such as people's names. It uses the number of matching characters and necessary transpositions to compute the string distance. The *Jaro–Winkler distance* [28] is a variation of the former, which assigns more weight to common prefixes.
- The *Levenshtein similarity* [12] counts the number of insertion, deletion, and substitution operations. Usually a unit cost is assigned to a single operation, and the sum of all costs is returned as the distance between strings. A variant of this is the *Damerau–Levenshtein distance*, which also allows transposition of two characters. Different cost values can be assigned to individual operations, leading to the weighted Levenshtein distance. By sacrificing the metric's properties, the Levenshtein distance measure can be turned into a ratio ($0 \leq r \leq 1$) such that higher ratio values indicate greater similarity.

According to the comparison studies, the Levenshtein similarity measure outperforms other edit-based methods in most cases [2,3]. Therefore it is widely used in many different application scenarios that require the computation of approximate string similarity measures ranging from plagiarism [24] to iris detection [26]. An ensemble approach that uses weighted compositions of Levenshtein and Jaccard similarity measures for company record linkage is described in [9].

We choose the Levenshtein similarity measure as the underlying method for the approximate matching of IT product descriptions because it allows for typos and small uncertainties of hand-written text. There are also fast implementations of the Levenshtein similarity algorithm. It is claimed in [1] that an approximation of Levenshtein similarity can be computed in near-linear time.

Token-based distance measures consider strings as multisets of characters:

- The *Jaccard coefficient* originally comes from biology and is used to compare finite sets. It is simply the quotient of the cardinalities of the intersection and the union of all characters or tokens in two strings.
- The *Cosine similarity* [25] for strings is usually computed on vocabulary vector encodings of a string.
- The *WHIRL* similarity [4,8] measures the distance of two strings in terms of cosine similarity of weighted tf–idf vectors of words. This introduces statistical weighting for the importance of terms in a set of documents.
- *Q-grams with tf–idf* [8] divide a string into q-grams instead of words and computes the weight of each word according to its tf–idf. The distance between two strings is computed as the cosine similarity of the weighted words.

For our task of linking the records of IT products, where the strings that describe the products can contain words (tokens) in arbitrary order, we also assess the performance of WHIRL, the frequency-weighted token-based distance measure. We do not assess the performance of more advanced token-based approaches, where tokens are smaller than words, as the word semantics in a product description are very strong, and we do not want to lose this by splitting the words into q-grams. Instead, to capture typos and small inconsistencies, we exploit the Levenshtein similarity for a token.

3   Hybrid Similarity Measure

In our case of matching IT product descriptions, we need a similarity measure that is independent of a token[4] order, resilient to minor typos and text inconsistencies, and assigns more weight to matching scores of discriminative tokens. On the one hand, the discriminative tokens can be defined in terms of tf–idf weighting captured by the WHIRL similarity, the efficiency of which is assessed in the experimental section (Sect. 4). On the other hand, with our customized similarity measure, we check the hypotheses of the domain experts in IT products that almost all tokens are important in the product descriptions from the query dataset and that the alphanumeric tokens should have more weight.

In this regard we propose a hybrid similarity measure (LT measure) based on the Levenshtein measure that is applied to tokenized product descriptions. Before applying the similarity measure, product descriptions from a query dataset are preprocessed by removing unnecessary punctuation, spaces and upper case, and short tokens are merged with consecutive numeric tokens, such as "DL 360" → "DL360". Vendor names are preprocessed further by eliminating uninformative stop-words such as "inc." or "corp.", and by using special mapping dictionaries for brand names and acronyms such as "hp" → "hewlett packard".

Next, a record from the query dataset q is split into tokens t_i, $i = 1, 2, ..., n$ which are compared with the tokenized records from the master dataset. For each token in the query record, we search for the closest token r_k, $k = 1, ..., m$ in the master record μ and obtain a similarity score of s_{t_i}.

$$s_{t_i} = \max_{r_k \in \mu} LevenshteinScore(t_i, r_k). \tag{1}$$

The query token scores are aggregated to yield the similarity score of the record pair. Let us note that $\mathbb{A}$ is a set of alphabetic[5] and $\mathbb{N}$ a set of alphanumeric tokens. The indicator function $\mathbb{1}_{t_j \in \mathbb{X}}$ outputs 1, if $t_j \in \mathbb{X}$, and 0 otherwise. The LT similarity score can be written as

$$LT(q, \mu) = \frac{\sum_{i=1}^{n} \alpha \cdot s_{t_i} \cdot \mathbb{1}_{t_i \in \mathbb{A}} + s_{t_i} \cdot \mathbb{1}_{t_i \in \mathbb{N}}}{\sum_{i=1}^{n} \alpha \mathbb{1}_{t_i \in \mathbb{A}} + \mathbb{1}_{t_i \in \mathbb{N}}}. \tag{2}$$

According to the assumption of the importance of alphanumeric tokens, we assign them a weight of $= 1$, whereas the alphabetic tokens are assigned a

---

[4] We refer to tokens as words.
[5] Alphabetic tokens contain only letters.

weight of $\alpha \in (0, 1]$. Thus, we ensure that alphabetic tokens are always assigned a smaller or equal weight. In the experimental section (Sect. 4) we verify the hypothesis regarding the importance of the alphanumeric tokens. We also evaluate the influence of $\alpha \geq 1$, when alphabetic tokens are assigned more weight. A pair with the highest LT similarity score is considered to be the best match.

As there are certain product records that should not be matched, there is one more parameter β that depends on the similarity score, $\beta \in (0, 1]$. If the closest similar record has a similarity score greater than β, we consider a product q from a query dataset to be matched to a product μ from the master dataset, otherwise q is considered to be unmatched:

$$q \sim \mu \iff sim_{score}(q, \mu) \geq \beta, \tag{3}$$

where sim_{score} is the score of a similarity function.

In this work, we consider only one best match with the similarity score greater than β. For other applications, top-k matches might be considered using a similar evaluation process. Parameters α and β are to be trained for the optimal LT similarity measure. For comparison approaches, namely Levenshtein and WHIRL similarity measures, only β is trained.

The proposed LT measure is similar to the Mongue–Elkan method [21] in that it also combines edit-based and token-based similarities. As discussed above, the LT measure additionally allows more impact for discriminative tokens.

4   Experimental Evaluation

In order to assess the performance of the promising similarity-matching techniques for our task of linking records, we use a labeled dataset containing 3570 records of IT products based on real examples from the datasets in our deployment. The manual labeling process is quite complicated because the master table contains 21 k products. A total of 544 records from the labeled dataset should be matched to particular records in the master dataset of IT products, whereas 3026 records should not be matched because they correspond to missing product entities in the master dataset. The query records contain certain variabilities of the product descriptions, which comes partially from human error and partially from variations of the product descriptions. One of the real examples combining two types of variance is shown in Fig. 1. We treat matched and unmatched products as two classes that should be correctly labeled: *class 1* stands for matched products and *class 2* for unmatched.

The similarity-matching algorithms we compare were chosen from the best performers in their classes and do not require extensive supervised training. These are the Levenshtein similarity measure as a representative of edit distance, WHIRL as a representative of token-based similarity measures and BERT-based similarity measure as a representative of contextual unsupervised neural network models. We compare these algorithms with the new customized similarity measure (LT measure) introduced in Sect. 3. Its accuracy is measured in terms of precision, recall and $F1$ score for each class. Precision means the ratio of correctly classified products among the retrieved products of a certain class. Recall means the ratio of correctly classified products among all the products of a certain class. $F1$ score is a combined accuracy measure that is the harmonic mean of precision and recall. We measure the $F1$ score for each class separately and the aggregated average $F1$ score for both classes.

The labeled dataset is split into training (60%) and test (40%) datasets. First, we tune the parameters of the similarity measure on the training dataset. The records from the training dataset are matched against the master dataset.

For our application it is often more important to retrieve all the objects from class 1 (matched products) correctly. Thus, we choose the best similarity measure and its parameters, such as β and α, for the LT measure (only β for other comparison measures) by maximizing the recall of the matched class in the first place, and then by maximizing the average $F1$ score. A grid search serves to optimize the parameters.

First, for the LT measure we choose the best α and β values in the training set to maximize the recall of the matched class. The plot of the $F1$ score together with the corresponding precision and recall values for various α levels is shown in Fig. 2. The accuracy values correspond to β with maximum recall.

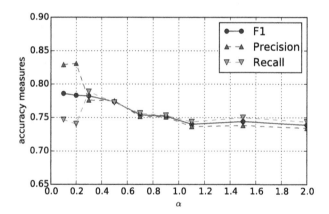

Fig. 2. Precision, recall and $F1$ score (training set) for class 1 as a function of the weight of alphanumeric tokens.

According to these results, $\alpha = 0.3$ provides the best retrieval of IT products from the matched class. This means that alphanumeric tokens should have a weight that is 3.3 times higher than that of alphabetical tokens. This supports the initial hypothesis that matching the alphanumeric tokens for IT product descriptions, such as "DL380" in "HPE ProLiant DL380" is much more important than matching alphabetical tokens, such as "HPE" or "Server".

Having chosen $\alpha = 0.3$, we compare the performance of LT with Levenshtein and WHIRL similarity measures for different values of β. The plot of performance measure variations for each similarity technique is shown in Fig. 3.

The maximum F1 scores for class 1, and overall, are reached at different threshold levels β for each similarity measure. For the LT measure, the optimal β (β^*) is 0.8, for the Levenshtein measure it is $\beta^* = 0.5$, for WHIRL it is $\beta^* = 0.3$ and for the BERT-based measure it is $\beta^* \geq 0.3$. This shows that all the distance measures indeed capture different characteristics of similarity between the IT product descriptions. Other statistics on the performance of these three methods as a function of β can be found in Table 1.

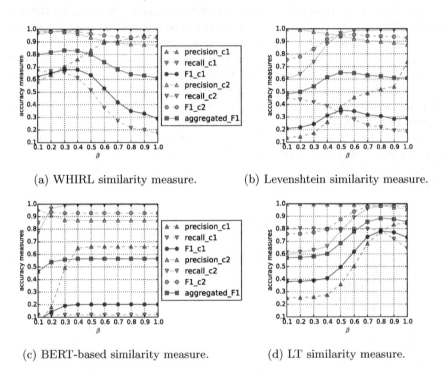

(a) WHIRL similarity measure.   (b) Levenshtein similarity measure.

(c) BERT-based similarity measure.   (d) LT similarity measure.

Fig. 3. Performance of three similarity measures. Label $c1$ corresponds to the records that should be matched and label $c2$ to those that should be unmatched.

Note that the maximum precision and recall values reported in Table 1 do not necessarily correspond to maximum $F1$ scores. For example, maximum precision for class 1 of the WHIRL algorithm comes with quite low recall and $F1\_1$ values, namely 0.19 and 0.31, respectively, which can also be verified in Fig. 3a. Thus, having a high precision for WHIRL means that, if WHIRL identifies a product as belonging to class 1, this is indeed a product of class 1 in 92% of cases. At the same time, owing to the low recall of 19%, WHIRL is able to identify only 19% of class 1 products among all the products belonging to class 1.

After using trained α and β parameters for these methods, we report the accuracy values on the test set for class 1 in Fig. 4. According to the results, the customized hybrid LT measure outperforms the best edit distance as well as BERT-based, token and tf–idf-based WHIRL distances.

Table 1. Maximum performance values of matching algorithms for various β values on the training set. Labels c1 and c2 correspond to class 1 and class 2, respectively.

Acc. measure	WHIRL	Levenshtein	BERT-based	LT
max precision_c1	**0.95**	0.74	0.66	0.84
max recall_c1	0.67	0.45	0.12	**0.80**
max F1_c1	0.68	0.35	0.20	**0.78**
max precision_c2	0.99	**1.00**	0.96	**1.00**
max recall_c2	**1.00**	0.99	0.99	0.99
max F1_c2	**0.98**	0.95	0.94	**0.98**
max aggregated F1	0.83	0.65	0.56	**0.88**
optimal trained β	0.3	0.5	0.4	0.8
optimal trained α	–	–	–	0.3

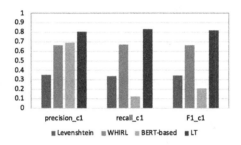

Fig. 4. Precision, recall and $F1$ score for test set.

A similar evaluation is performed to match vendor names of the products in query and master datasets. After vendors have been matched, product descriptions are matched within a vendor. In the case of vendor matching, the LT similarity measure also performs the best with $\beta^* = 0.85$. For lack of space, we do not describe the vendor matching process here. Vendor names are used as blocks within which the product records are matched in order to reduce the number of comparisons.

Although the BERT-based similarity measure did not outperform others, mainly having low recall for the products that should have been matched, we believe it can be adapted for our task in a better way. Its low recall is mainly associated with the fact that the large network (with many parameters) was solely retrained in the unsupervised manner with a limited amount of data. In addition, the task of linking IT product records is quite different from standard NLP language modeling tasks because the tokens in product descriptions are quite specific and therefore it is difficult to see them in usual texts. Moreover, typos and inconsistencies of encoding models for IT products create additional challenges for standard NLP modeling problems where vocabularies are fixed and typos are omitted.

5 Conclusions and Future Work

We introduced a customized similarity measure called the Levenshtein tokenized measure with the purpose to link the records of IT products. This similarity measure combines the benefits of edit and token-based measures and does not require extensive training. It also assigns higher weights to the alphanumeric tokens in product descriptions. This leads to higher matching accuracy for the tasks of record linkage and duplicate detection. We also evaluated the similarity measure based on contextual word embeddings. Although the LT measure did not outperform the proposed hybrid measure, we believe that neural unsupervised training of similar but easier neural network architectures combined with rule-based approaches such as our proposed LT measure might improve the accuracy of record linkage for IT products. We will investigate this in future work.

Acknowledgments. We thank the reviewers of this paper for their valuable comments, which greatly improved the paper.

References

1. Andoni, A., Krauthgamer, R., Onak, K.: Polylogarithmic approximation for edit distance and the asymmetric query complexity. In: Goldreich, O. (ed.) Property Testing. LNCS, vol. 6390, pp. 244–252. Springer, Heidelberg (2010). https://doi.org/10.1007/978-3-642-16367-8_16

2. Bilenko, M., Mooney, R., Cohen, W., Ravikumar, P., Fienberg, S.: Adaptive name matching in information integration. IEEE Intell. Syst. **18**(5), 16–23 (2003)

3. Cohen, W., Ravikumar, P., Fienberg, S.: A comparison of string metrics for matching names and records. In: KDD Workshop on Data Cleaning and Object Consolidation, vol. 3, pp. 73–78 (2003)

4. Cohen, W.W.: Integration of heterogeneous databases without common domains using queries based on textual similarity. In: ACM SIGMOD Record, vol. 27, pp. 201–212. ACM (1998)

5. Dallachiesa, M., Nushi, B., Mirylenka, K., Palpanas, T.: Similarity matching for uncertain time series: analytical and experimental comparison. In: Proceedings of the 2nd ACM SIGSPATIAL International Workshop on Querying and Mining Uncertain Spatio-Temporal Data, pp. 8–15. ACM (2011)

6. Dallachiesa, M., Nushi, B., Mirylenka, K., Palpanas, T.: Uncertain time-series similarity: return to the basics. Proc. VLDB Endow. **5**(11), 1662–1673 (2012)

7. Devlin, J., Chang, M., Lee, K., Toutanova, K.: BERT: pre-training of deep bidirectional transformers for language understanding. CoRR abs/1810.04805 (2018). http://arxiv.org/abs/1810.04805

8. Elmagarmid, A.K., Ipeirotis, P.G., Verykios, V.S.: Duplicate record detection: a survey. IEEE Trans. Knowl. Data Eng. **19**(1), 1–16 (2007)

9. Gschwind, T., Miksovic, C., Mirylenka, K., Scotton, P.: Fast record linkage for company entities (2019). http://arxiv.org/abs/1907.08667

10. Hettiarachchi, G.P., Hettiarachchi, N.N., Hettiarachchi, D.S., Ebisuya, A.: Next generation data classification and linkage: role of probabilistic models and artificial intelligence. In: IEEE Global Humanitarian Technology Conference (GHTC 2014), pp. 569–576 (2014)

11. Jaro, M.A.: Advances in record-linkage methodology as applied to matching the 1985 census of Tampa, Florida. J. Am. Stat. Assoc. **84**(406), 414–420 (1989)
12. Levenshtein, V.I.: Binary codes capable of correcting deletions, insertions and reversals. Soviet Physics Doklady **10**(8), 707–710 (1966)
13. Mirylenka, K.: Mining and learning in sequential data streams: interesting correlations and classification in noisy settings. Ph.D. thesis, University of Trento (2015)
14. Mirylenka, K., Cormode, G., Palpanas, T., Srivastava, D.: Conditional heavy hitters: detecting interesting correlations in data streams. VLDB J. Int. J. Very Large Data Bases **24**(3), 395–414 (2015)
15. Mirylenka, K., Dallachiesa, M., Palpanas, T.: Correlation-aware distance measures for data series. In: EDBT, pp. 502–505 (2017)
16. Mirylenka, K., Dallachiesa, M., Palpanas, T.: Data series similarity using correlation-aware measures. In: SSDBM 2017, pp. 11:1–11:12 (2017)
17. Mirylenka, K., Miksovic, C., Scotton, P.: Applicability of latent Dirichlet allocation for company modeling. In: Industrial Conference on Data Mining (ICDM 2016) (2016)
18. Mirylenka, K., Miksovic, C., Scotton, P.: Recurrent neural networks for modeling company-product time series. In: Proceedings of AALTD, pp. 29–36 (2016)
19. Mirylenka, K., Palpanas, T., Cormode, G., Srivastava, D.: Finding interesting correlations with conditional heavy hitters. In: 2013 IEEE 29th International Conference on Data Engineering (ICDE), pp. 1069–1080. IEEE (2013)
20. Mirylenka, K., Scotton, P., Miksovic, C., Dillon, J.: Hidden layer models for company representations and product recommendations. In: EDBT, pp. 468–476 (2019)
21. Monge, A.E., Elkan, C., et al.: The field matching problem: algorithms and applications. In: KDD, vol. 2, pp. 267–270 (1996)
22. Peters, M.E., et al.: Deep contextualized word representations. CoRR abs/1802.05365 (2018). http://arxiv.org/abs/1802.05365
23. Radford, A., Narasimhan, K., Salimans, T., Sutskever, I.: Improving language under-standing with unsupervised learning. Technical report, OpenAI (2018)
24. Su, Z., Ahn, B.R., Eom, K.Y., Kang, M.K., Kim, J.P., Kim, M.K.: Plagiarism detection using the levenshtein distance and Smith-Waterman algorithm. In: ICI-CIC 2008, pp. 569–569. IEEE (2008)
25. Tata, S., Patel, J.M.: Estimating the selectivity of tf-idf based cosine similarity predicates. SIGMOD Rec. **36**(2), 7–12 (2007)
26. Uhl, A., Wild, P.: Enhancing iris matching using Levenshtein Distance with alignment constraints. In: Bebis, G., et al. (eds.) ISVC 2010. LNCS, vol. 6453, pp. 469–478. Springer, Heidelberg (2010). https://doi.org/10.1007/978-3-642-17289-2_45
27. Wilson, D.R.: Beyond probabilistic record linkage: using neural networks and complex features to improve genealogical record linkage. In: The 2011 International Joint Conference on Neural Networks, pp. 9–14 (2011)
28. Winkler, W.E.: String comparator metrics and enhanced decision rules in the Fellegi-Sunter model of record linkage. In: Proceedings of the Section on Survey Research, pp. 354–359 (1990)

Pharos: Query-Driven Schema Inference for the Semantic Web

David Haller[(✉)] and Richard Lenz

Chair of Computer Science 6 (Data Management),
University of Erlangen, Erlangen, Germany
{david.haller,richard.lenz}@fau.de
https://www.cs6.tf.fau.eu

Abstract. The practical advantage of a data lake depends on the semantic understanding of its data. This knowledge is usually not externalized, but present in the minds of the data analysts who have used a great deal of cognitive effort to understand the semantic relationships of the heterogeneous data sources. The SQL queries they have written contain this hidden knowledge and should therefore serve as the foundation for a self-learning system. This paper proposes a methodology for extracting knowledge fragments from SQL queries and representing them in an RDF-based knowledge graph. The feasibility of this approach is demonstrated by a prototype implementation and evaluated using example data. It is shown that a query-driven knowledge graph is an appropriate tool to approximate the semantics of the data contained in a data lake and to incrementally provide interactive feedback to data analysts to help them with the formulation of queries.

Keywords: Data lake · Semantic Web · Knowledge graph · Schema inference · Query-driven · Data integration

1 Introduction

In the age of "Big Data" the requirements for data analysis have changed drastically. Whereas in the past we had to deal with manageable amounts of data and a limited number of well-defined data sources and structures, today we need access to many different heterogeneous data sources, which contain large amounts of data that have to be processed at high speed. New hardware and programming models help to achieve the necessary performance; however, a remaining challenge is to consolidate the semantic differences resulting from the heterogeneity of the data in order to understand their meaning. The various methods of current research to better understand data can be summarized under *Data Profiling* [10]. This paper presents a new methodology that derives semantics by analyzing SQL query logs and makes the results available in an RDF graph.

The traditional approach to providing data for analytical purposes is the data warehouse. All data is periodically extracted from its original data sources,

ⓒ Springer Nature Switzerland AG 2020
P. Cellier and K. Driessens (Eds.): ECML PKDD 2019 Workshops, CCIS 1168, pp. 112–124, 2020.
https://doi.org/10.1007/978-3-030-43887-6_10

transformed into a uniform schema and merged into a central database. This method has proven to be successful when working with a small and fixed number of internal data sources, whose structures and meanings are well-known. If, however, external and heterogeneous data sources shall be used, this method is not suitable because the cognitive effort for full data integration is too high, resulting in long delays until new data sources become available for analysis.

A modern approach is the data lake: Heterogeneous, non-integrated data made available via a uniform interface, usually SQL. The data is not merged into a common schema, instead, the structural and semantic data integration is performed as part of the analysis. The data analyst must therefore perform a partial data integration within the query. In that way, the analyst gradually gains a better understanding of the relationships between the heterogeneous data sources and their meaning. Unfortunately, this complex knowledge is not externalized, so other users cannot benefit from it. However, the SQL queries executed by the data analyst contain hints about the meaning of the data sources used, and this hidden knowledge could be externalized by analyzing the query log.

2   Approach

Any SQL query contains information about the schema of the tables or data sources referenced, which can be utilized to reconstruct the schema, at least partially. Each query can therefore be regarded as a partial schema definition. The more SQL queries are known, the more can be said about the interrelationships of their data sources. The following SQL query illustrates the idea.

```
select sum(salary), dep.id, dep.name
from person p join department dep on p.dep_id = dep.id
where dep.location = 'DE' and dep.est_year between 1994 and
    2019
group by dep.id
order by dep.name;
```

Listing 1.1. Example query for demonstrating schema inference

From this SQL query, the following information can be derived, both about the schema and about the values of possible tuples.

- person and department could have a foreign key relationship.
- location has "DE" as a possible value.
- est_year can take values between 1994 and 2019.
- salary is summable - possible data types integer or real.
- name is sortable - data type may be alphanumeric.

All these data points are then stored in a knowledge graph, which is amended each time an SQL query is analyzed. That graph can later help data analysts to formulate new SQL queries. In Fig. 1 a simplified knowledge graph can be seen. In order to process the collected knowledge uniformly, an ontology is defined that is able to partially describe the schema of a data source.

For this paper, a prototype was developed, named PHAROS, after the famous lighthouse of Alexandria. The prototype can analyze SQL query protocols using a set of predefined mapping rules and generate an RDF graph. This prototype was then used to assess the effectiveness of the rules set by comparing the reconstructed schema against the already known schema of the data source.

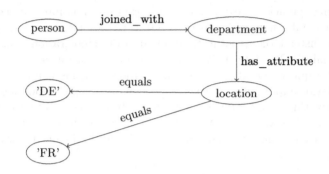

Fig. 1. Partial schema information stored in a knowledge graph

3   Related Work

In the OCEAN research project [12], SQL query log analysis was done using a graph-based approach [13]. PHAROS extends this approach by integrating derived schema information into an RDF-based knowledge graph. Little research has already been done on expressing partial schema information as an ontology. There are several approaches for the translation of a relational schema into an ontology using a mapping language. Two of these mapping languages are now a W3C recommendation. Those are *Direct Mapping* and *R2RML*. An overview of the different techniques for *RDB-RDF Mapping* can be found in [11] and [6]. The focus of most mapping languages is on the integration of relational schemata, as these are the most important in the real word. However, the existing mapping languages are only suitable for PHAROS to a limited extent, since they usually assume the existence of a complete schema on the one hand, and offer no means of expressing the unsharpness of a possibly contradictory partial schema definition on the other hand.

Apart from the way in which a mapping between the schemata of the data sources and the ontology of the knowledge graph can be expressed syntactically and semantically, the methods on how these mappings are created can also be distinguished. The idea behind the approaches which can be summarized under *Ontology-based Data Integration* is to use a domain-specific ontology to access a variety of heterogeneous data sources using a uniform method. The nodes and edges of the knowledge graph are created using the content of the different data sources. If there is only one data source, the approach is called *Ontology-based Data Access* [3]. The opposite way, automatically deriving a common ontology from the data sources, can be called *Database Reverse Engineering* [8] as it tries to understand the intentions of the database developers who designed the schema. If both ontologies and database schemata are already given that are required by existing applications, it is important to achieve interoperability between the platforms and synchronize the data. This can be done using *Mapping Discovery* [7] techniques which search for similarities between databases and ontologies. The approach taken by PHAROS does not fit into any of these categories: Although a predefined ontology exists, it is not domain-specific, but serves to describe the schema of a data source approximately, whereby this knowledge is derived from the requests of the users of the data sources.

4   Contribution

Schema inference is achieved in PHAROS by relying on a query-driven approach. The knowledge about the semantics of a data source that is available in the minds of data analysts is utilized by analyzing the SQL queries data analysts have written. At first glance, this may seem paradoxical. SQL is generally associated with relational database systems that require the definition of a schema which can trivially retrieved by any user.

In recent years, SQL has evolved into a general query language for heterogeneous data sources. This has already been implemented in some systems, such as Apache Drill. Therefore, despite the usage of SQL, the meaning and structure of a data source is often unknown, which is especially challenging when several heterogeneous data sources have to be integrated with each other. A data source can be completely unstructured (a sequence of bytes), it can be in a semistructured format, such as CSV, JSON or XML, or a relational database whose logical schema is known but the conceptual schema has been lost or is not documented.

Typically, data sources are used in some way or the other, like processed by other programs. This means that queries do exist, which were written with the knowledge of the meaning of the data sources. Each SQL query contains implicit knowledge about its data source, reflecting the user's idea of how the data source might be structured and what the stored data might mean. The idea behind this approach is that each SQL query provides another piece of the puzzle to iteratively reach a better understanding of the semantics.

4.1   Schema Inference

Any SQL query with a `select` statement can be considered as a partial schema definition. A subset of attributes is specified in the projection list; the use of type casts or aggregate functions allows inference about properties of these attributes, such as order or summability. The selection clause provides information about the value ranges of attributes, whether they are `not null`, about possible valid values, and about their data types. Join operations indicate possible primary key-foreign key relationships, a `group by` could indicate a hierarchical structure. The frequent use of `distinct` could also mean that an attribute is not `unique`.

PHAROS uses Semantic Web [2] techniques to iteratively construct a knowledge graph that approximates the meaning of heterogeneous data sources. This knowledge graph can then be used to help data analysts formulate better queries or discover new data sources that might fit their queries. As the knowledge graph is stored in an RDF database with a SPARQL [9] interface, it can then later easily be queried for information that is interesting for the respective application domain.

Fig. 2. Steps of the schema inference process

The more SQL queries are available for a data source, the more accurately the schema can be reconstructed. It is important to note that SQL queries can be "wrong", especially if they were executed in interactive mode. A user could experiment with the data and define completely meaningless joins or compare attributes with values that would never be valid anyway. The inferred schema is always only an approximation to the actual schema. Since it can be assumed that meaningful queries are executed more frequently, but wrong queries exactly only once, an obvious step is to weight schema information in relation to the frequency of the queries they were obtained from.

Figure 2 shows the steps that are performed when analyzing a single query. After the user has captured the query, the extraction rules discussed above are applied to discover the hidden knowledge. The information obtained is then inserted into the knowledge graph as a new fragment. If the same information has already been derived from another query, the weighting of the affected statement is adjusted accordingly. The query is then executed by the underlying database system.

4.2   Ontology

For enabling typesafe querying of the knowledge graph, an ontology has been defined. The namespace used by PHAROS is http://pharos.cs.fau.de. The two central resources that can be used as the subject of a statement are tables and attributes. A table URI is made up of http://pharos.cs.fau.de/table/name, an attribute URI is made up of http://pharos.cs.fau.de/table/name#attr. Attributes whose assignment to a table is unknown are assigned to the pseudo-table %unknown. This is necessary because identifiers are not always fully qualified. In that case, it can no longer be reliably concluded to which table an attribute belongs. The same applies to derived attributes such as the results of aggregate functions.

In addition to resources, there are a number of predicates. The first and most basic is `has_attribute`. It defines a table-attribute relationship, so the subject must always be a table and the object an attribute. Attribute types can be specified with `is_type`. Possible values are `boolean`, `string`, `char`, `decimal`, `long`, `double`, `time`, and `timestamp`. The type of an attribute is determined using assignments and comparisons with literals in the query. If an explicit type conversion takes place, the predicate `casted_to` is used. Comparison operations are represented by predicates like `greater_than`, `less_than` equal or `not_equal`. For mathematical expressions the predicates `add`, `subtract`, `multiply`, `divide` and `modulus` exist.

The predicate `has_order` is used to represent an `order by` clause. Possible values are the literals `ascending` and `descending`. Derived attributes are calculated during execution. Examples are the results of SQL functions such as `max`, `sum` or `replace`. Those attributes are described with `produced_by`, following the name of the producing function. Tables and attributes can have an assigned alias. Since the alias is chosen by the user, it can also contain a reference to the semantics of the data used. For this purpose, `aliased_by` is available. Groupings are represented by the predicate `grouped_by`. If several attributes are grouped, the `grouped_by` predicate appears several times in the knowledge graph. Join operations are initiated by the `joined_with` predicate, regardless of the syntax used for the join operation in the SQL query. The join condition is treated as a regular comparison operation.

All collected statements have to be weighted to express how often schema information could be derived from the analyzed SQL queries. Statements having a low weight could be a false positive, while a high weight indicates that the derived schema information is likely to be part of the actual schema. Unfortunately, the RDF data model corresponds to a directed, but unweighted graph. There is, however, an RDF concept allowing to formulate statements about statements. A statement itself is regarded as a resource, which in turn can participate in predicate relationships both as a subject and as an object. These statements are called *Reified Statements* in RDF jargon. The type used is called `rdf:Statement` with the predicates `rdf:subject`, `rdf:predicate` and `rdf:object`. Resources that are instances of this type can then have any additional predicates, making it possible to add meta-information to statements. In PHAROS, the two "metapredicates" `observed` and `source` are used. `observed` is a counter that indicates how often a statement was derived, while `source` refers to all the queries analyzed for a statement. Listing 1.2 contains a complete document in turtle syntax that uses Reified Statements.

```
[ a rdf:Statement ;
    rdf:subject <http://pharos.cs.fau.de/table/employee> ;
    rdf:predicate pharos:has_attribute ;
    rdf:object <http://pharos.cs.fau.de/table/employee#salary> ;
    pharos:observed "1"^^xsd:int ;
    pharos:source "select salary from employee" ] .
```

Listing 1.2. *Reified Statements* in PHAROS

5   Prototype

The prototype developed to demonstrate the feasibility of the proposed solution is written entirely in Java, utilizing the Apache Jena framework for RDF and SPARQL support. It can be used as a command line tool as well as a so called JDBC proxy driver.

In command line mode, PHAROS evaluates a set of SQL files and inserts the extracted knowledge into an RDF document. The RDF document may be empty, or may already contain statements from an earlier analysis run that are supplemented by the new statements. This mode does not require a running database system and works completely offline.

The alternative mode to run PHAROS is as a JDBC proxy driver. This allows PHAROS to latch into an existing SQL session, record the queries that have been sent, and analyze them in the background. The implementation concept of the JDBC proxy driver and parts of its source code can already be found in the TSUNAMI prototype, which was already presented in the bachelor thesis of the author [4] and a subsequent conference paper [5]. This section is therefore partly based on these publications.

The integration of PHAROS into database management systems takes place via a JDBC proxy driver, a database driver that implements the Java interface `java.sql.Driver`, but forwards all calls to a second JDBC driver, which in turn establishes the connection with the actual database management system. The driver makes it possible to intercept all calls and to manipulate them if necessary. This procedure is transparent for the actual database application. Each application can therefore be used together with PHAROS if it supports the manual configuration of a JDBC driver. The concept of the JDBC proxy driver has long been used in many applications for various purposes. For PHAROS, a generic driver was developed whose functionality can be easily changed by defining hooks. A Java application that wants to use PHAROS must load the `de.fau.cs.pharos.driver.ProxyDriver` driver. When establishing a connection to the database, a JDBC URL with the following format must be passed:

$$\texttt{jdbc:pharos:} \textit{original-driver} \texttt{:} \textit{original-arguments}$$

If, for example, an application should be used together with PHAROS, which was connected directly to Apache Drill up to now, the corresponding JDBC URL must be structured as follows:

```
jdbc:pharos:org.apache.drill.jdbc.Driver:jdbc:drill:zk=local
```

This string is interpreted by the proxy driver. It loads the specified original driver and passes its arguments to it. The original driver establishes the connection to the actual database management system and creates a new object which implements the `java.sql.Connection` interface. At this point the decorator pattern [1] is used. The `Connection` object is decorated with a wrapper object, which also implements the `Connection` interface. This is done using the `java.lang.reflect.Proxy` class, which allows you to intercept method calls and execute user-defined code. The code to be executed is defined in its own class, which must implement the interface `java.lang.reflect.InvocationHandler`. This interface defines the method `invoke()`, which is executed on every method call of the decorated object. This makes it possible to log or edit arguments or to change the return value of the called method.

As an intermediate step, the abstract class `ProxyInvocationHandler<T>` was created, which makes it easier to intercept method calls of objects of different types at a desired location in the code. For each type to be decorated, a subclass must be created which implements `beforeCall()`, `afterCall()`, `onMethodError()`, `onHandlerError()` and `onCleanup()`. PHAROS uses this interface to intercept all strings containing SQL queries so they can be analyzed.

The user of the proxy driver gets back a decorated `Connection` object after establishing the connection. A call to `createStatement()` would also create a decorated object of the type `java.sql.Statement`. With this object, it is possible to intercept calls from `execute()` and `executeQuery()` and to read the first argument of these methods, which contains the SQL query. This SQL query can now be analyzed by PHAROS and the resulting information can be stored in the knowledge graph. The SQL query is then passed on to the original driver.

This flexible architecture allows the query analysis process to be completely transparent to the database application, since the usual JDBC interface continues to be used. For the integration of an application into PHAROS, only the driver URL has to be adapted, as shown at the beginning of this section. It is also possible to nest several of these proxy drivers into each other or even manipulate SQL queries, which ensures easy extensibility of the prototype.

6   Evaluation

The PHAROS prototype described in the last section is now used to validate the feasibility of the given approach. A test data set consisting of seven large SQL queries and SQL schemata with eleven tables is used to check to what extent the prototype is able to correctly derive the real schemata of the underlying data sources from these exemplary queries. The data set is based on the SQL exercise sheets of our lecture *Conceptual Modeling*. By analyzing the SQL queries, a knowledge graph shall be created that describes the semantics of the database schema. Please note that the PHAROS prototype does not know the database schema, the input data consists exclusively of SQL queries. In the beginning, the knowledge graph is empty and gets filled with every SQL query that is analyzed. When the analysis phase is complete, the knowledge graph can be utilized by executing SPARQL queries.

First, the aliases found in the SQL queries are examined. Since these aliases were assigned by users, they could include a reference to the semantics of the respective data. The necessary SPARQL query, which determines the aliases, can be seen in Listing 1.3, the output in Table 1. Most aliases don't reveal anything interesting because they are just abbreviations of the table name. It's different with the aliases `LtrName` and `LtrVorn`, from which one could conclude that an

Table 1. Result of query from Listing 1.3

Attribute	Alias
<http://pharos.cs.fau.de/table/abteilung>	"ab"
<http://pharos.cs.fau.de/table/dienstwagen>	"dw"
<http://pharos.cs.fau.de/table/dienstwagen#fahrzeugtyp_id>	"Typ"
<http://pharos.cs.fau.de/table/dienstwagen#id>	"DIW"
<http://pharos.cs.fau.de/table/fahrzeug>	"fz"
<http://pharos.cs.fau.de/table/fahrzeughersteller>	"fh"
<http://pharos.cs.fau.de/table/fahrzeugtyp>	"ft"
<http://pharos.cs.fau.de/table/mitarbeiter>	"mi"
<http://pharos.cs.fau.de/table/mitarbeiter>	"mi1"
<http://pharos.cs.fau.de/table/mitarbeiter>	"mi2"
<http://pharos.cs.fau.de/table/mitarbeiter#abteilung_id>	"Abt"
<http://pharos.cs.fau.de/table/mitarbeiter#name>	"LtrName"
<http://pharos.cs.fau.de/table/mitarbeiter#vorname>	"LtrVorn"
<http://pharos.cs.fau.de/table/versicherungsnehmer>	"vn"
<http://pharos.cs.fau.de/table/versicherungsvertrag>	"vv"

employee can be a leader of something[1]. The two aliases `mi1` and `mi2` can also point to a *self-join*. This hypothesis can be tested by evaluating the `source` predicate of the RDF statement that points to the underlying SQL queries.

```
select ?attribute ?alias where
{
    ?x rdf:subject ?attribute .
    ?x rdf:predicate pharos:aliased_by .
    ?x rdf:object ?alias .
}
```

Listing 1.3. Schema Inference: Aliases

In the next step the data types derived from the queries are checked (see Listing 1.4). It is noticeable that only `string` and `long` are recognized, although for example `Abschlussdatum` (*inception date*) of the table `Versicherungsvertrag` (*insurance policy*) is declared as `date`. It is apparent that the prototype needs to be refined to better recognize strings that might represent a date (Table 2).

```
select ?attribute ?type where
{
    ?x rdf:subject ?attribute .
    ?x rdf:predicate pharos:is_type .
    ?x rdf:object ?type .
}
```

Listing 1.4. Schema Inference: Data types

---

[1] The German word *Leiter* (abbreviated as `Ltr`) is used for a person representing a department or group and can be translated as *director* or *manager* in English.

Table 2. Result of query from Listing 1.4

Attribute	Type
\<http://pharos.cs.fau.de/table/mitarbeiter#abteilung_id\>	long
\<http://pharos.cs.fau.de/table/abteilung#bezeichnung\>	string
\<http://pharos.cs.fau.de/table/mitarbeiter#ist_leiter\>	string
\<http://pharos.cs.fau.de/table/mitarbeiter#name\>	string
\<http://pharos.cs.fau.de/table/mitarbeiter#vorname\>	string
\<http://pharos.cs.fau.de/table/versicherungsnehmer#eigener_kunde\>	string
\<http://pharos.cs.fau.de/table/versicherungsvertrag#abschlussdatum\>	string
\<http://pharos.cs.fau.de/table/versicherungsvertrag#art\>	string

```
select ?attribute ?value where
{
    ?x rdf:subject ?attribute .
    ?x rdf:predicate pharos:equal .
    ?x rdf:object ?value .
    filter (isLiteral(?value))
}
```
Listing 1.5. Schema Inference: Attribute values

Now, possible values for attributes shall be found. This is done with the SPARQL request from Listing 1.5. The filter using the function isLiteral() ensures that no statements are included in the result set whose object is another RDF resource (Table 3).

Table 3. Result of query from Listing 1.5

Attribute	Value
\<http://pharos.cs.fau.de/table/abteilung#bezeichnung\>	"Vertrieb"
\<http://pharos.cs.fau.de/table/mitarbeiter#ist_leiter\>	"J"
\<http://pharos.cs.fau.de/table/mitarbeiter#ist_leiter\>	"N"
\<http://pharos.cs.fau.de/table/mitarbeiter#name\>	"Braun"
\<http://pharos.cs.fau.de/table/mitarbeiter#vorname\>	"Christian"
\<http://pharos.cs.fau.de/table/versicherungsnehmer#eigener_kunde\>	"J"
\<http://pharos.cs.fau.de/table/versicherungsvertrag#art\>	"HP"
\<http://pharos.cs.fau.de/table/versicherungsvertrag#art\>	"TK"
\<http://pharos.cs.fau.de/table/versicherungsvertrag#art\>	"VK"

In Listing 1.6, the previous SPARQL query is adjusted so that only values of the Ist_Leiter attribute of the Employees table are displayed. The result are the values "J" and "N", which correspond to the semantics of the database schema. It is of course also possible to search for attributes to which certain values have been assigned or which have been compared with certain values. To do this, it is needed to simply put the concrete value in place of ?value and use a variable such as ?attribute for the rdf:subject.

```
select ?value where
{
    ?x rdf:subject <http://pharos.cs.fau.de/table/mitarbeiter#
        ist_leiter> .
    ?x rdf:predicate pharos:equal .
    ?x rdf:object ?value .
}
```

Listing 1.6. Schema Inference: Possible Values for Ist_Leiter

SPARQL queries such as the ones shown above can be used to implement an automatic completion system within a Business Intelligence tool. The application could continuously send SPARQL queries with the current user input to an RDF Triple Store in the background to interactively support the user in formulating his SQL queries after evaluating the results, such as a list of possible values when entering the where clause.

Therefore, it can be observed that the PHAROS prototype can successfully reconstruct a partial schema of the original database and has the potential to provide valuable insights for data analysts, which helps them in writing better queries. The proxy driver architecture of PHAROS enables thereby a minimally invasive deployment in existing environments.

7   Future Work

There are multiple unanswered questions and possible extension points remaining. Until now, SQL queries were examined only in isolation, but not in the context of a complete session. This would open up additional possibilities for reconstructing the mental model from the user's mind. The weight of a node in the knowledge graph is currently determined by counting how often a statement could be derived from a query. Queries passed within a session or with a close *chronological* distance are often semantically related to each other, which should be considered when building the knowledge graph. A further approach would be the determination of the *semantic* distance between two queries, which is the difference between the two subsets of the knowledge graph formed by the two queries. By using these two distance measures and by also analyzing the result set that is returned, it could be possible to recognize trends. If the next query produces a less satisfying result, the user will likely undo his last change and try a new approach. In this way, the user's intention could be better understood and more targeted suggestions could be made.

8 Summary

The objective of this paper was to develop a system to better understand the meaning of heterogeneous data sources in data lakes. It was presumed that data analysts who have already explored the semantics behind a data source with great cognitive effort have already written down this hidden knowledge in the form of their SQL queries, but have not yet explicitly externalized it. Our approach therefore started with the logs of the SQL queries and tried to derive knowledge fragments from the individual clauses of an SQL query and stored them in a knowledge graph to be later used by data analysts to support them with the formulation of further SQL queries.

It was decided to use Semantic Web techniques to build and query the knowledge graph in order to profit from already available, mature tools and to simplify the possibility to use the collected data in other projects. The knowledge graph was therefore created using the *Resource Definition Format* (RDF); the queries were made using the *SPARQL Protocol And RDF Query Language*. A new ontology was defined for being able to express partial schema definitions in RDF. This ontology describes the tables and attributes of a data source as resources. The predicates correspond to the different types of knowledge fragments that can be extracted from an SQL query, such as "attribute `salary` has been casted to the data type `long`". Further conclusions can be drawn from this, for example that `salary` can be summed. Tables or attributes with certain desired properties can be determined using SPARQL queries. The possible contradictions between the different knowledge fragments was addressed with a weighting; if a fragment can be derived several times from different queries, it receives a higher weighting. Because an RDF graph is originally unweighted, the RDF language construct *Reified Statements* was used to formulate "statements about statements". Using this approach, the uncertainty of information can also be expressed in the knowledge graph.

To demonstrate the feasibility of this approach, the prototype PHAROS was developed, named after the famous lighthouse of Alexandria. The prototype was implemented as a JDBC proxy driver that makes it possible for PHAROS to connect to an existing SQL session of any program that has a JDBC interface, and thereby to record and analyze all queries in the background. The prototype was evaluated using a test database with a known schema and a set of test queries. The purpose of this evaluation was to check whether the prototype was capable of reconstructing the schema on the basis of the test queries. It has been demonstrated that with each request the prototype has come closer to the real schema; for example, the foreign key relationships between tables have been correctly recognized.

It has been shown that a query-driven knowledge graph is a promising tool for better understanding the importance of data sources and making this knowledge widely available, as well as providing a foundation on which to build further applications.

References

1. Gamma, E., Helm, R., Johnson, R., Vlissides, J.: Design Patterns: Elements of Reusable Object-Oriented Software. Addison-Wesley Professional Computing Series. Addison-Wesley, Boston (1995)
2. Antoniou, G., Groth, P., van Harmelen, F., Hoekstra, R.: A Semantic Web Primer, 3rd edn. MIT Press, Cambridge (2012)
3. Calvanese, D., De Giacomo, G., Lembo, D., Lenzerini, M., Rosati, R.: Ontology-based data access and integration. In: Liu, L., Özsu, M. (eds.) Encyclopedia of Database Systems, pp. 1–7. Springer, New York (2017). https://doi.org/10.1007/978-1-4899-7993-3
4. Haller, D.: Anfrage-getriebene Anbindung von Datenquellen an ein Datenmanagementsystem. Bachelorarbeit, Friedrich-Alexander-Universität, Erlangen (2017)
5. Haller, D.: Tsunami - Anfrage-getriebene Anbindung von Datenquellen an ein Datenmanagementsystem. In: Eibl, M., Gaedke, M. (eds.) Informatik 2017. LNI, vol. 275, pp. 2561–2566. Gesellschaft für Informatik, Bonn (2017). https://doi.org/10.18420/in2017_258
6. Hert, M., Reif, G., Gall, H.C.: A comparison of RDB-to-RDF mapping languages. In: Proceedings of the 7th International Conference on Semantic Systems - I-Semantics 2011, Graz, Austria, pp. 25–32. ACM Press (2011)
7. Hu, W., Qu, Y.: Discovering simple mappings between relational database schemas and ontologies. In: Aberer, K., et al. (eds.) ASWC/ISWC-2007. LNCS, vol. 4825, pp. 225–238. Springer, Heidelberg (2007). https://doi.org/10.1007/978-3-540-76298-0_17
8. Li, M., Du, X., Wang, S.: A semi-automatic ontology acquisition method for the semantic web. In: Fan, W., Wu, Z., Yang, J. (eds.) WAIM 2005. LNCS, vol. 3739, pp. 209–220. Springer, Heidelberg (2005). https://doi.org/10.1007/11563952_19
9. Mennicke, S., Kalo, J.C., Balke, W.T.: Using queries as schema-templates for graph databases. Datenbank-Spektrum **18**(2), 89–98 (2018). https://doi.org/10.1007/s13222-018-0286-9
10. Papenbrock, T., Bergmann, T., Finke, M., Zwiener, J., Naumann, F.: Data profiling with metanome. Proc. VLDB Endow. **8**(12), 1860–1863 (2015)
11. Sahoo, S.S., et al.: A Survey of Current Approaches for Mapping of Relational Databases to RDF. W3C RDB2RDF Incubator Group (2009)
12. Wahl, A.M.: A minimally-intrusive approach for query-driven data integration systems. In: 2016 IEEE 32nd International Conference on Data Engineering Workshops (ICDEW), Helsinki, Finland, pp. 231–235. IEEE, May 2016
13. Wahl, A.M., Endler, G., Schwab, P.K., Rith, J., Herbst, S., Lenz, R.: A graph-based framework for analyzing SQL query logs. In: Proceedings of the 1st ACM SIGMOD Joint International Workshop on Graph Data Management Experiences & Systems (GRADES) and Network Data Analytics (NDA) - GRADES-NDA 2018, Houston, Texas, pp. 1–5. ACM Press (2018)

Informativeness-Based Active Learning for Entity Resolution

Victor Christen[1]([✉]), Peter Christen[2], and Erhard Rahm[1]

[1] University of Leipzig, Leipzig, Germany
{christen,rahm}@informatik.uni-leipzig.de
[2] Research School of Computer Science, The Australian National University,
Canberra, Australia
peter.christen@anu.edu.au

Abstract. Entity Resolution is a crucial task to integrate data from different sources to identify records that represent the same entity. Entity resolution commonly employs supervised learning techniques based on training data of matching and non-matching pairs of records and their attribute similarities as represented by similarity vectors. To reduce the amount of manual labelling to generate suitable training data, we propose a novel active learning approach that does not require any prior knowledge about true matches and that is independent of the learning method used. Our approach successively identifies new training examples based on an informativeness measure for similarity vectors by considering their relationship to already classified vectors and the uncertainty in the similarity vector space covered by the current training set. Experiments on several data sets show that even for a small labelling effort our approach achieves comparable results to fully supervised approaches and it can outperform previous active learning approaches for entity resolution.

Keywords: Record linkage · Entropy · Uncertainty · Interactive labelling

1 Introduction

Entity Resolution (ER) is the task of identifying pairs of records from different data sources that refer to the same real-world entities [4]. ER is a crucial step for different application domains such as census analysis, national security, and the health, life, and social sciences. The quality and usefulness of any data analysis based on linked data highly depends upon how accurate ER was conducted.

To identify pairs of records that refer to the same entity, the attributes of records are generally compared using similarity functions such as approximate string comparators [4]. A crucial part of ER is the classification of two records as a *match* (same entity) or *non-match* (different entities) based on the calculated similarities between them. Machine learning approaches [13,23] can learn a classifier over sets of known matching and non-matching record pairs based on

© Springer Nature Switzerland AG 2020
P. Cellier and K. Driessens (Eds.): ECML PKDD 2019 Workshops, CCIS 1168, pp. 125–141, 2020.
https://doi.org/10.1007/978-3-030-43887-6_11

the similarities of their attributes as represented by a *similarity* or *weight vector*. For example, comparing first name, last name, street address, city and zipcode leads to a five-dimensional similarity vector per compared record pair [4].

To generate a classification model, labelled pairs of records are necessary. This however might require significant manual labelling efforts [26]. Moreover, the number of true matches (record pairs that refer to the same entity) is generally very small compared to the number of non-matching pairs because of the quadratic nature of the comparison space [4], and therefore the selection of labelled pairs is challenging if one wants to learn an unbiased classifier [6]. Active learning techniques promise to minimise the labelling effort as well as to select representative pairs that result in a good classifier.

Previous work in active learning for ER [1,2,19,26] has focused on selecting pairs based on a certain classification model and the resulting decision boundary of the learned classifier. In this paper, we propose a novel active learning approach for ER that considers the covered similarity vector space and the relationships between similarity vectors.

The main idea of our approach is to search for new unlabelled similarity vectors around *informative* similarity vectors that already are classified as matches or non-matches. In this process, we introduce an informativeness measure for a similarity vector based on the current training data set. The most informative vectors are then used to define a search space where new vectors are selected. We specifically make the following contributions:

- We propose an active learning technique for ER that iteratively selects new similarity vectors for manual classification by an oracle independent of any classifier using an informativeness measure. This measure is based on information entropy to characterise the relationship between vectors labelled as matches as well as non-matches. Moreover, the measure considers uncertainty so that new areas in the similarity vector space are queried.
- Our active learning technique is able to generate training data using a budget-limited human oracle [26], and it does not require any prior knowledge about true matches and non-matches.
- We evaluate our active learning technique on three data sets from different application domains. Our results show that our proposed approach outperforms a previous budget-limited active learning approach for ER [26] and achieves classification quality comparable to fully supervised approaches.

In the following we discuss work related to our approach. In Sect. 3 we formalise the problem that we aim to solve with our approach, which we describe in detail in Sect. 4. In Sect. 5 we then experimentally evaluate our approach and compare it with existing active learning as well as supervised methods for ER.

2   Related Work

ER is an essential part of data integration in various domains such as e-commerce, health and social science research, or national security. As a result,

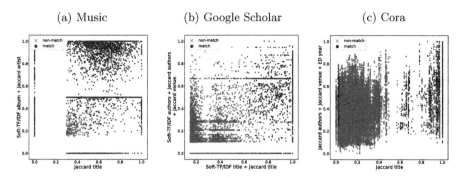

Fig. 1. Examples of similarity vectors where the monotonicity assumption does not hold. The three plots show similarity vectors of the data sets we use in our evaluation in Sect. 5. If an axis represents more than one similarity, they are summed and normalised into [0, 1].

ER has been intensively studied [4,11,17,18]. One challenge of ER is the quality of the data sources and their heterogeneity [20]. In order to overcome this problem, supervised as well as unsupervised approaches have been proposed [3,13,23]. Unsupervised approaches utilise clustering methods to identify groups of similar records that refer to the same entity. In contrast, supervised ER approaches require and use a training data set consisting of verified true matches and true non-matches to build a classifier. In general, unsupervised methods perform worse than supervised approaches as shown by extensive studies [12], where supervised approaches are able to achieve high ER quality for different domains such as consumer products, bibliographic records, and census data.

A crucial part of supervised approaches is the amount and quality of data available for training, because a non-informative or not representative training data set can result in biased, over-fitted, or inaccurate classifiers.

To overcome such issues, active learning techniques [1,2,19,26] have been applied to minimise the labelling effort and to select representative record pairs for manual classification. An active learning approach is an iterative process [5] where in each iteration a number of informative and unlabelled training instances are selected that are then manually classified by a human oracle. Many active learning approaches determine informative instances using the distance between instances [25] or their entropy [21] according to a certain classification model.

Previous work in active learning for ER [1,2] allows to specify a minimum required precision threshold, where the aim of these approaches is to then maximise the recall of the resulting classifier based on the selected record pairs. However, these approaches have the underlying assumption of monotonicity of precision which implies that a record pair with higher similarity is more likely to be a match than a pair with a lower similarity.

Recent work by Wang et al. [26] however has shown that the assumption of monotonicity does not generally hold. We validate this in Fig. 1 which shows the distribution of true matches and non-matches for three data sets according to their similarities. As can be seen, in each data set there are clear examples that violate the monotonicity assumption. Therefore, Wang et al. proposed a cluster based active learning approach that iteratively selects record pairs from a cluster. In each iteration, a cluster is processed by selecting a set of record pairs to be labelled by a human oracle. The labelled vectors are then added to the final training data set if the purity of the current cluster is above a user defined threshold. Otherwise, the cluster is split into two by classifying the unlabelled vectors of the current cluster based on the current classifier. The authors showed that their approach requires less examples than earlier active learning approaches for ER while achieving similar classification accuracy.

In comparison to our proposed approach, the selected examples by Wang et al. [26], and thus the resulting training data set, depend upon the applied classification model, and therefore the resulting ER quality can vary depending upon the classifier employed in this active learning approach.

Ngonga-Ngomo et al. [19] proposed a generation method of link specifications representing a complex match rule using genetic programming by iteratively improving a set of determined link specifications representing match rules. In each iteration, new examples are selected based on the disagreement according to the current link specification (for example, if 5 of 10 specifications classify a match for a record pair the disagreement is high). A disadvantage of this approach is that the generation of link specifications is not deterministic.

Related to active learning is crowd-sourced based ER [8,16,24,27], where ambiguous or controversial matches are resolved by evaluating votes from a crowd of human evaluators. Mozafari et al. [16] proposed two such approaches, named *Uncertainty* and *MinExpError*, being applicable for applications beyond ER. The main idea of these approaches is to use non-parametric bootstraping to estimate the uncertainty of classifiers. However, crowd-sourcing techniques that rely on a large number of human resources (often non-experts) cannot be used for sensitive data, such as personal health, financial, crime, or government records, where only a small number of experts have access to the data.

In contrast to previous work, our approach is independent of the classification model used to determine informative examples, because we characterise the informativeness of similarity vectors by considering the relationships between vectors within the vector space, as well as the relationships between unlabelled and already labelled vectors. Moreover, our work does not rely upon the monotonicity assumption that does not hold for many ER problems [26].

3   Problem Definition

Active learning approaches aim to reduce the manual efforts required for selecting training data, while keeping the quality of ER classification at a high level [1,2, 26]. In general, the goal of ER is to identify matches $m_i \in \mathbf{M}$ for a set of records $\mathbf{R}$ from one or multiple data sources, where each $m_i = (r_x, r_y)$, with $r_x, r_y \in \mathbf{R}$ and $r_x \neq r_y$. To determine a match for a record pair (r_x, r_y), the set of attributes $\mathbf{A} = \{A_1, ..., A_n\}$ characterising these records is used to calculate similarities

$s_1, ..., s_n$ between attribute values. Similarity functions $f_j(r_x.A_j, r_y.A_j)$, with $1 \leq j \leq n$, are used to measure how similar the values in attribute A_j are. We assume each similarity function f_j maps into $[0, 1]$, where 1 means two attribute values are the same and 0 means they are completely different [4].

A similarity or weight vector $\mathbf{w} \in [0, 1]^n$ consists of the calculated n similarities between the attributes in $\mathbf{A}$. For example, the two records r_1 and r_2 characterised by the attributes $\mathbf{A} = \{surname, address\}$ with $r_1.surname =$ "ashworth", $r_1.address =$ "fern hill" and $r_2.surname =$ "ashwort", $r_2.address =$ "fearn hill" might results in a similarity vector $\mathbf{w} = \langle 0.74, 0.78 \rangle$ when using approximate string comparison functions such as edit distance [4].

The goal of an active learning approach is to identify a set of classified similarity vectors $\mathbf{T} \subset \mathbf{W}$ for a given set of unclassified vectors $\mathbf{W}$, where $\mathbf{T}$ consists of *matches* and *non-matches* and is used as training data to learn a classifier. Our approach considers a predefined budget b of the total number of similarity vectors that can be labelled by a human oracle. The approach selects in each iteration a predefined number k of vectors where the selection depends on the informativeness of each vector in $\mathbf{T}$ and the vector space covered by $\mathbf{T}$.

As detailed below, to measure the informativeness $info(\mathbf{w}_i, \mathbf{T})$, of a vector $\mathbf{w}_i$, we consider the relationship of $\mathbf{w}_i$ to vectors $\mathbf{w}_k \in \mathbf{T} \backslash \{\mathbf{w}_i\}$, where we calculate the similarity between two vectors $\mathbf{w}_i$ and $\mathbf{w}_k$ using the Cosine similarity defined as $sim(\mathbf{w}_i, \mathbf{w}_k) = \frac{\mathbf{w}_i \cdot \mathbf{w}_k}{||\mathbf{w}_i|| \cdot ||\mathbf{w}_k||}$. We assume that the area around a vector $\mathbf{w}_i$ consists of more informative vectors than for a vector $\mathbf{w}_k$, if $info(\mathbf{w}_i, \mathbf{T}) > info(\mathbf{w}_k, \mathbf{T})$. The area $S(\mathbf{w}_i)$ around $\mathbf{w}_i$ represents the search space for selecting new unclassified vectors, where $S(\mathbf{w}_i)$ consists of similarity vectors $\mathbf{w} \in \mathbf{W}$ and where the similarity $sim(\mathbf{w}_i, \mathbf{w})$ is above a certain threshold that is dynamically calculated according to the current training data set $\mathbf{T}$.

4   Informativeness-Aware Active Learning

In this section, we describe our active learning approach beginning with a high-level description. Algorithm 1 describes our informativeness-aware active learning approach for generating a training data set $\mathbf{T}$. This training data set is generated by selecting a number of similarity vectors from the set of all similarity vectors $\mathbf{W}$, where a total budget b is available for manual labelling of selected similarity vectors. The set of all (unlabelled) vectors $\mathbf{W}$ is generated by comparing record pairs based on the set of attributes $\mathbf{A}$ and appropriate similarity functions [4]. Initially, we select a number of similarity vectors $k > 1$ from $\mathbf{W}$ based on selection strategies such as *stratified sampling* or *farthest first* (line 1).

Throughout the learning process, we identify in each iteration a set of informative vectors $\mathbf{I} \subseteq \mathbf{T}$ according to the current training data set $\mathbf{T}$. The vectors in $\mathbf{I}$ are used to determine a search space for selecting k new vectors from $\mathbf{W}$ that are to be labelled by the oracle in the current iteration.

To identify the set $\mathbf{I}$, we characterise the informativeness of a vector considering its relationship to all vectors already in $\mathbf{T}$ (line 4). In particular, the informativeness $info(\mathbf{w}, \mathbf{T})$ of a vector $\mathbf{w} \in \mathbf{T}$ is calculated using an entropy-based measure considering the similarities to vectors of both the same and the other class. Moreover, $info(\mathbf{w}, \mathbf{T})$ considers the potential search space around $\mathbf{w}$ with respect to the labelled vectors from $\mathbf{T}$. We describe the calculation of informativeness for similarity vectors and their selection in Sect. 4.2 below.

Algorithm 1. Informativeness-Aware Active Learning Approach

Input:
- **W**: Unlabelled similarity vectors
- b: Total manual labelling budget
- k: Number of similarity vectors to select in each iteration

Output:
- **T**: Training data set in the form of labelled similarity vectors

1 $\mathbf{T} \leftarrow \texttt{initialSelect} (\mathbf{W}, k)$ // Select initial training data set
2 **while** $|\mathbf{T}| < b$ **do**
3 | // Identify informative similarity vectors of the current training data set
4 | $\mathbf{I} \leftarrow \texttt{identifyInformativeVectors} (\mathbf{T})$
5 | // Select unlabelled similarity vectors around informative vectors
6 | $\mathbf{W}_o \leftarrow \texttt{selectVectors} (\mathbf{I}, \mathbf{W}, k, \mathbf{T})$
7 | $\mathbf{T}' \leftarrow \texttt{manualClassify} (\mathbf{W}_o)$ // Use oracle to classify selected vectors
8 | $\mathbf{T} \leftarrow \mathbf{T} \cup \mathbf{T}'$ // Add newly classified vectors to the overall training data set
9 |_ $\mathbf{W} \leftarrow \mathbf{W} \setminus \mathbf{W}_o$ // Remove classified vectors from set of unlabelled vectors
10 **return** T

For each similarity vector in **I**, we determine a search space based on its location in the similarity vector space and the location of the closest similarity vector in the opposite class as determined by the Cosine similarity. We consider each unlabelled vector contained in the search space as a candidate (line 6). The idea of the selection process is to identify similarity vectors in uncertain areas that are close to the boundary of matches and non-matches. The identified set of similarity vectors $\mathbf{W}_o$ is then manually classified by the oracle and added as $\mathbf{T}'$ to the total training data set $\mathbf{T}$ (lines 7 and 8). The approach terminates once the number of classified similarity vectors reaches the total budget b. In the following, we describe the initial selection strategies, the computation of informativeness, and the identification of new training vectors in more detail.

4.1   Initial Selection

Initially, we select a set of similarity vectors from the set of all unclassified vectors **W**. We propose two strategies: *stratified sampling* and *farthest first* [26].

Stratified sampling splits the set of similarity vectors **W** into several partitions $\{\mathbf{P}_1, .., \mathbf{P}_x\}$. To determine an appropriate number of partitions, x, we apply canopy clustering [15] on the unlabelled similarity vectors **W**. The generated partitions are used to determine the set of k initial similarity vectors. We iteratively select similarity vectors over the x partitions, where in each iteration we select the vector $\mathbf{w}_i$ of partition $\mathbf{P}_i$ that is the closest vector to its

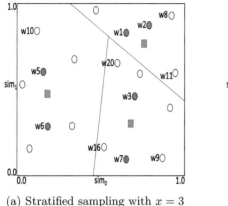 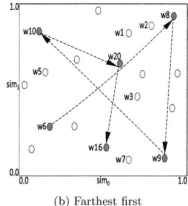

(a) Stratified sampling with $x = 3$ (b) Farthest first

Fig. 2. Examples of initial selection strategies for $k = 6$. The grey circles represent the selected similarity vectors while squares show the centroids of each partition. (Color figure online)

cluster centroid, and add $\mathbf{w}_i$ to $\mathbf{T}$. After that, we remove $\mathbf{w}_i$ from partition $\mathbf{P}_i$. The process terminates once the number of selected similarity vectors is k.

On the other hand, the farthest first method [26] initially selects a similarity vector at random from $\mathbf{W}$ and adds it to $\mathbf{T}$. After that, we iteratively add another similarity vector to $\mathbf{T}$ that has the maximum distance to all vectors already in $\mathbf{T}$. We repeat this process until $\mathbf{T}$ contains k similarity vectors.

For example, in Fig. 2a, stratified sampling selects the similarity vectors $w1$, $w2$, $w3$, $w5$, $w6$ and $w7$. The vector space is initially split into $x = 3$ partitions. After that, for each centroid (blue squares) of a partition we select the closest two similarity vectors. In Fig. 2b, the farthest first approach randomly selects, for example, $w6$ as the first similarity vector and adds it to $\mathbf{T}$. After that, $w8$ is selected since it is the vector farthest away from $w6$. The next selected vectors are $w9$, $w10$, $w20$, and $w16$, following the same process.

4.2   Informativeness of Similarity Vectors

In order to generate a representative training data set, we propose a selection approach that considers the informativeness of similarity vectors $\mathbf{w} \in \mathbf{T}$. The goal is to determine informative classified vectors that can be used to select unclassified vectors from $\mathbf{W}$. We describe the informativeness of a similarity vector by considering its location with respect to the vectors of the same as well as vectors from the other class in the vector space. The intuition is that we look for new vectors in the areas of classified vectors that are not outliers (i.e. are not surrounded only by vectors from the other class) but are also not easy to classify vectors (i.e. are not surrounded only by vectors from the same class).

To determine informative vectors of the current training data set $\mathbf{T}$, we define the following measure $info(\mathbf{w}_j, \mathbf{T})$, as shown in Eq. (1), for a classified vector

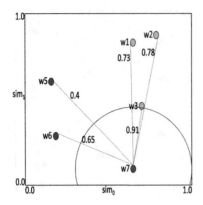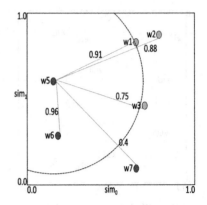

Fig. 3. Two examples for determining the informativeness of similarity vectors $w5$ and $w7$ of $\mathbf{T}=\{w1, w2, w3, w5, w6, w7\}$, based on the location in the vector space and the search spaces $S(w5)$ and $S(w7)$ for $w5$ and $w7$, as represented by the circles. Red coloured circles represent classified non-match similarity vectors while green coloured circles represent classified match vectors.(Color figure online)

$\mathbf{w}_j \in \mathbf{T}$, where sim is the Cosine similarity as described in Sect. 3. This measure is based on the entropy of a vector $\mathbf{w}_j$ according to all vectors in $\mathbf{T}$ and the uncertainty of a vector $\mathbf{w}_j$. Entropy and uncertainty are equally weighted when $\alpha = 0.5$.

$$info(\mathbf{w}_j, \mathbf{T}) = \alpha \cdot entropy(\mathbf{w}_j, \mathbf{T}) + (1 - \alpha) \cdot uncertainty(\mathbf{w}_j, \mathbf{T}) \quad (1)$$

Information entropy [22] can be used to describe how balanced a data set is. In our case, the entropy of a vector $\mathbf{w}_j$ is high if it is close to vectors representing both matches as well as non matches. To determine the entropy of $\mathbf{w}_j$, we compute the aggregated similarities between $\mathbf{w}_j$ and each vector $\mathbf{w}_k$ of $\mathbf{T}_S^{w_j}$ and $\mathbf{T}_O^{w_j}$, where $\mathbf{T}_S^{w_j}$ and $\mathbf{T}_O^{w_j}$ consist of vectors that are assigned to the same class and the other class, respectively, according to $\mathbf{w}_j$, as shown in Eq. (2).

$$entropy(\mathbf{w}_j, \mathbf{T}) = -\left[\frac{\sum_{\mathbf{w}_k \in \mathbf{T}_S^{w_j}} sim(\mathbf{w}_j, \mathbf{w}_k)}{|\mathbf{T}|-1} \cdot log(\frac{\sum_{\mathbf{w}_k \in \mathbf{T}_S^{w_j}} sim(\mathbf{w}_j, \mathbf{w}_k)}{|\mathbf{T}|-1}) \right.$$
$$\left. + \frac{\sum_{\mathbf{w}_k \in \mathbf{T}_O^{w_j}} sim(\mathbf{w}_j, \mathbf{w}_k)}{|\mathbf{T}|} \cdot log(\frac{\sum_{\mathbf{w}_k \in \mathbf{T}_O^{w_j}} sim(\mathbf{w}_j, \mathbf{w}_k)}{|\mathbf{T}|}) \right] \quad (2)$$

The uncertainty of a vector $\mathbf{w}_j$ is determined by the reciprocal of the intersection between the current training data set $\mathbf{T}$ and the search space determined as the area between $\mathbf{w}_j$ and the closest vector of the opposite class as shown in Eq. (3).

$$uncertainty(\mathbf{w}_j, \mathbf{T}) = \frac{1}{1 + |\mathbf{T} \cap S(\mathbf{w}_j)|} \quad (3)$$

For example, the entropy of $w7$ in Fig. 3 is 0.68 calculated by Eq. (2) utilising the aggregated similarity to vectors of the same class ($w6$ and $w5$) as $0.65 + 0.4 =$

Algorithm 2. Selection Method of New Similarity Vectors

Input:
- **I**:   Set of informative similarity vectors
- **T**   Current classified training data set
- **W**: Set of unlabelled similarity vectors
- k: Number of similarity vectors to be selected

Output:
- $\mathbf{W}_o$: Similarity vectors selected for manual classification by oracle

1   $\mathbf{C} = \emptyset$ // Initialise empty set of candidates
2   **foreach** $\mathbf{w}_j \in \mathbf{I}$ **do**
3 | // Determine vector being closest to w_j from the opposite class
4 | $\mathbf{w}_c \leftarrow$ getClosest $(\mathbf{w}_j, \mathbf{T})$
5 | $\delta \leftarrow sim(\mathbf{w}_j, \mathbf{w}_c)$ // Calculate threshold representing the search space of $\mathbf{w}_j$
6 | **foreach** $\mathbf{w}_u \in \mathbf{W}$ **do**
7 | | // Add unlabelled vector if its similarity is above the threshold δ
8 | | **if** $sim(\mathbf{w}_u, \mathbf{w}_j) > \delta$ **then**
9 | | | $\mathbf{C} \leftarrow \mathbf{C} \cup \{\mathbf{w}_u\}$

10 // Identify the k most diverse vectors from candidate set
11 $\mathbf{W}_o \leftarrow$ farthestFirstSelection $(\mathbf{C}, k)$
12 **return** $\mathbf{W}_o$

1.05, as well as to vectors of the other class ($w1$, $w3$ and $w2$) as $0.73 + 0.91 + 0.78 = 2.42$. The intersection between the search space $S(w7)$ and the current training data set **T** is empty and therefore $uncertainty(w7) = 1$. Consequently, $info(w7)$ is equal to $0.5 \cdot 0.68 + 0.5 \cdot 1 = 0.84$. The informativeness for $w5$ is calculated similarly where its entropy is 0.697 and its uncertainty is 0.5 since $S(\mathbf{w_5}) \cap T = \{w_6\}$, and therefore $info(w5, \mathbf{T}) = 0.6$.

We add a vector $\mathbf{w}_j$ to **I** if $info(\mathbf{w}_j, \mathbf{T})$ is above the mean according to the $info$ measure for the vectors of the current training data set **T**. In our running example, the mean of $info$ according to the current training data set is 0.61, and so we add $w7$ ($info = 0.84$) to **I**, but not $w5$. The set **I** of informative vectors is then used to select vectors of **W** to be manually classified and added to **T**.

4.3   Training Data Selection

The selection method shown in Algorithm 2 determines for each similarity vector of **I** a set of unlabelled vectors from **W**. For this, we identify for each vector $\mathbf{w}_j \in \mathbf{I}$ a search space $S(\mathbf{w}_j)$ determined by the closest vector $\mathbf{w}_c$ from the opposite class. For example, in Fig. 4 the closest vector from the other class for $w7$ is $w3$.

The objective is to identify new vectors in uncertain areas so that in each iteration an increasingly more representative training data set **T** is generated. A vector $\mathbf{w}_u \in \mathbf{W}$ is added to the set **C** of candidates if it is contained in the search space $S(\mathbf{w}_j)$ consisting of vectors $\mathbf{w}_u$ where the similarity $sim(\mathbf{w}_j, \mathbf{w}_u)$ is larger

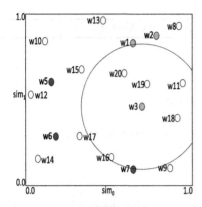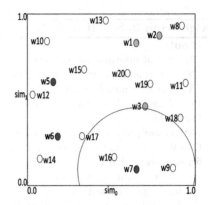

Fig. 4. Two examples of selecting new similarity vectors according to the search spaces $S(w3)$ and $S(w7)$ represented as circles, where $w3$ and $w7$ are the informative vectors. Red and green coloured circles represent classified vectors. (Color figure online)

than $sim(\mathbf{w}_j, \mathbf{w}_c)$ (line 9). At the end of the selection method, we determine the most k-diverse vectors of $\mathbf{C}$ by applying a farthest first approach (line 11).

Figure 4 shows an example for selecting vectors based on w_3 and w_7. The selection method selects all vectors as candidates into $\mathbf{C}$ that are in the search spaces $S(w7)$ and $S(w3)$, shown as circles around $w3$ and $w7$. Consequently, the combined candidate set, $\mathbf{C}$, based on $w7$ and $w3$ consists of the similarity vectors $w9$, $w11$, $w16$, $w18$, $w19$ and $w20$.

The identified set of similarity vectors $\mathbf{W}_o$ are then manually classified by an oracle and added to $\mathbf{T}$ (Algorithm 1, line 8). The updated training data set is used in the next iteration to identify a new set of informative vectors. This loop ends once the number of manually classified similarity vectors reaches the budget b.

4.4   Complexity Analysis

We now briefly discuss the complexity of our proposed approach. Because of the independence of our approach with regard to the actual classification model used, its complexity only depends upon the number of unlabelled similarity vectors, $\mathbf{W}$, the total budget $\mathbf{b}$, and the number k of similarity vectors to be selected in each iteration. In each iteration, we compute the similarities between all pairs of vectors in the current training data set, $\mathbf{T}$, resulting in a complexity of $O(|\mathbf{T}|^2)$. Moreover, we identify for each informative similarity vector of $\mathbf{I}$ the closest unlabelled similarity vectors in $\mathbf{W}$, a process which requires $|\mathbf{W}| \cdot |\mathbf{I}|$ comparisons where $|\mathbf{I}| \leq |\mathbf{T}|$ holds. At the end of each iteration, we determine the k most diverse similarity vectors of $\mathbf{C}$, where $|\mathbf{C}| \leq |\mathbf{W}|$, resulting in a complexity $O(k \cdot |\mathbf{C}|)$. Overall, the complexity to determine similarity vectors for one iteration is $O(|\mathbf{T}|^2 + |\mathbf{W}| \cdot |\mathbf{I}| + k \cdot |\mathbf{C}|)$, with $|\mathbf{I}| \leq |\mathbf{T}|$ and $|\mathbf{C}| \leq |\mathbf{W}|$. The number of iterations is bound by k and b as b/k.

5   Experiments and Results

We evaluated our active learning approach using three data sets as summarised in Table 1. The Cora and Google Scholar (GS) [12] data sets contain publication records that are to be linked, where the GS data set consists of matches between DBLP and GS. The Music data set contains records from the Music-Brainz database[1]. This data set is corrupted [10] and consists of five sources with duplicates for 50% of the original records. To avoid the comparison of the full Cartesian product of vectors, we applied blocking [4] and filtering [14].

Table 1. Overview of evaluated data sets.

Data set	Number of records	$|\mathbf{W}|$	Match: non-match	Attributes	$n = |\mathbf{w}|$
Cora	1,295	286,141	1:16	Title, authors, year, venue	4
Google Scholar	2,616/64,263	472,790	1:89	Title, authors, year, venue	6
Music	19,375	251,715	1:16	Title, artist, album, year, language, number	7

The ratios between matches and non-matches (with blocking and filtering applied) shown in Table 1 highlight the imbalance of these data sets and emphasise the challenges of selecting a representative training data set. The similarity vectors (of dimension n) were calculated using string comparison functions on the different attributes shown in Table 1, such as q-gram based Jaccard and Soft-TF/IDF [4]. To classify the similarity vectors as matches and non-matches, we used the decision tree classifier implemented in the Weka toolkit [7].

Our proposed active learning approach is implemented in Java 1.8 and we ran all experiments on a desktop machine equipped with an Intel Core i7-4470 CPU with 8×3.40 GHz CPUs, and 32 GBytes of main memory. To facilitate repeatability, both code and data sets are available from the authors.

We evaluated different parameter settings for our approach. As initialisation method we used *farthest first*, *stratified sampling* and *random selection*, set $\alpha = [0.3, 0.4, 0.5, 0.6, 0.7]$ to weight the *entropy* and *uncertainty* in Eq. (1) when determining informative similarity vectors, set the number of selected vectors in each iteration as $k = [30, 35, 40, 45, 50]$, and the total budget $b = [200, 500, 1000, 2000, 5000]$. We set default values as $\alpha = 0.5$, $k = 30$, $b = 1000$ and *farthest first* as the initialisation method, because we obtained good results with these settings for all three data sets based on preliminary experiments.

---

[1] Available at: https://musicbrainz.org.

We compared our approach with the two basic active learning approaches *Smallest Margin* [25] and *Entropy* [21], the *Uncertainty* selection approach [16], as well as the only budget limited active learning approach for ER we are aware of (named *Clu-AL*) [26]. We do not compare our approach with *MinExpError* [16] because this approach does not scale well for large budgets. Furthermore, we compared our approach with both fully supervised decision tree and support vector machine (using RBF and linear kernels) classifiers, as also used for comparison in previous work on active learning for ER [26].

To allow a comparative evaluation of our proposed approach with these earlier approaches we use the F-measure [9]. We acknowledge that there are issues when this measure is used to comparatively evaluate different ER classifiers, however there is currently no accepted alternative to the F-measure we are aware of.

5.1 Parameter Evaluation

Figure 5a shows the obtained ER classification quality for different initialisation methods averaged over different iteration sizes k. Farthest first slightly outperforms stratified sampling and random selection by 0.75% and 0.95%, respectively, for the Cora data set, and by 3.1% and 1.8% for Google Scholar. On the other hand, Farthest first achieves a lower F-Measure by 1.17% compared to stratified sampling for the Music data set. The small differences in F-measure results for the different initial selection strategies show that our main selection strategy based on the search space of informative vectors performs effectively independent of the initial set of similarity vectors.

As can be seen in Fig. 5b, changes for the weight parameter α only slightly influence the ER classification quality, between 2% to 4%, for the three data sets. For the Cora data set we observe a decreasing quality for $\alpha > 0.5$. With an α weight over 0.5 our approach prioritises the *entropy* of a vector more than the *uncertainty*, and therefore the approach mainly selects vectors as informative that are located in-between true matches and non-matches.

For all three data sets, the F-measure slightly decreases with a higher number of selected similarity vectors, k, per iteration as shown in Fig. 5c. This indicates that a higher number of selected similarity vectors increases the probability for selecting non-informative vectors. An increasing budget generally leads to an improvement of F-measure results as shown in Fig. 5d. Even for a small budget of $b = 200$, for all three data sets our approach achieves F-measure results of above 80%, with an increase up to 97% for the Music data set as more informative vectors are added to the training set. The runtime scales quadratically with respect to the total budget as shown in Fig. 5e, however, all runtimes are below 200 seconds for budgets up to $b = 1,000$.

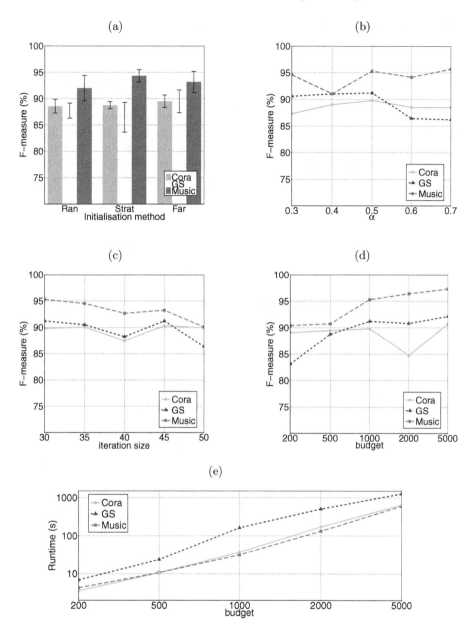

Fig. 5. Classification F-measure results for (a) different initialisation methods, (b) different values for weight parameter α of $info$, (c) different numbers of similarity vectors per iteration k, (d) different total budgets b, and (e) runtime for different total budgets b.

5.2   Comparison with Existing Approaches

We compare our active learning approach, named *InfoSpace-AL*, with the active learning approaches *Smallest Margin, Entropy*, and *Uncertainty*, as well as the clustering based active learning approach *Clu-AL* [26]. We also compare our approach with supervised approaches using fully supervised SVM and decision tree classifiers. To compare the different active learning approaches, we experimentally determined a suitable number of similarity vectors to select in each iteration, k, for each approach separately over all data sets. We use the following values for k: *Smallest Margin*: 45, *Entropy*: 50, *Uncertainty*: 45, and *InfoSpace-AL*: 30. The *Clu-AL* approach follows an adaptive strategy for determining the number of similarity vectors it selects in each iteration.

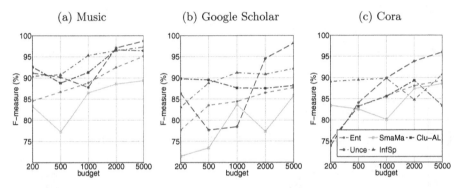

Fig. 6. F-measure results of our approach (named *InfoSpace-AL*, InfSp) as compared with the other active learning approaches *Entropy* (Entr) [21], *Smallest Margin* (SmaMa) [25], *Clu-AL* [26] and *Uncertainty* (Unce) [16].

Table 2. F-measure results of our approach (InfoSpace-AL) as compared with fully supervised classifiers (SVM and DTree) for a budget of $b = 1,000$.

Data set	Dtree	SVM	InfoSpace-AL
Google Scholar	88.63%	91.44%	91.21%
Cora	84.09%	82.22%	89.80%
Music	96.80%	96.90%	95.30%

Figure 6 shows the F-Measure of the considered approaches according to different budgets b. *InfoSpace-AL* is the only approach that, for a small budget, achieves an F-Measure above 80% for all three data sets. *Smallest Margin* and *Uncertainty* result in a high variance with an increasing budget, where the F-Measure achieved by *Uncertainty* is reduced by up to 8.7% from a budget of $b = 200$ to $b = 500$. In contrast, *InfoSpace-AL* achieves more stable F-Measure results compared to *Uncertainty* even for small budgets of $200 \leq b \leq 1,000$.

InfoSpace-AL and *Clu-AL* both achieve high F-Measure results for each data set for small budgets of $b = 500$ and $b = 1,000$. However, we observe that *Uncertainty* achieves high F-Measure values above 90% for each data set if the budget is above $b = 2,000$. To summarise, our approach achieves results comparable to *Clu-AL* and *Uncertainty*, and it is one of the best performing approaches for small budgets of up-to $b = 1,000$.

To evaluate the two supervised approaches, we applied 10-fold cross validation. Our approach achieves comparable results compared to the fully supervised approaches as shown in Table 2. Our informativeness-based active learning approach outperforms the supervised approaches by around 5.7% in F-Measure for the Cora data set. On the other hand, the supervised approaches achieve higher F-Measure results for the Google Scholar and Music data sets compared to our active learning approach. However, we emphasise that our approach achieves these comparable results with a moderate manual classification effort, so that the labelling effort is reduced by around 99% compared to a fully supervised classifier that requires much larger training data sets which are commonly not available in real-world ER applications.

6   Conclusions and Future Work

We have proposed an active learning approach for entity resolution (ER) that iteratively selects similarity vectors into a training data set based on the informativeness of vectors for a current training data set. Unlike with existing active learning approaches for ER, the main advantage of our approach is that it is independent of any intermediate classification results since it determines the search space for new vectors based on a defined informativeness measure considering the location of vectors in the vector space, as well as the uncertainty of the search space. In each iteration, our approach selects new vectors according to the most informative vectors. The evaluation showed that our approach can achieve results comparable to fully supervised approaches where much larger training data sets are required to achieve a high ER quality compared to our budget limited approach. Moreover, our approach outperforms a previous state-of-art active learning method for ER that is also based on a limited budget for the number of manual classifications possible. Furthermore, our approach does also not rely on the assumption of monotonicity of precision [26].

For future work we aim to investigate adaptive methods for determining an optimal number k of selected similarity vectors in each iteration such that the probability for selecting non-informative similarity vectors is minimised. We also plan to investigate filtering methods that initially reduce the set of vectors $\mathbf{W}$ to avoid the selection of non-informative vectors. Moreover, we like to integrate metric space approaches to improve the efficiency of the approach for determining new unlabelled similarity vectors.

Acknowledgements. This work was partially funded by the Australian Research Council (ARC) under Discovery Project DP160101934, and Universities Australia and the German Academic Exchange Service (DAAD).

References

1. Arasu, A., Götz, M., Kaushik, R.: On active learning of record matching packages. In: ACM SIGMOD, Indianapolis, pp. 783–794 (2010)
2. Bellare, K., Iyengar, S., Parameswaran, A.G., Rastogi, V.: Active sampling for entity matching. In: ACM SIGKDD, Beijing, pp. 1131–1139 (2012)
3. Christen, P.: Automatic record linkage using seeded nearest neighbour and support vector machine classification. In: ACM SIGKDD, Las Vegas, pp. 151–159 (2008)
4. Christen, P.: Data Matching - Concepts and Techniques for Record Linkage, Entity Resolution, and Duplicate Detection. Data-Centric Systems and Applications. Springer, Heidelberg (2012). https://doi.org/10.1007/978-3-642-31164-2
5. Dasgupta, S.: Two faces of active learning. Theoret. Comput. Sci. **412**(19), 1767–1781 (2011)
6. Ertekin, S., Huang, J., Bottou, L., Giles, L.: Learning on the border: active learning in imbalanced data classification. In: ACM CIKM, Lisbon, pp. 127–136 (2007)
7. Frank, E., et al.: Weka-a machine learning workbench for data mining. In: Maimon, O., Rokach, L. (eds.) Data Mining and Knowledge Discovery Handbook, pp. 1269–1277. Springer, Boston (2009). https://doi.org/10.1007/978-0-387-09823-4_66
8. Gokhale, C., et al.: Corleone: hands-off crowdsourcing for entity matching. In: ACM SIGMOD, Snowbird, Utah, pp. 601–612 (2014)
9. Hand, D., Christen, P.: A note on using the F-measure for evaluating record linkage algorithms. Stat. Comput. **28**(3), 539–547 (2017). https://doi.org/10.1007/s11222-017-9746-6
10. Hildebrandt, K., Panse, F., Wilcke, N., Ritter, N.: Large-scale data pollution with apache spark. IEEE Trans. Big Data 1 (2017). https://doi.org/10.1109/TBDATA.2016.2637378. ISSN 2372-2096
11. Köpcke, H., Rahm, E.: Frameworks for entity matching: a comparison. Data Knowl. Eng. **69**(2), 197–210 (2010)
12. Köpcke, H., Thor, A., Rahm, E.: Evaluation of entity resolution approaches on real-world match problems. PVLDB Endow. **3**(1–2), 484–493 (2010)
13. Köpcke, H., Rahm, E.: Training selection for tuning entity matching. In: QDB/MUD, Auckland, pp. 3–12 (2008)
14. Köpcke, H., Thor, A., Rahm, E.: Learning-based approaches for matching web data entities. IEEE Internet Comput. **14**(4), 23–31 (2010)
15. McCallum, A., Nigam, K., Ungar, L.H.: Efficient clustering of high-dimensional data sets with application to reference matching. In: ACM SIGKDD, Boston, pp. 169–178 (2000)
16. Mozafari, B., Sarkar, P., Franklin, M., Jordan, M., Madden, S.: Scaling up crowdsourcing to very large datasets: a case for active learning. PVLDB Endow. **8**(2), 125–136 (2014)
17. Naumann, F., Herschel, M.: An Introduction to Duplicate Detection. Synthesis Lectures on Data Management. Morgan and Claypool Publishers, San Rafael (2010)
18. Nentwig, M., Hartung, M., Ngonga Ngomo, A.C., Rahm, E.: A survey of current link discovery frameworks. Semant. Web **8**, 419–436 (2017)
19. Ngonga Ngomo, A.-C., Lyko, K.: EAGLE: efficient active learning of link specifications using genetic programming. In: Simperl, E., Cimiano, P., Polleres, A., Corcho, O., Presutti, V. (eds.) ESWC 2012. LNCS, vol. 7295, pp. 149–163. Springer, Heidelberg (2012). https://doi.org/10.1007/978-3-642-30284-8_17
20. Rahm, E., Do, H.H.: Data cleaning: problems and current approaches. IEEE Data Eng. Bull. **23**(4), 3–13 (2000)

21. Settles, B.: Active learning literature survey. Technical report, University of Wisconsin-Madison, Department of Computer Sciences (2009)
22. Shannon, C.E.: A mathematical theory of communication. Bell Syst. Tech. J. **27**, 379–423 (1948)
23. Sherif, M.A., Ngonga Ngomo, A.-C., Lehmann, J.: WOMBAT – a generalization approach for automatic link discovery. In: Blomqvist, E., Maynard, D., Gangemi, A., Hoekstra, R., Hitzler, P., Hartig, O. (eds.) ESWC 2017. LNCS, vol. 10249, pp. 103–119. Springer, Cham (2017). https://doi.org/10.1007/978-3-319-58068-5_7
24. Singh, R., et al.: Synthesizing entity matching rules by examples. PVLDB **11**(2), 189–202 (2017)
25. Tsai, M.H., Ho, C.H., Lin, C.J.: Active learning strategies using SVMs. In: The 2010 International Joint Conference on Neural Networks (IJCNN), Barcelona, pp. 1–8. IEEE (2010)
26. Wang, Q., Vatsalan, D., Christen, P.: Efficient interactive training selection for large-scale entity resolution. In: Cao, T., Lim, E.-P., Zhou, Z.-H., Ho, T.-B., Cheung, D., Motoda, H. (eds.) PAKDD 2015. LNCS (LNAI), vol. 9078, pp. 562–573. Springer, Cham (2015). https://doi.org/10.1007/978-3-319-18032-8_44
27. Wang, S., Xiao, X., Lee, C.H.: Crowd-based deduplication: an adaptive approach. In: ACM SIGMOD, Melbourne, pp. 1263–1277 (2015)

Encoding Hierarchical Classification Codes for Privacy-Preserving Record Linkage Using Bloom Filters

Rainer Schnell[(✉)] and Christian Borgs

German Record Linkage Center, University of Duisburg-Essen,
Lotharstr. 65, 47057 Duisburg, Germany
{rainer.schnell,christian.borgs}@uni-due.de

Abstract. Hierarchical classification codes are widely used in many scientific fields. Such codes might reveal sensitive personal information, for example medical conditions or occupations. This paper introduces a new encoding technique for encrypting sensitive codes, which preserves the hierarchical similarity of the codes. The encoding was developed for the use of hierarchical codes in Privacy-preserving Record Linkage (PPRL). The technique is demonstrated with real-world survey data containing occupational codes (ISCO codes). After describing the construction and its similarity preserving properties, Hierarchy Preserving Bloom Filters (HPBF) are compared with positional q-grams and standard Bloom filters in a PPRL context. The method presented here is similarity preserving for hierarchies, privacy-preserving and will increase linkage quality when used in Bloom filter-based PPRL.

Keywords: Positional Bloom filters · Hierarchy Preserving Bloom Filters · Entity resolution · ISCO codes · Hierarchical similarity · PPRL

1 Introduction

In many research settings, categorising elements with hierarchical categorical schemes is daily practice. Examples include taxonomies in biology, the classification of occupations [18], accident statistics [6,10], entity resolution of corporations using NACE-codes [20], or classifying diseases or causes of death with the ICD.

In data science, matching identifiers of different records (Record Linkage) is a central challenge [2]. In most applications, linkage is done on clear text identifiers. However, numerical attributes, dates and geographical information might also be used for linking. Hierarchical codes can be used for directly linking data as well [9].

For many applications, such hierarchical codes often relate to individuals, i.e. job classifications. Therefore, these codes can be sensitive, requiring special data protection, as would be the case for recording diseases of a person. Using these codes in a data linkage scenario would require encrypting them, which, while retaining discriminatory power, would lead to a loss of the hierarchical

© Springer Nature Switzerland AG 2020
P. Cellier and K. Driessens (Eds.): ECML PKDD 2019 Workshops, CCIS 1168, pp. 142–156, 2020.
https://doi.org/10.1007/978-3-030-43887-6_12

properties of the codes. However, a proper encoding technique for hierarchical codes should: (1) preserve the similarity of different codes if they agree on higher levels and disagree only on lower level details, (2) improve linkage quality by using the information contained in the hierarchy.

To the best of our knowledge, no other privacy-preserving method for the encryption of hierarchical codes has been published so far. Therefore, in this paper we suggest a new encoding technique for hierarchical codes. The new method is tested for use as an additional identifier in Privacy-preserving Record Linkage (PPRL) settings.

2   Methods

The newly suggested method for privacy-preserving hierarchy will be based on Bloom filters, which are commonly used for linking data privately [5, 26].

2.1   Bloom Filters

Bloom filters [1] (BF) are binary vectors with a length of l bits in which information is stored. They were first devised to rapidly check set membership [1], but have been used in other applications as well. For Privacy-preserving Record Linkage, they were first suggested [24] to be used by splitting strings into subsets of the length q (q-grams or n-grams). These q-grams determine a number k of bit positions $B_i \in \{1, \ldots, l\}$ to be set to one in a bit vector initially consisting of l zero bits. Currently, it is recommended to randomly select these bit positions by using the input q-gram together with a password as a seed for a PRNG that selects k random bit positions that are set to a value of one [23]. An example is shown in Fig. 1.

The attractive main property of Bloom filters is that they can be used to encrypt strings in a similarity-preserving way. One method to compute the similarity of two sets A and B of bigrams is the Dice coefficient, which is calculated as the doubled intersect of the two sets divided by the number of elements in both sets:

$$D = \frac{2|A \cap B|}{|A| + |B|}. \tag{1}$$

As can be seen in Fig. 1, the names Sahra and Sarah share three out of four bigrams (subsets of $q = 2$). This gives an unencrypted clear-text Dice similarity of $D = \frac{2*3}{4+4} = 0.75$.

For Bloom filters, the Dice similarity can be computed by comparing the sets of bit positions of two Bloom filters. Here, the Dice similarity of the Bloom filters is very close to the unencrypted bigram similarity, as both Bloom filters have 6 resp. 7 bits set to one, while sharing 5 bit positions. This gives a Dice coefficient for the encrypted names of $D_{BF} = \frac{2*5}{7+6} \approx 0.77$.

Bloom filters have been used for encoding numerical attributes, dates [29] and geographical information [7] as well. Up to now, no encoding technique

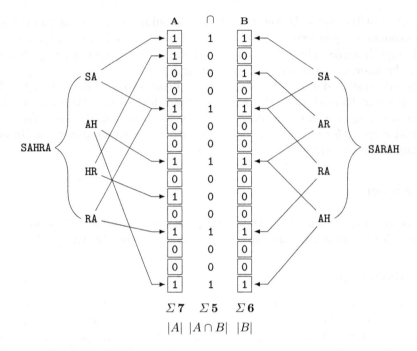

Fig. 1. Bloom filters constructed for two similar names using a length of $l = 15$ bits and $k = 2$ hash functions for each bigram.

for encoding hierarchical codes into Bloom filters has been suggested. A naive approach would use unigrams ($q = 1$). Of course, this approach would result in the loss of all hierarchical information, so using regular Bloom filters for hierarchical codes is not advised.

2.2 Positional BFs (PBFs)

In hierarchical codes, the positions of the code matter. For many applications, the first code position is the top of the hierarchy and important for all following positions, as they change the meaning of all following codes. To the best of our knowledge, no encoding of hierarchical codes into Bloom filters has been suggested before. However, the literature mentions some encodings of positional information of q-grams in strings. The concatenation of an index and the q-gram at the index position was first proposed by [21] as positional q-grams. They have, for example, been used in conjunction with Bloom filters in genome searches [11]. For PPRL settings, it has been used before [2,25]. The application of positional unigrams to hierarchical codes is straightforward: Taking the ISCO-88 code 3213 as an example, a Bloom filter using positional unigrams would hash the values 31, 22, 13, and 34 into the bit vector, where the second digit is the q-gram position.

In contrast, a standard Bloom filter would only map the elements 2, 1 and 3 of the ISCO code 3213 to the bit vector. Although positional unigrams identify the position of a code element, it does not reflect the relative importance of the code positions, as the first code positions in hierarchical codes are usually more important than the last bits. Therefore, we expect that Record Linkage using positional q-grams yields better results than naive representations of hierarchical information, but will still omit information on the relative importance given by the index position.

2.3   Hierarchy Preserving Bloom Filters (HPBFs)

As described in the last section, the code positions are crucial for preserving hierarchies, as the first position determines the meaning of all following codes. The same is true for all following positions. Since existing encoding methods ignore this information, a new method is proposed. It is shown in Fig. 2. The new encoding is based on two modifications of the standard procedure.

First, the code is split into unigrams. The first code position remains as is. All following code positions will contain all unigrams of the previous code positions. This will give more weight to the first positions.

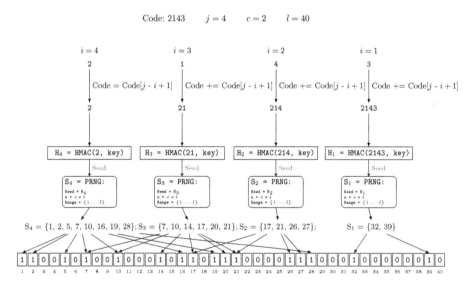

Fig. 2. Constructing Bloom filters for hierarchical codes from the ISCO 88-Code for electrical engineers (2143, code length $j = 4$). Bloom filter length is $l = 40$ with a stream length modifier of $c = 2$. This way, 16 bit positions are set to one. Four bit positions are set to one by more than one PRNG stream (S_i).

Second, the number of bit positions for coding is dependent on the position of the unigram within the code. To achieve this, the first positions receive more bits in the representation than the last positions. In the implementation shown here, the q-grams are hashed with an HMAC [15], such as SHA-3 [19], using a secret key. The numeric representation of the resulting hash is used as a seed for a PRNG, which draws bit positions in the range from 1 to l. The number of bit positions depends on the modifier c and the position within the code.

Since the first positions are more important than the last positions, for a code of length j, the variable $i \in \{j, j-1, \ldots, 1\}$ is multiplied by c, giving $c * i$ elements to draw. In Fig. 2, c is set to a value of 2, leading to eight bit positions to be drawn for the first code position, while the last code position sets only two bit positions to one. This way, the more important the code position in the hierarchy, the more bits are set to one.

Choice of c. In this particular application, the range of possible values for c is small since most codes are limited to three or four digits. This way, possible choices for c depend on the length and the frequency distribution of the actual codes in the space of all possible codes. Up to now, we have chosen c empirically. Finding an optimal value of c will require additional research, since it is highly domain-dependent.

Algorithm 1: HIERARCHYPRESERVINGBLOOMFILTERS($input, pwd = 42, l = 512, c = 1$)

$split \leftarrow$ STRSPLIT($input$)
$BF \leftarrow [0] * l$
local H
local S
for $i \leftarrow length(split)$ **downto** 1
do $\begin{cases} Code \leftarrow Code + split[length(split) - i + 1] \\ H[i] \leftarrow \text{SHA2}(Code, key = pwd) \\ \textbf{comment: } \text{Use numeric representation of hash as seed for PRNG} \\ S[i] \leftarrow \text{PRNG}(Seed = H[i], n = c * i, min = 1, max = l) \\ BF[S[i]] \leftarrow 1 \end{cases}$
return (BF)

The pseudocode for the suggested procedure is shown in Algorithm 1. We denote this encoding as Hierarchy Preserving Bloom Filters (HPBF) since the term Hierarchical Bloom Filters has been used for different data structures [4, 14,17], which are not intended do preserve hierarchies in codes. To empirically test this encoding, three datasets were used.

2.4   Data

Each of the three datasets used in the evaluation is described briefly.

Synthetic Data. Using ICD11-codes for classifying diseases, a sample of $n = 20,000$ ICD11-codes was drawn. A second sample of $n = 14,000$ codes was generated. By chance, both codes will agree or disagree on several code positions.

PASS ISCO-Codes. The PASS panel [28] is a longitudinal study on the effects of unemployment. A classification of the occupation is given by the ISCO-codes (ISCO-88).

Table 1. Exemplary jobs with their ISCO-88 major groups (M), sub-major groups (SM), minor groups (MI) and units (UN) as well as descriptions of them.

M	SM	MI	UN	Description
1				Legislators, senior officials and managers
		131		General managers
			1314	General managers in wholesale and retail trade
			1315	General managers of restaurants and hotels
3				Technicians and associate professionals
	34			Other associate professionals
		341		Finance and sales associate professionals
			3413	Estate agents

Exemplary ISCO codes can be seen in Table 1. The first positions are the major groups, where the largest differences between occupational groups are obvious. Every subsequent code position describes the occupation more precisely.

To study the reliability of the codes the occupation of each person was coded by two independent coding units. For our purpose here, we consider this as an example for the intended application of HPBFs: if the data collection has been done independently, the linkage between two datasets could be enhanced by using the encoded ISCO-codes. This allows a direct comparison of the true positive matches attained by HPBFs compared to exact matching and (positional) Bloom filters.

Synthetic Data for PPRL. Pairs of randomly selected ISCO-codes of the PASS study were randomly assigned to two real-life mortality datasets [25], for which a gold standard linkage solution existed. The personal information (first and last name and date of birth) were encrypted using CLKs with $k = 20$ hash functions and a length of $l = 1000$, while the ISCO-codes were encrypted as either standard Bloom filters, HPBFs or PBFs. These were then included in the CLKs. As the true match state was known, PPRL linkage quality using hierarchical information in different encodings can be evaluated.

2.5 Evaluation Methods

First, evaluation methods relating to the hierarchy-preserving properties of the encryption methods are reviewed. Next, evaluating linkage quality in a PPRL setting is discussed.

Hierarchical Recall and Precision. To map a hierarchical code into a tree structure, code positions form the tree leaves, where each position determines a level.
An example is shown in Fig. 3.

To calculate precision and recall for a tree-based classification, a modification
of the standard definitions of precision and recall is necessary [27]. This can best
be explained by an example (see Fig. 3). Here, the true value (dark green, denoted
as Y_{labels}) is compared to a second classification (dark blue, denoted as $\hat{Y}_{labels}$).
Let the true code be ZBC, while the second classification is the code ZBD.

The hierarchical precision is then calculated as the number of agreements on
the labels ($\hat{Y}_{labels} \cap Y_{labels}$) divided by the number of labels given by a classifier
($\hat{Y}_{labels}$).

The hierarchical recall is calculated as the number of agreements on the labels
($\hat{Y}_{labels} \cap Y_{labels}$) divided by the number of labels given by the true label (Y_{labels}).

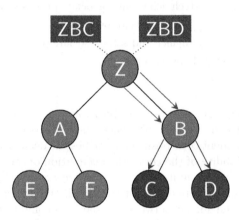

Fig. 3. Two example codes, Y_{labels} = ZBC (dark green) and $\hat{Y}_{labels}$ = ZBD (dark blue)
and their resulting tree structure. The path for querying both codes is drawn with
arrows. Both codes only differ on the final node, sharing two nodes (ancestors) on a
higher hierarchical level. Note that codes containing A, E and F are used in the full
classification, but not in the example codes ZBC and ZBD. (Color figure online)

In Fig. 3, the two exemplary codes (Y_{labels} = ZBC and $\hat{Y}_{labels}$ = ZBD) are
compared by tracing their sub-tree within the full tree. The final nodes for both
codes are C and D. The *ancestors* of a given node are all nodes on the path to
the root node (Z in this case) [12]. In this example, the sets of ancestors would
be $Y = \{Z, B, C\}$ and $\hat{Y} = \{Z, B, D\}$. For this particular set of codes, both
hierarchical precision and hierarchical recall can be calculated as:

$$\text{Hierarchical Recall} = R_h = \frac{|\hat{Y}_{labels} \cap Y_{labels}|}{|Y_{labels}|}, \tag{2}$$

and

$$\text{Hierarchical Precision} = P_h = \frac{|\hat{Y}_{labels} \cap Y_{labels}|}{|\hat{Y}_{labels}|}, \tag{3}$$

with the mean of both serving as the F-Score:

$$F_h = \frac{1}{2}(P_h + R_h).\qquad(4)$$

Both Y and $\hat{Y}$ agree on two ancestors ($\{Z, B\}$), giving a size of the elements in their intersect of $|\hat{Y}_{labels} \cap Y_{labels}| = 2$. Both have three ancestors ($|\hat{Y}_{labels}| = 3$ and $|Y_{labels}| = 3$). Therefore, $P_h = R_h = F_h = \frac{2}{3} \approx 0.667$.

Please note that hierarchical precision and recall require information on the actual position within a tree. Therefore, it can only be used if the hierarchical information is preserved in an encoding. In the following empirical study, we compare the pairwise similarity of hierarchical precision and recall in the clear-text with the pairwise Dice similarity of HPBFs.

Similarity by Linkage Category. If an encryption method is hierarchy-preserving, a full match of codes should yield a higher similarity than partial matches. In addition, a difference in the last characters of a code should result in a higher similarity than a difference in the first code positions, as these are usually more important for the code hierarchy. This idea is captured by the classification of partial matches by Klug et al. [13]. In their application, they used ICD codes and classified the type of agreement into six classes:

1. Full match (no code positions differ)
2. Subgroups differ (last two positions disagree)
3. Subgroups and fourth character differ (diagnostic subgroups differ)
4. Only the first two characters match (only diagnostic groups match)
5. Only the first character matches (only diagnostic chapter matches)
6. Full non-match (all code positions differ)

A full match should lead to similarities close to one, while a full non-match should give similarity values close to zero. Ideally, for all categories in between, the similarity values should not overlap, so that the range of similarity coefficients for each category is small. We compared Dice similarities of Bloom filters within each category given by the classification of Klug et al. [13].

Evaluation of Linkage Quality of PPRL Methods Using Hierarchical Codes. For linking all three Bloom filter-based PPRL methods, we used Multibit trees. Multibit trees were suggested for chemometrics by Kristensen et al. [16] and proposed for PPRL by Schnell [22].

The efficiency of Multibit trees for comparing Bloom filters is due to the fact that possible pairs below a pre-set similarity threshold are not evaluated. Therefore, Multibit trees are being used as an error-tolerant blocking method. The implementation of Multibit trees uses the Tanimoto similarity T as a similarity measure. T is defined as number of bits set to 1 in both vectors A and B divided by the total number of bits set to 1 in A and B:

$$T(A, B) = \frac{\Sigma_i(A_i \wedge B_i)}{\Sigma_i(A_i \vee B_i)}\qquad(5)$$

Lower thresholds will result in more pair comparisons and a higher number of false positive classifications. Conversely, the amount of true matches will increase as well.

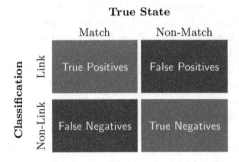

Fig. 4. Outcomes for linkage pairs by classification and true matching state.

For the empirical evaluation of the suggested method, we use the traditional evaluation criteria of precision and recall

$$\text{Recall} = \frac{\text{TP}}{\text{TP} + \text{FN}}, \tag{6}$$

$$\text{Precision} = \frac{\text{TP}}{\text{TP} + \text{FP}}, \tag{7}$$

$$\overline{F} = \frac{1}{2}(\text{Recall} + \text{Precision}), \tag{8}$$

as given by Fig. 4 and the corresponding Eqs. 6 to 8. Following the critique by [8], we use the unweighted arithmetic mean of precision and recall ($\overline{F}$) instead of the harmonic mean (F-Score).

3   Results

First, results concerning the hierarchy-preserving properties of the encryptions is reported, before linkage quality in PPRL settings is addressed.

Similarity by Linkage Category. Hierarchy-preserving encryptions should lead to similar Dice coefficients for each linkage category (as defined by [13]; see Sect. 2.5). To test this, the pairwise Dice similarities of the Bloom filters were computed for each encryption method (HPBF, PBF and standard BF).

The results are shown in Fig. 5, where the box plots for each category and encryption method are shown.

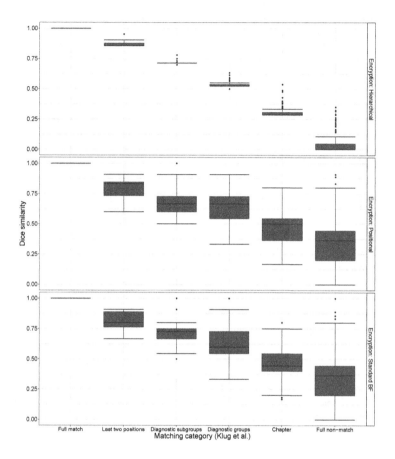

Fig. 5. Linkage categories and Dice similarity by method (HPBF with $c = 1$).

The HPBFs discriminate the categories very well and with little spread. In contrast, both the positional and standard BFs show a wide spread of Dice coefficients for all categories but the full match category. The plot demonstrates that Hierarchy Preserving Bloom filters have more discriminating power of encoded hierarchical codes than previous methods. To explore the properties of the HPBFs further, we examine the functional relationship between the Dice coefficient of pairs of Bloom filters and the corresponding hierarchical precision and recall of unencrypted codes.

Hierarchical Precision and Recall. For each pair of ISCO codes, the hierarchical precision and recall were computed [27]. Since ISCO occupational codes always have four characters, the number of ancestors will always be four. This way,

$R_h = P_h$, which is why only hierarchical recall (R_h) will be reported here. If the encryption is hierarchy-preserving, the hierarchical recall should be a monotone function of the Dice similarity of the two encrypted codes. By comparing the ISCO-codes of the same person generated by two different coding units, this relationship is shown for all methods in Fig. 6.

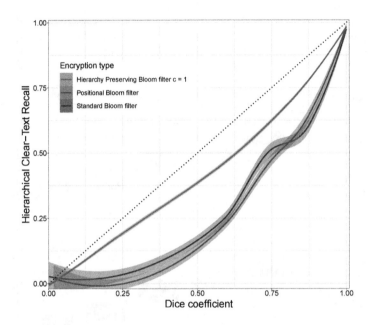

Fig. 6. Hierarchical recall plotted against the Dice similarity of the encrypted codes for Hierarchy Preserving Bloom filters (HPBF), positional Bloom filters (PBF) and Standard Bloom filters (BF). The lines shown are loess smoothers (based on $n = 5,033$ code pairs) with 2 standard errors (the shaded areas).

Given this dataset, standard and positional Bloom filters perform worse than Hierarchy Preserving Bloom filters. Furthermore, the standard errors are considerably larger at lower hierarchical recall and Dice coefficient values. In contrast, the HPBFs have smaller standard errors. Furthermore, the numeric value of the hierarchical recall is better approximated by the Dice coefficients of the HPBFs (the smoothed curve is much closer to the diagonal reference line).

3.1   Privacy-Preserving Record Linkage (PPRL)

In a PPRL setting, using as many stable identifiers as possible is recommended [23]. However, hierarchical codes can be unstable, especially when relying on humans to classify hierarchical codes. In the given dataset, two encoding units classified the persons' jobs independently. Of the resulting $n = 5,033$ ISCO code pairs, only $2,497$ matched exactly.

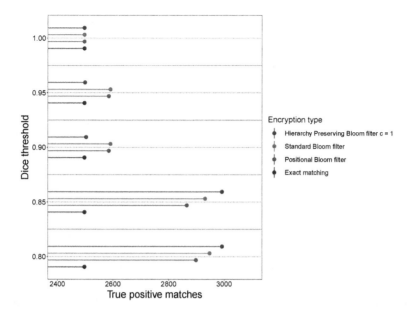

Fig. 7. True positive ISCO code matches by method and similarity threshold.

Increasing the number of matches by accepting differing codes at lower hierarchical levels (for example, 3122 and 3121: Computer equipment operators (3122) and computer assistants (3121)) is shown in Fig. 7. Here, the number of true positive matches increases substantially when lowering the level of accepted minimal similarity. HPBFs with a stream length modifier of $c = 1$ show the highest number of true positives at all similarity thresholds below 0.90.

However, even allowing for errors yields only about 3,000 matching code pairs. Obviously, using the ISCO code as single identifier, even when allowing for errors, is not sufficient for linking. Therefore, we studied the performance of HPBFs in a PPRL setting.

The bit vectors resulting from the standard, positional (PBF) and Hierarchy Preserving Bloom filter (HPBF) encryptions are inserted with an OR operation into standard CLKs (composite Bloom filters [23]) for names and dates of birth (with $k = 20$ hash functions and $l = 1000$ bits[1]). These CLKs are evaluated as described in Sect. 2.5.

The resulting mean of precision and recall is shown in Fig. 8. Although the same amount of information is encoded by all three methods, the combination of HBPFs with CLKs considerably improves the linkage quality compared to combining PBFs with CLKs and standard Bloom filters with CLKs.

---

[1] Above a certain minimal length, choices for l are arbitrary as long as k is adjusted accordingly.

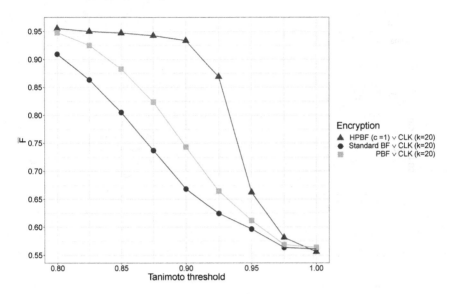

Fig. 8. Arithmetic mean of precision and recall $(\overline{F})$ of standard CLKs with $k = 20$ hash functions combined with ISCO codes encoded in Bloom filters (BF), positional Bloom filters (PBF) and Hierarchy Preserving Bloom filters (HPBF), linked using several Tanimoto thresholds.

4   Conclusion

In many research applications, codes representing a hierarchical relation are used. Preserving similarities of hierarchical codes with encrypted identifiers was shown to be possible with Hierarchy Preserving Bloom filters (HPBFs) presented in this paper. Using these encodings in Privacy-preserving Record Linkage (PPRL) settings, linkage quality will improve compared to previous methods for encoding hierarchical codes.

However, security of HPBFs remains an issue. As has been shown by Christen et al. [3], bit patterns within Bloom filters can be used as attack vectors in cryptographic attacks on Bloom filter encodings. Hence, reducing only the number of frequent Bloom filters is not sufficient to prevent attacks. The implications of these results for methods for encoding numerical and geographical information [7,29] have not been studied at all in the literature. Therefore, a study on the cryptographic properties and attack methods, as well as options to prevent these attacks for Hierarchy Preserving Bloom filters is subject of ongoing research.

Acknowledgements. The authors would like to thank the three anonymous reviewers for the comments, which improved the paper substantially.

References

1. Bloom, B.H.: Space/time trade-offs in hash coding with allowable errors. Commun. ACM **13**(7), 422–426 (1970)
2. Christen, P.: Data Matching - Concepts and Techniques for Record Linkage, Entity Resolution, and Duplicate Detection. Springer, Heidelberg (2012). https://doi.org/10.1007/978-3-642-31164-2
3. Christen, P., Vidanage, A., Ranbaduge, T., Schnell, R.: Pattern-mining based cryptanalysis of Bloom filters for privacy-preserving record linkage. In: Phung, D., Tseng, V.S., Webb, G.I., Ho, B., Ganji, M., Rashidi, L. (eds.) PAKDD 2018. LNCS (LNAI), vol. 10939, pp. 530–542. Springer, Cham (2018). https://doi.org/10.1007/978-3-319-93040-4_42
4. Crainiceanu, A.: Bloofi: a hierarchical Bloom filter index with applications to distributed data provenance. In: Darmont, J., Pedersen, T.B. (eds.) Proceedings of the 2nd International Workshop on Cloud Intelligence (Cloud-I 2013) Riva del Garda, Trento, Italy, 26 August. ACM, New York (2013)
5. Dantas Pita, R., et al.: On the accuracy and scalability of probabilistic data linkage over the Brazilian 114 million cohort. IEEE J. Biomed. Health Inform. **22**(2), 346–353 (2018)
6. European Commission: Eurostat: European Statistics on Accidents at Work (ESAW): Summary Methodology. Publications Office of the European Union, Luxembourg (2013)
7. Farrow, J.: Privacy preserving distance-comparable geohashing. In: International Health Data Linkage Conference, Vancouver, 28–30 April (2014)
8. Hand, D., Christen, P.: A note on using the F-measure for evaluating record linkage algorithms. Stat. Comput. **28**(3), 539–547 (2017). https://doi.org/10.1007/s11222-017-9746-6
9. Hejblum, B.P., et al.: Probabilistic record linkage of de-identified research datasets with discrepancies using diagnosis codes. Sci. Data **6**, 180298–180309 (2019). Article no. 180298
10. Jacinto, C., Santos, F.P., Soares, C.G., Silva, S.A.: Assessing the coding reliability of work accidents statistical data: how coders make a difference. J. Saf. Res. **59**, 9–21 (2016)
11. Kerschbaum, F., Beck, M., Schönfeld, D.: Inference control for privacy-preserving genome matching. CoRR abs/1405.0205 (2014)
12. Kiritchenko, S., Matwin, S., Nock, R., Famili, A.F.: Learning and evaluation in the presence of class hierarchies: application to text categorization. In: Lamontagne, L., Marchand, M. (eds.) AI 2006. LNCS (LNAI), vol. 4013, pp. 395–406. Springer, Heidelberg (2006). https://doi.org/10.1007/11766247_34
13. Klug, S.J., Bardehle, D., Ressing, M., Schmidtmann, I., Blettner, M.: Vergleich von ICD-Kodierungen zwischen Mortalitätsstatistik und studieninterner retrospektiver Nachkodierung. Gesundheitswesen **71**(4), 220–225 (2009)
14. Koloniari, G., Pitoura, E.: Bloom-based filters for hierarchical data. In: 5th Workshop on Distributed Data and Structures, Thessaloniki, 13–14 June 2003 (2003)
15. Krawczyk, H., Bellare, M., Canetti, R.: HMAC: Keyed-hashing for message authentication. Internet RFC 2104 (1997)
16. Kristensen, T.G., Nielsen, J., Pedersen, C.N.S.: A tree-based method for the rapid screening of chemical fingerprints. Algorithms Mol. Biol. **5**(1), 9–20 (2010)
17. Lillis, D., Breitinger, F., Scanlon, M.: Hierarchical Bloom filter trees for approximate matching. J. Digital Forensics Secur. Law **13**(1), 81–96 (2018)

18. McLean, D., et al.: Evaluation of the quality and comparability of job coding across seven countries in the INTEROCC study. Occup. Environ. Med. **68**(Suppl. 1), A61 (2011)
19. National Institute of Standards and Technology: Secure hash standard (SHS). FIPS PUB 180-4 (2012)
20. Peruzzi, M., Zachmann, G., Veugelers, R.: Remerge: regression-based record linkage with an application to PATSTAT. Technical report 2014/10iii, Bruegel Working Paper, Brussels (2014)
21. Riseman, E.M., Hanson, A.R.: A contextual postprocessing system for error correction using binary n-grams. IEEE Trans. Comput. **5**, 480–493 (1974)
22. Schnell, R.: An efficient privacy-preserving record linkage technique for administrative data and censuses. J. Int. Assoc. Off. Stat. **30**(3), 263–270 (2014)
23. Schnell, R.: Privacy-preserving record linkage. In: Harron, K., Goldstein, H., Dibben, C. (eds.) Methodological Developments in Data Linkage, pp. 201–225. Wiley, Hoboken (2016)
24. Schnell, R., Bachteler, T., Reiher, J.: Privacy-preserving record linkage using Bloom filters. BMC Med. Inform. Decis. Mak. **9**(1), 41–52 (2009)
25. Schnell, R., Richter, A., Borgs, C.: A comparison of statistical linkage keys with Bloom filter-based encryptions for privacy-preserving record linkage using real-world mammography data. In: Proceedings of the 10th International Joint Conference on Biomedical Engineering Systems and Technologies (BIOSTEC 2017), pp. 276–283 (2017)
26. Smith, D.: Secure pseudonymisation for privacy-preserving probabilistic record linkage. J. Inf. Secur. Appl. **34**, 271–279 (2017)
27. Sokolova, M., Lapalme, G.: A systematic analysis of performance measures for classification tasks. Inf. Process. Manage. **45**(4), 427–437 (2009)
28. Trappmann, M., Beste, J., Bethmann, A., Müller, G.: The PASS panel survey after six waves. J. Labour Market Res. **46**(4), 275–281 (2013)
29. Vatsalan, D., Christen, P.: Privacy-preserving matching of similar patients. J. Biomed. Inform. **59**, 285–298 (2016)

Machine Learning for Cybersecurity (MLCS)

Are Network Attacks Outliers?
A Study of Space Representations
and Unsupervised Algorithms

Félix Iglesias[1]([⊠]) [iD], Alexander Hartl[1] [iD], Tanja Zseby[1] [iD], and Arthur Zimek[2] [iD]

[1] Institute of Telecommunications, TU Wien, Vienna, Austria
{felix.vazquez,alexander.hartl,tanja.zseby}@tuwien.ac.at
[2] Departmaent of Mathematics and Computer Science (IMADA),
University of Southern Denmark (SDU), Odense, Denmark
zimek@imada.sdu.dk

Abstract. Among network analysts, "anomaly" and "outlier" are terms commonly associated to network attacks. Attacks are outliers (or anomalies) in the sense that they exploit communication protocols with novel infiltration techniques against which there are no defenses yet. But due to the dynamic and heterogeneous nature of network traffic, attacks may look like normal traffic variations. Also attackers try to make attacks indistinguishable from normal traffic. Then, are network attacks actual anomalies? This paper tries to answer this important question from analytical perspectives. To that end, we test the *outlierness* of attacks in a recent, complete dataset for evaluating Intrusion Detection by using five different feature vectors for network traffic representation and five different outlier ranking algorithms. In addition, we craft a new feature vector that maximizes the discrimination power of outlierness. Results show that attacks are significantly more outlier than legitimate traffic—specially in representations that profile network endpoints—, although attack and non-attack outlierness distributions strongly overlap. Given that network spaces are noisy and show density variations in non-attack spaces, algorithms that measure outlierness locally are less effective than algorithms that measure outlierness with global distance estimations. Our research confirms that unsupervised methods are suitable for attack detection, but also that they must be combined with methods that leverage pre-knowledge to prevent high false positive rates. Our findings expand the basis for using unsupervised methods in attack detection.

Keywords: Outlier detection · Network traffic analysis · Feature selection

1 Introduction

The study of previous research on network security analysis (NTA) at network level [13] discloses three main claimed goals: (a) attack detection, (b) anomaly

© Springer Nature Switzerland AG 2020
P. Cellier and K. Driessens (Eds.): ECML PKDD 2019 Workshops, CCIS 1168, pp. 159–175, 2020.
https://doi.org/10.1007/978-3-030-43887-6_13

detection, and (c) traffic classification. These three topics are not the same, but undoubtedly overlap. For instance, traffic classifications often include classes that are attacks. An anomaly might be an attack, but an attack does not necessarily show itself to be an anomaly. The traffic features selected for the analysis obviously play a determining role to see if a network attack expresses itself as an anomaly or not, but also the analysis perspective is relevant [33]. For example, Distributed Denial of Service (DDoS) attacks usually appear as anomalous peaks in network monitors that observe traffic as time series [12]; however, they are hardly anomalies from a spatial perspective, in which they can take a significant portion of the total captured traffic—note that DDoS attacks try to harm targets by bombarding them with false connection requests. Actually, DDoS and other types of illegitimate traffic (e.g., scanning activities) have become so common that they can rarely be considered anomalies in most networks any more [11,22].

When the term "outlier" comes into play, things become even more confusing. "Anomaly" and "outlier" are not smooth synonyms, and even the description of *outlier* can be ambiguous in practical implementations [39]. For instance, it is common to find small groups of close traffic instances that are distant from the data bulk. Together, they form an *outlying cluster*; individually, instances can be deemed as outliers or not. Even in spite of such ambiguities, in related research the meaning of *anomaly* is commonly assumed without discussion. Carefully reviewing such works (and excluding time series analysis), the empirical meaning of anomaly inferred from experiments habitually corresponds to *network attacks that show outlierness*. Some authors identify attacks as anomalies and perform their detection with outlier-based techniques [6,18,38]. Also many outlier-based detection proposals appear in other field surveys [5,8,25].

But, do network attacks actually show themselves as outliers or outlying clusters? This is the crux that will make unsupervised methods effective for attack detection or not. Related works take it for granted, but the question must be analytically answered, not in vain most attacks are designed to pass unperceived. As a starting point, we highly recommend that research works on *anomaly detection* in NTA clearly establish their definition of *anomaly*. Otherwise, whenever theoretical proposals are implemented into real scenarios—far from lab conditions—such methods are prone to trigger detection alarms in view of many harmless, meaningless, noisy instances. This discussion is critical because precisely unacceptable high false positives rates is what slows down the adoption of machine learning in real-world network attack detectors [15,17]. If this is true for supervised machine learning, it is even more severe for unsupervised methods, which are also commonly evaluated with the same Intrusion Detection System (IDS) datasets (e.g., [6,18,38]). Note that IDS datasets are usually not designed to match realistic ratios between normal and attack traffic, but to offer a variety of attack classes with sufficient representation in the dataset [16]. This is not ideal for unsupervised methods because they work by learning from the sample placement and space geometries drawn by the analyzed data. From here, and without considering irrelevant, easy-to-detect, illegitimate traffic that has become common, it naturally follows that the real-world ratio attack/non-attack is going to be considerably lower than in IDS datasets. Therefore, the probabilities for a detected anomaly to be an actual attack drop dramatically. How the base-rate fallacy affects IDS was already advised by Axelsson in [4].

The previous observations do not imply that unsupervised methods are not valid for attack detection; instead, they introduce the necessity for evaluating the outlierness of network attacks and to investigate if unsupervised methods suffice by themselves for the actual detection in real implementations. Note that signature-based detection or supervised approaches are limited in detecting novel threats and zero-day attacks; therefore, the contribution of unsupervised approaches is deemed highly valuable. A last challenge that unsupervised methods must additionally face is their traditional high computational complexity. Most popular outlier detection algorithms are naturally *instance-based* and show considerable time and memory overloads [9,30]. Network traffic analysis for attack detection must be fast and lightweight, since it must deal with ever-growing volumes of traffic (big data, streaming data) and is expected to promptly react when malicious instances are discovered.

The main contribution of this paper is answering the following questions:

- **Are network attacks outliers?** We study five popular and recent space representations used in NTA security applications and experiment with five popular and recent unsupervised outlier detection algorithms in order to elucidate if network traffic attacks show a distinguishable outlier nature.
- **What are the most suitable feature representations for attack detection?** We investigate which existing feature vectors perform best in conjunction with outlier detection.
- **Is the observed outlierness sufficient as indicator for implementing real-world attack detection?** We discuss if the detected outlierness suffices for implementing effective detectors in real environments. Additionally, we propose a new vector that maximizes attack/non-attack separation.

Unlike most papers that apply outlier detection in NTA, we do not use the KDD-Cup98', KDD-Cup99' or NSL-KDD datasets, which have not been representative any more for a long time. Moreover, such datasets use a set of ad-hoc features whose extraction is obscure, costly and unfeasible for modern lightweight detectors. Instead, our experiments are conducted on the CICIDS2017 dataset [34], which is one of the most complete, reliable IDS evaluation datasets to date. As for the selected features, we study five vector spaces created by the CAIA [36], CISCO-Joy [3], Consensus [14], TA [22] and AGM [21] formats. Outlierness ranks are obtained by using five different algorithms: k-nearest neighbours [32], LOF [7], HBOS [19], isolation Forest [27], and SDO [23]. Scripts and experiments are openly available for replication and reuse in [10].

2   Problem Spaces in NTA

When considering traffic at network level, the possibilities for extracting features are immense. Irrespective of the specific features, NTA is mostly faced by constructing homogeneous vectors from different perspectives, therefore leading to problem spaces where instances correspond to:

- *Packets*, meaning the contents of every datagram exchanged between two network devices. This type of analysis allows *dpi* (deep packet inspection), which has become obsolete due to high data rates, privacy concerns and encryption. Hence, packet based analysis is computationally too demanding and unable to explore modern network traffic with reasonable costs.
- *Flows.* The definition of a *traffic flow* given by IPFIX [1] is extremely flexible. A flow is defined "as a set of packets or frames passing an Observation Point in the network during a certain time interval. All packets belonging to a particular Flow have a set of common properties", which can vary depending on the use case. We underline two special cases: (a) *Application-based Flows.* For the last three decades flows have been principally defined with the 5-tuple key: [IP source, IP destination, source Port, destination Port, Protocol], which states the communication for a specific application between endpoints, e.g., a TCP connection. However, the use of the 5-tuple is not justified in terms of security, it is simply a reminiscence from network policies implemented in the 1990's that have become a *de facto* standard. A myriad of works in NTA for security assume the 5-tuple key (e.g., [24, 26, 35–37]). (b) *Endpoint-based Flows.* Yet rare in literature, a flow key can also be a 1-tuple, i.e., using the endpoint address as flow key (either IP source or IP destination). In such cases, a flow summarizes the behavior of a single device (in its role of source or destination) for a defined observation window.
- *Aggregated flows.* In this scenario, a set of features are timely aggregated to reveal the current status of the network as a whole (e.g., number of sent packets per second). Such approach usually analyzes *time series* and is effective to quickly detect attacks and events that have a strong impact in the network. However, this top perspective is useless to capture subtle, more selective attacks or threats that only aim at one or few destinations, or show a slow propagation. Also, tracking back attack sources is a challenging task.

We discard the analysis of *packets* and *aggregated flows* due to the reasons given above and focus on *flows* to capture application or device behavior (5-tuple and 1-tuple flow key respectively). We select a set of feature vectors that are popular in the NTA literature or have been recently proposed. They are:

- **CAIA vector.** As coined in [28], we use CAIA to refer to the feature vector originally proposed by Williams et al. [36]. The same vector has been commonly applied (as defined or with minimal variations) in the context of NTA, specifically when using machine learning-based solutions, e.g., [26, 35, 37]. The original CAIA vector stores bidirectional information and consists of 22 features. We extended it to 30 features as in [26].
- **Consensus vector.** In [14] a set of features for NTA are selected based on a meta-study including 71 of the most relevant, cited papers in NTA. This work concludes with 12 relevant features. We extend them based on the considerations discussed in [14] and [28], obtaining a final 20-feature vector.

Table 1. Studied NTA representations (feature vectors).

Object: Source hosts (unidirectional)			
Key: srcIP; **Obs.window:** 10sec			
Features (22 total):			

AGM

#dstIP	#dstPort	#TTL	#pktLength
mode_dstIP[1]	mode_dstPort	mode_TTL	mode_pktLength
pkts_mode_dstIP	pkts_mode_dstPort	pkts_mode_TTL	pkts_mode_pktLength
#srcPort	#protocol	#TCPflag	pkts
mode_srcPort	mode_protocol	mode_TCPflag	
pkts_mode_srcPort	pkts_mode_protocol	pkts_mode_TCPflag	

Object: Flows (unidirectional)
Key: srcIP, dstIP, protocol; **Obs.window/timeout:** 60sec
Features (13 total)

Time-Activity

srcPort	maxton	seconds-active
dstPort	minton	bytes_per_seconds-active
protocol	maxtoff	pkts_per_seconds-active
bytes	mintoff	
pkts	interval	

Object: Flows (bidirectional)
Key: srcIP, dstIP, srcPort, dstPort, protocol; **Idle/active timeout:** 300sec/1800sec
Features (30 total):

CAIA

protocol	max_srcPktLength	max_srcPktIAT	#srcTCPflag:fin
duration	stdev_srcPktLength	stdev_srcPktIAT	#srcTCPflag:cwr
srcPkts	min_dstPktLength	min_dstPktIAT	#dstTCPflag:syn
srcBytes	mean_dstPktLength	mean_dstPktIAT	#dstTCPflag:ack
dstPkts	max_dstPktLength	max_dstPktIAT	#dstTCPflag:fin
dstBytes	stdev_dstPktLength	stdev_dstPktIAT	#dstTCPflag:cwr
min_srcPktLength	min_srcPktIAT	#srcTCPflag:syn	
mean_srcPktLength	mean_srcPktIAT	#srcTCPflag:ack	

Object: Flows (bidirectional)
Key: srcIP, dstIP, srcPort, dstPort, protocol; **Idle/active timeout:** 300sec/1800sec
Features (20 total):

Consensus

srcBytes	dstPort	mode_dstPktLength	median_srcPktLength
srcPkts	protocol	min_srcPktLength	median_dstPktLength
dstBytes	duration	median_srcPktIAT	min_dstPktLength
dstPkts	max_srcPktLength	variance_srcPktIAT	median_dstPktIAT
srcPort	mode_srcPktLength	max_dstPktLength	variance_dstPktIAT

Object: Flows (bidirectional)
Key: srcIP, dstIP, srcPort, dstPort, protocol; **Idle/active timeout:** 300sec/1800sec
Features (650 total):

Cisco[2]

srcPort	#certificates
dstPort	#SAN
packet length sequence (100 features)	offered cipherSuites (139 features)
IAT sequence (100 features)	selected cipherSuites (26 features)
byte distribution (256 features)	offered TLSExtensions (13 features)
public key length	accepted TLS extensions (11)

1: Removed from the analysis.   2: The Cisco-Joy tool can extract more features. We removed features that did not contain usable information in the CICIDS2017 dataset.

– **Cisco-Joy vector.** Anderson et al. recently proposed this feature vector, which is able to discriminate attacks in supervised learning and is suitable for encrypted traffic [2,3]. It contains 650 features and can be easily extracted by using the Cisco/Joy open tool, https://github.com/cisco/joy.
– **Time-Activity vector (TA).** The Time-Activity vector [22] uses a 3/5-tuple key and is unidirectional. It was devised to profile flows from a time-behavioral perspective, allowing lightweight characterization of traffic by means of clustering methods. The final vector is formed by 13 features.

- **AGM vector.** Designed for the discovery of patterns in the Internet Background Radiation [21], this vector allows profiling traffic sources or destinations. The *basic AGM vector* contains 22 features, which are extended after removing nominal features or transforming them into dummy variables if distributions are concentrated on few values (e.g., more than 90% of traffic uses TCP, UDP or ICMP). The *extended AGM vector* is purely numerical.

The CAIA, Consensus, TA, and AGM vectors are compared in [28] for supervised attack detection with the UNSW-NB15 dataset [29]. Table 1 shows vector features in the format used here. We apply the nomenclature described in [28]. We refer the interested reader to the original papers for detailed descriptions.

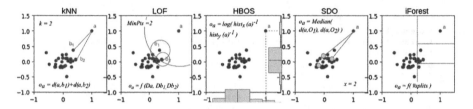

Fig. 1. A quick overview of how the studied algorithms estimate the outlierness (o_a) of a random point a.

3   Outlier Detection Algorithms

In this section we briefly introduce the used outlier detection algorithms (a visual overview of the different approaches is shown in Fig. 1):

kNN. The k-nearest neighbor distance (kNN) has been used for measuring object isolation in [32], where each instance outlierness is ranked based on the distance to its kth nearest neighbor. kNN is an instance-based method where estimations are locally approximated. It does not require training and the computational effort appears every time that a new instance must be evaluated and compared with the previous ones. kNN requires setting a k parameter.

LOF. The Local Outlier Factor algorithm (LOF) entailed a considerable enhancement in the task of measuring instance outlierness within data [7], generating a varied family of derived algorithms [33]. LOF compares the density estimate (D_i) based on the k-nearest neighbors with the density estimates for each of the k-nearest neighbors, thus adapting to different local densities. LOF is also an instance-based method and does not require training. In a recent comparison, LOF has shown to be a good benchmark solution, which, in general, has not been significantly outperformed by more recent methods in terms of accuracy [9]. LOF uses the *MinPts* parameter, which is equivalent to k in kNN.

HBOS. Histogram-Based Outlier Detection (HBOS) [19] is a simple, straightforward algorithm based on evaluating the feature empirical distributions of the analyzed dataset (i.e., histograms for continuous features and frequency tables for nominal features). Since it assumes feature independence, it sacrifices precision to achieve fast performances in linear times. Outlierness is calculated based on the relative position of the instance feature values with regard to the obtained empirical distributions (*hist*). HBOS does not require parameterization, but for the histogram binning, which allows static bin-widths (k equal width bins) or dynamic bin-widths (in every bin falls N/k instances, being N the total number of instances). In our experiments, bins-widths are "static".

iForest. Isolation Forest (iForest) [27] is a model-based outlier ranking method that shows linear time complexity with low memory requirements even in front of large datasets. The operation principle is as follows: for a given instance, features and splits are randomly selected in a procedure that progressively reduces the range of feature values until the instance is isolated (i.e., the only instance in the remaining subspace). The number of splits defines the outlierness value of the instance, since outliers are expected to be easier to isolate (less splits) than inliers (more splits). The partitioning procedure can be abstracted as a tree (an iTree), therefore an iForest provides the weighted evaluations of a set of iTrees. During training, iTrees are built using the training dataset; in application phases, instances pass through iTrees to obtain outlierness scores. iForest parameters are: t, the number of estimators or iTrees; ψ, the sample size to train every iTree; and f, number of features passed to each iTree.

SDO. The Sparse Data Observers (SDO) algorithm is a model-based unsupervised outlier ranking method that has been designed to provide fast evaluations and be embedded in autonomous frameworks [23]. SDO is conceived to avoid the common bottleneck problems implied by traditional instance-based outlier detection when a continuous evaluation of incoming instances is demanded. SDO creates a low density data model by sampling a training population. During training, model instances—called *observers* (O)—are evaluated in a way that low-active observers are removed. Thus, the low density model becomes free of potential outliers. In application phases, observers provide instance outlierness based on joined distance estimations. SDO is light, easy to tune, and makes the most of pre-knowledge. SDO parameters are intuitive and stable, *rule of thumb* parameterization works well in most applications. Parameters are: k, the number of observers; x, the number of closest observers that evaluate every instance; and q (or qv), which establishes the threshold for the removal of low-active observers.

4   Dataset

The CICIDS2017 dataset [34] was recently published by the Canadian Institute for Cybersecurity (CIC). The CIC has developed some of the most widely-used IDS and IPS (Intrusion Prevention Systems) datasets in research for the last decades, including NSL-KDD and ISCX series. The CICIDS2017 dataset accomplishes the quality criteria that IDS/IPS datasets must meet in order to provide suitable evaluation tests. These criteria [16] control that data is complete, realistic, representative, diverse, and heterogeneous in terms of protocols, attacks, and legitimate uses, as well as in formats and supporting metadata. In CICIDS2017, attack families are implemented according to the most common security threats reported by McAfee in 2016, including: Web based, Brute force, DoS, DDoS, Infiltration, Heart-bleed, Bot, and Scan. The CICIDS2017 dataset is available at: https://www.unb.ca/cic/datasets/ids-2017.html.

Table 2. Used parameters in the experiments.

	Consensus	CAIA	AGM	TA	Cisco-Joy
kNN	$k = 2$	$k = 2$	$k = 15$	$k = 3$	$k = 15$
LOF	$MinPts = 5$	$MinPts = 5$	$MinPts = 18$	$MinPts = 5$	$MinPts = 39$
HBOS	$k = 20$	$k = 22$	$k = 992$	$k = 21$	$k = 20$
iForest	$t = 50, f = 37$	$t = 95, f = 26$	$t = 96, f = 2$	$t = 64, f = 1$	$t = 73, f = 428$
	$\psi = 860$	$\psi = 873$	$\psi = 696$	$\psi = 529$	$\psi = 281$
SDO	$k = 553, x = 9$	$k = 396, x = 5$	$k = 823, x = 11$	$k = 926, x = 23$	$k = 281, x = 11$
	$qv = 0.2$	$qv = 0.25$	$qv = 0.2$	$qv = 0.5$	$qv = 0.2$

5   Experiments

This section describes the conducted experiments. Henceforth, we refer to the feature formats as the subset F and the used algorithms as the subset A:

$$F = \{\text{CAIA, Consensus, TA, Cisco-Joy, AGM}\} \tag{1}$$

$$A = \{\text{kNN, LOF, HBOS, iForest, SDO}\} \tag{2}$$

We describe the experiments step-by-step:

1. Flow extraction
From CICIDS2017 *pcaps*, we extracted features to match the studied representations. Therefore, for each vector format we obtained a structured dataset, $D_i = M_i \times (N_i + 2)$, where $i \in F$, M_i is the respective number of instances (flows), and $N_i + 2$ is the respective number of features plus a binary label (attack, non-attack) and a multiclass label (attack family). Feature vectors were extracted with a feature extractor based on Golang[1].
2. Cleaning and normalization
We removed nominal features from preprocessed datasets (see Table 1), except for the "Protocol", which was transformed into the dummies "TCP", "UDP", "ICMP" and "others". Datasets were min-max normalized:

$$Z_i = normalize\big(remove\_nominal(D_i)\big) \tag{3}$$

---

[1] https://golang.org/.

3. Stratified sampling

Datasets were sampled and a 5% subset was drawn for hyperparameter search and tuning: $Z_i' = strat\_sample_{.05}(Z_i)$, where $i \in F$. The sampling process was stratified with respect to the multiclass labels to keep balanced distributions.

4. Hyperparameter search

For each vector format ($i \in F$) and algorithm ($j \in A$), hyperparameter search was conducted by means of evolutionary algorithms[2]:

$$param_{i,j} = hyperparam\_search(Z_i', j) \tag{4}$$

Obtained hyperparameters are shown in Table 2.

5. Univariate analysis of outlierness ranks

We split each Z_i dataset into a *non-attack* (Z_{in}) and *attack* subsets (Z_{ia}). Later, measures of central tendency and histograms over Z_{in} and Z_{ia} were extracted with each algorithm j.

6. Analysis with outlier ranking metrics

For each dataset Z_i, the performance of each algorithm j was evaluated with the metrics defined in Sect. 6.

7. Feature selection for maximizing outlierness

Finally, CAIA, Consensus and AGM formats (i.e., the best ones in previous experiments) were joined, vectors were extracted from pcaps, and a 5% stratified sample was drawn, obtaining the final Z_F' dataset. By means of a forward wrapper with SDO as nested algorithm, features were gradually selected to find a set that maximizes the separation between attack and non-attack outlierness. ROC-AUC (Sect. 6) was selected as optimization criterion. The obtained vector was named "OptOut" (from Optimized Outlierness), it is shown in Table 4. Steps 4, 5 and 6 were repeated for the OptOut vector.

6 Outlier Detection Evaluation Indices

For evaluating algorithms, we have used the same metrics applied by Campos et al. in their recent outlier detection algorithm comparison [9]. We refer the reader to this paper for further explanations about the performance indices. They are: *P@n*, precision at the top n ranks; *AdjP@n*, P@n adjusted for chance; *AP*, average precision; *AdjAP*, AP adjusted for chance; *MaxF1*, Maximum F1 score [31]; *AdjMF1*, MaxF1 score adjusted for chance; *ROC-AUC*, area under the ROC curve. Indices named *adjusted* are based on the recommendations given by Hubert et al. [20]. Following Campos et al. [9], in our experiments $P@n$ and $adjP@n$ define n as the number of instances of labeled outliers in the dataset.

---

[2] https://github.com/rsteca/sklearn-deap.

7   Results and Discussion

We proceed to show results and discuss the questions raised in the Introduction.

7.1   Are Network Attacks Outliers?

Figure 2 shows box plots obtained from the *univariate analysis of outlierness ranks* step. For the sake of visibility, extreme values (top outliers) have been removed and outlierness ranks have been normalized. Upper and lower box boundaries correspond to 75th and 25th percentiles respectively, whereas upper and lower whiskers correspond to 95th and 5th percentiles. Additionally, we show some histograms in Fig. 3 (attack and non-attack empirical densities are equalized by normalizing histograms). There are four immediate evidences that stand out from the statistics:

(a) The differences between attack and non-attack instances in terms of outlierness for the Cisco-Joy are useless for discriminating attacks. Note that boxplots and distributions overlap or non-attacks show higher values.
(b) Regardless of the used algorithm, as a general rule attacks show higher outlierness than non-attack instances when using the CAIA, TA or AGM vectors, being AGM the format that shows major differences.
(c) Attack and non-attack outlierness ranges significantly overlap.
(d) SDO shows the best performances, followed by HBOS.

The inability of the Cisco-Joy format for discriminating attacks based on outlierness (a) was expected since this vector uses a considerably high dimensional space with a majority of binary features, drawing an input space highly unsuitable for methods based on Euclidean metrics. On the other hand, the preponderance of

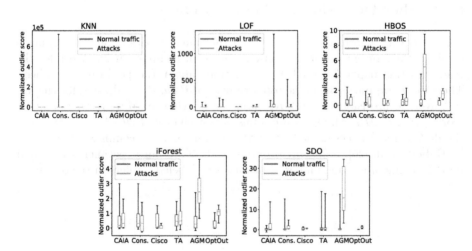

Fig. 2. Box plots for outlierness ranks.

SDO and HBOS (d), when considered together with observations (b) and (c), suggests that network attacks tend to be global, but clustered outliers, and not local outliers. The spaces drawn by the feature vectors are highly noisy and rich in density variations, and such noise and multiple densities are mainly generated by legitimate traffic. Network attacks tend to set small clusters relatively far from the data bulk. Such conditions favor non-local distance-based methods like HBOS and SDO. In any case, the significant range overlap (c) makes detection solely based on outlier ranking algorithms hardly suitable for real applications, in which high false positive rates would be unacceptable.

7.2   What Are the Best Feature Vectors for the Task?

Table 3 shows the performance of algorithms for each feature vector with the indices defined in Sect. 6. As for the algorithms, the evaluation measures corroborate the findings discussed in Sect. 7.1, confirming the prevalence of HBOS and SDO. On the other hand, noteworthy is the fact that the AGM vector shows high ROC-AUC and low values of other indices, whereas CAIA and Consensus show low ROC-AUC but higher values for the other indices when compared with AGM. This fact suggest that, in the AGM case, most attacks show higher outlierness than most non-attack instances, but still top outlierness values correspond to legitimate traffic. Contrarily, in the CAIA and Consensus cases most attacks and most non-attacks show similar outlierness, but top outlier positions are considerably taken by attacks (note that attacks in the dataset are negligible compared to normal instances). Such circumstance favors the use of the AGM vector to build a general purpose detector, but CAIA or Consensus as a support detector for evaluating only extreme outlierness cases; more interestingly, it suggests that vector formats are complementary and a new feature vector that maximizes attack outlierness can be built from them.

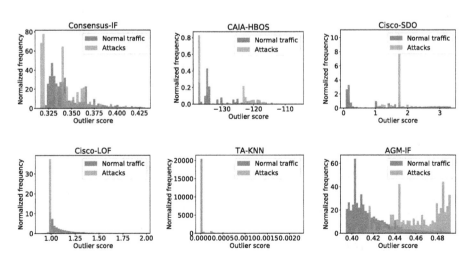

Fig. 3. Normalized histograms (top 5% outliers removed for a better visualization).

7.3 Can We Improve Vectors and Use Them in Real Detection?

Results in Table 3 show that the studied vectors would generate many false positives in real-world applications. As described in Sect. 5, we constructed a feature vector OptOut that maximizes the separation between attack and non-attack outlierness. OptOut uses the 5-tuple key, but enriched with features that describe the behavior of the network device as information source, therefore instances profile application-based and endpoint-based behavior at the same time. Table 4 shows the included features in the OptOut vector and Fig. 4 the forward selection process. We performed hyperparameter search also for this vector and obtained the following values: kNN, $k = 15$; LOF, $MinPts = 50$; HBOS, $k = 22$; iForest, $t = 50, f = 4, \psi = 456$; SDO, $k = 241, x = 25, qv = 0.35$. Some histograms are shown in Fig. 5.

Table 3. Algorithm performances.

		P@n	Adj. P@n	AP	Adj. AP	Max. F1	Adj. mF1	ROC-AUC
Cons.	HBOS	0.40	0.20	0.26	0.01	0.44	0.25	0.42
	LOF	0.22	−0.04	0.20	−0.07	0.41	0.21	0.47
	kNN	0.18	−0.10	0.06	−0.26	0.41	0.21	0.47
	iForest	0.20	−0.07	0.12	−0.18	0.41	0.21	0.40
	SDO	0.58	0.44	0.40	0.20	0.72	0.62	0.82
CAIA	HBOS	0.45	0.27	0.27	0.02	0.47	0.29	0.45
	LOF	0.21	−0.05	0.18	−0.10	0.41	0.20	0.47
	kNN	0.18	−0.10	0.06	−0.26	0.41	0.22	0.47
	iForest	0.31	0.08	0.21	−0.06	0.47	0.30	0.56
	SDO	0.32	0.09	0.45	0.26	0.52	0.36	0.60
AGM	HBOS	0.03	0.03	0.04	0.03	0.10	0.09	0.92
	LOF	0.01	0.01	0.03	0.02	0.02	0.02	0.63
	kNN	0.13	0.13	0.20	0.20	0.13	0.13	0.81
	iForest	0.04	0.04	0.05	0.04	0.09	0.08	0.91
	SDO	0.00	−0.00	0.00	−0.00	0.09	0.09	0.95
TA	HBOS	0.03	0.03	0.04	0.04	0.03	0.03	0.53
	LOF	0.00	0.00	0.00	−0.00	0.01	0.00	0.53
	kNN	0.04	0.03	0.05	0.05	0.04	0.03	0.58
	iForest	0.03	0.03	0.03	0.02	0.04	0.03	0.54
	SDO	0.04	0.04	0.07	0.07	0.05	0.05	0.54
Cisco	HBOS	0.02	−0.20	0.01	−0.21	0.32	0.17	0.26
	LOF	0.09	−0.12	0.15	−0.04	0.32	0.17	0.28
	kNN	0.01	−0.21	0.01	−0.21	0.31	0.15	0.12
	iForest	0.02	−0.20	0.01	−0.21	0.33	0.18	0.27
	SDO	0.02	−0.20	0.01	−0.21	0.52	0.41	0.65

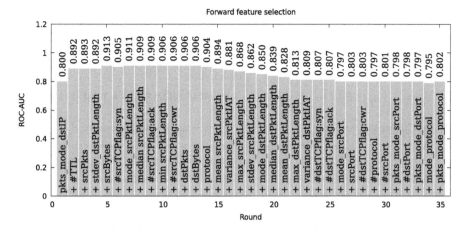

Fig. 4. OptOut forward selection process.

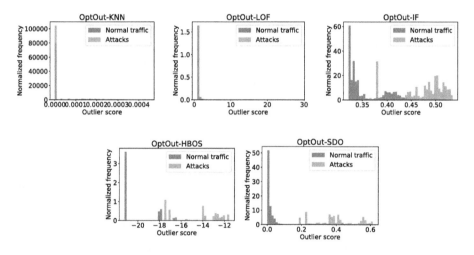

Fig. 5. OptOut vector. Normalized histograms (top 5% outliers removed for a better visualization).

Obtained outlierness box plots are shown in Fig. 2, histograms in Fig. 5, and performance indices in Table 5. Results disclose that the OptOut vector considerably increases performances and, therefore, the capability of algorithms to discriminate attacks based on outlierness (particularly when using SDO). However, real-world detection demands high accuracy to minimize the proliferation of false positives. Attack detection based on unsupervised algorithms can hardly solve the problem alone, but its combination with supervised methods and techniques that leverage pre-knowledge is expected to build detection frameworks with highly effective performances.

Table 4. OptOut feature vector after forward selection (SDO nested).

Original vector	Feature	Description
AGM	pkts_mode_dstIP	Packets received by the most common IP destination
AGM	#TTL	Number of different TTL used by the IP source
CAIA/Consensus	srcPkts	Packets sent by the IP source
CAIA	stdev_dstPktLength	Standard deviation of the length of the packets sent by the IP destination
CAIA/Consensus	srcBytes	Bytes sent by the IP source

Table 5. Algorithm performances for the OptOut feature vector.

	P@n	Adj. P@n	AP	Adj. AP	Max. F1	Adj. mF1	ROC-AUC
HBOS	0.74	0.65	0.93	0.90	0.87	0.83	0.96
LOF	0.20	−0.07	0.20	−0.07	0.41	0.21	0.46
KNN	0.09	−0.22	0.17	−0.11	0.40	0.20	0.43
iForest	0.78	0.70	0.72	0.63	0.79	0.72	0.92
SDO	0.90	0.87	0.97	0.96	0.92	0.90	0.98

8   Conclusions

In this work we have faced three relevant aspects of network attacks, namely: (a) if they are actually outliers, (b) what are the most suitable algorithms and feature vectors for implementing outlierness-based detectors, and (b) if the attack outlierness is enough for implementing real-world detection. We have studied these questions from analytical perspectives by evaluating five different feature vectors used in the literature with five different outlier ranking algorithms. For our experiments we have used a dataset for intrusion detection evaluation that reflects modern attacks as well as legitimate behavior profiles.

The conducted experiments reveal that, as a general rule, network attacks have higher global distance-based outlierness averages than normal traffic. Given the characteristics of network feature spaces—noisy, highly varied, with normal instances covering a broad spectrum and drawing subspaces with many density differences—local algorithms show low performances for attack detection. Algorithms with a more global space interpretation—like SDO or HBOS—tend to perform better, specially when representation spaces capture the behavior of network devices and hosts (e.g., the AGM format). We have proposed a feature space that maximizes the separation of attacks and non-attacks in terms of outlierness; however, the risk of high false positive rates still prevails due to the base-rate fallacy problem inherent to network security spaces.

Outlier detection algorithms can be a powerful tool for detecting known and novel attacks, but leveraging pre-knowledge with supervised methods should not be omitted, since supervised and unsupervised methods are complementary and, together, can build highly refined solutions.

References

1. RFC 7011 - Specification of the IP Flow Information Export (IPFIX) Protocol for the Exchange of Flow Information. Technical report Internet Engineering Task Force (IETF), September 2013. https://www.ietf.org/rfc/rfc7011.txt
2. Anderson, B., McGrew, D.: Machine learning for encrypted malware traffic classification: accounting for noisy labels and non-stationarity. In: Proceedings of the 23rd ACM SIGKDD International Conference on Knowledge Discovery and Data Mining, pp. 1723–1732 (2017)
3. Anderson, B., McGrew, D.A.: Identifying encrypted malware traffic with contextual flow data. In: Proceedings of the 2016 ACM Workshop on Artificial Intelligence and Security, AISec@CCS 2016, Vienna, Austria, 28 October 2016, pp. 35–46 (2016)
4. Axelsson, S.: The base-rate fallacy and its implications for the difficulty of intrusion detection. In: Proceedings of the 6th ACM Conference on Computer and Communications Security, CCS 1999, pp. 1–7 (1999)
5. Bhuyan, M., Bhattacharyya, D., Kalita, J.: Network anomaly detection: methods, systems and tools. IEEE Commun. Surv. Tutorials **PP**(99), 1–34 (2013)
6. Bhuyan, M., Bhattacharyya, D., Kalita, J.: A multi-step outlier-based anomaly detection approach to network-wide traffic. Inf. Sci. **348**, 243–271 (2016)
7. Breunig, M.M., Kriegel, H.P., Ng, R., Sander, J.: LOF: identifying density-based local outliers. In: Proceedings of the ACM International Conference on Management of Data (SIGMOD), Dallas, TX, pp. 93–104 (2000)
8. Buczak, A.L., Guven, E.: A survey of data mining and machine learning methods for cyber security intrusion detection. IEEE Commun. Surv. Tutorials **18**(2), 1153–1176 (2016)
9. Campos, G.O., et al.: On the evaluation of unsupervised outlier detection: measures, datasets, and an empirical study. Data Min. Knowl. Disc. **30**(4), 891–927 (2016)
10. CN-Group TUWien: network-attack-outlierness (2019). https://github.com/CN-TU/network-attack-outlierness
11. Durumeric, Z., Bailey, M., Halderman, J.A.: An internet-wide view of internet-wide scanning. In: Proceedings of the 23rd USENIX Conference on Security Symposium, SEC 2014, pp. 65–78. USENIX Association, Berkeley (2014)
12. Fachkha, C., Bou-Harb, E., Debbabi, M.: Towards a forecasting model for distributed denial of service activities. In: 2013 IEEE 12th International Symposium on Network Computing and Applications, pp. 110–117, August 2013
13. Ferreira, D.C., Bachl, M., Vormayr, G., Iglesias, F., Zseby, T.: Curated research on network traffic analysis, November 2018. https://doi.org/10.5281/zenodo.1243050
14. Ferreira, D.C., Vázquez, F.I., Vormayr, G., Bachl, M., Zseby, T.: A meta-analysis approach for feature selection in network traffic research. In: Proceedings of the Reproducibility Workshop, Reproducibility 2017, pp. 17–20. ACM (2017)
15. García-Teodoro, P., Díaz-Verdejo, J., Maciá-Fernández, G., Vázquez, E.: Anomaly-based network intrusion detection: techniques, systems and challenges. Comput. Secur. **28**(1), 18–28 (2009)

16. Gharib, A., Sharafaldin, I., Habibi Lashkari, A., Ghorbani, A.: An evaluation framework for intrusion detection dataset. In: International Conference on Information Science and Security (ICISS), pp. 1–6, December 2016
17. Goeschel, K.: Reducing false positives in intrusion detection systems using data-mining techniques utilizing support vector machines, decision trees, and naive bayes for off-line analysis. In: SoutheastCon 2016, pp. 1–6 (2016)
18. Gogoi, P., Bhattacharyya, D.K., Borah, B., Kalita, J.K.: A survey of outlier detection methods in network anomaly identification. Comput. J. **54**(4), 570–588 (2011)
19. Goldstein, M., Dengel, A.: Histogram-based outlier score (HBOS): a fast unsupervised anomaly detection algorithm. In: KI 2012: Advances in artificial intelligence: 35th Annual German Conference on AI, pp. 59–63 (2012)
20. Hubert, L., Arabie, P.: Comparing partitions. J. Classif. **2**(1), 193–218 (1985)
21. Iglesias, F., Zseby, T.: Pattern discovery in internet background radiation. IEEE Tran. Big Data **PP**(99), 1 (2017)
22. Iglesias, F., Zseby, T.: Time-activity footprints in IP traffic. Comput. Netw. **107**(P1), 64–75 (2016)
23. Iglesias Vázquez, F., Zseby, T., Zimek, A.: Outlier detection based on low density models. In: 2018 IEEE International Conference on Data Mining Workshops, ICDM Workshops, Singapore, Singapore, 17–20 November 2018, pp. 970–979 (2018)
24. Karagiannis, T., Papagiannaki, K., Faloutsos, M.: Blinc: multilevel traffic classification in the dark. SIGCOMM Comput. Commun. Rev. **35**(4), 229–240 (2005)
25. Kaur, R., Singh, M.: A survey on zero-day polymorphic worm detection techniques. IEEE Commun. Surv. Tutorials **16**(3), 1520–1549 (2014)
26. Lim, Y.S., Kim, H.C., Jeong, J., Kim, C.K., Kwon, T.T., Choi, Y.: Internet traffic classification demystified: on the sources of the discriminative power. In: Proceedings of the 6th International Conference, Co-NEXT 2010, pp. 9:1–9:12. ACM, New York (2010)
27. Liu, F.T., Ting, K.M., Zhou, Z.H.: Isolation-based anomaly detection. ACM Trans. Knowl. Discovery Data (TKDD) **6**(1), 3:1–3:39 (2012). https://doi.org/10.1145/2133360.2133363
28. Meghdouri, F., Zseby, T., Iglesias Vazquez, F.: Analysis of lightweight feature vectors for attack detection in network traffic. Appl. Sci. **8**(11), 2196 (2018)
29. Moustafa, N., Slay, J.: The evaluation of network anomaly detection systems: statistical analysis of the UNSW-NB15 data set and the comparison with the KDD99 data set. Inf. Sec. J. Global Perspect. **25**(1–3), 18–31 (2016)
30. Petrovskiy, M.I.: Outlier detection algorithms in data mining systems. Program. Comput. Softw. **29**(4), 228–237 (2003)
31. Powers, D.M.W.: Evaluation: from precision, recall and F-measure to ROC, informedness, markedness and correlation. J. Mach. Learn. Technol. **2**(1), 37–63 (2011)
32. Ramaswamy, S., Rastogi, R., Shim, K.: Efficient algorithms for mining outliers from large data sets. In: Proceedings of the ACM International Conference on Management of Data (SIGMOD), Dallas, TX, pp. 427–438 (2000)
33. Schubert, E., Zimek, A., Kriegel, H.P.: Local outlier detection reconsidered: a generalized view on locality with applications to spatial, video, and network outlier detection. Data Min. Knowl. Disc. **28**(1), 190–237 (2014). https://doi.org/10.1007/s10618-012-0300-z
34. Sharafaldin, I., Habibi Lashkari, A., Ghorbani, A.: Toward generating a new intrusion detection dataset and intrusion traffic characterization. In: 4th International Conference on Information Systems Security and Privacy, pp. 108–116, January 2018

35. Vlădutu, A., Comăneci, D., Dobre, C.: Internet traffic classification based on flows' statistical properties with machine learning. Int. J. Netw. Manag. 27(3), e1929-n/a (2017)
36. Williams, N., Zander, S., Armitage, G.: A preliminary performance comparison of five machine learning algorithms for practical ip traffic flow classification. SIGCOMM Comput. Commun. Rev. 36(5), 5–16 (2006)
37. Zhang, J., Chen, X., Xiang, Y., Zhou, W., Wu, J.: Robust network traffic classification. IEEE/ACM Trans. Networking 23(4), 1257–1270 (2015)
38. Zhang, J., Zulkernine, M.: Anomaly based network intrusion detection with unsupervised outlier detection. In: 2006 IEEE International Conference on Communications, vol. 5, pp. 2388–2393 (2006)
39. Zimek, A., Filzmoser, P.: There and back again: outlier detection between statistical reasoning and data mining algorithms. Wiley Interdisc. Rev. Data Min. Knowl. Discov. 8(6), e1280 (2018). https://doi.org/10.1002/widm.1280

Auto Semi-supervised Outlier Detection for Malicious Authentication Events

Georgios Kaiafas[1]([✉]) [ID], Christian Hammerschmidt[2] [ID], Sofiane Lagraa[1] [ID], and Radu State[1] [ID]

[1] SnT, University of Luxembourg, Luxembourg City, Luxembourg
{georgios.kaiafas,sofiane.lagraa,radu.state}@uni.lu
[2] TU Delft, Delft, The Netherlands
c.a.hammerschmidt@tudelft.nl

Abstract. Cyber-attacks become more sophisticated and complex especially when adversaries steal user credentials to traverse the network of an organization. Detecting a breach is extremely difficult and this is confirmed by the findings of studies related to cyber-attacks on organizations. A study conducted last year by IBM found that it takes 206 days on average to US companies to detect a data breach. As a consequence, the effectiveness of existing defensive tools is in question. In this work we deal with the detection of malicious authentication events, which are responsible for effective execution of the stealthy attack, called *lateral movement*. Authentication event logs produce a pure categorical feature space which creates methodological challenges for developing outlier detection algorithms. We propose an auto semi-supervised outlier ensemble detector that does not leverage the ground truth to learn the normal behavior. The automatic nature of our methodology is supported by established unsupervised outlier ensemble theory. We test the performance of our detector on a real-world cyber security dataset provided publicly by the Los Alamos National Lab. Overall, our experiments show that our proposed detector outperforms existing algorithms and produces a 0 *False Negative Rate* without missing any malicious login event and a *False Positive Rate* which improves the state-of-the-art. In addition, by detecting malicious authentication events, compared to the majority of the existing works which focus solely on detecting malicious users or computers, we are able to provide insights regarding when and at which systems malicious login events happened. Beyond the application on a public dataset we are working with our industry partner, POST Luxembourg, to employ the proposed detector on their network.

Keywords: Outlier detection · Ensemble learning · Cybersecurity · Embedding · Semi-supervised learning

1 Introduction

Lateral movement attack is a stealth and well orchestrated attack where the adversaries gain shell access without necessarily creating abnormal network traffic. They make use of legitimate credentials to log into systems, escalate privileges

© Springer Nature Switzerland AG 2020
P. Cellier and K. Driessens (Eds.): ECML PKDD 2019 Workshops, CCIS 1168, pp. 176–190, 2020.
https://doi.org/10.1007/978-3-030-43887-6_14

using lateral movements and subsequently manage to traverse a network without any detection. The JP Morgan Chase [36] and Target hacks [23] are two well known examples of attacks where the adversaries stayed undetected while they traversed network.

Researchers have addressed malicious logins detection by evaluating their methods on a real-world cyber security dataset provided freely by the Los Alamos National Lab [19]. Existing works focus on detecting malicious users or computers which leads to classifying all the generated events from a user or computer as malicious or legit. As a result, it fails to detect which specific events are malicious and does not provide any information regarding when the adversaries manage to impersonate benign users. Additionally, most of the existing approaches on this dataset are questionable and the authors in [32] provide further details of their study.

A common characteristic of *login logs* or *authentication events* is being comprised of multidimensional categorical variables. Categorical variables stem from discrete entities and their properties, e.g. source user, destination computer, or protocol type. The underlying values of this type of variables are inherently unordered and as a consequence it is often hard to define similarity between different values of the same variable. As such, detecting anomalies on discrete data is challenging and is not a well studied topic in academia; the primary focus is on continuous data. Moreover, the prominent challenge in the defensive cyber world is to develop effective approaches which are realistic.

A possible solution to this point comes from the semi-supervised approaches [22] that do not require anomalous instances in the training phase. These approaches model the normal class and identify anomalies as the instances that diverge from the normal model. In real-world problems where the amount of unlabeled data is immense, identifying events that are not suspicious needs a lot of manual work and underlies the risk of miss-labeling true anomalous events. Hence, our motivation to develop our auto approach is to alleviate analysts from time expensive and monotonous tasks that include a significant amount of uncertainty.

In this work, we analyze authentication events using the Los Alamos authentication dataset [19] and we aim at detecting unauthorized events to services or computers in contrast to the majority of the existing works. We propose an embedding based and automatic semi-supervised outlier detector to reduce the false positives produced by an unsupervised outlier ensemble. In particular, our approach is an ensemble approach where we develop an unsupervised outlier ensemble to identify the most confident normal data points which will feed the semi-supervised detector to ultimately detect outliers. Our technique could be considered as a sequential outlier ensemble approach where two dependent components are developed for an outlier detection task. We refer to the authors of [1] for the details of outlier ensembles categories.

The contributions of our proposed approach are:

- We produce an embedding space via the Logistic PCA [25] algorithm that has potentiality of better representing the normal behavior.

- We develop the Restricted Principal Bagging (*RPB*) technique, an improved variant of the well established feature bagging technique [27], that works on the principal components space.
- We introduce a new unsupervised combination function, *Vertical Horizontal Procedure* (*VHP*), that leverages gradually the predictions of individual and smaller scale ensemble members.
- We automatically build an automatic semi-supervised ensemble by combining the aforementioned novel components to effectively detect malicious events.

Overall, our approach improves current state-of-the-art by achieving a 0.0017 *FPR* and 0 *FNR*; without missing any malicious login event. It is tested on an extremely imbalanced data sample of the real-world authentication log dataset provided by Los Alamos. In this challenging data sample the percentage of malicious events is 0.0066% which is 1348 times lower than the average outlier percentage in datasets used for outlier detection [33].

These improvements enhance our understanding of anomalous patterns since existing state of the art methods fail to capture all the anomalous patterns. It is particularly important for the practical implementation to keep the base rate fallacy in mind: Reducing the number of the false positives by 150 compared to state of the art means that we enable cyber analysts spending less time on monotonous tasks of pruning false alerts.

Detecting malicious events instead of users or computers provides actionable insights to analysts by answering questions related to when exactly and at which systems a malicious event happened. Our work could also be used to extend existing methodologies which detect malicious users to further detect malicious events. To the best of our knowledge, this work is the first automatic semi-supervised attempt that aims at detecting anomalous authentication events.

The rest of the paper is organized as follows. We briefly review related work in Sect. 2. Then, we continue by describing extensively how we develop each component of our approach in Sect. 3. In Sect. 4 we explain in detail the dataset and we present the experimental settings and results. We close in Sect. 5, where we conclude with remarks and future research directions.

2   Related Work

Anomaly Detection in Categorical Data. In [17], the authors proposed a distance based semi-supervised anomaly detection method. In particular, the distance between two values of a categorical attribute is determined by the co-occurrence of the values of other attributes in the dataset. In [30], the authors proposed an unsupervised anomaly detector based on subspaces. It examines only a small number of low dimensional subspaces randomly selected to identify anomalies. In [7], the authors proposed an anomaly detection method on heterogeneous categorical event data. The method maximizes the likelihood of the data by embedding different events into a common latent space and then assessing the compatibility of events. Furthermore, approaches that are based on pattern mining techniques have been developed. For instance, in [2], the authors proposed to identify anomalies using pattern-based compression, and [14] detects patterns in short sequences of categorical data.

Malicious Logins Detection. The Los Alamos National Lab provides a publicly available dataset [19] which is the most used and is related to authentication login events. There is a non-exhaustive list of papers analyzing this dataset for detecting abnormal authentication activities. The majority of the related work of this dataset focuses on detecting anomalous entities, users or computers [4,13,15,16,21,39]. On the other hand, only few works [18,29,35] detect anomalous events. The most used approach among all the existing works is the bipartite graph.

This work effectively detects malicious authentication events instead of malicious entities which gives the opportunity to analysts to correlate identified malicious authentication events with malicious events on other data sources. In addition, detecting anomalous entities could be considered as a subset of detecting malicious events because from the latter we can derive the former but not vice versa. Furthermore, our work is the first automatic semi-supervised outlier ensemble approach that is developed with the aid of established theory on outlier ensembles [1,44]. It is composed of novel and existed methods never tested for outlier detection on categorical data and especially on authentication logs.

3 Methodology

We propose a novel outlier ensemble detector for categorical data which automatically creates the "non-polluted" by outliers training set of a semi-supervised ensemble. More specifically, first it builds in an unsupervised way an outlier ensemble on all data points to identify with a relative confidence data points that are normal Secondly, it develops a semi-supervised ensemble detector which is trained only on the (normal) data points derived from the first phase. Finally, the semi-supervised ensemble classifies new observations (data points not in the training set) as belonging to the learned normal class or not. Figure 1 illustrates the sequential and automatic nature of our approach. Throughout this work we use *outliers* and *anomalies* interchangeably.

3.1 Phase 1

Unsupervised outier detection algorithms detect outliers based on their algorithmic design [45]. In this work, we reverse the problem of unsupervised outlier detection to unsupervised normal detection by using established oulier ensemble theory. The aim of the this phase is to create the training dataset of the semi-supervised model; normal data points. In particular, we independently employ two unsupervised detectors to build an outlier ensemble on bagged subspaces and finally identify the most confident normal data points.

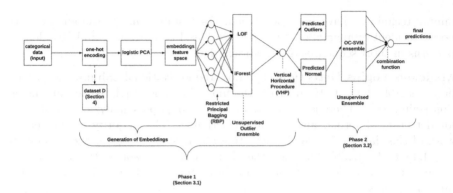

Fig. 1. Auto Semi-supervised Outlier Detector

Generation of Embeddings. Our dataset is a pure categorical dataset and we produce the embeddings of our proposed detector via the Logistic PCA algorithm [25]. This algorithm produces principal components and our aim is to find principal components that explain at least 90% of the total variance. We suggest a high percentage of explained variance because it means that we represent an amount of information very close to the information included in the original variables. We could have selected a different number of principal components that explain more than 90% of the total variance but we leave this sensitivity analysis for the future. Additionally, according to Theorem 2 of [25] we select columns to decrease the deviance the most. This Theorem states that for Logistic PCA the standard basis vector which decreases deviance the most is the one corresponding to column with mean closest to 1/2.

Restricted Principal Bagging. Our motivation for developing the *RPB - Restricted Principal Bagging* technique is to upper bound the sample space of the principal components and then add randomness in a similar way like the Feature Bagging technique [27]; randomness is a key ingredient of outlier ensemble techniques. Our technique aims at capturing the individual contribution of each principal component to the total explained variance. As such, we adjust the Feature Bagging technique [27] to work for principal components and find subspaces to detect ouliers more effectively. We explain in detail the *RPB* technique in Algorithm 1.

Firstly, *RPB* creates multiple random subsets of the first p principal components and each of these subsets is denoted by S_j. We denote by PCs the principal components that we keep after we have applied the Theorem 2 and we also call as V the set of all the S_j. Hence, $V = \{S_1, S_2, S_3, S_3, S_4, S_5\} = \{0.04 * \mid PCs \mid, 0.1 * \mid PCs \mid, 0.2 * \mid PCs \mid, 0.3 * \mid PCs \mid, 0.4 * \mid PCs \mid, 1.0 * \mid PCs \mid\}$. Then for a S_j and for *Iter* iterations it samples from a uniform distribution $U(d/2, d-1)$ without replacement, where d is the dimensionality of S_j. Hence, for each *Iter* iteration N_j principal components are sampled out and create a dataset F_j. Finally, an unsupervised outlier detector with random parameters is applied to F_j.

Algorithm 1. Restricted Principal Bagging

Input:
- **V** the set of all the S_j
- **OD** is an unsupervised Outlier Detection Algorithm which outputs numeric outlier scores for each data point
- **Iter** represents how many times we perform feature sampling

Output
- **E** is a vector composed of oulier scores for each data point

Procedure:

1: **for all** S_j in V **do**
2:    **for** $i = 1, 2, 3, 4, ...Iter$ **do**
3:       Randomly sample from a uniform distribution between $\lceil d/2 \rceil$ and $(d-1)$, where d is the number of the principal components in S
4:       Randomly pick, without replacement, N_i principal components to create a subset F_i
5:       Apply OD on F_i feature space
6:    **end for**
7: **end for**

Unsupervised Outlier Detectors. We employ two well performing and established unsupervised detectors to combine them and identify the most confident normal points that will feed afterwards the semi-supervised learner. We intentionally select heterogeneous detectors in order to increase the probability that they capture different patterns of anomalies. Also, we could have selected more than two heterogeneous unsupervised detectors to build the ensemble but for the current experiments we showcase the promising performance of the most straightforward version of our approach.

Firstly, we select iForest [28] which is a tree-based and state-of-the-art detector which performs the best across many datasets [11] and applications [9,41]. Secondly, we select LOF [5] which is a proximity-based method and designed to detect local outliers (see [1] for details in local and global outliers). It is also a state-of-the-art outlier detection algorithm and there is a large body of research on this detector [3,12,27,45].

The procedure that we follow at this phase is of running a detector over a range of parameters without leveraging the ground truth to tune the detectors. This procedure is interpreted as an ensemblar approach and we refer to [1] where the authors discuss the topic extensively. As such, we run LOF with different random values for the neighborhood parameter. Also, we run iForest with the Cartesian product of parameters $IF = \{(Number\ Of\ Estimators \times Maximum\ Samples \times Maximum\ Features)\}$.

LOF and iForest independently apply RPB on set V to build the ensemble version of LOF and iForest. Henceforth, we call LOF - RPB $scores_j$ and $iForest$ - RPB $scores_j$ the outlier scores that are produced by applying the RPB technique on a subset S_j and employing the LOF and iForest respectively. The final step is to combine these results in an unsupervised way to find the most confident normal data points. We introduce later the *VHP* combination function to combine these results. Finally, we call as W the most confident normal data points that will feed the semi-supervised algorithm to learn the normal behavior and as O the least confident normal data points.

VHP Combination Function. As we discussed before, the *RPB* algorithm builds a couple of LOF and iForest ensembles on each subset S_j. Hence, we propose a strategy to effectively combine and gradually take advantage of these couples of ensembles instead of applying a global combination function across all the *LOF - RPB scores$_j$* and *iForest - RPB scores$_j$*. The authors in [43] develop a novel local combination function and highlight the effectiveness of this type of combination functions.

In our strategy we utilize the *Averaging* combination function to calculate the average scores of ensemble members. The reason why we select this function is that the average score is the most widely used in outlier ensemble literature and performs the best in most cases [8]. It is worth noting that combining effectively outlier ensemble members without leveraging the ground truth is challenging and the authors in [1,24,44] extensively discuss the topic.

In particular, firstly we normalize all the *LOF - RPB scores$_j$* and *iForest - RPB scores$_j$* and then apply the *Averaging* function to get the average scores on each S_j. As such for each subset S_j we build an ensemble produced by these combined outlier scores. We refer to this ensemble as *LOF Ens & iForest Ens*. Afterwards, we convert the numeric outlier scores of each *LOF Ens & iForest Ens* ensemble to binary values based on a threshold. Finally, we combine these binary values by utilizing the unweighted majority voting [40] technique to produce the output of *Phase 1*.

The conversion to binary values is referred as the *Vertical Strategy* and the combination of the binary values as the *Horizontal Strategy*. Henceforth, we call this combination function as *VHP*, Vertical Horizontal Procedure. All the outlier scores are normalized with the Z-score normalization scheme which is the most commonly used in outlier detection literature (see [1] for details in different normalization schemes).

3.2   Phase 2

At this phase we leverage the produced W dataset of *Phase 1* to build the semi-supervised ensemble. The W dataset is composed of the most confident normal class data points and via this dataset we learn the normal class patterns. As a result this procedure of our analysis makes our approach sequential and automatic at the same time. The desired outcome of this sequential approach is to reduce significantly the number of false positives of O dataset after we have learnt the contour of the normal class.

Hence, we employ the OCSVM - One-Class SVM algorithm [34] which is a well performing algorithm that is applied to several problems such as, fraud detection [37] and network intrusion detection [26]. OCSVM is a boundary method that attempts to define a boundary around the training data (normal class), such that new observations that fall outside of this boundary are classified as outliers [38].

Our proposed approach is developed on a pure unsupervised setup and as a result we do not seek for the best performing parameters. Hence, without any loss of generality we select as parameters of the OCSVM algorithm the Cartesian product $B = \{(Type\ of\ kernel \times Upper\ bound\ of\ training\ errors \times Kernel\ coefficient)\}$. The procedure that we follow at this phase is analogous to *Phase 1* where we execute each detector over a range of parameters without leveraging the ground truth to tune the performance.

In particular, we independently execute several training runs of the OCSVM on W with different parameter values from set B. The number of training executions is equal to the carnality of set B. Next, for each execution of OCSVM an outlier score vector is produced which has length equal to the number of observations of O dataset. Finally, we combine these outlier score vectors, without leveraging the ground truth, to ultimately produce the final outlier score for each data point. It is worth noting that we could have selected any other set of parameters as input for the OCSVM algorithm. The procedure of running a detector over a range of parameters without the use of labels is interpreted as an ensemblar approach (see [1] for details).

4   Experiments and Evaluation

The major objective of our experiments is to demonstrate the effectiveness of our proposed auto semi-supervised detector by comparing it with works which detect malicious login events. On the one hand, we do not leverage the ground truth to tune any component of our methodology on the other hand, we use the ground truth to present the performance of *Phase 1* as well as *Phase 2*.

4.1   Dataset

The Los Alamos National Laboratory provides a freely available and comprehensive dataset[1] [19]. It includes 58 consecutive days of credential-based login events, of which days the 3 to 29 are labelled as malicious or normal via a RedTeam table. This dataset consists of 1 billion events and is an excessively imbalanced dataset; the percentage of the malicious login events is 0.000071%.

Each authentication event contains the attributes: time, source user, destination user, identifier per domain, source computer, destination computer, authentication type, logon type, authentication orientation, and authentication result. In addition, the authentication events are Windows-based authentication events from both individual computers and centralised Active Directory domain controller servers [20]. We also create a new attribute for each authentication event based on if source computer and destination computer are the same or different.

---

[1] https://csr.lanl.gov/data/cyber1/.

This new boolean feature quantifies the Local or Remote rule respectively. In our analysis, the time variable is excluded and as a result a purely categorical feature space is produced.

Developing a data mining methodology on 1 billion events would require a big data infrastructure but our work is not on proposing a computer engineering tool. Hence, we use a data sample to develop and evaluate our methodology. Sampling from a such an excessively imbalanced dataset usually produces samples composed of zero malicious login events which makes the evaluation of both classes impossible.

Hence, we seek for a random sample of 150, 000 consecutive authentication events that contains at least 5 malicious events in order to thoroughly evaluate our approach. In other words, the percentage of malicious events has to be at least 0.0033%. Consequently, our randomly selected sample contains 10 malicious events and its percentage of malicious events is 0.0066%. Then, on the sampled categorical space we apply the one-hot technique to produce the input binary space of the Logistic PCA algorithm. The dimension of the this binary space is $150, 000 \times 2700$ and we refer to this dataset as D.

4.2   Experiment Environment

We used the logisticPCA [25] R package for the implementation of the Logistic PCA algorithm and the data.table [10] R package for fast data manipulation. The iForest, LOF and OCSVM algorithms were executed using the Python *Scikit-learn* library [31].

4.3   Experimental Settings

Phase 1. We apply the Logistic PCA on the D dataset $(150, 000 \times 2700)$ and we keep 900 principal components which explain 93% of the total variance. Afterwards, we apply the Theorem 2 we explained in Sect. 3.1 and we return 500 principal components which will be the embeddings feature space denoted by PCs.

The exact parameters of LOF and iForest are presented in Table 1. LOF is employed with different number of neighbors as input whereas the input parameter set of iForest is the Cartesian product $IF = \{(Number\ Of\ Estimators \times Maximum\ Samples \times Maximum\ Features)\}$.

Phase 2. Table 2 presents the parameters at *Phase 2*. In Sect. 3.2 we defined the set B which is the Cartesian product of the input parameter values of OCSVM. In addition, the *Averaging* combination function is utilized to unify the outlier scores of all OCSVM executions.

Table 1. Setting parameters

	Subsets S	Parameters
LOF	$V = \{4\%, 10\%, 20\%,$ $30\%, 40\%, 100\%\}$	$Neighbors = \{5, 10,$ $15, 20, 30, 40, 50, 60,$ $70, 80, 90, 100\}$
iForest	$V = \{4\%, 10\%, 20\%,$ $30\%, 40\%, 100\%\}$	$NumberOfEstimators$ $= \{100, 200, 300, 400\}$
		$MaximumFeatures =$ $\{10\%, 20\%, 40\%, 60\%\}$
		$MaximumSamples =$ $\{10\%, 30\%, 50\%\}$

Table 2. Setting parameters

nu	$\{0.0001, 0.0005, 0.001, 0.005\}$
$gamma$	$\{0.01, 0.05, 0.09, 0.001\}$
$kernel$	$\{\text{"rbf"}, \text{"sigmoid"}\}$

Settings for Comparisons: We develop the *VHP-Ensemble* with our proposed *VHP* combination function accompanied with the *RPB* algorithm by leveraging different subsets of principal components as we have discussed earlier. Also, we develop the *Vanilla-Ensemble* to compare our proposed ensemble with. It employs the iForest and LOF detector on the whole *PCs* embeddings space, the feature bagging technique by Lazarevic [27] and the *Averaging* combination function. The components of the developed ensembles and their corresponding names are presented in Table 3.

Table 3. Ensembles of *Phase 1*

	Detector		Principal components of subsets S							Combination		Bagging	
Ensmbles	LOF	iForest	20	50	100	150	150	200	500	VHP	Avg.	RPB	Lazarevic
VHP	Yes	Yes	Yes	Yes	Yes	Yes	Yes	Yes	No	Yes	No	Yes	No
Vanilla	Yes	Yes	No	No	No	No	No	No	Yes	No	Yes	No	Yes

4.4   Evaluation

Phase 1

In Table 4, we summarize the performance of the ensembles discussed previously and presented in Table 3. Since the output of *Phase 1* is two sets, W and O, we evaluate our detectors using the precision and recall measures. We also showcase the sensitivity of the ensembles by reporting the presicion and recall scores based on different thresholds m; number of reported outliers. In our analysis m plays the role of the confidence of finding normal data points.

We denote by $P_{@m}$ and $R_{@m}$ respectively, the precision and recall score produced with m ranked data points which are considered as outliers. Table 4 is a typical example of the trade off between precision and recall. In our proposed approach the cost of higher precision is less than the cost of higher recall.

Table 4. Precision and Recall of the output of *Phase 1*

Enembles	$P_{@1500}$	$R_{@1500}$	$P_{@5000}$	$R_{@5000}$	$P_{@7000}$	$R_{@7000}$
VHP	**0.015**	**1.0**	**0.008**	**1.0**	**0.007**	**1.0**
Vanilla	0.005	0.8	0.0016	0.8	0.0011	0.8

Phase 2.
Since all the components of this work are developed in a pure unsupervised setup it is important to investigate the sensitivity of our approach, *VHP-Ensemble*. As such, we test multiple variants of this ensemble detector based on different numbers of reported outliers at *Phase 1*. In this way, we investigate the effect of *Phase 1* on building the semi-supervised ensemble detector.

Hence, we denote by *Detector-1500* the semi-supervised detector which is developed when a threshold rank $m = 1500$ is chosen for the *VHP-Ensemble*. The most outlier point among the $m = 1500$ reported outliers has a rank of 1. In the same fashion, we develop *Detector-5000* and *Detector-7000* where $m = 5000$ and $m = 7000$ respectively. Our motivation for selecting so large m is that we want to feed the semi-supervised detector with the most confident normal data points. We identify them based on our intuition for the outliers percentage in our dataset. In our case, m is at least 150 times greater than the number of true malicious authentication events.

We compare our methodology with works that are developed on the same level of granularity; detecting malicious authentication events. Detecting malicious users or computers means a huge amounts of events have to be further analyzed to identify which specific events are malicious. Since the existing works on malicious events is limited we compare our proposed detector with any kind of machine learning(supervised, semi-supervised, unsupervised) approach that is tested on authentication events. Hence, we evaluate all variants of our detector with (i) Siadati et al. [35], (ii) Lopez et al. [29], (iii) Kaiafas et al. [18].

In Fig. 2 we present a summary of the FPR and TPR scores of all the competitors. Amongst the competitors, Siadati et al. [35] achieves the lowest FPR whereas Kaiafas et al. and all the variants achieve the highest TPR; they do not miss any malicious login. In addition, *Detector-1500* achieves the lowest FPR among all the competitors. Ultimately, *Detector-1500* improves FPR of the Kaiafas et al. supervised detector by 10% (150 login events) and more than doubles Siadati's TPR. Siadati et al. detector is based on integrating security analysts knowledge into the detection system in the form of rules that define login patterns. In other words, this detector does not improve the existing knowledge of the cyber analysts for anomalous patterns but instead relies on known rules to detect anomalies. As a consequence, the Siadati's rule based visualization detector misses 53% of the malicious logins.

In addition, each of the aforementioned approaches outperform the logisitic classifier of Lopez et al. [29] which achieves AUC 82.79%. We do not plot their reported FPR and TPR scores in Fig. 2 because their FPR scores are at least 5 times worse than the maximum FPR value in Fig. 2. Consequently, we avoid presenting a figure that is less readable and informative for the majority of the competitors.

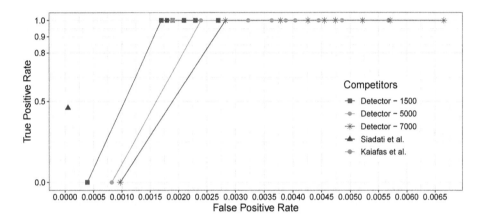

Fig. 2. Comparison of the Auto Semi-supervised Outlier Detector

5   Conclusion and Future Work

Our proposed automatic semi-supervised detector for malicious authentication detection outperforms the existent supervised algorithms and tools with the human in the loop. It is capable of capturing underlying mechanisms that produce anomalous authentication events. Our evaluation on a real-world authentication log dataset shows that we do not miss any malicious login events and improve the current state-of-the-art methods. Also, the sensitivity analysis showed that the rank threshold at *Phase 1* does not affect at all the TPR. On the other hand, the effect of the threshold on the FPR is not so noticeable. The semi-supervised ensemble detector improves the FPR of the unsupervised ensemble almost 9 times while all the developed variants did not miss any true malicious login events.

In the future we would like to extend this work by building an ensemble with multiple heterogeneous one-class classification algorithms [38]. Also, we want to model the authentication logs as graphs to produce embeddings with deep learning models [6]. Additionally, we intend to extend the existing work with network representation learning techniques [42] instead of embeddings. Finally, an extensive comparative evaluation will follow based on the above improvements on many cyber-security datasets.

Acknowledgement. Georgios Kaiafas is supported by the National Research Fund of Luxembourg (AFR-PPP Project ID 11824564). Additionally, the authors would like to thank POST Luxembourg, the industrial partner of this project.

References

1. Aggarwal, C.C., Sathe, S.: Outlier Ensembles. Springer, Cham (2017). https://doi.org/10.1007/978-3-319-54765-7
2. Akoglu, L., Tong, H., Vreeken, J., Faloutsos, C.: Fast and reliable anomaly detection in categorical data. In: Proceedings of the 21st ACM international conference on Information and knowledge management. pp. 415–424. ACM (2012)
3. Alshawabkeh, M., Jang, B., Kaeli, D.: Accelerating the local outlier factor algorithm on a GPU for intrusion detection systems. In: Proceedings of the 3rd Workshop on General-Purpose Computation on Graphics Processing Units, pp. 104–110. ACM (2010)
4. Bohara, A., Noureddine, M.A., Fawaz, A., Sanders, W.H.: An unsupervised multidetector approach for identifying malicious lateral movement. In: 2017 IEEE 36th Symposium on Reliable Distributed Systems (SRDS), pp. 224–233. IEEE (2017)
5. Breunig, M.M., Kriegel, H.P., Ng, R.T., Sander, J.: LOF: identifying density-based local outliers. In: ACM SIGMOD Record, vol. 29, pp. 93–104. ACM (2000)
6. Cai, H., Zheng, V.W., Chang, K.C.C.: A comprehensive survey of graph embedding: problems, techniques, and applications. IEEE Trans. Knowl. Data Eng. **30**(9), 1616–1637 (2018)
7. Chen, T., Tang, L.A., Sun, Y., Chen, Z., Zhang, K.: Entity embedding-based anomaly detection for heterogeneous categorical events. In: Proceedings of the Twenty-Fifth International Joint Conference on Artificial Intelligence, IJCAI 2016, pp. 1396–1403 (2016)
8. Chiang, A., Yeh, Y.R.: Anomaly detection ensembles: In defense of the average. In: 2015 IEEE/WIC/ACM International Conference on Web Intelligence and Intelligent Agent Technology (WI-IAT), vol. 3, pp. 207–210. IEEE (2015)
9. Ding, Z., Fei, M.: An anomaly detection approach based on isolation forest algorithm for streaming data using sliding window. IFAC Proc. Volumes **46**(20), 12–17 (2013)
10. Dowle, M., Srinivasan, A.: data.table: extension of 'data.frame' (2019). https://CRAN.R-project.org/package=data.table, r package version 1.12.2
11. Emmott, A.F., Das, S., Dietterich, T., Fern, A., Wong, W.K.: Systematic construction of anomaly detection benchmarks from real data. In: Proceedings of the ACM SIGKDD workshop on outlier detection and description, pp. 16–21. ACM (2013)
12. Goldstein, M., Uchida, S.: A comparative evaluation of unsupervised anomaly detection algorithms for multivariate data. PloS one **11**(4), e0152173 (2016)
13. Goodman, E., Ingram, J., Martin, S., Grunwald, D.: Using bipartite anomaly features for cyber security applications. In: 2015 IEEE 14th International Conference on Machine Learning and Applications (ICMLA), pp. 301–306. IEEE (2015)
14. Hammerschmidt, C., Marchal, S., State, R., Verwer, S.: Behavioral clustering of non-stationary IP flow record data. In: 2016 12th International Conference on Network and Service Management (CNSM), pp. 297–301. IEEE (2016)

15. Heard, N., Rubin-Delanchy, P.: Network-wide anomaly detection via the dirichlet process. In: 2016 IEEE Conference on Intelligence and Security Informatics (ISI), pp. 220–224 (2016)
16. Heard, N., Rubin-Delanchy, P.: Network-wide anomaly detection via the dirichlet process. In: 2016 IEEE Conference on Intelligence and Security Informatics (ISI), pp. 220–224. IEEE (2016)
17. Ienco, D., Pensa, R.G., Meo, R.: A semisupervised approach to the detection and characterization of outliers in categorical data. IEEE Trans. Neural Netw. Learn. Syst. **28**(5), 1017–1029 (2017)
18. Kaiafas, G., Varisteas, G., Lagraa, S., State, R., Nguyen, C.D., Ries, T., Ourdane, M.: Detecting malicious authentication events trustfully. In: NOMS 2018–2018 IEEE/IFIP Network Operations and Management Symposium. IEEE, April 2018
19. Kent, A.D.: Comprehensive, Multi-Source Cyber-Security Events. Los Alamos National Laboratory, New Mexico (2015). https://doi.org/10.17021/1179829
20. Kent, A.D.: Cyber security data sources for dynamic network research. In: Dynamic Networks and Cyber-Security, pp. 37–65. World Scientific, Singapore (2016)
21. Kent, A.D., Liebrock, L.M., Neil, J.C.: Authentication graphs: analyzing user behavior within an enterprise network. Comput. Secur. **48**, 150–166 (2015)
22. Khan, S.S., Madden, M.G.: One-class classification: taxonomy of study and review of techniques. Knowl. Eng. Rev. **29**(3), 345–374 (2014)
23. Krebs, B.: Target hackers broke in via HVAC company. Krebs on Security (2014)
24. Kriegel, H.P., Kroger, P., Schubert, E., Zimek, A.: Interpreting and unifying outlier scores. In: Proceedings of the 2011 SIAM International Conference on Data Mining, pp. 13–24. SIAM (2011)
25. Landgraf, A.J., Lee, Y.: Dimensionality reduction for binary data through the projection of natural parameters. arXiv preprint arXiv:1510.06112 (2015)
26. Lazarevic, A., Ertoz, L., Kumar, V., Ozgur, A., Srivastava, J.: A comparative study of anomaly detection schemes in network intrusion detection. In: Proceedings of the 2003 SIAM International Conference on Data Mining, pp. 25–36. SIAM (2003)
27. Lazarevic, A., Kumar, V.: Feature bagging for outlier detection. In: Proceedings of the Eleventh ACM SIGKDD International Conference on Knowledge Discovery in Data Mining, pp. 157–166. ACM (2005)
28. Liu, F.T., Ting, K.M., Zhou, Z.H.: Isolation forest. In: 2008 Eighth IEEE International Conference on Data Mining, pp. 413–422. IEEE (2008)
29. Lopez, E., Sartipi, K.: Feature engineering in big data for detection of information systems misuse. In: Proceedings of the 28th Annual International Conference on Computer Science and Software Engineering, pp. 145–156. IBM Corp. (2018)
30. Pang, G., Ting, K.M., Albrecht, D., Jin, H.: Zero++: harnessing the power of zero appearances to detect anomalies in large-scale data sets. J. Artif. Intell. Res. **57**, 593–620 (2016)
31. Pedregosa, F., et al.: Scikit-learn: machine learning in Python. J. Mach. Learn. Res. **12**, 2825–2830 (2011)
32. Pritom, M.M.A., Li, C., Chu, B., Niu, X.: A study on log analysis approaches using sandia dataset. In: 26th ICCCN, pp. 1–6 (2017)
33. Rayana, S.: Odds library. Stony Brook,-2016. Department of Computer Science, Stony Brook University, NY (2016). http://odds.cs.stonybrook.edu (2017)
34. Schölkopf, B., Platt, J.C., Shawe-Taylor, J., Smola, A.J., Williamson, R.C.: Estimating the support of a high-dimensional distribution. Neural Comput. **13**(7), 1443–1471 (2001)

35. Siadati, H., Saket, B., Memon, N.: Detecting malicious logins in enterprise networks using visualization. In: 2016 IEEE Symposium on Visualization for Cyber Security (VizSec), pp. 1–8. IEEE (2016)
36. Silver-Greenberg, J., Goldstein, M., Perlroth, N.: JPMorgan chase hack affects 76 million households (2014)
37. Sundarkumar, G.G., Ravi, V., Siddeshwar, V.: One-class support vector machine based undersampling: application to churn prediction and insurance fraud detection. In: 2015 IEEE International Conference on Computational Intelligence and Computing Research (ICCIC), pp. 1–7. IEEE (2015)
38. Swersky, L., Marques, H.O., Sander, J., Campello, R.J., Zimek, A.: On the evaluation of outlier detection and one-class classification methods. In: 2016 IEEE International Conference on Data Science and Advanced Analytics (DSAA), pp. 1–10. IEEE (2016)
39. Turcotte, M., Moore, J., Heard, N., McPhall, A.: Poisson factorization for peer-based anomaly detection. In: 2016 IEEE Conference on Intelligence and Security Informatics (ISI), pp. 208–210. IEEE (2016)
40. Van Erp, M., Vuurpijl, L., Schomaker, L.: An overview and comparison of voting methods for pattern recognition. In: Proceedings. Eighth International Workshop on Frontiers in Handwriting Recognition, 2002, pp. 195–200. IEEE (2002)
41. Wu, K., Zhang, K., Fan, W., Edwards, A., Philip, S.Y.: RS-forest: a rapid density estimator for streaming anomaly detection. In: 2014 IEEE International Conference on Data Mining, pp. 600–609. IEEE (2014)
42. Zhang, D., Yin, J., Zhu, X., Zhang, C.: Network representation learning: A survey. IEEE transactions on Big Data (2018)
43. Zhao, Y., Nasrullah, Z., Hryniewicki, M.K., Li, Z.: LSCP: Locally selective combination in parallel outlier ensembles. In: Proceedings of the 2019 SIAM International Conference on Data Mining, pp. 585–593. SIAM (2019)
44. Zimek, A., Campello, R.J., Sander, J.: Ensembles for unsupervised outlier detection: challenges and research questions a position paper. ACM SIGKDD Explor. Newsl. **15**(1), 11–22 (2014)
45. Zimek, A., Schubert, E.: Outlier detection. In: Liu, L., Özsu, M. (eds.) Encyclopedia of Database Systems, pp. 1–5. Springer, New York (2017). https://doi.org/10.1007/978-1-4899-7993-3

Defense-VAE: A Fast and Accurate Defense Against Adversarial Attacks

Xiang Li and Shihao Ji[(⊠)]

Department of Computer Science, Georgia State University, Atlanta, USA
xli62@student.gsu.edu, sji@gsu.edu

Abstract. Deep neural networks (DNNs) have been enormously successful across a variety of prediction tasks. However, recent research shows that DNNs are particularly vulnerable to adversarial attacks, which poses a serious threat to their applications in security-sensitive systems. In this paper, we propose a simple yet effective defense algorithm Defense-VAE that uses variational autoencoder (VAE) to purge adversarial perturbations from contaminated images. The proposed method is generic and can defend white-box and black-box attacks without the need of retraining the original CNN classifiers, and can further strengthen the defense by retraining CNN or end-to-end finetuning the whole pipeline. In addition, the proposed method is very efficient compared to the optimization-based alternatives, such as Defense-GAN, since no iterative optimization is needed for online prediction. Extensive experiments on MNIST, Fashion-MNIST, CelebA and CIFAR-10 demonstrate the superior defense accuracy of Defense-VAE compared to Defense-GAN, while being 50x faster than the latter. This makes Defense-VAE widely deployable in real-time security-sensitive systems. Our source code can be found at https://github.com/lxuniverse/defense-vae.

1 Introduction

Deep neural networks (DNNs) have demonstrated remarkable success in solving complex prediction tasks. However, recent studies show that they are particularly vulnerable to adversarial attacks [2,22,29] in the form of small perturbations to inputs that lead DNNs to predict incorrect outputs. For images, such perturbations are often almost imperceptible to human vision system, while being very effective at fooling DNN-based systems. Both white-box attacks [24] and black-box attacks [23] have been proposed to attack DNNs, and they can often fool the network with high probabilities. These attacks pose a serious threat to the applications of DNNs in security-sensitive systems, e.g., identity authentication surveillance, self-driving cars, malware detection, and voice command recognition. As a result, it is critical to develop effective and efficient defense mechanisms to counter adversarial attacks.

In this paper, we propose a simple yet effective defense mechanism called Defense-VAE that uses Variational AutoEncoder (VAE) [10,26] to purge the adversarial perturbations from contaminated images before feeding the images

© Springer Nature Switzerland AG 2020
P. Cellier and K. Driessens (Eds.): ECML PKDD 2019 Workshops, CCIS 1168, pp. 191–207, 2020.
https://doi.org/10.1007/978-3-030-43887-6_15

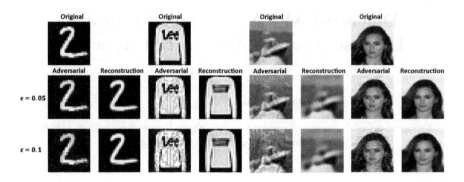

Fig. 1. Defense-VAE purges adversarial perturbations from contaminated images. Example images are from MNIST, F-MNIST, CIFAR-10, and CelebA, respectively. FGSM [7] with $\epsilon = 0.05$ and $\epsilon = 0.1$ are used to generate the adversarial attacks.

to the downstream CNN classifiers. To illustrate the idea, we generate some adversarial images based on the FGSM attack [7] with $\epsilon = 0.05$ and $\epsilon = 0.1$ on four popular image classification benchmarks: MNIST [13], Fashion-MNIST [31], CIFAR-10 [11] and CelebA [14]. These adversarial images are then fed into Defense-VAE for reconstruction. Figure 1 illustrates some of the typical examples from Defense-VAE. As we can see, the Defense-VAE generated images are the faithful reconstructions from the underlying clean images, with the majority of adversarial perturbations removed. As we will demonstrate later, such reconstructed images can recover almost all the accuracy losses due to adversarial attacks, without introducing much computation overhead compared to Defense-GAN [28], a closely related state-of-the-art defense algorithm that is based on Generative Adversarial Networks (GAN) [6].

Compared with the state-of-the-art defense algorithms, our method has the following properties:

- Defense-VAE is very generic and can defend white-box attacks and black-box attacks without the need of retraining the original CNN classifiers, and can further strengthen the defense by retraining or end-to-end finetuning;
- Defense-VAE achieves much higher accuracy than the state-of-the-art defense algorithms on white-box and black-box attacks. Especially, it outperforms Defense-GAN by about 30% in defending black-box attacks on F-MNIST;
- Defense-VAE is very efficient compared to the optimization-based alternatives, such as Defense-GAN, as no iterative optimization is needed for online prediction. From our experiments, it shows that Defense-VAE is about 50x faster than Defense-GAN. This makes our method widely deployable in real-time security-sensitive applications.

2   Defense-VAE: The Proposed Algorithm

At a high level, Defense-VAE is a defense algorithm that is based on deep generative models for image reconstruction. That is, given an adversarial image as input, the generative model attempts to produce a denoised image that is closely related to the underlying clean image, with the adversarial perturbations removed. As the name suggested, Defense-VAE is built upon Variational AutoEncoder (VAE) [10, 26]. Therefore, we first give a brief introduction to VAE.

2.1 Variational Auto-Encoder

Variational Autoencoder (VAE) [10, 26] is one of the most powerful deep generative models that is based on latent variable models. It consists of an encoder network to encode an input image to the latent variable z and a decoder network to decode the latent variable z back to the image domain:

$$z \sim \text{Enc}(\boldsymbol{x}) = q(\boldsymbol{z}|\boldsymbol{x}), \quad \boldsymbol{x} \sim \text{Dec}(\boldsymbol{z}) = p(\boldsymbol{x}|\boldsymbol{z}). \tag{1}$$

Since the maximum likelihood (ML) estimate of this latent variable model is intractable, a variational lower bound (ELBO) is optimized instead:

$$\begin{aligned} \mathcal{L}_{\text{VAE}} &= -\mathbb{E}_{q(\boldsymbol{z}|\boldsymbol{x})} \left[\log \frac{p(\boldsymbol{x}|\boldsymbol{z})p(\boldsymbol{z})}{q(\boldsymbol{z}|\boldsymbol{x})} \right] \\ &= -\mathbb{E}_{q(\boldsymbol{z}|\boldsymbol{x})} [\log p(\boldsymbol{x}|\boldsymbol{z})] + D_{\text{KL}}(q(\boldsymbol{z}|\boldsymbol{x}) \| p(\boldsymbol{z})) \end{aligned} \tag{2}$$

where the first term is the reconstruction error and the second term is a regularization that prefers the posterior to be close to the prior. Typically, a simple unit Gaussian prior is assumed in VAE. To facilitate efficient computation, a diagonal covariance Gaussian posterior is further assumed, which enables the use of the reparameterization trick to reduce the variance of Monte-Carlo sampling [10].

As a generative model, VAE can generate high quality images that follow the similar distribution of the training images.

2.2 Defense-VAE

VAE is typically trained to reproduce the same image from an input image. As for adversarial defense, reproducing the same adversarial images is an undesirable task as the adversarial perturbations may be preserved during the image reconstruction. Instead, in Defense-VAE, we modify the encoder and the decoder of the latent variable model as follows:

$$z \sim \text{Enc}(\hat{\boldsymbol{x}}) = q(\boldsymbol{z}|\hat{\boldsymbol{x}}), \quad \boldsymbol{x} \sim \text{Dec}(\boldsymbol{z}) = p(\boldsymbol{x}|\boldsymbol{z}), \tag{3}$$

where $\hat{\boldsymbol{x}} = \boldsymbol{x} + \boldsymbol{\delta}$ is an adversarial image with the perturbation $\boldsymbol{\delta}$ added on top of a clean image $\boldsymbol{x}$. This adversarial image is encoded to a latent variable z, which is decoded to the underlying clean image $\boldsymbol{x}$. Accordingly, the training loss of Defense-VAE is updated as follows:

$$\begin{aligned} \mathcal{L}_{\text{Defense-VAE}} &= -\mathbb{E}_{q(\boldsymbol{z}|\hat{\boldsymbol{x}})} \left[\log \frac{p(\boldsymbol{x}|\boldsymbol{z})p(\boldsymbol{z})}{q(\boldsymbol{z}|\hat{\boldsymbol{x}})} \right] \\ &= -\mathbb{E}_{q(\boldsymbol{z}|\hat{\boldsymbol{x}})} [\log p(\boldsymbol{x}|\boldsymbol{z})] + D_{\text{KL}}(q(\boldsymbol{z}|\hat{\boldsymbol{x}}) \| p(\boldsymbol{z})), \end{aligned} \tag{4}$$

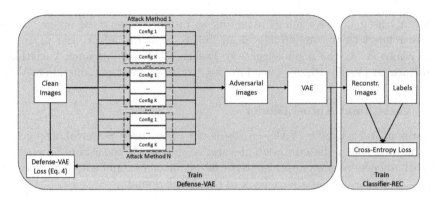

Fig. 2. Training pipeline of Defense-VAE. Defense-VAE (left) and Classifier-REC (right) can be trained separately, or jointly end-to-end (from scratch or by fine-tuning). See text for more details.

where the input to Defense-VAE is an adversarial image $\hat{x} = x + \delta$, and the expected output is the underlying clean image x. The compatibility between input and output pair is measured by the loss function 4.

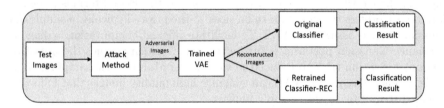

Fig. 3. Test pipeline of Defense-VAE

To train the Defense-VAE model, we can generate adversarial images given any clean image from a training set. Since there are many different adversarial attack algorithms and for each attack algorithm we can generate multiple adversarial images with different configurations, we can in principle generate an unlimited amount of training pairs for Defense-VAE, i.e., multiple adversarial images can be mapped to one clean image. The detailed training pipeline is demonstrated in Fig. 2 (left). Being an effective approach of generating sufficient training pairs for Defense-VAE, using multiple attack algorithms to produce adversarial training examples will also boost the capability of Defense-VAE to counter an ensemble of adversarial attacks and make Defense-VAE a generic defense algorithm that is robust to a wide range of attacks. As we will discuss later, this ensemble training strategy entails Defense-VAE superior defense capability over Defense-GAN.

Once the Defense-VAE model is trained, we can also use the reconstructed images from Defense-VAE to retrain the downstream CNN classifiers Fig. 2

(right). As we will see later, the retrained CNN classifier can further boost the defense accuracy over the original CNN classifier.

We can also train the whole pipeline end to end from scratch or finetuning from pre-trained VAE and CNN classifier by optimizing the joint loss function:

$$\mathcal{L}_{\text{End-to-End}} = \mathcal{L}_{\text{Defense-VAE}} + \lambda \mathcal{L}_{\text{Cross-Entropy}}. \tag{5}$$

As we will see from the experiments, this end to end training can boost the defense accuracy even further.

After training the Defense-VAE model and potentially retraining CNN classifiers or end-to-end finetuning the whole pipeline, we can use the trained Defense-VAE to purge the adversarial perturbations from any contaminated images, and the reconstructed images are then fed to the original CNN classifier or retrained CNN classifier for the final image classification. This test pipeline is shown in Fig. 3.

3   Related Work

Adversarial attacks and defenses is one of the active research areas in deep learning, with tens of different attack and defense algorithms developed in the past few years. For a general introduction to this exciting research area and the related terminologies, we refer the readers to [28,30,33] for more details. Here we will focus on the defense algorithms that are most closely related to Defense-VAE.

Defending against adversarial attacks is a challenging task. Different types of defense algorithms [18,25] have been proposed in the past few years. The first type of defense algorithms [4,9,15] augments the training data to make the DNN model resilient to the trained adversarial attacks. The second type of defense algorithms [5,8,16,20,21,27] modifies the training process by introducing regularization to the objective functions. The third type of defense algorithms [1, 9,32] attempts to remove the adversarial perturbations via input transformations before feeding the image to the classifier. According to this categorization, our Defense-VAE belongs to the input transformation based defense approach. In the following, we will therefore review the defense algorithms that are closely related to our work.

Adversarial training [7,12] is a popular and well investigated defense approach against adversarial attacks. It attempts to use adversarial images as data augmentation to train a robust classifier. It shows that this method can improve the defense accuracy effectively and sometimes it can even improve the accuracy upon the model trained only on the original clean training set. However, this defense mechanism is more effective in white-box attacks than in black-box attacks due to the gradient masking problem. In Defense-VAE, we also use adversarial examples to improve the robustness of the defense model. However, instead of improving the targeted CNN classifiers directly, adversarial training is used to train a Defense-VAE model to purge adversarial perturbations for the downstream CNN classifiers.

Magnet proposed by Meng and Chen [17] is another effective strategy to defend adversarial attacks. Magnet has two phases for defense: detector network and reformer network. Detector network learns the manifold of the normal clean images so that it can detect if an input image is an adversarial. If an image is detected as an adversarial, it will be forwarded to the reformer network, which will modify the adversarial image to the manifold of normal images. In Magnet, the reformer network is trained only on clean images with the goal of reconstructing the same clean input images, while Defense-VAE is trained on adversarial and clean image pairs with the goal of removing the adversarial perturbations from the contaminated images.

Another closely related work is Defense-GAN that is proposed by Samangouei et al. in [28], where a Generative Adversarial Network (GAN) [6] is used to reconstruct a clean image from an adversarial image. Defense-GAN firstly trains a GAN model purely on a training set of clean images, and as such it learns the distribution of the normal images. Then given an adversarial image, multiple iterations of back-propagations are used to identify a proper z from the clean image latent space, such that after decoded through the GAN generator, the reconstructed image is expected to be as close as possible to the adversarial image. Given the non-convex loss function of the GAN generator model, multiple random z's are used to initialize the back-propagation image search. Typically, given an adversarial image, Defense-GAN needs to perform L iterations of back-propagation for each of R random initializations, with the typical values of $L = 200$ and $R = 10$. As a comparison, to reconstruction a clean image, Defense-VAE can directly identify a proper z by forward-propagating an adversarial image through the VAE encoder network, and the z is subsequently used to reconstruct a clean image through the VAE-decoder network. No expensive iterative online optimization is needed in Defense-VAE. As we will discuss later, such reconstructed images are not only more accurate, but the whole process is much faster than Defense-GAN.

4   Experiments

We validate our algorithm on four popular image classification benchmarks: MNIST [13], F-MNIST [31], CelebA [14] and CIFAR-10 [11]. MNIST and F-MNIST are two gray-level image datasets, each containing 60,000 training images and 10,000 test images with the size of 28 × 28. While MNIST consists of 10 hand-written digits, F-MNIST contains 10 different articles, e.g., shoes, shirts, etc. CelebA contains 202,599 RGB images of human faces, split into training and test sets. We use this dataset for binary classification to distinguish if a face image is from a male or a female. CIFAR-10 contains 10 classes of RGB images of the size of 32 × 32, in which 50,000 images are for training and 10,000 images are for test.

We consider both the white-box attacks and the black-box attacks to test the defense performance of our algorithm. For the white-box attacks, FGSM [7], Randomized FGSM [12], and CW [3] attacks are used. For the black-box attacks, we train a substitute model to generate adversarial images to attack the targeted CNN classifiers. For a fair comparison, our experimental setups closely follow those of Defense-GAN[1].

To demonstrate the generalization of our algorithm, we test our algorithms with the targeted CNN classifiers of different architectures: different number of convolutional or full-connected layers, different convolution parameters, and with/without dropout or batch normalization. For the black-box attacks, different architectures are also considered for the substitute models. When we present results, we denote the targeted model as A, B, C, D and the substitute model as B, E. Detailed network architectures of the VAE model, the targeted CNN classifiers and their substitutes are summarized in Appendix A.

For the defense algorithms, we compare our algorithm with Adversarial Training [7,12], MagNet [17] and Defense-GAN [28]. All of our experiments are performed on NVIDIA Titan-Xp GPUs. Our source code can be found at https://github.com/lxuniverse/defense-vae.

4.1   Results on White-Box Attacks

First, we test our algorithm on three types of white-box attacks: FGSM, RAND-FGSM and CW attacks. The targeted CNN models are trained on the original training dataset for 10 epochs until convergence. Then for each clean training image we generate 12 different adversarial images by using 3 different white-box attack algorithms, each with 4 different configurations. For FGSM and RAND-FGSM, 4 different $\epsilon = 0.25, 0.3, 0.35$ and 0.4 are used. For the CW attack, 4 different learning rates $lr = 6, 8, 10$ and 12 are used. We combine these adversarial images and the original clean images to form the input and output pairs to train the Defense-VAE model. We initialize the weights of VAE with the normal distribution of $\mathcal{N}(0, 0.02)$ for the convolutional layers and $\mathcal{N}(1, 0.02)$ for the batch normalization layers. We note that usually 5 epochs are required for the Defense-VAE models to converge.

Additionally, we use the reconstructed images of Defense-VAE to retrain the CNN classifiers to improve the classification accuracy. Although the original CNN classifiers have already yielded very competitive performance compared with Defense-GAN, we note that retraining CNN classifiers for Defense-VAE can further strengthen the defense accuracy notably. Interestingly, the authors of Defense-GAN reported that for Defense-GAN retraining of CNN classifiers has negligible impact to the defense accuracy, while this is not true for Defense-VAE.

As discussed in Sect. 2.2, we can also train the whole pipeline end to end by optimizing the joint loss function Eq. 5 directly. This can be done through two approaches: (1) randomly initialize the VAE and CNN classifier model parameters and train the whole pipeline from scratch, and (2) pretrain VAE and CNN classifier separately and finetune the whole pipeline. Our experiments show that both approaches are almost equally effective, with the finetuning yielding slightly better results. We therefore only report the finetuning results in the following.

---

[1] https://github.com/kabkabm/defensegan.

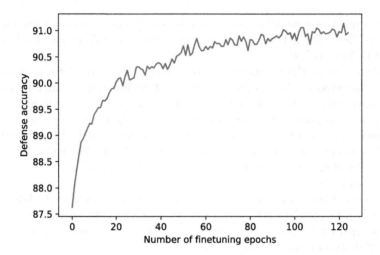

Fig. 4. The end-to-end finetuning can boost the defense accuracy even further, and yields the strongest defense model.

To demonstrate the effectiveness of this end-to-end finetuning approach, we provide one typical learning curve of the finetuning process in Fig. 4, where the adversarial attacks are generated by FGSM with $\epsilon = 0.3$. Starting from separately pretrained VAE model and CNN classifier (a.k.a., Defense-VAE-REC), we finetune the whole pipeline by optimizing the joint loss function 5. As we can see, the end-to-end finetuning boosts the defense accuracy by about 4% over the Defense-VAE model.

Table 1 reports the defense accuracies of Defense-VAE on three different white-box attacks: FGSM, RAND-FGSM and CW attacks. As a comparison, we also include the results of Defense-GAN, MagNet and Adversarial Training under the same experimental setups; for those results, we import them directly from the Defense-GAN paper [28]. As we can see, Defense-VAE and Defense-GAN are very competitive to each other, and outperform all the other defense algorithms by significant margins on all four benchmarks. Defense-VAE achieves superior performance over Defense-GAN, and can recover almost all the accuracy losses due to the adversarial attacks. We also note that retraining CNN classifiers (Defense-VAE-REC) and finetuning (Defense-VAE-E2E) can further improve the defense accuracies beyond the original CNN classifiers (Defense-VAE) by a notable margin, with the finetuning yielding the strongest defense against adversarial attacks.

Table 1. Classification accuracies of different defense methods under FGSM, RAND-FGSM and CW white-box attacks on the (top) F-MNIST and (bottom) MNIST image classification benchmarks. The defense accuracies of Defense-GAN, MagNet, and Adversarial Training are from Defense-GAN [28]. Results on CelebA and CIFAR-10 have the same pattern as above. Details can be found in Appendix B.

Attack	Classifier model	No attack	No defense	Defense VAE	Defense VAE-REC	Defense VAE-E2E	Defense GAN	MagNet	Adv. Tr. $\epsilon = 0.3$
FGSM $\epsilon = 0.3$	A	90.85	9.18	86.9	89.03	91.02	87.9	8.9	79.7
	B	71.62	15.89	70.88	74.41	77.86	62.9	16.8	13.6
	C	90.78	8.68	85.8	89.72	90.85	89.6	11.0	80.4
	D	86.94	8.51	85.36	87.09	89.26	87.5	9.9	69.8
RAND FGSM $\epsilon = 0.3$ $\alpha = 0.05$	A	90.85	7.91	86.42	88.91	90.57	88.8	9.6	44.7
	B	71.62	13.14	71.12	73.91	77.09	66.1	16.1	11.9
	C	90.78	5.48	86.42	89.38	90.28	89.3	11.2	69.9
	D	86.94	7.79	85.77	87.18	88.97	86.2	10.4	62.6
CW l_2 norm	A	90.85	11.67	81.81	86.99	88.54	89.6	6.0	15.7
	B	71.62	18.74	67.43	73.69	74.72	65.6	13.1	11.8
	C	90.78	7.70	78.64	87.47	88.69	89.6	8.4	10.7
	D	86.94	9.35	64.38	86.21	87.83	87.5	6.9	14.9
Average		84.05	10.34	79.24	84.50	**86.31**	82.55	10.69	40.48

Attack	Classifier model	No attack	No defense	Defense VAE	Defense VAE-REC	Defense VAE-E2E	Defense GAN	MagNet	Adv. Tr. $\epsilon = 0.3$
FGSM $\epsilon = 0.3$	A	99.15	14.65	98.29	98.98	99.28	98.8	19.1	65.1
	B	96.10	1.81	95.92	95.97	96.91	95.6	8.2	6.0
	C	99.08	29.53	98.41	98.91	99.24	98.9	16.3	78.6
	D	97.87	4.33	97.56	98.16	98.05	98.0	9.4	73.2
RAND FGSM $\epsilon = 0.3$ $\alpha = 0.05$	A	99.15	8.65	98.40	99.08	99.34	98.8	17.1	77.4
	B	96.10	1.65	95.83	96.04	96.87	94.4	9.1	13.8
	C	99.08	5.99	98.33	98.87	99.35	98.5	15.1	90.7
	D	97.87	3.25	97.81	98.3	98.05	98.0	11.5	53.9
CW l_2 norm	A	99.15	8.45	92.69	95.12	96.95	98.9	3.8	7.7
	B	96.10	3.00	87.66	88.56	95.08	91.6	3.4	28.0
	C	99.08	5.53	94.46	96.05	96.44	98.9	2.5	3.1
	D	97.87	3.92	83.42	89.46	95.71	98.3	2.1	1.0
Average		98.05	7.56	94.90	96.13	**97.61**	97.39	9.80	27.38

4.2   Robustness Under Untrained Attacks

In principle we can train Defense-VAE on all known adversarial attacks to best counter possible attacks in test. However, in reality new attacks are constantly invented; it's almost certain that after the deployment of Defense-VAE, some new adversarial attacks will emerge and Defense-VAE has never been trained on those attacks. To investigate the robustness of Defense-VAE in this circumstance, in this part of the experiments we train Defense-VAE on two attacks and test its defense capability against the third untrained attack. Again, three adversarial attacks are considered: FGSM, RAND-FGSM and CW, which gives three possible combinations that are shown in Table 2. As we can see, Defense-VAE is very robust for the first two attacks: FGSM and RAND-FGSM as the defense accuracies largely remain the same as it's trained on all three attacks. But for the CW attack, Defense-VAE is less robust, manifested by the significant accuracy loss compared to the Defense-VAE trained on all three attacks. Indeed, the CW attack is considered a much stronger attack and could have a very distinct attack pattern to that of FGSM and RAND-FGSM. We therefore incorporate Deepfool [18] to the training of Defense-VAE to counter the untrained CW attack since DeepFool and CW have very similar attack patterns. The results in parentheses show that this is indeed the case and Defense-VAE again can recover the most accuracy losses under untrained CW attack.

Table 2. Defense accuracy of Defense-VAE when it's trained on two attacks but is used to defend another attack. The results in parentheses are the accuracies after incorporating DeepFool [19] as additional adversarial training examples for Defense-VAE.

Attack	Classifier	Trained on other 2	Trained on 3
FGSM	A	87.34	89.03
	B	73.38	74.41
	C	88.03	89.72
	D	86.49	87.09
RAND FGSM	A	87.30	88.91
	B	73.59	73.91
	C	88.19	89.38
	D	86.73	87.18
CW	A	43.48 (85.06)	86.99
	B	34.52 (71.64)	73.69
	C	44.45 (85.22)	87.47
	D	30.77 (84.69)	86.21

4.3   Results on Black-Box Attacks

Next, we test the defense capability of Defense-VAE under black-box attacks on the MNIST and F-MNIST datasets. We train the targeted CNN model on the training set for 10 epochs with the batch size of 100 and the learning rate of 10^{-3} until convergence. Then the substitute model is trained with 150 images from the test set with the labels predicted by the targeted CNN classifier.

In the black-box attacks, Defense-VAE, as a defender, has no prior knowledge of the trained substitute model. Thus, we can only train Defense-VAE on the white-box attacks. Therefore, the same Defense-VAE model trained from

Table 3. Classification accuracies of different defense methods under FGSM black-box attacks on different image classification benchmarks: (top) F-MNIST, (bottom) MNIST. The defense accuracies of Defense-GAN, MagNet, and Adversarial Training are from the Defense-GAN paper [28]. Results on CIFAR-10 have the same pattern as above. Details can be found in Appendix B.

Classifier/ substitute	No attack	No defense	Defense-VAE	Defense-VAE-REC	Defense-VAE-E2E	Defense-GAN	MagNet	Adv. Tr. $\epsilon = 0.3$
A/B	90.85	37.92	83.69	86.64	86.39	58.60	54.04	73.93
A/E	90.85	24.94	76.97	83.02	83.61	47.90	33.11	69.45
B/B	71.62	17.61	73.66	72.42	75.22	49.40	38.12	31.77
B/E	71.62	13.44	69.29	69.36	71.78	37.20	31.19	26.17
C/B	90.78	39.14	83.64	86.88	87.67	52.89	46.64	77.91
C/E	90.78	22.89	76.27	80.16	80.32	48.71	30.16	75.04
D/B	86.94	32.87	80.31	85.80	84.78	57.79	54.78	61.72
D/E	86.94	23.51	70.66	79.48	77.53	40.07	33.96	50.93
Average	85.05	26.54	76.81	80.47	**80.91**	49.07	40.25	58.37

Classifier/ substitute	No attack	No defense	Defense-VAE	Defense-VAE-REC	Defense-VAE-E2E	Defense-GAN	MagNet	Adv. Tr. $\epsilon = 0.3$
A/B	99.15	65.89	98.68	98.71	99.16	93.12	69.37	96.54
A/E	99.15	76.32	98.64	98.92	99.19	91.39	67.10	96.68
B/B	96.10	14.40	95.89	95.95	96.71	90.57	56.87	20.92
B/E	96.10	26.48	96.26	95.81	97.09	88.41	46.27	11.20
C/B	99.08	60.74	97.91	98.02	99.15	93.57	75.71	98.34
C/E	99.08	72.73	98.30	98.59	99.28	92.23	67.60	98.43
D/B	97.87	33.36	97.68	98.22	97.85	92.72	68.17	76.67
D/E	97.87	39.95	97.72	98.22	97.69	91.64	60.73	76.76
Average	98.05	48.73	97.63	97.81	**98.27**	91.71	63.98	71.92

the experiments of white-box attacks is used to defend the black-box attacks.[2] In this experiment, 4 targeted CNN classifiers: A, B, C, and D, and 2 substitute models: B and E are considered, and this produces 8 possible Classifier/Substitute combinations. In this part of experiments, only the black-box FGSM attack is considered, with the results on MNIST and F-MNIST reported in Table 3. As a comparison, we also include the results of Defense-GAN, MagNet and Adversarial Training under the same experimental setups; again, for this set of results, we import them directly from the Defense-GAN paper [28]. As we can see, on both datasets Defense-VAE outperforms Defense-GAN and all other defense algorithms by significant margins. In particular, on F-MNIST, Defense-VAE improves the accuracy over Defense-GAN by about 30%. Also, as in the white-box attack experiments, retrained CNN classifiers

---

[2] In other words, we just need to train one Defense-VAE to defend both white-box and black-box attacks.

(Defense-VAE-REC) and finetuning (Defense-VAE-E2E) can further boost the defense accuracies over the original CNN classifiers (Defense-VAE) by a notable margin, with the end-to-end finetuning yielding the best defense accuracies among all the methods.

4.4 Why Is Defense-VAE so Effective?

The results above demonstrated superior performance of Defense-VAE over Defense-GAN. For the black-box FGSM attack, the former even outperforms the latter by about 30%. To understand why Defense-VAE can have such a large leap, we investigate the reconstructed images by Defense-VAE and Defense-GAN in this experimental setup, i.e., the black-box FGSM attack on F-MNIST. Figure 5 shows some typical examples from this experiment. As can be seen, the reconstructed images from Defense-VAE often preserve the correct class information of their underlying clean images, while Defense-GAN has a harder time to identify a correct reconstruction even though it searches for the right z from R random initializations and optimizes in L back-propagations, with typical $R = 10$ and $L = 200$. As we discussed in Sect. 3, Defense-VAE identifies a proper z directly by forward-propagating the input adversarial image through the VAE-encoder, and reconstructs a high quality denoised image through the VAE-decoder, and no online iterative optimization is involved.

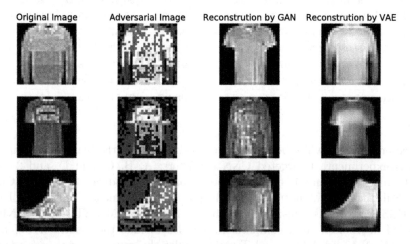

Fig. 5. The example reconstructions by Defense-VAE and Defense-GAN from the black-box FGSM attacks on F-MNIST: (a) original images; (b) adversarial images; (c) reconstruction by Defense-GAN; (d) reconstruction by Defense-VAE.

Table 4. Run time comparison between Defense-VAE and Defense-GAN, where *
denotes Defense-GAN recommended configuration.

Defense method		Run time on 1000 Images (s)
Defense-VAE		9.03
Defense-GAN	$L^* = 200$, $R^* = 10$	441.81
	$L = 400$, $R = 10$	875.48
	$L = 200$, $R = 20$	876.10
	$L = 400$, $R = 20$	1720.13

4.5 Defense Speed

Besides the superior defense accuracy of Defense-VAE, another advantage of
Defense-VAE is its superior defense speed over Defense-GAN. As discussed
above, to identify a high quality reconstruction, Defense-VAE doesn't need
expensive online iterative optimizations, while Defense-GAN requires L iterative back-propagations with R random restarts. To have a quantitative speed
comparison between Defense-VAE and Defense-GAN, we calculate their reconstruction times on 1000 adversarial images from F-MNIST, with the results
reported in Table 4, where different R and L configurations are considered.

As we can see, compared to the default Defense-GAN configuration, i.e.,
$L = 200$ and $R = 10$, Defense-VAE is about 50x faster than Defense-GAN.
Moreover, as L and R increase, Defense-GAN generally has a slightly better
defense accuracy, but the run time also increases linearly as $\mathcal{O}(L \times R)$. The
constant run-time complexity of Defense-VAE makes it widely deployable in
real-time security-sensitive systems.

5 Conclusion

In this paper, we propose Defense-VAE, a fast and accurate defense algorithm
against adversarial attacks. The algorithm is generic and can defense both white-
box and black-box attacks without the need of retraining the original CNN
classifier, and can further boost the defense strength by retraining or end-to-
end finetuning. Compared with the state-of-the-art algorithms, in particular,
Defense-GAN, our algorithm outperforms them in almost all white-box and
black-box defense benchmarks. In addition, Defense-VAE is very efficient as compared to the optimization-based defense alternatives, such as Defense-GAN, as
no expensive iterative online optimizations is needed. Speed test shows that
Defense-VAE is about 50x faster than Defense-VAE. Given the superior defense
accuracy and speed, we believe Defense-VAE is widely deployable in real-time
security-sensitive systems.

A   Network Architectures

The details of network architectures used in our experiments are described. Table 5 shows the architectures of the CNN classifiers and their substitute models, which are identical to those used in Defense-GAN [28] for a fair comparison.

Table 5. The architectures of the classifiers and the substitute models used in the white-box and black-box attacks.

A	B	C	D	E
Conv(*, 64, 5, 1, 2)	Dropout(0.2)	Conv(*, 128, 3, 1, 1)	FC(200)	FC(200)
ReLU	Conv(*, 64, 8, 2, 5)	ReLU	ReLU	ReLU
Conv(64, 64, 5, 2, 0)	ReLU	Conv(128, 64, 5, 2, 0)	Dropout(0.5)	FC(200)
ReLU	Conv(64, 128, 6, 2, 0)	ReLU	FC(200)	ReLU
Dropout(0.25)	ReLU	Dropout(0.25)	ReLU	FC(10) + Softmax
FC(128)	Conv(128, 128, 5, 1, 0)	FC(128)	Dropout(0.25)	
ReLU	ReLU	ReLU	FC(10) + Softmax	
Dropout(0.5)	Dropout(0.5)	Dropout(0.5)		
FC(10) + Softmax	FC(10) + Softmax	FC(10) + Softmax		

Table 6 shows the architecture of the Defense-VAE model used in the experiments on MNIST and F-MNIST. The architectures used for CelebA [14] and CIFAR-10 [11] are largely the same except that they are 1 or 2 layers deeper.

Table 6. The encoder and decoder of Defense-VAE used in the experiments.

Encoder	Decoder
Conv(*, 64, 5, 1, 2) + BN + ReLU	FC(128, 4096) + ReLU
Conv(64, 64, 4, 2, 3) + BN + ReLU	ConvT(256, 128, 4, 2, 1) + BN + ReLU
Conv(64, 128, 4, 2, 1) + BN + ReLU	ConvT(128, 64, 4, 2, 1) + BN + ReLU
Conv(128, 256, 4, 2, 1) + BN + ReLU	ConvT(64, 64, 4, 2, 3) + BN + ReLU
FC1(4096, 128), FC2(4096, 128)	ConvT(64, 64, 5, 1, 2) + BN + ReLU

B   Experiments on CelebA and CIFAR-10

We perform the white-box and black-box attacks on CelebA [14] and CIFAR-10 [11] datasets, with the results provided in Tables 7 and 8. Since Defense-GAN didn't provide results on CIFAR-10, we run their code on it and make sure the experimental settings for both algorithms are the same. We didn't provide the results related to the classifier model B due to its improper configuration for CIFAR-10, e.g., model B has much more parameters due to the large convolutional kernel size (e.g., 8×8) and 3 input channels.

Table 7. Classification accuracies of different defense methods under FGSM, RAND-FGSM and CW white-box attacks on CelebA and CIFAR-10. Since the Defense-GAN paper didn't provide the white-box attack results on CIFAR-10, we run their code and provide the results in the table.

Attack	Classifier model	No attack	No defense	Defense VAE	Defense VAE-REC	Defense VAE-E2E	Defense GAN	MagNet	Adv. Tr. $\epsilon = 0.3$
FGSM $\epsilon = 0.3$	A	96.55	3.94	92.40	94.89	95.10	92.55	9.85	12.25
	B	93.69	5.20	90.05	92.45	92.85	91.40	9.20	23.45
	C	95.62	4.45	92.47	94.46	95.25	92.55	10.85	11.30
	D	94.89	5.92	90.05	93.66	93.91	92.05	9.75	77.55
RAND FGSM $\epsilon = 0.3$ $\alpha = 0.05$	A	96.55	4.04	92.11	94.56	95.34	92.80	11.05	7.00
	B	93.69	4.76	90.55	92.57	93.07	90.30	10.15	45.15
	C	95.62	5.12	91.70	93.76	94.15	92.00	10.45	10.55
	D	94.89	6.15	91.42	93.53	93.87	91.65	11.05	6.96
CW l_2 norm	A	96.55	4.94	93.70	95.07	95.90	82.10	9.85	56.90
	B	93.69	4.90	90.65	92.40	93.55	74.65	9.55	7.25
	C	95.62	8.00	93.28	94.57	95.92	79.85	9.85	26.35
	D	94.89	6.47	91.15	93.12	93.39	77.40	10.40	50.10
Average		95.19	5.32	91.63	93.75	**94.36**	87.44	10.17	27.90

Attack	Classifier model	No attack	No defense	Defense VAE	Defense VAE-REC	Defense VAE-E2E	Defense GAN
FGSM $\epsilon = 0.3$	A	86.52	2.44	44.86	48.52	50.72	51.92
	C	87.62	5.05	43.92	47.29	47.39	47.84
	D	61.76	8.24	47.75	50.69	53.36	33.80
RAND FGSM $\epsilon = 0.3$ $\alpha = 0.05$	A	86.52	3.71	39.84	47.80	50.51	50.36
	C	87.62	3.87	41.28	46.16	47.91	48.52
	D	61.76	7.94	47.88	50.67	51.18	26.78
CW l_2 norm	A	86.52	2.34	38.41	45.91	49.44	45.62
	C	87.62	7.13	41.21	46.26	46.19	43.87
	D	61.76	7.78	53.32	55.81	57.21	20.35
Average		78.63	5.39	44.27	48.79	**50.43**	41.01

Table 8. Classification accuracies under FGSM black-box attacks on CIFAR-10. Since the Defense-GAN paper didn't provide the black-box attack results on CIFAR-10, we run their code and provide the results in the table.

Classifier/Substitute	No attack	No defense	Defense-VAE	Defense-VAE-REC	Defense-VAE-E2E	Defense-GAN
C/E	87.62	14.13	37.22	42.68	45.72	20.24
D/E	61.76	10.39	32.60	38.10	37.18	11.68
Average	74.69	12.16	34.91	40.39	**41.45**	16.32

References

1. Akhtar, N., Liu, J., Mian, A.: Defense against universal adversarial perturbations. arXiv preprint arXiv:1711.05929 (2017)
2. Biggio, B., et al.: Evasion attacks against machine learning at test time. In: Blockeel, H., Kersting, K., Nijssen, S., Železný, F. (eds.) ECML PKDD 2013. LNCS (LNAI), vol. 8190, pp. 387–402. Springer, Heidelberg (2013). https://doi.org/10.1007/978-3-642-40994-3_25
3. Carlini, N., Wagner, D.: Towards evaluating the robustness of neural networks. In: 2017 IEEE Symposium on Security and Privacy (SP), pp. 39–57 (2017)
4. Dziugaite, G.K., Ghahramani, Z., Roy, D.M.: A study of the effect of JPG compression on adversarial images. arXiv preprint arXiv:1608.00853 (2016)
5. Gao, J., Wang, B., Lin, Z., Xu, W., Qi, Y.: DeepCloak: masking deep neural network models for robustness against adversarial samples. arXiv preprint arXiv:1702.06763 (2017)
6. Goodfellow, I.J., et al.: Generative adversarial networks. In: NIPS (2014)
7. Goodfellow, I.J., Shlens, J., Szegedy, C.: Explaining and harnessing adversarial examples (2014). arXiv preprint arXiv:1412.6572 (2014)
8. Gu, S., Rigazio, L.: Towards deep neural network architectures robust to adversarial examples. arXiv preprint arXiv:1412.5068 (2014)
9. Guo, C., Rana, M., Cisse, M., van der Maaten, L.: Countering adversarial images using input transformations. arXiv preprint arXiv:1711.00117 (2017)
10. Kingma, D.P., Welling, M.: Auto-encoding variational Bayes. arXiv preprint arXiv:1312.6114 (2013)
11. Krizhevsky, A.: Learning multiple layers of features from tiny images. Technical report (2009)
12. Kurakin, A., Goodfellow, I., Bengio, S.: Adversarial machine learning at scale. arXiv preprint arXiv:1611.01236 (2016)
13. Lecun, Y., Bottou, L., Bengio, Y., Haffner, P.: Gradient-based learning applied to document recognition. In: Proceedings of the IEEE, pp. 2278–2324 (1998)
14. Liu, Z., Luo, P., Wang, X., Tang, X.: Deep learning face attributes in the wild. In: Proceedings of International Conference on Computer Vision (ICCV) (2015)
15. Luo, Y., Boix, X., Roig, G., Poggio, T., Zhao, Q.: Foveation-based mechanisms alleviate adversarial examples. arXiv preprint arXiv:1511.06292 (2015)
16. Lyu, C., Huang, K., Liang, H.N.: A unified gradient regularization family for adversarial examples. In: 2015 IEEE International Conference on Data Mining, November 2015
17. Meng, D., Chen, H.: MagNet: a two-pronged defense against adversarial examples. In: Proceedings of the Conference on Computer and Communications Security (2017)
18. Moosavi-Dezfooli, S.M., Fawzi, A., Frossard, P.: DeepFool: a simple and accurate method to fool deep neural networks. In: Proceedings of the IEEE Conference on Computer Vision and Pattern Recognition, pp. 2574–2582 (2016)
19. Moosavi-Dezfooli, S.M., Fawzi, A., Frossard, P.: DeepFool: a simple and accurate method to fool deep neural networks. In: CVPR (2016)
20. Nayebi, A., Ganguli, S.: Biologically inspired protection of deep networks from adversarial attacks. arXiv preprint arXiv:1703.09202 (2017)
21. Nguyen, L., Wang, S., Sinha, A.: A learning and masking approach to secure learning. arXiv preprint arXiv:1709.04447 (2017)

22. Papernot, N., McDaniel, P., Goodfellow, I.: Transferability in machine learning: from phenomena to black-box attacks using adversarial samples. arXiv preprint arXiv:1605.07277 (2016)
23. Papernot, N., McDaniel, P., Goodfellow, I., Jha, S., Celik, Z.B., Swami, A.: Practical black-box attacks against machine learning. In: Proceedings of the 2017 ACM on Asia Conference on Computer and Communications Security (2017)
24. Papernot, N., McDaniel, P., Jha, S., Fredrikson, M., Celik, Z.B., Swami, A.: The limitations of deep learning in adversarial settings. In: 2016 IEEE European Symposium on Security and Privacy (EuroSP), March 2016
25. Papernot, N., McDaniel, P., Wu, X., Jha, S., Swami, A.: Distillation as a defense to adversarial perturbations against deep neural networks. In: 2016 IEEE Symposium on Security and Privacy (SP), pp. 582–597 (2016)
26. Rezende, D.J., Mohamed, S., Wierstra, D.: Stochastic backpropagation and approximate inference in deep generative models. In: ICML (2014)
27. Ross, A.S., Doshi-Velez, F.: Improving the adversarial robustness and interpretability of deep neural networks by regularizing their input gradients. arXiv preprint arXiv:1711.09404 (2017)
28. Samangouei, P., Kabkab, M., Chellappa, R.: Defense-GAN: protecting classifiers against adversarial attacks using generative models. In: ICLR (2018)
29. Szegedy, C., et al.: Intriguing properties of neural networks. arXiv preprint arXiv:1312.6199 (2013)
30. Vorobeychik, Y., Kantarcioglu, M.: Adversarial Machine Learning. Morgan & Claypool, San Rafael (2018)
31. Xiao, H., Rasul, K., Vollgraf, R.: Fashion-MNIST: a novel image dataset for benchmarking machine learning algorithms (2017)
32. Xu, W., Evans, D., Qi, Y.: Feature squeezing: detecting adversarial examples in deep neural networks. In: Proceedings 2018 Network and Distributed System Security Symposium (2018)
33. Yuan, X., He, P., Zhu, Q., Li, X.: Adversarial examples: attacks and defenses for deep learning. arXiv preprint arXiv:1712.07107 (2017)

Analyzing and Storing Network Intrusion Detection Data Using Bayesian Coresets: A Preliminary Study in Offline and Streaming Settings

Fabio Massimo Zennaro$^{(\boxtimes)}$ (iD)

Department of Informatics, University of Oslo,
Blindern, PO Box 1080, 0316 Oslo, Norway
fabiomz@ifi.uio.no

Abstract. In this paper we offer a preliminary study of the application of Bayesian coresets to network security data. Network intrusion detection is a field that could take advantage of Bayesian machine learning in modelling uncertainty and managing streaming data; however, the large size of the data sets often hinders the use of Bayesian learning methods based on MCMC. Limiting the amount of useful data is a central problem in a field like network traffic analysis, where large amount of redundant data can be generated very quickly via packet collection. Reducing the number of samples would not only make learning more feasible, but would also contribute to reduce the need for memory and storage. We explore here the use of Bayesian coresets, a technique that reduces the amount of data samples while guaranteeing the learning of an accurate posterior distribution using Bayesian learning. We analyze how Bayesian coresets affect the accuracy of learned models, and how time-space requirements are traded-off, both in a static scenario and in a streaming scenario.

Keywords: Network intrusion data · Bayesian machine learning · Bayesian coresets · Logistic models

1 Introduction

Securing modern networks is a non-trivial challenge that requires high throughput (to evaluate in real-time the behavior of the network) and expert knowledge (to decide whether suspicious or malicious activity is taking place on the network). Network intrusion detection, in particular, is concerned with the early detection of attempts of breaking into and/or comprising a network. Packet collection techniques allows the continuous monitoring of networks and the collection of large amounts of traffic data.

Modern network intrusion detection data sets quickly exceeds the processing capacity of human reviewers, and thus automatic processing techniques for

© Springer Nature Switzerland AG 2020
P. Cellier and K. Driessens (Eds.): ECML PKDD 2019 Workshops, CCIS 1168, pp. 208–222, 2020.
https://doi.org/10.1007/978-3-030-43887-6_16

filtering and analyzing the data become necessary. A simple solution consists in eliciting knowledge from experts and encoding it into logical rules. Such rule-matching or signature-matching algorithms can process the data quickly, but their effectiveness is limited by the sort and the range of knowledge provided by the experts; in particular, these algorithms lack any sort of generalization and are unable to deal with novel data that do not perfectly match the encoded rules [8]. Machine learning has been put forward as a potential solution to this shortcoming. By learning directly from the data, and not from the particular knowledge provided by experts, machine learning algorithms aim at learning patterns that may hold not only for historical collected data, but also for future unforeseen data.

Many current machine learning techniques rely on large amount of data to learn useful patterns. The success of *deep learning*, in particular, has been explained by, among other factors, the availability of big data sets [13]. However, large data sets have also significant drawbacks. Large collections of data are problematic to archive and to store on drives; they are challenging to manipulate and load, requiring either a high amount of memory or frequent swapping operations; they are often redundant, to the point that this redundancy contributes little or nothing to the learning process.

Beyond deep learning, other machine learning algorithms may be severely affected in a negative way by large redundant data sets. This is the case, for instance, of *Bayesian machine learning*. Bayesian machine learning provides a rigorous framework for performing inference over data. It allows the estimation of complete probability distributions, a precise evaluation of uncertainty, the possibility of neatly integrating prior expert knowledge in the learned model, and the ability to update the learned model when provided with new data. However, all these possibilities come at a high computational cost, as standard Bayesian learning relying on *Markov chain Monte Carlo* (MCMC) algorithms do not to scale well with respect to the size of the data.

In presence of large redundant data sets, a possible solution to make inference via MCMC feasible consist in the reduction of the number of samples used to learn. Simple solutions include statistical techniques like *random sampling* or unsupervised learning algorithms such as *clustering via k-means* [2]. A more exact approach is based on the idea of creating *coresets*: instead of learning on the whole redundant data sets, it may be possible to define a (weighted) subset of samples which is probabilistically guaranteed to return a result close to the one that would be obtained by processing the whole data. In Bayesian machine learning, [3,4] recently proposed an efficient and promising algorithm to learn *Bayesian coresets in Hilbert space* (BCH) that computes a weighted subset of samples by smartly exploiting the structure of that space.

As many other applications, network intrusion detection could take advantage of Bayesian data analysis. A careful estimation of uncertainty when evaluating the possibility of a threat on the network is critical in order to take decisions. The possibility of integrating expert knowledge and update models are also important features in the complex and constantly changing environment of networks. Unfortunately though, the amount of data collected by capturing packets on a network quickly exceeds the feasibility of performing Bayesian machine learning via MCMC. By filtering redundant data and reducing the amount of samples via BCH, network traffic data sets could be reduced in size (thus offering a concrete benefit for storing and management) and processed using Bayesian techniques (thus producing more complete and versatile results).

In this paper we conduct a preliminary study of the possibility of applying BCH to network intrusion detection data and perform Bayesian machine learning via MCMC. We consider two main problems. (i) *How effective is the use of BCH to learn models of network intrusion?* We address this question considering (subsets of) realistic network traffic data and evaluating the effectiveness of BCH on a simple supervised learning problem. While an evaluation of BCH on computer security data (phishing data sets) is already provided in [3,4] in terms of metrics of posterior quality, we offer here an analysis in terms of accuracy, which is more relevant to the field of cyber-security. In particular we consider our results in light of the trade-off between time-space and accuracy and with respect to the sensitivity of BCH to its hyper-parameters. (ii) *How effective is the use of BCH to reduce the amount of data in a streaming environment?* We answer this question by considering the same realistic network data, but setting up a more challenging scenario in which the data samples are received sequentially. In this context, we analyze, once again in terms of accuracy and time-space savings, the advantages that BCH may bring when processing the data in real-time, upon arrival. Our results confirms that the trade-off between accuracy and time-space savings when using BCH is mainly regulated by one of the free hyper-parameters of BCH. Moreover, we show that the algorithm could be successfully used in a streaming environment, where it succeeds in sensibly reducing the computational time over several iterations and in ensuring good performances by aggregating coresets over the same iterations.

On the side, while tackling these questions, we also offer a practical contribution in the form of a porting of BCH algorithms[1] [3] into the framework of the probabilistic programming library Edward [19]. Specifically, we adapt the original code for coreset computation to work with Edward models, thus exploiting the probabilistic programming features of Edward[2] and the automatic differentiation feature of Tensorflow[3]. Code for this implementation is available online[4].

The rest of the paper is organized as follows. Section 2 briefly describes Bayesian machine learning and BCH. Section 3 outlines the problem of network intrusion detection and references previous work. Section 4 introduces our experimental setup. Section 5 tackles our first research question by analyzing the use of BCH on network traffic data. Section 6 deals with our second research

---

[1] https://github.com/trevorcampbell/bayesian-coresets.
[2] http://edwardlib.org/.
[3] https://www.tensorflow.org/.
[4] https://github.com/FMZennaro/BayesianCoresets-Edward.

question by evaluating the use of BCH in a streaming environment. Finally, Sect. 7 summarizes our results and presents some of the several avenues available for further development of this work.

2   Background

In this section we first introduce our general notation for the learning problem. We review the Bayesian approach to learning and its limitations. We then explain how Bayesian coresets deal with the problem of scalability. Finally, we review alternative approaches to work around the computational challenges of Bayesian learning.

2.1   Notation

In the following, we will deal with standard *supervised learning problems*. We consider a *data matrix* $\mathbf{X}$ of dimension $N \times F$, containing N samples described by F features; a sample $\mathbf{x}_i$ is a vector of dimension $1 \times F$. We also assume we are given a *label vector* $\mathbf{y}$ of dimension N, such that for each sample $\mathbf{x}_i$ we have a label y_i. Our aim is to learn a model mapping samples to labels: $f_\theta : \mathbf{x}_i \mapsto y_i$, where θ is a set of parameters defining the mapping function f.

The standard approach of machine learning is to convert this learning problem in an optimization problem as a function of the parameters θ. The optimal solution is found by computing the *point* estimate $\hat{\theta}$ of the parameters. For each input sample $\mathbf{x}_i$ we can then compute the output as $y_i = f_{\hat{\theta}}(\mathbf{x}_i)$. The result y_i (which can be interpreted probabilistically if calibrated [16]) is the output of the single model $f_{\hat{\theta}}$ on which we invested all our trust.

2.2   Bayesian Machine Learning

In Bayesian machine learning we tackle the problem of supervised learning with the aim of computing a full *distributional* estimation $P(\theta)$ of the parameters θ, instead of a point estimation. In this way, for each input sample $\mathbf{x}_i$ we can compute a distribution over the possible outputs $P(y_i|\mathbf{x}_i; \theta)$. This result represents the probability distribution of the output, computed considering all possible values of the parameters θ scaled by the trust assigned to them.

More formally, in Bayesian machine learning we estimate the *posterior distribution* $\pi(\theta) = P(\theta|\mathbf{X})$ of the parameters given the data using *Bayes' formula*:

$$P(\theta|\mathbf{X}) = \frac{P(\mathbf{X}|\theta)\,P(\theta)}{P(\mathbf{X})} = \frac{P(\mathbf{X}|\theta)\,P(\theta)}{\int P(\mathbf{X}|\theta)\,P(\theta)d\theta} = \frac{\mathcal{L}(\mathbf{X};\theta)\,\pi_0(\theta)}{\int \mathcal{L}(\mathbf{X};\theta)\,\pi_0(\theta)\,d\theta},$$

where $\pi_0(\theta) = P(\theta)$ is the *prior probability distribution* over the parameters, $\mathcal{L}(\mathbf{X};\theta) = P(\mathbf{X}|\theta)$ is the *likelihood function* of the data with respect to the parameters, and $P(\mathbf{X})$ is the *evidence*.

Computing the posterior distribution is a challenging task that requires the evaluation of the product of likelihood function $\mathcal{L}(\mathbf{X};\theta)$ and prior distribution

$\pi_0(\theta)$, and the evaluation of the evidence integral. Monte Carlo Markov chain (MCMC) algorithms are a practical solution to this problem based on the idea of sampling from the posterior distribution [10]. The main drawback of this approach is the computational scalability as the complexity of sampling a posterior point grows linearly with the size of the data [3].

2.3   Bayesian Coresets

Evaluating the posterior distribution via MCMC sampling requires the computation of the likelihood $\mathcal{L}(\mathbf{X};\theta)$. Under the assumption of independent and identically distributed data, the likelihood for the whole data set $\mathcal{L}(\mathbf{X};\theta)$ may be factorized in the product of the likelihoods of individual data points $\mathcal{L}(\mathbf{x}_i;\theta)$:

$$\mathcal{L}(\mathbf{X};\theta) = \prod_{i|x_i \in \mathbf{X}} \mathcal{L}(\mathbf{x}_i;\theta),$$

or, equivalently, in the product of *log-likelihoods*:

$$\log \mathcal{L}(\mathbf{X};\theta) = \sum_{i|x_i \in \mathbf{X}} \log \mathcal{L}(\mathbf{x}_i;\theta).$$

Bayesian coresets compute a small weighted subset of the original data $\mathbf{T} \subseteq \mathbf{X}$ such that the log-likelihood computed on $\mathbf{T}$ approximates the log-likelihood computed on $\mathbf{X}$:

$$\log \mathcal{L}(\mathbf{X};\theta) = \sum_{i|x_i \in \mathbf{X}} \log \mathcal{L}(\mathbf{x}_i;\theta) \approx \sum_{n|x_n \in \mathbf{T}} w_n \log \mathcal{L}(\mathbf{x}_n;\theta) = \log \mathcal{L}(\mathbf{T};\mathbf{w},\theta),$$

where $\mathbf{x}_n$ are samples belonging to the coreset and w_n are the associated weights. The degree of approximation may be evaluated in terms of *epsilon*-distance between the original log-likelihood and the coreset likelihood:

$$|\log \mathcal{L}(\mathbf{X};\theta) - \log \mathcal{L}(\mathbf{T};\mathbf{w},\theta)| \leq \epsilon \cdot |\log \mathcal{L}(\mathbf{X};\theta)| \quad \forall \theta. \tag{1}$$

Estimating this distance is challenging, and an approximation is offered by *Huggin's algorithm* [11].

Bayesian Coresets in Hilbert Spaces. A refinement of this solutions has been proposed by [3], with the suggestion of embedding log-likelihoods in a Hilbert function space. This reformulation has several advantages. First, by taking the objects of this space to be functions of the form $g : \Theta \rightarrow \mathbb{R}$, log-likelihoods $\log \mathcal{L}(\mathbf{x}_i;\theta)$ or $\log \mathcal{L}(\mathbf{x}_n;w_n,\theta)$ become vectors of this space; consequently, the total likelihood over the whole data set $\log \mathcal{L}(\mathbf{X};\theta)$ or the coreset $\log \mathcal{L}(\mathbf{T};\mathbf{w},\theta)$ can be expressed in terms of vector sum. Second, by taking as a norm of the space a bounded sup norm $\|g\| = \sup_\theta \left| \frac{g(\theta)}{\log \mathcal{L}(\mathbf{X};\theta)} \right|$, we can restate the constrained problem in Eq. 1 as a sparse quadratic minimization problem:

$$\min_{\mathbf{w}} \|\log \mathcal{L}(\mathbf{X};\mathbf{w}) - \log \mathcal{L}(\mathbf{X})\|^2$$

under the constraints:

$$w_n \geq 0$$
$$\sum 1 \left[w_n > 0 \right] \leq M,$$

where $1\left[c\right]$ is the identity function that returns 1 if c holds or 0 otherwise, and M is a maximum number of coreset samples $\mathbf{x}_n$ that we allow selecting. This fomalization turns the problem of constructing a coreset into an optimization problem aimed at finding the minimal set of samples $\mathbf{x}_n$ that approximates the log-likelihood on the data set. Finally, the structure of the Hilbert space allows us to exploit the directionality of the space in order to account for residual errors between $\log \mathcal{L}\left(\mathbf{X}; \mathbf{w}\right)$ and $\log \mathcal{L}\left(\mathbf{X}\right)$ and to better select samples $\mathbf{x}_n$ that would improve the approximation. Algorithms for coreset construction that exploit the properties of the Hilbert space include the *coreset construction based on Frank-Wolfe algorithm* [3] and the *GIGA algorithm* [4].

Model Dependency of Bayesian Coresets. As it has been underlined by [5], it is important to remark that a coreset computed by a BCH algorithm is tightly connected to a specific family of models. Such a coreset does not constitute a generic weighted non-redundant distillation of the original data set; it is a subset of the original data optimized with respect to a specific family of models in order to produce a posterior distribution as close as possible to the one that we would learn from the original data set. In sum, *a BCH is actually a tuple made up by a family of models and a weighted set of samples.*

2.4   Alternative Approaches to BCH for Bayesian Learning

BCH is just one of the possible approaches to make Bayesian machine learning feasible on large data sets. Other approaches, which do not involve reducing the number of samples, include *variational Bayes, parallel MCMC* and *approximate MCMC*. Variational Bayes algorithms forgo the idea of using MCMC algorithms to perform inference, and rely instead on variational approximations of the posterior [2]. The variational approach allows learning in presence of large data sets, but the method does not provide guarantees on the degree of approximation of the uncertainty of the posterior [9]. Parallel MCMC algorithms rely on parallelization: large data sets are divided among multiple clusters; each cluster runs locally Bayesian inference via MCMC and produces a posterior distribution; finally, all the posteriors are aggregated by finding a unique posterior in the metric space of the posterior distributions [14,18]. The parallel approach allows to deal with large data sets, but it still requires a high computational budget and does not address the problem of storing redundant data. Finally, approximate MCMC aims at speeding up existing algorithms by replacing costly transition in the Markov chain process with approximations [12]. Again, this approach is effective when we have to process large data sets, but it requires analyzing the execution of the MC algorithm, and, once again, it does not consider the problem of storing redundant data.

3   Network Security

Network security is one of the main challenges in the management of online systems. Network administrators try to monitor and prevent malicious activity through the deployment of intrusion detection systems, the collection of network traffic, and the analysis of this data [15]. Processing these data in a timely manner in order to detect suspicious activity as early as possible is a crucial problem. The ease with which large amount of data can be collected on a network poses severe scalability problems, both in terms of storage and in terms of processing [8]. Given this constraint, computationally-cheap algorithms, such as *signature-matching*, *random forests* and *support vector machines*, have been favored; for a review of pattern recognition and classical machine learning algorithms applied to network security, see, for instance, [8] and [7].

4   Experimental Setup

In this section we provide a formal description of our study by defining the exact learning problem we considered, by presenting the data sets and the transformations we applied to them, and, finally, by discussing the models we implemented.

Problem Definition. Given a large data set for network intrusion detection, we express our learning problem as a supervised learning problem in which we try to discover a function that maps network flows to an output defining whether a flow is malicious or not. More precisely, we try to infer an optimal set of parameters θ that define the mapping function $f_\theta : \mathbf{x}_i \mapsto y_i$. In a first static scenario, we process our data with and without BCH, running the simulations multiple times ad observing the contribution of the BCH algorithm. In a second scenario, we simulate the progressive collection of large chunks of data. All the data samples are taken to be independent and identically distributed. In this case, we observe what the contribution of BCH would be if we were to filter out data as soon as they are collected.

Network Data Set. To run our experiments we use the network traffic data collected in the CICIDS2017 data set [17]. Processing real-world network data presents challenges from a privacy perspective; for this reason, the CICIDS2017 was collected running a simulated network designed to behave in a realistic fashion. Five days of simulated traffic were collected; during each day, different types of attacks and malicious behaviors were enacted. Network packets are gathered and aggregated in network flows. In total, the data set contains more than 2.5 million sample flows. Each sample is defined by a 78-dimensional vector reporting features such as packet flags and packet lengths (see [17] for a complete description of the data). Finally, a binary label has been assigned to each network flow, denoting whether a flow is legitimate or not. We restrict our attention to the second day, Tuesday[5], which is made up by 445708 data samples, of which 13835 constitute instances of brute force attacks.

Network Data Preprocessing. The CICIDS2017 data set contains a very limited number of samples (201) with missing values for the day of Tuesday. Given their limited number we simply assume that they are *missing completely at random* [1] and we just drop them. Also, before each experiment we always standardize the data to zero mean and unit variance by feature. Standardization parameters are always computed exclusively on the training data being processed.

Data Set Subsampling. For the purpose of this preliminary study, we evaluate our algorithms only on limited subsets of the large CICIDS2017 data set. Studying smaller data sets guarantees some advantages: (i) it allows us to compare Bayesian learning on coresets with the "ground truth" of learning on the whole data sets, which would not be feasible if we were to consider the entire data set; (ii) manipulating the data set allows us to simulate scenarios in which we receive streaming data. While we plan to extend our evaluation to the full data set in order to assess the true potential of BCH for network intrusion detection, we were still able to get useful insights by studying its application on more modest-sized subsets of the original data set. Thus, from the large original pool of data, we programmatically create subsets by random sampling. In order to preserve the imbalance between positive and negative instances in the training data, we always select ten times as many positive samples as negative samples. The prototypical training data set $\mathbf{X}_i^{tr}$ consists of 800 positive instances and 80 negative instances. For a test data set, instead, we selected an even number of positive and negative instances, thus simplifying the interpretation of the results. Our prototypical test data set $\mathbf{X}_i^{te}$ consists of 200 positive an 200 negative samples.

Sample Reduction techniques. Given a training data set $\mathbf{X}_i^{tr}$, in order to reduce the amount of data samples, we apply BCH using the GIGA algorithm [4]. This algorithm has two free hyper-parameters: (i) the number of random dimension on which to project the samples; and (ii) the number of computational iteration M, which implicitly limits the maximum number of coreset samples that can be selected.

Network Models. For our simulations, we consider two models:

1. *Bayesian logistic regression (BLR):* a discriminative generalized-linear Bayesian machine learning algorithm [2]. We define our weighted BLR model as:

$$\theta = \mathtt{N}\,(0,1)$$
$$y_i = \mathtt{Bern}\,((\sigma\,(\theta \mathbf{x}_i))^{w_i})$$

---

[5] Notice that we did not consider the data on Monday because no malicious activity takes place on this day, thus providing us only with positive instances.

where $N(0,1)$ is a Gaussian prior, and the likelihood $\mathcal{L}(\mathbf{X};\theta)$ is the likelihood under a Bernoulli pdf $\mathrm{Bern}(\sigma(\theta\mathbf{x}_i))$ scaled by the weight w_i associated to the sample $\mathbf{x}_i$. If the data samples are not weighted, then $w_i = 1$ for all $\mathbf{x}_i$, and the model reduces to a standard BLR. Notice that, in general, if all the samples $\mathbf{x}_i$ are scaled by a constant value $w_i = k$ the inference process will not change[6].

Given a new sample $\mathbf{x}_i$, its probability of belonging to a class y_i can be obtained by integrating over all the models under the posterior: $P(y_i|\mathbf{x}_i) = \int P(y_i|\mathbf{x}_i,\theta)P(\theta)d\theta$.

2. *Support Vector Machine (SVM):* as a baseline and comparison, we consider support vector machine, a linear maximum-margin discriminator [6]. We train a SVM model to find the slope θ of a discriminating hyper-plane between samples[7].

Given a new samples $\mathbf{x}_i$, its class is computed as $y_i = \mathrm{sign}(\theta\mathbf{x}_i)$, where $\mathrm{sign}(z)$ is the sign function, $\mathrm{sign}(z) = 1$ if $z > 0$, 0 otherwise.

5 Simulation 1: BCH Applied to Network Intrusion Detection Data

In this simulation we analyze the use of BCH applied to network intrusion detection data. We evaluate the contribution they provide both from the point of view of the performance they achieve and the time-space they save. We compare these results to the baseline offered by SVM and by a BLR computed over the whole data set.

Protocol. We generate five subsets of training data $\mathbf{X}_i^{tr}$ and test data $\mathbf{X}_i^{te}$, $1 \leq i \leq 5$, using the methodology presented in Sect. 4.

We apply BCH to each training data set $\mathbf{X}_i^{tr}$. We set the free hyper-parameters of BCH as follows: (i) we fix the number of random dimension to 500, following the experimental evaluation in [3]; (ii) we consider three values for the number of computational iteration M, that is $M = 1000$, following again the evaluation in [3], $M = 500$, and an aggressively lower value of $M = 100$, which is expected to guarantee a higher saving in terms of space and time.

For each subset i, we train and test an SVM model, a BLR model trained on the whole data set $\mathbf{X}_i^{tr}$, and a BLR model trained on the coreset computed from the training data $\mathbf{X}_i^{tr}$.[8] The SVM model is trained with default parameters from the scikit[9] library. The BLR models are trained using the Hamiltonian Monte

---

[6] We take advantage of this constant scaling to prevent overflowing errors in the simulations.

[7] Notice that for a fair comparison with the BLR model, we compute only the slope of the discriminating hyperplane and not its intercept.

[8] Notice that we do not train the SVM model on the coresets because, as discussed in Sect. 2.3, coresets are not generic non-redundant sub-data sets, but they are sub-selections optimized for a specific statistical model.

[9] https://scikit-learn.org/stable/.

Carlo algorithm offered in the Edward library with the following settings: sampling 10000 points, using a burn-in period of half of the samples, thinning every second sample, and adjusting the step size manually to guarantee an acceptance rate around 0.8. When doing prediction, we use 1000 posterior samples. We repeat each training and testing 10 times and we average the results.

We evaluate the results in terms of classification accuracy and wall-clock time required for the training of the model (all the models are run on a non-dedicated mid-range laptop machine with no GPU support).

Results. First of all, the data processing via BCH with different hyper-parameters produced different coresets. Table 1 reports the number of data points selected, specifying the number of samples in the minority class that have been preserved, and the wall-clock computation time as a function of the hyper-parameter M. Notice that the number of iterations M does not correspond to the number of coreset samples selected; in all the cases the algorithm selects a number of samples well below this threshold. With respect to the original number of samples, the amount of data points selected by BCH ranges from around one tenth, when using a low $M = 100$, to one third, when using a high $M = 1000$.

Table 1. Coresets computed on each training data set $\mathbf{X}_i^{tr}$. The table reports the number of data points selected to the left of the slash $(/)$, and the number of these points belonging to the minority class to the right of the slash $(/)$. The last column reports average and standard deviation of the wallclock time to compute the coresets.

Hyper-parameter	$\mathbf{X}_1^{tr}$	$\mathbf{X}_2^{tr}$	$\mathbf{X}_3^{tr}$	$\mathbf{X}_4^{tr}$	$\mathbf{X}_5^{tr}$	Time(s)
$M = 100$	84/1	82/3	82/1	88/1	87/3	260.78 ± 2.37
$M = 500$	184/4	191/10	186/7	189/9	200/10	251.82 ± 3.05
$M = 1000$	259/5	270/15	239/4	252/8	252/6	268.63 ± 2.51

Figure 1 shows the accuracy of our models on the different data sets we considered. Consistently with our expectations, the two linear models, SVM and BLR on the whole data set, perform similarly; BLR models trained on coresets show, in general, decreasing performances as we decreased the hyper-parameter M of the BCH algorithm.

Figure 2 compares the wall-clock time of each algorithm. The highly optimized SVM algorithm shows some variability, but in general terminates in tenths of seconds. On the other hand, BLR takes up to two orders of magnitude longer. BLR on coresets is faster, even if on the same time scale; surprisingly using coresets with $M = 500$ took the shortest time, which may be due to the particularly good sub-selection of points, or, more likely, to other contingent processes running on the same machine.

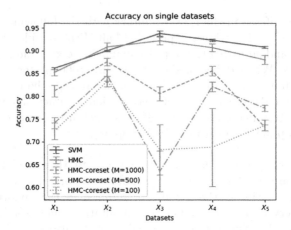

Fig. 1. Mean and standard deviation of accuracy of the models on each data set $\mathbf{X}_i^{te}$.

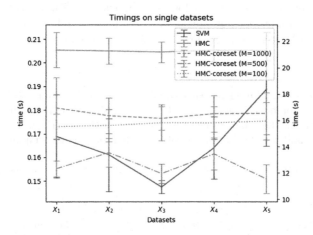

Fig. 2. Mean and standard deviation of wall-clock time required for training the models on each data set $\mathbf{X}_i^{tr}$. Notice the different y-scale for SVM on the left side, and the BLR model on the right side.

Discussion. These basic experiments highlight that the time-space savings offered by BCH inevitably come at the cost of the accuracy of the final model. The number of iterations M provides a key hyper-parameter to manage such a trade-off, as it exchanges the dimension of the data set for the accuracy of the model.

Notice that from these experiments, the time saving offered by coresets does not appear particularly remarkable. It is worth, though, to underline that such an improvement is relevant when related to the small data sets we are processing. The time savings when using larger data set are discussed in detail in [3].

Interestingly, the computation of coresets has a subsampling effect with respect to the minority class, as shown in Table 1: while in the original data set the ratio between the two classes was set to 1:10, this ratio has sensibly decreased. This may seem undesirable if we were expecting BCH to produce a more balanced data set; in reality, though, the algorithm selects only samples useful for a proper reconstruction of the likelihood function, and the result seems to suggest that the instances of malicious behaviours may actually be quite redundant, probably due to the fact that we are considering only one specific form of attack (brute force).

6   Simulation 2: BCH in a Streaming Environment

In this simulation we try to setup a more interesting and realistic scenario. We simulate the collection of batches of data in real-time and we learn from the cumulative set of collected samples. The aim is to evaluate how the learning process would be affected if the sets of collected data were to be downsized using BCH before being processed and stored. Such a scenario seems particularly interesting because BCH would immediately discard redundant data, thus solving at once the problem of making Bayesian inference feasible and reducing the required amount of memory and storage space.

Protocol. As before, we generate five subsets of training data $\mathbf{X}_i^{tr}$ and test data $\mathbf{X}_i^{te}$, $1 \leq i \leq 5$. Now, however, instead of processing each data set independently, we simulate the arrival of a data set $\mathbf{X}_i^{tr}$ at time steps t_i. At each time step t_i, we want to learn from all the collected data sets and so we pool together all the data $\mathbf{X}_j^{tr}$, for $j \leq i$. Notice that all the samples are independent and identically distributed.

When using coresets, we apply BCH to each training data set $\mathbf{X}_i^{tr}$ as soon as it is collected. At each time step t_i, instead of aggregating together all the previously collected data, we just aggregate the coresets. This operation is theoretically justified by the possibility of aggregating coreset computed in parallel [3]. We use the same hyper-parameters for BCH used in the previous simulation.

We also run the same models as before, and we repeat each simulation 10 times.

Results. We work with the same coresets computed in the previous simulation and we refer back the reader to Table 1 for their details.

Figure 3 shows the accuracy of the models on the different data sets we generated. The starting values of accuracy computed on a single data set (X_1) are consistent with the values computed in the previous experiments and shown in Fig. 1. When we start aggregating more data sets, we notice that the performance of SVM and HMC on the whole data set is only slightly improved; on the other hand, the performance of BLR on coresets shows a consistent improvement. Even aggregating only two coresets $(X_1 + X_2)$ the performance gap between BLR on coresets and SVM or BLR on the whole data set is significantly reduced. This improvement is expected when aggregating two coresets computed with the hyper-parameter $M = 500$; in this case, the final amount of selected data points would be close to the amount obtained computing a single coreset with the hyper-parameter $M = 1000$; and we know from the previous simulations that the performance of BLR trained on a coreset computed with the hyper-parameter $M = 1000$ is very close to the performance of BLR on the whole data set. More surprising is the improvement registered by aggregating only two coresets computed with hyper-parameter $M = 100$.

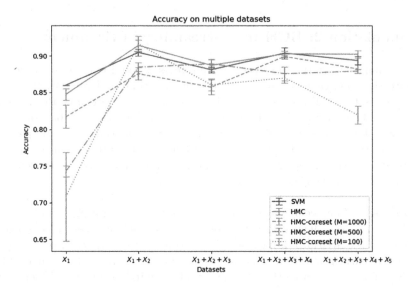

Fig. 3. Mean and standard deviation of accuracy of the models on each data set $\mathbf{X}_i^{te}$.

Figure 4 compares the wall-clock time of each algorithm. Again, the time scale of the two family of algorithms, SVM and BLR, are very different. However, notice that while the computational time for SVM and BLR on the whole data sets tend to grow in a linear fashion, the growth in the required computational time when running BLR on coresets is almost flat.

Discussion. This simulation showed the potential advantages that could be obtained by deploying BCH in a streaming scenario. In such an instance, the aggregation of two or more coresets can provide a performance very close to SVM or BLR trained on the whole data. One of the most significant advantages, though, is that BCH reduces the amount of data in real-time before learning, thus limiting the amount of memory necessary for processing; guaranteeing a sub-linear growth in the time required for learning as more data are gathered may prove especially advantageous when data is collected in real-world environments in which batches of data are generated continuously over multiple timesteps

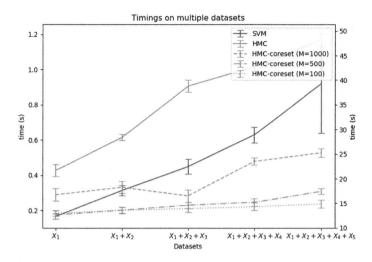

Fig. 4. Wall-clock time required for training the models on each data set $\mathbf{X}_i^{tr}$. Notice the different y-scale for SVM on the left side, and the BLR model on the right side.

7   Conclusion and Future Work

This preliminary study showed the feasibility of applying BCH to network data. Network intrusion detection could take great advantage by employing fully probabilistic descriptions of network traffic, and BCH may prove to be an enabler for such an approach. Moreover, we showed that the same algorithm is also effective in reducing the number of samples to be stored; this issue is particularly relevant when collecting network packets, as the amount of data may quickly grow and cause severe challenges in their management.

Our experiments first confirmed the concrete trade-off between model accuracy and time-space saving when adopting BCH. More interestingly, we also investigated how BCH may be deployed in a dynamic streaming environment in which data samples would be filtered in real-time before processing. This last scenario returned particularly good and interesting results, showing that BCH can be effectively used to sub-select relevant data samples at different time steps and aggregate together only the coresets. We demonstrated that in a streaming scenario the use of BCH may guarantee a better scalability by ensuring that the computational time for learning grows in a strongly sub-linear fashion.

Of course this study is just a preliminary evaluation of the potential of BCH applied to the challenging problem of processing the large data sets for network security. Further investigation is clearly necessary to assess more precisely the role that BCH may serve. Immediate directions of further study that we consider are the following: applying our protocol to bigger and more realistic data sets; include other typologies of attacks; compare BCH to other data reduction techniques, such as random sampling or k-means. More interesting questions concern also the recursive application of BCH in a streaming scenario and its effectiveness when used to process streaming data that do not conform to the assumption of independent and identically distributed data anymore.

References

1. Barber, D.: Bayesian Reasoning and Machine Learning. Cambridge University Press, New York (2012)
2. Bishop, C.M.: Pattern Recognition and Machine Learning. Springer, Boston (2006). https://doi.org/10.1007/978-1-4615-7566-5
3. Campbell, T., Broderick, T.: Automated scalable Bayesian inference via Hilbert coresets. arXiv preprint arXiv:1710.05053 (2017)
4. Campbell, T., Broderick, T.: Bayesian coreset construction via greedy iterative geodesic ascent. arXiv preprint arXiv:1802.01737 (2018)
5. Coleman, C., et al.: Select via proxy: efficient data selection for training deep networks (2018)
6. Cortes, C., Vapnik, V.: Support-vector networks. Mach. Learn. **20**(3), 273–297 (1995). https://doi.org/10.1007/BF00994018
7. Garcia-Teodoro, P., Diaz-Verdejo, J., Maciá-Fernández, G., Vázquez, E.: Anomaly-based network intrusion detection: techniques, systems and challenges. Comput. Secur. **28**(1–2), 18–28 (2009)
8. Gardiner, J., Nagaraja, S.: On the security of machine learning in malware C&C detection: a survey. ACM Comput. Surv. (CSUR) **49**(3), 59 (2016)
9. Giordano, R.J., Broderick, T., Jordan, M.I.: Linear response methods for accurate covariance estimates from mean field variational Bayes. In: Advances in Neural Information Processing Systems, pp. 1441–1449 (2015)
10. Givens, G.H., Hoeting, J.A.: Computational Statistics, vol. 710. Wiley, Hoboken (2012)
11. Huggins, J., Campbell, T., Broderick, T.: Coresets for scalable Bayesian logistic regression. In: Advances in Neural Information Processing Systems, pp. 4080–4088 (2016)
12. Johndrow, J.E., Mattingly, J.C., Mukherjee, S., Dunson, D.: Approximations of Markov chains and high-dimensional Bayesian inference. arXiv preprint arXiv:1508.03387 (2015)
13. LeCun, Y., Bengio, Y., Hinton, G.: Deep learning. Nature **521**(7553), 436 (2015)
14. Neiswanger, W., Wang, C., Xing, E.: Asymptotically exact, embarrassingly parallel MCMC. arXiv preprint arXiv:1311.4780 (2013)
15. Northcutt, S., Novak, J.: Network Intrusion Detection. Sams Publishing, Indianapolis (2002)
16. Shalizi, C.: Advanced data analysis from an elementary point of view (2013)
17. Sharafaldin, I., Lashkari, A.H., Ghorbani, A.A.: Toward generating a new intrusion detection dataset and intrusion traffic characterization. In: ICISSP, pp. 108–116 (2018)
18. Srivastava, S., Cevher, V., Dinh, Q., Dunson, D.: WASP: scalable Bayes via barycenters of subset posteriors. In: Artificial Intelligence and Statistics, pp. 912–920 (2015)
19. Tran, D., Kucukelbir, A., Dieng, A.B., Rudolph, M., Liang, D., Blei, D.M.: Edward: a library for probabilistic modeling, inference, and criticism. arXiv preprint arXiv:1610.09787 (2016)

6th Workshop on Sports Analytics: Machine Learning and Data Mining for Sports Analytics (MLSA)

Analyzing Soccer Players' Skill Ratings Over Time Using Tensor-Based Methods

Kenneth Verstraete[1](✉)🆔, Tom Decroos[2], Bruno Coussement[2],
Nick Vannieuwenhoven[2]🆔, and Jesse Davis[2]🆔

[1] Department of Chronic Diseases, Metabolism and Ageing, KU Leuven,
3000 Leuven, Belgium
kenneth.verstraete@kuleuven.be

[2] Department of Computer Science, KU Leuven, 3001 Leuven, Belgium
{tom.decroos,nick.vannieuwenhoven,jesse.davis}@cs.kuleuven.be

Abstract. Soccer players have a variety of skills such as passing, tackling, shooting and dribbling. However, their abilities are not fixed and evolve over time. Understanding this evolution could be interesting from many perspectives. We analyze player skill data from the FIFA video game series by EA Sports using tensor methods. This data can be organized as a tensor over three dimensions, namely players, skills, and age, which we explore in two different ways. First, we use a polyadic decomposition to uncover hidden structures among skills and see how these structures evolve over time. Second, we use a Tucker decomposition to predict how a specific player's skills will evolve over time.

1 Introduction

Playing soccer involves multiple technical skills such as passing, tackling, shooting and dribbling as well as physical skills like acceleration and endurance. Over time, a player's ability will change; with practice skills can improve, but as a player ages certain skills will inevitably decline. Understanding this evolution is interesting from several perspectives. First, gaining intuition about the relationship between different skills and how they change over time can yield insights into the game. Second, predicting how a player will evolve can help a club with roster decisions such as identifying potential transfer targets or deciding which current players to retain and for how long.

One potential source of skill data comes from the FIFA video game series, where EA Sports models all soccer players using ratings for in-game skills with the aim of reflecting their real-life soccer skills.[1] In this paper, we explore modeling temporal evolutions in this skill data set. These data can be naturally modeled using a tensor with three *modes* or axes: players, skills and age. We make three contributions. First, we highlight a variety of challenges that arose while analyzing the SoFIFA data. Second, we explored the SoFIFA data tensor using the *canonical polyadic decomposition* (CPD) in order to extract interpretable

---

[1] These skill ratings are accessible online at https://www.sofifa.com.

P. Cellier and K. Driessens (Eds.): ECML PKDD 2019 Workshops, CCIS 1168, pp. 225–234, 2020.
https://doi.org/10.1007/978-3-030-43887-6_17

latent structures. Moreover, this decomposition can yield some insights into how these structures evolve over time. Third, we employ the *Tucker decomposition* of a tensor in order to project how a specific player's skills will evolve over time. That is, we can predict a player's ratings for each skill at a future age.

2   Tensors and the SoFIFA Data

We provide a brief background on tensors, describe our data set and highlight a number of challenges encountered when analyzing this data.

2.1   Tensors

Tensors are a generalization of vectors and matrices that allow us to model the interactions between different variables such as players, skills, and age. An order-d tensor $\mathcal{A} \in \mathbb{R}^{n_1 \times n_2 \times \cdots \times n_d}$ can be identified with a d-array. The different dimensions of the tensors are called the modes. This paper focuses on order-3 tensors, but all presented theory can be generalized to higher orders as well. See the survey by Kolda and Bader [4] for more details.

2.2   Data Description

We used all available data from the FIFA 07 game up to the FIFA 18 game to track players' skill evolution. Human experts employed by EA Sports assign each soccer player ratings between 0 and 100 for 24 different skills (e.g., crossing, finishing, dribbling, etc; see Fig. 1). Which experts and how many assigned ratings to each player is not known. Updating the skill ratings is done in an open-source fashion; a community of 8,000 coaches, scouts and season ticket holders can submit inconsistencies which are then checked and fixed by a small team of 25 EA producers [10]. SoFIFA.com details each player's biological data, wage, club, positions and skills in player cards (Fig. 1). These player cards are updated through time. From 2007 to 2012, the player cards were updated biannually, and thereafter EA Sports started updating them weekly.

For each player, we can capture the evolution of his skills as he ages in a matrix by scraping the skill values on his player card as they were on January 1st of each year from 2007 to 2018. We then stack these player matrices in a three-dimensional tensor of the form: *players* × *skills* × *ages*.

2.3   Data Challenges

Analyzing real-world data is challenging. In this paper, most of the challenges are related to the quality of the skill ratings. We discuss five challenges.

Cold start. Players have played few matches at the start of their career. Hence it is hard for experts to construct a player card with accurate skill ratings for new players based on a small amount of data.

Lionel Messi (ID: 158023)
Age 31 (Jun 24, 1987) 170cm 72kg
CF RW ST
Value €110.5M   Wage €565K

94 Overall Rating   94 Potential

Attacking		Skill		Movement		Goalkeeping	
86	Crossing	97	Dribbling	91	Acceleration	6	GK Diving
95	Finishing	94	FK Accuracy	86	Sprint Speed	11	GK Handling
70	Heading Accuracy	89	Long Passing	95	Reactions	15	GK Kicking
92	Short Passing	96	Ball Control			14	GK Positioning
						8	GK Reflexes

Power		Mentality		Defending	
85	Shot Power	48	Aggression	33	Marking
72	Stamina	75	Penalties	28	Standing Tackle
66	Strength				
94	Long Shots				

Fig. 1. Lionel Messi's player card with his ratings for the 24 different skills [2]. Each skill rating is a number between 0 and 100 that quantifies how proficient the player is at that skill estimated by human experts. The overall rating is a weighted average of the skill ratings. The exact weighting scheme employed by FIFA is unknown, as is the way they predict the potential of a player, which is the highest overall rating a player can attain in the future.

Human subjectivity. Rating players' skills is inherently a subjective task. A difference in experience between human experts can lead to different opinions about the same player.

Ratings for irrelevant skills. Each player receives a rating for every skill available, including those that are not relevant for their position (e.g., field players receive ratings for goalkeeping skills).

Artificial boosts. EA Sports is known to manually boost the ratings for well-known players if they disagree with the experts' opinion [5], which disrupts the quality and consistency of the ratings in the data set.

Disruptive skill corrections. Three types of corrections can occur that are unrelated to actual changes in player skill. First, due to the cold start problem, early-career players can receive wildly inaccurate initial skill ratings, which are later (when the player has played more matches) corrected by a large value, causing a disruptive jump in the skill evolution. Second, a collective correction can happen where rating for a specific skill is dramatically altered for a large group of players. For example, all field players had their rating for the *GK Kicking* skill significantly lowered in the FIFA 11 edition of the game. Finally, the newer FIFA games receive updates on a weekly basis and hence, rating adjustments have become more event-related now compared to older FIFA games, when updates only happened once or twice a year.

For example, Messi's penalty skill was lowered from 78 to 75 in the update from 21 December 2017 after he missed a penalty kick on 17 December 2017 [6,7].

3   Exposing Hidden Structure with CPD

Since we are dealing with temporal data that describes the evolution of players over time, tensor decompositions provide a straightforward and natural way to discover patterns and predict evolution. A decomposition that allows us to discover intelligible hidden, underlying patterns in the player data is the canonical polyadic decomposition [3]. The CPD approximates a tensor $\mathcal{A} \in \mathbb{R}^{n_1 \times n_2 \times n_3}$ as a sum of r *pure tensors* $\mathcal{A}_i$, where each $\mathcal{A}_i$ can be written as the tensor product of the vectors $\mathbf{p}_i, \mathbf{s}_i$, and $\mathbf{a}_i$, living in respectively $\mathbb{R}^{n_1}, \mathbb{R}^{n_2}$ and $\mathbb{R}^{n_3}$:

$$\mathcal{A} \approx \sum_{i=1}^{r} \mathcal{A}_i = \sum_{i=1}^{r} \mathbf{p}_i \otimes \mathbf{s}_i \otimes \mathbf{a}_i. \qquad (1)$$

Recall that for $\mathbf{x} \in \mathbb{R}^{n_1}, \mathbf{y} \in \mathbb{R}^{n_2}$ and $\mathbf{z} \in \mathbb{R}^{n_3}$ we have

$$\mathbf{x} \otimes \mathbf{y} \otimes \mathbf{z} := [x_i y_j z_k]_{(i,j,k)=(1,1,1)}^{(n_1,n_2,n_3)} \in \mathbb{R}^{n_1 \times n_2 \times n_3}.$$

We then define the player factor matrix as $P := [\mathbf{p}_1\ \mathbf{p}_2\ ...\ \mathbf{p}_r]$, the skill factor matrix as $S := [\mathbf{s}_1\ \mathbf{s}_2\ ...\ \mathbf{s}_r]$ and the age factor matrix as $A := [\mathbf{a}_1\ \mathbf{a}_2\ ...\ \mathbf{a}_r]$. If the CPD is of rank r and all the elements in the rank-1 terms are non-negative, then the rank-1 terms can be seen as the r dominant building blocks of our tensor, which is conceptually similar to non-negative matrix factorization. A crucial feature of CPD that distinguishes it from many matrix-based approaches is that the factorization into the pure tensors $\mathcal{A}_i$ is unique up to the order of the summands; see [1] for details.

To illustrate a CPD-based analysis, we take the rank-7 CPD of a tensor containing skill ratings of 17,859 players for all 24 skills over the age range [23..29]. Each term can be interpreted as a collection of correlated skills that make up a common characteristic or trait observed in different soccer players. Figure 2a illustrates each term's skill vector containing the weights of all 24 skills. Each weight in such a vector represents how important the skill is in the term. Each term's evolution over player age is visualized in Fig. 2b. For most of the terms the evolution is near-constant. We now briefly discuss the terms.

1. **Mean.** The first term of the decomposition represents the mean of the data as it is the most dominant part of the data.
2. **Evolution term.** The second term describes the overall positive evolution of skills up to the age of 28–29.
3. **Defender trait.** The third term represents the trait of a defender, showing high weights for the *Marking* and *Standing Tackle* skills.
4. **Heading trait.** The fourth term represents the trait of players that are good at heading, showing high weights for *Strength* and *Aggression*. Interestingly, this trait is also correlated with *Acceleration* and *Sprint Speed* according to our decomposition.

5. **Goalkeeper trait.** The fourth term represents the trait of goalkeepers, with high weights for all goalkeeper skills (e.g., *GK Diving, GK Handling, GK Positioning*) and more general skills also associated with goalkeepers (e.g., *Strength, Reactions*).

6. **Correction term.** Prior to FIFA 11, all players had a high rating for *GK Kicking*. Starting from the FIFA 11 edition, all field players' ratings for this skill were significantly lowered. This term accounts for this correction and illustrates how CPD can expose hidden structure in the data as we were initially unaware of this correction in the data.

7. **Striker trait.** The seventh term is the trait of players who are good at skills typically associated with strikers such as *Finishing, Dribbling, Shot Power, Long Shots.* and *Penalties.*

Player Examples. We illustrate how each player is essentially a linear combination of these traits. The coefficients for each player can be found in the player factor matrix. We give an example for a famous player in each position. Their coefficients can be found in Table 1.

Table 1. Coefficients for Messi, Modrić, Chiellini and Buffon.

Player	Position	Mean	Evolution	Defender	Heading	Goalkeeper	Correction	Striker
Messi	ATT	811.82	27.42	0.0	0.0	0.0	6.03	46.30
Modrić	MID	766.61	34.63	26.55	0.0	0.26	9.70	14.88
Chiellini	DEF	572.59	14.92	56.53	21.05	5.55	17.24	7.81
Buffon	GK	98.32	12.17	7.62	4.52	75.19	2.06	29.51

Lionel Messi. His coefficients for the mean term (811.82) and the striker trait (46.30) are extremely high. This matches his reputation as one of the best offensive soccer players ever. His coefficients for the defender, heading and goalkeeper traits are equal to zero, which also matches his playing style and small stature.

Luka Modrić. With midfielders being a combination of attackers and defenders, there is no real midfielder trait in the CPD. The Croatian player does not possess particularly high coefficients for any of the traits compared to other players, but is reasonably strong in all of them, except for heading.

Giorgio Chiellini. A typical classic defender. With his tall physical presence, he is good at heading, as illustrated by his high coefficient for the heading trait (21.05). Compared to Messi, whose coefficient for the heading trait is zero, Chiellini shows significantly higher ratings than Messi for heading skills such as *Heading Accuracy* (83 vs 70) and *Strength* (89 vs 59).

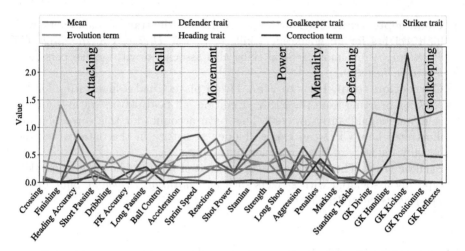

(a) The skill vectors of the seven terms in the CPD of our data tensor.

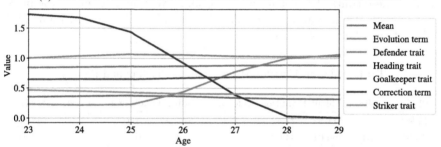

(b) The age vectors of the seven terms in the CPD of our data tensor. The orange evolution term describes the positive evolution of skills up to the age of 28-29. The brown correction term accounts for the reduction in *GK Kicking* all players received at the introduction of FIFA 11. The other five terms are relatively stable over all ages.

Fig. 2. The vectors of the rank-7 CPD visualized. (Color figure online)

Gianluigi Buffon. Buffon has a low coefficient for the mean (98.32) because goalkeepers have low ratings for most skills. Naturally, his coefficient for the goalkeeper trait (75.19) is extremely high compared to the field players. Having a low value for the coefficient of the mean can cause other coefficients to have illogical values; a small peak in any non-goalkeeper skill has to be fitted by a non-goalkeeper trait with a large coefficient. As an example, Buffon has a strong value for the striker trait (29.51) but that does not make him a good striker. Interpreting the values of non-goalkeeper traits for goalkeepers is often not useful.

4 Predicting Skill Ratings Using Tucker Decomposition

Discovering underlying patterns in player data is certainly interesting from a research perspective. However, a more interesting direction from an application perspective is predicting a player's evolution, as this can directly influence decision making on player acquisition and player retainment. Our task is then the following:

Given: Skill ratings of players for ages $[h_1..h_k]$
Predict: Skill ratings of the same players for ages $[h_{k+1}..h_N]$

We address this task by learning latent structures in a full tensor with training data for ages $[h_1..h_N]$ using the Tucker decomposition. Unlike the CPD, the factor matrices of this decomposition are less intuitive as they are not unique.

4.1 Tucker Decomposition Theory

The multilinear multiplication is a tensor multiplication in which a tensor is multiplied with a matrix in each mode [4]. The multilinear multiplication in which a tensor $\mathcal{A} \in \mathbb{R}^{n_1 \times n_2 \times n_3}$ is multiplied in each mode i by matrix $M_i \in \mathbb{R}^{m_i \times n_i}$, is denoted by

$$(M_1, M_2, M_3) \cdot \mathcal{A} \in \mathbb{R}^{m_1 \times m_2 \times m_3}.$$

The Tucker decomposition can be seen as a higher-order analogue of the principal component analysis [8]. In this decomposition, a tensor $\mathcal{A} \in \mathbb{R}^{n_1 \times n_2 \times n_3}$ is factorized in a core tensor $\mathcal{S} \in \mathbb{R}^{r_1 \times r_2 \times r_3}$ multiplied with factor matrices along each mode. In the three-dimensional case, the factorization is

$$\mathcal{A} = (U_1, U_2, U_3) \cdot \mathcal{S},$$

with $\mathcal{A} \in \mathbb{R}^{n_1 \times n_2 \times n_3}$ the original tensor, $\mathcal{S} \in \mathbb{R}^{r_1 \times r_2 \times r_3}$ the core tensor and $U_1 \in \mathbb{R}^{n_1 \times r_1}, U_2 \in \mathbb{R}^{n_2 \times r_2}, U_3 \in \mathbb{R}^{n_3 \times r_3}$ the factor matrices.

If $n_i > r_i, i = 1, 2, 3$, the core tensor $\mathcal{S}$ can be seen as a compressed version of the tensor $\mathcal{A}$. It describes how and to which extent the elements of the tensor interact with each other using the factor matrices.

4.2 Predicting Skill Ratings

Given the full data tensor $\mathcal{A} \in \mathbb{R}^{p_{\text{total}} \times M \times N}$ with p_{total} players, M skills and the data from age h_1 until h_N, the tensor is split into a training tensor $\mathcal{A}_{\text{train}} \in \mathbb{R}^{p_{\text{train}} \times M \times N}$ with p_{train} players and a test tensor $\mathcal{A}_{\text{test}} \in \mathbb{R}^{p_{\text{test}} \times M \times N}$ with p_{test} players, such that $p_{\text{total}} = p_{\text{train}} + p_{\text{test}}$. From the test set, only the first h_k ages are used. The predictions are made for the ages $[h_{k+1}..h_N]$.

In order to make predictions, a model has to be trained first. Unlike most learning algorithms, there are no iterative steps in which a model is trained. The latent structures are extracted by computing a single decomposition. Our approach consists of the following three steps (see Fig. 3).

1. **Extracting latent structures.** The latent structures of the data are extracted using the Tucker decomposition on the training tensor $\mathcal{A}_{\text{train}}$. These structures are represented by the factor matrices of the decomposition; $\hat{U}_1$ for the player factor matrix, $\hat{U}_2$ for the skill factor matrix and $\hat{U}_3$ for the age factor matrix (Fig. 3a).

2. **Finding the player factor matrix.** We find a player factor matrix $\hat{U}_1^{\text{test}}$ such that the test tensor $\mathcal{A}_{\text{test}}$ is best approximated by a Tucker decomposition using the core $\hat{S}$, the skill factor matrix $\hat{U}_2$ and the truncated age factor matrix $\hat{U}_3^{\text{trunc}}$ using only the rows of ages $[h_1..h_k]$ from the train tensor $\mathcal{A}_{\text{train}}$ (Fig. 3b). Finding this player factor matrix can be reduced to solving a least-squares problem.

3. **Completing the test tensor.** We can now complete the missing data in $\mathcal{A}_{\text{test}}$ by multiplying the player factor matrix $\hat{U}_1^{\text{test}}$ with the core $\hat{S}$, the skill factor matrix $\hat{U}_2$, and the other part of the age factor matrix $\hat{U}_3^{\text{other}}$ corresponding to the ages $[h_{k+1}..h_N]$ (Fig. 3c).

4.3 Experiments

To evaluate how well our tensor-based method can predict future skill ratings, we address three different instances of our prediction task. $h_k : [h_{k+1}..h_N]$ denotes the task of using data of the players at age h_k to predict their skill ratings at the ages $[h_{k+1}..h_N]$. We predict the evolution of young players $(18 : [19..26])$, mid-career players $(23 : [24..31])$, and older players $(26 : [27..34])$. The goalkeeper skills *GK Diving*, *GK Handling*, *GK Kicking*, *GK Positioning*, and *GK Reflexes* are disregarded, as these skills are irrelevant for most players.

For each task, we report the mean absolute error (MAE) over all players over all skills over ages $[h_{k+1}..h_N]$ and compare our tensor-based method against two baseline models. The first baseline model predicts players to have no evolution at all, i.e., we use a player's skill ratings at age h_k as the predictions for his skill ratings at ages $[h_{k+1}..h_N]$. The second baseline model uses the well-known k-nearest neighbors algorithm (KNN) in which the predictions are made using the k closest neighbors in the data set. The results can be found in Table 2.

Table 2. The mean absolute error of the models for the three prediction tasks.

Prediction method	$18 : [19..26]$	$23 : [24..31]$	$26 : [27..34]$
No evolution baseline	10.45	8.23	6.90
KNN baseline $(k = 10)$	**8.31**	7.82	7.67
Tensor-based method	8.57	**7.71**	**6.74**

If we focus on short-term predictions (1–2 years into the future), then the no evolution model shows the best performance. This is expected, as soccer players tend to improve gradually and thus show little evolution over 1–2 years.

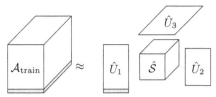

(a) The Tucker decomposition of the training tensor $\mathcal{A}_{\text{train}}$ is computed. The ith row of the player factor matrix, $\hat{U}_1$, corresponds to the data of the ith player in the training tensor.

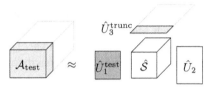

(b) Given $\mathcal{A}_{\text{test}}$ containing the data for ages $[h_1..h_k]$ of the test players, core tensor $\hat{\mathcal{S}}$, skill factor matrix $\hat{U}_2$ and the truncated age factor matrix $\hat{U}_3^{\text{trunc}}$, the coefficients for all players in the test set, $\hat{U}_1^{\text{test}}$, are computed.

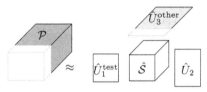

(c) With $\hat{U}_1^{\text{test}}$ computed, the predictions $\mathcal{P}$ are now calculated by using the multilinear multiplication with $\hat{U}_3^{\text{other}}$: $\mathcal{P} = (\hat{U}_1^{\text{test}}, \hat{U}_2, \hat{U}_3^{\text{other}}) \cdot \hat{\mathcal{S}}$.

Fig. 3. The three steps of our tensor-based method to predict skill ratings.

If we focus on long-term predictions (5–7 years into the future), then KNN shows the best performance. This matches the observations of Vroonen et al. who showed the superiority of KNN over the no evolution baseline for long-term predictions [9]. Our tensor-based method seems to hit a sweet spot between these two prediction methods, as evidenced by its lowest overall MAE for two of the three prediction tasks in Table 2.

Note that the predictions in our experiments are based on only one year of data. The performance of the tensor-based method degrades if we use more than one year of data. We can think of two reasons for this. First, the algorithm to decompose a tensor aims to optimally compress the entire data set and not necessarily to optimally predict missing data for certain ages. There is thus a technical misalignment between our prediction task and the task that the tensor decomposition is solving. Fixing this technical misalignment could be interesting future work. Second, skill ratings barely change in one or two years. Hence, taking three years instead of one year, for example, does not add much more information to the prediction model. Vroonen et al. [9] also came to the conclusion that increasing the number of years of given data to make predictions, has a limited effect on performance.

5 Conclusions

In this paper, we made three contributions to the field of soccer analytics. First, we highlighted a variety of challenges that arose while processing the SoFIFA data. Second, we explored the SoFIFA data tensor using the canonical polyadic decomposition (CPD) in order to extract interpretable latent structures. We showed how these latent structures group together related skills, how they evolve over players' ages, and how each player can be summarized as a linear combination of these latent structures. Third, we employed the Tucker decomposition of a tensor in order to project how a specific player's skills will evolve as he ages.

Acknowledgments. Tom Decroos is supported by the Research Foundation Flanders (FWO). Nick Vannieuwenhoven is supported by a Postdoctoral Fellowship of the Research Foundation Flanders with project 12E8119N. Jesse Davis is partially supported by KU Leuven Research Fund (C14/17/070 and C32/17/036), Interreg V A project NANO4Sports and the Flemish Government under the "Onderzoeksprogramma Artificiële Intelligentie (AI) Vlaanderen" programme.

References

1. Chiantini, L., Ottaviani, G., Vannieuwenhoven, N.: Effective criteria for specific identifiability of tensors and forms. SIAM J. Matrix Anal. Appl. **38**, 656–681 (2017). https://doi.org/10.1137/16M1090132
2. EA Vancouver: FIFA 19, video game (2018)
3. Hitchcock, F.L.: The expression of a tensor or a polyadic as a sum of products. J. Math. Phys. **6**(1–4), 164–189 (1927). https://doi.org/10.1002/sapm192761164
4. Kolda, T., Bader, B.: Tensor decompositions and applications. SIAM Rev. **51**(3), 455–500 (2009). https://doi.org/10.1137/07070111X
5. Mulani, D.: The embarrassing story behind Thomas Muller's FIFA rating (2017). https://www.sportskeeda.com/esports/the-embarrassing-story-behind-thomas-mullers-fifa-rating. Accessed 16 Dec 2018
6. SoFIFA.com. https://sofifa.com/player/158023/lionel-messi/changeLog. Accessed 27 Dec 2018
7. transfermarkt.com. https://www.transfermarkt.com/lionel-messi/elfmetertore/spieler/28003. Accessed 27 Dec 2018
8. Tucker, L.R.: Some mathematical notes on three-mode factor analysis. Psychometrika **31**(3), 279–311 (1966). https://doi.org/10.1007/bf02289464
9. Vroonen, R., Decroos, T., Haaren, J.V., Davis, J.: Predicting the potential of professional soccer players. In: Davis, J., Kaytoue, M., Zimmermann, A. (eds.) 4th Workshop on Machine Learning and Data Mining for Sports Analytics (MLSA), CEUR Workshop Proceedings, Aachen, vol. 1971, pp. 1–10 (2017). http://ceur-ws.org/Vol-1971/paper-02.pdf
10. WIRED: Meet the data master behind EA Sports' popular FIFA franchise. https://datamakespossible.westerndigital.com/meet-data-master-ea-sports-fifa/. Accessed 15 Dec 2018

Exploring Successful Team Tactics in Soccer Tracking Data

L. A. Meerhoff[1]($\boxtimes$), F. R. Goes[2], A-.W. De Leeuw[1], and A. Knobbe[1]

[1] Universiteit Leiden, Leiden, The Netherlands
l.a.meerhoff@liacs.leidenuniv.nl
[2] Rijksuniversiteit Groningen, Groningen, The Netherlands

Abstract. In recent years, professional soccer leagues have started collecting tracking data of players on the pitch during all matches of the league. This tracking data might provide an important addition to existing tactical analyses (e.g., video analysis and annotated events). By characterizing the spatial relations between players over time, the dynamic context in which success takes place can be determined. Tactical analysis of events can be enriched with spatial relations between the players during these events. Here, we demonstrate our automatized methodological approach where we use tracking data of 48 matches to (1) identify key events, (2) construct interpretable spatial relations between the players, (3) systematically examine the spatial relations over time, (4) define the success of an event, and (5) discover interpretable and actionable patterns in the spatial relations to report back to the coaching staff. With our approach, future analyses of tactics can be less tedious and more data-driven. Moreover, the context-of-play can be assessed in more detail when implementing tracking data.

Keywords: Subgroup Discovery · Tracking data · Association football

1 Introduction

Data is having an increasing impact on the world around us, also on sports such as soccer due to developments in sensor technology and optical tracking [18]. Recently, in addition to the annotated event data of soccer match-play, competition-wide high-quality tracking data of the players and the ball on the pitch during match-play has become available. This spatio-temporal data is rich and complex and offers many opportunities for analyzing and optimizing tactics in soccer by applying modern data science techniques [22]. We will demonstrate that it is possible to analyze tactical behavior in soccer without having to strictly define beforehand which specific metrics may construe this tactical behavior by adopting an exploratory data mining technique.

This work is part of the research programme *Data-driven research on Sports & Healthy Living* with project number 629.004.012, which is (partly) financed by the Netherlands Organization for Scientific Research (NWO).

© Springer Nature Switzerland AG 2020
P. Cellier and K. Driessens (Eds.): ECML PKDD 2019 Workshops, CCIS 1168, pp. 235–246, 2020.
https://doi.org/10.1007/978-3-030-43887-6_18

1.1 Tactical Analysis in Practice

In technical terms, tactics concern how teams and individuals manage space and time, and adapt to the opponent and conditions of play [10]. The coach, often supported by their staff, is the 'tactical mastermind' who designs the game plan. Essentially, the coach has to make a head-to-head comparison: which strategy works best specifically *for* us, *against* a specific opponent.

Currently, it is common practice to have a video analyst and often also an embedded scientist to provide additional insights for the coaching staff. Video analysts laboriously go through video footage of match-play to highlight specific game situations (e.g., typical strengths and weaknesses). This kind of qualitative analysis is highly tuned to the coaching staff's philosophy, but often relies a great deal on the 'expert eye', making it prone to bias.

An embedded scientist is more focused on quantitative analyses, for example using annotated event data. This data varies from straightforward performance indicators such as the number of successful passes per player, to more complicated analyses such as passing networks among players highlighting who is well-connected to whom [6]. The type of event is recorded (e.g., a pass), but also the estimated location and the players involved [4]. Interestingly, these events are available for almost every professional league worldwide. As such, this data is even used to inform clubs about potential new acquisitions on the transfer market [2,20] and can certainly be used for tactical analysis. However, the manually coded data does not provide all the context in which a play took place (i.e., what did all the players do up until the event).

Given the subjectivity of video analysis and the lack of context in event data, the coaching staff could benefit from more systematic analyses of tactics that allow for an objective comparison between the success rates of different playing styles.

1.2 Positional Tracking Data

In one form or another, 'data' already plays a role in the decision-making process of the coaching staff, however, a new type of data will make the role of data in sports even more important. The latest development in soccer data is the semi-automatic tracking of the positions of the players on the pitch by professional leagues such as the German Bundesliga and the Dutch Eredivisie. This spatio-temporal data has the potential to allow for the systematic analysis of tactical behavior in invasion-based team sports (e.g., soccer, hockey, rugby) [18,22].

Although event data also contains information about space and time, event data is much more superficial as only the locations of some players are known. With positional tracking data on the other hand, the positions of the players (and the ball) leading up to an event can be assessed (e.g., a pass to a closely guarded or an open player), thus providing the necessary context[1] as to why a sequence of actions may have been successful.

By coupling the events and the positioning of the players, novel and insightful patterns can be uncovered in the tactics of an invasion-based game such as soccer. In fact, the positional tracking data is so rich and complex, that numerous hand-crafted metrics could be conceived. Indeed, in recent years metrics have been developed that describe –for example– how threatening a player is on offence (e.g., Dangerousity) [15], how well a player positioned itself off-the-ball [26], or how effective a pass was, based on the displacement in the defensive team it triggered [9,12]. Gudmundsson and Horton [11] provide a clear overview of the pioneering work on tracking data in sports. They highlight that one of the open problems is that not many spatially informed metrics for player and team performance have been rigorously tested, often because only limited match data is available. It is thus pertinent to carefully consider which (variation of a) metric is the most indicative of success.

1.3   Aims

In the current paper, we aim to demonstrate a methodological approach that deals with two challenges for tactical analysis in an invasion-based team sport. First, current analyses either could be more objective and lack scalability (video analysis), or could be more reliable and lack context (annotated event data). We will demonstrate that with tracking data, the subjective constructs deemed important by experts can be operationalized algorithmically. The second challenge is that with tracking data so many (slightly) different metrics can be derived, that it is difficult to assess which is the most informative. Fortunately, now that tracking data is abundantly available, it is possible to discover the metrics that are most informative of success using a descriptive data mining technique.

2   Methodological Approach

To put this spatio-temporal data information in an actionable and interpretable context, we adopt an event-based approach. For a chosen type of event, we compute many of the metrics that already exist in scientific literature. By formulating a qualification of success, we can then use a descriptive data mining technique to uncover actionable and interpretable patterns. Here, a pattern refers to the features and their value ranges that best classify success. For our experiment, we selected the event *Turnovers*, which we classified based on the location (in- or outside the opponent's penalty box). We used Subgroup Discovery to identify patterns in the data.

---

[1] Although sometimes 'context' refers to the state of the match with respect to for example goal differential or time remaining [16,23], we refer to context as the locations (and thus actions) of all players on the pitch around a key event.

2.1 Key Events

During a match, discrete events occur that can be the key to understanding successful performance. Taking a shot on goal, for example, is a key occurrence that is directly related to winning a match. The outcome of an event can be classified: if the shot on goal led to a goal, it was a successful event. However, even with the increase of available data, goals and even shots on goal occur so infrequently that other more frequently occurring key can be analyzed more productively. Although dealing with sparse successful events is a key challenge in soccer analytics, we here circumvent this issue by looking at a more frequently occurring event such as the moment that a team loses possession of the ball.

Such changes in possession are an important part of soccer match-play as they can reveal successful match-play, without relying on infrequent events such as a scored goal. For our methodology, we define a turnover as the instant that the opposing team gains possession of the ball.[2] Typically, any turnover would be considered as unsuccessful from the perspective of the team losing the ball. However, our definition of *Turnovers* includes *any* change in possession, also as a consequence of a goal or some other proxy of success (e.g., an intercepted cross pass, a shot on- or off target). Based on the tracking data, our automatized methodological approach identifies the location of the ball at the instant that the possession changes team. Successful events are then the events where the defending team gained possession inside their penalty box. That is, if an attacking team managed to get the ball inside the opponent's penalty box at the instant the possession ended, the turnover was classified as successful. From here onward, we refer somewhat counter-intuitively to possession sequences that ended inside the opponents' penalty box as successful *Turnovers* and all other possession sequences as unsuccessful *Turnovers*. In other words, an attacking sequence where the team in possession got very close to (scoring) a goal.

2.2 Feature Construction

The features that describe the key events are constructed from a range of theory-driven metrics [1,7,8,15,19,22]. Here, we briefly explain the metrics conceptually, but for algorithmic details we refer the reader to the literature. All metrics require some form of spatial aggregation: a distance-based interpretation of what is happening on the pitch. This could for example be the *Width* of the team, which is the distance between the player closest to one and the player closest to the other sideline. The metrics are always considered with respect to a team (i.e., the *Width* of the team with the ball). Additionally, it is possible to formulate

---

[2] A possession starts the instant a player gains control over the ball and ends the instant a player of the opposing team gains control over the ball. Control of the ball is established when a player has been the closest player to the ball for at least 0.5 s, with the limitation that the distance has to be less than 1.5 m. The ball carrier loses control of the ball the instant that another player fits these criteria, or when the distance to the ball is more than 4 m. Additionally, one-touch passes were identified based on a direction change of the ball.

slight variations (see A-D in Table 1) by for example excluding the goalkeeper or looking at other subsets of players (e.g., the defenders, midfielders or attackers).

Distance-Based Metrics. In addition to the *Width*, we incorporated some other distance-based metrics of what happens on the pitch. The *Centroid* refers to the average positioning of the players on a team [7,8]. Similarly, the *Spread* is the standard deviation of the distances between each player and its team's *Centroid* [1,19]. The *Surface* refers to the area covered by the Convex Hull, of which we also take the *Circumference* [22], that can be drawn around different subsets of players (e.g., the defenders) [8,22]. Finally, the shape ratio is the ratio between the *Width* and distance between the player closest to- and farthest away from- the goalkeeper [7].

Potential Danger. Moreover, the context of the players relative to each other can be taken into account. Link and colleagues [15], for example, developed a measure called *Dangerousity* which captures how threatening a ball carrier is. It is a combination of the *Pressure* exerted by the defending team, the *Zone* the player is in (i.e., closer to the goal and inside the penalty box is more threatening), the *Control* the player has over the ball and the *Density* of the players around the ball carrier. *Pressure* is based on the position of the defender(s) with respect to the ball carrier. The closer a defender is to the ball carrier, the higher the pressure. Additionally, pressure is scaled based on the defender's position with respect to the goal and the ball carrier. The pressure of a defender in the 'head-on' zone (between the goal and the ball carrier) is weighted higher than a defender in the 'hind' zone (i.e., the ball carrier is in-between the defender and the goal). *Zone* is a value assigned to each location in the final third of the pitch. The Zone-values increase as the distance to the goal gets smaller, with an additional increase for zones that Link and colleagues deemed threatening (e.g., inside the penalty box). *Control* is based on the difference in velocity between the ball and the ball carrier, where a small difference indicates high control. Finally, the *Density* is based on the number of players and how crowded they are on the line between the ball carrier and the goal. For more details on how Dangerousity and its components are defined, see Link and colleagues [15].

Temporal Aggregation. From the available metrics, the event-based features are generated by reducing them to scalar values by systematically compressing the temporal dimension. As can be seen in Table 1, from every metric we construct multiple features. Multiple windows could be examined, but for the sake of simplicity we limit ourselves to one specific window. We aggregate the metrics from 10 until 5 s preceding each event. We opted for a window of 5 s as it captures a relatively short term process, as many decisive moments have a rather immediate effect. By excluding the time directly preceding the event, we force a more predictive analysis that captures what happens preceding, rather than at the instant of, the event. We aggregate over time by taking the average and the standard deviation of the metrics. Furthermore, each metric is aggregated for each team separately and for some of the metrics there are some more specific variations as can bee seen in Table 1.

Table 1. An overview of all metrics and the features that were constructed from them.

Metric	Team	Method	Variations	Features (#)
Centroid [7,8]	2	2	2^A	8
Circumference [22]	2	2		4
Control [15]	2	2		4
Dangerousity [15]	2	2		4
Density [15]	2	2		4
Pressure [15]	2	2	4^B	16
Spread [1,19]	2	2	2^C	8
Surface [8,22]	2	2	3^D	12
Shape Ratio [7]	2	2		4
Width [7]	2	2		4
Zone [15]	2	2		4
Total number of constructed features				**72**

^{A} With and without goalkeeper
^{B} The maximum and average pressure of all defenders, the pressure of the closest defender and the average pressure of all defenders within 9 m.
^{C} *Spread* (average distance to *Centroid*) and *Spread Uniformity* (standard deviation of distance to *Centroid*)
^{D} For the whole team, the midfielders & attackers, and the midfielders & attackers per player

2.3   Discovering Patterns

Once the rich and complex positional data has been reduced to a tabular format, the data can be explored for patterns. We will use Subgroup Discovery with the tool *Cortana* [17]. Subgroup discovery is an exploratory, descriptive data mining technique, targeted at labeled examples. It has previously been shown to be informative in a sports-related setting [13,14]. A subgroup is a part of the dataset that has a distribution of the target attribute that stands out compared to that of the rest of the dataset.

Take the following example: of a dataset with shot attempts, each shot is labeled as on target or not on target. In the whole dataset, the percentage of shots on targets may be rather low. A subgroup, identified by a (set of) condition(s), of the dataset might have a larger percentage of successful events. It could be, for example, that the percentage of successful events increases when the distance of the ball carrier to the goal is small.

3   Experiment

In this experiment, we show an implementation of our methodology for the event *Turnovers*. We defined success based on whether the possession ended inside the opponent's penalty box. We generated the features as presented in Table 1, which we explored using Subgroup Discovery.

3.1   Data

We used a database with 48 matches from the seasons 2014–2018 from two top-level soccer clubs in the Dutch premier division ('Eredivisie'). The data was collected by the clubs for performance analysis. The database included matches from the regular competition, the national cup and the Europa League. The clubs obtained written consent from the players to collect, share and store their data. We, in turn, obtained written informed consent from the clubs, to allow us to use the data for scientific purposes. All personal data was anonymized and the principles of the Declaration of Helsinki were adhered to throughout the research project. The X and Y coordinates of all players and the ball were recorded at 10 Hz with a video-based tracking system (SportsVU, STATS LLC, Chicago, IL, USA). For our experiment, we used tracking data only; the ball possession and key events were all computed algorithmically.

3.2   Subgroup Discovery

Our tabular data contained 6729 examples (i.e., *Turnovers*) and 72 features (see Table 1). The prior was 13.5%, that is, 910 *Turnovers* took place inside the opponent's penalty box. To assess the quality of the subgroups, we will use the Weighted Relative Accuracy (*WRAcc*):

$$WRAcc(S, T) = p(S) * (p(T|S) - p(T)),$$

where S is the subgroup indicator variable (a binary function that decides for each example whether it is covered by the subgroup) and T the target variable. Additionally, we compute the Area Under the Convex Hull of all of the subgroups' True- and False- Positive Rates (*ROC AUC*). Subgroups that have no correlation with the target (i.e., based on a random subset) will lie on the diagonal of the ROC-curve, yielding an *ROC AUC* of 0.5 (i.e., the naive baseline). We searched at depth 1, meaning that the exploration was restricted to one condition per subgroup. We adopted the *intervals* strategy, which means that conditions for subgroups could be formulated both as a range as well as a cut-off. Through swap-randomization with 100 repetitions we determined that subgroups with a *WRAcc* of at least 0.076 were not found by chance ($p < 0.05$).

3.3   Subgroups

We found 24 significant ($p < 0.05$) subgroups with an ROC AUC of 0.627 (see Table 2), wh With a prior of 13.5%, the percentage-point increase varied from

Table 2. Overview of significant subgroups ($p < 0.05$) ranked based on the *WRAcc*. The coverage and posterior indicate how large and successful a subgroup is. The condition of the subgroup is specified by the interval of a feature constructed from a metric referring to a specific Team (attacking or defending) and aggregation method (average or standard deviation).

Rank	Coverage	WRAcc	Posterior	Metric	Team	Method	Lower	Upper
Related to the team's dispersion on the pitch								
1	2794	0.21	19.4%	*Circumference*	DEF	AVG	-inf	144
2	3050	0.20	18.7%	*Circumference*	ATT	AVG	161	213
3	3630	0.19	17.7%	*Spread[A]*	DEF	AVG	-inf	17.82
4	3687	0.19	17.6%	*Spread[B]*	ATT	AVG	8.60	15.19
5	3740	0.18	17.3%	*Spread[A]*	ATT	AVG	17.87	24.27
6	2595	0.18	19.0%	*Spread[B]*	DEF	AVG	-inf	8.10
8	3582	0.17	17.3%	*Surface[C]*	DEF	AVG	-inf	1307
10	3477	0.17	17.3%	*Surface[C]*	ATT	AVG	1352	2249
11	3363	0.15	17.0%	*Surface[D]*	DEF	AVG	-inf	12.21
14	3144	0.13	16.8%	*Surface[E]*	ATT	AVG	118	166
15	3794	0.13	16.2%	*Width*	DEF	AVG	-inf	40.11
16	3188	0.13	16.7%	*Width*	ATT	AVG	41.53	62.88
17	3250	0.12	16.4%	*Surface[D]*	ATT	AVG	12.40	16.62
18	4373	0.12	15.6%	*Surface[E]*	DEF	AVG	-inf	121
Related to the team's potential danger								
7	3249	0.17	17.7%	*Control*	ATT	AVG	0.09	1.00
9	3482	0.17	17.4%	*Control*	ATT	STD	0.00	0.50
12	4776	0.14	15.8%	*Control*	DEF	AVG	-inf	0.06
13	4704	0.14	15.8%	*Control*	DEF	STD	-inf	0.00
19	951	0.10	22.0%	*Dangerousity*	ATT	STD	0.00	0.33
20	937	0.10	22.1%	*Dangerousity*	ATT	AVG	0.00	0.24
21	948	0.10	21.8%	*Zone*	ATT	AVG	0.00	0.26
22	968	0.10	21.6%	*Zone*	ATT	STD	0.00	0.38
23	606	0.08	24.3%	*Density*	ATT	STD	0.10	0.44
24	766	0.08	21.5%	*Density*	ATT	AVG	0.02	inf

[A] Average player distance to team centroid
[B] Variability of player distance to team centroid
[C] Of the whole team
[D] Of the midfielders and attackers
[E] Of the midfielders and attackers, divided by the number of players involved

1.7 to 10.6% for the different subgroups. The subgroups ranked highest had the best combination of an increase in percent point successful events whilst still covering many examples. Given the similarity of some of the constructed features, we present the similar subgroups together.

Dispersion. Many of the subgroups concern a measure that captures the dispersion of the players on the pitch. Subgroups of offensive sequences where the ball ended inside the penalty box were either characterized by a relatively compact defending team, or a relatively spread out attacking team. Note that although there might be overlap between these subgroups, they do not necessarily concern the same subsets of *Turnovers*. A compact defending team could refer to specific game situations where all defenders are bunched together, such as a corner or a free kick on the attacking half. A spread-out attacking team might in practice correspond to a counter attack situation, where the attacking players are unorganized and thus spread out.

The width-related subgroups show us that an offensive success is slightly more likely (increase from prior 13.5 to posterior 16.2%) if the defending team is rather narrowly positioned (less than 40.11 m). In contrast, the attacking team should be rather broadly positioned (given that the width of the pitch is 70 m).

Potential Danger. The various components of Dangerousity, and Dangerousity itself, are all normalized between 0 and 1 to denote more (closer to 1) and less (closer to 0) threatening situations. In terms of potential *Control*, the subgroups indicate that the attacking team should and the defending team should not be in control of the ball. Furthermore, the more threatening the *Zone*, the more likely it is that success follows 5 s later. The *Density* reflects on how many players there were around the ball carrier. The interval of the related subgroups indicates that it should not be too crowded around the ball carrier. The subgroup based on the standard deviation of the compound measure *Dangerousity* (rank 19) tells us that there must have been a stark increase of Dangerousity.

4   Discussion

The focus of the current paper was on demonstrating the potential value of the relatively new positional tracking data which could be employed to enrich event data. First of all, the scalability and objectivity of current daily practice can be improved by using tracking data. We demonstrated that key events can be identified automatically, making it easier to analyze many matches at the same time and reducing the variable errors. Secondly, the numerous features that can be generated from tracking data can be dealt with by using an exploratory data mining technique. We demonstrated that the most prominent patterns in the data can be discovered among many features by using Subgroup Discovery.

Admittedly, the discovery that counter attacks lead to situations where the ball is likely to end in the opponent's penalty box will not revolutionize soccer. Nevertheless, being able to quantify the importance of specific game situations -regardless of how obvious these situations are- is a step forward in objectifying soccer analyses. Moreover, our methodology can be tuned to a coaching staff's specific interests in many ways. The most difficult parameter choice is the label of success. In our case, reaching the opponent's penalty box, it is safe to assume that there is some correlation with success. However, by itself reaching the penalty box will never result in winning a match. In our current approach, we simplified the setting by reducing success to a Boolean. Future implementations of this approach should consider other (numeric) targets as well.

There are also some other notable parameters that could be tuned. It might be that a coach is interested in an entirely different type of key event. One could apply specific conditions to an event (e.g., turnovers on the opponent's half), but also consider other familiar events such as *Passes* and *Shots on goal*. Another aspect that could be examined further is how to compress the temporal dimension. Specifically, the window within which the metrics are aggregated could be further explored. Windows could be chosen to reflect specific short- and/or long-term processes. Currently, how the spatial relations develop over time is often neglected by taking either an arbitrarily chosen window [8,9] or sometimes an instantaneous value [15,19]. Our methodology allows for the systematic comparison of various windows, which could yield interesting insights on short- and long-term processes during a match. Moreover, there are other aggregation function that could be considered in addition to the average and standard deviation that we included in the current analysis. Particularly the minimum and maximum could be interesting, as in soccer success can be the result of seizing a small window of opportunity. For example, Link and colleagues [15] aggregate their Dangerousity measures over time by looking at the 'peak danger' during specific periods of time. Finally, it is possible to extend the metrics that are included in the analysis. Although we implemented a broad range of metrics, the list definitely not exhaustive. There are many more existing and yet-to-be-formulated metrics that could be incorporated in our methodology. Most notably, there are many more ways to quantify how an area on the pitch is controlled [3,5,21,24,25]. Each of these features could extend our methodology to cover more grounds in finding the key tactics that lead to success.

Furthermore, to link the findings from our methodology to practice, it is pertinent to create a tool that demonstrates the metrics and their subgroups. Therefore, future work would extend the practical value of our modelling approach by creating an interactive and dynamic plotting tool where the (sometimes rather abstract) features are plotted over time, for example in combination with the positions of the players on the pitch in a two-dimensional bird's eye view. Also, the discovery of actionable patterns should be aimed at identifying strengths and weaknesses of specific teams with respect to each other. By contrasting two teams, a head-to-head comparison could be made that informs about what works against whom, and vice versa (i.e., which strategy is typically successful for a specific team).

From a scientific point of view, it is important to note that findings from data mining are not the same as scientific facts in the traditional sense. For the more applied sports science domain, these findings can still be highly informative as a basis for generating new data-driven hypotheses. The findings from our approach, could be used to inform about a fingerprint of tactical behavior. With such a fingerprint, pertinent questions in the sports science domain can be further examined, such as how tactics develops with age, or the differences between countries. With better tools to analyze tactics in soccer, the next step will be to test the outcome of a specific intervention. When this can be reliably done, this kind of analysis could be used in a tool for coaches. Our long-term vision is that practitioners can experiment with their tactics in a 'Cockpit' that would help them come up with the best intervention for the specific situation at hand.

4.1 Conclusions

Tracking data has the clear potential to add context of the goings-on on the pitch to key events. Our methodological approach shows that interpretable and actionable results could be obtained by systematically exploring many metrics and determining how well they represent success and with which thresholds. Our work also shows that care must be taken in generating features, as it is inevitable that many arbitrary decisions have to be made. With more data available, data mining techniques should be employed to critically assess which metrics best represent success. Future work should focus on further developing the metrics that represent the context-of-play. Moreover, future analyses should take the playing style of specific teams, and maybe even players, into account. By making a head-to-head comparison between teams, the contrasts could best highlight the strengths and weaknesses for a specific team against another specific team.

Our proposed modelling approach can be used to further understand tactics in invasion-based team sports by comparing specific targets, teams and styles of play. When the full potential of tracking data is captured, it will affect the way soccer and possibly other team sports are played.

References

1. Bourbousson, J., Sève, C., McGarry, T.: Space-time coordination dynamics in basketball: part 2. The interaction between the two teams. J. Sports Sci. **28**(3), 349–358 (2010)
2. Bransen, L., van Haaren, J., van de Velden, M.: Measuring soccer players' contributions to chance creation by valuing their passes. J. Quant. Anal. Sports **15**(2), 97–116 (2019)
3. Brefeld, U., Lasek, J., Mair, S.: Probabilistic movement models and zones of control. Mach. Learn. **108**(1), 127–147 (2018). https://doi.org/10.1007/s10994-018-5725-1
4. Decroos, T., van Haaren, J., Davis, J.: Automatic discovery of tactics in spatio-temporal soccer match data. In: Proceedings of the 24th ACM SIGKDD International Conference on Knowledge Discovery & Data Mining, KDD 2018, London, United Kingdom (2018)
5. Dick, U., Brefeld, U.: Learning to rate player positioning in soccer. Big Data **7**(1), 71–82 (2019)
6. Duch, J., Waitzman, J.S., Amaral, L.A.N.: Quantifying the performance of individual players in a team activity. PLoS ONE **5**(6), 1–7 (2010)
7. Folgado, H., Lemmink, K.A.P.M., Frencken, W., Sampaio, J.: Length, width and centroid distance as measures of teams tactical performance in youth football. Eur. J. Sport Sci. **14**, 487–492 (2014)

8. Frencken, W., Lemmink, K., Delleman, N., Visscher, C.: Oscillations of centroid position and surface area of soccer teams in small-sided games. Eur. J. Sport Sci. **11**(4), 215–223 (2011)
9. Goes, F.R., Kempe, M., Meerhoff, L.A., Lemmink, K.A.P.M.: Not every pass can be an assist: a data-driven model to measure pass effectiveness in professional soccer matches. Big Data **7**(1), 57–70 (2019)
10. Grehaigne, J.F., Godbout, P., Bouthier, D.: The foundations of tactics and strategy in team sports. J. Teach. Phys. Educ. **18**, 159–174 (1999)
11. Gudmundsson, J., Horton, M.: Spatio-temporal analysis of team sports. ACM Comput. Surv. **50**(2), 22:1–22:34 (2017)
12. Kempe, M., Goes, F.R.: Move it or lose it: exploring the relation of defensive disruptiveness and team success. EasyChair Preprint, no. 989 (2019)
13. Knobbe, A., Orie, J., Hofman, N., van der Burgh, B., Cachucho, R.: Sports analytics for professional speed skating. Data Min. Knowl. Discov. **31**(6), 1872–1902 (2017). https://doi.org/10.1007/s10618-017-0512-3
14. de Leeuw, A.W., Meerhoff, L.A., Knobbe, A.: Effects of pacing properties on performance in long-distance running. Big Data **6**(4), 248–261 (2018)
15. Link, D., Lang, S., Seidenschwarz, P.: Real time quantification of dangerousity in football using spatiotemporal tracking data. PLoS ONE **11**(12), 1–16 (2016)
16. Ljung, D., Carlsson, N., Lambrix, P.: Player pairs valuation in ice hockey. In: MLSA@PKDD/ECML (2018)
17. Meeng, M., Knobbe, A.: Flexible enrichment with cortana-software demo. In: Proceedings of BeneLearn, pp. 117–119 (2011)
18. Memmert, D., Lemmink, K.A., Sampaio, J.: Current approaches to tactical performance analyses in soccer using position data. Sports Med. **47**, 1–10 (2016)
19. Moura, F., Martins, L.E.B., Anido, R., Barros, R., Cunha, S.: Quantitative analysis of brazilian football players' organisation on the pitch. Sports Biomech. **11**, 85–96 (2012)
20. Payyappalli, V.M., Zhuang, J.: A data-driven integer programming model for soccer clubs' decision making on player transfers. Environ. Syst. Decis. **39**(4), 466–481 (2019). https://doi.org/10.1007/s10669-019-09721-7
21. Power, P., Ruiz, H., Wei, X., Lucey, P.: Not all passes are created equal: Objectively measuring the risk and reward of passes in soccer from tracking data, pp. 1605–1613 (2017)
22. Rein, R., Memmert, D.: Big data and tactical analysis in elite soccer: future challenges and opportunities for sports science. SpringerPlus **5**(1), 1–13 (2016). https://doi.org/10.1186/s40064-016-3108-2
23. Routley, K., Schulte, O.: A markov game model for valuing player actions in ice hockey. In: Proceedings of the 31st Conference on Uncertainty in Artificial Intelligence, UAI 2015, pp. 782–791. AUAI Press, Arlington (2015)
24. Spearman, W.: Beyond expected goals. In: Proceedings of the 12th MIT Sloan Sports Analytics Conference, pp. 1–17 (2018)
25. Spearman, W., Basye, A., Dick, G., Hotovy, R., Pop, P.: Physics-based modeling of pass probabilities in soccer. In: Proceeding of the 11th MIT Sloan Sports Analytics Conference (2017)
26. Steiner, S.: Passing decisions in football: Introducing an empirical approach to estimating the effects of perceptual information and associative knowledge. Front. Psychol. **9**, 361 (2018)

Soccer Team Vectors

Robert Müller$^{(\boxtimes)}$, Stefan Langer, Fabian Ritz, Christoph Roch, Steffen Illium, and Claudia Linnhoff-Popien

Mobile and Distributed Systems Group, LMU Munich, Munich, Germany
{robert.mueller,stefan.langer,fabian.ritz,christoph.roch,
steffen.illium,linnhoff}@ifi.lmu.de

Abstract. In this work we present STEVE - Soccer **TE**am **VE**ctors, a principled approach for learning real valued vectors for soccer teams where similar teams are close to each other in the resulting vector space. *STEVE* only relies on freely available information about the matches teams played in the past. These vectors can serve as input to various machine learning tasks. Evaluating on the task of team market value estimation, *STEVE* outperforms all its competitors. Moreover, we use *STEVE* for similarity search and to rank soccer teams.

Keywords: Sports analytics · Representation learning · Self-supervised learning · Soccer · Football

1 Introduction

The field of soccer analytics suffers from poor availability of free and affordable data. While Northern American sports have already been the subject of data analytics for a long time, soccer analytics has only started to gain traction in the recent years.

Feature vectors usually serve as an input to machine learning models. They provide a numeric description of an objects characteristics. However, in the case of soccer analytics these features are hard to obtain. For example, collecting non-trivial features for a soccer player or a team involves buying data from a sports analysis company which employs experts to gather data.

In this work we propose STEVE - Soccer **TE**am **VE**ctors, a method to automatically learn feature vectors of soccer teams. *STEVE* is designed to only use freely available match results from different soccer leagues and competitions. Thus, we alleviate the problem of poor data availability in soccer analytics. Automatically extracted feature vectors are usually referred to as representations in the literature. These representations can conveniently serve as input to various machine learning tasks like classification, clustering and regression. In the resulting vector space, similar teams are close to each other. We base the notion of similarity between soccer teams on four solid assumptions (Sect. 3). The most important one is that two teams are similar if they often win against the same opponents. Hence, *STEVE* can be used to find similar teams by computing the distance between representations and to rank a self chosen list of teams according to their strengths.

© Springer Nature Switzerland AG 2020
P. Cellier and K. Driessens (Eds.): ECML PKDD 2019 Workshops, CCIS 1168, pp. 247–257, 2020.
https://doi.org/10.1007/978-3-030-43887-6_19

This paper is organized as follows: In Sect. 2 we review related work. It consists of an in depth discussion of the process of learning real valued vectors for elements in a set and its applications. We also briefly review a recent approach to team ranking. In Sect. 3 we introduce *STEVE*, our approach to learn meaningful representations for soccer teams. After giving an overview of the underlying idea, we introduce the problem with more rigor and conclude the section by formulating an algorithm for the task. In Sect. 4 we conduct various experiments to evaluate the approach. Finally, Sect. 5 closes out with conclusions and outlines future work.

2   Related Work

Learning real valued vectors for elements in a set has been been of particular interest in the field of natural language processing. Elements are usually words or sentences and their representation is computed in such a way, that they entail their meaning. Modern approaches typically learn a distributed representation for words [3] based on the distributional hypothesis [22], which states that words with similar meanings often occur in the same contexts.

Mikolov et al. [16,17] introduced *word2vec*, a neural language model which uses the skip-gram architecture to train word representations. Given a center word *word2vec* by iteratively maximizes the probability of observing the surrounding window of context words. The resulting representations can be used to measure semantic similarity between words. According to *word2vec* the most similar word to *soccer* is *football*. Moreover, vector arithmetic can be used to compute analogies. Although having recently been put in question [2,12], a very famous example is the following: *king - man + woman = queen*. The concept has since then been extended to graph structured data to learn a representation for each node in a graph. Perozzi et al. [21] and Dong et al. [7] treat random walks as the equivalent of sentences. This is based on the assumption that these walks can be interpreted as sampling from a language graph. The resulting sentences are fed to *word2vec*. Building upon graph based representation learning approaches, *LinNet* [20] builds a weighted directed match-up network where nodes represent lineups from NBA basketball teams. An edge from node i to node j is inserted if lineup i outperformed lineup j. The edge weight is set to the performance margin of the corresponding match-up. Lineup representations are computed by deploying *node2vec* [9] on the resulting network. Afterwards, a logistic regression model based on the previously computed lineup representations is learned to model the probability of lineup λ_i outperforming lineup λ_i.

More recently, the aforementioned findings have also been applied to sports analytics. (batter|pitcher)2vec [1] computes representations of Major League Baseball players through a supervised learning task that predicts the outcome of an at-bat given the context of a specific batter and pitcher. The resulting representations are used to cluster pitchers who rely on pitches with dramatic movement and predict future at-bat outcomes. Further, by performing simple arithmetic in the learned vector space they identify opposite-handed doppelgangers.

Le et. al [14] introduce a data-driven ghosting model based on tracking data of a season from a professional soccer league to generate the defensive motion patterns of a *league average* team. To fine-tune the *league average* model to account for a team's structural and stylistic elements, each team is associated with a team identity vector.

Our approach aims to learn representations for soccer teams and is thus closely related to the presented approaches. As we use the representations to rank teams, we briefly review related work on the topic.

Neumann et al. [18] propose an alternative to classical ELO and Pi rating based team ranking approaches [6,11]. A graph based on the match results and a generalized version of agony [10] is used to uncover hierarchies. The approach is used to categorize teams into a few discrete levels of playing quality. General match-up modeling is addressed by the *blade-chest* model [5]. Each player is represented by two d-dimensional vectors, the *blade* and *chest* vectors. Team a won if its blade is closer to team b's chest than vice versa. Intransitivity is explicitly modeled by using both blade and chest vectors, something that cannot be accounted for by approaches that associate a single scalar value with each team [4].

3   Soccer Team Vectors

In this section we present *STEVE - Soccer Team Vectors*. We first give an overview of the underlying idea and the goal of this work. Afterwards we discuss the problem definition and introduce an algorithm to learn useful latent representations for soccer teams.

3.1   Overview

STEVE aims to learn meaningful representations for soccer teams where representations come in the form of low dimensional vectors. If two teams are similar, their representations should be close in vector space while dissimilar teams should exhibit a large distance. Furthermore, these learned latent representations can be used as feature vectors for various machine learning tasks like clustering, classification and regression. Due to the fact that there is no clear definition of similarity for soccer teams, we base our approach on the following four assumptions:

1. The similarity between two teams can be determined by accounting for the matches they played in the past.
2. Frequent draws between two teams indicate that they are of approximately equal strength. Hence, both teams are similar.
3. Two teams are similar if they often win against the same opponents.
4. More recent matches have a higher influence on the similarity than those a long time ago.

Since data acquisition is expensive and time-consuming, especially in the field of sports analytics, *STEVE* is designed to learn from minimal information. More precisely, we only use data about which teams played against each other, during which season a match took place and whether the home team won, lost or the match resulted in a draw. Note that the assumptions mentioned above do not require any further information and are therefore well suited for this setting.

3.2 Problem Definition

To simplify definitions, let $M = \{1, 2, \ldots, m\}$, we assume that each of the m soccer teams we want to learn a representation for is associated with an identification number $i \in M$. Further, let $\Phi \in \mathbb{R}^{m \times \delta}$, where each row Φ_i represents team i's δ dimensional latent representation. The goal of this work is to find Φ in such a way, that $dist(\Phi_i, \Phi_j)$ is small for similar teams i and j and $dist(\Phi_i, \Phi_k)$ is large for dissimilar teams i and k. $dist(\cdot, \cdot)$ is some distance metric and similarity between teams is determined according to the assumptions made in Sect. 3.1. To solve this task, data is given in the following form: $\mathcal{D} = \{(a, b, s, d) \in M \times M \times \{1, \ldots, x_{max}\} \times \{0, 1\}\}$. The quadruple (a, b, s, d) represents a single match between teams a and b, s is an integer indicating during which of the x_{max} seasons the match took place and d is a flag set to 1 if the match resulted in a draw and 0 otherwise. If $d = 0$, the quadruple is arranged such that team a won against team b.

3.3 Algorithm

According to the first assumption, we can loop over the dataset $\mathcal{D}$ while adjusting Φ. If $d = 1$, we minimize the distance between Φ_a and Φ_b, thereby accounting for the second assumption. The third assumption addresses a higher order relationship, where teams that often win against the same teams should be similar. We introduce a second matrix $\Psi \in \mathbb{R}^{m \times \delta}$ and call each row Ψ_i team i's loser representation. Further, we call Φ_i the winner representation of team i. Both matrices Φ and Ψ are initialized according to a normal distribution with zero mean and unit variance. When team a wins against team b we minimize the distance between Φ_a and Ψ_b, bringing b's loser representation and a's winner representation closer together. That is, the loser representations of all teams a often wins against, will be in close proximity to team a's winner representation. Consequently, if other teams also often win against these teams, their winner representations must be close in order to minimize the distance to the loser representations. Parameters Φ and Ψ are estimated using stochastic gradient descent where the objective we aim to minimize is given as follows:

$$\underset{\Phi, \Psi}{\arg \min} \sum_{(a,b,s,d) \in \mathcal{D}} d * dist(\Phi_a, \Phi_b) + (1 - d) * dist(\Phi_a, \Psi_b)$$

We minimize the distance between Φ_a and Φ_b directly when both teams draw ($d = 1$). Otherwise ($d = 0$) we minimize the distance between Φ_a (winner representation) and Ψ_b (loser representation). With the squared euclidean distance as the distance metric, the expression can be rewritten as illustrated below.

$$\underset{\Phi,\Psi}{\arg\min} \sum_{(a,b,s,d)\in\mathcal{D}} d * \|\Phi_a - \Phi_b\|^2 + (1 - d) * \|\Phi_a - \Psi_b\|^2$$

In its current form, matches played in long past seasons contribute as much to the loss as more recent matches. We alleviate this problem by down-weighting matches from older seasons using the linear weighting scheme $\frac{s}{x_{xmax}}$, thereby completing the formulation of the objective:

$$\underset{\Phi,\Psi}{\arg\min} \sum_{(a,b,s,d)\in\mathcal{D}} \frac{s}{x_{max}} \left[d * \|\Phi_a - \Phi_b\|^2 + (1 - d) * \|\Phi_a - \Psi_b\|^2 \right]$$

This approach has the advantage of no complex statistics having to be gathered. All our assumptions are captured in the teams's representations. We describe the algorithm in more detail in Algorithm 1. Note that here gradients are computed after observing a single data point and the regularization term is omitted. This is done for illustration purposes only. In our implementation, we train the algorithm in a batch-wise fashion. For In lines 9, 12 and 15 the representations are normalized as we have found this to speed up training. It also helps to keep distances within a meaningful range.

Algorithm 1. STEVE($\mathcal{D}$, m, δ, α, x_{max},e)

1:   $\Phi \sim \mathcal{N}(0,1)^{m\times\delta}$ ▷ Initialize Φ
2:   $\Psi \sim \mathcal{N}(0,1)^{m\times\delta}$ ▷ Initialize Ψ
3:   **for** i in $\{1,\ldots,e\}$ **do**
4:       $\mathcal{D} = \text{shuffle}(\mathcal{D})$ ▷ Shuffle dataset
5:       **for** each (a,b,s,d) in $\mathcal{D}$ **do**
6:          $L(\Phi,\Psi) = \frac{s}{x_{max}}\left[d * \|\Phi_a - \Phi_b\|^2 + (1 - d) * \|\Phi_a - \Psi_b\|^2 \right]$ ▷ Compute loss
7:         **if** $d = 0$ **then**   ▷ a won the match
8:             $\Psi_b = \Psi_b - \alpha * \frac{\partial L}{\partial \Psi_b}$ ▷ Gradient descent on b's loser representation
9:             $\Psi_b = \Psi_b/\|\Psi_b\|_2$ ▷ Normalize b's loser representation
10:       **else**   ▷ Match is a draw
11:            $\Phi_b = \Phi_a - \alpha * \frac{\partial L}{\partial \Phi_b}$ ▷ Gradient descent on b's winner representation
12:            $\Phi_b = \Phi_b/\|\Phi_b\|_2$ ▷ Normalize b's winner representation
13:         **end if**
14:         $\Phi_a = \Phi_a - \alpha * \frac{\partial L}{\partial \Phi_a}$ ▷ Gradient descent on a's winner representation
15:         $\Phi_a = \Phi_a/\|\Phi_a\|_2$ ▷ Normalize a's winner representation
16:       **end for**
17:   **end for**
18:   **return** Φ, Ψ

---

4   Experiments

In this section, we provide an overview of the dataset. We also conduct various experiments to investigate the expressiveness and efficacy of our approach.

4.1 Dataset and Experimental Setup

The dataset consists of all the matches from the Bundesliga (Germany), Premier League (Great Britain), Serie A (Italy), La Liga (Spain), Eredivisie (Netherlands), League 1 (France), Sper Lig (Turkey), Pro League (Belgium), Liga NOS (Portugal), Europa League and the Champions League played from 2010 until 2019. A total of 29529 matches between 378 different teams was carried out where approximately 25% ended in a draw. Unless stated otherwise, for all experiments we set $\delta = 16$ and batch size $= 128$. We use Adam [13] with a learning rate $\alpha = 0.0001$ for parameter estimation and train for $e = 40$ epochs. Additionally, we add a small L_2 weight penalty of 10^{-6}.

4.2 Similarity Search

We select five European top teams and run *STEVE* on all the matches from season 2010 until 2019 in the corresponding league. Since we are dealing with small datasets, we set $\delta = 10$ and batch size $= 32$. For each team, we note the five most similar teams (smallest distance) in Table 1. Note that we use the distance between the corresponding winner representations. As expected, we clearly observe that top teams are similar to other top teams. For example, the team most similar to Barcelona is Real Madrid. Both teams often compete for supremacy in *La Liga*. In general, we observe that similarities in Table 1 roughly reflect the average placement in the respective league.

Table 1. Five most similar teams for five European top teams.

Top soccer team per league chosen for similarity search				
Bayern München	Barcelona	Paris SG	Manchester U	Juve. Turin
Five most similar teams chosen by *STEVE*				
RB Leipzig	Real Madrid	Lyon	Liverpool	SSC Napoli
Dortmund	Valencia	Marseille	Manchester C.	AS Roma
Mönchengladbach	Atletico Madrid	Monaco	Chelsea	AC Milan
Leverkusen	Sevilla	St Etienne	Tottenham	Inter. Milano
Hoffenheim	Villarreal	Lille	Arsenal	SS Lazio

4.3 Ranking Soccer Teams

To retrieve a ranked list of soccer teams, one could simply use a league table. However, the list will only reflect the team's constitution accumulated over a single season. The ranking will not take past successes into account. One might alleviate this problem by averaging the league table over multiple seasons. Nevertheless, another problem arises: the list will only consist of teams from a single league. Combining league tables from different countries and competitions to obtain a more diverse ranking is considerably less straightforward. It gets even more complicated when we wish to rank a list of self chosen teams, possibly from many different countries. *STEVE* provides a simple yet effective way to generate rankings for the use case mentioned above. Given a list of teams, we simulate a tournament where each team plays against all other teams. The list is then sorted according to the number of victories. To compute the outcome of a single match (victory or defeat) between team a and b, let $\alpha = \|\Phi_a - \Psi_b\|^2$ and $\beta = \|\Phi_b - \Psi_a\|^2$. If $\alpha < \beta$, then team b's loser representation is closer to team a's winner representation than team a's loser representation is to team b's winner representation. Thus, team a is stronger than team b and we increase team a's victory counter. The same line of reasoning is applied to the case where $\alpha > \beta$.

In Fig. 1 we generated two rankings using the aforementioned approach. Each list consists of twelve teams from different European countries of different strengths. Our approach produces reasonable rankings: Highly successful international top teams like *Real Madrid, FC Bayern Munich, FC Barcelona*, and *AS Roma* are placed at the top of the list while mediocre teams like *Espanyol Barcelona* and *Werder Bremen* are placed further back in the list. The least successful teams like *FC Toulouse, Cardiff City, Fortuna Düsseldorf* and *Parma Calcio* occupy the tail of the list. *STEVE* can be seen as an alternative to previous soccer team ranking methods [11,15] based on the ELO rating [8].

[1] Real Madrid, FC Bayern Munich, Inter Milano, Liverpool FC, Borussia Dortmund, Ajax Amsterdam, FC Porto, Club Brugge KV, Werder Bremen, 1.FC Nuremberg, FC Toulouse, Cardiff City

[2] FC Barcelona, AS Roma, Atlético Madrid, Paris SG, Tottenham, PSV Eindhoven, Arsenal London, SL Benfica, Espanyol Barcelona, VFB Stuttgart, Fortuna Dusseldorf, Parma Calcio

Fig. 1. Team rankings generated by *STEVE*. Each row[1,2] depicts one ranked list from the strongest (left) to the weakest team (right). Numbers represent a team's relative strength - the number hypothetical matches won.

4.4 Team Market Value Estimation

The goal of this work is to learn representations that are well suited for various downstream machine learning tasks. We validate this property by evaluating *STEVE* with respect to regression and classification performance. We argue that a meaningful representation should carry enough information to reliably predict a team's market value. Therefore, both tasks involve predicting the value of a team given its representation. We obtained current market values for all teams in the dataset from season 2018/2019. A team's market value is determined by the sum of the market values of all its players. On average, a team is worth €183.7 million with a standard deviation of €241.2M. The least valuable team is *BV De Graafschap* (€10.15M) and the most valuable team is *FC Barcelona* (€1180M). The first, second and third quartiles are €25M, €93.7M and €232.5M, respectively. For regression and classification, we use the following team representations as an input to a multi layer perceptron (MLP) with two hidden layers. The first hidden layer has a size of 50, the second one 20. Apart from changing hidden layer sizes, we use default parameters provided by [19] for all further analyses.

- **STEVE.** Representations are computed using *STEVE* with $\delta \in \{8, 16, 32\}$. A team's winner and loser representation is concatenated to form its team vector. The resulting feature vectors are of size 16, 32, 64.
- **Season-stats.** We extract count based features for each team in the dataset to mimic traditional feature extraction. For season 2018/2019 we collect the following statistics: number of victories, draws, defeats as well as goals scored and goals conceded. Each feature is computed for matches in the Champions League, Euro League and the respective national league. Additionally, we use goals per match, goals per national and international match. This results in a 18 dimensional feature vector (representation) for each team.
- **Season-stats (CAT-x).** Season-stats for the last x seasons are concatenated. The resulting feature vectors are of size $x * 18$.
- **Season-stats (SUM-x).** Season-stats for the last x seasons are summed together. The resulting feature vectors are of size 18.

Comparability between the different representations mentioned above is ensured due to the fact that none of them requires information absent in the dataset.

Season-Stats has many features that are intuitively well suited for team value estimation. For example, a large proportion of teams that participate in international competitions are more valuable than those who don't. Statistics about goals and match results are helpful for assessing a team's strength which is in turn correlated to the team's market value.

Regression. Team value estimation naturally lends itself to be cast as a regression problem. During training we standardize team values (targets) and *Season-Stats* features by subtracting the mean and dividing by the standard deviation. Evaluation is carried out using 5-fold cross-validation and results are reported in Table 2.

Classification. By grouping team values into bins, we frame the task as classification problem. Teams are assigned classes according to the quartile their value lies in. Consequently, each team is associated with one of four classes. We apply the same standardization procedure as in the case of regression and use 5-fold cross-validation. Results are reported in Table 3.

Results. *STEVE* clearly outperforms the other representations both in terms of regression and classification performance. While $\delta = 64$ generally yields the best results, even $\delta = 16$ produces superior results compared to *Season-Stats*. In terms

Table 2. Results for the regression task of team value estimation. To quantify the quality of prediction, we use root mean squared error (RMSE), mean absolute error (MAE) and the mean median absolute error (MMAE), all reported in million €.

	RMSE	MAE	MMAE
STEVE-16	142.12 ± 75.22	88.37 ± 25.69	52.01 ± 13.42
STEVE-32	131.51 ± 40.15	83.20 ± 24.51	46.87 ± 21.89
STEVE-64	**111.27** ± 48.58	**67.14** ± 30.51	**32.80** ± 10.42
Season-Stats	173.75 ± 119.35	110.32 ± 63.61	69.96 ± 15.43
Season-Stats (CAT-3)	200.77 ± 157.55	138.15 ± 87.06	86.98 ± 39.83
Season-Stats (CAT-5)	172.05 ± 70.83	119.81 ± 43.18	80.74 ± 18.96
Season-Stats (CAT-9)	151.09 ± 80.37	105.98 ± 41.96	68.82 ± 23.15
Season-Stats (SUM-3)	158.44 ± 108.50	105.65 ± 53.95	69.16 ± 11.39
Season-Stats (SUM-5)	154.71 ± 115.76	104.04 ± 59.34	69.81 ± 15.69
Season-Stats (SUM-9)	158.33 ± 120.90	106.67 ± 62.61	67.74 ± 17.75

Table 3. Results for the classification task of team value estimation, measured with micro F_1 score and macro F_1 score.

	Micro F_1	Macro F_1
STEVE-16	0.67 ± 0.10	0.64 ± 0.10
STEVE-32	**0.74** ± 0.11	0.71 ± 0.14
STEVE-64	**0.74** ± 0.10	**0.72** ± 0.09
Season-Stats	0.52 ± 0.14	0.45 ± 0.19
Season-Stats (CAT-3)	0.50 ± 0.12	0.44 ± 0.15
Season-Stats (CAT-5)	0.55 ± 0.14	0.51 ± 0.16
Season-Stats (CAT-9)	0.60 ± 0.13	0.56 ± 0.11
Season-Stats (SUM-3)	0.49 ± 0.09	0.40 ± 0.10
Season-Stats (SUM-5)	0.48 ± 0.08	0.39 ± 0.07
Season-Stats (SUM-9)	0.47 ± 0.09	0.37 ± 0.15

of regression performance, we observe that *Season-Stats* is most competitive when using information from multiple seasons (CAT-x and SUM-x). All forms of representation manage to estimate the general tendency of a team's market value but *STEVE's* predictions are far more precise. Similar conclusions can be drawn when inspecting classification performance. The best competing representation is *Season-Stats (CAT-9)* which is 162 dimensional, 92 dimensions more than *STEVE (δ = 64)*. Still, *STEVE (δ = 64)* provides ≈20% better performance than *Season-Stats (CAT-9)*. It can therefore be concluded that *STEVE* is able to compress information needed for the task and succeeds to provide high efficacy representations.

5   Conclusion

In this work we introduced *STEVE*, a simple yet effective way to compute meaningful representations for soccer teams. We provided qualitative analysis using soccer team vectors for team ranking and similarity search. Quantitative analysis was carried out by investigating the usefulness of the approach by estimating the market values of soccer teams. In both cases, *STEVE* succeeds to provide meaningful and effective representations. Future work might investigate further upon different weighting schemes for the season during which a match took place. For example instead one can use the exponential distribution to weigh down past seasons. Moreover, including the number of goals scored during a match and accounting for the home advantage might help to capture more subtleties.

References

1. Alcorn, M.A.: (batter| pitcher)2vec: statistic-free talent modeling with neural player embeddings. In: MIT Sloan Sports Analytics Conference (2016)
2. Allen, C., Hospedales, T.M.: Analogies explained: towards understanding word embeddings. In: Proceedings of the 36th International Conference on Machine Learning, ICML 2019, 9–15 June 2019, Long Beach, California, USA, pp. 223–231 (2019)
3. Bengio, Y., Ducharme, R., Vincent, P., Jauvin, C.: A neural probabilistic language model. J. Mach. Learn. Res. **3**(Feb), 1137–1155 (2003)
4. Bradley, R.A., Terry, M.E.: Rank analysis of incomplete block designs: I. the method of paired comparisons. Biometrika **39**(3/4), 324–345 (1952)
5. Chen, S., Joachims, T.: Predicting matchups and preferences in context. In: Proceedings of the 22nd ACM SIGKDD International Conference on Knowledge Discovery and Data Mining, pp. 775–784. ACM (2016)
6. Constantinou, A., Fenton, N.: Determining the level of ability of football teams by dynamic ratings based on the relative discrepancies in scores between adversaries. J. Quant. Anal. Sports **9**, 37–50 (2013)
7. Dong, Y., Chawla, N.V., Swami, A.: Metapath2vec: scalable representation learning for heterogeneous networks. In: Proceedings of the 23rd ACM SIGKDD International Conference on Knowledge Discovery and Data Mining, KDD 2017, pp. 135–144. ACM, New York (2017)
8. Elo, A.E.: The Rating of Chessplayers, Past and Present. Arco Pub., New York (1978)
9. Grover, A., Leskovec, J.: Node2vec: scalable feature learning for networks. In: Proceedings of the 22Nd ACM SIGKDD International Conference on Knowledge Discovery and Data Mining, KDD 2016, pp. 855–864. ACM, New York (2016)
10. Gupte, M., Shankar, P., Li, J., Muthukrishnan, S., Iftode, L.: Finding hierarchy in directed online social networks. In: Proceedings of the 20th International Conference on World Wide Web, WWW 2011, pp. 557–566. ACM, New York (2011)
11. Hvattum, L.M., Arntzen, H.: Using ELO ratings for match result prediction in association football. Int. J. Forecast. **26**, 460–470 (2010)
12. Kalidindi, K.V.: Deconstructing word embeddings. CoRR abs/1902.00551 (2019)
13. Kingma, D.P., Ba, J.: Adam: a method for stochastic optimization. In: 3rd International Conference on Learning Representations, ICLR 2015, San Diego, CA, USA, 7–9 May 2015, Conference Track Proceedings (2015)

14. Le, H.M., Peter, C., Yue, Y.: Data-driven ghosting using deep imitation learning. In: MIT Sloan Sports Analytics Conference, pp. 1–15 (2017)
15. Leitner, C., Zeileis, A., Hornik, K.: Forecasting sports tournaments by ratings of (prob)abilities: a comparison for the EURO 2008. Int. J. Forecast. **26**, 471–481 (2010)
16. Mikolov, T., Chen, K., Corrado, G., Dean, J.: Efficient estimation of word representations in vector space. In: 1st International Conference on Learning Representations, ICLR 2013, Scottsdale, Arizona, USA, 2–4 May 2013, Workshop Track Proceedings (2013)
17. Mikolov, T., Sutskever, I., Chen, K., Corrado, G.S., Dean, J.: Distributed representations of words and phrases and their compositionality. In: Burges, C.J.C., Bottou, L., Welling, M., Ghahramani, Z., Weinberger, K.Q. (eds.) Advances in Neural Information Processing Systems, vol. 26, pp. 3111–3119. Curran Associates, Inc. (2013)
18. Neumann, S., Ritter, J., Budhathoki, K.: Ranking the teams in european football leagues with agony. In: Brefeld, U., Davis, J., Van Haaren, J., Zimmermann, A. (eds.) MLSA 2018. LNCS (LNAI), vol. 11330, pp. 55–66. Springer, Cham (2019). https://doi.org/10.1007/978-3-030-17274-9_5
19. Pedregosa, F., et al.: Scikit-learn: machine learning in Python. J. Mach. Learn. Res. **12**, 2825–2830 (2011)
20. Pelechrinis, K.: LinNet: probabilistic lineup evaluation through network embedding. In: Brefeld, U., et al. (eds.) ECML PKDD 2018, Part III. LNCS (LNAI), vol. 11053, pp. 20–36. Springer, Cham (2019). https://doi.org/10.1007/978-3-030-10997-4_2
21. Perozzi, B., Al-Rfou, R., Skiena, S.: Deepwalk: online learning of social representations. In: Proceedings of the 20th ACM SIGKDD International Conference on Knowledge Discovery and Data Mining, KDD 2014, pp. 701–710. ACM, New York (2014)
22. Harris, Z.S.: Distributional structure. Word **10**, 146–162 (1954)

Tactical Analyses in Professional Tennis

Arie-Willem de Leeuw[1]([⊠]), Aldo Hoekstra[2], Laurentius Meerhoff[1],
and Arno Knobbe[1]

[1] Leiden Institute of Advanced Computer Science (LIACS), Leiden, The Netherlands
a.de.leeuw@liacs.leidenuniv.nl
[2] Koninklijke Nederlandse Lawn Tennis Bond (KNLTB),
Amstelveen, The Netherlands

Abstract. In tennis, applying a proper game strategy is an important
aspect in performance optimization. In this work, we perform tactical
analyses for a specific professional tennis player by using a manually
annotated data collection of 4,593 points. Primarily, we will apply Subgroup Discovery to find generic characteristics of successful points in
tennis and descriptions that are specific for our player of interest. To
demonstrate that easily understandable patterns can be gleaned from
our method, that are relatively simple to put into practice, we will focus
on finding the most important descriptions of won service points. In general, the most profound characterisation of successful service points in
tennis are points that last maximally two strokes. For our specific player,
we have found that more service points are won if the player avoids hitting a backhand.

Keywords: Data Mining · Subgroup Discovery · Tennis

1 Introduction

Many sports develop over the years. While some advancements are caused by
changes in the rules, others are a consequence of technological improvements.
For example, in tennis the developments in the materials and physical condition
of the players have increased the dynamics in matches [6]. As a consequence, it
has become more difficult for the players to improvise during points. Therefore,
there are patterns that frequently occur in professional tennis matches [5]. Since
most players adopt a specific style, the details of these patterns depend on the
players that are involved in the match. For coaches, it is a valuable asset to have
detailed information about these patterns before the start of a match. In this
way, the strengths and weaknesses of both players are made explicit, which can
be used for upon deciding on a proper match strategy and thereby optimizing
the chances of being successful [4].

This work is part of the research programme *Citius Altius Sanius*, which is (partly)
financed by the Netherlands Organisation for Scientific Research (NWO).

© Springer Nature Switzerland AG 2020
P. Cellier and K. Driessens (Eds.): ECML PKDD 2019 Workshops, CCIS 1168, pp. 258–269, 2020.
https://doi.org/10.1007/978-3-030-43887-6_20

There are multiple data-driven approaches for identifying the main characteristics of playing styles in professional tennis. First, computer vision techniques have been applied to extract data from video sequences and analyse tennis tactics [3,19,20]. Second, hand-crafted analyses are used to record certain key aspects in tennis matches [10,11]. These data sets are mostly used to investigate dependencies between single variables, e.g., first serve strategy versus winning percentage [9]. Third, the introduction of Hawk-eye system [1] opened up the possibility to collect spatiotemporal data. By using multiple cameras, the (x, y, z) position of the ball is tracked. This information for example can be used to predict serve directions [21] or analyse the efficiency of tennis services [18]. More recently, also deep learning is applied on tracking data in tennis [8,22].

In this study, we will apply Subgroup Discovery [12,17] on a manually annotated data collection of tennis points. The advantage of this approach is that we can investigate dependencies between controlled combinations of the variables. Furthermore, this technique has already proved its value in several other sport-specific settings [13,14]. Since our data collection only consists of points that feature a specific tennis player, we consider the specific task of *personalization*, i.e., we will find player-specific characteristics. More specifically, we we will focus on finding the main characteristics of *successful service points*. By using the data set of service points of the opponents as a benchmark, we will make a distinction between generic characteristics of successful service points and results that are specific for our player of interest.

This article is structured as follows. First, we describe the data at hand and the preprocession that we have applied. Hereafter, we discuss the feature engineering procedure, which is an important part in our approach. Subsequently, we discuss the methods that we have used and report on the results of our experiments. Finally, we end with conclusions and mention possible directions for future research.

2   Materials

To understand the content of the data and the performed analyses, it is important to know the basic concepts in tennis. Readers that are not familiar with tennis can find an overview of the rules and regulations in [2]. In the remainder of this section, we will discuss the data that is available for our analyses and also mention the preprocession that we have applied.

2.1   Data

We have a data collection consisting of 31 official tennis singles matches of which twenty matches are played on hardcourt, nine matches on clay and two matches on grass. The anonymous professional tennis player in case, that is the subject of this study and from now on will be named as player X, is always one of the two players.

Of every match, we have a point-by-point description. In this data set, all points of the matches are annotated manually. For every point, there is a substantial amount of information recorded. In the remainder of the section, we will touch upon the most relevant attributes. First, we have general information about the point. More specifically, we know

1. **Match and game score** The score in the match, set and game.
2. **Number of strokes** The number of strokes of the point. Points that consist of more than 9 strokes are in the same category.
3. **Start of the point** The serve can be either hit from the advantage or deuce side of the court.
4. **Score type** A variable that describes the end of the point.
5. **Winner** The player that has won the point.

In addition to this general information, our data collection contains information about some strokes of the points. As recording the information of all strokes would be extremely time-consuming, the data set captures only the characteristics of the most important strokes. Therefore, we only have detailed information about the first four and final two strokes of each point.

For every point, we can write the information about the strokes as a tuple $t = (A_1, ..., A_j, ..., A_n)$, where n characterises the length of the point. For points shorter than 6 strokes, the length of t is equal to the number of strokes in the point. For the other points, the length of t is equal to 6. Moreover, A_j is a collection of stroke-specific information with j specifying the stroke of the point, e.g., A_3 corresponds to the third stroke of the point.

The content of A_j depends on the stroke that is considered. In our data collection, the first two strokes of the point, the serve and return respectively, are treated with special care. For the service, we have

1. **Service type** The specific details about the serve.
2. **Direction** The direction of the serve.
3. **Service volley** Indicates if the server played a specific service strategy, i.e., service volley.

There are two specific properties of the return recorded

1. **Type return** The movements the receiving player had to make to return the ball.
2. **Block** If the player blocked the ball.

In addition to these return-specific features, there is even more information gathered about the return. For the return, we also have the following attributes

1. **Intention** The kind of intention with which the player hit the ball.
2. **Situation** The situation how the ball arrives at a player.
3. **Direction** The part of the court to which the ball is directed.
4. **Stroke type** The kind of stroke with which the player hits the ball.

The attributes *intention* and *situation* can take several values, such as 'opbouwend' or 'VvS'. The former indicates a constructive stroke and the latter correspond to a stroke that prevents the opponent from scoring. These attributes can be difficult to infer and are sensitive to human interpretation. This problem is tackled by carefully instructing the people that are responsible for the annotation. In case there were doubts, the values of the attributes were recorded after consulting the embedded scientist of the tennis association.

The aforementioned four attributes are not only collected for the return, but also for the two strokes after the return and the final two strokes. Finally, for the third, fourth and last two strokes of the rally, we also have

1. **Slice** Indicates whether a type of effect, i.e., slice, is applied to the ball.

In total, we have a collection of binary, numeric and categorical attributes. Moreover, the number of distinct categories varies between the different categorical attributes.

2.2   Data Preparation

As human annotation is sensitive to errors, we need to investigate the quality of the data thoroughly. This includes checking that the change in game score corresponds with the outcome of the previous point, there is no specific information about a certain stroke if the point is already finished, and many other checks. After performing these analyses, we had to reduce our data collection by almost 20%. Finally, we are left with a collection of 4,593 points consisting of 2,260 points in which player X is serving and 2,333 instances where the opponent is starting the point with a serve.

In our data collection, there is information about the strokes of point but it is not specified which of the two players hits the ball. This implies that the raw data can not directly be used to analyse the playing style of player X. Thus, we first make a distinction between the strokes of player X and the other strokes of the point. Hence, instead of having a single tuple t that contains all information about all strokes of the point, we introduce a player-specific tuple t_P that only contains the information of the strokes of player P. Note that the length of the point determines the number of strokes that is part of t_P. In this case, t_P can contain the attributes of the first stroke, the first two strokes or the first two strokes and the last stroke of the player.

Finally, in our data collection the attribute *stroke type* of the fourth stroke of the point contains many missing values. Therefore, this attribute is excluded from the analyses. This implies that the player-specific tuples do not include the stroke type of the player's second stroke.

3   Feature Engineering

Before we can perform our analyses, we first use the data to construct a broad range of different features that capture the most relevant information. We have divided the features into several categories, which we will discuss separately.

3.1 Point Characteristics (4 Features)

First, we include some specific information about the point. We take into account the total number of strokes, the end of the point, e.g., winner or error, and from which side of the court the server started the point. Recall that we could not include the stroke type of the second stroke of the player in the player-specific tuples due to the many missing values. Therefore, we have included the stroke type of the third stroke in the point as a separate feature.

3.2 Match Situation (13 Features)

In the second category, we consider the match situation. First, the match situation is taken into account by the set and game number at the start of the point, i.e., an integer $i \in [1,5]$ or $i \in [1,13]$, respectively. Note that we describe the tiebreak as the thirteenth game. Second, we separately consider the scores in the game, set and match, and we construct three categorical features that divide the situation into three different situations. The value of these features is equal to 1 if the player is in the lead, 0 if both players are on equal footing and -1 if the player is behind.

In our analyses, we also want to include temporal effects. First, we include a numerical feature that describes the streak of points that the player is serving or returning. Second, we also incorporate mental aspects by introducing features that take into account the outcome of previous points. As a start, we have a binary feature that describes if the previous point was won or lost. Moreover, we separately consider the previous five, four, three and two points and determine the fraction of points the player won. This gives four additional features. So, suppose the outcome of the previous 5 points is described by $m = (\text{won, won, lost, lost, won})$, where the last element of m is the most recent point. In this case, for example three of the last five points are won. Therefore, the feature that quantifies the outcome of the previous five points is equal to 3/5. Similarly, the other features that describe the mental aspects take on the values 2/4, 1/3, 1/2 and 1.

3.3 Stroke Characteristics (27 Features)

We have 27 features that describe characteristics of the different strokes. First, there are three features that characterise the serve. We have the direction of the service, the type of the service and whether the player applied service volley. Here, the type is a categorical feature that combines the intention of the service and whether it was a first or second serve. In this feature, the double fault, i.e., the second serve resulted in a fault, is treated as a separate category.

For the second stroke of the point, i.e., the return, we have six features. Finally, we also have features about the second and final stroke of both players. For the second stroke, we have four features per player and thus eight features in total. The last stroke of the player is described by five distinct features. For the return, there are some specific features, such as position where the return was hit. Other stroke characteristics, such as the direction or intention of the stroke are features for the return, second and last strokes of both players.

In our data collection, the short points need some additional attention. Here, short points have maximally five strokes. In this case, some of the aforementioned strokes are not part of the point. As the features for the different strokes are categorical, we introduce a separate category for these cases. Moreover, in this case there is also some overlap between the strokes. For example, a point of three strokes implies that for one of the players the second stroke is equal to the final stroke. In this case, the values of the features for the two strokes are also equal.

3.4 Rally Features (61 Features)

In addition to features of specific strokes of both players separately, e.g., the second or last stroke, we also consider features that describe the sequence of shots between the tennis players, i.e., rally features. Therefore, for every point we consider the collection of all strokes together and the strokes of both players separately. Of these distributions, we first construct several binary features that describe if certain categories of the aforementioned stroke-specific features are present or not. As an example, consider a point of four strokes. Suppose the directions of the strokes are 'wide', 'down the line', 'cross' and 'cross'. In this case, the rally feature that indicates if the direction 'cross' is present is equal to one. However, the direction 'middle' is not present and therefore the feature that specifies whether this direction is present is zero. By considering all different categories, we have constructed 57 binary features.

Finally, we also introduce features that characterise the entire point. For these features, we consider the values of the stroke-specific features of all strokes and select the most common category. If there is no difference between the occurrence of the two most frequent categories, we describe this with another label. We apply this to the direction, intention, situation and stroke type. Therefore, this leads to 4 additional features.

4 Experiments

In this work, we are interested in finding generic patterns in both successful and unsuccessful points in tennis and characteristics that are specific for player X. To find these descriptions, we have performed several experiments. In this section, we will describe the experiments and elaborate on the results.

4.1 Methods

To find the main patterns in our data collection, we have used Subgroup Discovery. Subgroup Discovery is an exploratory supervised data mining technique that aims to find subsets of the data where the distribution in the target variable is different from the distribution of the target in the entire data set.

Table 1. Overview of the best subgroups of successful service points for player X. This data collection consists of 2,260 service points from which 1,429 points, i.e., 63.2%, are won by player X. For each subgroup, we give the size, the winning percentage and the value for the Cortana Quality measure [13]. The threshold for this quality measure for finding statistical significant results at $\alpha = 0.05$ is roughly 0.06 and therefore the listed subgroups are statistical significant. Finally, we specify the condition that characterise the points that are part of the subgroups. The intentions 'opbouwend' and 'VvS' corresponds to a stroke with a constructive intention or a stroke with the intention to prevent the opponent from scoring, respectively.

	Coverage	Winning percentage	Quality	Condition
1	894	85.1%	0.372	Point is maximally two strokes
2	659	92.9%	0.372	Last stroke of player X is 'first serve' or 'double fault'
3	1164	77.2%	0.310	The point contains no stroke with intention 'opbouwend'
4	1378	74.7%	0.302	Player X is not hitting a backhand
5	1524	73.6%	0.301	Player X hits no stroke with intention 'VvS'

In this work, we want to find characteristics of points that are often won or lost by a specific player. Hence, we are in a classification setting as the point outcome is characterised by a binary variable that takes on the value 1 if the point is won, and 0 otherwise. To perform our classification experiments, we have used the freely available online tool Cortana [16] and selected the Cortana Quality measure that is introduced in [13].

In tennis, there are two distinct situations, i.e., the tennis player can be either serving or receiving. In the first case, the first stroke of the player is the service. For the second option, the player starts with a return. As both situations are fundamentally different, we have performed our experiments for both cases separately. In total, we therefore have four different cases. Namely, our tennis player can have two different roles, i.e., serving or returning, and the point can be either won or lost.

The data collection that consists of return points of player X is equivalent to the data set of service points of a collection of several different tennis players. Thus, this data set can be used for setting a benchmark for general characteristics of winning service points in tennis. By comparing the results for this data set with the results for the collection of service points of player X, we can find service strengths of this specific player. Similarly, we can also find player-specific strengths in return games as well as his weaknesses while serving or returning. In the remainder of this work, for the sake of brevity we will restrict ourselves to discussing strengths and weaknesses in *service points*.

4.2   Serve Characteristics

In Tables 1 and 2, we show the characteristics of the best subgroups at search depth 1 for successful service points of player X and for players serving against player X, respectively. We observe that there are many qualitative similarities between the subgroups. For example, we find that tennis players are more successful on their serve if the point last maximally two strokes. Apart from these similarities, we also find some differences between player X and the other players. The other players are more successful on their first serve and player X is more successful if player X does not hit a backhand.

Table 2. Overview of the best subgroups of successful service points for the data collection of service points of tennis players in matches against player X. In total, this set contains 2,333 service points of which 1,505 points, i.e., 64.5%, are won by the serving player. The same characteristics of the subgroups as in Table 1 are displayed. The subgroups are statistical significant as the threshold for this quality measure for finding statistical significant results at $\alpha = 0.05$ is equal to approximately 0.06. Recall that the intention 'opbouwend' corresponds to a stroke with a constructive intention and a stroke that is hit with the intention to prevent the opponent from scoring is denoted by the intention 'VvS'.

	Coverage	Winning percentage	Quality	Condition
1	562	94.0%	0.310	Last stroke of player X is 'first serve' or 'double fault'
2	797	82.6%	0.269	Length of point maximally two strokes
3	1452	73.4%	0.242	Server hits a first serve
4	1239	74.7%	0.235	The point contains no stroke with intention 'opbouwend'
5	1761	70.8%	0.206	Server has no stroke with intention 'VvS'

We have also performed experiments at search depth 2. For player X and the other players, the best subgroup consists of service points that maximally last two strokes and are started with a first serve. The values for the Cortana Quality are 0.405 and 0.341 for player X and the other players, respectively. Also at depth 2, there are some subgroups that are only of high quality for player X. In this case, we specifically have found that player X is more successful if the point maximally last two strokes and the point contains no stroke with the intention of finishing the point.

We have used the *distribution of false discoveries* to determine a threshold for the quality measure of the subgroups that indicates whether the results are statistically significant [7]. For both data collections, the thresholds at confidence level $\alpha = 0.05$ are roughly 0.06 and 0.09 for search depths 1 and 2, respectively.

As the quality of the presented subgroups is much larger than these thresholds, we obtain results that are statistically significant.

The subgroups in Tables 1 and 2 only show the subgroups with highest value for the quality measure. However, there are many more subgroups at both search depth 1 and 2 that are statistically significant. These subgroups are described by conditions on numerous different features. Therefore, there are also more player-specific characteristics of successful service points. For example, if we only focus on subgroups that are described by a condition on match situation features, we find at search depth 1 that the opponents are more successful on service points if they are in front in the game. However, for player X, there are no statistical significant subgroups that are described by a condition on match situation features. This suggests that players in general win more service points if the are in front in the game, but that this is not the case for player X.

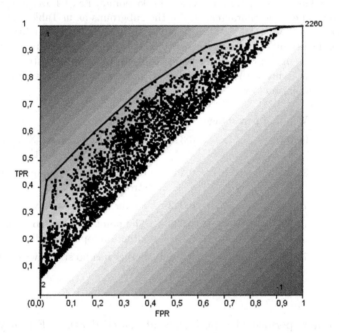

Fig. 1. The ROC Curve that illustrates the quality of the classification of successful service points of player X. Every point in this figure, represents a subgroup that is obtained from the Subgroup Discovery experiments at search depth 2. The Area under the ROC Curve is equal to 0.79.

To assess the quality of the classification model, we consider the Receiving Operation Characteristics (ROC) Curve. For every subgroup, we determine the false positive and true positive rates. In Fig. 1, we show the ROC curve for the data collection of service points of player X. By constructing the convex hull of all subgroups, we can determine the Area Under the ROC Curve (AUC). This measure quantifies the overall performance of the classification [15]. At search depth 2, we find values of 0.79 and 0.74 for the AUC of the data collection that consists of service points of player X and the service points of the opponents in matches against player X, respectively. This implies that the classification potential is reasonable good.

In addition to considering all service points together, we can also look into specific subsets to obtain characteristics of a specific class of service points. For example, we considered the collection of points that started with a serve of player X that last maximally three strokes. Interestingly, we have found a statistically significant subgroup at search depth 1, that indicates that player X is more successful if the direction of the service is the 'T' direction, i.e., a service towards the center of the tennis court. For this subset, the quality of the subgroups that consists of points with serves in the other directions is not above the significance threshold. Thus, this suggest that player X is most successful with a service in the 'T' direction. This is another example of a characterization for the playing style of player X.

5   Conclusion

In this paper, we have performed tactical analyses in tennis. We have used a manually annotated data set of 4,593 points that features a professional player X. We have used Subgroup Discovery to find characteristics of successful service points. By applying this technique to all service points of player X and comparing the results with the outcomes for the service points of all opponents, we have made a distinction between generic descriptions of successful service points and characterizations that are specific for player X.

Primarily, we have focused on the descriptions of successful service points that are most significant. At search depth 1, we have found the general description that service points are more often won if the point is maximally two strokes long. From our experiments at search depth 2, we have obtained that the serving player is even more successful if in addition the aforementioned condition the point starts with a first serve. Moreover, specifically for player X we have obtained that player X is less successful on his first serve and is more vulnerable with his backhand. Additionally, we have considered the subset of service points that are maximally three strokes long and we have demonstrated that player X is more successful if the service is directed towards the middle of the court. Moreover, by determining the AUC for the classification of all service points of player X and the collection of points where the opponent of player X is serving, we have obtained that the quality of our classification is reasonably good.

In addition to the considered subset of short service points, it is also interesting to look into other specific subsets. For example, we could compare the results in different match situations to investigate whether this affects the playing style. Moreover, the methods we have presented in this work also have many applications in other sports with a tactical component. For example, our approach can be used for comparing playing styles of different football teams. As there are many interesting sport-specific questions that can be addressed, it is undesirable to compare the results for all different settings manually. Therefore, a possible direction of future research is to develop an automatised procedure for applying Contrast Data Mining with Subgroup Discovery.

References

1. www.hawkeyeinnovations.co.uk
2. www.usta.com/content/dam/usta/sections/pacific-northwest/pdfs/play/leaguenew/ITFRuleswheader.pdf
3. Calvo, C., Micarelli, A., Sangineto, E.: Automatic annotation of tennis video sequences. In: Van Gool, L. (ed.) DAGM 2002. LNCS, vol. 2449, pp. 540–547. Springer, Heidelberg (2002). https://doi.org/10.1007/3-540-45783-6_65
4. Crespo, M., Piles, J.: Tactics for elite level men's tennis - part 1. ITF Coach. Sport Sci. Rev. **56**(20), 9–10 (2012)
5. Crespo, M., Reid, M.: Introduction to modern tactics. ITF Coach. Sport Sci. Rev. **27**, 2 (2002)
6. Cross, R., Pollard, G.: Grand slam men's singles tennis 1991–2009 serve speeds and other related data. ITF Coach. Sport Sci. Rev. **16**(49), 8–10 (2009)
7. Duivestreijn, W., Knobbe, A.: Exploiting false discoveries statistical validation of patterns and quality measures in subgroup discovery. In: Proceedings of International Conference on Data Mining, pp. 151–160 (2011)
8. Fernando, T., Denman, S., Sridharan, S., Fookes, C.: Memory augmented deep generative models for forecasting the next shot location in tennis. IEEE Trans. Knowl. Data Eng. 1 (2019, PrePrints)
9. Gillet, E., Leroy, D., Thouvarecq, R., Stein, J.F.: A notational analysis of elite tennis serve and serve-return strategies on slow surface. J. Strength Cond. Res. **23**(2), 532–539 (2009)
10. Hughes, M., Franks, I.: Notational Analysis of Sport: Systems for Better Coaching and Performance in Sport. Routledge Publishing, London (2004)
11. Hughes, M., Hughes, M., Behan, H.: The evolution of computerised notational analysis through the example of racket sports. Int. J. Sports Sci. Eng. **1**(1), 3–28 (2007)
12. Klösgen, W., Zytkow, J. (eds.): Handbook of Data Mining and Knowledge Discovery. Oxford University Press Inc., New York (2002)
13. Knobbe, A., Orie, J., Hofman, N., van der Burgh, B., Cachucho, R.: Sports analytics for professional speed skating. Data Min. Knowl. Disc. **31**(6), 1872–1902 (2017). https://doi.org/10.1007/s10618-017-0512-3
14. de Leeuw, A.W., Meerhoff, L., Knobbe, A.: Effects of pacing properties on performance in long-distance running. Big Data **6**(4), 248–261 (2018)
15. Mason, S., Graham, N.: Areas beneath the relative operating characteristics (ROC) and relative operating levels (ROL) curves: statistical significance and interpretation. Q. J. R. Meteorol. Soc. **128**(584), 2145–2166 (2002)
16. Meeng, M., Knobbe, A.: Flexible enrichment with cortana - software demo. In: Proceedings of BeneLearn, the Annual Belgian-Dutch Conference on Machine Learning, pp. 117–119 (2011)
17. Novak, P., Lavrač, N., Webb, G.: Supervised descriptive rule discovery: a unifying survey of contrast set, emerging pattern and subgroup mining. J. Mach. Learn. Res. **10**, 377–403 (2009)

18. Rioult, F., Mecheri, S.M.B., Kauffmann, F., Benguigui, N.: What can hawk-eye data reveal about serve performance in tennis? In: Machine Learning and Data Mining for Sports Analytics ECML/PKDD Workshop (2015)
19. Terroba, A., Kosters, W., Varona, J., Manresa-Yee, C.: Finding optimal strategies in tennis from video sequences. Int. J. Pattern Recogn. Artif. Intell. **27**(6), 1–31 (2013)
20. Wang, P., Cai, R., Yang, S.-Q.: A tennis video indexing approach through pattern discovery in interactive process. In: Aizawa, K., Nakamura, Y., Satoh, S. (eds.) PCM 2004. LNCS, vol. 3331, pp. 49–56. Springer, Heidelberg (2004). https://doi.org/10.1007/978-3-540-30541-5_7
21. Wei, X., Lucey, P., Morgan, S., Carr, P., Reid, M., Sridharan, S.: Predicting serves in tennis using style priors. In: Proceedings of the 21th ACM SIGKDD International Conference on Knowledge Discovery and Data Mining, pp. 2207–2215 (2015)
22. Wei, X., Lucey, P., Morgan, S., Sridharan, S.: Forecasting the next shot location in tennis using fine-grained spatiotemporal tracking data. IEEE Trans. Knowl. Data Eng. **28**(11), 2988–2997 (2016)

Difficulty Classification of Mountainbike Downhill Trails Utilizing Deep Neural Networks

Stefan Langer$^{(\boxtimes)}$, Robert Müller$^{(\boxtimes)}$, Kyrill Schmid$^{(\boxtimes)}$,
and Claudia Linnhoff-Popien$^{(\boxtimes)}$

Mobile and Distributed Systems Group, LMU Munich, Munich, Germany
{stefan.langer,robert.mueller,kyrill.schmid,linnhoff}@ifi.lmu.de

Abstract. The difficulty of mountainbike downhill trails is a subjective perception. However, sports-associations and mountainbike park operators attempt to group trails into different levels of difficulty with scales like the *Singletrail-Skala* (S0-S5) or colored scales (blue, red, black, ...) as proposed by *The International Mountain Bicycling Association*. Inconsistencies in difficulty grading occur due to the various scales, different people grading the trails, differences in topography, and more. We propose an end-to-end deep learning approach to classify trails into three difficulties easy, medium, and hard by using sensor data. With mbientlab Meta Motion r0.2 sensor units, we record accelerometer- and gyroscope data of one rider on multiple trail segments. A 2D convolutional neural network is trained with a stacked and concatenated representation of the aforementioned data as its input. We run experiments with five different sample- and five different kernel sizes and achieve a maximum Sparse Categorical Accuracy of 0.9097. To the best of our knowledge, this is the first work targeting computational difficulty classification of mountainbike downhill trails.

Keywords: Sports analytics · Deep neural networks · Mountainbike · Accelerometer · Gyroscope · Convolutional neural networks

1 Introduction

Mountainbiking is a popular sport amongst outdoor enthusiasts, comprising many different styles. There are styles like cross country riding, characterized by long endurance rides, styles like downhill riding, characterized by short, intense rides down trails, and more [1]. Mountainbiking, as it is known today, originated in the US in the 1970s and since then went through various levels of popularity [2]. Official, competitive riding started in the 1980s with the foundation of the *Union Cycliste Internationale (UCI)*, followed by the first World Championship in 1990 [3]. In this work, we focus on the difficulty classification of mountainbike downhill trails and do not take into account uphill or flat sections of trails. There are multiple approaches in trail difficulty classification, whereby a color-inspired grading is most commonly used [4–6]. The *International Mountain*

© Springer Nature Switzerland AG 2020
P. Cellier and K. Driessens (Eds.): ECML PKDD 2019 Workshops, CCIS 1168, pp. 270–280, 2020.
https://doi.org/10.1007/978-3-030-43887-6_21

Bicycling Association (IMBA) proposes a trail difficulty rating system comprised of five grades, ranging from a green circle (easiest) to a double black diamond (extremely difficult) [4]. In addition, the *IMBA Canada* offers a guideline on how to apply those gradings to mountainbike trails [7]. *British Cycling* also propose a colored difficulty scale, including four basic grades from green (easy) to black (severe) with an additional orange for bike park trails [5]. Inspired by rock climbing difficulty grading, as well as ski resort gradings, Schymik et al. created the *Singletrail-Skala*, containing three main difficulty classes (blue, red, black) and a more fine granular six grades ranging from S0 to S5 [6]. Trails on *Openstreetmap* [8] are rated with respect to the IMBA grading as well as the *Singletrail-Skala*, whereas the latter also describes tracks which are not specifically made for mountainbiking [9]. Due to factors like the various scales, different people grading the trails or differences in topography, estimating the difficulty of mountainbike trails consistently is not an easy task. This work aims to make mountainbike track difficulty assessment less subjective and more measurable. In order to do so, we collect acceleration-, as well as gyroscope-data from multiple sensor units that are connected to the mountainbike frame as well as the rider. Because we do not collect data in dedicated mountainbike parks, but on open trails (hiking paths among others), we decided to use the three main difficulties given by the *Singletrail-Skala* as the set of labels.

Table 1. Mapping of the *Singletrail-Skala* grades to the labels used in our data set. Label 0 describes easy trails, label 1 medium trails, and label 2 hard trails.

Colored grading	Fine grading	Label	Description
Blue	S0, S1	0	Easy
			Mostly solid and non-slip surface
			Slight to moderate gradient
			No switchbacks
			Basic skills needed
Red	S2	1	Medium
			Loose surface, bigger roots and stones
			Moderate steps and drops
			Moderate switchbacks
			Advanced skills needed
Black	S3+	2	Hard
			Loose surface, slippery, big roots and stones
			High drops
			Tight switchbacks
			Very good skills needed

Table 1 gives an overview of the three grades blue, red and black. Schymik et al. [6] define the difficulties as follows: Blue describes easy trails, comprising the grades S0 and S1. Red describes medium trails and is equal to the grade S2. Black describes all difficulties above and can be considered hard. *Openstreetmap* provides difficulty classifications for all trails on which this dataset is collected [9]. We then train a 2D convolutional neural network with a stacked and concatenated representation of the aforementioned data as its input. Thereby we can grade sections of downhill trails regarding their difficulty.

1.1   Related Work

For training purposes, mountainbikes of professional athletes get set up with telemetry technology, such as BYB Telemetry's sensors [10]. Their sensors are connected to the suspension fork as well as the suspension shock and measure the movement of each. Stendec Data extends those capabilities and adds sensors for measuring brake pressure and acceleration in order to capture braking points, wheel movements, and more [11]. However, the two systems mentioned above are expensive and hard to get. Therefore, we use mbientlab Meta Motion sensor units to capture acceleration and gyroscope data. Ebert et al. [12] automatically recognized the difficulty of boulder routes with mbientlab sensor units. To the best of our knowledge there is no scientific work regarding the difficulty classification of mountainbike trails using accelerometers or gyroscopes yet. However, there has been a great amount of work done in the field of activity recognition with acceleration data [13–19]. Many of those approaches make use of classical machine learning methods [13–16,20]. Preece et al. [20] compare feature extraction methods for activity recognition in accelerometer data. Bao et al. [14] classify activities using custom algorithms and five biaxial acceleration sensors worn simultaneously on different parts of the body. Furthermore, there has been a noticeable shift towards deep learning approaches in recent years [17–19]. Moya Rueda et al. [21] use multiple convolutional neural networks which they concatenate in a later stage with fully connected layers. Zeng et al. utilize a 1D convolutional neural network, treating each axis of the accelerometer as one channel of the initial convolutional layer [18]. In a survey by Wang et al. the authors give an overview of state-of-the art deep learning methods in activity recognition [22]. The authors claim that deep learning outperforms traditional machine learning methods and has been widely adopted for sensor-based activity recognition tasks.

2   The Dataset

2.1   Collecting and Labeling Data

Instead of working with dedicated mountainbike telemetry systems, we use mbientlab Meta Motion r0.2 sensor units to record data [23]. Those units contain multiple sensors, including an accelerometer as well as a gyroscope. Mbientlab sensors offer a Bluetooth Low Energy interface to which an Android or iOS application can be connected. The rider is equipped with two sensor units. Figure 1

Fig. 1. Mounting point on the helmet (left), mounting point on the downtube of the frame (right)

visualizes the mounting points of the mbientlab sensor units. One unit is connected to the downtube of the mountainbike, the other one to the back of the rider's helmet. For each recording, the sensors are facing the same direction to keep the axes layout consistent. The accelerometer creates datapoints in three axes (x, y, z) in the unit g (equals 9.80665 m/s^2) with a frequency of 12.50 Hz. The gyroscope creates datapoints in three axes (x, y, z) in the unit deg/s with a frequency of 25.00 Hz. We synchronize the starting points of the recordings and linearly interpolate missing datapoints to reach a constant frequency of 25.00 Hz for all sensors.

Labeling of the data happens after the actual data collection process. We record every downhill ride with an action camera (mounted to the rider's chest), synchronize the video with the data recordings, and manually label subsections of the trail. For the majority of subsections on open trails, we use the difficulty grading provided by *Openstreetmap*. Those gradings are made visible in mountainbike specific *Openstreetmap* variants and can also be found in the (XML-like) *.OSM* exports of an area. One "way" node (which describes a trail) then includes another node "tag", comprising the difficulty description. For subsections that the *Singletrail Skala* would consider to not represent this difficulty (as per their description), we up- or downgrade the difficulty label. Downgrading mostly occurs for fireroads or other very easy sections, upgrading for particularly steep or tight sections.

2.2 Input Data Representation

For each ride, we collect data with two sensor units. Every unit provides data for the accelerometer and the gyroscope sensors. Each sensor generates datapoints for three axes (x, y, z) with an additional timestamp value. Zeng et al. [18] interpret each axis of a sensor as a filter of the input to a 1D convolutional layer. We keep the same procedure but additionally stack each of the four sensors (two

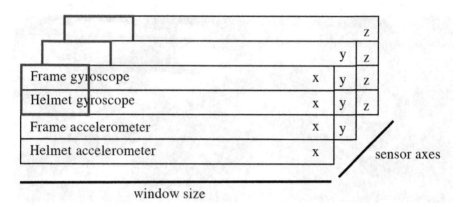

Fig. 2. Input shape of one sample comprising four sensors with three dimensions (axes) each

accelerometers, two gyroscopes) vertically to create an image-like representation. Figure 2 visualizes the shape of our input data. Height and width of the image-like representation are represented by four sensors and n datapoints. RGB-like channels are represented by the three axes x, y, and z. The square in the top left corner visualizes the kernel sliding across the input data. We split each recording into smaller samples utilizing a sliding window with an overlap of 75%. This allows us to create many examples from few data recordings. In our experiments we test five different window sizes, namely 1000 ms, 2000 ms, 5000 ms, 10000 ms, and 20000 ms resulting in 25, 50, 125, 250, and 500 data points per example. This leads to 5937, 2971, 1150, 575, and 286 samples respectively. For each experiment, we use a 80/20 test/train split in order to evaluate the network's performance on unseen data.

3 Classification Through a 2D Convolutional Neural Network

In order to classify mountainbike downhill trails regarding their difficulty, we apply a convolutional neural network. Figure 3 visualizes the network's architecture. The input to the first block is of shape $(n, 4, 3)$, with n being the amount of data points per sample. One sample consists of data of four sensors (vertically stacked), with each three axes (filters), and a sample size of n data points. We chain three convolutional blocks followed by two Dense Layers. Each convolutional block consists of one Conv2D [24], a Batch Normalization [25], a ReLU Activation [26], a Max Pool [27] and a Dropout Layer [28]. The convolutional layers use a kernel of shape $(m, 2)$ and a stride of $(1, 1)$, with m being the length of the kernel. Multiple values for n and m are tested in the experiments. All convolutional layers use the padding 'same' [29]. With this setting, the width and height dimensions of the in- and output of a convolutional layer stay the same. Furthermore, we add L2 regularization to each convolutional layer [30].

L2 regularization shifts outlier weights closer to 0. Max Pool layers use a pool size of (2, 1), which reduces the shape by approximately half in length. The dropout rate of each Dropout Layer is 0.3. The Conv2D Layers of the second and third convolutional block have 8 and 16 filters respectively. After the convolutional blocks, we add two Dense Layers. The first layer has 128 units and a ReLU Activation. The second and final layer has three Softmax activated units, which represent the predicted label. The network uses the Adam optimizer [31] with a learning rate of 0.001 and a Sparse Categorical Crossentropy as it's loss function. This configuration proofed to be the best in our experiments.

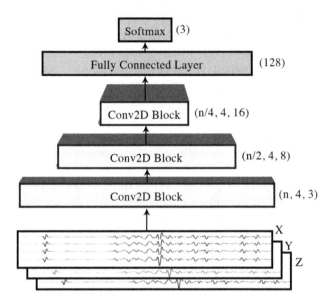

Fig. 3. Convolutional Neural Network for trail difficulty classification on stacked accelerometer and gyroscope data. The input data is of shape $(n, 4, 3)$, with n being the amount of data points per sample. After the input layer, three convolutional blocks and two fully connected layers follow.

3.1 Experiments

Due to the fact that there are no established neural network configurations for trail difficulty classification, we evaluate 25 combinations of window- and kernel sizes. We test five window sizes (1000 ms, 2000 ms, 5000 ms, 10000 ms, 20000 ms) and five kernel sizes ((5,2), (10,2), (20, 2), (40,2), (60,2)) (see Table 2). For empty result cells, the amount of datapoints per sample is smaller than the kernel length. The dataset includes approximately 32% of samples of label 0, 56% of samples of label 1 and 12% of samples of label 2. This uneven distribution led the model to rarely predict the labels 0 and 2. Therefore we decided to compensate the inequality by copying existing examples of the underrepresented

Table 2. Sparse categorical accuracy and amount of epochs of the experiments. A window size of 10000 ms with a kernel size of (60,2) creates the highest accuracy (0.9097) after 781 epochs on the test dataset.

		Kernel size					Samples	Samples after over-sampling
		(5,2)	(10,2)	(20,2)	(40,2)	(60,2)		
Window size	1000 ms	0.4990 (271)	0.5313 (124)	0.5165 (249)	-	-	5937	10368
	2000 ms	0.5155 (259)	0.5585 (300)	0.5734 (247)	0.5599 (183)	-	2971	5073
	5000 ms	0.6181 (300)	0.6632 (515)	0.7743 (726)	0.6632 (63)	0.7778 (79)	1150	2019
	10000 ms	0.7222 (425)	0.6875 (473)	0.7639 (548)	0.8681 (607)	**0.9097 (781)**	575	978
	20000 ms	0.6250 (456)	0.6111 (5)	0.6111 (162)	0.6250 (8)	0.6806 (545)	286	498

classes within the training set (so that the classes are balanced equally). In order to reduce overfitting, we add an early stopping callback to the network, which stops the training process when there is no improvement for 250 epochs (the patience value). With smaller patience values the network stopped learning too early in some cases. The maximum amount of epochs for training is 1500. We run a batch size of 32 and a steady learning rate of 0.001. In Table 2 we show the resulting Sparse Categorical Accuracy, the amount of epochs before training was stopped and the amount of samples in the train set. The Sparse Categorical Accuracy measures the accuracy of the result of sparse multiclass classification problems [32]. For every experiment, we use a sliding window with an overlap of 75%. To not have many highly similar examples in one batch, we shuffle the data before training. Short window sizes (1000 ms, 2000 ms) show lower accuracy than the larger samples across all kernel sizes. This could be attributed to the low amount of datapoints within a sample (25) as well as the short sample not representing the subsection of the trail.

The lowest accuracy (0.4990) was reached with window size 1000 ms and kernel size (5,2). With a window size of 10000 ms and a kernel size of (60,2), we achieve a high sparse categorical accuracy of 0.9097. This leads to the conclusion, that a window length of 10000 ms is necessary to represent a downhill trail subsection appropriately. Longer sequential dependencies (by using larger kernel lengths) show a positive effect on the difficulty classification as well.

Figure 4 shows the curves of the Sparse Categorical Accuracy on the train as well on the test dataset across 1000 epochs. Both values increase early on and level out with no major overfitting visible in the plot. The highest accuracy on the test dataset was achieved after 781 epochs.

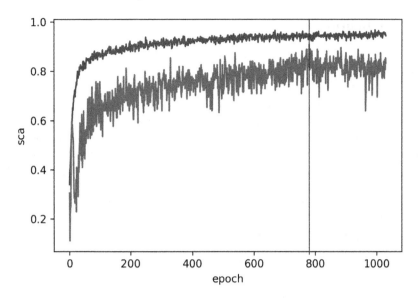

Fig. 4. Plot of the Sparse Categorical Accuracy (sca) on the train set (blue, upper) as well as on the test set (orange, lower). (Color figure online)

Figure 5 shows the confusion matrix of the best resulting configuration, namely a window size of 10000 ms and a kernel size of (60,2). Good results for all three classes are shown, with only few false positives in neighbored areas. The matrix also highlights the fact, that the label 2 (hard) is underrepresented. However, the distribution of correctly predicted labels matches the distribution of the raw dataset well.

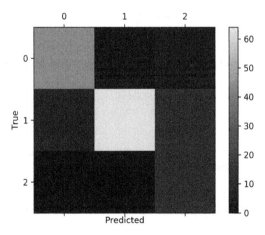

Fig. 5. The confusion matrix for a window size of 10000 ms and a kernel size of (60,2).

278 S. Langer et al.

4 Conclusion

In this work, we proposed an end-to-end deep learning approach to classify mountainbike downhill trails regarding their difficulty. We gave an introduction to multiple official difficulty scales and decided to use the *Singletrail-Skala* for this work. Using mbientlab Meta Motion r0.2 sensor units, we recorded multiple rides on multiple trail segments, resulting in 2971 training examples for the best window-size/kernel-size combination. The sensor units provided us with accelerometer and gyroscope data in each three axes, which we concatenated to create an image-like representation of the data. Downhill trails were labeled according to their *Singletrail-Skala* rating and a subjective up- or downgrading for subsections, that strongly diverge from their rating. We implemented a 2D convolutional neural network with two dense layers at the end for the classification process. We ran experiments with five different window sizes (1000 ms, 2000 ms, 5000 ms, 10000 ms, 20000 ms) and five different kernel sizes ((5,2), (10,2), (20,2), (40,2), (60,2)). The best result could be observed with a sample size of 10000 ms and a kernel size of (60,2), resulting in a Sparse Categorical Accuracy of 0.9097 on a 80/20 train/test split. In future work, one could think of a non-supervised clustering method to avoid subjective input. Additionally, the dataset could possibly be improved by using more sensors, like high-resolution barometers or heartrate sensors. As can be seen in Fig. 5 more examples for hard sections (label 2) are needed. This category is underrepresented in the data we collected.

With this work, we hope to reduce the amount of subjective rating of mountainbike trails and make their difficulty measurable. An automated recognition of downhill trail difficulty could be advantageous in diverse scenarios. For unlabeled, or mislabeled trails, our sensor analysis architecture can generate a fitting label. This can help tourist areas or mountainbike park operators describe the difficulty of new or existing trails consistently across areas, topographies, or countries. For social fitness networks like e.g. Strava [33] one could think of an automated difficulty grading of rides (or subsections of rides). This would extend the existing performance comparison factors, like speed or distance, by a value for downhill trail difficulty. Furthermore, we hope to promote data analytics in the sport of mountainbiking by releasing a bigger and improved version of our dataset soon.

References

1. U. C. Internationale: The evolution of mountain bike and its many formats, June 2019. https://www.uci.org/news/2019/the-evolution-of-mountain-bike-and-its-many-formats
2. Gaulrapp, H., Weber, A., Rosemeyer, B.: Injuries in mountain biking. Knee Surg. Sports Traumatol. Arthrosc. **9**(1), 48–53 (2000). https://doi.org/10.1007/s001670000145
3. Impellizzeri, F.M., Marcora, S.M.: The physiology of mountain biking. Sports Med. **37**(1), 59–71 (2007). https://doi.org/10.2165/00007256-200737010-00005

4. I. M. B. Association: Trail difficulty rating system, June 2019. https://www.imba.com/resource/trail-difficulty-rating-system
5. B. Cycling: MTB trail grading system, June 2019. https://www.britishcycling.org.uk/search/article/mtbst20100615-MTB-Trail-Grading-System-0
6. Schymik, C., Philipp, H., Werner, D.: Singletrail-skala (STS) version1. 4. Einstufung in technische Schwierigkeitsgrade. Zugriff am **15**, 2015 (2008)
7. I. M. B. Association: Trail rating guidelines, June 2019. https://imbacanada.com/trail-rating-guidelines/
8. openstreetmap.org, Openstreetmap, June 2019. https://wiki.openstreetmap.org
9. openstreetmap.org: Mountain biking: June 2019. https://wiki.openstreetmap.org/wiki/Mountain_biking
10. B. Telemetry: Byb telemetry, June 2019. https://www.bybtelemetry.com/
11. S. Racing: Stendec racing, June 2019. https://stendecracing.com/about/
12. Ebert, A., Schmid, K., Marouane, C., Linnhoff-Popien, C.: Automated recognition and difficulty assessment of boulder routes. In: Ahmed, M.U., Begum, S., Bastel, J.-B. (eds.) HealthyIoT 2017. LNICST, vol. 225, pp. 62–68. Springer, Cham (2018). https://doi.org/10.1007/978-3-319-76213-5_9
13. Ravi, N., Dandekar, N., Mysore, P., Littman, M.L.: Activity recognition from accelerometer data. In: AAAI, vol. 5, issue 2005, pp. 1541–1546 (2005)
14. Bao, L., Intille, S.S.: Activity recognition from user-annotated acceleration data. In: Ferscha, A., Mattern, F. (eds.) Pervasive 2004. LNCS, vol. 3001, pp. 1–17. Springer, Heidelberg (2004). https://doi.org/10.1007/978-3-540-24646-6_1
15. Kwapisz, J.R., Weiss, G.M., Moore, S.A.: Activity recognition using cell phone accelerometers. ACM SigKDD Explor. Newslett. **12**(2), 74–82 (2011)
16. Lara, O.D., Labrador, M.A.: A survey on human activity recognition using wearable sensors. IEEE Commun. Surv. Tutor. **15**(3), 1192–1209 (2012)
17. Yang, J., Nguyen, M.N., San, P.P., Li, X.L., Krishnaswamy, S.: Deep convolutional neural networks on multichannel time series for human activity recognition. In: Twenty-Fourth International Joint Conference on Artificial Intelligence (2015)
18. Zeng, M., Nguyen, L.T., Yu, B., Mengshoel, O.J., Zhu, J., Wu, P., Zhang, J.: Convolutional neural networks for human activity recognition using mobile sensors. In: 6th International Conference on Mobile Computing, Applications and Services, pp. 197–205. IEEE (2014)
19. Ronao, C.A., Cho, S.-B.: Human activity recognition with smartphone sensors using deep learning neural networks. Expert Syst. Appl. **59**, 235–244 (2016)
20. Preece, S.J., Goulermas, J.Y., Kenney, L.P., Howard, D.: A comparison of feature extraction methods for the classification of dynamic activities from accelerometer data. IEEE Trans. Biomed. Eng. **56**(3), 871–879 (2008)
21. Moya Rueda, F., Grzeszick, R., Fink, G., Feldhorst, S., ten Hompel, M.: Convolutional neural networks for human activity recognition using body-worn sensors. In: Informatics, vol. 5, issue 2, p. 26. Multidisciplinary Digital Publishing Institute (2018)
22. Wang, J., Chen, Y., Hao, S., Peng, X., Hu, L.: Deep learning for sensor-based activity recognition: a survey. Pattern Recogn. Lett. **119**, 3–11 (2019)
23. Mbientlab: Meta motion c, June 2019. https://mbientlab.com/metamotionc/
24. Krizhevsky, A., Sutskever, I., Hinton, G.E.: Imagenet classification with deep convolutional neural networks. In: Advances in Neural Information Processing Systems, pp. 1097–1105 (2012)
25. Ioffe, S., Szegedy, C.: Batch normalization: Accelerating deep network training by reducing internal covariate shift, arXiv preprint arXiv:1502.03167 (2015)

26. Hahnloser, R.H., Sarpeshkar, R., Mahowald, M.A., Douglas, R.J., Seung, H.S.: Digital selection and analogue amplification coexist in a cortex-inspired silicon circuit. Nature **405**(6789), 947 (2000)

27. Matsugu, M., Mori, K., Mitari, Y., Kaneda, Y.: Subject independent facial expression recognition with robust face detection using a convolutional neural network. Neural Netw. **16**(5–6), 555–559 (2003)

28. Srivastava, N., Hinton, G., Krizhevsky, A., Sutskever, I., Salakhutdinov, R.: Dropout: a simple way to prevent neural networks from overfitting. J. Mach. Learn. Res. **15**(1), 1929–1958 (2014)

29. Keras: Convolutional layers, June 2019. https://keras.io/layers/convolutional/

30. Ng, A.Y.: Feature selection, L1 vs. L2 regularization, and rotational invariance. In: Proceedings of the Twenty-First International Conference on Machine Learning, p. 78. ACM (2004)

31. Kingma, D.P., Ba, J.: Adam: A method for stochastic optimization, arXiv preprint arXiv:1412.6980 (2014)

32. Moolayil, J.: Keras in Action. Learn Keras for Deep Neural Networks, pp. 17–52. Springer, Berkeley (2019). https://doi.org/10.1007/978-1-4842-4240-7_2

33. Strava: Die beste app für läufer und radfahrer, June 2019. https://www.strava.com/?hl=de

First Workshop on Categorizing Different Types of Online Harassment Languages in Social Media

Categorizing Online Harassment
on Twitter

Mozhgan Saeidi[1(✉)], Samuel Bruno da S. Sousa[1,2(✉)], Evangelos Milios[1(✉)],
Norbert Zeh[1(✉)], and Lilian Berton[2(✉)]

[1] Dalhousie University, Halifax, Canada
{mozhgan.saeidi,samuelsgousa}@dal.ca, {eem,nzeh}@cs.dal.ca
[2] Federal University of São Paulo, São José dos Campos, Brazil
lberton@unifesp.br

Abstract. Harassment on social media is a hard problem to tackle since
those platforms are virtual spaces in which people enjoy the liberty to
express themselves with no restrictions. Furthermore, a large amount of
users generating publications on online media like Twitter contributes
to the hardness of controlling sexism and sexual harassment content,
requesting robust methods of Machine Learning (ML) to be applied in
this task. To do so, this work aims at comparing the performance of
supervised ML algorithms to categorize online harassment in Twitter
posts. We tested Logistic Regression, Gaussian Naïve Bayes, Decision
Trees, Random Forest, Linear SVM, Gaussian SVM, Polynomial SVM,
Multi-Layer Perceptron, and AdaBoost methods on the SIMAH Competition benchmark data, using TF-IDF vectors and Word2Vec embeddings
as features. As results, we reached scores above 0.80% of accuracy for
all the harassment types in the data. We also showed that, when using
TF-IDF vectors, Linear and Gaussian SVM are the best methods to predict harassment content, while Decision Trees and Random Forest better categorize physical and sexual harassment. Overall, by using TF-IDF
vectors presented higher performance on these data, suggesting that the
training corpus for Word2Vec influenced negatively on the classification
task outcomes.

Keywords: Online harassment · Text classification · Twitter

1   Introduction

Language reveals the values of people and their perspectives [21]. Sexism in
language has been discussed in different communication media, such as adver-
tisements, newspapers, TV, and more recently in online social networks. Sexism
can be defined as an aggregate of hostile stereotypes towards women [22] manifested on language, behaviors, and cultural traits. In social media platforms,
sexist comments present different categories, according to the intent they are

Supported by the organization of ECML PKDD 2019.

written [21]. It is a widely known problem to detect sexist content online automatically [19–21] since text is an unstructured data type, besides the intrinsic
ambiguity of language. To explore contents on social platforms, such as Twitter,
employing ML techniques, the most significant role is with the data [19], because
several processes have to be done to balance, collect, label, or even measure the
overlap of datasets' classes.

Detecting harassment on online social media is a challenge in the field of Natural Language Processing (NLP) so recent as the major online social networks.
For instance, Facebook and Twitter were created in 2004 and 2006, respectively. The discussion concerning online harassment types, in turn, only gained
large audiences in 2017 due the #MeToo movement, which encourages women
to denounce offensive content towards them in real life and online [20]. Since this
time, tools to control the spreading of offensive posts against women and other
groups were released, despite of not being able to detect 100% of the content in
the networks. Therefore, this is a promising research area [19,21] whose results
have an influential role to promote an educational culture among social network
users, as well as combating harassment online.

In this study, we focused on the sexual harassment tweets gathered by Sharifirad and Matwin [19]. By using NLP and supervised classifiers, we classify tweets
into two groups: harassment versus no harassment in the first task. In the second task, we categorized different types of online harassment tweets into three
categories: "indirect harassment", "physical harassment" and "sexual harassment". The ML algorithms which we used for both tasks are included of: Logistic Regression, Naïve Bayes, Decision Tree, Support Vector Machines, Random
Forest, Multi-layer Perceptron, and AdaBoost. The choice of a robust classifier
to detect this kind of content justifies testing of all of those different methods.

The rest of the paper is organized as follows: Sect. 2 presents some related
work to the problem. Section 3 includes the methodology, the dataset, algorithms
and word embeddings employed in the work. In Sect. 4, we will go through the
experiments and the required steps for them. In Sect. 5 we show the results based
on our experiments. Finally, Sect. 6 presents the conclusion and future works.

2   Related Work

Automatically detecting content containing sexual harassment could be the basis
for removing it, or flagging it for human evaluation. Sexism as a classification
task was introduced by [23], in which they collected tweets around the famous
Australian TV show My Kitchen Rules and the hashtag #mkr, annotated 16
thousand tweets, and categorized them as racist, sexist, or neither. In the next
step, different methods were performed in the task, such as character level grams
and word grams, and logistic regression with 10-fold cross-validation was run
to classify the tweets. In 2017, in the other work [11], they categorized tweets
exhibiting sexism, using several classifiers if they have one of the three following features: "protective paternalism", "complementary gender differentiation"
and "heterosexual intimacy". Taking advantage of deep neural networks models happened in [1], which used the dataset collected in [23], input into different

types of convolutional and recurrent neural networks, combining long-short term memory (LSTM) with random embedding. Among the combinations, gradient boosted decision Tree had the best performance.

The other work on this problem is considering the tweets towards top feminists and collects a range of online abuse and categorize them into classes such as sexual harassment, physical threats, flaming and trolling, stalking, electronic sabotage, impersonation and defamation [14]. In the same manner, [21] came up with new categories, with name, definitions, points, hints, and examples. They ran a pilot study of 50 tweets labeled by one male and 12 women non-activist, calculate the Kappa score. The application of affect classification on different sexist has been showed in [20] and they categories by focusing on the sexist dataset included of four categories of: Indirect harassment, Information threat, Sexual harassment, and Physical harassment. This study aims to understand the emotion type and intensity of emotion in each category of sexual harassment, which shows there are some similarities in the physical and sexual harassment categories.

Aware of pre-processing step influence in the methods' final performance, [4] gathered messages from Twitter and proposed a new method based on person identification combined to word normalization leading to gain of performance. Other common technique to boost the harassment detection in Twitter is text augmentation, as performed by [19], who added more information to enrich the dataset and compensate the lack of training data for some specific kinds of harassment, as indirect harassment, which is more difficult to predict since it is commonly presented as sarcasm or jokes against women [20]. This method has been used in different applications like bioinformatics [13], image processing [12], computer vision [7], video and audio processing [9,15]. However, in this work, no text augmentation is performed since our main goal is to measure how robust supervised classifiers are when running on imbalanced, non-structured, and ambiguous textual data.

3 Methodology

In this section, the Twitter harassment dataset is described, as well as the nine classifiers trained to classify sexist content on social media and the Word2Vec model [16].

3.1 Data

The dataset is composed of 10,622 posts collected from Twitter in the English language, whose statistics are presented in Table 1. In this benchmark for harassment content classification, 6,374 data instances are available for training, 2,125 data instances were released for validation, while 2,123 are provided for the test. The number of those tweets which present harassment content is 3,956, and sexual harassment is the most numerous kind of hate speech in this set with 3,419 data instances. Besides sexual, there is also indirect and physical harassment

(a) Harassment content.

(b) Indirect harassment.

(c) Physical harassment.

(d) Sexual harassment.

Fig. 1. Word clouds for the tweets which present each kind of harassment content.

content in the data. The former concerns sexism and indirect offenses, involving conduct that is not directed at a particular individual but results in overall poisoned content, like the invasion of personal space, suggestive remarks or sounds, offensive jokes, ridicule, innuendo; while the latter refers to violent threats on the social media. By analyzing the table, it is possible to notice that the classes are imbalanced. So special attention is required in the pre-processing step for balancing the classes of the dataset.

Table 1. Dataset Statistics. 'Indirect H.' stands for indirect harassment; 'Physical H.' refers to physical harassment; and 'Sexual H.' means sexual harassment.

Subset	Data instances	Harassment	Indirect H.	Physical H.	Sexual H.
Training	6,374	2,713	55	76	2.582
Validation	2,125	632	71	36	525
Test	2,123	611	197	100	312
Total	10,622	3,956	323	297	3,419

The content of the Tweets in this dataset comprises coarse language and swear words, as shown in Fig. 1. Each kind of harassment presents specific language constructions. For instance, the words in tweets with indirect harassment content are related to sexism and offensive jokes (Fig. 1b), whereas physical harassment tweets present violent threats (Fig. 1c), and sexual harassment posts are composed of terms and expressions with a sexual connotation (Fig. 1d).

3.2 Algorithms

The contribution of this work is on both binary and multi-class classification tasks. The goal of binary classification is to learn a function $F(x)$ that minimizes the misclassification probability $P(yF(x) < 0)$, where y is the class label with +1 for positive and −1 for negative [17]. On the other hand, multi-class classification aims at assigning one of several classes to each data instance. This task finds a model which maps an input vector x to binary vectors y, where $y \in \{0, 1\}$. Since the detection of harassment is a brand-new task of NLP, we proposed to compare supervised classifiers performance for this problem. Therefore, different algorithms were chosen to be tested in our study. Among the methods for classification, we used the supervised models, choosing based on [25], below:

- Logistic regression [17];
- Gaussian Naïve Bayes [17];
- Decision Trees (DTs) [17];
- linear, gaussian, and polynomial Support Vector Machines (SVM) [2];
- Random forest (RF) [3];
- Multi-Layer Perceptron (MLP) [17];
- AdaBoost [8].

3.3 Word Embeddings

Word embeddings are continuous representations of words and their semantic features in low-dimensional vector spaces [5,10]. It is capable of capturing the context of a word in a document, semantic and syntactic similarity, and relation with other words. Word2Vec [16] is one of the most popular techniques to learn word embeddings, whose representations can be obtained using two methods (both involving Neural Networks): Common Bag-of-Words (CBOW) and Skip-gram, using either hierarchical softmax or negative sampling.

CBOW takes the context of each word as the input and tries to predict the word corresponding to the context [5], minimizing the values for the following loss function:

$$loss = -\log(p(\overrightarrow{w_t}|\overrightarrow{W_t}))\qquad(1)$$

in which w_t is the target word in the sequence of words W_t.

Skip-gram, otherwise, usually tries to achieve the reverse of what the CBOW model does [16]. It tries to predict the source context words (surrounding words) given a target word (the center word).

288   M. Saeidi et al.

4 Experiments

The experiments pipeline started with the dataset pre-processing, which comprises tokenization, stop-words removal, stemming or lemmatization, and word vectors extraction. We have tested different approaches to come up with features for the classifiers and, after that, we performed the classification tasks, validating the models on a 10-fold cross-validation setup.

4.1 Pre-processing

The start point working with tweets dataset is preprocessing, which in this task includes removing hyperlinks, hashtags, numbers, and punctuation marks, including hashtags and at signs. For the pre-processing step, we used the functions provided by NLTK library[1]. They are: tokenization with 'word_tokenize'; stop-words removal for the English language; lemmatization with Wordnet Lemmatizer; or stemming with SnowBall stemmer. In addition to the standard English stop words from Scikit-learn[2] package, we have removed from the text the Twitter acronyms and HTML tags showed in Table 2 since they play a role as noise in the dataset. As we tested two different representations for the words in the tweets, the lemmatized the words were used to extract their representations in the Word2Vec model, whereas the stemmed forms of the same words were used to yield their term frequency-inverse document frequency (TF-IDF) [18] representation with the 'TfidfVectorizer' function from Scikit-learn.

Table 2. Acronyms and HTML tags assigned as stop words in the tweets dataset pre-processing.

Additional stop words
'don', 'http', 'amp', 'cc', 'rt', 'x89', 'x8f', 'x95', 'x9d', 'na', 'im', 'co', 'id'

After that, the TF-IDF vectors were computed for the posts to classify. The original dataset (prior training and test sets) had 19,945 words, which lead to a large and sparse matrix. Thus, it was needed to decrease the dimension of this structure to reduce the time and computational complexity in the classification. To do so, we pruned the number of terms, selecting the most relevant ones. We started selecting the 25 most relevant according to TF-IDF scores, and increased this number up to 50, running the 9 supervised classifiers described in Sect. 3.2, with 10-fold cross-validation, and measuring the performance metric of accuracy for each model. On the validation set, the achieved accuracies for the supervised classifiers while varying the number of features from 25 to 50 with a stride of 5 suggest that 45 is the best number of features extracted from the tweets, as demonstrated in the Table 4. Table 3, additionally, shows the 45 most relevant words in the data when using TF-IDF scores to extract features from the text.

---

[1] https://www.nltk.org/.
[2] https://scikit-learn.org/.

Table 3. The 45 most relevant words in the dataset after stemming with SnowBall stemmer.

Words selected according to TF-IDF scores
'alway', 'ass', 'ava', 'becaus', 'bitch', 'black', 'chop', 'cumshot', 'friend',
'fuck', 'girl', 'good', 'got', 'guy', 'horni', 'just', 'know', 'like',
'littl', 'look', 'love', 'make', 'nake', 'nude', 'peopl', 'porn', 'pussi',
'realli', 'right', 'sassi', 'say', 'sex', 'shame', 'shit', 'slut', 'think',
'time', 'today', 'video', 'want', 'watch', 'whore', 'women', 'xxx', 'year'

As Table 1 showed, this dataset is imbalanced. Hence to avoid the dominant influence of the major class on the outcomes of the algorithms, we have employed SMOTE (Synthetic Minority Over-sampling Technique) [6] before running the models. SMOTE makes the minor classes equal to the major class by yielding synthetic samples of the instances with less recurrent labels. It selects similar data points and randomly changes the values one column at a time in the range of the difference to the nearest points.

4.2   Task A – Binary Classification

As mentioned in Sect. 4.1, we tested different numbers of features extracted from text, according to TF-IDF scores. For each set of features, we have run the models to classify whether a tweet has harassment content or not. The accuracy scores for each model on the validation set are presented in Table 4. Among the classifiers, LR RF, Linear SVM, Gaussian SVM, MLP, and Adaboost reached performances above 0.90 of accuracy. GNB and DT did not perform as good as the other ones since they are based on probabilities. So as many words overlap in both classes, they lead to lower performance on assigning the correct label to the data points. Polynomial SVM with degree 2 has also not performed well since the polynomial kernel is likely to overfitting on classification.

Table 4. Accuracy values for the supervised methods on the validation set using features extracted by TF-IDF scores.

Classifier	20 F.	25 F.	30 F.	35 F	40 F.	45 F.	50 F.
Logistic Regression	0.896	0.904	0.905	**0.907**	**0.907**	**0.907**	0.906
Gaussian Naïve Bayes	0.838	0.840	**0.844**	0.842	0.831	0.825	0.822
Decision Trees	0.887	0.888	**0.894**	0.888	0.891	0.888	0.890
Random Forest	0.889	0.894	0.896	0.896	0.899	**0.900**	0.899
Linear SVM	0.896	0.904	0.904	**0.908**	0.906	0.906	0.904
Gaussian SVM	0.887	0.897	0.896	0.901	0.802	**0.903**	0.895
Polynomial SVM	**0.788**	0.702	0.702	0.702	0.702	0.702	0.702
MLP	0.896	0.702	0.903	0.702	**0.906**	0.901	0.903
AdaBoost	0.891	0.903	0.904	0.905	0.904	**0.906**	0.898
Average	0.874	0.848	0.872	0.850	0.861	0.871	0.869

We validated all the models on 10-fold cross-validation before classifying the validation set. Figure 2 depicts the learning curves for each model on k-fold cross-validation with $k = \{2, 4, 6, 8, 10\}$. It is possible to notice that Linear and Gaussian SVMs hit the highest scores, whereas GNB and Polynomial SVM hit the two least. The remaining classifiers surpassed 85% of accuracy.

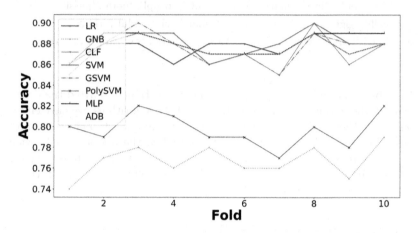

Fig. 2. Classifiers learning curve on a 10-fold cross-validation over the validation set with 45 TF-IDF features.

4.3   Task B – Multi-class Classification

This part is a multi-class classification of online harassment Tweets into three categories of "indirect harassment", "sexual harassment", and "physical harassment". We have run the same algorithms over the dataset using the 45 features extracted in Sect. 4.1. Physical harassment was the hardest content to classify since it is the less frequent one among the classes in the dataset. On the other hand, sexual harassment was easier to detect in the tweets, as shown in Table 5, because sexual treats are explicitly written in the posts. RF results surpassed the remaining methods' on the three labels, hitting 0.935 of average accuracy. As noticed in Sect. 4.2, Gaussian Naïve Bayes and Polynomial SVM were the worst-performing models.

4.4   Classification with Word Embeddings

To verify how good word embeddings are to detect harassment content on social media, we have trained a 50-dimensional CBOW model on the English Wikipedia corpus, which is the largest text collection freely available to collect and train models on. The Table 6 presents the results on the validation set.

Table 5. Accuracy values on the three classes of harassment over the validation set using 45 features extracted by TF-IDF scores.

Classifier	Indirect H.	Physical H.	Sexual H.	Average
Logistic Regression	0.913	0.913	**0.949**	**0.925**
Gaussian Naïve Bayes	0.827	0.807	0.944	0.859
Decision Trees	0.899	**0.921**	0.948	0.923
Random Forest	**0.923**	**0.931**	**0.952**	**0.935**
Linear SVM	0.912	0.872	**0.949**	0.911
Gaussian SVM	0.893	0.854	**0.949**	0.899
Polynomial SVM	0.779	0.736	**0.949**	0.821
MLP	**0.918**	0.877	**0.949**	0.915
AdaBoost	0.913	0.897	0.948	0.919
Average	0.886	0.868	0.949	0.901

Table 6. Accuracy values for the Word2Vec embeddings on the Validation set.

Classifier	Harassment	Indirect H.	Physical H.	Sexual H.	Average
Logistic Regression	**0.822**	0.834	0.846	**0.949**	0.863
Gaussian Naïve Bayes	0.750	0.834	0.814	**0.949**	0.837
Decision Trees	0.706	0.957	0.957	0.939	0.890
Random Forest	0.787	**0.963**	**0.973**	**0.950**	0.918
Linear SVM	0.731	0.711	0.892	0.948	0.821
Gaussian SVM	0.727	**0.966**	**0.982**	0.776	0.863
Polynomial SVM	0.822	0.825	0.849	**0.949**	0.861
MLP	**0.819**	0.913	0.870	0.813	0.854
AdaBoost	0.786	0.935	0.938	0.948	**0.902**
Average	0.772	0.882	0.902	0.913	0.868

5   Results and Discussion

The results of the test data are presented in this section. The final experiments were performed on the test set, which is the same data from the challenge proposed by the SIMAH Competition. We have compared the performance of TF-IDF feature vectors to Word2Vec embeddings and reported the best scoring methods on both set of features.

5.1   Classification with TF-IDF Vectors

We performed both classification tasks on the 45 features extracted by TF-IDF values and measured the accuracy for each running on the test set. For the binary classification, LR, Linear SVM, and Adaboost surpassed the remaining algorithms.

Otherwise, DT and RF presented the highest results when ran over the three kinds of harassment content. The complete results are shown in the Table 7.

Table 7. Accuracy values for the TF-IDF vectors.

Classifier	Harassment	Indirect H.	Physical H.	Sexual H.	Average
Logistic Regression	**0.814**	0.891	0.834	0.859	0.849
Gaussian Naïve Bayes	0.776	0.871	0.818	**0.876**	0.835
Decision Trees	0.799	0.879	**0.887**	**0.876**	**0.860**
Random Forest	0.813	0.886	**0.895**	**0.876**	**0.867**
Linear SVM	**0.814**	**0.896**	0.835	0.859	0.851
Gaussian SVM	0.810	0.870	0.817	0.859	0.839
Polynomial SVM	0.782	0.809	0.759	0.859	0.802
MLP	0.811	**0.900**	0.861	0.859	0.857
AdaBoost	**0.814**	0.887	0.867	0.862	0.857
Average	0.803	0.876	0.841	0.865	0.846

In order to evaluate the outcomes of the algorithms, we also have computed the macro F1 scores for each class. This performance measure is the harmonic mean of precision and recall and it aims to evaluate the classification performance on all the classes, without taking into account whether the data is balanced or not. As Table 8 shows, physical harassment is the hardest kind of offensive content to detect. DT and RF hit better results on physical and sexual harassment, but on average LR and Linear SVM were the best scoring methods. The results in the Table 8 are lower than the ones in the Table 7, since accuracy scores are sensitive to the presence of the major classes [17].

Table 8. Macro F1 values for the TF-IDF vectors on all the classes.

Classifier	Harassment	Indirect H.	Physical H.	Sexual H.	Average
Logistic Regression	0.726	**0.775**	0.598	0.793	**0.723**
Gaussian Naïve Bayes	0.691	0.738	0.593	0.798	0.705
Decision Trees	0.697	0.673	**0.613**	**0.802**	0.696
Random Forest	0.717	0.686	**0.623**	**0.813**	0.709
Linear SVM	**0.728**	**0.773**	0.596	0.793	**0.722**
Gaussian SVM	**0.737**	0.745	0.597	0.793	0.718
Polynomial SVM	0.668	0.687	0.568	0.793	0.679
MLP	0.726	0.763	0.595	0.793	0.719
AdaBoost	0.724	0.726	**0.613**	0.798	0.715
Average	0.712	0.729	0.599	0.797	0.709

5.2   Classification with Word Embeddings

When we ran the supervised methods on the dataset represented by the word vectors yielded by the Word2Vec model, the accuracy values for the binary classification were lower than the ones hit on the TF-IDF features, as the Table 9 shows. The main reason for this difference of performance is the nature of the training data input to the Word2Vec model. As Wikipedia, in general, presents only formal language the swear words in the tweets do not have meaningful representations in the model. Furthermore, if we have trained Word2Vec on the data used in this study, the word representations could lead to overfitting. Among the supervised algorithms, RF and Gaussian SVM hit best results in most of the tries.

Table 9. Accuracy values for Word2Vec Embeddings.

Classifier	Harassment	Indirect H.	Physical H.	Sexual H.	Average
Logistic Regression	**0.781**	0.837	0.822	0.858	0.825
Gaussian Naïve Bayes	0.735	0.831	0.809	**0.860**	0.809
Decision Trees	0.660	0.899	0.933	0.857	0.837
Random Forest	0.760	**0.909**	**0.948**	0.858	**0.869**
Linear SVM	0.727	0.894	0.847	**0.859**	0.832
Gaussian SVM	0.707	**0.907**	**0.952**	0.847	0.853
Polynomial SVM	**0.781**	0.843	0.836	**0.859**	0.830
MLP	0.768	0.869	0.848	0.669	0.789
AdaBoost	0.752	0.899	0.912	0.857	**0.855**
Average	0.741	0.876	0.879	0.836	0.833

Table 10. Macro F1 values when using Word2Vec Embeddings.

Classifier	Harassment	Indirect H.	Physical H.	Sexual H.	Average
Logistic Regression	**0.731**	0.666	**0.599**	**0.791**	**0.696**
Gaussian Naïve Bayes	0.692	0.665	0.589	0.794	0.685
Decision Trees	0.608	0.579	0.575	0.782	0.636
Random Forest	0.681	0.527	0.528	0.785	0.630
Linear SVM	0.533	**0.667**	0.595	0.793	0.647
Gaussian SVM	0.414	0.475	0.487	0.458	0.458
Polynomial SVM	**0.729**	**0.670**	**0.604**	0.793	**0.699**
MLP	0.716	0.513	0.523	0.583	0.583
AdaBoost	0.697	0.652	0.589	0.787	0.681
Average	0.644	0.601	0.565	0.729	0.635

By computing the macro F1 scores (Table 10), we can see that LR and Polynomial SVM presented the highest average values of F1 score, as well as the top scores on most of the classes. The difference between accuracy and macro F1 scores is due F1 score penalizes low recall or low precision, at the same time as accuracy values keep high when the model hits good results in the major classes. Concerning the harassment content in the tweets, sexual harassment was the easiest one to detect, mostly because the dataset presents it more frequently.

To assure the performance of our approaches for harassment content on social media, we computed the average accuracies for the three kinds of offensive tweets using the scores from the Table 7 and compared to the results reported by [19]. The 4 lines at the top and at the bottom of the Table 11 were extracted from their work. In the top 4 rows of the table, AWR stands for a text augmentation technique. In our experiments, we did not perform this kind of data enrichment. As it is possible to notice, our results on the 45 TF-IDF vectors surpassed the results reported by Sharifirad et al., 2018, without text augmentation. However, when compared to their augmented dataset results, our approaches still have to be improved. Anyway, the rise on accuracy compared to the results on literature which do not make use of text augmentation shows that pre-processing tasks have a huge influence on the results, and our approach is stable and robust to detect harassment content on Twitter.

Table 11. Average accuracy values compared to the results in the literature.

Classifier	Average
CNN + AWR	**0.980**
LSTM + AWR	**0.980**
SVM + AWR	0.920
Naïve Bayes + AWR	0.940
Logistic Regression	0.861
Gaussian Naïve Bayes	0.855
Decision Trees	0.880
Random Forest	0.885
Linear SVM	0.863
Gaussian SVM	0.848
Polynomial SVM	0.809
MLP	0.873
AdaBoost	0.872
CNN	0.750
LSTM	0.740
SVM	0.680
Naïve Bayes	0.600

6   Conclusion and Future Works

In this paper, we aim to improve the classification performance of different models on Twitter data. By considering different types of online harassment on this social media, we used 9 supervised algorithms to categorize them. We also performed an empirical comparison between TF-IDF feature vectors and Word2Vec embeddings trained on the Wikipedia English corpus. We have validated all the runnings of the algorithms on a 10-fold cross-validation process, applied them over a validation set, and after this step, we have classified the posts collected from Twitter.

Among the 9 models to categorize offensive content online, DT, RF, and Linear SVM showed the best results. DT and RF classify instances according to information gain, whereas Linear SVM finds the hyperplane which maximizes the boundary decision between the classes. In social media content, the information gain of the first two algorithms is influenced by the frequency that each word appears associated to some label, while in the last one the hyperplane is calculated according to optimization functions (see [2] for more details). We also have noticed that embeddings trained on textual corpus whose domain is different from the target data tend to decrease the classification performance. It also shows that these representations are not robust to perform well regardless of the domain of the data in which the predictions are performed over.

This automatic classification into three categories of "indirect harassment", "sexual harassment", and "physical harassment" will significantly improve the process of detecting these types of speech on social media by reducing the time and effort required by human beings. We also expect this word to leverage the discussions on harassment detection on Twitter, as well as other social networks. Furthermore, as future work, we plan to test deep architectures on our set of features, as well as testing different strategies of data augmentation, especially the ones based on pseudo-labeling [24].

Acknowledgements. This study was financed in part by the Coordenação de Aperfeiçoamento de Pessoal de Nível Superior – Brasil (CAPES)– Finance Code 001 –; São Paulo Research Foundation (FAPESP) under grant number 2018/09465-0; Natural Sciences and Engineering Research Council of Canada (NSERC); and Dalhousie University.

References

1. Badjatiya, P., Gupta, S., Gupta, M., Varma, V.: Deep learning for hate speech detection in tweets. In: Proceedings of the 26th International Conference on World Wide Web Companion, pp. 759–760. International World Wide Web Conferences Steering Committee (2017)
2. Boser, B.E., Guyon, I.M., Vapnik, V.N.: A training algorithm for optimal margin classifiers. In: Proceedings of the Fifth Annual Workshop on Computational Learning Theory, COLT 1992, pp. 144–152. ACM, New York (1992). https://doi.org/10.1145/130385.130401, http://doi.acm.org/10.1145/130385.130401

3. Breiman, L.: Random forests. Mach. Learn. **45**(1), 5–32 (2001). https://doi.org/10.1023/A:1010933404324
4. Bretschneider, U., Wöhner, T., Peters, R.: Detecting online harassment in social networks. In: Proceedings of the International Conference on Information Systems - Building a Better World through Information Systems, ICIS 2014, Auckland, New Zealand, 14–17 December 2014 (2014). http://aisel.aisnet.org/icis2014/proceedings/ConferenceTheme/2
5. Camacho-Collados, J., Pilehvar, M.T.: From word to sense embeddings: a survey on vector representations of meaning. J. Artif. Intell. Res. **63**, 743–788 (2018)
6. Chawla, N.V., Bowyer, K.W., Hall, L.O., Kegelmeyer, W.P.: Smote: synthetic minority over-sampling technique. J. Artif. Intell. Res. **16**, 321–357 (2002)
7. Chen, C.h.: Handbook of Pattern Recognition and Computer Vision. World Scientific, Singapore (2015)
8. Freund, Y., Schapire, R.E.: Experiments with a new boosting algorithm. In: Proceedings of the Thirteenth International Conference on International Conference on Machine Learning, ICML 1996, pp. 148–156. Morgan Kaufmann Publishers Inc., San Francisco (1996). http://dl.acm.org/citation.cfm?id=3091696.3091715
9. Gambäck, B., Sikdar, U.K.: Using convolutional neural networks to classify hate-speech. In: Proceedings of the first workshop on abusive language online, pp. 85–90 (2017)
10. Huang, C.H., Yin, J., Hou, F.: A text similarity measurement combining word semantic information with TF-IDF method. Jisuanji Xuebao (Chin. J. Comput.) **34**(5), 856–864 (2011)
11. Jha, A., Mamidi, R.: When does a compliment become sexist? Analysis and classification of ambivalent sexism using twitter data. In: Proceedings of the Second Workshop on NLP and Computational Social Science, pp. 7–16 (2017)
12. Kamavisdar, P., Saluja, S., Agrawal, S.: A survey on image classification approaches and techniques. Int. J. Adv. Res. Comput. Commun. Eng. **2**(1), 1005–1009 (2013)
13. Larranaga, P., et al.: Machine learning in bioinformatics. Briefings in Bioinform. **7**(1), 86–112 (2006)
14. Lewis, R., Rowe, M., Wiper, C.: Online abuse of feminists as an emerging form of violence against women and girls. Br. J. Criminol. **57**(6), 1462–1481 (2016)
15. Lu, X., Zheng, B., Velivelli, A., Zhai, C.: Enhancing text categorization with semantic-enriched representation and training data augmentation. J. Am. Med. Inform. Assoc. **13**(5), 526–535 (2006)
16. Mikolov, T., Chen, K., Corrado, G.S., Dean, J.: Efficient estimation of word representations in vector space. In: Proceedings of ICLR (2013)
17. Mitchell, T.M.: Machine Learning, 1st edn. McGraw-Hill Inc., New York (1997)
18. Salton, G., McGill, M.J.: Introduction to Modern Information Retrieval. McGraw-Hill Inc., New York (1986)
19. Sharifirad, S., Jafarpour, B., Matwin, S.: Boosting text classification performance on sexist tweets by text augmentation and text generation using a combination of knowledge graphs. In: Proceedings of the 2nd Workshop on Abusive Language Online (ALW2), pp. 107–114. Association for Computational Linguistics, Brussels, October 2018. https://www.aclweb.org/anthology/W18-5114
20. Sharifirad, S., Jafarpour, B., Matwin, S.: How is your mood when writing sexist tweets? Detecting the emotion type and intensity of emotion using natural language processing techniques. arXiv preprint arXiv:1902.03089 (2019)
21. Sharifirad, S., Matwin, S.: When a tweet is actually sexist. a more comprehensive classification of different online harassment categories and the challenges in NLP. arXiv preprint arXiv:1902.10584 (2019)

22. Vandenbossche, L., Spruyt, B., Keppens, G.: Young, innocent and sexist? Social differences in benevolent and hostile sexist attitudes towards women amongst flemish adolescents. Young **26**(1), 51–69 (2018). https://doi.org/10.1177/1103308817697240
23. Waseem, Z., Hovy, D.: Hateful symbols or hateful people? predictive features for hate speech detection on twitter. In: Proceedings of the NAACL student research workshop, pp. 88–93 (2016)
24. Weston, J., Ratle, F., Mobahi, H., Collobert, R.: Deep learning via semi-supervised embedding. In: Montavon, G., Orr, G.B., Müller, K.-R. (eds.) Neural Networks: Tricks of the Trade. LNCS, vol. 7700, pp. 639–655. Springer, Heidelberg (2012). https://doi.org/10.1007/978-3-642-35289-8_34
25. White, C.: Atlantic geoscience society abstracts: 44th annual colloquium and general meeting 2018. Atlantic Geol. J. Atlantic Geosci. Soc./Atlantic Geol. Revue de la Société Géoscientifique de l'Atlantique **54**, 81–132 (2018)

Learning to Detect Online Harassment on Twitter with the Transformer

Margarita Bugueño$^{(\boxtimes)}$ and Marcelo Mendoza

Instituto Milenio Fundamentos de Los Datos, Departamento de Informática,
Universidad Técnica Federico Santa María, Santiago, Chile
margarita.bugueno.13@sansano.usm.cl, mmendoza@inf.utfsm.cl

Abstract. This paper describes our submission to the SIMAH challenge (SocIaL Media And Harassment). The proposed competition addresses the challenge of harassment detection on Twitter posts as well as the identification of a harassment category. Automatically detecting content containing harassment could be the basis for removing it. Accordingly, the task is considered to be an essential step to distinguishing different types of harassment provides the means to control such a mechanism in a fine-grained way. In this work, we classify a set of Twitter posts into non-harassment or harassment tweets where the last ones are classified as indirect harassment, sexual harassment, or physical harassment. We explore how to use self-attention models for harassment classification in order to combine different baselines' outputs. For a given post, we use the transformer architecture to encode each baseline output exploiting relationships between baselines and posts. Then, the transformer learns how to combine the outputs of these methods with a BERT representation of the post, reaching a macro-averaged F-score of 0.481 on the SIMAH test set.

Keywords: Harassment detection · Self-attention models · Social media

1 Introduction

Social networks have been the battlefield of users for many years. Natural language reveals values, perspectives, and emotions. Among all types of hate and abusive language, harassment tweets have been very dominant on social media platforms such as Twitter and Facebook. The Canadian Human Rights Commission[1] defines harassment as a form of discrimination which includes any unwanted physical or verbal behavior that offends or humiliates someone.

Harassment can also be a way to silence the speech of others, especially women [8]. The extensive debate about the use of social media has allowed identifying that hate language is a catalyst for discrimination and social segregation[2]. Thus, the concepts of sexism and harassment are very related.

---

[1] https://www.chrc-ccdp.gc.ca/eng/content/what-harassment-1.
[2] The Washington Post: http://tiny.cc/sltgcz.

© Springer Nature Switzerland AG 2020
P. Cellier and K. Driessens (Eds.): ECML PKDD 2019 Workshops, CCIS 1168, pp. 298–306, 2020.
https://doi.org/10.1007/978-3-030-43887-6_23

There are many definitions of sexism. For example, sexism is a form of discrimination against women [9]. However, sexism is not just about discrimination, and what is happening on social networks is far from this definition. This observation drives the need to conceptualize sexism and harassment on social media. The categorization of sexism in social media into hostile or benevolent sexism has changed over the years, giving way to a more specific vision in terms of the types of harassments.

Waseem and Hovy [18] collected hateful tweets, categorizing them into sexism, racism, or neither. Later, Jha and Mamidi [9] focused on sexist tweets and proposed two categories of hostile and benevolent sexism. These categories mutated to a finer granularity in [15]. That work proposed a distinct category of sexism, including indirect harassment, information threat, sexual harassment, and physical harassment. The work proposes a more comprehensive and in-depth categorization of online harassment in social media. From that work, due to the significant problem to apply automatic methods to strongly unbalanced data, techniques such as text augmentation and text generation [13] have been applied to achieve performance improvements.

In this paper, we focus on the categories proposed in [15]. Our approach applies self-attention models for harassment classification in order to combine different baselines outputs with a BERT-based representation of each tweet [5]. To accomplish this task, we use the transformer [16], a successful deep learning architecture used for translation in natural language processing. The transformer can detect which part of the data ingested is useful to solve a given task. As a consequence, the encoding learned by the transformer can consistently produce a better prediction of the harassment label than the ones provided by the baselines. Experiments on the proposed dataset show that our proposal reaches a macro-averaged F-score of 0.481 with an accuracy of 0.764.

This work is organized as follows. In Sect. 2 we present a literature review. In Sect. 3 we introduce our proposal. Sections 4 and 5 present the experimental configurations with the results. We conclude in Sect. 6 providing remarks and outlining future work.

2   Related Work

The massiveness with which users interact on social networks has driven new analytical tasks. Among them, the detection of hate speech in social media has captured the interest of the scientific community. Due to the massive volume of social media data, the need for automatic hate speech detection methods has become increasingly urgent.

Several works have approached the problem from a classic machine learning perspective [3,4,17]. These jobs generally combine features extracted from messages with features retrieved from user profiles, using a feature-engineering strategy. Combining both sources of information, several of these methods train supervised learning algorithms like support vector machines or random forests. A limitation of many of these works is that they are sensitive to the imbalance of labeled data. In practice, many of these methods fail to generalize well to other datasets, which limits their use in real environments. A thorough review of these types of methods is addressed in [12].

More sophisticated models, such as those studied in deep learning, have also been applied to the problem of hate speech detection. One of the advantages of deep learning architectures is that they allow the neural network to learn an adequate representation for the problem. The use of text encoders has offered advantages to these types of models over conventional machine learning models. For example, convolutional networks [7] have shown good results in the Wasem and Hovy dataset [18]. Recurrent neural networks have also shown good results in this dataset, based on the GRU architecture [20]. Nearly perfect results in this dataset were also reported using deep learning by Badjatiya et al. [2]. Unfortunately, many of these models have overfitting problems, and then, they are not transferable to production. Recently, Arango et al. [1] showed that there are also problems in the generation of these datasets considered as standard for the evaluation of this type of tasks. Among these problems, the most worrying is the population bias used to generate the samples that make up the dataset. These works show that the hate speech detection problem is far from being solved.

A significant problem that these datasets have is the imbalance between classes. The hate speech detection must be carried out in scenarios where most of the conversations are mostly neutral, and the harassment is exceptional. However, not being exceptional is less critical. The consequences that harassment and hate language produce on social network users is fierce. To address the problem of imbalance, in [13], the authors use techniques to increase and generate texts that allow generating training data with balanced classes. In this same line, Sharifirad et al. [15] show that a promising way to address the problem is to define a finer type of harassment. Based on this latest work, the Simah challenge defines a dataset with three types of harassment, which will be addressed by our work.

Hate speech has many variants and has been at the center of attention of many researchers in recent years. Recently, the relationship between hate speech and mood detection has shown to be a promising way of research [14], which would allow linking two tasks that apparently might seem unconnected, sentiment analysis, and hate speech detection. The advances shown in the consolidation of hate speech lexicons have also been impressive [4], which would allow the flourishing of unsupervised techniques to address this task such as graph-based techniques [10].

Far from showing itself as a task with mature and robust solutions, this task shows many challenges. For more details on all hate speech detection variants, the reader is recommended to review the Fortuna and Nunes survey [6].

3   Proposal

The SIMAH competition defines two sub-tasks. The first task is a binary classification to separate online harassment tweets versus non-harassment tweets.

The second task is a multi-class classification of online harassment tweets into indirect harassment, sexual harassment, and physical harassment. Our proposal jointly faces both tasks without the need to have a phase for each sub-task. The trained system employs an adaptation of the transformer [16] in order to exploit its self-attention modules.

3.1   Applying Self-attention

For a given tweet, each baseline provides a label from the set of possible tags, i.e., Non-Harassment, Indirect Harassment, Sexual Harassment, or Physical Harassment (Non-H, IH, SH, PH). To ingest these outputs into the transformer, we encode each label using a one-hot encoding vector of the class, producing orthogonal vectors for different categories. To ingest the text of the tweet, we use its BERT vector (Bidirectional Encoder Representations from Transformers) [5] which are computed at sentence-level. BERT vectors are provided by Google research in a library as a service[3]. Our model uses four baselines. We concatenate these four vectors with the BERT vector of the tweet, which is ingested into the transformer. We use the encoder of the transformer using two layers, each one with four attention heads. Each layer has an attention module and a position-wise feed-forward layer. The position-wise layer is a crucial module that allows to code from which baseline the data is encoded. The outputs of the encoders are concatenated, and then by applying a Hadamard product between them, we obtain a state vector that represents what the transformer learned from the baselines and the tweet. Then, the vector is ingested into a softmax layer, who is in charge of producing an output. The model is depicted in Fig. 1.

Despite the original architecture [16] was proposed as a sequence transduction model based on an encoder-decoder structure, we only use the encoder of the transformer with its attention mechanisms. Hence, the transformer uses stacked self-attention and fully connected layers. The encoder used is composed of a stack of two layers. The transformer uses a residual connection between each module and a normalization. The links inside the transformer are produced by inputs and outputs of the same dimension. The attention mechanism of the transformer is wired using a scaled dot-product operator. Then, multi-head attention consists of several attention layers running in parallel. After the attention module, a position-wise feedforward module is applied to each position, consisting of two linear transformations with ReLU activations in between. The output sequence produced by the encoder gives five vectors, one for each input ingested into the encoder, which are then combined using the Hadamard product.

3.2   Baselines

One of the baseline models is based on convolutional neural networks (CNN) and the other three on recurrent neural networks (RNN). One RNN used one recurrent layer while the others used two layers (as the CNN). Note that in the

---

[3] https://github.com/google-research/bert#pre-trained-models.

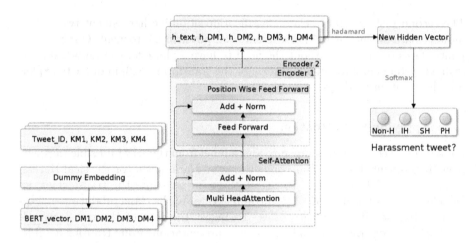

Fig. 1. To apply self-attention, each baseline is encoded and concatenated with the BERT vector of the tweet. This encoding is ingested into the transformer using two levels of encodings, each one with a self-attention module and four attention heads. At the output of the encoder, we apply a Hadamard product. The resulting vector is ingested in a softmax layer producing the output.

case of RNNs, we used GRU layers. For the RNN baselines, the output was produced using a softmax and focal loss as loss function while for CNN, we used categorical cross-entropy. Table 1 shows the parameters of each architecture.

4   Experiments

4.1   Data

The dataset provided for this challenge contains 10622 annotated tweets, split into training, validation, and testing partitions, as it is shown in Table 2. The competition has two related tasks: the first one is a binary classification (harassment or non-harassment tweet) and the second task is a multi-class classification of online harassment tweets into three categories: indirect harassment, sexual harassment, and physical harassment.

As social media data sources are unstructured and noisy, we need to do some transforms of the irregular input text. Accordingly, we considered stopwords removal, punctuation marks, digits removal, and text transform to lowercase. Furthermore, it is worth mentioning that we leave important question marks and exclamation marks since have proven to be helpful [19]. To process jargon, we removed emojis. In addition, HTML marks were replaced by the term <*url*>, #*word* with the term <*hashtag*>, @*word* terms by the term <*user*>, and numerical terms with <*number*>.

Table 1. Architecture of our baseline models. CNN1_cce: each convolutional layer is followed by a max_pool (stride = 3) and a batch normalization layer

Model configuration			
CNN1_cce	RNN1_focal	RNN2_focal	RNN3_focal
BatchNorm	GRU-64	GRU-128	GRU-128
conv5-128 (relu)	BatchNorm	drop-0.3	drop-0.45
drop-0.65	drop-0.3	GRU-64	BatchNorm
conv5-64 (relu)		drop-0.2	GRU-64
drop-0.65			dropout-0.2
F-100 (relu)			BatchNorm
BatchNorm			

Table 2. Distribution of tweets into training, validation, and testing data partitions across each class. *Two tweets were labeled as harassment posts but they were not classified into any valid category for the second task.

	Non-H	IH	SH	PH	Total
Training	3661	55	2582	76	6374
Validation	1493	71	525	36	2125
Testing	1512	197	312	100	2123*

4.2 Traning

Baselines. Once each tweet was preprocessed, we used GloVe [11] word embeddings (pre-trained on a Twitter corpus) to represent each word in each tweet. These 100 dimension vectors were used in the four baselines and were ingested one-at-a-time as a sequence of word vectors per tweet.

Transformer. Once the baselines' outputs were computed, we encoded each output using class vectors with 768 dimensions, to be consistent with the dimensionality of BERT. Each tweet was encoded using BERT as service[4], a library that maps a variable length-sentence to a fixed-length vector with 768 dimensions. In the transformer, we used gradient descent for parameter update. The size of the hidden units was set to 256 with a dropout of 0.3. We varied the learning rate throughout training for 100 epochs, according to recommendations provided by the transformer's authors using a warmup of 500 and a factor of 3. We used focal loss with class weights inversely proportional to each class size as a loss function.

---

[4] https://github.com/hanxiao/bert-as-service.

5 Results and Discussion

The performance of our model on the validation and testing sets is shown in Tables 3 and 4. Together with the accuracy, in Table 3, we show macro-averaged F1-score and per-class macro-averaged F1-scores as these metrics account for the class imbalance. In this way, it is easy to appreciate that accuracy accuses values that could be considered quite good considering the number of classes. However, when observing the F-score, we note that the metric is much lower than expected. The imbalance between classes explains this fact. Minority classes are the most complicated classes to detect. This fact explains why theses classes reach an F1-score of about 20% (16.7% for indirect harassment tweets).

Table 3. Results on the development and testing sets. Accuracy and F1-scores: macro-averaged and per class

	Accuracy	Macro F	Non-H	IH	SH	PH
Validation	**0.872**	**0.544**	0.926	0.202	0.853	0.195
Testing	**0.764**	**0.481**	0.880	0.167	0.681	0.196

Table 4. Results on the development and testing sets. Precision and Recall macro-averaged and per class

	Precision					Recall				
	Non-H	IH	SH	PH	Macro	Non-H	IH	SH	PH	Macro
Validation	0.936	0.224	0.827	0.174	**0.540**	0.916	0.183	0.882	0.222	**0.551**
Testing	0.868	0.214	0.637	0.214	**0.483**	0.892	0.137	0.731	0.180	**0.485**

Table 4 shows the accuracy and recall per class (Non-H: Non-Harassment, IH: Indirect Harassment, SH: Sexual Harassment, PH: Physical Harassment) as well as the macro-averaged metrics. Indeed, the precision and recall metrics are comparable between the validation and testing partitions. However, IH and PH classes have a low recall, although they have good enough precision, given the complexity of the task. This fact indicates that, of the total of examples classified as indirect or physical harassment tweets, an acceptable portion is correctly labeled, but the amount of recovered examples is insignificant. In other words, the predictions for both classes are poorly contaminated but somewhat incomplete, especially for the IH class with only 13.7%. This fact occurs due to the difference in the distribution of the examples in the training partition versus the evaluation partitions, which made it hard to recognize and extract patterns correctly.

6   Conclusions

We have presented a method based on the transformer architecture for harassment detection and classification. Experimental results show that our model can detect a substantial proportion of the hardest classes of this challenging task. Our architecture achieves a macro-averaged F1-score of 0.481 in the Simah competition dataset.

We are currently extending this work to improve its performance. One change we are making is to replace the one-hot encoders of the baselines with their confidence vectors. Another promising line is to use data augmentation techniques to handle the imbalance in minority classes. The use of SMOTE techniques is promising in this line of work.

Acknowledgements. Authors acknowledge funding from the Millennium Institute for Foundational Research on Data. Mr. Mendoza was partially funded by the project BASAL FB0821 while Ms. Bugueño was partially funded by the *Programa de Iniciación Científica* PIIC-DGIP of Universidad Técnica Federico Santa María.

References

1. Arango, A., Pérez, J., Poblete, B.: Hate speech detection is not as easy as you may think: a closer look at model validation. In: SIGIR 2019, pp. 45–54 (2019)
2. Badjatiya, P., Gupta, S., Gupta, M., Varma, V.: Deep learning for hate speech detection in tweets. In: WWW 2017, pp. 759–760 (2017)
3. Chatzakou, D., Kourtellis, N., Blackburn, J., de Cristo-Faro, E., Stringhini, G., Vakali, A.: Mean birds: detecting aggression and bullying on Twitter. In: WebSci 2017, pp. 13–22 (2017)
4. Davidson, T., Warmsley, D., Macy, M., Weber, I.: Automated hate speech detection and the problem of offensive language. In: ICWSM 2017, pp. 512–515 (2017)
5. Devlin, J., Chang, M.-W., Lee, K., Toutanova, K.: BERT: pre-training of deep bidirectional transformers for language understanding. arXiv preprint arXiv:1810.04805 (2018)
6. Fortuna, P., Nunes, S.: A survey on automatic detection of hate speech in text. ACM Comput. Surv. **51**(4), 1–30 (2018). Article no. 85
7. Gambäck, B., Sikdar, U.K.: Using convolutional neural networks to classify hate-speech. In: 1st Workshop on ACL, pp. 85–90 (2017)
8. Hess, A.: Why women aren't welcome on the internet. Pac. Stand. (2014). https://psmag.com/social-justice/women-arent-welcome-internet-72170
9. Jha, A., Mamidi, R.: When does a compliment become sexist? Analysis and classification of ambivalent sexism using Twitter data. In: 2nd Workshop on NLP and Computational Social Science, pp. 7–16 (2017)
10. Papegnies, E., Labatut, V., Dufour, R., Linarès, G.: Graph-based features for automatic online abuse detection. In: Camelin, N., Estève, Y., Martín-Vide, C. (eds.) SLSP 2017. LNCS (LNAI), vol. 10583, pp. 70–81. Springer, Cham (2017). https://doi.org/10.1007/978-3-319-68456-7_6
11. Pennington, J., Socher, R., Manning, C.D.: GloVe: global vectors for word representation. In: EMNLP 2014, pp. 1532–1543 (2014)

12. Schmidt, A., Wiegand, M.: A survey on hate speech detection using natural language processing. In: 5th Workshop on NLP for Social Media, pp. 1–10 (2017)
13. Sharifirad, S., Jafarpour, B., Matwin, S.: Boosting text classification performance on sexist tweets by text augmentation and text generation using a combination of knowledge graphs. In: 2nd Workshop on Abusive Language Online (ALW2), pp. 107–114 (2018)
14. Sharifirad, S., Jafarpour, B., Matwin, S.: How is your mood when writing sexist tweets? Detecting the emotion type and intensity of emotion using natural language processing techniques. arXiv preprint arXiv:1902.03089 (2019)
15. Sharifirad, S., Matwin, S.: When a tweet is actually sexist. A more comprehensive classification of different online harassment categories and the challenges in NLP. arXiv preprint arXiv:1902.10584 (2019)
16. Vaswani, A., et al.: Attention is all you need. In: NIPS 2017, pp. 6000–6010 (2017)
17. Waseem, Z.: Are you a racist or am i seeing things? Annotator influence on hate speech detection on Twitter. In: 1st Workshop on NLP and Computational Social Science, pp. 138–142 (2016)
18. Waseem, Z., Hovy, D.: Hateful symbols or hateful people? Predictive features for hate speech detection on Twitter. In: Proceedings NAACL, pp. 88–93 (2016)
19. Zhao, Z., Resnick, P., Mei, Q.: Enquiring minds: early detection of rumors in social media from enquiry posts. In: WWW 2015, pp. 1395–1405 (2015)
20. Zhang, Z., Robinson, D., Tepper, J.: Detecting hate speech on Twitter using a convolution-GRU based deep neural network. In: Gangemi, A., et al. (eds.) ESWC 2018. LNCS, vol. 10843, pp. 745–760. Springer, Cham (2018). https://doi.org/10.1007/978-3-319-93417-4_48

Detection of Harassment on Twitter with Deep Learning Techniques

Ignacio Espinoza[(✉)] and Fernanda Weiss

Millennium Institute for Foundational Research on Data, Universidad Técnica
Federico Santa María, Valparaíso, Chile
{ignacio.espinozav,fernanda.weiss.13}@sansano.usm.cl

Abstract. Online harassment is a common issue since the beginning of
social networks and it's still present nowadays, causing serious consequences to victims because their gender, race, sexuality, among others.
We have seen efforts to fight these behaviors creating automated systems
to detect and report this kind of bad conduct. However, these solutions
tend to perform well only on a specific type of data without generalizing
well. In this paper, we present a new dataset of harassment detection on
Twitter with four classes, presented for the SIMAH competition. Then
we apply three different deep learning architectures (CNN, LSTM, and
BiGRU) to classify these tweets showing that it is a hard problem to solve
especially because of the lack of annotated data within some classes. The
results only on the test set reach 46% in f1-score and using all data to
train gives 55% using the same metric.

Keywords: Harassment detection · Text classification · Deep learning

1 Introduction

We are currently living in an era where hate language becomes more present in
social networks, where users can comment by hiding themselves behind a profile
without fear of reprisals. This hate language is generally used to attack other
people about their sexuality, ethnicity, political affiliation, among others, and
can cause great harm to them. This problem is so serious that studies[1] indicate
that 41% of the adults in the United States have had personal experiences of
harassment or abusive behavior online.

Efforts have been made to combat harassment in which they intend to publicly publicize these reprehensible attitudes. Within these we find campaigns
like #MeToo[2] on Twitter that has fought sexual harassment and assault against
women, making public denunciations of men with high positions in Hollywood
who used their power to abuse many women. These campaigns also seek to
encourage victims to report harassment. However, it is difficult for them to

---

[1] https://www.pewinternet.org/2017/07/11/online-harassment-2017/.
[2] https://twitter.com/hashtag/metoo.

© Springer Nature Switzerland AG 2020
P. Cellier and K. Driessens (Eds.): ECML PKDD 2019 Workshops, CCIS 1168, pp. 307–313, 2020.
https://doi.org/10.1007/978-3-030-43887-6_24

make complaints for fear of repercussions. In social networks this kind of event are public and every user can see the user's interactions, so it would be very useful to have automatic harassment detection systems in social networks which trigger alarms when these situations occur in order to report bad user behaviors and take action regarding these users.

Studies related to this problem have been carried out but the results indicate that it is hard to make an accurate detection since the works of the state-of-the-art show strong overfitting to the data used for their studies. Therefore, this paper proposes different ways of detecting harassment on Twitter, showing that this is still a difficult problem to solve.

This paper is organized as follows. In Sect. 2 we start with a review of investigations in harassment classification, what approaches on harassment detection have been researched. In Sect. 3 we discuss in detail the methodological approaches taken to study the harassment on Twitter. Then, in Sect. 4 we explain the data that we use in this study. Then we explain the processing over dataset for the classification. And then we discuss experiments and results. Finally, we conclude in Sect. 5 giving our conclusion about our findings outlining future work.

2   Related Work

The harassment problem has been studied recently by [1, 2]. In [1], the authors made an in-depth characterization of the types of harassment that are more common in social networks like Twitter. This more comprehensive categorization will be used later in this work. In [2], the authors made a study focused on the mental and effective state of people who comment with some intention of harassment in their messages on Twitter. This work characterized each type of harassment with the emotions that users convey in each of them.

In previous works, the harassment was encapsulated within another problem:the hate-speech detection [3–5]. This problem typically is treated as the classification of racism, sexism or neither of them. And the harassment is considered inside the sexual class and it doesn't have the details of the meaning of each class presented in [1]. In the literature, works of the hate-speech problem found different focuses to solve the problem. Some of them take advantage of the input that they use, where they create different features embeddings to fed another classifier, from machine learning techniques [3] and deep learning [4] to adding more data than just the text [5].

Although it seems that the problem of detecting harassment and hate speech in social networks is solved, a recent study [6] have shown otherwise. There is a strong data overfitting in current state-of-the-art works and they don't have the ability to generalize with other data, as well as dataset bias.

We must also take into account that there are other social problems that have a similar approach to the problem of harassment but that affects other groups of people such as teenagers [7] and black people [8]. This could add another difficulty layer to the detection problem making harder solving this problem. Therefore it is important to continue with these studies.

3   Method

For the harassment classification task, we will use neural networks and data augmentation techniques.

We have the text of tweets and its labels, so our work is focus in the text. We take the embedding for each word in the dataset from a pre-trained embedding model GloVe [9]. After that, we take each tweet and create its embedding having one embedding for each document/tweet. To create it, we take the average of the embedding of each word of the tweet.

Once we have all embeddings, we feed a neural network with these documents embeddings. Our election was Convolutional neural network [10] and Long short-term memory [11]. The first, CNNs have recently been applied to various NLP tasks and the results were promising. They give us a good tool for pattern recognition over the text. The second, LSTM is a common neural network for NLP tasks and give us a good point of comparison with CNN.

With the setup explained above, we can try to predict the harassment classification, however we treat with an unbalanced dataset. To face this problem, we use a data augmentation technique. The technique is called SMOTE [12]. SMOTE allows to generate new artificial instances of the minority class by interpolating the real instances that belong to the minority class.

4   Experiments and Results

4.1   Dataset and Evaluation

For the harassment detection problem, we used the dataset published in [1]. This dataset allows us to make a granularity classification of three classes on the harassment case, in which each is a different approach to online harassment in social networks. The description of each class is shown below:

- **Harassment**: Indicates if a tweet contains harassment or not.
- **Indirect harassment**: If a tweet contains harassment and it's indirect. They are not directly violent or sexist and doesn't have any swear words.
- **Physical harassment**: If a tweet contains harassment and it's physical. They directly refer to the biologic, physical or mental ability of an individual.
- **Sexual harassment**: If a tweet contains harassment and it's sexual. They are violent and allude to the use of force toward sex.

In this way, we take a configuration of four classes for the classification of our proposal: "No harassment", "Indirect harassment", "Physical harassment" and "Sexual harassment". In Table 1 we show the distribution of each class in the dataset.

To evaluate the performance of our proposal we used three basic measures of statistical analysis: precision, recall and F1 score. For these measures, we used the implementation provided by the Scikit-learn python library[3]. Due to the unbalance of the classes shown above we don't use the accuracy score.

---

[3] https://scikit-learn.org/stable/modules/model_evaluation.html.

Table 1. Distribution of classes in the dataset.

Dataset	No harassment	Indirect h.	Physical h.	Sexual h.	**Total examples**
Train	3661	55	76	2582	6374
Validation	1494	71	36	525	2126
Test	1512	198	100	313	2123

4.2   Data Processing

One important thing to notice is that the tweets already came without punctuation, that means words like "don't" appear as "don t" in the dataset. This is the same for URLs where "w3.url.com" appears as "w3 url com". Having said that, we took the tweets and applied a basic preprocessing to the text: symbols and numbers were removed.

After preprocessing the text embeddings were created for each word using GloVe [9], a pre-trained embedding model. We use the implementation off Spacy library for Python[4] with the pre-trained model called 'en_vectors_web_lg', which has 300 dimensions and it's trained over common crawl texts. With Spacy we have the embeddings for each word, but we worked at tweet level which means there is one embedding for each tweet. So to pass from a word embedding to tweet embedding we averaged the word embeddings per tweet. In this way, we create an average word embedding for each tweet in the dataset. This decision was made due both ways gave similar results but the average word embeddings models took less time to train.

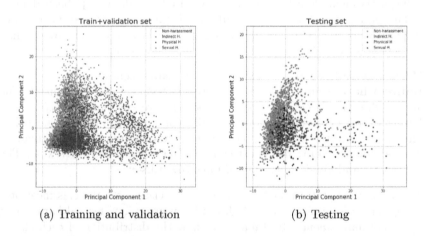

(a) Training and validation   (b) Testing

Fig. 1. Visualization of the GloVe embeddings using PCA.

To visualize the embeddings of the tweets we apply the PCA technique to bring the created 300-dimensional embeddings to two dimensions, as shown in

---

[4] https://spacy.io/models/en.

Fig. 1. The image shows that, for both training and testing data, there is no clear division between the four classes.

Due to the imbalance between classes, we opted to use the data augmentation technique SMOTE [12]. This oversampling strategy uses the local neighborhood of the real samples to create synthetic data based on the mean of near data in the space, allowing to have an equal number of samples per class. As embeddings are n-dimensional vectors we passed the collection of embeddings to SMOTE to generate the missing data in the unbalanced classes. In this new scenario, it should be easier for the model to learn the minority classes. Finally, we had a dataset of 20.616 examples as a result of applying over-sampling to both training and validation dataset.

To compare the results of applying SMOTE, we handled a version of the data with oversampling and a version without it.

4.3   Harassment Detection

The harassment detection tests were performed on three different deep learning architectures: CNN, LSTM, and Bi-GRU. Those three networks were implemented with 50 epochs, batch size of 32 examples and learning rate of 0,001. The training dataset was divided into 80% to train and 20% to validate the models. And the final results get of a test set with different data of training set. With the aforementioned configuration, the results are shown in Table 2.

Table 2. Result of three models: CNN, LSTM and Bi-GRU with (+) and without (w/o) smote.

Configuration	F1 macro average	Precision	Recall
CNN base + smote	**0.46**	0.48	0.46
CNN w/o smote	0.43	0.49	0.45
LSTM + smote	**0.46**	0.46	0.47
LSTM w/o smote	0.41	0.54	0.44
Bi-GRU + smote	**0.46**	0.48	0.47
Bi-GRU w/o smote	0.44	0.44	0.46

To see how well the classifier sare working we show in Fig. 2 the confusion matrix of the best architecture from Table 2. We can see the most difficult classes for detecting are the minorities, the indirect and physical harassment, where we just can detect it 12% and 4% of the examples respectively.

It is important to note that this dataset presents an unbalanced problem, so we use a data augmentation technique, SMOTE, to compensate this. However, it was not very useful since it only helped in a small percentage, although it improves any base technique.

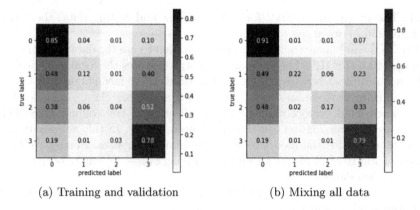

(a) Training and validation   (b) Mixing all data

Fig. 2. Confusion matrix for the CNN model + SMOTE. (0: non-harassment, 1: Indirect, 2: Physical, 3: Sexual). Figure (a) shows the result of a train with training and validation and figure (b) is using all data for training.

On the other hand, mixing the training and test set and taking it 80% to train and 20% to test we get better results, reaching 55% of F1 macro with a CNN with the same configuration mentioned above. This shows that the test data could have examples that are not representative of the training set, and doing a review of this set we verify that many examples look synthetic since several of them are the same sentence and only change one word and keeping the context. This could add noise to the classifier.

5 Conclusions

This work takes the problem of detecting harassment in social networks, specifically on Twitter. We work with three neural networks of the state-of-the-art in natural language processing, including CNN, LSTM, and Bi-GRU. Although the results are not very encouraging, they show the difficulty of solving this problem. However, this is not why this problem should be set aside. We must continue working on this since the detection of harassment in social networks can be very helpful for people who suffer from this harassment. Future work could explore sophisticated architectures who put attention on minority classes.

Acknowledgments. The authors acknowledge funding support from the Millennium Institute for Foundational Research on Data. The authors also would like to thank Global Affairs Canada and Dalhousie University for funding the research reported in this paper through the Emerging Leaders in the Americas Program (ELAP).

References

1. Sharifirad, S., Matwin, S.: When a tweet is actually sexist. a more comprehensive classification of different online harassment categories and the challenges in NLP. arXiv preprint arXiv:1902.10584 (2019)

2. Sharifirad, S., Jafarpour. B., Matwin, S.: How is your mood when writing sexist tweets? detecting the emotion type and intensity of emotion using natural language processing techniques. arXiv preprint arXiv:1902.03089 (2019)
3. Badjatiya, P., et al.: Deep learning for hate speech detection in tweets. In: Proceedings of the 26th International Conference on World Wide Web Companion. International World Wide Web Conferences Steering Committee, pp. 759–760 (2017)
4. Gambäck, B., Sikdar, U.K.: Using convolutional neural networks to classify hate-speech. In: Proceedings of the First Workshop on Abusive Language Online, pp. 85–90 (2017)
5. Pitsilis, G.K., Ramampiaro, H., Langseth, H.: Detecting offensive language in tweets using deep learning. arXiv preprint arXiv:1801.04433 (2018)
6. Arango, A., Pérez, J., Poblete, B.: Hate speech detection is not as easy as you may think: a closer look at model validation. In: Proceedings of the 42nd International ACM SIGIR Conference on Research and Development in Information Retrieval, pp. 45–54. ACM (2019)
7. Chen, Y., et al.: Detecting offensive language in social media to protect adolescent online safety. In: 2012 International Conference on Privacy, Security, Risk and Trust and 2012 International Conference on Social Computing, pp. 71–80. IEEE (2012)
8. Kwok, I., Wang, Y.: Locate the hate: detecting tweets against blacks. In: Twenty-Seventh AAAI Conference on Artificial Intelligence (2013)
9. Pennington, J., Socher, R., Manning, C.: GloVe: global vectors for word representation. In: Proceedings of the 2014 Conference on Empirical Methods in Natural Language Processing (EMNLP), pp. 1532–1543 (2014)
10. Kalchbrenner, N., Grefenstette, E., Blunsom, P.: A convolutional neural network for modelling sentences. arXiv preprint arXiv: 1404.2188 (2014)
11. Hochreiter, S., Schmidhuber, J.: Long short-term memory. Neural Comput. **9**(8), 1735–1780 (1997)
12. Chawla, N.V., et al.: SMOTE: synthetic minority over-sampling technique. J. Artif. Intell. Res. **16**, 321–357 (2002)

Gradient Boosting Machine and LSTM Network for Online Harassment Detection and Categorization in Social Media

Fabíola S. F. Pereira[1(✉)], Thiago Andrade[2],
and André C. P. L. F. de Carvalho[1]

[1] ICMC University of São Paulo, São Carlos, Brazil
{fabiola.pereira,andre}@usp.br
[2] INESC TEC, Porto, Portugal
thiago.a.silva@inesctec.pt

Abstract. We present a solution submitted to the Social Media and Harassment Competition held in collaboration with ECML PKDD 2019 Conference. The dataset used is as set of tweets and the first task was on the detection of harassment tweets. To deal with this problem, we proposed a solution based on a gradient tree-boosting algorithm. The second task was categorization harassment tweets according to the type of harassment, a multiclass classification problem. For this problem we proposed a LSTM network model. The solutions proposed for these tasks presented good predictive accuracy.

Keywords: Harassment detection · Twitter mining · Sexism analysis

1 Introduction

In this paper, we present our solution for the SIMAH (Social Media And Harassment) Discovery Challenge co-located with ECML/PKDD 2019. This competition focus on online harassment in Twitter.

According to [11], online harassment is emerging as a specific communication type in Twitter messages. As consequence, automatically monitoring these messages becomes a very important task, with high social impact. State-of-the-art studies address simple and broad categorization of media posts with sexism, such as "sexist/non-sexist" content, "more negative than positive" sexist tweet or a "hostile/benevolent" sexism categorization [5,9,10]. However, these studies presented preliminary experimental results. Additionally, they have not focused on the sentiment of sexist posts. An in-depth understanding and categorization of online harassment in social media can reveal, for instance, the mental state of the author [10], and can either avoid or reduce harmful consequences.

[10] proposes a comprehensive categorization of online harassment in social media into the following categories: "indirect harassment", "sexual harassment", "physical harassment" and "not sexist". The SIMAH discovery challenge is essentially based on [10]'s categorization and has two related tasks:

© Springer Nature Switzerland AG 2020
P. Cellier and K. Driessens (Eds.): ECML PKDD 2019 Workshops, CCIS 1168, pp. 314–320, 2020.
https://doi.org/10.1007/978-3-030-43887-6_25

- the first (Task A) is a binary classification task, in which we are asked to classify "harassment" tweets versus "non-harassment" tweets.
- The second (Task B) is a multi-class classification task in which we have to classify harassment tweets into one of four categories "indirect harassment", "sexual harassment" or "physical harassment".

The proposed solution is based on (i) a basic text pre-processing pipeline, (ii) the Extreme Gradient Boosting (XGBoost) [1] algorithm for the first task and (iii) the Long short-term memory (LSTM) [12] neural network algorithm for the second task. Section 2 briefly describes the dataset used. Section 3 presents the text pre-processing pipeline and the steps followed to get insights from the dataset. Section 4 details our proposals for the two tasks. Experimental results are in Sects. 5 and 6 concludes the paper.

2   Dataset Description

The data used has a set of tweets in English language without punctuation characters. Competition provided the training and validation subsets to be used. Table 1 describes the main characteristics of the dataset. While the dataset is roughly balanced for the binary classification task, more the multiclass task there is a highly imbalanced.

Table 1. Distribution of dataset classes.

# tweets	Dataset		
	Train	Validation	Test (gold labels)
Harassment	2713	632	611*
Indirect	55	71	197
Physical	76	36	100
Sexual	2582	525	312
Non-harassment	3661	1493	1512
Total	6374	2125	2123

*2 harassment tweets without type label

3   Feature Engineering

We performed a basic text pre-processing pipeline for both tasks. Afterwards, from tweets text data, we derived one more feature, also used in both tasks. Finally, we tried a data visualization based on similarity network analysis in order to get insights from the challenge dataset. Next, we describe such feature engineering steps.

3.1   Text Pre-processing Pipeline

The pre-processing of the tweets involved the following tokenizer steps: transformation to lower case, keeping of only alphanumeric characters, removal of stop words using standard NLTK[1] english language list and stemming with Porter Stemmer, also from NLTK. It is important to mention that for task B the standard stop words list was extended with the 'RT' term, in order to remove retweets referrals. After the token-level processing, we built a tf-idf matrix keeping very frequent terms (present in more than 90% of tweets) and removing very infrequent terms (present in less than 5% of the tweets). We considered as tokens [1–3]-grams. Finally, we applied the TruncatedSVD algorithm [4] to reduce the number of predictive attributes to 300. Figure 1 illustrates this pipeline.

3.2   Feature Extraction

Using as primary feature the tweet text, the unique feature we derived was "number of words". It is worth mentioning that we applied a standard scaler to standardize this feature to mean 0 and variance 1.

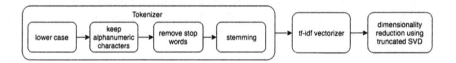

Fig. 1. Text pre-processing pipeline.

3.3   Visualizing Tweets as a Similarity Network

To extract insights from the Twitter dataset, we built a similarity network of tweets (Fig. 2). Given that a tweet is represented by a vector from a corresponding row of a tf-idf matrix, we computed the cosine similarity for each pair of tweets. After, considering each tweet as a node, we link two tweets if they have a similarity degree >0.7, building the similarity network. From Fig. 2 we can infer that (i) non-harassment and sexual tweets are very similar and (ii) there is no similarity between indirect and physical tweets. The similarity network ratifies that the categorization of harassment tweets is a complex problem, not addressed by a simple similarity measure.

4   Proposed Approaches

We experimentally investigated several algorithms and strategies, selecting the best according to the training set results. Using the *random forest* algorithm as baseline, we selected the *tree-based boosting* algorithm (task A) and the *LSTM* neural network based algorithm (task B). The best results from the two tasks are presented next.

---

[1] NLTK Python package: https://www.nltk.org/.

4.1   Task A: Online Harassment Detection

In order to solve the binary classification task, we explored two ensemble tree-based models and a recurrent neural network.

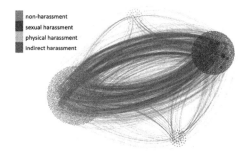

non-harassment
sexual harassment
physical harassment
indirect harassment

Fig. 2. Tweets cosine similarity network. Nodes are tweets. An edge (u, v) indicates that u and v are similar. An edge weight indicates the degree of similarity of its nodes. The color of the nodes indicates its tweet type. Only similarity relations with degree >0.7 are presented.

Random Forest (RF). A random forest is an ensemble bagging meta estimator that fits a number of independent decision tree classifiers on various sub-samples of the dataset and uses averaging to improve the predictive accuracy and control over-fitting. The `RandomForestClassifier` from scikit-learn library [7] was used with default parameters.

Extreme Gradient Boosting (XGBoost). Boosting is an ensemble technique in which the predictors are not made independently, but sequentially. Tree-based gradient boosting trains many decision trees models in a gradual, additive and sequential manner, identifying the shortcomings of weak learners by using gradients in the loss function [3]. XGBoost [1] is an implementation of gradient boosting machines, sparsity-aware for sparse data, built to scale in a distributed environment. The `XGBClassifier` from XGBoost library [1] was used with parametrization of 300 estimators, max depth = 2 and learning rate = 0.1.

Long Short-Term Memory Network (LSTM). This algorithm is a recurrent neural network that uses internal memory to deal with different sequences of inputs. Applied to text data, in natural language, it can capture dependencies in a sequence of words, modeling the context very well [12]. The used model implementation was from Keras library [2], with parametrization considering a `Sequential` model, a fixed number of 5000 most frequent words, a sequence length of 15 and embedded dimension of 100. LSTM layer with 200 units, dropout = 0.2 and recurrent dropout = 0.2. Softmax as activation function. Categorical cross-entropy as loss function and Adam optimizer [6].

4.2   Task B: Categorizing Types of Online Harassment

Second task is a multiclass classification problem in which we explored approaches varying both the multiclass strategy and the classifier algorithms. Regarding multiclass strategy, the *one-vs-one* strategy consists in fitting one classifier per class pair. At prediction time, the class which received the most votes is selected. The *output-code* based strategies consist in representing each class with a binary code (an array of 0s and 1s). At fitting time, one binary classifier per bit in the code book is fitted. At prediction time, the classifiers are used to project new points in the class space and the class closest to the points is chosen. As classifier algorithms we explored *random forest* to serve as baseline, *gradient boosting* and *catboost* as boosting tree-based algorithms and *LSTM* for a neural network-based model.

One-vs-One Random Forest (OORF). This is our baseline approach, run with `RandomForestClassifier` and `OneVsOneClassifier` scikit-learn implementations over default parameter setup.

Output Code Random Forest (OCRF). Random forest algorithm with output code multiclass strategy implemented using `OutputCodeClassifier` from scikit-learn, with code size $= 2$.

Output Code with Gradient Boosting (OCGB). `OutputCodeClassifier` with code size $= 15$ and `GradientBoostingClassifier` with 20 estimators, learning rate $= 1$ and max depth $= 7$, both from scikit-learn library.

One-vs-One Catboost Classifier (OOCB). Catboost is a machine learning algorithm also based on gradient boosting on decision trees [8]. We used the `CatBoostClassifier` implementation from catboost library [8], tunned with 5 iterations, depth $= 10$, learning rate $= 1$, and logloss as loss function.

Long-Short-Term-Memory (LSTM). The same setup used for Task A.

5   Experimental Results

The SIMAH competition provided three datasets: train, validation and final test. For our final solution, we trained all the models over the union of train and validation datasets. The results here presented refer to application of our trained models over final test dataset. Training models only over train dataset resulted in lower performances for all approaches and we do not present the results in this paper.

Figure 3 presents the results for Task A. XGBoost model achieved the best performance for the binary classification problem. LSTM had a poor performance, even below our RF baseline. Task B results are depicted in Fig. 4. Clearly, LSTM reached the best performance, followed by the gradient boosting model OCGB.

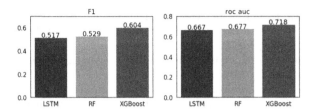

Fig. 3. Experimental results for Task A harassment detection over final test dataset.

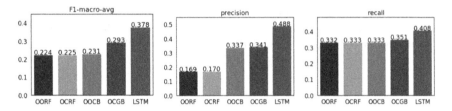

Fig. 4. Experimental results for Task B harassment categorization over final test data.

6   Conclusion

This paper presents a solution for the SIMAH discovery challenge. The competition focus on posts with harassment content in Twitter. The solution for the first task, a binary classification of "harassment/non-harassment" tweets, is based on a XGBoost model which achieved $F1 = 0.604$. For the second task, a multiclass classification problem ("indirect harassment", "sexual harassment" or "physical harassment"), LSTM neural network is our best model with F1 macro-average $= 0.378$.

References

1. Chen, T., Guestrin, C.: XGBoost: a scalable tree boosting system. In: Proceedings of the 22nd ACM SIGKDD International Conference on Knowledge Discovery and Data Mining, KDD 2016, pp. 785–794 (2016)
2. Chollet, F., et al.: Keras (2015). https://github.com/fchollet/keras
3. Friedman, J.H.: Greedy function approximation: a gradient boosting machine. Ann. Stat. **29**(5), 1189–1232 (2001)
4. Hansen, P.C.: The truncated SVD as a method for regularization. BIT Numer. Math. **27**(4), 534–553 (1987)
5. Jha, A., Mamidi, R.: When does a compliment become sexist? Analysis and classification of ambivalent sexism using Twitter data. In: Proceedings of the Second Workshop on NLP and Computational Social Science, pp. 7–16, August 2017
6. Kingma, D.P., Ba, J.: Adam: a method for stochastic optimization (2014)
7. Pedregosa, F.: Scikit-learn: machine learning in Python. J. Mach. Learn. Res. **12**, 2825–2830 (2011)

8. Prokhorenkova, L., Gusev, G., Vorobev, A., Dorogush, A.V., Gulin, A.: CatBoost: unbiased boosting with categorical features. In: NIPS, pp. 6639–6649 (2018)
9. Sharifirad, S., Jafarpour, B., Matwin, S.: Boosting text classification performance on sexist tweets by text augmentation and text generation using a combination of knowledge graphs. In: ALW (2018)
10. Sharifirad, S., Jafarpour, B., Matwin, S.: How is your mood when writing sexist tweets? Detecting the emotion type and intensity of emotion using natural language processing techniques. CoRR abs/1902.03089 (2019)
11. Sharifirad, S., Matwin, S.: When a tweet is actually sexist. A more comprehensive classification of different online harassment categories and the challenges in NLP. CoRR abs/1902.10584 (2019)
12. Sherstinsky, A.: Fundamentals of recurrent neural network (RNN) and long short-term memory (LSTM) network. CoRR abs/1808.03314 (2018)

Attention-Based Method for Categorizing Different Types of Online Harassment Language

Christos Karatsalos[(✉)] and Yannis Panagiotakis

Athens University of Economics and Business, Athens, Greece
ckarats@gmail.com, giannispanagiwtakis@gmail.com

Abstract. In the era of social media and networking platforms, Twitter has been doomed for abuse and harassment toward users specifically women. Monitoring the contents including sexism and sexual harassment in traditional media is easier than monitoring on the online social media platforms like Twitter, because of the large amount of user generated content in these media. So, the research about the automated detection of content containing sexual or racist harassment is an important issue and could be the basis for removing that content or flagging it for human evaluation. Previous studies have been focused on collecting data about sexism and racism in very broad terms. However, there is no much study focusing on different types of online harassment attracting natural language processing techniques. In this work, we present an multi-attention based approach for the detection of different types of harassment in tweets. Our approach is based on the Recurrent Neural Networks and particularly we are using a deep, classification specific multi-attention mechanism. Moreover, we tackle the problem of imbalanced data, using a back-translation method. Finally, we present a comparison between different approaches based on the Recurrent Neural Networks.

Keywords: Text classification · Twitter · Hate speech · Deep learning · Attention mechanism

1 Introduction

In the era of social media and networking platforms, Twitter has been doomed for abuse and harassment toward users specifically women. In fact, online harassment becomes very common in Twitter and there have been a lot of critics that Twitter has become the platform for many racists, misogynists and hate groups which can express themselves openly. Online harassment is usually in the form of verbal or graphical formats and is considered harassment, because it is neither invited nor has the consent of the receipt. Monitoring the contents including sexism and sexual harassment in traditional media is easier than monitoring on the online social media platforms like Twitter. The main reason is because of the large amount of user generated content in these media. So, the research

P. Cellier and K. Driessens (Eds.): ECML PKDD 2019 Workshops, CCIS 1168, pp. 321–330, 2020.
https://doi.org/10.1007/978-3-030-43887-6_26

about the automated detection of content containing sexual harassment is an important issue and could be the basis for removing that content or flagging it for human evaluation. The basic goal of this automatic classification is that it will significantly improve the process of detecting these types of hate speech on social media by reducing the time and effort required by human beings.

Previous studies have been focused on collecting data about sexism and racism in very broad terms or have proposed two categories of sexism as benevolent or hostile sexism [1], which undermines other types of online harassment. However, there is no much study focusing on different types online harassment alone attracting natural language processing techniques.

In this paper we present our work, which is a part of the SociaL Media And Harassment Competition of the ECML PKDD 2019 Conference. The topic of the competition is the classification of different types of harassment and it is divided in two tasks. The first one is the classification of the tweets in harassment and non-harassment categories, while the second one is the classification in specific harassment categories like indirect harassment, physical and sexual harassment as well. We are using the dataset of the competition, which includes text from tweets having the aforementioned categories. Our approach is based on the Recurrent Neural Networks and particularly we are using a deep, classification specific attention mechanism. Moreover, we present a comparison between different variations of this attention-based approach like multi-attention and single attention models. The next Section includes a short description of the related work, while the third Section includes a description of the dataset. After that, we describe our methodology. Finally, we describe the experiments and we present the results and our conclusion.

2   Related Work

Waseem et al. [2] were the first who collected hateful tweets and categorized them into being sexist, racist or neither. However, they did not provide specific definitions for each category. Jha and Mamidi [1] focused on just sexist tweets and proposed two categories of hostile and benevolent sexism. However, these categories were general as they ignored other types of sexism happening in social media. Sharifirad and Matwin [3] proposed complimentary categories of sexist language inspired from social science work. They categorized the sexist tweets into the categories of indirect harassment, information threat, sexual harassment and physical harassment. In the next year the same authors proposed [4] a more comprehensive categorization of online harassment in social media e.g. twitter into the following categories, indirect harassment, information threat, sexual harassment, physical harassment and not sexist.

For the detection of hate speech in social media like twitter, many approaches have been proposed. Jha and Mamidi [1] tested support vector machine, bidirectional RNN encoder-decoder and FastText on hostile and benevolent sexist tweets. They also used SentiWordNet and subjectivity lexicon on the extracted phrases to show the polarity of the tweets. Sharifirad et al. [5] trained, tested

Table 1. Class distribution of the dataset.

Dataset	Tweets	Harassment	Harassment (%)	Indirect (%)	Sexual (%)	Physical (%)
Train	6374	2713	42.56	0.86	40.50	1.19
Validation	2125	632	29.74	3.34	24.76	1.69
Test	2123	611	28.78	9.28	14.69	4.71

and evaluated different classification methods on the SemEval2018 dataset and chose the classifier with the highest accuracy for testing on each category of sexist tweets to know the mental state and the affectual state of the user who tweets in each category. To overcome the limitations of small data sets on sexist speech detection, Sharifirad et al. [6] have applied text augmentation and text generation with certain success. They have generated new tweets by replacing words in order to increase the size of our training set. Moreover, in the presented text augmentation approach, the number of tweets in each class remains the same, but their words are augmented with words extracted from their ConceptNet relations and their description extracted from Wikidata. Zhang et al. [7] combined convolutional and gated recurrent networks to detect hate speech in tweets. Others have proposed different methods, which are not based on deep learning. Burnap and Williams [8] used Support Vector Machines, Random Forests and a meta-classifier to distinguish between hateful and non-hateful messages. A survey of recent research in the field is presented in [9]. For the problem of the hate speech detection a few approaches have been proposed that are based on the Attention mechanism. Pavlopoulos et al. [10] have proposed a novel, classification-specific attention mechanism that improves the performance of the RNN further for the detection of abusive content in the web. Xie et al. [11] for emotion intensity prediction, which is a similar problem to ours, have proposed a novel attention mechanism for CNN model that associates attention-based weights for every convolution window. Park and Fung [14] transformed the classification into a 2-step problem, where abusive text first is distinguished from the non-abusive, and then the class of abuse (Sexism or Racism) is determined. However, while the first part of the two step classification performs quite well, it falls short in detecting the particular class the abusive text belongs to. Pitsilis et al. [15] have proposed a detection scheme that is an ensemble of RNN classifiers, which incorporates various features associated with user related information, such as the users' tendency towards racism or sexism.

3   Dataset Description

The dataset from Twitter that we are using in our work, consists of a train set, a validation set and a test set. It was published for the "First workshop on categorizing different types of online harassment languages in social media". The whole dataset is divided into two categories, which are harassment and non-harassment tweets. Moreover, considering the type of the harassment, the

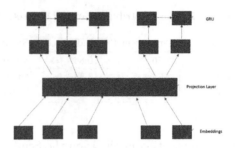

Fig. 1. Projection layer

tweets are divided into three sub-categories which are indirect harassment, sexual and physical harassment. We can see in Table 1 the class distribution of our dataset. One important issue here is that the categories of indirect and physical harassment seem to be more rare in the train set than in the validation and test sets. To tackle this issue, as we describe in the next section, we are performing data augmentation techniques. However, the dataset is imbalanced and this has a significant impact in our results.

4 Proposed Methodology

4.1 Data Augmentation

As described before one crucial issue that we are trying to tackle in this work is that the given dataset is imbalanced. Particularly, there are only a few instances from indirect and physical harassment categories respectively in the train set, while there are much more in the validation and test sets for these categories. To tackle this issue we applying a back-translation method [16], where we translate indirect and physical harassment tweets of the train set from english to german, french and greek. After that, we translate them back to english in order to achieve data augmentation. These "noisy" data that have been translated back, increase the number of indirect and physical harassment tweets and boost significantly the performance of our models.

Another way to enrich our models is the use of pre-trained word embeddings from 2B Twitter data [17] having 27B tokens, for the initialization of the embedding layer.

4.2 Text Processing

Before training our models we are processing the given tweets using a tweet pre-processor[1]. The scope here is the cleaning and tokenization of the dataset.

---

[1] https://pypi.org/project/tweet-preprocessor/

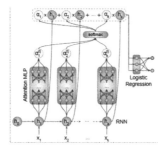

Fig. 2. Attention mechanism, MLP with l layers

4.3   RNN Model and Attention Mechanism

We are presenting an attention-based approach for the problem of the harassment detection in tweets. In this section, we describe the basic approach of our work. We are using RNN models because of their ability to deal with sequence information. The RNN model is a chain of GRU cells [18] that transforms the tokens $w_1, w_2, ..., w_k$ of each tweet to the hidden states $h_1, h_2, ..., h_k$, followed by an LR Layer that uses h_k to classify the tweet as harassment or non-harassment (similarly for the other categories). Given the vocabulary V and a matrix $E \in R^{d \times |V|}$ containing d-dimensional word embeddings, an initial h_0 and a tweet $w = < w_1, .., w_k >$, the RNN computes $h_1, h_2, ..., h_k$, with $h_t \in R^m$, as follows:

$$h'_t = \tanh(W_h x_t + U_h(r_t \odot h_{t-1}) + b_h)$$

$$h_t = (1 - z_t) \odot h_{t-1} + z_t \odot h'_t$$

$$z_t = \sigma(W_z x_t + U_z h_{t-1} + b_z)$$

$$r_t = \sigma(W_r x_t + U_r h_{t-1} + b_r)$$

where $h'_t \in R^m$ is the proposed hidden state at position t, obtained using the word embedding x_t of token w_t and the previous hidden state h_{t-1}, $\odot$ represents the element-wise multiplication, $r_t \in R^m$ is the reset gate, $z_t \in R^m$ is the update gate, σ is the sigmoid function. Also $W_h, W_z, W_r \in R^{m \times d}$ and $U_h, U_z, U_r \in R^{m \times m}$, $b_h, b_z, b_r \in R^m$. After the computation of state h_k the LR Layer estimates the probability that tweet w should be considered as harassment, with $W_p \in R^{1 \times m}, b_p \in R$:

$$P_{RNN}(harassment|w) = \sigma(W_p h_k + b_p).$$

We would like to add an attention mechanism similar to the one presented in [10], so that the LR Layer will consider the weighted sum h_{sum} of all the hidden states instead of h_k:

$$h_{sum} = \sum_{t=1}^{k} \alpha_t h_t$$
$$P_{attentionRNN} = \sigma(W_p h_{sum} + b_p)$$

Alternatively, we could pass h_{sum} through an MLP with k layers and then the LR layer will estimate the corresponding probability. More formally,

$$P_{attentionRNN} = \sigma(W_p h_* + b_p)$$

where h_* is the state that comes out from the MLP. The weights α_t are produced by an attention mechanism presented in [10] (see Fig. 2), which is an MLP with l layers. This attention mechanism differs from most previous ones [19,20], because it is used in a classification setting, where there is no previously generated output sub-sequence to drive the attention. It assigns larger weights α_t to hidden states h_t corresponding to positions, where there is more evidence that the tweet should be harassment (or any other specific type of harassment) or not. In our work we are using four attention mechanisms instead of one that is presented in [10]. Particularly, we are using one attention mechanism per category. Another element that differentiates our approach from Pavlopoulos et al. [10] is that we are using a projection layer for the word embeddings (see Fig. 1). In the next subsection we describe the Model Architecture of our approach.

4.4   Model Architecture

The Embedding Layer is initialized using pre-trained word embeddings of dimension 200 from Twitter data that have been described in a previous sub-section. After the Embedding Layer, we are applying a Spatial Dropout Layer, which drops a certain percentage of dimensions from each word vector in the training sample. The role of Dropout is to improve generalization performance by preventing activations from becoming strongly correlated [13]. Spatial Dropout, which has been proposed in [12], is an alternative way to use dropout with convolutional neural networks as it is able to dropout entire feature maps from the convolutional layer which are then not used during pooling. After that, the word embeddings are passing through a one-layer MLP, which has tanh as activation function and 128 hidden units, in order to project them in the vector space of our problem considering that they have been pre-trained using text that has a different subject. In the next step the embeddings are fed in a unidirectional GRU having 1 Stacked Layer and size 128. We prefer GRU than LSTM, because it is more efficient computationally. Also the basic advantage of LSTM which is the ability to keep in memory large text documents, does not hold here, because tweets supposed to be not too large text documents. The output states of the GRU are passing through four self-attentions like the one described above [10], because we are using one attention per category (see Fig. 2). Finally, a one-layer MLP having 128 nodes and ReLU as activation function computes the final score for each category. At this final stage we have avoided using a softmax function to decide the harassment type considering that the tweet is a harassment, otherwise we had to train our models taking into account only the harassment tweets and this might have been a problem as the dataset is not large enough.

5   Experiments

5.1   Training Models

In this subsection we are giving the details of the training process of our models. Moreover, we are describing the different models that we compare in our experiments.

Table 2. The results considering F1 Score.

Model	sexual_f1	indirect_f1	physical_f1	harassment_f1	f1_macro
attentionRNN	0.674975	0.296320	0.087764	0.709539	0.442150
MultiAttentionRNN	0.693460	0.325338	0.145369	0.700354	0.466130
MultiProjectedAttentionRNN	0.714094	0.355600	0.126848	0.686694	**0.470809**
ProjectedAttentionRNN	0.692316	0.315336	0.019372	0.694082	0.430276
AvgRNN	0.637822	0.175182	0.125596	0.688122	0.40668
LastStateRNN	0.699117	0.258402	0.117258	0.710071	0.446212
ProjectedAvgRNN	0.655676	0.270162	0.155946	0.675745	0.439382
ProjectedLastStateRNN	0.696184	0.334655	0.072691	0.707994	0.452881

Batch size which pertains to the amount of training samples to consider at a time for updating our network weights, is set to 32, because our dataset is not large and small batches might help to generalize better. Also, we set other hyperparameters as: epochs $= 20$, patience $= 10$. As early stopping criterion we choose the average AUC, because our dataset is imbalanced.

The training process is based on the optimization of the loss function mentioned below and it is carried out with the Adam optimizer [21], which is known for yielding quicker convergence. We set the learning rate equal to 0.001:

$$L = \frac{1}{2}BCE(harassment) + \frac{1}{2}(\frac{1}{5}BCE(sexualH) + \frac{2}{5}BCE(indirectH)$$
$$+ \frac{2}{5}BCE(physicalH))$$

where BCE is the binary cross-entropy loss function,

$$BCE = -\frac{1}{n}\sum_{i=1}^{n}[y_i log(y_i') + (1 - y_i)log(1 - y_i'))]$$

i denotes the ith training sample, y is the binary representation of true harassment label, and y' is the predicted probability. In the loss function we have applied equal weight to both tasks. However, in the second task (type of harassment classification) we have applied higher weight in the categories that it is harder to predict due to the problem of the class imbalance between the training, validation and test sets respectively.

5.2   Evaluation and Results

Each model produces four scores and each score is the probability that a tweet includes harassment language, indirect, physical and sexual harassment language respectively. For any tweet, we first check the score of the harassment language and if it is less than a specified threshold, then the harassment label is zero, so the other three labels are zero as well. If it is greater than or equal to that threshold, then the harassment label is one and the type of harassment is the one among these three having that has the greatest score (highest probability). We set this threshold equal to 0.33.

We compare eight different models in our experiments. Four of them have a Projected Layer (see Fig. 1), while the others do not have, and this is the only difference between these two groups of our models. So, we actually include four models in our experiments (having a projected layer or not). Firstly, Last-StateRNN is the classic RNN model, where the last state passes through an MLP and then the LR Layer estimates the corresponding probability. In contrast, in the AvgRNN model we consider the average vector of all states that come out of the cells. The AttentionRNN model is the one that it has been presented in [10]. Moreover, we introduce the MultiAttentionRNN model for the harassment language detection, which instead of one attention, it includes four attentions, one for each category.

We have evaluated our models considering the F1 Score, which is the harmonic mean of precision and recall. We have run ten times the experiment for each model and considered the average F1 Score. The results are mentioned in Table 2. Considering F1 Macro the models that include the multi-attention mechanism outperform the others and particularly the one with the Projected Layer has the highest performance. In three out of four pairs of models, the ones with the Projected Layer achieved better performance, so in most cases the addition of the Projected Layer had a significant enhancement.

6   Conclusion - Future Work

We present an attention-based approach for the detection of harassment language in tweets and the detection of different types of harassment as well. Our approach is based on the Recurrent Neural Networks and particularly we are using a deep, classification specific attention mechanism. Moreover, we present a comparison between different variations of this attention-based approach and a few baseline methods. According to the results of our experiments and considering the F1 Score, the multi-attention method having a projected layer, achieved the highest performance. Also, we tackled the problem of the imbalance between the training, validation and test sets performing the technique of back-translation.

In the future, we would like to perform more experiments with this dataset applying different models using BERT [22]. Also, we would like to apply the models presented in this work, in other datasets about hate speech in social media.

References

1. Jha, A., Mamidi, R.: When does a compliment become sexist: analysis and classification of ambivalent sexism using Twitter data. In: Proceedings of the Second Workshop on Natural Language Processing and Computational Social Science (2017)
2. Waseem, Z., Hovy, D.: Hateful symbols or hateful people: predictive features for hate speech detection on Twitter. In: Proceedings of NAACL-HLT, pp. 88–93 (2016)
3. Sharifirad, S., Matwin, S.: Classification of different types of sexist languages on Twitter and the gender footprint on each of the classes. In: CICLing 2018 (2018)
4. Sharifirad, S., Matwin, S.: When a tweet is actually sexist. A more comprehensive classification of different online harassment categories and the challenges in NLP (2019). https://arxiv.org/abs/1902.10584
5. Sharifirad, S., Matwin, S., Jafarpour, B.: How is your mood when writing sexist tweets? Detecting the emotion type and intensity of emotion using natural language processing techniques (2019). https://arxiv.org/abs/1902.03089
6. Sharifirad, S., Matwin, S., Jafarpour, B.: Boosting text classification performance on sexist tweets by text augmentation and text generation using a combination of knowledge graphs (2018). http://aclweb.org/anthology/W18-5114
7. Zhang, Z., Robinson, D., Tepper, J.: Detecting hate speech on Twitter using a convolution-GRU based deep neural network. In: Gangemi, A., et al. (eds.) ESWC 2018. LNCS, vol. 10843, pp. 745–760. Springer, Cham (2018). https://doi.org/10.1007/978-3-319-93417-4_48
8. Burnap, P., Williams, M.: Cyber hate speech on Twitter: an application of machine classification and statistical modeling for policy and decision making. Policy Internet **7**(2), 223–242 (2015)
9. Schmidt, A., Wiegand, M.: A survey on hate speech detection using natural language processing. In: Proceedings of the Fifth International Workshop on Natural Language Processing for Social Media, Valencia, Spain, pp. 1–10. Association for Computational Linguistics (2017)
10. Pavlopoulos, J., Malakasiotis, P., Androutsopoulos, I.: Deeper attention to abusive user content moderation. In: Proceedings of the 2017 Conference on Empirical Methods in Natural Language Processing, pp. 1125–1135. Association for Computational Linguistics (2017)
11. Xie, H., Feng, S., Wang, D., Zhang, Y.: A novel attention based CNN model for emotion intensity prediction. In: Zhang, M., Ng, V., Zhao, D., Li, S., Zan, H. (eds.) NLPCC 2018. LNCS (LNAI), vol. 11108, pp. 365–377. Springer, Cham (2018). https://doi.org/10.1007/978-3-319-99495-6_31
12. Tompson, J., Goroshin, R., Jain, A., LeCun, Y., Bregler, C.: Efficient object localization using convolutional networks. In: The IEEE Conference on Computer Vision and Pattern Recognition (CVPR), June 2015 (2015)
13. Hinton, G.E., Srivastava, N., Krizhevsky, A., Sutskever, I., Salakhutdinov, R.R.: Improving neural networks by preventing co-adaptation of feature detectors. arXiv preprint arXiv:1207.0580 (2012)
14. Park, J.H., Fung, P.: One-step and two-step classification for abusive language detection on Twitter. In: 1st Workshop on Abusive Language Online, ACL 2017, Vancouver, Canada, 4th August 2017 (2017)
15. Pitsilis, G., Ramampiaro, H., Langseth, H.: Detecting offensive language in tweets using deep learning. arXiv preprint arXiv:1801.04433 (2018)

16. Sennrich, R., Haddow, B., Birch, A.: Improving neural machine translation models with monolingual data. In: Proceedings of the 54th Annual Meeting of the Association for Computational Linguistics (Volume 1: Long Papers), Berlin, Germany, August 2016, pp. 86–96. Association for Computational Linguistics (2016)

17. Pennington, J., Socher, R., Manning, C.D.: GloVe: global vectors for word representation. In: Proceedings of the 2014 Conference on Empirical Methods in Natural Language Processing (EMNLP), Doha, Qatar, pp. 1532–1543 (2014)

18. Cho, K., et al.: Learning phrase representations using RNN encoder-decoder for statistical machine translation. In: Proceedings of the 2014 Conference on Empirical Methods in Natural Language Processing, Doha, Qatar, pp. 1724–1734 (2014)

19. Luong, T., Pham, H., Manning, C.D.: Effective approaches to attention-based neural machine translation. In: Proceedings of the 2015 Conference on Empirical Methods in Natural Language Processing, Lisbon, Portugal, pp. 1412–1421 (2015)

20. Bahdanau, D., Cho, K., Bengio, Y.: Neural machine translation by jointly learning to align and translate. In: Proceedings of the 3rd International Conference on Learning Representations, San Diego, CA, USA (2015)

21. Kingma, D.P., Ba, J.: Adam: a method for stochastic optimization. CoRR, abs/1412.6980 (2014)

22. Devlin, J., Chang, M.W., Lee, K., Toutanova, K.: BERT: pretraining of deep bidirectional transformers for language understanding. arXiv preprint arXiv:1810.04805 (2018)

IoT Stream for Data Driven Predictive Maintenance

SPICE: Streaming PCA Fault Identification and Classification Engine in Predictive Maintenance

Cristian Axenie$^{(\boxtimes)}$, Radu Tudoran, Stefano Bortoli, Mohamad Al Hajj Hassan, Alexander Wieder, and Goetz Brasche

Huawei German Research Center, Riesstrasse 25, 80992 Munich, Germany
{cristian.axenie,radu.tudoran,stefano.bortoli,mohamad.alhajjhassan, alexander.wieder,goetz.brasche}@huawei.com

Abstract. Data-driven predictive maintenance needs to understand high-dimensional "in-motion" data, for which fundamental machine learning tools, such as Principal Component Analysis (PCA), require computation-efficient algorithms that operate near-real-time. Despite the different streaming PCA flavors, there is no algorithm that precisely recovers the principal components as the batch PCA algorithm does, while maintaining low-latency and high-throughput processing. This work introduces a novel processing framework, employing temporal accumulate/retract learning for streaming PCA. The framework is instantiated with several competitive PCA algorithms with proven theoretical advantages. We benchmark the framework in a real-world predictive maintenance scenario (i.e. fault classification in a coal coke production line) and prove its low-latency (millisecond level) and high-throughput (thousands events/second) processing guarantees.

Keywords: Stream processing · PCA · Online learning · Predictive maintenance · Fault identification · Fault classification

1 Introduction

Given today's industrial IoT sensory streams, predictive maintenance systems need to process streams of data under tight computational constraints for decision making (e.g. fault / normal operation). Processing data streams is quite different from querying static data, as data might be transient and follow a non-stationary distribution. These complications impose significant constraints on the problem of streaming PCA, which is an essential building block for many inference and decision making tasks. Due to its practical relevance, there is renewed interest in this problem [17]. PCA is good at maintaining data structure in reduced subspaces in an unsupervised way. It is also useful in updating the decision boundaries by adding discriminately informative features with newly added samples through updating the feature vectors by incremental eigenvector estimates [8]. This is highly relevant in multi-class classification problems, such

© Springer Nature Switzerland AG 2020
P. Cellier and K. Driessens (Eds.): ECML PKDD 2019 Workshops, CCIS 1168, pp. 333–344, 2020.
https://doi.org/10.1007/978-3-030-43887-6_27

as fault identification in predictive maintenance [6]. However, there is no algorithm that recovers the principal components in the same precision regime as the batch PCA algorithm does, employing low-latency high-throughput processing. The core contribution of the paper resides in exploiting a novel streaming learning paradigm, termed accumulate-retract learning [2], that leverage state-of-the-art PCA algorithms to achieve low-latency high-throughput processing, in real-world predictive maintenance fault identification and classification.

2   Related Work

Recent focus in predictive maintenance is on understanding streaming high-dimensional data, where the dimensionality of the data can potentially scale together with the number of available sample points [12,19]. Current trends in predictive maintenance have led to an exploration of the complexity of covariance estimation underlying PCA [3]. Such algorithms have provable complexity guarantees [7], but either store all samples (i.e. for looping through samples) or explicitly maintain the covariance matrix. In high-dimensional applications, such as sensory-rich production lines, storing all data is prohibitive. Different from previous approaches, our work brings the focus on two critical quantities: latency and throughput. In the streaming, data-driven, setting many approaches for incremental or online PCA have been developed, some focusing on replacing the inefficient steps in the traditional PCA algorithm [4]. Despite the multitude of successful dedicated algorithms, such as [18], there is no algorithm that brings the focus on the two critical quantities of focus in streaming predictive maintenance. Our work utilizes a new paradigm for stream processing and learning and proposes the implementation of several competitive streaming PCA algorithms for efficient computation with low-latency and high-throughput. Our accumulate/ retract learning framework [2] offers a solution for incrementally computing combinations of statistics and learning, as needed in fault identification. The proposed framework is among the first attempts towards this paradigm shift initially set by Massive Online Analytics (MOA) and algorithms like Adaptive Windowing (ADWIN) [1]. Despite the fact that ADWIN implementations have theoretical guarantees, they do not guarantee latency and throughput, the goal of our system. Our system finds a trade-off between these two guarantees to be able to perform inference (i.e. fault identification) with low-latency and high-throughput. Similar to ADWIN, a main advantage of our approach is that it does not require any prior about how fast or how often the stream will drift, as it continuously estimates that while updating the models.

3   Methods

This section covers the design, implementation details, and the motivation to tackle the inherent problems in traditional PCA impeding it to achieve low-latency and high-throughput processing. We identified three aspects in existing approaches which impact the streaming PCA formulation: (a) the continuous calculation of the mean and other descriptive statistics (i.e. covariance) on the datastream; (b) sorting the dominant eigenvalues in the rank update of the QR decomposition and the ordering of lower/upper triangular sub-matrices; (c) the complexity of computations performed at each training step. Dissecting the theory of PCA, we found those bottlenecks that kept PCA away from streaming applications. Employing a novel method and system for online machine learning [2], capable of incrementally computing machine learning models on data streams, we successfully instantiated multiple streaming PCA approaches in a predictive maintenance scenario (i.e. fault identification in coal coke production) and prove processing performance (i.e. low-latency, high-throughput). The underlying computational mechanism in our approach is the accumulate-retract framework. Such a framework allows for dual model updates as soon as new data comes into the system and leaves the system, respectively, as the stream progresses. The technique builds on top of the sliding window paradigm, and provides a novel, incremental computation model for closed-form learning rules used in streaming PCA. For example, the average calculation in the eigenvalue update, in incremental form, can be visualized in Fig. 1. Such incremental computation tackles successfully the first bottleneck, namely the incremental calculation of the mean and other descriptive statistics on the datastream. As shown in Fig. 1, the average computation is split in dual operations that keep subquantities in the update of the mean consistent as the stream progresses (i.e. window slide). Going further, the second problem that the accumulate-retract framework solves, is sorting the dominant eigenvalues in the rank update of the QR decomposition. For this, let's assume we need to sort the current list of dominant eigenvalues, as shown in Fig. 2. In this case, the caches (i.e. fixed memory areas) are used to store content (i.e. in buckets) on updates depending on counts (i.e. histogram). Updates are done in buckets, which contain sorted eigenvalues - which in follow a histogram sorting. Each time new eigenvalues are computed sorting is triggered (Step 1). In such an instantiation the retraction cache stores the last calculated eigenvalues (in time) in each bucket, whereas when the accumulation cache moves according to the sliding convention, new buckets are brought and the entire structure is sorted (Steps 2, 3, 4). Moreover, in this instantiation, the buckets stored on disk or 3rd party storage devices contain data organized based on value/indexes. The last eigenvalue (time-wise) in each bucket has a reference in the retraction cache, as shown in Fig. 2. There are many models for learning PCA efficiently: from stochastic approximation models (i.e. Hebbian and Oja's Learning Rules [9]), to subspace learning, and nonlinear PCA denoising autoencoders [13]. Yet, in order to explore the potential that the accumulate/retract learning framework offers, we selected three Streaming PCA algorithms, which employ only local learning rules and have proven theoretical advantages. Their dual, accumulate/retract implementation is one novel aspect of this work, enabling such theoretical model to perform in real-world predictive maintenance scenarios.

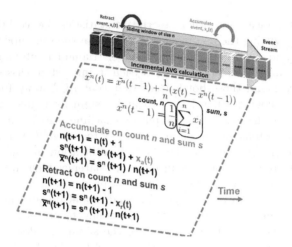

Fig. 1. Streaming PCA incremental average update using accumulate/retract.

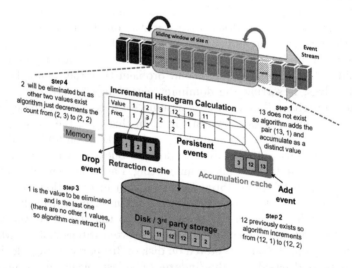

Fig. 2. Streaming PCA eigenvalues sorting on streams: histogram update using accumulate/retract.

4   Materials

The first model considered in our experiments is a single-layer neural network based on the Stochastic Gradient Ascent (SGA) [9]. We chose this model because it efficiently provides a description of the covariance matrix, which is typically too expensive to be estimated online. The accumulate/retract learning maps easily on the structure of the network and the corresponding learning rule in

Eq. 1 due to its dual operation when forward propagating from the input $x(k)$ to the hidden layer $y(k)$ by updating the weight matrix W.

$$\Delta w_j(t-1) = \gamma(t)y_j(t)(x(t) - y_j(t)w_j(t-1) - 2\sum_{i<j} y_i(t)w_i(t-1)) \quad (1)$$

where $\gamma(t)$ is the learning rate. The second model employed in our experiments is the Generalized Hebbian Algorithm (GHA) [10]. The model uses Hebbian learning to achieve optimal unsupervised extraction of the eigenvectors of the autocorrelation matrix of the input distribution, given samples from that distribution. Each output of such a trained network represents the response to one eigenvector, whereas outputs are ordered by decreasing eigenvalues. Considering the same single-layer feed-forward neural network, the GHA-based Streaming PCA learning rule is given by:

$$\Delta w_{ij}(t) = \gamma(t)(y_j(t)x_i(t) - y_j(t)\sum_{k<i} w_{kj}(t)y_k(t)) \quad (2)$$

GHA requires only the computation of the outer products yx^T and yy^T so that if the number of outputs is small the computational and storage requirements can be correspondingly decreased.

Finally, the third Streaming PCA model employed in our experiments is the Candid Covariance-free Incremental PCA (CCIPCA) [14]. This Streaming PCA model is able to compute the principal components of a sequence of samples, incrementally, without estimating the covariance matrix. The method is based on the concept of statistical efficiency (i.e. the estimate has the smallest variance given the observed data). In order to achieve this, CCIPCA keeps the scale of observations and computes the mean of observations incrementally. Assuming that we have a $d - dimensional$ input stream $u(n), n = 1, 2, ...$ with covariance matrix A and given the definition, $\lambda x = Ax$, and $v = \lambda x$, then the learning rule for CCIPCA is given by:

$$v(n) = \frac{n-1-l}{n}v(n-1) + \frac{1+l}{n}u(n)u^T(n)\frac{v(n-1)}{||v(n-1)||} \quad (3)$$

where l is the "amnesic parameter" (i.e. a forgetting parameter). Independent of the learning rule the weights w_j converge to the eigenvectors c_i as the streaming PCA model finds the unique set of weights which is both optimal and gives uncorrelated outputs. This is also true for the $v(n)$, in 3, analog to w. In the accumulate/retract framework such an update is performed as new elements are available from the stream. The accumulate event (i.e. incoming) triggers an incremental update of the weight matrix, whereas the retraction event (i.e. outgoing) triggers a decremental update of the weight matrix. This ensures the weight matrix stays consistent as the stream progresses. For each of the considered models, we derived the closed form incremental learning rules in the accumulate-retract framework. As shown previously, the eigenvectors $z_1, z_2, z_3, ..., z_p$ are given by $Cz_i = \lambda_i z_i$, where λ_i are the eigenvalues of the covariance matrix, $C[c_{jk}]$.

We can then rewrite C as $c_{jk} = \frac{1}{n-1}\sum_{i=1}^{n}(z_{ij}-\bar{z}_j)(z_{ik}-\bar{z}_k)$ where $\bar{z} = \frac{1}{n}\sum_{i=1}^{n} z_i$ is the incremental average. Hence, we can rewrite the covariance matrix C as $C = \frac{1}{n-1}\sum_{i=1}^{n}(z_i-\bar{z})(z_i-\bar{z})^T$, which is, in fact, the autocorrelation. The problem is that these measures are not robust statistics and hence not resistant to outliers. We replace z with its estimate at time t $z(t)$ so that $v = \lambda z$ estimate at time t is $v(t) = \frac{1}{n}\sum_{i=1}^{n} x(t)x^T(t)z(t)$. We can now calculate the eigenvalues and eigenvectors given v as we know that $\lambda = \|v\|$. If we consider $z = \frac{v}{\|v\|}$ we can rewrite $v(t) = \frac{1}{n}\sum_{i=1}^{n} x(t)x^T(t)\frac{v(t-1)}{\|v(t-1)\|}$. Such computation steps provide a closed form learning rule in the accumulate-retract framework [2]. Independent of the considered streaming PCA model, the system converges from an initially random set of weights to the eigenvectors of the input autocorrelation in the eigenvalues order. The optimal weights are found by minimizing the linear reconstruction error $E\{(x-\hat{x})^2\}$ when the rows of W span the first p eigenvectors of C and the Linear Least Squares (LLS) estimate of x given y is $\hat{x} = CW^T(WCW^T)^{-1}y$. If the rows of W are the first eigenvectors then $WW^T = I$ and $C = W^T\Lambda W$ where Λ is the diagonal matrix of C in descending order. Then, $y = Wx$ is the Karhunen-Loève Transformation (KLT). In the LLS optimization routine, if we have an unknown function f the best estimator of y is $\min_{f}\sum_{i=1}^{n}(y_i-f(x_i))^2$. Moreover, if f is linear in x and $y = ax+b$ then the best estimator is the search for the best a, b that minimize $\min_{f}\sum_{i=1}^{n}(y_i-ax_i-b)^2$. In the accumulate-retract framework such a problem is incrementally solved as the datastream progresses, using simple updates shown in Fig. 3 Given that covariance can be incrementally calculated as $cov_{xy}(t) = \frac{n-2}{n-1}cov_{xy}(t-1)+\frac{1}{n}(x^n(t)-\bar{x}^{n-1}(t))(y^n(t)-\bar{y}^{n-1}(t))$, the problem in closed form assumes calculating incrementally a and b as $a(t) = \frac{cov_{xy}(t)}{m_2(t)}$, $b(t) = \bar{y}(t) - a\bar{x}(t)$ where $m_2(t)$ is the 2nd statistical moment in incremental form. Yet, up to now we made the assumption that only y values contain errors while x are known

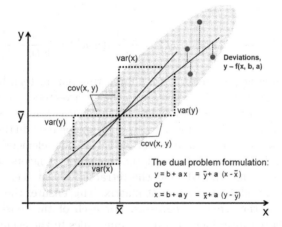

Fig. 3. Streaming PCA optimization using LLS in accumulate/retract framework.

accurately. This is not true in practical applications, thus we are seeking the values of a and b that minimize $\min\limits_{a,b} \sum_{i=1} \frac{(y_i - ax_i - b)^2}{1+a^2}$, which amounts for the Total Least Squares (TLS). It has been shown that the TLS problem can be solved by performing a Minor Component Analysis (MCA) [16], thus finding the linear combinations (or directions) which contain the minimum variance. Since there might be errors in both x and y, we incorporate y into x and reformulate the problem by finding the direction d in $E_{TLS} = \min\limits_{d} \frac{(dx+b)^2}{d^2}$ over all inputs x (including y). We can rewrite the same constraint as $E_{TLS} = n\min\limits_{d} \frac{d^t Rd + 2bd^T E(x) + b^2}{d^T d}$ where $R = \frac{1}{n}\sum_{i=1}^{n} x_i x_i^T$ is the autocorrelation of the input and $E(x) = \frac{1}{n}\sum_{i=1}^{n} x_i$ is the average of the input. At convergence ($\frac{dE_{TLS}}{dz} = 0$) we must have $Rd + bE(x) - \theta d = 0$ where $\theta = \frac{d^T Rd + 2bd^T E(x) + b^2}{d^T d}$. We need to find a hyperplane $dx + b = 0$. Taking the expectation we have $C = -dE(x)$ which we can substitute in $Cd - \theta d = 0$ where now $\theta = \frac{d^T Cd}{d^T d}$ and $C = R - E(xx^T)$ is the covariance matrix. We can now see that every eigenvector is a solution of the minimization of E_{TLS}. Given the analytical walk-through of the methods, in the following section, we instantiate the considered streaming PCA models within the accumulate-retract framework for a multi-class classification problem. In such a problem, PCA acts as a preprocessing step and due to its incremental nature, preserves the discriminant information within the data and can provide classification boundaries [15].

5   Experiments and Discussion

This section introduces the results and the analysis of the three streaming PCA models instantiated in our framework using Apache Flink [5]. Flink is an open source system for parallel scalable processing on real-time streaming data. At its core, Flink builds on an optimized distributed dataflow runtime that supports our accumulate-retract framework, crucial in obtaining low-latency

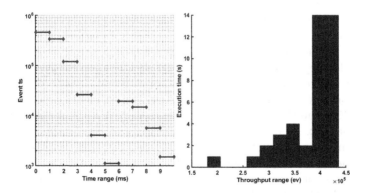

Fig. 4. Analysis of QR-based PCA without accumulate/retract (left latency, right throughput).

high-throughput online learning. The experimental setup for our tests used 4 machines, each with 24 CPU cores and 196 GB RAM, and Flink for cluster management. During the experiments we consider a fixed sliding window, but the system can support also adaptive windowing. At the same time the caches (i.e. in RAM) mechanism allows to maintain new and old data in order to allow the retraction of individual stream events when sliding. This allows our system to learn from continuous data in a single pass. We used a real-world stream with sensory readings from a coal coke prediction production line (i.e. data from 1 preheater temperature sensor, 2 briquetting temperature sensors, 2 cooker temperature sensors, 2 coke quencher temperature sensors, 2 coke transport system temperature sensors and 2 blast furnace temperature sensors). We addressed the problem of identifying faults in the production line and queried the eigenvalues and eigenvectors to extract the normal and faulty operation configuration prior to a multi-class classifier. The datastream contained 2M incoming events at 40 kHz. Moreover the datastream had the property that the eigenvalues of the input X are close to the class labels (i.e. $1, 2, ..., d$) and the corresponding eigenvectors are close to the canonical basis of R^d, where d is the number of principal components to extract and the class number for the multi-class classification task (i.e. various types of faults and normal operation - in our scenario, we consider 10 classes, 9 faults and 1 normal). Basically, the system: (a) supports the automatic generation of a concise, reliable, low-dimensional model which describes the operation mode of the various sensors in the coal coke production line; (b) identifies different alarm type conditions (i.e. the eigenvectors configuration, given the 10 different classes); and (c) conjectures the most likely cause of failure (i.e. the eigenvalues configuration). In order to evaluate the three streaming PCA models, we also implemented an efficient QR-based PCA of [11] using Householder transformation as ground-truth and ran it in the accumulate-retract framework on the same experimental setup. Important to note that the three streaming PCA models do not need to compute the correlation matrix in advance, since the eigenvectors are derived directly from the data. This is an important feature of streaming PCA, particularly if the number of inputs is large. The scope of our analysis is to emphasize that using simple incremental operations and exploiting an efficient data orchestration, the accumulate/retract framework can leverage low-latency high-throughput streaming PCA. Such a platform allows the three streaming PCA models to learn from datastreams in a single pass. In our evaluation, the latency measure refers to the single stream event processing time, whereas the throughput refers to the number of stream events processed in a second. In order to provide a baseline, we performed initial experiments with streaming PCA and the, ground-truth, QR-based PCA without the accumulate-retract framework. As one can see in both Fig. 4 and Fig. 5, respectively, without the accumulate-retract framework, the system can only process up to ~21k events independent of the chosen model. The latency distribution, is centered on values ~1 ms, as one can see in the left panel of both Figs. 4 and 5, respectively. There is an advantage that the streaming PCA holds in the overall event processing latency, with no event processed in over 8 ms. Important to mention that in

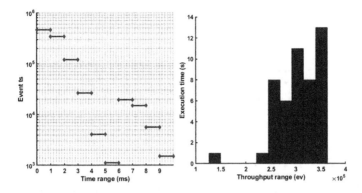

Fig. 5. Analysis of streaming PCA (SGA model) without accumulate/retract (left latency, right throughput).

both experiments the PCA model computed the eigenvectors and the eigenvalues simultaneously for each new incoming stream event. The central experiments of this paper are meant to emphasize the advantages of the accumulate-retract learning framework [2]. We now analyze a series of large-scale experiments meant at extracting the eigenfeatures for the multi-class fault identification scenario (i.e. 2M events streamed at 40 kHz, a coal coke production line in China). In terms of latency, one can observe that the streaming PCA models outperform the QR-based PCA model, with a substantial distribution of events processed at 1 ms and just a limited number of events processed at over 8 ms (less than 1000), as shown in Fig. 6. This is supported by the gain of $\sim$1k events throughput, as shown in Fig. 7 in the accumulate/retract framework and up to $\sim$10k more than in the baseline experiments. This is also visible in the core distribution of throughput ranges peaking at around 40k events/s (Table 1). To offer an understanding of the actual estimation performance of the system, the next table shows the eigenvalues of the input and how close they are to the class labels (i.e. $1, 2, ..., d = 10$) and the corresponding eigenvectors variance with respect to the canonical basis of R^d. This analysis didn't address a model comparison, rather an emphasis on the capabilities of the accumulate/retract learning framework to leverage streaming PCA models to reach guarantees of performance. Pushing real-world performance constraints, the accumulate-retract framework instantiation of various streaming PCA models stands out as good candidate for low-latency high-throughput systems for dimensionality reduction, in critical applications such as fault identification in predictive maintenance. The code repository for the Streaming PCA benchmarking is available at [1].

---

[1] https://github.com/omlstreaming/iotstream2019.

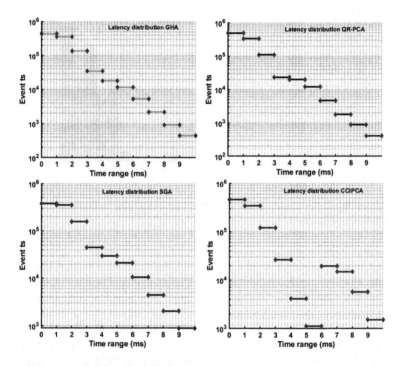

Fig. 6. Comparison of streaming PCA models latency in the accumulate/retract.

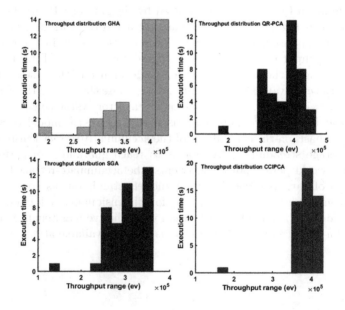

Fig. 7. Comparison of streaming PCA models throughput in the accumulate/retract.

Table 1. Eigenvalues and Eigenvectors estimates analysis

Eigenvalue	Eigenvalue estimate	Eigenvector variance
1	0.994071965	0.033719458
2	1.99658601	0.023661145
3	3.00600192	0.013884741
4	4.00420688	0.025106736
5	5.04173253	0.022354039
6	5.95475267	0.007637369
7	6.88985141	0.011129644
8	7.87972194	0.015864081
9	8.90795326	0.007244545
10	10.0642228	0.014663302

6 Conclusion

Tackling the theoretical bottlenecks in traditional PCA algorithms and focusing on two critical quantities, namely latency and throughput, the current work supports the renewed interest and performance improvements in streaming machine learning. As our experiments show, streaming PCA models can be leveraged by the accumulate/retract framework. Such a system offers flexibility, allowing for arbitrary combinations of multiple functions underlying complex machine learning models (i.e. average, least squares regression, histogram sorting) to be calculated on the stream, with no time- and resource-penalty, by exploiting the underlying hardware, data processing and data management for true low-latency, high-throughput stream processing. In our predictive maintenance benchmark scenario, our system could enable immediate alarming of conditions outside the normal mode of operation (i.e. the system will provide the class label corresponding to normal/fault operation). Failure identification would be based on the capability of the system to "fingerprint" potential failure branch types, based on the underlying eigenvectors/eigenvalues configurations. More experiments are planned to prove the competitiveness of the approach.

References

1. Bifet, A., Gavaldà, R., Holmes, G., Pfahringer, B.: Machine Learning for Data Streams with Practical Examples in MOA. MIT Press, Cambridge (2018)
2. Axenie, C., Tudoran, R., Bortoli, S., Hassan, M.A.H., Foroni, D., Brasche, G.: STARLORD: sliding window temporal accumulate-retract learning for online reasoning on datastreams. In: 17th IEEE International Conference on Machine Learning and Applications, ICMLA 2018, Orlando, FL, USA, 17–20 December 2018, pp. 1115–1122 (2018)

3. Baker, C.G., Gallivan, K.A., Van Dooren, P.: Low-rank incremental methods for computing dominant singular subspaces. Linear Algebra Appl. **436**(8), 2866–2888 (2012)
4. Boutsidis, C., Garber, D., Karnin, Z., Liberty, E.: Online principal components analysis. In: Proceedings of the Twenty-Sixth Annual ACM-SIAM Symposium on Discrete Algorithms, pp. 887–901 (2015)
5. Carbone, P., Katsifodimos, A., Ewen, S., Markl, V., Haridi, S., Tzoumas, K.: Apache Flink^{TM}: stream and batch processing in a single engine. IEEE Data Eng. Bull. **38**, 28–38 (2015). https://flink.apache.org/introduction.html
6. Gajjar, S., Kulahci, M., Palazoglu, A.: Real-time fault detection and diagnosis using sparse principal component analysis. J. Process Control **67**, 112–128 (2018)
7. Hallgren, F., Northrop, P.: Incremental kernel PCA and the Nyström method. arXiv preprint arXiv:1802.00043 (2018)
8. Lois, B., Vaswani, N.: A correctness result for online robust PCA. In: 2015 IEEE International Conference on Acoustics, Speech and Signal Processing (ICASSP), pp. 3791–3795. IEEE (2015)
9. Oja, E.: Principal components, minor components, and linear neural networks. Neural Netw. **5**(6), 927–935 (1992)
10. Sanger, T.D.: Optimal unsupervised learning in a single-layer linear feedforward neural network. Neural Netw. **2**(6), 459–473 (1989). http://www.sciencedirect.com/science/article/pii/0893608089900440
11. Sharma, A., Paliwal, K.K., Imoto, S., Miyano, S.: Principal component analysis using QR decomposition. Int. J. Mach. Learn. Cybern. **4**(6), 679–683 (2013). https://doi.org/10.1007/s13042-012-0131-7
12. Susto, G.A., Schirru, A., Pampuri, S., McLoone, S., Beghi, A.: Machine learning for predictive maintenance: a multiple classifier approach. IEEE Trans. Ind. Inform. **11**(3), 812–820 (2014)
13. Valpola, H.: From neural PCA to deep unsupervised learning. In: Bingham, E. (ed.) Advances in Independent Component Analysis and Learning Machines, pp. 143–171. Elsevier, Amsterdam (2015)
14. Weng, J., Zhang, Y., Hwang, W.S.: Candid covariance-free incremental principal component analysis. IEEE Trans. Pattern Anal. Mach. Intell. **25**(8), 1034–1040 (2003)
15. Woo, S., Lee, C.: Incremental feature extraction based on decision boundaries. Pattern Recogn. **77**, 65–74 (2018). http://www.sciencedirect.com/science/article/pii/S003132031730496X
16. Xu, L., Oja, E., Suen, C.Y.: Modified Hebbian learning for curve and surface fitting. Neural Netw. **5**(3), 441–457 (1992)
17. Zhan, J., Lois, B., Guo, H., Vaswani, N.: Online (and offline) robust PCA: novel algorithms and performance guarantees. In: Artificial Intelligence and Statistics, pp. 1488–1496 (2016)
18. Zhao, F., Rekik, I., Lee, S.W., Liu, J., Zhang, J., Shen, D.: Two-phase incremental kernel PCA for learning massive or online datasets. Complexity **2019**, 1–17 (2019)
19. Zheng, H., Wang, R., Yang, Y., Li, Y., Xu, M.: Intelligent fault identification based on multi-source domain generalization towards actual diagnosis scenario. IEEE Trans. Ind. Electron. **67**(2), 1293–1304 (2019)

Event-Based Predictive Maintenance on Top of Sensor Data in a Real Industry 4.0 Case Study

Athanasios Naskos[1]([✉]), Georgia Kougka[1,2], Theodoros Toliopoulos[1,2], Anastasios Gounaris[2], Cosmas Vamvalis[1,3], and Daniel Caljouw[4]

[1] Atlantis Engineering, Thessaloniki, Greece
{naskos,kougka,vamvalis}@abe.gr
[2] Aristotle University of Thessaloniki, Thessaloniki, Greece
{georkoug,tatoliop,gounaria}@csd.auth.gr
[3] European Federation of National Maintenance Societies, Brussels, Belgium
chairman@efnms.org
[4] Philips Consumer Lifestyle, Drachten, The Netherlands
Daniel.Caljouw@philips.com

Abstract. We deal with a real predictive maintenance case study encountered in modern Industry 4.0 settings: based on logs of past failures, we train a model to predict critical failures of equipment without any other domain expert knowledge well in advance. We propose a novel methodology that combines and extends the state-of-the-art in event-based predictions with advanced time-series analytics. This renders our technique applicable directly onto the sensor data, as it is produced in a modern factory setting. Further, we show that employing unsupervised learning techniques, such as continuous outlier monitoring, is a competitive approach. Although our techniques are developed and tested in a specific case study, they can be transferred to arbitrary settings.

1 Introduction

One of the key characteristics of the 4th Industrial Evolution, broadly known as Industry 4.0, is the wide-spread of Predictive Maintenance (PdM), which is reported to be capable of yielding *"tremendous"* benefits.[1] The goal of PdM is to eliminate machinery downtime and operational costs. Broadly, PdM capitalizes on the benefits stemming from mature techniques in the fields of data analytics, machine learning and big data management in distributed and IoT/edge-computing settings [11].

In this work, we aim to devise PdM techniques, when there is no prior domain expert knowledge, e.g., rules or models that can be used to predict events. Event-based PdM approaches train models with logged events to recognize patterns

---

[1] https://www.pwc.nl/nl/assets/documents/pwc-predictive-maintenance-beyond-the-hype-40.pdf.

© Springer Nature Switzerland AG 2020
P. Cellier and K. Driessens (Eds.): ECML PKDD 2019 Workshops, CCIS 1168, pp. 345–356, 2020.
https://doi.org/10.1007/978-3-030-43887-6_28

which precede a failure incident (called *target events*). We explore both supervised and unsupervised learning techniques. For the former, a novel technique for timeseries discretization, which is based on [16], is utilized in order to leverage event-based prediction approaches, especially in cases where the provided maintenance logs are not sufficient to be used for predictions (which is a commonplace situation). In this way, the timeseries is transformed to a sequence of *artificial* events, which may not have meaning in the physical world but are shown capable of capturing preceding hidden patterns before failures. Then, an event-based PdM approach borrowed from the field of aviation industry is adapted to process such *artificial* events and to feed a regression analysis model that predicts prominent machinery failures.

In addition, we apply anomaly detection on streaming data [1] to detect early signs of failure. Anomaly detection naturally lends itself to failure detection in Industry 4.0; here we provide concrete evidence that it can also be used for prediction per se and thus become part of an advanced PdM solution. The key strength of this approach is that it does not rely on model training.

The approaches are tested in a real case study of a cold forming press in the Philips Consumer Lifestyle plant in the Netherlands, but we ensure that the approach is applicable to arbitrary settings, provided that sensor measurements are available and, for the supervised learning approach, there exists information about the time a failure occurred. We have provided the implementation of the core parts of both the supervised and unsupervised learning techniques as open source[2], so that third parties can easily adapt our solution to their problems. In summary, our contribution is threefold:

1. We present a novel methodology that is directly applicable to sensor measurements. Our key rationale is to transform timeseries to a series of events, which allow for state-of-the-art event-based PdM techniques to apply.
2. We present how streaming outlier detection can be leveraged to predict critical equipment failures.
3. We evaluate our solutions to a real setting, and the results are particularly encouraging. The supervised learning solution managed to achieve 61% precision and 61% recall (0.61 F1-score) when predicting 1–8 h ahead. The unsupervised learning one managed to achieve different but equally interesting trade-offs. The performance is further improved with combinations of predictors, e.g., exhibiting 59% recall and 82% precision (0.67 F1-score).

Paper structure: the next section presents the case study we consider. In Sect. 3, we present our novel methodology, which is based on combination of timeseries and event-based PdM techniques. Section 4 discusses the application of outlier detection directly on streaming data for PdM purposes. In Sect. 5, the obtained results are discussed. We conclude with the related work and the final remarks in Sects. 6 and 7, respectively.

---

[2] http://interlab.csd.auth.gr/anaskos/pdm-solution.

2   The Case Study

High-Level Description. The focus of the current use case is the production line of a Philips factory and more specifically, the cold forming press that is a part of the line. During a production run the input material passes through many form-shaping and quality check stages to reach its final form. The cold forming press is one such form-shaping stage and a complex and expensive piece of equipment, where a metal strip goes in and cold formed products come out. The press contains various modules that cut, bend and flatten the input strip. The main problem of the specific machinery is that it is acting as a black box, which means that there is no option to monitor its status or the status of each individual module that is contained within it, without completely stopping the machine. The modules within the press are arranged in a specific order to produce the correct output shape. The different modules form a pipeline, where the output of one module upstream forms the input of the next module downstream. The metal strip that comes into the press passes through all of the modules in the specified order with its shape constantly changing. The metal strip comes from an input reel and passes through the first quality check. The strip then enters the press and is processed by each of the six modules. After the strip takes its final form, it exits the press and passes a second quality check.

The modules are subject to breakage induced by age and other reasons, which cascade from one module to the next in line and so forth. Hence, planned preventive maintenance checks are applied removing the whole press from the production line. The cost for stopping the production run and especially the cost for repairing the faulty parts of the machinery can be very high for the company. This means that there is a need for some type of prediction or early detection of failures in the cold forming press.

Failure Types and Sensors Available. There are multiple data sources in place. The first one considers the profile of the raw material (i.e. metal strip), such as part number, while the second one considers the thickness and the temperature of the metal strip entering the press. The quality check at the end product of the press contain measurements on the thickness, the shape and other proprietary quality measurements. The final data source comes from the press itself. To monitor the opaqueness of the cold forming process, an acoustic emission sensor is placed in the press; overall, the different acoustic sensors are comprise 6 channels, and at each step, a measurement for each angle and channel is monitored (the dimensionality of each acoustic channel is at the orders of several hundreds). The sensor detects and emits the acoustic waves that radiate from the material when it is processed by the press. The difference on the emission can detect possible faults on the parts of the press that cannot be detected through other means. In the following sections, acoustic emission measurements are used to predict and/or detect prominent machine stops due to machinery failures.

3   A Supervised Learning Technique Based on a Timeseries Discretization Methodology

Our rationale is summarized as follows. We take for granted that event-based machine learning for PdM is a rather mature field, in the sense that PdM techniques tailored to Industry 4.0 setting have been developed, e.g., [8,12,14]. The key characteristics of a PdM task compared to traditional classification is that events are very rare and the feature set very sparse. Then, the main challenge is to fill the gap from the initial measurements from sensors to event generation. Our key novelty is that, instead of classifying timeseries as a discrete set of events, we map timeseries to a sequence of *artificial* events thus placing no burden to engineers to annotate the sensor measurements. We only require information about the time of several critical failures in order to train our methods.

The key part is the discretization of the timeseries to a series of events. To this end, we employ the *Matrix Profile* (*MP*), which is a data structure that annotates a time-series [16][3].

3.1   Data Pre-processing and MP-Based Timeseries Analytics

For the purpose of the Philips's case study, we focus on subsequences of a predefined pattern-length of size PL. The subsequences's length corresponds to individual acoustic measurements. Accoustic measurements are processed per channel after taking the maximum value among all angles for each time point; therefore, all acoustic channels are transformed to 1-dimensional series. For a timeseries T of length n, we estimate MP, which is a vector of length $n - PL + 1$. $MP(i)$ denotes the distance of the sub-sequence starting at the ith position in T to its nearest neighbor. Any distance metric can be used, but as explained in [16], the default option is the *z-normalized Euclidean Distance*. The MP vector is accompanied by the *Matrix Profile Index (MPI)*, which is of same size as MP. $MPI(i)$ keeps the pointer to the position of the closest neighbor of the subsequence of length PL starting at $T(i)$. The lower the values in the MP, the higher the similarity of the PL-size pattern beginning at the corresponding point to its closest neighbor.

3.2   The MP-based Algorithm to Extract Artificial Events

We introduce an algorithm that is based on the estimation of MP in order to create the artificial, yet significant, events through similarity estimates. Then, on top of this, a complete technique for extracting hidden patterns to predicting or early detect failures is developed. The technique for artificial event extraction is summarized in Algorithm 1.

The proposed technique does not require any logs apart from raw measurements, not even information about past failures. Given the pattern-length parameter (PL), we apply the MP methodology in order to compute the MP based on

---

[3] Open source implementations are provided from https://www.cs.ucr.edu/~eamonn/MatrixProfile.html.

Algorithm 1. Mapping timeseries values to artificial events

Require: A time-series of press incidents $T = t_1, \ldots, t_n$, the pattern-length PL.
1: MP,MPI ← computeMatrixProfile(T,pl)
2: X ← 10 //default value for the X threshold
3: Y ← 2 //default value for the Y threshold
4: Z ← 2 //default value for the Z threshold
5: **for** $i = 1$ to $n - PL + 1$ **do**
6:    **if** MP(i) > X*MP(MPI(i)) **then**
7:       Remove the corresponding edge from MPI
8:    **end if**
9:    **if** MP(i) > Y*mean(MP) **then**
10:       Remove the corresponding edge from MPI
11:    **end if**
12: **end for**
13: $conComp$ ← Calculate the weakly connected components of MPI
14: Filter the distinct components with less than Z members
15: **for** $i = 1$ to $n - PL + 1$ **do**
16:    **if** MPI(i) belongs to a connected component **then**
17:       map T(i) to the connected component id
18:    **end if**
19: **end for**

---

subsequences of length PL and thus also generate the MPI. MPI is essentially a directed graph, where each edge points to the most similar subsequence. We consider the MPI as a graph $G = (V, E)$, where the $V = v_1, \ldots, v_n$ denotes a set of nodes and $E = e_1, \ldots, e_z$ defines the edges of the graph G weighted by the values in MP. Some edges have either globally or locally very high weights. Therefore, we apply a set of thresholds in order to eliminate the nodes that probably are noise and are connected to other nodes with low similarity. More specifically, we filter out the edges of the graph that connect two nodes when their distance is X times greater than the distance of the edge connecting the sink node to its nearest neighbor (local rule). A global rule is that we prune all edges with a weight more than Y times the mean MP value. As a following step, we estimate the weakly connected components (sub-graphs) of the MPI graph and map each such component to a distinct artificial event, i.e., in this step we disregard edge directions. We prune small components with less than Z members. Finally, every point of the timeseries that is part of a connected component is labeled by the id of that component. In Algorithm 1, we provide default values for the three thresholds employed, based on our experiments in the real dataset (we omit sensitivity analysis due to space constraints).

Variants. Similarly to the version of the algorithm that we described above, we have also implemented and tested a community detection algorithm, proposed in [4], for finding communities of graph nodes instead of connected components. Then all the members of a detected community are labeled with the community id.

3.3 Supervised Event-Based PdM Approach Using Artificial Events

Utilizing the events generation technique presented in Sect. 3.2, we are able to use any data-driven event-based prediction approach to tackle the PdM problem. In this work, we showcase the usage of a PdM approach applied on aviation data [8], adapted to the Philips use case. Adapting the selected PdM approach to the press use case, the *artificial* events obtained by the MP-based algorithm are subject to intensive preprocessing in order to expose the patterns of machine failure and then leverage such patterns to train a model to predict prominent machine failures. The proposed approach penalizes both rare and frequent events (implicitly performing feature selection) and amplifies the strength of the events closer to machine failure incidents, applying a Multi-Instance Learning (MIL) technique to over-sample the aforementioned events. Such preprocessed log data form the training set, which is then fed into a regression analysis algorithm for the prediction of the machine failures. Next, we further elaborate on this approach as a key representative of the state-of-the-art.

3.4 Event-Based PdM Solution Details

The artificial events are mapped to actual timestamps based on the origin time-series and partitioned in ranges defined by the occurrences of the fault that PdM targets. These ranges are further partitioned into time segments, the size of which (i.e. minutes, hours, days) correspond to the time granularity of the analysis. In the press use case, hourly segments are used, based on the knowledge acquired from the maintenance engineers. More specifically, engineers want to be warned at least one hour and at most several hours before the occurrence of a machine failure. The rationale behind the time segmentation is that the segments that are closer to the end of the range may contain fault events that are potentially indicative of the main event. The goal is to learn a function that quantifies the risk of the targeted failure occurring in the near future, given the events that precede it. Hence, a sigmoid function is proposed, which maps higher values to the segments that are closer to a machine failure. The steepness and shift of the sigmoid function are configured to better map the expectation of the time before the failure at which correlated events will start occurring. The segmented data in combination with the risk quantification values are fed into a Random Forests algorithm as a training set to form a regression problem.

In practice, the event types are hundreds if not thousands. Each event type is essentially a dimension. Therefore, to increase the effectiveness of the approach standard preprocessing techniques can be applied: (i) Multiple occurrences (*MO*) of the same event in the same segment can either be noise or may not provide useful information. Hence, multiple occurrences can be collapsed into a single one. (ii) Standard feature selection (*FS*) techniques (like [5]) can also be used in order to further reduce the dimensionality of the data.

Finally, to deal with the imbalance of the labels (given that the fault events are rare) and as several events appear shortly before the occurrence of the fault events, but only a small subset of them is related to them, Multiple Instance Learning (*MIL*) can be used for bagging the events and automatically detecting the events that can act as predictors. A single bag contains events of a single hour. Also, the data closer to the fault events (according to a specified threshold) are over-sampled, so that training is improved.

3.5 Experiments and Results

The experiments were done using historical timeseries and converting them to a continuous stream. The ground truth used for the measurements is the information of the timestamps that the machine stopped working due to technical reasons, e.g. damage of a module on the press; this information has been provided by the engineers responsible for the examined machinery. Each machine stop represents a failure mode, each prediction represents an alarm and the detected stops are the ones that have at least one preceding alarm within a fixed period before the fault. We assess the efficiency of the technique using the *recall* and *precision* metrics adapted to the PdM context, measured according to the following definitions: *precision* is the ratio of the successfully predicted stops to the number of total alarms, and *recall* is the ratio of the predicted stops to the number of total stops, where a stop is considered as successfully predicted if there is any prediction made in a specified time gap before a machine stop. Multiple alarms inside the specified time gap for the same machine stop are counted as a single alarm, while the false alarms (i.e. before the time gap) are counted individually. The rationale is that the maintenance engineers are prompted to respond to the first alarm for a specific machine stop, while in the case of the false alarms, they are called to respond to every one of them.

The data used for the assessment of the supervised learning approach are the acoustic emission measurements. The acoustic emission sensors are placed in 6 different spatial positions on the cold forming press, generating data in 6 distinct channels providing measurements in hundreds of different angles per channel. We perform dimensionality reduction through maintaining only the maximum value across all angles per channel per time point. I.e., finally, for each timestamp, we consider a single measurement per acoustic channel.

The experiments share a common parametrization and fine-tuning is beyond the scope of this work (due to space limitations). Three values of pattern-length (PL) for the Matrix Profile are used (i.e. 5, 10 and 50). The number of distinct artificial events are depicted in Table 1. MO is enabled in all the experiments and over-sampling is applied. The steepness and the shift parameters of the sigmoid function are set to 0.8 and 4, respectively; the threshold for the value of the sigmoid function to set an alarm is set to 0.3, while the time gap for true alarm consideration is set between 1 and 8 h before a machine stop incident. All measurements refer to 10-fold cross validation. As there are lots of event types, FS preprocessing step is also tested. For partitioning the dataset into 10 folds, we use the number of incidents and not the number of time segments.

Table 2 presents the recall and precision values, of the results that achieved the best F1-score per channel. The second column depicts the pattern-length used in the Matrix Profile algorithm, while the third one indicates the usage of the FS preprocessing step. The Table also presents the results in two of the channels (i.e. 1st and 2nd) where the community detection (CD) algorithm

Table 1. Number of distinct artificial event types generated per pattern length (PL).

Source	PL-5	PL-10	PL-50
Channel 1	6411	5615	1794
Channel 2	6401	5800	5234
Channel 3	6374	5722	3591
Channel 4	6334	5148	1305
Channel 5	6353	5837	3445
Channel 6	6396	5724	4216

Table 2. Experimental results on all the acoustic channels using the supervised learning technique (CD: community detection).

Source	PL	FS	Recall	Precision	F1-score
Channel 1	5	x	0.61	0.61	0.61
Channel 1 CD	5	x	0.57	0.63	0.59
Channel 2	10		0.63	0.53	0.55
Channel 2 CD	5		0.49	0.45	0.47
Channel 3	10	x	0.55	0.61	0.57
Channel 4	50		0.49	0.76	0.59
Channel 5	50	x	0.45	0.80	0.54
Channel 6	50	x	0.48	0.68	0.50

Table 3. Experimental results on all the acoustic channels using an ensemble of the supervised learning technique.

Source	Strategy	Recall	Precision	F1-score
Ch.4 - Ch.1	AND	0.62	0.5	0.55
Ch.1 CD - Ch.1	OR	0.59	0.82	0.67

is used in place of the connected component (CC) algorithm utilized in the MP-based *artificial* event generation approach. As we observe, Channel 1 and Channel 4 achieved the highest F1-score (0.61 and 0.59 resp.). There is no clear winner between the different pattern-lengths and whether feature selection has been applied or not. Regarding the application of the CD, the results are inferior to those achieved by CC, despite the fact that the number of the generated artificial event types is almost the same in both the cases.

Next, we employ two simple ensemble strategies with two predictors each: the AND strategy, where two predictors need to raise an alarm, and OR strategy, where an alarm is raised whenever at least one of the predictors votes for it. We have computed the precision, recall and F1-score of all the possible pairs between all the previous experiments. The results with the highest F1-score per strategy are shown in Table 3. As we observe, the OR strategy was able to enhance the previous results, achieving 0.67 F1-score combining two cases with low recall but high precision. Note that in this scenario, a random predictor achieved F1-score of 0.31; moreover, a dummy predictor with recall 1 through raising an alarm every 7 h cannot exceed F1-score of 0.58.

Table 4. Best results in terms of F1-score of the outlier detection technique for each of the 6 acoustic channels. *AND* and *OR* refer to the ensemble experiments.

Source	Recall	Precision	F1 score
Channel 1	0.57	0.41	0.48
Channel 2	0.83	0.48	0.61
Channel 3	0.74	0.49	0.59
Channel 4	0.38	0.82	0.52
Channel 5	0.73	0.48	0.58
Channel 6	0.84	0.43	0.58
Ch.5 - Ch.6 *AND*	0.72	0.75	0.74
Ch.2 - Ch.4 *OR*	0.84	0.57	0.68

4   An Unsupervised Learning Technique

In this section, we present the streaming distance-based outlier detection algorithm, namely MCOD [7], that was used for early detection of failure on the dataset. We employ sliding windows. Given a set of objects $\mathbb{O}$ and the threshold parameters R and k, we report all the objects $o_i \in \mathbb{O}$ for which the number of objects o_j, $j \neq i$ for which $dist(o_i, o_j) \leq R$ is less than k. The report should be updated after each window slide. Note that according to this definition, outliers may be reported during any time they belong to the window and not necessarily when they are first inserted into it.

The experimental setting is the same as the one used in the supervised approach with the difference that the *precision* is the ratio of all the true alarms to the number of total alarms, where true alarm is any prediction (in the form of warning outlier detection) made in a specified time gap before a machine stop. To avoid the problem of fine-tuning, we experimented with the combinations of 3 values of R and 3 values of k, i.e., 9 combinations of parameters. The window contains the last 3600 measurements and the window slide is fixed to 10% of the window size. As previously, we aim to predict faults 1 to 8 h ahead.

In almost all of the experiments, there is a trade-off between recall and precision. Based on the algorithm parameters chosen, this trade-off can be configured in favor of either of the measures depending on the output needed or preferred. Table 4 shows the best results per channel in terms of the F1 score. An observation that can be drawn from the table is the difference in the results among the data sources, which means that each source can have a different impact on the PdM technique. It seems that especially regarding the acoustic emission sensors, specific positions in the press can yield better understanding of the faults in the press. In general, the unsupervised learning one managed to achieve different but equally interesting trade-offs compared to the supervised one, e.g., 83% recall and 48% precision (0.61 F1-score). We also tested an ensemble of the different parameterizations of the acoustic emission channels. Running more than one parameterization helps to improve the results and either exacerbate or mitigate the trade-off. Certain combinations can greatly increase one measure while slightly decreasing the other one. The last two lines of Table 4 show indicative results from this experiment.

5   Discussion

In the previous sections, we have presented a supervised predictive approach based on a novel timeseries discretization approach and an unsupervised detection approach used as predictive mechanism. Both the approaches are applied on the same use case and dataset. The results of the two approaches suggest that there actually exist preceding hidden patterns or indications for most of the machine stops and prove that the novel MP-based timeseries discretization approach has managed to successfully reveal those previously unknown patterns. In addition, outlier detection can also enhance PdM.

Some further comments on the logs are as follows. The logs with the machine stops contained 82 records of 8 machine stop categories. These categories are quite different in their nature. We did not preprocess the machine stop logs before the execution of the experiments in order to assess the robustness of the approaches. Cleaning the logs is expected to decrease the precision and affect the recall of the unsupervised approach, however it will potentially increase the metrics of the supervised approach. Also, to stress-test the supervised learning approach, we have considered all the different machine stop categories as a single failure category, thus training a single model. The results of the supervised approach will potentially improve if separate models are used for each individual machine stop category.

Finally, we clarify that the purpose of this research work is not to compare the supervised against the unsupervised learning approach and promote the most efficient or appropriate one. On the contrary, the goal is to promote the strengths of each approach and to provide the basis for an heterogeneous ensemble solution utilizing multiple instances of both the supervised and unsupervised approaches. Overall, we envisage a multi-layer ensemble solution, where different instances of the same predictor type are combined at a lower layer, while, at a higher-level, different types of predictors form an ensemble.

6   Related Work

Data-driven techniques, where the data refer to past events, commonly in the form of log entries, are widely used in PdM. [8] is a key representative of the state-of-the-art. Another event-based approach is presented in [12], where historical and service data from a ticketing system are combined with domain knowledge to train a binary classifier for the prediction of a failure. As in the previous work, a feature selection [3] and an event amplification technique is used to enhance the effectiveness of the SVM-based classifier. Event-based analysis, based on event and failure logs, is also performed in [14], where it is assumed that the system is capable of generating exceptions and error log entries that are inherently relevant to significant failures. This work relies on pattern extraction and similarity between patterns preceding a failure, while emphasis is posed on feature selection.

The work in [17] proposes a correlation-driven approach between different sensor signals and fault events to guide the PdM process. This approach tries to identify correlations between detected anomalies in different sensor signals, which are mapped to specific faults. Here, we focus on event processing, where events are generated from the sensors artificially.

Data-driven PdM is also related to online frequent episodes mining; research works [2] and [10] propose techniques in this topic. The key strength of [2] is that it can apply to an online (i.e. streaming) scenario. [10] further improves upon it through providing solutions for the case where the event types is unbounded. Complex-event processing (CEP) [6] is also a technology that enables PdM enforcement after warning sequential patterns have been extracted. A good overview of the data-driven PdM is presented in [9].

Motif-detection in timeseries can also be used in prediction scenarios. The authors in [15] propose a tool that is able to predict outcomes based on weakly labeled time series of millions of data points. Finally, outlier detection is a vivid research field that has developed broad and multifaceted algorithmic solutions. The comparative study [13] presents a wide range of distance-based outlier detection algorithms and suggests that the MCOD algorithm, which is employed in this work, is considered as a state-of-the-art solution in the streaming data processing for distance-based outlier detection.

7 Conclusions and Future Work

In this work, three state-of-the-art techniques in timeseries analysis, event-based PdM and streaming outlier detection are leveraged in order to provide effective PdM solutions operating directly on the output of sensors. The key strength is that no domain expert knowledge is required and one of the techniques does not require model building at all. More specifically, a novel timeseries discretization approach is proposed for the generation of *artificial* events, in order to enable the utilization of event-based predictive approaches. In parallel, distance-based outlier detection is shown to be effective in capturing early signs of abnormal equipment behavior. The solution is evaluated in a real setting, and the results are particularly encouraging achieving high F1 scores and useful trade-offs between the recall and precision metrics. As future work, we intend to work towards an ensemble solution and derive an efficient manner to tune the various parameters involved automatically. However, the most important next step is to build on these early insights into the benefits of our proposal and proceed to a more thorough experimentation and testing.

Acknowledgement. This research work is funded by the Z-BRE4K project (funded by European Union's Horizon 2020 research and innovation program under grant agreement No 768869).

References

1. Aggarwal, C.C.: Outlier analysis. Data Mining, pp. 237–263. Springer, Cham (2015). https://doi.org/10.1007/978-3-319-14142-8_8
2. Ao, X., Luo, P., Li, C., Zhuang, F., He, Q.: Online frequent episode mining. In: IEEE 31st International Conference on Data Engineering (ICDE), pp. 891–902 (2015)
3. Bach, F.R.: Bolasso: model consistent Lasso estimation through the bootstrap. In: Proceedings of the 25th International Conference on Machine learning, pp. 33–40. ACM (2008)
4. Ghosh, S., et al.: Distributed Louvain algorithm for graph community detection. In: IPDPS. pp. 885–895 (2018)
5. Kira, K., Rendell, L.A.: The feature selection problem: traditional methods and a new algorithm. AAAI **2**, 129–134 (1992)
6. Kolchinsky, I., Schuster, A.: Efficient adaptive detection of complex event patterns. PVLDB **11**(11), 1346–1359 (2018)
7. Kontaki, M., Gounaris, A., Papadopoulos, A.N., Tsichlas, K., Manolopoulos, Y.: Efficient and flexible algorithms for monitoring distance-based outliers over data streams. Inf. Syst. **55**, 37–53 (2016)
8. Korvesis, P., Besseau, S., Vazirgiannis, M.: Predictive maintenance in aviation: failure prediction from post flight reports. In: IEEE International Conference on Data Engineering (ICDE), pp. 1414–1422 (2018)
9. Kovalev, D., Shanin, I., Stupnikov, S., Zakharov, V.: Data mining methods and techniques for fault detection and predictive maintenance in housing and utility infrastructure. In: 2018 International Conference on Engineering Technologies and Computer Science (EnT), pp. 47–52 (2018)
10. Li, H., Peng, S., Li, J., Li, J., Cui, J., Ma, J.: ONCE and ONCE+: counting the frequency of time-constrained serial episodes in a streaming sequence. arXiv preprint arXiv:1801.09639 (2018)
11. Lu, Y.: Industry 4.0: a survey on technologies, applications and open research issues. J. Ind. Inf. Integr. **6**, 1–10 (2017)
12. Sipos, R., Fradkin, D., Moerchen, F., Wang, Z.: Log-based predictive maintenance. In: Proceedings of the 20th ACM SIGKDD International Conference on Knowledge Discovery and Data Mining, pp. 1867–1876. ACM (2014)
13. Tran, L., Fan, L., Shahabi, C.: Distance-based outlier detection in data streams. Proc. VLDB Endowment **9**(12), 1089–1100 (2016)
14. Wang, J., Li, C., Han, S., Sarkar, S., Zhou, X.: Predictive maintenance based on event-log analysis: a case study. IBM J. Res. Dev. **61**(1), 11–121 (2017)
15. Yeh, C.C.M., Kavantzas, N., Keogh, E.: Matrix profile IV: using weakly labeled time series to predict outcomes. Proc. VLDB Endowment **10**(12), 1802–1812 (2017)
16. Yeh, C.M., et al.: Matrix profile I: all pairs similarity joins for time series: a unifying view that includes motifs, discords and shapelets. In: IEEE ICDM, pp. 1317–1322 (2016)
17. Zhu, M., Liu, C.: A correlation driven approach with edge services for predictive industrial maintenance. Sensors (Basel, Switzerland) **18**(6), 1844 (2018)

Forecasting of Product Quality Through Anomaly Detection

Mehmet Dinç[1(✉)], Şeyda Ertekin[1], Hadi Özkan[2], Can Meydanlı[2],
and Volkan Atalay[1]

[1] Department of Computer Engineering, Middle East Technical University,
Ankara, Turkey
{dinc.mehmet,sertekin,vatalay}@metu.edu.tr
[2] Arçelik, Istanbul, Turkey
{hadi.ozkan,can.meydanli}@arcelik.com

Abstract. Forecasting of product quality by means of anomaly detection is crucial in real-world applications such as manufacturing systems. In manufacturing systems, the quality is assured through tests performed on sample units randomly chosen from a batch of manufactured units. One of the major issues is to detect defective units among the sample test units as early as possible in terms of test time and of course as accurate as possible. Traditional way of detecting defective units is to make use of human experts during test. However, human intervention is prone to errors and it is time consuming. On the other hand, automated systems are efficient alternatives and of assistance to human experts. There are on-line and off-line approaches for automated systems. Our ultimate aim is to design a system that automates the detection of defective units among the sampled freezer units manufactured in high volumes in a factory of one of the leading home appliances manufacturers. We start by analyzing the data of the test units sampled from the batches of freezer units. For analysis, we first embedded data in two-dimensional space to observe if there are any structures exist in the data. Clustering was then applied to see if the data can be grouped into two classes without their labels. As off-line approaches, state-of-the-art classifier methods including one-class-classifier are employed. Finally, a deep learning method for time-series analysis combined with a classifier is applied as an on-line method.

Keywords: Product quality · Anomaly detection · Forecasting · Manufacturing systems · On-line analysis · Off-line analysis

1 Introduction

Freezers are durable consumer goods that are manufactured in mass volumes. In order to assure the quality during the manufacturing process, a certain number of freezer units are randomly selected from each production batch and the selected freezer units are tested for various types of defects including cosmetic defects

© Springer Nature Switzerland AG 2020
P. Cellier and K. Driessens (Eds.): ECML PKDD 2019 Workshops, CCIS 1168, pp. 357–366, 2020.
https://doi.org/10.1007/978-3-030-43887-6_29

and functional defects. In general, a human expert conducts tests, interprets data and results, and concludes a decision. The way of detecting defective units by means of human experts is prone to errors and takes time. It is highly desirable to automate the process such that forecasting of product quality is performed by means of anomaly detection preferably by applying machine learning and data-driven methods.

Our ultimate aim is to design a system that automates the detection of defective units during cooling tests of freezer units manufactured in high volumes in a factory of one of the leading home appliances manufacturers, Arçelik (Beko). In the design of such a system, extra attention should be paid to the sensitivity (accuracy on detecting defective units), since missing a defective freezer unit might potentially lead to a totally defective batch that may be delivered to the market. However, false alarms (false positives) would only lead to an extra manual test which is a small drawback comparing to a miss (false negative). We start by analyzing the data of the test units sampled from the batches of freezer units. Data is then embedded onto two-dimensions to visualize its distribution. Such a visualization may yield particularly structures and outliers existing in the data. Clustering is then applied to see if the data can be grouped into two classes. As off-line approaches, state-of-the-art classifier methods including one-class-classifier are employed. Finally, a deep learning method for time-series analysis combined with a classifier is applied as an on-line approach.

2   Related Work

Anomaly detection is a major forecasting method used in assessing the product quality in real-world applications. Several methods have been proposed to detect anomalies in data [1,3]. Traditionally, statistical methods such as cumulative sum (CUSUM) and exponentially weighted moving average (EWMA) were employed [2]. There exist also methods based on Support Vector Machines (SVM) [7,11]. When anomaly detection is defined as outlier detection, the solution may come from one-class support vector machine as well [13]. Technology companies such as Twitter and Netflix have also proposed their own solutions for this problem [5,6]. With its reemergence, Long Short Term Memory Networks (LSTM) [4] became the most popular method for time series modeling and forecasting. Various methods exist to incorporate LSTMs; both stacked LSTMs and an LSTM-based encoder-decoder for detecting anomalies in time series data have been described [9,10]. However, a set of rules must be set in order to decide, if the predicted points are indeed anomalies; Shipomon et al. compared different rules for anomaly detection [12].

3   Data

The results of every freezer unit that has been tested between 2016 and 2018 at Arçelik (Beko) Refrigerator Plant are available as data. During the test of a randomly selected unit, two sensors measure the temperature inside the freezer unit and another sensor measures the ambient temperature. Finally, a fourth sensor measures the power consumed by the compressor. The freezer compressor is both a motor and a pump that move the refrigerant through the system. These measurements are recorded at each minute. Since the two sensors inside the freezer unit give out the same temperature, we make use of only one of them. The ambient temperature stays constant during the test. Therefore, only one temperature sensor data is considered in the rest of this study.

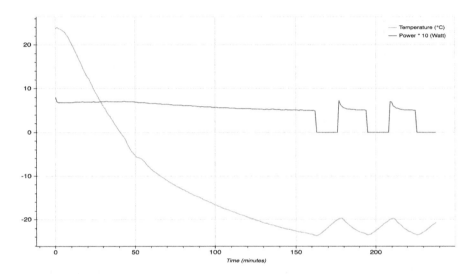

Fig. 1. Sensor reading data of a non-defective freezer unit during cooling test. (Color figure online)

Expected behavior of a non-defective freezer unit is given in Fig. 1 and it can be described as follows. Temperature sensors start from ambient temperature and drops down until around $-20\,°C$, while power is consumed steadily. After reaching the target temperature, freezer unit starts its cycling phase. During this phase, compressor stops and thus power sensor starts reading zero. Simultaneously, temperature starts rising for a few degrees. Compressor then turns on starts consuming power and also cooling the freezer unit. In Fig. 1, data drawn with red color correspond to the temperature inside the freezer unit while the consumed power data is given by purple color. A freezer unit is labeled as defective if it is unable to reach its target temperature in a few hours. The test goes on until the human expert decides whether the unit is defective or not. This means that the tests vary in terms of time depending on the experience of the expert. Even though most tests are concluded in 120 min, there are tests that last more than 220 min. Especially, the defective units are tested for several hours so that the cause of failure becomes clear. In addition, freezer unit model also affects the test time. Different models may take different lengths of test time to reach their target temperatures. Similarly, there are some events that disrupt the test process. During the test, a unit might be of subject for further examination or it might be tested for extreme cases. As a result, we have data of units for which the human expert indicates to be OK with potentially anomalous measurements. If a unit is labeled to be defective, a metadata is recorded including model, product identification number, batch number, test date, and error code. Furthermore, defective units are grouped according to their types of defect and labeled with respective error codes.

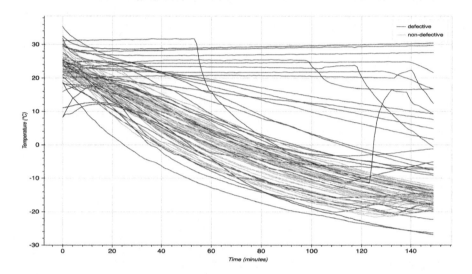

Fig. 2. Plot of temperature sensor measurements of a sub-set of the original dataset for 150 min of the test. (Color figure online)

Figure 2 presents the plot of temperature sensor measurements of a subset of the original dataset. Data drawn with red color represent the non-defective freezer units whereas data drawn with blue color refer to the defective freezer units.

4   Initial Analyses of Data

4.1   Embedding the Data in 2D Space

150 min temperature sensor data can be considered as a 150 dimensional feature vector. We have embedded 150 dimensional feature vectors in a lower dimensional space using t-Distributed Stochastic Neighbor Embedding (t-SNE) [8] with its perplexity parameter set to 50. The embedding of data from 150-dimensional space into 2-dimensional space is shown in Fig. 3; the data points with red color correspond to defective freezer units while data points with black color correspond to non-defective freezer units. Most of the data points corresponding to defective freezer units are grouped in a cluster while the data points corresponding to non-defective units are spread out. It is interesting to individually analyze

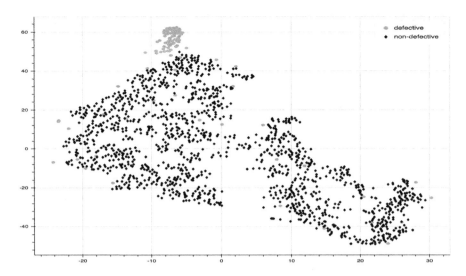

Fig. 3. Embedding of temperature sensor data from 150-dimensional space onto 2-dimensional space using t-SNE. (Color figure online)

each of the defective-labeled freezer units which are embedded among or nearby non-defective units. About 95% of the defective freezer units can be separated from non-defective units by a classifier based on a Support Vector Machine with linear kernel.

4.2   Clustering the Data

We wanted to see if the data could be clustered with respect to the temperature sensor measurement values into two groups as properly working and defective freezer units. 150 dimensional feature vectors are used for a subset of freezer units in the dataset. The result of applying k-means clustering algorithm with $k = 2$ is shown in Fig. 4. Even though the data looks clustered neatly, resulting cluster labels do not match with the original labels shown in Fig. 2.

4.3   Applying Classifiers

We have applied state-of-the-art classification algorithms to the original dataset in order to have a baseline for further improvements. Since the dataset is imbalanced, we have applied under-sampling to the data corresponding to non-defective freezer units. We have applied the classifiers with 10-fold cross validation and Table 1 shows the average accuracy and sensitivity values for the tests.

4.4   One-Class Classification

One-class classification is a very common method employed in outlier detection. We have used One-Class Support Vector Machines in order to be able to define

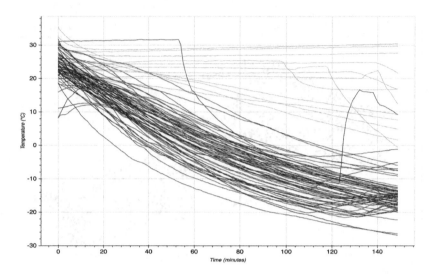

Fig. 4. Result of k-means clustering based on temperature sensor measurement values for 150 min of test.

a hyper-plane containing of all data points corresponding to non-defective units. Any data point that is outside this hyper-plane is considered to be an anomaly (or defective). The results of this method greatly depend on the strictness of the hyper-plane. When we select a hyper-plane that contains all of the data points corresponding to non-defective units, only 70% of the data points corresponding to defective units remain outside the hyper-plane whereas the other 30% are inside the hyper-plane. This gives an accuracy score of 85% and a sensitivity score of 70%. By changing the hyper-plane, these scores may change. However, the best scores reached with this method are 85% of accuracy and 80% of sensitivity.

Table 1. Average accuracy and sensitivity values for different classifiers.

	Decision tree	Random forest	SVM (RBF kernel)	Multilayer perceptron
Accuracy (%)	85	85	78	84
Sensitivity (%)	79	80	84	78

4.5   Applying Long Short-Term Memory Network Model

In order to make use of the time-series property of the data, we consider applying Long Short-Term Memory (LSTM) network models. The behavior of non-defective freezer units under the cooling test is modeled by training a LSTM network model with only data items corresponding to non-defective freezer units. A fixed-length sliding window of a data item is taken as input in order first to train for and then eventually to predict the value at the subsequent time step.

We then calculate the error between the predicted value and the real measurement value at that particular time step. We hypothesize that the error would be higher for data items corresponding to defective freezer units compared to those corresponding to non-defective units.

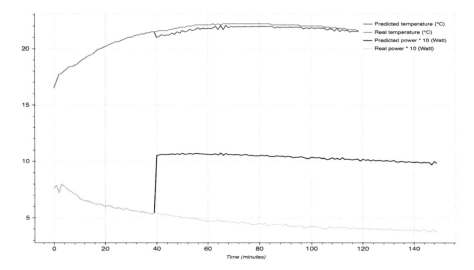

Fig. 5. Plot of the data of the two sensors (temperature sensor and power sensor) of a defective unit along with the prediction values of the trained LSTM network model. (Color figure online)

Figure 5 gives the plot of a data item of the two sensors (temperature sensor and power sensor) for a defective unit along with the predictions of trained LSTM network model. The actual temperature sensor values are shown by red, temperature prediction values by blue, actual consumed power values by cyan and finally power prediction values by black. Power values are multiplied by 10 in order to be able to show the details. Figure 6 gives the plot of the data of the two sensors of a non-defective unit. It can be observed that prediction values and the real values are very close in the non-defective unit graph. However, in the defective case, the error between the prediction and actual values is high.

A threshold value for the error should be set in order to be able to decide if the data item corresponding to a freezer unit is defective or not. There is again a trade-off between the accuracy and the sensitivity score. We have tried several window sizes; starting from a window size of 1 min and going up to 90 min. In addition, different test duration values from 70 min up to 150 min and different threshold values are tested. The best result is achieved by using a 40 min window length: 91% of accuracy and 88% of sensitivity. Figure 7 shows the effect of the threshold level on the accuracy. By lowering the threshold, sensitivity can be increased. However, this also increases the number of false positives. An ideal threshold should be determined according to the needs of the manufacturer.

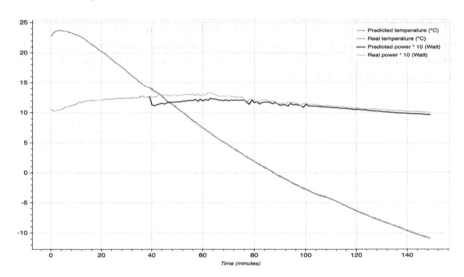

Fig. 6. Plot of the data of the two sensors (temperature sensor and power sensor) of a non-defective unit along with the prediction values of the trained LSTM network model.

Reducing the test time is one of the objectives of this study. We have picked the best two performing window sizes and tested them with shorter test duration. Table 2 shows the accuracy and sensitivity values for these settings. As expected, as the test time reduces accuracy scores drop as well.

Table 2. Effect of test duration on the accuracy and sensitivity.

	150 min	140 min	130 min	110 min	90 min	70 min
40 min window accuracy (%)	91	90	89	87	87	86
60 min window accuracy (%)	92	91	90	88	88	87
40 min window sensitivity (%)	88	86	84	79	79	78
60 min window sensitivity (%)	87	86	83	79	79	78

Aside from the increase in accuracy scores, LSTM network model also has the advantage of being an on-line algorithm. This means that, the model can be run simultaneously with the product test.

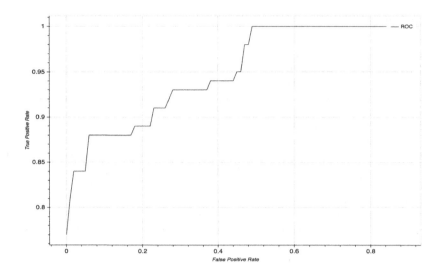

Fig. 7. ROC curve of the best model.

5 Conclusion

Anomaly detection is an essential method in order to forecast the product quality in manufacturing plants. The quality can be assured through tests performed on sample units randomly chosen from a batch of manufactured units in such a plant. With the aim to build an automated test system in Arçelik's freezer factory manufacturing in high volumes, our ultimate goal is to detect defective units during the cooling test among the sample units as early as possible in terms of test time and as accurate as possible. For this purpose, we analyzed the cooling test data of the units sampled from the batches of manufactured freezer units. The first steps of the analysis were composed of the embedding and clustering of the data. Traditional classification algorithms were applied and their performances were assessed. Finally, a deep learning method for time-series analysis combined with a classifier was applied. Our analysis results showed the feasibility of such an automated system. However, the classifier models described in this study should be further studied and customized to be deployed under the factory conditions.

An automated test system can be initially deployed to assist the human expert performing the test. By this way, more data can be collected. Ultimately, the automated test system should perform continual learning so that it would learn and adapt itself in real-time. Various models of freezer units are produced in the factory and therefore, building individual and customized test systems for freezer models may yield better results. Since the products are sold to the customers and problems occur at the customer side when the products are operational in the field, it would be very beneficial to find possible links between the problems in the field and the results of quality tests performed in the factory.

References

1. Andrade, T., Gama, J., Ribeiro, R., Sousa Lima, W., de Carvalho, A.: Anomaly Detection in Sequential Data: Principles and Case Studies, pp. 1–14 (2019). https://doi.org/10.1002/047134608x.w8382
2. Basseville, M., Nikiforov, I.V.: Detection of Abrupt Changes - Theory and Application. Prentice Hall, Upper Saddle River (1993)
3. Chandola, V., Banarjee, A., Kumar, V.: Anomaly detection: a survey. ACM Comput. Surv. **41**(3), 795–825 (2009)
4. Hochreiter, S., Schmidhuber, J.: Long short-term memory. Neural Comput. **9**, 1735–1780 (1997). https://doi.org/10.1162/neco.1997.9.8.1735
5. Wong, J., Colburn, C., Meeks, E., Vedaraman, S.: RAD - outlier detection on big data, February 2015. https://medium.com/netflix-techblog/rad-outlier-detection-on-big-data-d6b0494371cc. Accessed 07 Apr 2019
6. Kejariwal, A.: Introducing practical and robust anomaly detection in a time series, January 2015. https://blog.twitter.com/engineering/en_us/a/2015/introducing-practical-and-robust-anomaly-detection-in-a-time-series.html. Accessed 07 Apr 2019
7. Ma, J., Perkins, S.: Online novelty detection on temporal sequences. In: Proceedings of the Ninth ACM SIGKDD International Conference on Knowledge Discovery and Data Mining, KDD 2003, pp. 613–618. ACM, New York (2003). https://doi.org/10.1145/956750.956828
8. van der Maaten, L., Hinton, G.E.: Visualizing data using t-SNE. J. Mach. Learn. Res. **9**, 2579–2605 (2008). http://www.jmlr.org/papers/v9/vandermaaten08a.html
9. Malhotra, P., Ramakrishnan, A., Anand, G., Vig, L., Agarwal, P., Shroff, G.: LSTM-based encoder-decoder for multi-sensor anomaly detection. CoRR abs/1607.00148 (2016)
10. Malhotra, P., Vig, L., Shroff, G., Agarwal, P.: Long short term memory networks for anomaly detection in time series. In: ESANN, April 2015
11. Polat, K., Gunes, S.: Detection of ECG Arrhythmia using a differential expert system approach based on principal component analysis and least square support vector machine. Appl. Math. Comput. **186**(1), 898–906 (2007). https://doi.org/10.1016/j.amc.2006.08.020
12. Shipmon, D.T., Gurevitch, J.M., Piselli, P.M., Edwards, S.T.: Time series anomaly detection; detection of anomalous drops with limited features and sparse examples in noisy highly periodic data. CoRR abs/1708.03665 (2017)
13. Tax, D., Duin, R.: Support vector data description. Mach. Learn. **54**, 45–66 (2004). https://doi.org/10.1023/B:MACH.0000008084.60811.49

Data Preprocessing and Dynamic Ensemble Selection for Imbalanced Data Stream Classification

Paweł Zyblewski[1]($\boxtimes$) (iD), Robert Sabourin[2], and Michał Woźniak[1] (iD)

[1] Department of Systems and Computer Networks, Faculty of Electronics,
Wrocław University of Science and Technology, Wrocław, Poland
`{pawel.zyblewski,michal.wozniak}@pwr.edu.pl`
[2] LIVIA, École de Technologie Supérieure, University of Quebec,
Montreal, Quebec, Canada
`robert.sabourin@etsmtl.ca`

Abstract. Learning from the non-stationary imbalanced data stream is a serious challenge to the machine learning community. There is a significant number of works addressing the issue of classifying non-stationary data stream, but most of them do not take into consideration that the real-life data streams may exhibit high and changing class imbalance ratio, which may complicate the classification task. This work attempts to connect two important, yet rarely combined, research trends in data analysis, i.e., non-stationary data stream classification and imbalanced data classification. We propose a novel framework for training base classifiers and preparing the dynamic selection dataset (DSEL) to integrate data preprocessing and dynamic ensemble selection (DES) methods for imbalanced data stream classification. The proposed approach has been evaluated on the basis of computer experiments carried out on 72 artificially generated data streams with various imbalance ratios, levels of label noise and types of concept drift. In addition, we consider six variations of preprocessing methods and four DES methods. Experimentation results showed that dynamic ensemble selection, even without the use of any data preprocessing, can outperform a naive combination of the whole pool generated with the use of preprocessing methods. Combining DES with preprocessing further improves the obtained results.

Keywords: Imbalanced data · Data stream · Dynamic ensemble selection · Data preprocessing · Concept drift

1 Introduction

This work is an attempt to connect two of the important research directions, i.e., data stream classification employing ensemble approach [12] as well as data analysis with imbalanced data distribution [11]. Although the real data streams may be characterized by high class imbalance ratio, which could further inhibit

© Springer Nature Switzerland AG 2020
P. Cellier and K. Driessens (Eds.): ECML PKDD 2019 Workshops, CCIS 1168, pp. 367–379, 2020.
https://doi.org/10.1007/978-3-030-43887-6_30

the classification task, there are not many state-of-art methods which take this fact into account. There are only a few works that distinguish the differences between imbalanced data stream classification problem and a scenario where the prior knowledge about the entire data set is given. This divergence is a result of the lack of knowledge about the class distribution and it is notably present in the initial stages of the data stream classification. Additionally, during designing classifiers for streaming data we have to take into consideration a few important issues, which are usually ignored by the traditional classifier learning algorithms:

- Limited computational resources as memory and time are available.
- Usually short time limit to make a decision for each incoming sample.
- Possibility of *concept drift* appearance, i.e., changes in data distribution.
- Impossibility or delay in data labeling.

One of the very promising directions of the stream data analysis is classifier ensemble, where a plethora of methods have been proposed, but this approach still remains the focus of intense research and its high flexibility and accuracy in many real-life decision problems brought its popularity. Nevertheless, obtaining high-quality classifier ensemble depends on addressing the most important problems on how to ensure a high diversity of the ensemble and how to produce the final decision of the pool of individual models [20].

Kuncheva analysed different approaches of applying ensemble techniques to data stream classification [13] and distinguished the following basic strategies of their adaptation to new incoming data:

- Dynamic combiners, where base models are trained in advance, but the classifier ensemble is changing the combination rule (e.g., changing weights for weighted voting or aggregation).
- Updating training data – recent training examples are used to online-update base classifiers (e.g. in on-line bagging or its further generalizations [2]).
- Updating base classifiers.
- Update the classifier ensemble line-up, e.g., by replacing the worst performing base classifiers by a new classifier trained on the newest data.

In this work we focus on the last strategy, precisely on the classifier (or classifier ensemble) selection methods. They employ the idea of *overproduce-and-select*, where for a given classification task we have more classifiers at our disposal than we are going to use, but for each sample being recognized the local competencies of individual models should be detected. Basically, there are two approaches:

- *Static selection*, where a feature space is divided into several partitions and for each of them one classification model is assigned. The decision about the new instance is made by a classifier assigned to a partition where the example belongs to.
- *Dynamic Selection*, in where the features space is not partitioned in advance, but during the classification of a given example, the competencies of each available classification model are evaluated and the final decision is made

according to the most competent classifier. One of the important variation of this domain is Dynamic Ensemble Selection (DES), which is recognized as very promising direction in classifier ensemble learning [6]. DES uses the notion of competence to select the best models to classify a given test instance. Usually, the competence of a base classifier is estimated on the basis of its immediate vicinity, called the local region of competence. It is formed using a set of labeled samples from either the training or validation set, which is called the *dynamic selection dataset* (DSEL).

This work is focusing on dynamic ensemble selection used to mitigate the difficulties related to skewed class distribution embedded in non-stationary data streams using data preprocessing approach. In a nutshell, the main contributions of this work are as follows:

- Presentation of several strategies used for forming *dynamic selection dataset* (DSEL), i.e., set of neighbouring examples of a recognizing sample.
- Proposition of a novel framework for training base classifiers and preparing DSEL for using the dynamic selection process during imbalanced data stream classification.
- Experimental evaluation of the discussed approaches on the basis of diverse data streams and a detailed comparison with the state-of-art method.

2 Dynamic Ensemble Selection and Data Preprocessing

Let us shortly discuss some of the most popular and recent approaches to DES. Woloszynski and Kurzynski proposed *Randomized Reference Classifier* [19], which produces supports for each class that are realizations of random variables with the beta probability distributions. Lysiak et al. [15] discussed how to enhance the selection step using diversity measures. Cruz et al. proposed META-DES.Oracle [5], which employs meta-learning over multiple datasets and feature selection to improve the selection process. Zyblewski et al. [21] proposed the *Minority Driven Ensemble* algorithm, which employs dynamic classifier selection approach to exploit local data characteristics and combat with data imbalance.

In this paper, we consider four different DES strategies. Two of those strategies (KNOR A-Eliminate and KNORA-Union) are based on oracle information, while DES-KNN and DES-Clustering select classifiers on the basis of their local competence but they also take into consideration an ensemble diversity. Let's briefly describe the methods we used during experiments.

- KNORA-Eliminate (KNORA-E) [10] selects only the local oracles - classifiers which can correctly classify all samples within the local region of competence. If no classifier is selected, the size of competence region is reduced by removing the farthest neighbor,
- KNORA-Union (KNORA-U) [10] makes the decision based on weighted voting, where the weight assigned to each base classifier equals to the number of correctly classified samples in the competence region,

- DES-KNN [18] ranks the base models in decreasing order of accuracy and increasing order of diversity and select the most accurate and diverse ones to form the ensemble.
- DES-Clustering [18] uses the *K-means* algorithm for defining the region of competence, then the most accurate and diverse classifiers ale selected for the ensemble.

As we mentioned above, this work deals with DES application to imbalanced data classification. One of the most promising direction of imbalanced data analysis is data preprocessing. Such methods focus on changing the data distribution by reducing the number of majority class objects (*undersampling*) or generating new minority class objects (*oversampling*), e.g., Random undersampling (RUS) [1] removes random instances from the majority class, while Random over-sampling (ROS) replicates minority class examples. Nevertheless, theses methods have the serious disadvantages, as RUS may remove biased samples, what may deteriorate the classification performance, while ROS may increase the likelihood of overfitting. Therefore methods which are able to generate synthetic minority examples have been developed. Chawla et al. proposed SMOTE [4], which generates new instances from existing minority samples using its nearest neighbors. Regular SMOTE does not impose any rule in selecting existing instances. SVM SMOTE [16] uses an SVM classifier to find support vectors and generate new samples based on them. Borderline SMOTE [8] selects only those existing instances of which at least half of the neighbors are from the same class. Borderline-1 SMOTE select neighbors from the same class as the existing sample and Borderline-2 SMOTE considers neighbors from any class. Safe-level SMOTE [3] samples minority instances along the same line with different weight degree, computed by using nearest neighbour minority instances. ADASYN [9] is similar to SMOTE but the number of generated samples is proportional to the number of samples which are not from the same class as the selected existing instance in a given neighborhood.

3   The Proposed Framework

To deal with the imbalanced data streams classification we propose the following framework for classifier ensemble forming and preparing the *dynamic selection dataset* (DSEL) for the dynamic selection process.

Let's assume that the data stream consists of fixed-size data chunks $\mathcal{DS}_k$, where k is the chunk index and Ψ_k denotes the classifier trained on the basis of the kth chunk. Each based model Ψ_k learns from the T_k training set which is obtained by preprocessing $\mathcal{DS}_k$. DSEL$_k$ denotes *dynamic selection dataset* for the kth data chunk and it is considered as previously preprocessed $\mathcal{DS}_{k-1}$. On the beginning each new trained classifier (one per each incoming chunk) is added to the ensemble until the maximum size of the ensemble (ES) is achieved. Then if new model is added, we evaluate each classifier in the ensemble (according to BAC score) and the worst one is removed. Additionally, at each step we remove from the ensemble all models which BAC scores are lower than a given threshold α.

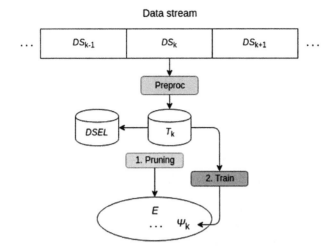

Fig. 1. The framework for training base classifiers and to prepare a DSEL for dynamic selection process. Here, T_k is the training data produced by preprocessing (*Preproc*) data chunk DS_k and Ψ_k is the base classifier trained on the kth data chunk. E denotes the classifier pool.

Pruning process is performed before adding kth classifier to the ensemble. The concept behind the proposed framework is presented in Fig. 1 and the pseudocode is shown in Algorithm 1.

In the beginning the classifier pool E is empty. We train our first classifier (Ψ_0) on the preprocessed zero chunk (steps 4, 5 and 6). When the first chunk arrives, we use (Ψ_0) to classify it and then we use it to train our second model (steps 8, 9 and 10). We also store the T_1 training set as the DSEL for future (step 12). Then, with the arrival of each chunk, the following steps are performed:

- In step 14, we use previously stored training set as DSEL for the dynamic selection process to create the list of ensembles for classifying each test instance in DS_k.
- In step 15, we use the ensembles selected by DES method to classify instances the current data chunk.
- In step 16, based on the current chunk, we evaluate BAC scores of all models in the ensemble in order to use this information for pruning in the next steps.
- In steps 17 and 18, we remove from the ensemble all models with BAC scores lower than a given threshold α.
- In steps 19 and 20, we remove the worst rated base model in the ensemble.
- In steps 21, 22 and 23 we use the data preprocessing method on the current chunk to generate training set T_k, on the basis of which we train a new classifier and add it to the pool E.
- Finally, in step 24, we store the current training set T_k in order to use it as DSEL when the next chunk arrives.

Algorithm 1. Pseudocode for the proposed framework

Input:
    Fixed ensemble size (ES),
    Pruning threshold (α)
    $E \leftarrow \varnothing$
    $k \leftarrow -1$
 1: **while** *Stream* **do**
 2:    $k \leftarrow k + 1$
 3:    **if** $k == 0$ **then**
 4:        $T_k \leftarrow Preprocess(\mathcal{DS}_k)$
 5:        $\Psi_k \leftarrow trainNewClassifier(T_k)$
 6:        $E \leftarrow \Psi_k$
 7:    **else if** $k == 1$ **then**
 8:        $\mathcal{DS}_{k\_class} \leftarrow classify(\mathcal{DS}_k, E)$
 9:        $T_k \leftarrow Preprocess(\mathcal{DS}_k)$
10:        $\Psi_k \leftarrow trainNewClassifier(T_k)$
11:        $E \leftarrow \Psi_k$
12:        $\mathcal{DSEL} \leftarrow T_k$
13:    **else**
14:        $DE_k \leftarrow dynamicSelection(E, DS_k, \mathcal{DSEL})$
15:        $\mathcal{DS}_{k\_class} \leftarrow classify(\mathcal{DS}_k, DE_k)$
16:        $S \leftarrow scoreBaseModels(\mathcal{DS}_k)$
17:        **if** $len(E) > 1$ **then**
18:            $E \leftarrow pruneThreshold(E, S, \alpha)$
19:        **if** $len(E) > ES - 1$ **then**
20:            $E \leftarrow pruneWorstClassifier(E, S)$
21:        $T_k \leftarrow Preprocess(\mathcal{DS}_k)$
22:        $\Psi_k \leftarrow trainNewClassifier(T_k)$
23:        $E \leftarrow \Psi_k$
24:        $\mathcal{DSEL} \leftarrow T_k$
25: **end while**

4   Experimental Evaluation

The goal of experiments is to show how the combination of dynamic ensemble selection methods and preprocessing perform in terms of classifying imbalanced data streams with various imbalance ratios and different types of concept drift.

4.1   Experimental Setup

As the experimental protocol TEST AND TRAIN framework [12] was used, i.e., every classification model is trained on a recent data chunk, but it is evaluated on the basis of the following one. Evaluation of the proposed framework was based on the *balanced accuracy measure* (BAC) according to scikit-learn implementation [17], which for the binary case is equal to the arithmetic mean of sensitivity (the true positive rate) and specificity (the true negative rate) and *geometric mean measure* (G-mean) according to imbalanced-learn implementation [14] defined as $g = \sqrt{a^+ a^-}$, where a^+ denotes sensitivity and a^- denotes specificity. As the base classifier we used *classification and regression tree* (CART). For ensemble pruning (see line 18 in Algorithm 1) we used $\alpha = 0.55$, i.e., all base classifiers which BAC scores are lower than α were removed from ensemble. The choice of this value was motivated to get the classifiers a slightly better than the random ones. The maximum size of the classifier pool was $ES = 20$.

Experiments were implemented in Python programming language and may be repeated according to source code published on *Github*[1].

Performance of the naive aggregation of the whole classifier pool (Naive) and the dynamic ensemble methods (KNORA-E, KNORA-U, DES-KNN and DES-Clustering) is evaluated depending on the data preprocessing methods that they were coupled with. Neighborhood size for DES methods is $k = 7$.

Six preprocessing methods chosen for the experiments are: SMOTE, SVM-SMOTE, two variants of Borderline-SMOTE (B1-SMOTE and B2-SMOTE), Safe-level SMOTE (SL-SMOTE) and ADASYN. We also check how the ensemble methods behave without the use of any data preprocessing.

The proposed framework was evaluated using 72 artificially generated data streams. Each stream is composed of one hundred thousand instances divided into 200 chunks of 500 objects described by 8 features, and contains five concept drifts. The base concepts were generated according to procedure of creating the Madelon [7] synthetic classification dataset, the used stream generator is available at *Github*[2]. Each combination was generated three times, based on the determined seeds. The variety of streams was ensured by generating three streams for each combination of the following parameters:

- *the imbalance ratio*—successively 10, 20, 30 and 40% of the minority class.
- *the level of label noise*—successively 0, 10 and 20%.
- *the type of concept drift*—sudden or incremental.

The results of experiments for two measures: BAC (a) and G-mean (b) for different IR values and drift types are presented in Figs. 2 and 3. The radar charts present how each data preprocessing technique influenced the performance of a given DES method and are followed by the classification results for the best performing dynamic selection methods coupled with the most effective data preprocessing techniques. Presented methods were selected based on the statistical evaluation and are compared to the aggregation of probabilities of the whole classifier pool and to the results obtained only with the use of dynamic selection or preprocessing. The Figs. 2 and 3 show results related to different imbalance

---

[1] https://github.com/w4k2/ECML19-IoT-DES-preproc.
[2] https://github.com/w4k2/ECML19-IoT-DES-preproc/blob/master/csm/StreamGenerator.py.

ratios and drift types. The complete statistical evaluation of all methods, which was the basis for all figures shown, is available on *Github*[3] in PDF format.

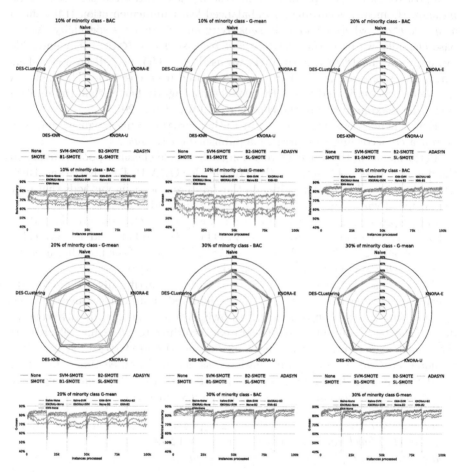

Fig. 2. Comparison of different sampling approaches for different classifier ensembles with respect to performance measures (BAC and G-mean) for imbalance ratio 1 : 9, 2 : 8 and 3 : 7.

During evaluation of the proposed framework in the case of different types (sudden or incremental) of concept drift, we focus on the streams with high imbalance ratios (i.e., 1 : 9 and 2 : 8), typical for the real-life decision tasks. The comparison is shown in Fig. 3.

---

[3] https://github.com/w4k2/ECML19-IoT-DES-preproc/blob/master/ paper_appendix/ECML19_Statistical_appendix.pdf.

4.2 Lessons Learned

Based on the statistical analysis, which is available in its entirety on *Github* (see footnote 3) repository, we can see that for the 1 : 9 imbalance ratio, according to BAC, DES-KNN was the best performing method without the use of any preprocessing. In cases where we coupled DES with preprocessing methods, KNORA-U performed best except for the use of SL-SMOTE, where it was not statistically better than DES-KNN. According to G-mean for 1 : 9 IR DES-KNN was statistically the best dynamic ensemble selection method. For the Borderline2-SMOTE preprocessing method, both DES-KNN and KNORA-U performed statistically similar. The best preprocessing methods were SVM-SMOTE and Borderline2-SMOTE.

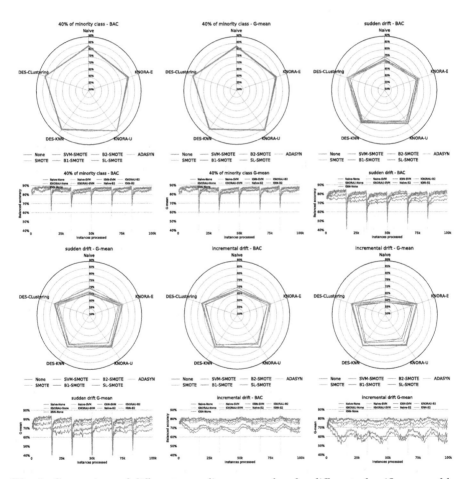

Fig. 3. Comparison of different sampling approaches for different classifier ensembles with respect to performance measures (BAC and G-mean) for imbalance ratio 4 : 6 and for sudden drift and incremental drift.

For the 2 : 8 IR, both in terms of BAC and G-mean, KNORA-U performed the best when paired with any preprocessing method. In case of not using any data preprocessing, DES-KNN was statistically the best. As for preprocessing techniques, in most cases the SVM-SMOTE was statistically significant, Borderline2-SMOTE performed the best for naive aggregation of the whole classifier pool.

For the 3 : 7 imbalance ratio, again, KNORA-U turned out to be statistically the best DES method. Only exception (according to G-mean) was the case when we do not use preprocessing, then DES-KNN works best. Best performing data preprocessing method for DES, according to both measures, was the SVM-SMOTE. Borderline2-SMOTE again performed the best for naive aggregation.

In case of 4 : 6 IR according to both BAC and G-mean KNORA-U was the statistically significant DES method in every case. Borderline2-SMOTE worked best for naive aggregation and in the remaining cases SVM-SMOTE was statistically the best preprocessing method.

For sudden drift, in terms of both measures, DES-KNN was statistically the best without the use of any preprocessing method and KNORA-U was statistically leading when paired with every oversampling method. Borderline2-SMOTE was the best for naive aggregation and for KNORA-U according to G-mean, for the rest of DES methods SVM-SMOTE performed the best.

Finally, for incremental drift, according to BAC, DES-KNN performed statistically the best without the use of preprocessing and for the SL-SMOTE while KNORA-U was the best for other oversampling techniques. SVM-SMOTE was the best preprocessing method for KNORA-E, DES-KNN and DES-Clustering and Borderline2-SMOTE performed the best coupled with naive aggregation and KNORA-U. According to G-mean, KNORA-U was statistically leading DES method for Borderline2-SMOTE and ADASYN while DES-KNN was statistically significant for all other preprocessing techniques. SVM-SMOTE worked best with KNORA-E and DES-KNN, Borderline2-SMOTE proved to be statistically significant for naive aggregation, KNORA-U and DES-Clustering.

In general, the order of the presented approaches in terms of performance, starting with the worst, is as follows: (1) naive aggregation without the use of any preprocessing methods → (2) naive aggregation combined with preprocessing → (3) dynamic ensemble selection methods without preprocessing → (4) DES methods coupled with preprocessing methods. The lower the imbalance ratio, the smaller the differences between approaches, but the order is maintained. The conducted experiments showed that the best performing DES method among the considered strategies across all tested imbalance ratios is the KNORA-U, which uses the weighted voting scheme. As the KNORA-Union method selects all base models that are able to correctly classify at least one instance in the local region of competence and then it combines them based on the weighted voting, where the number of votes is equal to the number of correctly recognized samples, it allows us to select both accurate and diverse ensemble. As both of these characteristics are determinants of a good classifier ensemble model, they may be the reason for high results of this DES method. Worth mentioning is also the DES-KNN, which is doing well for high imbalance ratios, especially for the

10% of minority class and for incremental drift in terms of G-mean. DES-KNN performs the best for high IR (10 and 20% of minority class) in case of not using any preprocessing method. The worst performing DES method, for low IR (30 and 40%) worse even than naive aggregation, was KNORA-E. This may be due to the fact, that the local oracles are found only for regions of competence with a significantly reduced size, which negatively affects the performance.

Based on the results achieved by DES-KNN and DES-Clustering methods we may suspect that the *K-Nearest Neighbors* technique is better suited for defining the local region of competence in case of imbalanced data streams than the clustering technique. Despite the higher computational cost, KNN allows for more precise estimation of the region of competence which leads to more possible ensemble configurations for classifying new instances.

On the other hand, SVM-SMOTE and Borderline2-SMOTE have proven to be the preferred preprocessing strategies for the used dynamic ensemble selection methods. The combination of KNORA-U or DES-KNN with one of those preprocessing methods always leads to the best classification performance.

5 Conclusions

The main goal of this work was to propose a novel framework for training base classifiers and preparing the *dynamic selection dataset* (DSEL) for the dynamic selection process during imbalanced data stream classification. We proposed the self-updating ensemble model employing data preprocessing techniques. The computer experiments confirmed the effectiveness of the proposed framework and based on the statistical analysis we can conclude that dynamic ensemble methods coupled with data preprocessing techniques are statistically significantly better than the approaches that do not combine both of these concepts.

The results presented in this paper are quite promising therefore they encourage us to continue our work on employing dynamic ensemble selection for imbalanced data stream classification. Future research may consist of analysing different methods of defining the local region of competence (e.g. applying various distance metrics) and the development of a weighted voting combination approach based on the KNORA-Union, specialized in dealing with imbalanced data. Analysing the impact of pruning threshold α value on the performance of DES methods for the proposed framework might be another idea worth noting.

Acknowledgment. This work was supported by the Polish National Science Centre under the grant No. 2017/27/B/ST6/01325 as well as by the statutory funds of the Department of Systems and Computer Networks, Faculty of Electronics, Wroclaw University of Science and Technology.

References

1. Barandela, R., Valdovinos, R., Sánchez, J.: New applications of ensembles of classifiers. Pattern Anal. Appl. **6**(3), 245–256 (2003)

2. Bifet, A., Holmes, G., Pfahringer, B.: Leveraging bagging for evolving data streams. In: Balcázar, J.L., Bonchi, F., Gionis, A., Sebag, M. (eds.) ECML PKDD 2010. LNCS (LNAI), vol. 6321, pp. 135–150. Springer, Heidelberg (2010). https://doi.org/10.1007/978-3-642-15880-3_15

3. Bunkhumpornpat, C., Sinapiromsaran, K., Lursinsap, C.: Safe-level-SMOTE: safe-level-synthetic minority over-sampling technique for handling the class imbalanced problem. In: Theeramunkong, T., Kijsirikul, B., Cercone, N., Ho, T.-B. (eds.) PAKDD 2009. LNCS (LNAI), vol. 5476, pp. 475–482. Springer, Heidelberg (2009). https://doi.org/10.1007/978-3-642-01307-2_43

4. Chawla, N.V., Bowyer, K.W., Hall, L.O., Kegelmeyer, W.P.: SMOTE: synthetic minority over-sampling technique. J. Artif. Int. Res. **16**(1), 321–357 (2002)

5. Cruz, R.M.O., Sabourin, R., Cavalcanti, G.D.C.: META-DES.Oracle: meta-learning and feature selection for dynamic ensemble selection. Inform. Fusion **38**, 84–103 (2017)

6. Cruz, R.M., Sabourin, R., Cavalcanti, G.D.: Dynamic classifier selection: recent advances and perspectives. Inf. Fusion **41**(C), 195–216 (2018)

7. Guyon, I.: Design of experiments of the NIPS 2003 variable selection benchmark. In: NIPS 2003 Workshop on Feature Extraction and Feature Selection (2003)

8. Han, H., Wang, W.-Y., Mao, B.-H.: Borderline-SMOTE: a new over-sampling method in imbalanced data sets learning. In: Huang, D.-S., Zhang, X.-P., Huang, G.-B. (eds.) ICIC 2005. LNCS, vol. 3644, pp. 878–887. Springer, Heidelberg (2005). https://doi.org/10.1007/11538059_91

9. He, H., Bai, Y., Garcia, E.A., Li, S.: ADASYN: adaptive synthetic sampling approach for imbalanced learning. In: 2008 IEEE International Joint Conference on Neural Networks, IEEE World Congress on Computational Intelligence, pp. 1322–1328 (2008)

10. Ko, A.H., Sabourin, R., Britto Jr., A.S.: From dynamic classifier selection to dynamic ensemble selection. Pattern Recogn. **41**(5), 1718–1731 (2008)

11. Krawczyk, B.: Learning from imbalanced data: open challenges and future directions. Prog. Artif. Intell. **5**(4), 221–232 (2016). https://doi.org/10.1007/s13748-016-0094-0

12. Krawczyk, B., Minku, L.L., Gama, J., Stefanowski, J., Woźniak, M.: Ensemble learning for data stream analysis: a survey. Inform. Fusion **37**, 132–156 (2017)

13. Kuncheva, L.I.: Classifier ensembles for changing environments. In: Roli, F., Kittler, J., Windeatt, T. (eds.) MCS 2004. LNCS, vol. 3077, pp. 1–15. Springer, Heidelberg (2004). https://doi.org/10.1007/978-3-540-25966-4_1

14. Lemaître, G., Nogueira, F., Aridas, C.K.: Imbalanced-learn: a Python toolbox to tackle the curse of imbalanced datasets in machine learning. J. Mach. Learn. Res. **18**(17), 1–5 (2017)

15. Lysiak, R., Kurzynski, M., Woloszynski, T.: Optimal selection of ensemble classifiers using measures of competence and diversity of base classifiers. Neurocomputing **126**, 29–35 (2014)

16. Nguyen, H.M., Cooper, E.W., Kamei, K.: Borderline over-sampling for imbalanced data classification. IJKESDP **3**, 4–21 (2011)

17. Pedregosa, F., et al.: Scikit-learn: machine learning in Python. J. Mach. Learn. Res. **12**, 2825–2830 (2011)

18. Soares, R.G.F., Santana, A., Canuto, A.M.P., de Souto, M.C.P.: Using accuracy and diversity to select classifiers to build ensembles. In: The 2006 IEEE International Joint Conference on Neural Network Proceedings, pp. 1310–1316, July 2006

19. Woloszynski, T., Kurzynski, M.: A probabilistic model of classifier competence for dynamic ensemble selection. Pattern Recogn. **44**(10–11), 2656–2668 (2011)
20. Woźniak, M., Graña, M., Corchado, E.: A survey of multiple classifier systems as hybrid systems. Inform. Fusion **16**, 3–17 (2014)
21. Zyblewski, P., Ksieniewicz, P., Woźniak, M.: Classifier selection for highly imbalanced data streams with *minority driven ensemble*. In: Rutkowski, L., Scherer, R., Korytkowski, M., Pedrycz, W., Tadeusiewicz, R., Zurada, J.M. (eds.) ICAISC 2019. LNCS (LNAI), vol. 11508, pp. 626–635. Springer, Cham (2019). https://doi.org/10.1007/978-3-030-20912-4_57

A Study on Imbalanced Data Streams

Ehsan Aminian[1], Rita P. Ribeiro[1,2], and João Gama[1,3(✉)]

[1] LIAAD - INESC TEC, Porto, Portugal
ehaminian@gmail.com
[2] Faculty of Science, University of Porto, Porto, Portugal
rpribeiro@dcc.fc.up.pt
[3] School of Economics, University of Porto, Porto, Portugal
jgama@fep.up.pt

Abstract. Data are growing fast in today's world and great portion of that is in the form of stream. In many situations, data streams are imbalanced making it difficult to use with classical data mining methods. However, mining these special kinds of streams is one of the most attractive research area. In this paper, we propose two algorithms for learning from imbalanced regression data streams. Both methods are based on Chebychev's inequality but in a different way. The first method, undersamples from the frequent target value examples while the second method over-samples the rare and extreme target value examples. This way, the learner will focus in the rare and more difficult cases. We applied our methods to train regression models using two benchmark datasets and two well-known regression algorithms: Perceptron and FIMT-DD. Our obtained results from the simulations indicate the usefulness of our proposed methods.

Keywords: Data streams · Imbalanced data streams · Chebyshev's inequality

1 Introduction

Data streams are a large volume of data originated from one or more sources arriving in a sequential manner. There are various applications that generate streaming data, like network traffic, financial transactions, telecommunication calling records, remote surveillance system, share market data, remote sensor etc. The data coming from these sources must be processed before they become outdated and the decisions must be taken in real time. The big volume and speed of data streams are the two main challenges in their mining process. Data stream mining can be investigated in classification and regression problems depending whether the target variable is discrete or continuous. In both domains, data stream can be skewed meaning that the distribution of the target variable is imbalanced and its most important values are scarcely represented [3].

Take for example, the problem of predicting the probability of debtor's delinquency from a credit scoring data stream. In such data stream, only small percentage of the observations are delinquent but the goal is to have a prediction for

© Springer Nature Switzerland AG 2020
P. Cellier and K. Driessens (Eds.): ECML PKDD 2019 Workshops, CCIS 1168, pp. 380–389, 2020.
https://doi.org/10.1007/978-3-030-43887-6_31

these cases as much correct as possible. Similar situations are observed in other areas, such as detecting network intrusions [5], managing risk and predicting failures of technical equipment.

In these kind of problems, classical learning algorithms lose their effectiveness because they concentrate in the frequent cases, and tend to ignore rare cases. The learned model will be biased towards the frequent cases. In other words, classical learning algorithms have problems in dealing with infrequent and rare cases.

Performance evaluation may also be affected by imbalanced data sets. Imagine in the above-mentioned example, we use accuracy as the metric for evaluation the model. In such situation, the accuracy of the model may be high but the predicted values of the probability for delinquent debtors can be disappointing. That is because the model is trained mostly over the frequent cases, the accuracy of prediction for these cases are high affecting the overall accuracy. However, in this case, the accuracy of the prediction for our desired cases (i.e. delinquent debtors) may be disappointing. This example shows that conventional metrics based on the average behavior, such as accuracy or mean square error, are no longer effective for assessing the performance of a given model.

The focus of this paper are regression problems. We propose two novel under/over sampling methods for handling imbalanced data streams. In the next section, we have a brief review over past works in this area. We present our methods in Sect. 3 which is followed by illustrating our obtained results in Sect. 4. Finally, we draw our conclusions in Sect. 5.

2   Related Work

One of the solutions for tackling imbalanced distribution problems is modifying the original collection of training set in order to reduce or eliminate the extent of imbalance in datasets. Generating new examples for the rare cases (over-sampling) [1], getting rid of objects from frequent cases (under-sampling) [7] and combining both methods [9] to change the distribution balance of original data are the techniques called data-level solutions.

Nitesh et al. [4] proposed a method for classification in imbalanced context called SMOTE (Synthetic Minority Over-sampling TEchnique), that has gained intensive attention. In their approach, synthetic examples are generated from the minority class examples. Each synthetic example is generated from the combination of a minority class example and any or all of its k nearest neighbours. Their experiments showed that the combination of SMOTE and random under-sampling surpassed their previous methods.

SmoteR [11] was introduced for addressing imbalanced regression tasks, which has been inspired from SMOTE. SmoteR partitioned samples into normal and interesting cases by using a user defined relevance function. Cases with a relevance value higher than a user defined threshold are considered interesting, and the others are considered normal cases. To get ride of imbalanced distribution, both under-sampling and over-sampling is applied by SmoteR. The normal cases are randomly under-sampled while interesting cases are over-sampled in a more informed way. The over-sampling of SmoteR is an adaption of that from SMOTE in which two interesting examples are picked up and a new synthetic example is build by their interpolation. Also, weighted average of the target values of the two examples yields the target value for the new created example.

Random Over-Sampling (RO) and introduction of Gaussian Noise (GN), which was originally developed for classification tasks, were adopted recently for regression problems in [3]. The range of the target variable is partitioned into several bins where each bin contains only the interesting examples or the normal examples. To specify whether an example is interesting or normal, a user defined relevance function and threshold is used as in SmoteR. Starting from the smallest target value, each example is labeled as normal or interesting. At the end, consecutive examples with the same label construct a bin in which all examples are normal or interesting. In the RO method, replicas of the examples in the interesting bins are added to make the dataset balanced. In the GN method, normal examples are randomly under-sampled but for over-sampling, they have adopted the classification method described in [8] for regression tasks. To create a new example, they chose an example from interesting bin and add a random noise to both its features and its target value. Another method introduced in [3] is WEighted Relevance-based Combination Strategy (WERCS) in which examples are given a weight based on their relevance value of the target variable. Each example can be over or under sampled but with a different probability. Examples with high relevance value (weight) have small chance to be selected for removing from the dataset (under-sampling) but have a great chance to be candidate for contributing in an over-sampling procedure.

3 Problem Definition

We consider the online learning framework. In this framework, when an example becomes available, the current decision model makes a prediction. Only after the prediction has been taken, the environment reacts providing feedback, e.g. the true value of the target variable.

In this paper we address regression problems in which the output variable is continuous. Suppose a sequence of instances in the form of pairs (x_t, y_{t-1}), arriving one at a time. x_t is a p-dimensional vector belonging to an instance space X observed at time t, and y_{t-1} is the true value for the output corresponding to x_{t-1}. Prediction of y_{t-1} has been done over x_{t-1} at time $t-1$. Instead of minority class examples, we use rare extreme values, as the target variable is continuous. Let $H(x)$ be the online learner, updating its hypothesis $H : X \rightarrow Y$ sequentially with the example at the current time step. When a target instance x_t arrives, the task of $H(x)$ is to predict its output value i.e. y_t, such that this prediction minimize the prediction error on cases in which the target variable has a rare extreme value, during the learning process.

4   Proposed Method

In the following sections, we propose two methods for learning from imbalanced data stream. Both methods are based on the Chebyshev inequality. This inequality is derived from Markov inequality and can be used to bound the tail probabilities of a random variable Y.

The Chebyshev inequality guarantees that in any probability distribution, 'nearly all' values are close to the mean. More precisely no more than $\frac{1}{t^2}$ of the distribution's values can be more than t standard deviations away from the mean. Although conservative, the inequality can be applied to completely arbitrary distributions (unknown except for mean and variance). Let Y be a random variable with finite expected value $\bar{y}$ and finite non-zero variance σ^2. Then for any real number $t > 0$:

$$\Pr(|y - \bar{y}| \geq t\sigma) \leq \frac{1}{t^2} \tag{1}$$

Only the case $t > 1$ provides useful information. When $t < 1$ the right-hand side is greater than one, so the inequality becomes useless, as the probability of any event cannot be greater than one. When $t = 1$ it just says the probability is less than or equal to one, which is always true. Therefore, for $t = \frac{|y - \bar{y}|}{\sigma}$ and $t > 1$, the rareness score of an observation Y is:

$$P(|\bar{y} - y| = t) = \frac{1}{(\frac{|y - \bar{y}|}{\sigma})^2} \tag{2}$$

The above definition states that the probability of observing y far from the mean is small and it decreases as we get farther away from the mean. In imbalanced data streams, rare cases are more likely to occur far from the mean while frequent examples are closer to the mean. So, given mean and variance of a random variable, the Chebyshev's inequality is an indication of the degree of rareness of an observation.

Figure 1a shows the probability values calculated from Eq. 2 for ailerons dataset[1], which is an imbalanced dataset. Box plot of the output variable for the mentioned dataset is also illustrated in Fig. 1b. As can be seen from these figures, Chebyshev's probability value for examples near to the mean is close to one and the value decreases as we are getting far from the mean until it gets close to zero for examples in the farthest distance. As a result, the output value from Eq. 2 can indicate if an example is a rare case or a frequent one.

After being able to specify if an example in data stream is a rare extreme value case or a frequent one, the next step would be training the learner using this extra information. In this paper, we use this extra information and propose an under-sampling as well as an over-sampling method.

---

[1] https://www.dcc.fc.up.pt/~ltorgo/Regression/ailerons.html.

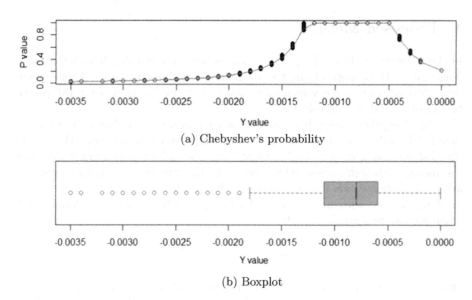

(a) Chebyshev's probability

(b) Boxplot

Fig. 1. Chebyshev's probability and Boxplot for the target variable Y in ailerons dataset

4.1 Under-Sampling

As it is mentioned above, Chebyshev's probability value for the frequent target value examples are high while the value is low for the rare extreme target value cases. In order to do under-sampling over data, first, Chebyshev's probabilities value of the examples are reversed. Then the learner is trained over each example based on the corresponding inverted probability of the example. The probabilities value are reversed by subtracting them from the max value that is one, as follows:

$$inv(P(|\ y - \bar{y}\ |= t)) = 1 - P(|\ y - \bar{y}\ |= t) = \begin{cases} 1 - \frac{\sigma^2}{|y-\bar{y}|^2}, & t > 1 \\ 0, & t \leq 1 \end{cases} \tag{3}$$

Figure 2 shows the output of Eq. 3 for the ailerons dataset, which is the inverse of Fig. 3b. Our proposed under-sampling method is illustrated in Algorithm 1.

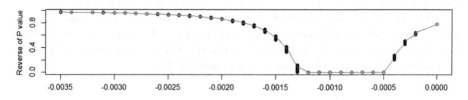

Fig. 2. Inverted probability values for the Ailerons dataset using Eq. 3

In this algorithm, first, a learner is created with random value for its parameters. Then, it calculates the mean and the variance from the target variable examples through recursive methods. That is, they can be estimated for a dataset including $n + 1$ examples if we just have access to the mean and the variance values for n examples and we access to the $(n+1)$-th example itself. The greater n we have, the more accurate estimation will be. For the first examples, mean and variance are not accurate and, therefore, Chebyshev's probability will not be accurate enough. That means that, we can not specify if the incoming example is a rare or a frequent one. In this situation the learner is trained over the example but we use the example for calculating mean and variance. We define this phase of the algorithm as the *warming phase*. Using the warming phase, our estimates of the used statistics (i.e. mean and variance) become stable, which enables the algorithm to detect rare and frequent examples. The *warming phase size* (WS) parameter in the algorithm determines the number of examples considered for making the statistics value stable. After the warming phase, for each incoming example, inverted Chebyshev's probability is computed and the learner is trained over the example only if a random number is less than that value. So, it is expected the learner to be trained over large portion of rare cases and a small percentage of frequent examples. This results in training the learner over a nearly balance dataset.

Algorithm 1. The proposed Under-sampling Algorithm

1: **procedure** UNDER-SAMPLING(S:DATASTREAM, WS: WARMING PHASE SIZE)
2:     $H \leftarrow CreateEmptyModel()$
3:     $i \leftarrow 0$
4:     **while** *true* **do**
5:         $\langle x, y \rangle \leftarrow GetExample()$
6:         $\hat{y} \leftarrow H(x)$
7:         $\overline{y}, \sigma \leftarrow UpdateStatistics(y)$
8:         **if** $i < WS$ **then**
9:             $H \leftarrow TrainModel(H, \langle x, y \rangle)$
10:         **else**
11:             $p \leftarrow ComputeInvProbability(y, \overline{y}, \sigma)$ ▷ Equation 3
12:             **if** $RandomNumber < p$ **then**
13:                 $H \leftarrow TrainModel(H, \langle x, y \rangle)$
14:             **end if**
15:         **end if**
16:         $i \leftarrow i + 1$
17:     **end while**
18: **end procedure**

4.2   Over-Sampling

Instead of training the learner over a small portion of frequent examples, we can re-train the learner several times over rare extreme value cases. We compute the value of $t = \frac{|y - \overline{y}|}{\sigma}$, where t can be any number in $[0 \quad + \infty)$. Like the previous case, the value for frequent examples is small, while it is large for the rare cases. We limit the function to produce only natural numbers as follows:

$$k = \lceil t = \frac{|y - \overline{y}|}{\sigma} \rceil, t > 1 \tag{4}$$

For each observation, the output value of Eq. 4 corresponds to the number of times that the learner is trained with that observation. Algorithm 2 describes our over-sampling proposed method.

Algorithm 2. The proposed Over-sampling Algorithm

1: **procedure** OVERSAMPLING(S: DATASTREAM, WS: WARMING PHASE SIZE)
2:     $H \leftarrow CreateEmptyModel()$
3:     $i \leftarrow 0$
4:     **while** *true* **do**
5:         $\langle x, y \rangle \leftarrow GetExample()$
6:         $\hat{y} \leftarrow H(x)$
7:         $\overline{y}, \sigma \leftarrow UpdateStatistics(y)$
8:         **if** $i < WS$ **then**
9:             $H \leftarrow TrainModel(H, \langle x, y \rangle)$
10:        **else**
11:            $k \leftarrow ComputeK(y, \overline{y}, \sigma)$ ▷ Equation 4
12:            **for** $i \leftarrow 1$ **to** k **do**
13:                $H \leftarrow TrainModel(H, \langle x, y \rangle)$
14:            **end for**
15:        **end if**
16:        $i \leftarrow i + 1$
17:    **end while**
18: **end procedure**

Like in the previous method, we use a warming phase to stabilize the statistics parameters. After that, the learner is trained one or multiple times over each example based on the output of Eq. 4 for that example.

5   Experimental Results

As we mentioned before, in real world, the cost of the errors over rare cases is more than those over the frequent examples. To simulate this, we have used a relevance function $\phi()$ [10] giving weight to each example of dataset. Figure 3 shows the weights given to the examples in the aileron dataset, according to the distribution of the target variable. This relevance function takes values between 0 and 1 and smoothly interpolates boxplot statistics of the target variable. The goal is that 0 is assigned to the median of the target variable and 1 to the most extreme cases. As can be seen in Fig. 3a, value of the weights for frequent

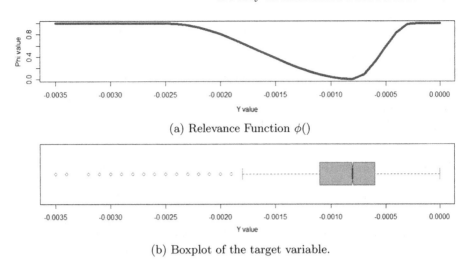

(a) Relevance Function $\phi()$

(b) Boxplot of the target variable.

Fig. 3. $\phi()$ relevance function and boxplot of the target variable Y in ailerons dataset

examples is small but it grows as we go far from the mean until it reaches its maximum over the rare cases.

As evaluation metric, we use modified RMSE function formulated in Eq. 5 where the prediction error for each example is multiplied by the relevance $\phi()$ value assigned to that example, according to its target variable value.

$$RMSE_\phi = \sqrt{\frac{1}{n}\sum_{i=1}^{n}\phi(y_i)\times(y_i-\hat{y}_i)^2} \tag{5}$$

In the above equation, $y_i, \hat{y}_i$ and $\phi()$ refer to the true value, the predicted value and relevance value for i-th example in the dataset, respectively.

To demonstrate the effectiveness of our proposed methods, we perform experiments over two real world datasets which are summarized in Table 1. We use a combination of Java (using MOA package) and R programming language in a laptop equipped with an Intel Core i7-45510U CPU @ 2.00 GHz processor and 8.00 GB of RAM for implementations. We report the mean and variance of $RMSE_\phi$ averaged over 10 runs.

Table 1. Datasets characterization.

Dataset	# Attributes	# Examples
Ailerons	40	13750
Elevator	19	16559

We have used two learner models for prediction over each dataset: perceptron [2] and fast incremental model trees with drift detection (FIMT-DD) [6].

E. Aminian et al.

In each experiment two instances from one type of a learner model are created with the same random initial parameters. One of the instances is trained using our methods and the other is trained over all of the examples in the dataset. In all experiments, warming size parameter has been set to 100. We use prequential evaluation scheme in which we first make prediction for each example and then the example is considered as a candidate for training. The obtained results for the proposed methods are reported in Tables 2 and 3.

Table 2. $RMSE_\phi$ for the Baseline and the Under-sampling method (Algorithm 1)

Dataset	Perceptron		FIMT-DD	
	Baseline	Under-sampling	Baseline	Under-sampling
Ailerons	0.0079 ± 0.0052	0.0022 ± 0.0007	0.0063 ± 0.0000	0.0006 ± 0.0000
Elevator	0.0257 ± 0.0189	0.0096 ± 0.0004	0.0317 ± 0.0129	0.0100 ± 0.0013

Table 3. $RMSE_\phi$ for the Baseline and the Over-sampling method (Algorithm 2)

Dataset	Perceptron		FIMT-DD	
	Baseline	Over-sampling	Baseline	Over-sampling
Ailerons	0.0079 ± 0.0052	0.0043 ± 0.0010	0.0063 ± 0.0000	0.0060 ± 0.0000
Elevator	0.0257 ± 0.0189	0.0210 ± 0.0012	0.0317 ± 0.0129	0.0241 ± 0.0027

As can be seen from the above tables, our method yields to smaller $RMSE_\phi$ in all cases. Also, the proposed under-sampling method seems to be more effective than our over-sampling method.

6 Conclusions and Future Work

In this paper, we introduce two data-level algorithms for learning from imbalanced regression data streams. Both methods are independent of the learning algorithm. Both methods are based on Chebychev's inequality but in a different way. The first method, under-samples from the frequent target value examples while the second method over-samples the rare and extreme target value examples to create a more balanced dataset for the learner. We applied our methods to train regression models using two benchmark datasets and two well-known regression algorithms: Perceptron and FIMT-DD. For comparative purposes, we use, as baseline, the same regression algorithms, without any re-sampling technique. The experimental results, show that our methods reduce prediction errors, in both datasets The impact of the re-sampling strategies is a reduction of the largest errors of the baseline.

This improvement encourages us to further pursue this work. The methods needs to be compare with other state-of-art algorithms from data stream mining community. There are still some issues that need to be addressed. For example, the optimum value for the warming size parameter should be better explored. In the future works, we will apply our methods over more datasets and using other learning models.

Acknowledgments. This research was carried out in the context of the project Fail-Stopper (DSAIPA/DS /0086/2018).

References

1. Batista, G.E., Prati, R.C., Monard, M.C.: A study of the behavior of several methods for balancing machine learning training data. ACM SIGKDD Explor. Newsl. **6**(1), 20–29 (2004)
2. Block, H.D.: The perceptron: a model for brain functioning. In: Neurocomputing: Foundations of research, vol. I, pp. 135–150. MIT Press, Cambridge (1988). http://dl.acm.org/citation.cfm?id=65669.104392
3. Branco, P., Torgo, L., Ribeiro, R.P.: Pre-processing approaches for imbalanced distributions in regression. Neurocomputing **343**, 76–99 (2019)
4. Chawla, N.V., Bowyer, K.W., Hall, L.O., Kegelmeyer, W.P.: SMOTE: synthetic minority over-sampling technique. J. Artif. Intell. Res. **16**, 321–357 (2002)
5. Galar, M., Fernandez, A., Barrenechea, E., Bustince, H., Herrera, F.: A review on ensembles for the class imbalance problem: bagging-, boosting-, and hybrid-based approaches. IEEE Trans. Syst. Man Cybern. Part C Appl. Rev. **42**(4), 463–484 (2012)
6. Ikonomovska, E., Gama, J., Dzeroski, S.: Learning model trees from evolving data streams. Data Min. Knowl. Discov. **23**(1), 128–168 (2011). https://doi.org/10.1007/s10618-010-0201-y
7. Kubat, M., Matwin, S., et al.: Addressing the curse of imbalanced training sets: one-sided selection. In: ICML, Nashville, USA, vol. 97, pp. 179–186 (1997)
8. Lee, S.S.: Noisy replication in skewed binary classification. Comput. Stat. Data Anal. **34**(2), 165–191 (2000)
9. Moniz, N., Ribeiro, R.P., Cerqueira, V., Chawla, N.: Smoteboost for regression: improving the prediction of extreme values. In: 5th IEEE International Conference on Data Science and Advanced Analytics, DSAA 2018, Turin, Italy, 1–3 October 2018, pp. 150–159 (2018). https://doi.org/10.1109/DSAA.2018.00025
10. Ribeiro, R.P.: Utility-based regression. Ph.D. thesis, Department Computer Science, Faculty of Sciences, University of Porto (2011)
11. Torgo, L., Ribeiro, R.P., Pfahringer, B., Branco, P.: SMOTE for regression. In: Correia, L., Reis, L.P., Cascalho, J. (eds.) EPIA 2013. LNCS (LNAI), vol. 8154, pp. 378–389. Springer, Heidelberg (2013). https://doi.org/10.1007/978-3-642-40669-0_33

Mining Human Mobility Data to Discover Locations and Habits

Thiago Andrade[1,2(✉)], Brais Cancela[1,3], and João Gama[1,2]

[1] INESC TEC, Porto, Portugal
`thiago.a.silva@inesctec.pt`
[2] University of Porto, Porto, Portugal
[3] Universidade da Coruña, A Coruña, Spain

Abstract. Many aspects of life are associated with places of human mobility patterns and nowadays we are facing an increase in the pervasiveness of mobile devices these individuals carry. Positioning technologies that serve these devices such as the cellular antenna (GSM networks), global navigation satellite systems (GPS), and more recently the WiFi positioning system (WPS) provide large amounts of spatio-temporal data in a continuous way. Therefore, detecting significant places and the frequency of movements between them is fundamental to understand human behavior. In this paper, we propose a method for discovering user habits without any a priori or external knowledge by introducing a density-based clustering for spatio-temporal data to identify meaningful places and by applying a Gaussian Mixture Model (GMM) over the set of meaningful places to identify the representations of individual habits. To evaluate the proposed method we use two real-world datasets. One dataset contains high-density GPS data and the other one contains GSM mobile phone data in a coarse representation. The results show that the proposed method is suitable for this task as many unique habits were identified. This can be used for understanding users' behavior and to draw their characterizing profiles having a panorama of the mobility patterns from the data.

Keywords: Habits · Meaningful places · Gaussian Mixture Model · Pattern · Mobility · Spatio-temporal clustering

1 Introduction

Understanding human mobility patterns can help in the exploration of the underlying driving factors of society as many aspects of life are associated with them. The first efforts to learn human mobility patterns were associated with classic social sciences. Since the nineteenth century, sociologists in what are called time-use or time-budget studies have been measuring the time people spend doing different activities throughout the day [13]. In contrast, methods for human mobility data collection have shifted over time as now both developed and developing countries are facing the increase of the pervasiveness of mobile devices [3,6].

© Springer Nature Switzerland AG 2020
P. Cellier and K. Driessens (Eds.): ECML PKDD 2019 Workshops, CCIS 1168, pp. 390–401, 2020.
https://doi.org/10.1007/978-3-030-43887-6_32

Positioning technologies that serve these devices such as the cellular antenna (GSM networks), global navigation satellite systems (GPS), and more recently the WiFi positioning system (WPS) provide large amounts of spatio-temporal data in a continuous way at low costs [9]. When dealing with raw data, final users cannot make sense of it without processing and applying techniques to extract meaningful information from its content. Many researchers have made efforts in exploring these data in order to find places, locations, and regions [7,17,18]. Hence, individuals can state a place as something with a meaning such as work, home, university while a pair of numbers like "39.98450, 116.29929" has no useful meaning to them. Therefore, detecting significant places and the frequency of movements between them is fundamental to understand human behavior.

Several studies confirmed the intuition that human mobility is highly predictable, centered on a small number of base locations [5]. This opens a wide range of opportunities for more intelligent recommendations and support for routine activities. Still, empirical studies on individual mobility patterns are scarce.

The main contributions of this paper are related as follows: we introduce a new dataset acquired from a Telecom company that comprises many different cities in Brazil. We also present a new density-based clustering for spatio-temporal data to identify meaningful places. Moreover, in the last step, we apply a Gaussian Mixture Model (GMM) over the Origin × Destination matrix of trips between meaningful places to automatically separate the trajectories for identification of user habits.

The following section presents the literature review and the most important related works. The remainder of the paper describes the methodology and the data sets utilized to assert the validity of the methods in Sect. 3, in Sect. 4 we discuss the experiments and results obtained. Finally, the conclusions and future work are presented in Sect. 5.

2   Related Work

Many researchers have been proposing methods to identify meaningful locations and habits from users for diverse goals. In this section, we review some relevant works which leverage the information contained in GPS and mobile phone data (GSM) for a multitude of different applications.

According to [7], several methods based on density have been proposed in order to discover regions of interest although most of these methods are used to aggregate spatial point objects. Some authors were more interested in the semantic movement trajectories. [8] introduced a model that makes use of movement datasets which has trajectories defined as sequences of time-stamped stops and moves between locations. In order to discover personalized visited-POIs, [11] proposed a method to estimate fine-grained and pre-defined locations. In [1] the authors explore raw GPS data to identify meaningful places in a region and describe user's profiles and similarities among them.

Many researchers were also interested in mobility patterns. Most location-based services provide recommendations based on a user's current location or a given route or destination. Even though there are indications that human movement is highly predictable, daily and weekly routines of individual users constitute a largely unexplored and unexploited area. [12] used more than 800 million of CDR data to identify weekly patterns of human mobility through mobile phone data. In [10], the authors present a methodology based on density-based clustering, clustering-based sequential mining and Apriori algorithm for analyzing user location information in order to identify user habits.

3   Problem Statement and Methodology

The objective of this work is to propose a methodology to identify user habits from GPS and GSM data without any apriori or external information. We propose a variation of DBSCAN clustering technique that is able to perform cluster of locations like buildings and squares in a better way and apply a GMM in order to separate the days and hours a given user moves between the clustered locations.

Before entering in details of the methodology, we introduce the definition of points and trajectories:

A point is a triple of the form p = *(latitude, longitude, time)* that represents a latitude-longitude location and a time-stamp. A trajectory is a sequence of ordered points triples $Tr = (p_1, p_2, \ldots, p_n)$ where p_i is a point and $p_1.time < p_2.time < \ldots < p_n.time$.

The first step of the methodology is the preprocessing task that is including among other activities, the data cleaning process where we perform outliers and noise removal. The second is the feature engineering to derive new information from the original data (in the form of latitude, longitude, and a time-stamp) to calculate key features such as time delta of the transitions, traveled distance between points, velocity, start and stop positions, time and day of the week, length and duration of a trajectory. In this work, we denote a new trajectory every time an individual stop moving or the time delta between points is more than 30 min.

3.1   User Stay Points Detection

Stay points are regions where a given user has stayed for a while within a defined radius. The algorithm is a hybrid density and time-based proposed in [15] that calculates the distance between two sets of points p_1 and p_0 in order to find those that are below a distance threshold. Next, it checks for how long the user stayed in that radius by looking at time threshold. At last step, it calculates the stay points centroid by getting the mean of the coordinates of the set of points. For this experiment, we set the parameters Distance-threshold as 200 m and the Time-threshold to 20 min as suggested in [18].

3.2   Meaningful Locations

A meaningful place is defined as a frequent location visited by an individual and does not need to be related to any other person or group like in the case of the POIs. Taking into account we already have the user's stay points, now we need to look for those places (stay points) a person visits repeatedly in order to form the so-called users' meaningful places.

Location detection techniques commonly make use of density-based methods. This is because the mechanism of density-based clustering is able to detect clusters of arbitrary shapes without specifying the number of the clusters in the data a priori and is also tolerant of outliers (noise).

The Location Clustering method proposed by [2], operates attributing in a way that once it forms a cluster, these points are eliminated from the neighbourhood and avoid new points to overlap to them. In this way, the remaining observations are available to form new clusters surrounding the previous center that could maybe be part of it. Our method, on the other hand, keeps a short memory for those points revisiting and maybe reclassifying them to the new cluster as the density of the new class turns to be more relevant.

One main advantage over the classical DBSCAN [4] implementation is that given the arbitrary shape of the trajectories, sometimes the clusters form straight chains which may not be a good representation of a location as normally buildings are in a squared or circular shape. Our method is robust to these situations as it classifies as noise those points that fall out of the neighbour's radius which is away from the centroid of the cluster. Another drawback of this original DBSCAN approach is that it does not return a centroid for each cluster. As we are looking for meaningful places over the set of stay points (Sect. 3.1), we need to find the centroid for each of the returned labels of the DBSCAN.

To overcome these issues, we propose a variation of the clustering algorithm DBSCAN [4] and Location Clustering [2] methods. The method starts searching for a given p point neighbours ($MinPts$) in as Eps radius. While the set of neighbours still changing, it keeps on looping through the data points. Once it stops changing, it checks if the number of items in the class is greater than the minimum points to form a cluster. If this condition holds, we set all the points into this given neighbourhood to noise and move the centroid of the list of points to iterate over again. The algorithm proposed is described the pseudo-code 1.

3.3   Identification of Habits

Individuals have a remarkable propensity to return to their frequently visited places. Hence, the interactions between individuals and these places are likely to represent the individual's characteristics. After clustering the user stay points into meaningful places as described in Sect. 3.2 we ended up with: trajectories connecting non-meaningful places (those who start and end in places classified as noise), trajectories connecting one meaningful place at the end or at the start and trajectories connecting two meaningful places. For the habits study purpose,

---
Algorithm 1. DBMeans Algorithm

```
1:  function DBMEANS(P, eps, MinPts)
2:      % P: a set of points (lat, long)
3:      % eps: the radius of the cluster > 0
4:      % MinPts: the minimum size of a cluster > 0
5:      % C: the label of each point in P
6:      C ← NOT VISITED
7:      Centroids ← []
8:      while ∃ p_c ∈ P|C(p_c) = NOT VISITED do:
9:          M ← {p_c}
10:         M_bak ← {∅}
11:         C(p_c) ← NOISE                                    ▷ Mark p_c as noise
12:         c = p_c                              ▷ Random choice from shuffled input P
13:         while M ≠ M_bak do
14:             M_bak ← M
15:             M ← {∅}
16:             C_M ← []
17:             for each p_x ∈ P do:
18:                 if distance(c, p_x) < eps then
19:                     M ← M ∪ {p_x}
20:                     if C(p_x) ∉ {NOTVISITED, NOISE} then
21:                         C_M ← C_M ∪ {C(p_x)}
22:             c ← Mean(M)
23:             for each c_m ∈ set(C_M) do:
24:                 if |c_m| ≥ MinPts then
25:                     C_M(C_M = c_m) ← NOISE                      ▷ Mark c_m as noise
26:                     Centroids ← Centroids \ c_m
27:         C(M) ← c_c              ▷ Mark all neighbour points with the same class
28:         Centroids(c_c) ← c
29:     C ← predict(P, Centroids, eps)
30:     return C
```
---

we will focus on the last item as we are interested in discovering frequent movements across meaningful places.

From this list of grouped trajectories is possible to identify the most important places of a given user as we can perform a count on the occurrences of trips connecting two locations. Groups with very low values, close to zero, means that there are no habits connecting those places or the *eps* parameter used to perform the clustering in step 3.2 is too small. For this study, we are considering only the two locations that have at least 5 (five) trajectories connecting them.

3.4   Gaussian Mixture Model to Classify the Different Habits

In order to discover user habits, we need to analyze the features that are emerging from the discovery process. One way we can utilize to separate the trips into habits is by the time they happen. To tackle this issue we create two new features, deriving a sine and cosine transform from the start hour.

In Fig. 1 we show an example of the transformation based on the start hours all trajectories of a user to show the new representation of time.

The cyclical representation of the time is not enough when dealing with individuals that use to go to a certain place in a non-strict way. The distribution of the data may be non-normal resulting in more than one peak along the day. Here we propose utilizing a Gaussian Mixture Model to handle these cyclical data and segment it into habits in a dynamic way. Figure 2 shows the starting hours of

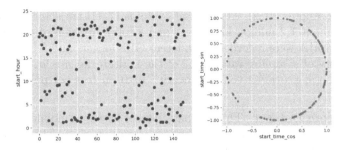

Fig. 1. Transformation of the start hour based on the Sin and Cos. The left image represents the hours in a plain representation (X axis is the trajectory order), the right is a circular where two or more points can fall over the same region no mater the trajectory order

a given user habit. One can notice that this user has 37 different starting hours for the same Origin × Destination pair and is possible to verify the segmentation made by the multiple Gaussians in the start hour distribution. Note that there are blue dots on the top and the bottom of the left image representing the same class of trajectories that occur close to 23:00 pm and 02:00 am.

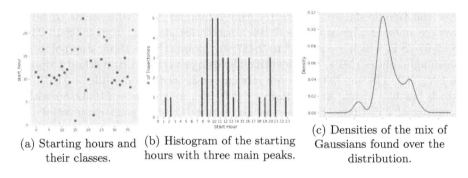

(a) Starting hours and their classes.

(b) Histogram of the starting hours with three main peaks.

(c) Densities of the mix of Gaussians found over the distribution.

Fig. 2. GMM model over the start hours of the trajectories. (Color figure online)

4   Experiments and Results

In this section, several experiments with the two real-world datasets are performed to evaluate our proposed method. The datasets description and their preparation are described in Subsect. 4.1. Subsection 4.2 corresponds to clustering results and Subsect. 4.3 presents the results regarding the habits extraction.

4.1   Datasets

Geolife GPS Dataset. This GPS trajectory dataset was collected in (Microsoft Research Asia) Geolife project by 182 users in a period of over three years (from

April 2007 to August 2012). The dataset contains 17,621 trajectories with a total distance of about 1,2 million kilometers and a total duration of 48,000+ hours. These trajectories were recorded by different GPS loggers and GPS-phones, and have a variety of sampling rates. 90% of the trajectories are logged in a dense representation, e.g. every 1 to 5 s or every 5 to 10 meters per point [16–18].

GSM Telecom Dataset. This is a new dataset based on mobile phone (GSM) data. The dataset contains 526,894 instances from a period of 12 months or 350 days starting on September 2017 and finishing in September 2018 consisting of 4,545 different individuals. After cleaning and removing the duplicates, it was reduced to 461,778 instances. The points were recorded in many cities in Brazil with a coarse granularity of one point at every 15 min. No information about the users is derived from these data, as the entire dataset is anonymized. Each point consists of a user sequential identification number, a pair of (latitude, longitude), and a timestamp. All the data was delivered in a single file that is available in the project folder on the web page.[1]

4.2   Clustering Results

Following we present the results of the experiments over the two datasets with respect to the identification of Meaningful Places. To conduct the experiments over the Geolife dataset we elected the individual '004' who seems to be an average person. This user has 1.100 trajectories starting from 2008-10-23 to 2009-07-28 in which are related to 2.437 stay points. From those stay points, 50 meaningful places (MPs) were identified by using the clustering method proposed in Sect. 3.2. The top two MPs are latitude = 39.99993, longitude = 116.32730 which has 659 visits, and latitude = 40.01086, longitude = 116.32186 with a counting of 235 times. Here we set the Home (Qinghuayuan Residential District) and Work (Tsinghua University Northwest) locations respectively based on the frequency of these observations as many other works propose [2,5,10,14]. To perform a visual inspection of the formed clusters, Fig. 3 illustrates the differences obtained using each one of the methods. Notice that our approach results in clusters that are more robust and handle the noise with more efficiency.

Regarding the GSM Mobile Telecom dataset, as in this dataset the granularity is coarse, the results are quite different from the ones shown in the GPS dataset as we have one observation at every 15 min. Although Brazil is a very large and populated country, the latitudes and longitudes encountered in this dataset fall into some very small up to medium cities with traffic conditions very different from Beijing. One can notice that in this case, a 15-min interval can lead to the transportation of the individual to a very different location without any details of the trajectory taken. Basically, we end up with the start and end of the trajectory only. To conduct the experiments over this dataset we elected the individual '10837'. This user has 19 trajectories starting from 2018-05-10 to 2018-07-01 in which are related to 135 stay points from 4 meaningful places

---

[1]   https://bit.ly/2ZVERKO.

(a) Our method returns more concise clusters as it doesn't erase the nearby points after forming a cluster. The surrounding points are set to noise when the mean of the points inside the radius stops changing.

(b) Location Clustering: returns satellite clusters over the main location as it works erasing those points who form a cluster after the mean stops changing. Noisy points can be set to a possible new cluster as can be seen in detail.

(c) DBSCAN: forms very large clusters from chaining points which are density-connected. This is of the main disadvantage in this context as the shape of the locations is usually in squared or circular different from ellipses.

Fig. 3. The dense region in the top shows the clear difference among the methods: while our approach (a) returns only two clusters, the Location Clustering (b) returns 9 and DBSCAN (c) returns only one large cluster. The X symbol stands for noise

(MPs). The top two MPs are latitude $= -18.96081$, longitude -48.32141 which has 38 visits, and latitude $= -18.94969$, longitude $= -48.31219$ with a counting of 6 times. In Fig. 4 we can see the locations over the map of Uberlândia/Brazil.

4.3 Habits Results

The knowledge discovery process over raw location data has led us to a panorama of the mobility patterns of the given community. Main factors that characterize habits are related to the start hour, length and duration of the trajectories that follow an Origin × Destination pattern. The Fig. 5 shows the three different habits returned from the Gaussians for the trajectories of the user connecting the two locations. The Fig. 6 illustrates the hourly distribution of the trajectories between the two main groups of meaningful places for the Geolife dataset user '004'.

The length of the trajectories is also a discriminant feature, as users tend to follow the same path to go from places according to evaluated conditions such as day of the week, the hour of the day, weather. In rush hours is more reasonable to avoid areas with too many people and traffic as the time taken to run the same path can be completely distinct. The Fig. 7 shows the length of the trajectories with respect to their groups of meaningful places. As we can see,

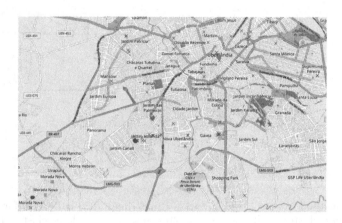

Fig. 4. Meaningful locations (colored circles) of the user 10837 over the Uberlândia/Brazil map. The x symbol stands for noise (Color figure online)

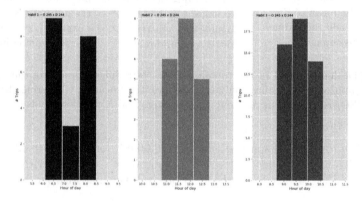

Fig. 5. Three main habits returned from the start hours connecting the top two locations of the user '004'

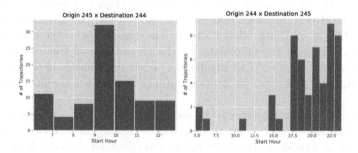

Fig. 6. Distribution of starting hours from the top two meaningful locations of the user '004'

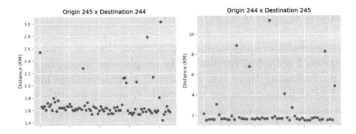

Fig. 7. Distribution of trajectory distance from the top two meaningful locations of the user '004'

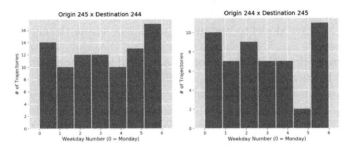

Fig. 8. Distribution of the trajectories according to the day of week from the top two meaningful locations of the user '004'

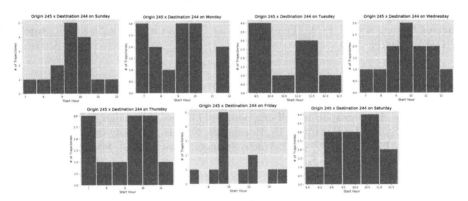

Fig. 9. Distribution of the trajectories according to the hour and day of week from the top two meaningful locations of the user '004'

some few trajectories have distance greater than 2 km. Those can be justified as non-habitual paths.

Another relevant way to analyze habits is looking for the day of the week a trip was taken. Routines are very common in human patterns and some of them may occur less often than the others. In Figs. 8 and 9 we show the distribution of the trips in a weekly view.

5   Conclusions and Future Work

A historical record of the daily mobility pattern of the users hides an unexpectedly high degree of potential predictability despite the apparent randomness of human nature. Following this idea, we show that most people have a relatively regular schedule of moments when they travel from one location to another.

In this research, we present a new density-based clustering method to filter mobility data finding the most frequent places of a given individual and compare our method with two other proposals and show that this approach provides more informative results for this context. We also explore a new GSM dataset of diverse cities in Brazil showing the usefulness of the proposed clustering method to identify meaningful places over data with different granularity. We also introduce a Gaussian Mixture Model to find individuals' habits from the clustered data in a dynamic way.

For future work, we intend to propose a method to find the patterns of people visiting and leaving different places at different times in an order (weekly basis, daily basis) similar to sequential pattern mining methods. Also includes some map matching tasks including external information in order to find the semantic meaning of the individuals' movements. We also intend to apply the method in other datasets to verify its usefulness generalizing in other scenarios. Location prediction is also a field that is considered the results of this paper are strongly related to it.

Acknowledgement. This work is financed by National Funds through the Portuguese funding agency, FCT - Fundação para a Ciência e a Tecnologia within project: UID/EEA/50014/2019.

References

1. Andrade, T., Gama, J.: Identifying points of interest and similar individuals from raw GPS data. In: Cagáñová, D., Horňáková, N. (eds.) Mobility Internet of Things 2018. EAISICC, vol. 39, pp. 293–305. Springer, Cham (2020). https://doi.org/10.1007/978-3-030-30911-4_21
2. Ashbrook, D., Starner, T.: Using GPS to learn significant locations and predict movement across multiple users. Pers. Ubiquit. Comput. **7**(5), 275–286 (2003)
3. Berry, D.: The computational turn: thinking about the digital humanities. Cult. Mach. **12**, 1–22 (2011)
4. Ester, M., Kriegel, H.P., Sander, J., Xu, X., et al.: A density-based algorithm for discovering clusters in large spatial databases with noise. In: KDD, vol. 96, pp. 226–231 (1996)
5. Herder, E., Siehndel, P.: Daily and weekly patterns in human mobility. In: UMAP Workshops. Citeseer (2012)
6. Lazer, D., et al.: Computational social science. Science **323**(5915), 721–723 (2009)
7. Lee, I., Cai, G., Lee, K.: Mining points-of-interest association rules from geo-tagged photos. In: 2013 46th Hawaii International Conference on System Sciences, pp. 1580–1588. IEEE (2013)

8. Li, Q., Zheng, Y., Xie, X., Chen, Y., Liu, W., Ma, W.Y.: Mining user similarity based on location history. In: Proceedings of the 16th ACM SIGSPATIAL, p. 34. ACM (2008)

9. Liu, H., Darabi, H., Banerjee, P., Liu, J.: Survey of wireless indoor positioning techniques and systems. IEEE Trans. Syst. Man Cybern. B Cybern. Part C (Appl. Rev.) **37**(6), 1067–1080 (2007)

10. Sardianos, C., Varlamis, I., Bouras, G.: Extracting user habits from google maps history logs. In: 2018 IEEE/ACM International Conference on Advances in Social Networks Analysis and Mining (ASONAM), pp. 690–697. IEEE (2018)

11. Suzuki, J., Suhara, Y., Toda, H., Nishida, K.: Personalized visited-poi assignment to individual raw gps trajectories. arXiv preprint arXiv:1901.06257 (2019)

12. Thuillier, E., Moalic, L., Lamrous, S., Caminada, A.: Clustering weekly patterns of human mobility through mobile phone data. IEEE Trans. Mob. Comput. **17**(4), 817–830 (2018)

13. Toch, E., Lerner, B., Ben-Zion, E., Ben-Gal, I.: Analyzing large-scale human mobility data: a survey of machine learning methods and applications. Knowl. Inf. Syst. **58**(3), 501–523 (2018). https://doi.org/10.1007/s10115-018-1186-x

14. Yang, M., Cheng, C., Chen, B.: Mining individual similarity by assessing interactions with personally significant places from GPS trajectories. ISPRS Int. J. Geo-Inf. **7**(3), 126 (2018)

15. Ye, Y., Zheng, Y., Chen, Y., Feng, J., Xie, X.: Mining individual life pattern based on location history. In: Mobile Data Management: Systems, Services and Middleware, MDM 2009, pp. 1–10. IEEE (2009)

16. Zheng, Y., Li, Q., Chen, Y., Xie, X., Ma, W.Y.: Understanding mobility based on GPS data. In: Proceedings of the 10th International Conference on Ubiquitous Computing, pp. 312–321. ACM (2008)

17. Zheng, Y., Xie, X., Ma, W.Y.: GeoLife: a collaborative social networking service among user, location and trajectory. IEEE Data Eng. Bull. **33**(2), 32–39 (2010)

18. Zheng, Y., Zhang, L., Xie, X., Ma, W.Y.: Mining interesting locations and travel sequences from GPS trajectories. In: Proceedings of the 18th International Conference on World Wide Web, pp. 791–800. ACM (2009)

Imbalanced Data Stream Classification Using Hybrid Data Preprocessing

Barbara Bobowska$^{(\boxtimes)}$, Jakub Klikowski$^{(\boxtimes)}$, and Michał Woźniak$^{(\boxtimes)}$

Department of Systems and Computer Networks,
Wrocław University of Science and Technology, Wrocław, Poland
{barbara.bobowska,jakub.klikowski,michal.wozniak}@pwr.edu.pl

Abstract. Imbalanced data streams have gained significant popularity among the researchers in recent years. This area of research is not only still greatly underdeveloped, but there are also numerous inherent difficulties that need to be addressed when creating algorithms that could be utilized in such dynamic environment and achieve satisfactory results when it comes to their predictive abilities. In this paper, a novel algorithm that combines both over- and under-sampling techniques in order to create a more robust classifier dedicated to imbalanced data streams is proposed. The efficiency and high predictive quality of the proposed method have been confirmed on the basis of extensive experimental research carried out on the real and the computer-generated data streams.

Keywords: Imbalanced data · Data stream classification · Data preprocessing

1 Introduction

In the last couple of years, a sharp rise in products and systems using machine learning to enhance their performance is observed. Many of the applications such as predicting user behavior on social platforms like Twitter, or client activity on online stores fall into the category of imbalanced data stream classification [24]. When designing methods for data stream classification one has to take into account the characteristics of a data stream such as the sequential manner that the data arrives, over which one has no control when it comes to the order of the arriving samples, as well as the fact that the size of the stream could be possibly infinite. Due to that requirement, it is impossible to process the upcoming data in multiple passes and such the samples can be processed once [26]. Furthermore, one has to consider the rapid rate at which the data arrives, at the same time ensuring that the processing of the data stream is done in a timely fashion such that the delay in the performance of the algorithm is minimal. Data streams can exhibit a change in data and target concepts over time (so-called Non-stationary data streams) [16,26]. Such a phenomenon is called *concept drift* [12] and it is quite common i.e. the change of popular topics on Twitter.

© Springer Nature Switzerland AG 2020
P. Cellier and K. Driessens (Eds.): ECML PKDD 2019 Workshops, CCIS 1168, pp. 402–413, 2020.
https://doi.org/10.1007/978-3-030-43887-6_33

Due to the *concept drift* the performance of the classifier can degrade over time and as such the classifier has to be trained incrementally to accommodate the changes of concepts of non-stationary data streams. Moreover, the proportion between classes is often skewed with one class being over-represented. In cases where the imbalance ratio is present traditional accuracy driven methods are not applicable especially when misclassification of the minority class examples is much more costly, as is often the case i.e. fraud detection [24]. It is worth mentioning that not only the imbalance ratio can influence the performance of the classifier. Some examples can be easy to classify even when the IR is high if the classes are separated from each other the decision boundary and be determined with ease. However, it has been observed that instances of the minority class have a tendency to create sparsely spread throughout the object space clusters, often surrounded by majority class examples [4]. The presence of noise and outliers is another difficulty factor that needs to be addressed. In [3,15] authors created preprocessing methods with those issues in mind.

Data streams may be processed either in blocks or one instance at a time. One of the most important issues in learning from the data stream is when to update the classifier [22]. Most researchers distinguish between two approaches: active and passive. In the former, the update is performed only if drift is detected while the later updates the classifier continuously regardless if the drift was detected or not [9]. In order to satisfy the time and memory requirements, a forgetting or data management mechanism must be used. One of the most popular approaches to forgetting is using sliding windows, which can be either sequence based, where the size of the window is defined by a number of instances and time stamp based where the window is defined by a certain duration time. In the simplest example sliding windows are of fixed size, and include only the most recent examples. The oldest samples in a window are discarded in favor of new ones. Some methods implement sliding windows of varying size depending on the response from drift detectors [2].

The main contributions of this work are as follows:

– Proposition of the two novel imbalanced data stream classifiers (DSC-R and DSC-S) which employ under- and oversampling techniques for balancing data.
– Experimental evaluation of the proposed algorithms and their comparison with state-of-art methods.

The article is organized as follows. Sections 1 and 2 present a brief introduction to the problem of imbalance data stream classification and a quick overview of the state-of-the-art algorithms dedicated to it. Section 3 offers an in-depth explanation of the proposed solution. Section 4 showcases the results of the computer experiments, comparing the proposed algorithm to different techniques for imbalanced data classification, proving the usefulness of the developed algorithm. Section 5 presents the conclusions and describes possible future improvements to the proposed method.

2   Related Works

Studies over the years presented algorithms dedicated to data stream analysis. Very fast decision tree (VFDT) proposed by Domingos and Hulten [13] was among the first methods for stream analysis, that to this day has been a basis for many modifications. VFDT utilizes the Hoeffding bound in order to calculate the proper number of examples needed to select the split-node. The algorithm incrementally creates a tree form from a data stream ensuring that once the examples were used to update the tree they are negligible and can be removed. The aforementioned modifications include ideas such as pruning mechanisms or utilizing sliding windows and drift detectors in order to better the algorithms in case of non-stationary streams [10]. Worth noting are several methods using ensembles of classifiers. Weighted Majority Algorithm [18] adjusts the weights of the classifiers in the ensemble so that the weight of an expert that misclassified an instance is decreased accordingly to the user-specified value. A modification of the method with an added procedure which adds new classifiers to the ensemble when the overall performance is unsatisfactory called Dynamic Weighted Majority (DWM) was introduced in [14]. In Accuracy Weighted Ensemble (AWE) a new classifier is added only if the ensemble's size is not exceeded [25] while in Learn++.NSE [8] such a constraint is not applied. In Learn++.CDS Ditzler and Polikar combine their previous work Learn++.NSE with SMOTE sampling in order to better address the data imbalance and later replacing SMOTE with an original bagging-based method of data balancing [7]. In SEA [23] a new classifier candidate is evaluated to determine whether or not it is worth including into the ensemble at the cost of replacing some other classifier already in the ensemble. Other approaches such as OUSEnsemble (Over Under Sampling Ensemble) [11] make use of sampling techniques. The stream is divided into blocks that consist of examples from both majority and minority class. The idea is to propagate all the instances of the minority class from the previous block and under-sample the majority examples in the current block such that the desired imbalance ratio is acquired. Afterwards, from the resultant subset, datasets later used to build component classifiers for the ensemble, are created by propagating all instances of the minority class to each of the datasets while each example from the majority class is propagated to only one dataset. Proposed by Chen and He the Selectively Recursive Approach (SERA) [5] uses a Mahalanobis distance to determine which of the examples from the minority class in the previous block are most similar to the minority examples in the current block. Based on that a limited number of minority class examples is selected and added to the majority class examples in the current block. Chen and He later designed a Recursive Ensemble Approach (REA) [6]. In REA minority class examples from the previous block that are nearest neighbors of minority class examples from the current block are added in order to balance the given training block. Both REA and SERA proved to make more accurate predictions than the method proposed by [19]. A Chunk-based ensemble approach, proposed by Wang et al. called KMean-Clustering [25] utilizes k-mean clustering in order to under-sample the majority class, by using the centroids created in the clustering process to resample the majority instances.

3   The Deterministic Sampling Classifier

The proposed method, called *Deterministic Sampling Classifier* (DSC), for data stream classification, processes the upcoming data in chunks. Each chunk is used in two operations. Firstly, the instances of the majority class present in the currently processed block are under-sampled in order to produce a balanced class representation in a data chunk (Fig. 1).

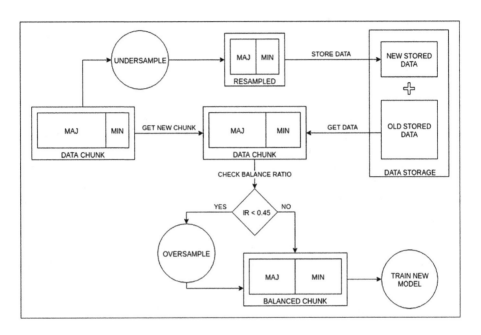

Fig. 1. Proposed method flow diagram

   The resulting data (referred to in the Fig. 1 as NEW STORED DATA) is then stored in a memory buffer (DATA STORAGE). Secondly, that same block of data is combined with a part of the data from the buffer, called OLD STORED DATA, using GET NEW CHUNK, which copies the data from the currently processed block and the GET DATA method, which takes OLD STORED DATA from the DATA STORAGE buffer. OLD STORED DATA, consists of all the previously accumulated under-sampled chunks. When a new chunk of data arrives the data from NEW STORED DATA are moved to the OLD STORED DATA part of the buffer. The DATA STORAGE is of fixed size. When the buffer is full, the oldest examples are removed from it. Afterward, the imbalance ratio of the data block created as a result of the GET NEW CHUNK and GET DATA is calculated, and if the value is lower than 0.45, an oversampling of the minority class is performed, and then a classification model is trained. Otherwise, the algorithm accepts the chunk as properly balanced and uses it to train the model right away. The implementation allows one to choose sampling algorithms of their liking. In this paper, the authors created two versions of the method DSC-S (Deterministic Sampling Classifier-SMOTE) and DSC-R (Deterministic Sampling Classifier-Random). For the DSC-R method the chosen sampling methods were: random over- and under-sampling and for the DSC-S: SMOTE and NearMiss (implementation from the imbalanced-learn library [17]) for over- and under-sampling accordingly.

4   Experimental Evaluation

The quality of the proposed algorithms was evaluated on the basis of computer experiments, using 26 real and 60 synthetic data streams. The evaluation procedure used in order to assess the predictive performance of a data stream classifier was conducted by interleaving testing with training (test-then-train) [16]. Each block is first used to test the classifier and afterward it is used for training. As a measure of comparison, the following methods were used: OUSEnsemble, KMeanClustering, REA, Learn++.CDS, Learn++.NIE and MLPClassifier (Multi-layer Perceptron classifier), using a k-NN as a base classifier. The algorithms were implemented in Python using Scikit-learn [20] and imbalanced-learn [17] libraries[1]. The selected real streams were downloaded from the KEEL [1] and PROMISE Software Engineering Repository [21]. The chosen datasets consisted of multidimensional binary classification problems with the imbalance ratio ranging from 1 to 39. The datasets were described in Table 1. The results were analyzed using the KEEL software evaluation tool [1]. Non-parametrical statistical tests were performed namely the Friedman Test as well as a Nemenyi's Post-Hoc Procedure. The metrics chosen were F-score, Gmean and AUC score. Tables 2, 3 and 4 present the obtained results. The table presents the obtained results as the mean value for each of the metrics, as well as, the information on those methods that performed poorly in comparison with the method named in the column, placed directly below the score. For instance, given the abalone-17-vs-7-8-9-10 dataset, the DSC-R algorithm performed statistically better than the 3rd, 5th, the 7th and 8th algorithm in the table (read from left to right). The obtained results prove the usefulness of the proposed algorithms. For the F-score the proposed DSC-R and DSC-S algorithms along with the REA algorithm have the best results. What is interesting the MLPC algorithm performed consistently the worst. For the Gmean the results are similar. The methods introduced in the paper perform favorably in comparison with other algorithms, the Learn++.CDS and Learn++.NIE techniques, as well as REA, have comparable results to the DSC-R and DSC-S methods. Lastly, the results in Table 4 representing the results for AUC score indicate the proposed algorithm obtained satisfactory results,

---

[1] Repository link: https://github.com/w4k2/iot-ecml2019.

with only the LCDS algorithm performing marginally better. It is worth mentioning, that the created methods are robust enough, so that imbalance ratio (whether low or high) does not negatively impact their performance.

Table 1. Overview of datasets selected for experimental evaluation (source: KEEL and PROMISE Software Engineering Repository).

Dataset	IMB. RATIO	SAMPLES	FEATURES
abalone-17_vs_7-8-9-10	39	2338	8
australian	1	690	14
glass-0-1-2-3-vs-4-5-6	3	214	9
glass0	2	214	9
glass1	2	214	9
heart	1	270	13
jm1	4	10885	22
kc1	5	2109	22
kc2	4	522	22
kr-vs-k-three_vs_eleven	35	2935	6
kr-vs-k-zero-one_vs_draw	26	2901	6
page-blocks0	9	5472	10
pima	2	768	8
segment0	6	2308	19
shuttle-c0-vs-c4	14	1829	9
vehicle0	3	846	18
vowel0	10	988	13
wisconsin	2	683	9
yeast-0-2-5-6-vs-3-7-8-9	9	1004	8
yeast-0-2-5-7-9-vs-3-6-8	9	1004	8
yeast-0-3-5-9-vs-7-8	9	506	8
yeast-0-5-6-7-9-vs-4	9	528	8
yeast-2-vs-4	9	514	8
yeast1	2	1484	8
yeast3	8	1484	8

Table 2. Overview of the results for F-score.

Dataset	1 DSC-R	2 DSC-S	3 KMC	4 LCDS	5 LNIE	6 REA	7 OUSE	8 MLPC
abalone-17-vs-7-8-9-10	0.237	0.208	0.060	0.250	0.046	0.176	0.072	0.000
	3,5,7,8	3,5,7,8	8	3,5,7,8	–	–	8	–
australian	0.571	0.601	0.444	0.514	0.604	0.518	0.602	0.000
	3,8	3,6,8	8	8	3,4,6,8	8	3,4,6,8	–
electricity-normalized	0.631	0.623	0.580	0.588	0.598	0.575	0.593	0.324
	3,4,5,6,7,8	3,4,6,7,8	8	8	8	8	8	–
glass-0-1-2-3-vs-4-5-6	0.765	0.801	0.781	0.814	0.768	0.782	0.452	0.000
	7,8	7,8	7,8	7,8	7,8	7,8	8	–
glass0	0.662	0.673	0.592	0.640	0.516	0.597	0.473	0.000
	5,7,8	5,7,8	7,8	7,8	8	8	8	–
glass1	0.687	0.649	0.576	0.517	0.558	0.412	0.573	0.000
	4,5,6,8	6,8	6,8	8	6,8	8	6,8	–
heart	0.583	0.569	0.489	0.569	0.641	0.538	0.580	0.144
	8	8	8	8	3,8	8	8	–
jm1	0.395	0.390	0.182	0.399	0.372	0.331	0.317	0.170
	3,7,8	3,7,8	–	3,6,7,8	3,8	3,8	3,8	–
kc1	0.387	0.394	0.186	0.377	0.394	0.331	0.311	0.127
	3,7,8	3,6,7,8	–	3,7,8	3,7,8	3,8	3,8	–
kc2	0.532	0.560	0.198	0.511	0.509	0.467	0.380	0.389
	3	3,7,8	–	3	3	3	3	3
kr-vs-k-three-vs-eleven	0.701	0.847	0.256	0.730	0.408	0.836	0.335	0.000
	3,5,7,8	1,3,4,5,7,8	8	3,5,7,8	3,8	1,3,4,5,7,8	8	–
kr-vs-k-zero-one-vs-draw	0.673	0.804	0.395	0.773	0.591	0.785	0.367	0.000
	3,7,8	1,3,5,7,8	8	1,3,5,7,8	3,7,8	1,3,5,7,8	8	–
page-blocks0	0.541	0.576	0.252	0.359	0.335	0.533	0.352	0.323
	3	3,4,5,7,8	–	–	–	3,7	–	–
pima	0.601	0.578	0.508	0.539	0.536	0.581	0.518	0.370
	3,5,7,8	8	8	8	8	8	8	–
segment0	0.671	0.792	0.405	0.587	0.339	0.580	0.294	0.123
	3,4,5,6,7,8	1,3,4,5,6,7,8	7,8	3,5,7,8	8	3,5,7,8	8	–
shuttle-c0-vs-c4	0.995	0.995	0.923	0.960	0.931	0.995	0.955	0.202
	3,5,7,8	3,5,7,8	8	8	8	3,5,7,8	8	–
vehicle0	0.812	0.824	0.653	0.744	0.563	0.758	0.563	0.257
	3,5,7,8	3,4,5,7,8	8	5,7,8	8	5,7,8	8	–
vowel0	0.682	0.753	0.274	0.553	0.272	0.553	0.251	0.059
	3,5,7,8	3,4,5,6,7,8	8	3,5,7,8	8	3,5,7,8	8	–
wisconsin	0.966	0.951	0.708	0.942	0.941	0.949	0.862	0.019
	3,8	3,8	8	3,8	3,8	3,8	3,8	–
yeast-0-2-5-6-vs-3-7-8-9	0.446	0.393	0.358	0.362	0.389	0.514	0.257	0.073
	7,8	8	8	8	8	3,4,7,8	8	–
yeast-0-2-5-7-9-vs-3-6-8	0.667	0.710	0.580	0.675	0.590	0.722	0.366	0.000
	7,8	3,5,7,8	7,8	7,8	7,8	3,5,7,8	8	–
yeast-0-3-5-9-vs-7-8	0.251	0.288	0.222	0.265	0.145	0.403	0.185	0.018
	8	8	8	8	–	3,5,7,8	8	–
yeast-0-5-6-7-9-vs-4	0.350	0.432	0.327	0.428	0.257	0.379	0.176	0.000
	7,8	7,8	7,8	7,8	8	7,8	8	–
yeast-2-vs-4	0.615	0.670	0.581	0.569	0.571	0.702	0.389	0.000
	7,8	7,8	8	8	8	7,8	8	–
yeast1	0.527	0.495	0.520	0.504	0.430	0.476	0.498	0.000
	8	8	8	8	8	8	8	–
yeast3	0.566	0.589	0.429	0.573	0.436	0.644	0.314	0.100
	7,8	7,8	8	7,8	8	7,8	8	–

Table 3. Overview of the results for Gmean score.

Dataset	1 DSC-R	2 DSC-S	3 KMC	4 LCDS	5 LNIE	6 REA	7 OUSE	8 MLPC
abalone-17-vs-7-8-9-10	0.782	0.818	0.541	0.684	0.102	0.282	0.536	0.000
	3,5,6,7,8	3,5,6,7,8	5,8	5,6,8	–	–	5,8	–
australian	0.622	0.640	0.433	0.557	0.187	0.580	0.100	0.000
	3,5,7,8	3,4,5,6,7,8	5,7,8	3,5,7,8	8	3,5,7,8	–	–
electricity-normalized	0.682	0.676	0.639	0.644	0.510	0.635	0.284	0.393
	3,4,5,6,7,8	3,4,5,6,7,8	5,7,8	5,7,8	7,8	5,7,8	–	7
glass-0-1-2-3-vs-4-5-6	0.865	0.892	0.836	0.905	0.869	0.869	0.165	0.000
	7,8	7,8	7,8	7,8	7,8	7,8	–	–
glass0	0.739	0.756	0.588	0.720	0.287	0.697	0.071	0.000
	3,5,7,8	3,5,7,8	5,7,8	5,7,8	8	5,7,8	–	–
glass1	0.750	0.717	0.560	0.600	0.254	0.526	0.127	0.000
	3,4,5,6,7,8	3,4,5,6,7,8	5,7,8	5,7,8	8	5,7,8	–	–
heart	0.623	0.618	0.518	0.589	0.346	0.566	0.284	0.205
	3,8	3,8	8	8	–	8	–	–
jm1	0.660	0.666	0.275	0.675	0.630	0.557	0.369	0.291
	3,6,7,8	3,6,7,8	–	3,6,7,8	3,7,8	3,7,8	–	–
kc1	0.651	0.670	0.278	0.655	0.658	0.557	0.391	0.204
	3,6,7,8	3,6,7,8	–	3,6,7,8	3,6,7,8	3,7,8	8	–
kc2	0.717	0.772	0.230	0.711	0.715	0.618	0.259	0.347
	3,7,8	3,7,8	–	3,7,8	3,7,8	3,7	–	–
kr-vs-k-three-vs-eleven	0.988	0.985	0.911	0.981	0.836	0.985	0.933	0.000
	3,7,8	3,7,8	8	3,7,8	8	3,7,8	8	–
kr-vs-k-zero-one-vs-draw	0.970	0.965	0.898	0.932	0.873	0.933	0.935	0.000
	3,5,7,8	3,5,8	8	8	8	8	8	–
page-blocks0	0.850	0.848	0.543	0.586	0.649	0.812	0.805	0.447
	3,4,5,8	3,4,5,8	–	–	–	3,8	3,8	–
pima	0.690	0.673	0.590	0.634	0.400	0.670	0.197	0.482
	3,4,5,7,8	3,5,7,8	5,7,8	5,7,8	7	3,5,7,8	–	7
segment0	0.911	0.934	0.719	0.848	0.585	0.851	0.415	0.242
	3,4,5,6,7,8	3,4,5,6,7,8	5,7,8	3,5,7,8	7,8	3,5,7,8	8	–
shuttle-c0-vs-c4	0.995	0.995	0.949	0.963	0.950	0.995	0.996	0.304
	8	8	8	8	8	8	3,5,8	–
vehicle0	0.916	0.901	0.819	0.881	0.723	0.882	0.708	0.380
	3,5,7,8	3,5,7,8	5,7,8	5,7,8	8	5,7,8	–	–
vowel0	0.937	0.938	0.523	0.797	0.634	0.790	0.634	0.224
	3,4,5,7,8	3,4,5,7,8	8	3,5,7,8	8	3,8	8	–
wisconsin	0.974	0.960	0.747	0.953	0.950	0.957	0.781	0.042
	3,8	3,8	8	3,8	3,8	3,8	8	–
yeast-0-2-5-6-vs-3-7-8-9	0.761	0.661	0.685	0.709	0.683	0.745	0.511	0.178
	7,8	8	7,8	7,8	8	7,8	8	–
yeast-0-2-5-7-9-vs-3-6-8	0.862	0.848	0.827	0.879	0.839	0.875	0.681	0.000
	7,8	7,8	7,8	7,8	7,8	7,8	8	–
yeast-0-3-5-9-vs-7-8	0.617	0.650	0.517	0.596	0.339	0.651	0.470	0.069
	8	7,8	8	8	–	8	8	–
yeast-0-5-6-7-9-vs-4	0.750	0.822	0.777	0.783	0.612	0.694	0.485	0.000
	7,8	7,8	7,8	7,8	8	8	8	–
yeast-2-vs-4	0.827	0.826	0.801	0.830	0.723	0.847	0.731	0.000
	8	8	8	8	8	8	8	–
yeast1	0.634	0.616	0.599	0.612	0.330	0.593	0.344	0.000
	5,7,8	5,7,8	5,7,8	5,7,8	8	5,7,8	8	–
yeast3	0.871	0.855	0.815	0.841	0.773	0.875	0.698	0.197
	7,8	7,8	8	7,8	8	5,7,8	8	–

Table 4. Overview of the results for AUC score.

Dataset	1 DSC-R	2 DSC-S	3 KMC	4 LCDS	5 LNIE	6 REA	7 OUSE	8 MLPC
abalone-17-vs-7-8-9-10	0.997	0.989	0.684	0.999	0.828	0.967	0.789	0.329
	3,5,6,7,8	3,5,6,7,8	8	3,5,6,7,8	3,8	3,5,7,8	3,8	–
australian	0.827	0.829	0.551	0.819	0.662	0.803	0.578	0.448
	3,5,7,8	3,5,7,8	8	3,5,7,8	3,7,8	3,5,7,8	8	–
electricity-normalized	0.961	0.961	0.927	0.974	0.871	0.971	0.551	0.671
	3,5,7,8	3,5,7,8	5,7,8	1,2,3,5,6,7,8	7,8	1,2,3,5,7,8	–	7
glass-0-1-2-3-vs-4-5-6	0.988	0.989	0.894	0.989	0.879	0.984	0.731	0.722
	7,8	7,8	8	7,8	8	7,8	–	–
glass0	0.937	0.929	0.792	0.930	0.724	0.881	0.628	0.675
	3,5,7,8	3,5,7,8	5,7	3,5,7,8	–	5,7,8	–	–
glass1	0.913	0.918	0.681	0.830	0.668	0.790	0.546	0.669
	3,5,6,7,8	3,5,6,7,8	–	5,7	–	7	–	–
heart	0.846	0.855	0.728	0.877	0.662	0.860	0.632	0.374
	3,5,7,8	3,5,7,8	8	3,5,7,8	8	3,5,7,8	8	–
jm1	0.866	0.862	0.398	0.906	0.788	0.858	0.720	0.467
	3,5,7,8	3,5,7,8	–	1,2,3,5,6,7,8	3,7,8	3,5,7,8	3,8	–
kc1	0.865	0.860	0.405	0.902	0.786	0.858	0.734	0.399
	3,5,7,8	3,5,7,8	–	1,2,3,5,6,7,8	3,7,8	3,5,7,8	3,8	–
kc2	0.877	0.889	0.487	0.908	0.832	0.882	0.769	0.757
	3,8	3,7,8	–	3,7,8	3	3,7,8	3	3
kr-vs-k-three-vs-eleven	1.000	1.000	0.981	1.000	0.987	1.000	0.978	0.968
	3,5,7,8	3,5,7,8	–	3,5,7,8	7,8	3,5,7,8	–	–
kr-vs-k-zero-one-vs-draw	1.000	1.000	0.972	1.000	0.982	0.999	0.981	0.691
	3,5,7,8	3,5,6,7,8	8	3,5,6,7,8	8	3,5,7,8	8	–
page-blocks0	0.994	0.994	0.850	0.999	0.962	0.992	0.913	0.734
	3,5,7,8	3,5,7,8	–	1,2,3,5,6,7,8	3,7,8	3,5,7,8	–	–
pima	0.870	0.871	0.736	0.890	0.725	0.850	0.645	0.508
	3,5,7,8	3,5,7,8	7,8	3,5,7,8	7,8	3,5,7,8	8	–
segment0	0.996	0.997	0.859	0.993	0.831	0.975	0.794	0.628
	3,5,6,7,8	3,4,5,6,7,8	7,8	3,5,6,7,8	7,8	3,5,7,8	8	–
shuttle-c0-vs-c4	1.000	1.000	0.999	1.000	0.500	1.000	0.999	0.480
	5,8	5,8	5,8	5,8	–	5,8	5,8	–
vehicle0	0.988	0.990	0.881	0.987	0.834	0.970	0.841	0.767
	3,5,6,7,8	3,5,6,7,8	8	3,5,7,8	–	3,5,7,8	–	–
vowel0	0.997	0.999	0.852	0.999	0.862	0.989	0.870	0.260
	3,5,6,7,8	3,5,6,7,8	8	3,5,6,7,8	8	3,5,7,8	8	–
wisconsin	0.998	0.997	0.929	0.997	0.716	0.995	0.958	0.081
	7,8	7,8	8	7,8	8	7,8	8	–
yeast-0-2-5-6-vs-3-7-8-9	0.951	0.943	0.789	0.989	0.807	0.929	0.803	0.444
	3,5,7,8	3,5,7,8	8	1,2,3,5,6,7,8	8	3,5,7,8	8	–
yeast-0-2-5-7-9-vs-3-6-8	0.983	0.979	0.896	0.997	0.925	0.978	0.900	0.252
	3,5,7,8	3,5,7,8	8	2,3,5,6,7,8	8	3,5,7,8	8	–
yeast-0-3-5-9-vs-7-8	0.947	0.940	0.659	0.989	0.741	0.889	0.730	0.270
	3,5,7,8	3,5,7,8	8	3,5,6,7,8	8	3,5,7,8	8	–
yeast-0-5-6-7-9-vs-4	0.984	0.975	0.867	0.994	0.891	0.965	0.828	0.089
	3,5,7,8	3,5,7,8	8	3,5,6,7,8	8	3,5,7,8	8	–
yeast-2-vs-4	0.987	0.986	0.906	1.000	0.911	0.982	0.891	0.326
	3,7,8	3,7,8	8	1,2,3,5,6,7,8	8	3,7,8	8	–
yeast1	0.887	0.885	0.768	0.922	0.754	0.877	0.655	0.339
	3,5,7,8	3,5,7,8	7,8	3,5,7,8	7,8	3,5,7,8	8	–
yeast3	0.987	0.984	0.919	0.996	0.953	0.982	0.904	0.380
	3,5,7,8	3,5,7,8	8	1,2,3,5,6,7,8	7,8	3,5,7,8	8	–

5   Conclusions and Future Directions

The proposed in this paper methods for imbalanced stream classification DSC-R and DSC-S performed favorably in comparison with other dedicated algorithms. The evaluation of the predictive abilities of the techniques was conducted on the basis of computer experiments. The obtained results were analyzed using statistical tests and for all the chosen metrics F-score, Gmean and AUC score, the proposed methods obtained satisfactory results, comparable to algorithms such as REA or Learn++.CDS or Learn++.NIE. The algorithm utilizes memory buffer in order to propagate the instances from the previous block that were chosen as the representatives. Since the buffer is of fixed size, after it is full some instances must be removed from it. In the current implementation, the oldest examples are deleted. A more advanced "forgetting" mechanism, that could favor the instances from the minority class and only the instances from the majority that are the best representatives could be introduced in order to further improve the performance of the classifier. Additionally testing other sampling methods for under- and over-sampling may prove to produce better results.

Acknowledgement. This work was supported by the Polish National Science Centre under the grant No. 2017/27/B/ST6/01325 as well as by the statutory funds of the Department of Systems and Computer Networks, Faculty of Electronics, Wroclaw University of Science and Technology.

References

1. Alcalá-Fdez, J., Fernández, A., Luengo, J., Derrac, J., García, S.: Keel data-mining software tool: data set repository, integration of algorithms and experimental analysis framework. Mult. Valued Log. Soft Comput. **17**(2–3), 255–287 (2011). http://dblp.uni-trier.de/db/journals/mvl/mvl17.html
2. Bifet, A., Gavaldà, R.: Learning from time-changing data with adaptive windowing. In: In SIAM International Conference on Data Mining (2007)
3. Bobowska, B., Woźniak, M.: Experimental study on modified radial-based over-sampling. In: Graña, M., et al. (eds.) SOCO'18-CISIS'18-ICEUTE'18 2018. AISC, vol. 771, pp. 110–119. Springer, Cham (2019). https://doi.org/10.1007/978-3-319-94120-2_11
4. Brzezinski, D., Stefanowski, J.: Ensemble classifiers for imbalanced and evolving data streams, pp. 44–68, March 2018. https://doi.org/10.1142/9789813228047_0003
5. Chen, S., He, H.: SERA: selectively recursive approach towards nonstationary imbalanced stream data mining. In: 2009 International Joint Conference on Neural Networks, pp. 522–529, June 2009. https://doi.org/10.1109/IJCNN.2009.5178874
6. Chen, S., He, H.: Towards incremental learning of nonstationary imbalanced data stream: a multiple selectively recursive approach. Evol. Syst. **2**(1), 35–50 (2011). https://doi.org/10.1007/s12530-010-9021-y
7. Ditzler, G., Polikar, R.: Incremental learning of concept drift from streaming imbalanced data. IEEE Trans. Knowl. Data Eng. **25**(10), 2283–2301 (2013)

8. Ditzler, G., Roveri, M., Alippi, C., Polikar, R.: Learning in nonstationary environments: a survey. IEEE Comput. Intell. Mag. **10**, 12–25 (2015). https://doi.org/10.1109/MCI.2015.2471196
9. Gama, J., Žliobaitė, I., Bifet, A., Pechenizkiy, M., Bouchachia, A.: A survey on concept drift adaptation. ACM Comput. Surv. **46**(4), 44:1–44:37 (2014). https://doi.org/10.1145/2523813. http://doi.acm.org/10.1145/2523813
10. Gama, J.: Knowledge Discovery from Data Streams, 1st edn. Chapman & Hall/CRC, Boca Raton (2010)
11. Gao, J., Ding, B., Fan, W., Han, J., Philip, S.Y.: Classifying data streams with skewed class distributions and concept drifts. IEEE Internet Comput. **12**(6), 37–49 (2008)
12. Hoens, T.R., Polikar, R., Chawla, N.V.: Learning from streaming data with concept drift and imbalance: an overview. Prog. Artif. Intell. **1**(1), 89–101 (2012). https://doi.org/10.1007/s13748-011-0008-0
13. Hulten, G., Spencer, L., Domingos, P.: Mining time-changing data streams. In: Proceedings of the Seventh ACM SIGKDD International Conference on Knowledge Discovery and Data Mining, KDD 2001, pp. 97–106. ACM, New York (2001). https://doi.org/10.1145/502512.502529. http://doi.acm.org/10.1145/502512.502529
14. Kolter, J.Z., Maloof, M.A.: Dynamic weighted majority: a new ensemble method for tracking concept drift. In: Proceedings of the Third IEEE International Conference on Data Mining, ICDM 2003, p. 123. IEEE Computer Society, Washington, D.C. (2003). http://dl.acm.org/citation.cfm?id=951949.952136
15. Koziarski, M., Krawczyk, B., Woźniak, M.: Radial-based approach to imbalanced data oversampling. In: Martínez de Pisón, F.J., Urraca, R., Quintián, H., Corchado, E. (eds.) HAIS 2017. LNCS (LNAI), vol. 10334, pp. 318–327. Springer, Cham (2017). https://doi.org/10.1007/978-3-319-59650-1_27
16. Krawczyk, B., Minku, L.L., Gama, J., Stefanowski, J., Woniak, M.: Ensemble learning for data stream analysis. Inf. Fusion **37**(C), 132–156 (2017). https://doi.org/10.1016/j.inffus.2017.02.004
17. Lemaître, G., Nogueira, F., Aridas, C.K.: Imbalanced-learn: a Python toolbox to tackle the curse of imbalanced datasets in machine learning. J. Mach. Learn. Res. **18**(17), 1–5 (2017). http://jmlr.org/papers/v18/16-365.html
18. Littlestone, N., Warmuth, M.K.: The weighted majority algorithm. Inf. Comput. **108**(2), 212–261 (1994). https://doi.org/10.1006/inco.1994.1009
19. Masud, M., Gao, J., Khan, L., Han, J., Thuraisingham, B.: A practical approach to classify evolving data streams: training with limited amount of labeled data, pp. 929–934, December 2008. https://doi.org/10.1109/ICDM.2008.152
20. Pedregosa, F., et al.: Scikit-learn: Machine learning in Python. J. Mach. Learn. Res. **12**, 2825–2830 (2011). http://www.jmlr.org/papers/volume12/pedregosa11a/pedregosa11a.pdf
21. Sayyad Shirabad, J., Menzies, T.: The PROMISE repository of software engineering databases. School of Information Technology and Engineering, University of Ottawa, Canada (2005). http://promise.site.uottawa.ca/SERepository
22. Stefanowski, J., Brzezinski, D.: Stream classification. In: Sammut, C., Webb, G.I. (eds.) Encyclopedia of Machine Learning and Data Mining, pp. 1191–1199. Springer, Boston (2017). https://doi.org/10.1007/978-1-4899-7687-1_908

23. Street, W.N., Kim, Y.: A streaming ensemble algorithm (SEA) for large-scale classification. In: Proceedings of the Seventh ACM SIGKDD International Conference on Knowledge Discovery and Data Mining, KDD 2001, pp. 377–382. ACM, New York (2001). https://doi.org/10.1145/502512.502568. http://doi.acm.org/10.1145/502512.502568
24. Sun, Y., Wong, A.K.C., Kamel, M.S.: Classification of imbalanced data: a review. IJPRAI **23**, 687–719 (2009)
25. Wang, Y., Zhang, Y., Wang, Y.: Mining data streams with skewed distribution by static classifier ensemble. In: Chien, B.C., Hong, T.P. (eds.) Opportunities and Challenges for Next-Generation Applied Intelligence. SCI, vol. 214, pp. 65–71. Springer, Heidelberg (2009). https://doi.org/10.1007/978-3-540-92814-0_11
26. Woźniak, M., Graña, M., Corchado, E.: A survey of multiple classifier systems as hybrid systems. Inf. Fusion **16**, 3–17 (2014)

A Machine Learning-Based Approach for Predicting Tool Wear in Industrial Milling Processes

Mathias Van Herreweghe[1], Mathias Verbeke[2]([⊠]), Wannes Meert[1], and Tom Jacobs[3]

[1] Department of Computer Science, KU Leuven, Leuven, Belgium
mathias.vanherreweghe@student.kuleuven.be, wannes.meert@cs.kuleuven.be
[2] Data and AI Competence Lab, Sirris, Brussels, Belgium
mathias.verbeke@sirris.be
[3] Precision Manufacturing Lab, Sirris, Diepenbeek, Belgium
tom.jacobs@sirris.be

Abstract. In industrial machining processes, the wear of a tool has a significant influence on the quality of the produced part. Therefore, predicting wear upfront can result in significant improvements of machining processes. This paper investigates the applicability of machine learning approaches for predicting tool wear in industrial milling processes based on real-world sensor data on exerted cutting forces, acoustic emission and acceleration. We show that both Gradient Boosting Machines and Temporal Convolutional Networks prove particularly useful to this end. The validation was performed using the PHM 2010 tool wear prediction dataset as a benchmark, as well as using a proper dataset gathered from an industrial milling machine. The results show that the approach is able to predict the tool wear within an industrially-relevant error margin.

Keywords: Tool wear prediction · Industrial milling processes · Temporal Convolutional Network · Gradient Boosting Machine

1 Introduction

In industrial machining processes, the wear of a tool has a significant influence on the quality of the produced part. When the amount of tool wear is too great, the process quality diminishes and unacceptable risks will occur (such as tool breakage, collision of machines, etc.). Consequently, being able to predict the wear of a tool upfront can lead to significant benefits for the production process, such as a reduction of the machine downtime or a reduction of the scrap that is generated during production with a tool that has worn too much. Therefore, tool wear prediction can contribute to increase the automation of production processes and is an important step towards Industry 4.0 for manufacturing companies.

© Springer Nature Switzerland AG 2020
P. Cellier and K. Driessens (Eds.): ECML PKDD 2019 Workshops, CCIS 1168, pp. 414–425, 2020.
https://doi.org/10.1007/978-3-030-43887-6_34

However, the prediction of tool life, i.e., the time a tool can work with an acceptable amount of tool wear, is a non-trivial challenge. In most cases the theoretical lifetime of a tool is estimated by the tool suppliers at certain conditions (e.g., speeds, amount of material to be machined within a certain time), based on which safety zones can be defined in which the tool will be replaced. However, these theoretical estimates are often too conservative and the actual operating conditions are taken into account only to a limited extent. More accurate estimation models can lead to a very important gain in production processes. This problem can be solved in two ways: either the current amount of wear can be estimated based on historical data and the actual operating conditions, or the tool's *remaining useful life* (RUL, i.e., the remaining time during which the tool can be used with an acceptable amount of wear) can be predicted at a certain point. When production companies know that the RUL is larger than the time needed for a certain product, they can safely use that tool for producing another product.

However, predicting tool wear is challenging, since wear is influenced by a lot of factors such as friction, heat generation and cutting forces. Traditionally, model- or physics-based approaches are used to predict the tool wear. However, developing these kinds of mathematical models typically requires a thorough understanding of the physical properties of the system. This prior knowledge on the properties of a system is not always available, especially in the case of complex manufacturing systems and processes. Due to the increasing availability of sensors and infrastructure to monitor these machines and the respective manufacturing processes, there is a growing interest in data-driven techniques to perform tool wear prediction.

A number of data-driven approaches for the purpose of tool wear prediction have already been proposed. Whereas they perform well, these methods are typically only validated on benchmark data, which is typically generated using optimal conditions. In this paper, the applicability of machine learning for tool wear prediction is examined on real-world industrial data. We will focus on milling, which is a machining operation where material from a workpiece is removed by a rotary cutter. During this process, the milling condition is a crucial factor for the quality of the workpiece. If the wear is too large, the milling process becomes unstable and errors may occur. Examples of errors are a workpiece with too rough of a surface, a cutter that breaks off or a collision of machine parts. The main motivation to focus on milling operations is that the failure of a milling tool can be responsible for 20% of the time that a milling machine is out of service or not operational [8].

2   Related Work

Early research on tool wear prediction mostly focused on model-based approaches to form mathematical formulas representing the accumulation of wear [2,7,9,16], which requires expert knowledge of the underlying physical dynamics of the milling process. This knowledge is costly and not always available.

In contrast, data-driven models try to derive wear based on historical data gathered by sensors on working machines. This problem got increased attention since it was the subject of the PHM 2010 Data Challenge [14]. More specifically, the challenge focused on RUL estimation for a high-speed CNC milling machine cutter using dynamometer, accelerometer, and acoustic emission data. Since then, most recent papers use the resulting dataset as a benchmark for this problem setting.

Several regression models have been tested on their ability to predict tool wear using sensory input. Convolutional neural networks (CNN) [10] are typically used to address problems involving sequential data, and have already been used for tool wear prediction [15,19]. However, generic CNNs are only able to capture spatial information from input data, which limits their predictive accuracy on temporal data. Long short-term memory networks (LSTM) are capable of capturing this temporal information. They typically outperform less complex models such as support vector machines, random forests and multilayer perceptrons [20]. In an effort to combine the strengths of LSTMs and CNNs, combinations of both network types have emerged, such as the convolutional LSTM [18]. Different variants of these networks have been used to predict tool wear. Zhao et al. [19] compare two types of auto-encoders, a deep belief network, a regular and a bi-directional LSTM and a convolution network for this problem. The current state-of-the-art results on the PHM 2010 Challenge Dataset were obtained by Qiao et al. [15] using a time-distributed convolutional LSTM, which also exploits the temporal properties of the milling process to estimate the wear. Besides LSTMs, gradient boosting machines (GBMs) have also been shown to accurately predict tool wear [13].

During our research, we observed however that a number of papers apply randomized cross-validation on a time series data set. As pointed out by Bergmeir, Hyndman and Koo [3], this is an incorrect approach when dealing with time series data. For example, Wu et al. [17] first combine 3 time series datasets, randomly shuffle them and subsequently split them into training and test sets. By doing so, one overfits on the test set as the model gains knowledge on the process in each of the sets due to the random shuffling of the data.

3   Model Selection

As the goal of this paper is to test the industrial applicability of machine learning for tool wear prediction in industrial milling processes, the methods were selected based on their computation speed (to enable near-real time prediction) as well as their accuracy, were we put a threshold error margin of $20\,\mu m$ in order to be industrially relevant. Gradient Boosting Machine was selected due to its accuracy in predicting the tool wear as well as its computation speed for predictions on new input data. As a second model, Temporal Convolutional Network (TCN) was selected due to its ability to exploit the temporal properties of the data, which voids the need for manual feature engineering. While performance should be comparable to LSTMs, in contrast TCNs are shown to be more computationally efficient at prediction time. In this section, the most important elements of both methods are briefly discussed.

3.1 Gradient Boosting Machine

A gradient boosting machine for regression is an ensemble learning method developed by Friedman [4], consisting of multiple weak learners, typically decision trees. Gradient boosting is a technique where weak learners are iteratively added to the current model in a stage-forward fashion. These weak learners are fitted to predict the negative gradient of any differentiable loss function. If the loss function $L(y, F(x)) = (y - F(x))^2/2$ with y the target value, $F(x)$ the prediction of the current model for input example x and $J = \sum_i L(y_i, F(x_i))$ the function to minimize, it follows that

$$\frac{\partial J}{\partial F(x_i)} = \frac{\partial \sum_i L(y_i, F(x_i))}{\partial F(x_i)} = \frac{\partial L(y_i, F(x_i))}{\partial F(x_i)} = F(x_i) - y_i. \tag{1}$$

This can be rearranged to

$$y_i - F(x_i) = -\frac{\partial J}{\partial F(x_i)}, \tag{2}$$

which shows that we can interpret the residuals as the negative gradients. Decision trees in which the response is continuous are called regression trees. This iterative process is repeated until the model consists of a predefined amount of regression trees.

The negative gradient is the direction of a step towards the minimum of the loss function, but it does not determine the size of the step. The size of a step is determined using the *line search* technique. This technique attempts to optimize the loss function in function of the step size. For a regression tree with T leaf nodes, the negative gradient will be calculated T times, once for every leaf node. For every such computation, line search will be applied.

The *shrinkage parameter* is used to shrink the step size by a fixed factor. The decrease of the step size should lead to better generalization capacities of the gradient boosting machine [5].

3.2 Temporal Convolutional Network

The convolutional neural network variant suggested to capture spatio-temporal information from sequential data is called a temporal convolutional network. A temporal convolutional network is simply put a combination of a one-dimensional fully-convolutional network and (dilated) causal convolutions. This network also makes use of residual blocks.

Dilated Causal Convolutions. Since the TCN is a one-dimensional fully–convolutional network (FCN) [12], the network consists solely of learnable filters in one dimension, including the final layer of the network. This means that decisions are only based on local spatial input. Causal convolutions guarantee that no information from the future leaks to the past. Regular convolutions can only reach a history with size linear in the depth of the network. Dilated convolutions are used to circumvent this by enabling a receptive field that is exponentially large in relation to the amount of layers in the network. The dilation depends on a dilation factor d and a filter size k. The dilation factor indicates per layer a fixed step between every two adjacent filter taps. This factor d is increased exponentially with the depth of the network, which ensures that there is a filter that hits each input within the effective history. The filter size k determines the length of the filter and thus the amount of input elements that are used for a convolution.

Residual Blocks. The receptive field of the TCN depends on the chosen values for the dilation factor and filter size. Depending on these values, the network can become too deep or large to be able to predict simple functions, such as the identity function for example. To be able to overcome this, the TCN is made of residual blocks [6] which aims to increase the stability of the network. A residual block adds the input to the output of the transformations at the end block, which makes the network learn modifications to the entity mapping rather than the entire transformation. An 1×1 convolution is used to alleviate possible discrepancies between the width of input and output. Within the used implementation, a residual block consists of two dilated convolutional layers, each followed by weight normalization, a rectified linear unit (ReLU) for non-linearity and a spatial dropout layer. This setup is shown on Fig. 1.

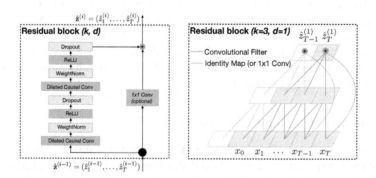

Fig. 1. Residual block and residual connection where blue lines represent filters and green lines are identity mappings [1] (Color figure online)

4 Experiments

To validate their industrial applicability, both the GBM and the TCN setup were tested on both a benchmark dataset as well as on a dataset gathered from an industrial setup. In this section, we will briefly discuss the datasets, the experimental setup, the preprocessing steps and the model parametrization, as well as the achieved results.

4.1 Data and Experimental Setup

The results will be evaluated on two different datasets, namely the PHM 2010 Challenge Dataset [14], which we will use as a benchmark dataset, as well as a dataset gathered from an industrial setup in the Precision Manufacturing Lab of Sirris. Both datasets contain time series data on forces (X, Y, Z), accelerations (X, Y, Z) and the root mean square of the acoustic emission (AE), resulting in a set of 7 parameters.

The benchmark dataset consists of six sets. In this paper, we will only use subsets $c1$, $c4$ and $c6$, as these are the most frequently used sets in state-of-the-art comparisons. Each of the sets consists of 315 cuts and for each cut around 220k measurements are available. The industrial dataset consists of 2 groups of data, which each contain 3 sets. The groups differ in terms of cutting speed and feeding rate, only the first group of sets will be used in this paper. Each set consists of approximately 100 cuts, with around 7k measurements for the forces and the AE and 80 measurements for the accelerometer data (due to a lower sampling rate of the available sensor). The milling operations in the industrial dataset were performed with a Haas Super Minimill 3-axis milling machine, using a Sandvik milling tool with a diameter of 16 mm. A similar setup as Li et al. [11] was used (as depicted in Fig. 2), yet in contrast to the benchmark dataset, the AE sensor is placed on the spindle instead of on the part itself. This was done to keep the distance between the sensor and the cutting point constant. The cutting speed and feed rate was set to 200 m/min and 159 mm/min respectively. The cutting diameter was 16 mm, the radial and axial cutting depth were both set to 1 mm.

4.2 Preprocessing

Benchmark Data. This dataset is already fairly preprocessed in the sense that it is ready to use the sensory data to generate features (for the GBM) or to reduce the amount of time steps (for the TCN). Some force signals, however, have very high peaks throughout the signals, which may be due to sensor faults. The affected cuts have one or more of their force signals shifted, as shown in Fig. 3. These peaks are flattened by linear interpolation between the beginning and end of such peaks to verify the impact of these peaks on the performance of both the GBM and TCN. The peak-flattening can occur on any force signal and on one or more of its axes independently.

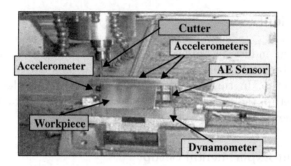

Fig. 2. Sensors on milling machine (taken from [11])

To this end, first the mean for every force signal for every cut is calculated. We iterate over these mean values until we encounter an absolute change in value that exceeds a set threshold. The current index in the iteration is then defined as the end of the peak. From this index we iterate in a backwards fashion until we encounter an index of which the corresponding absolute value is less than the absolute value of the end of the peak. This index is the start of the peak. We do a linear interpolation for all indices between the start and end of the peak and continue the iteration from the end of the peak, continuing the search for more peaks. When the end of the list of mean values is reached, we subtract this flattened list from the original list. The result is then a list with the differences between the flattened and original values at indices where a peak is present, and zero on all other indices. For each cut in a set, the differences are added to all values of the corresponding force signal of that cut, which undoes the shift shown on Fig. 3.

Industrial Data. The data gathered from the industrial process is raw and thus requires extensive preprocessing. Only the main preprocessing steps will be explained. First, the measured acoustic emission ranges over multiple frequencies. These measurements will be converted into a one-dimensional signal per cut by calculating the root mean square (RMS) over all frequencies per time step.

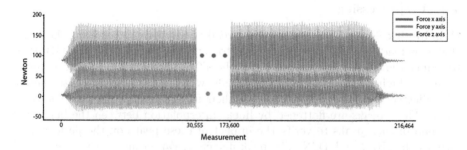

Fig. 3. Shifted force signal on x axis

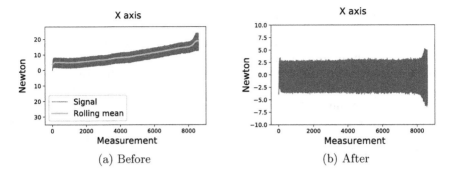

Fig. 4. Drift correction on force signal

Second, since measurements start before and after a complete cut, only the part of the signal during which the tool is in physical contact with the workpiece is retained. This results in the synchronization of the signals. Third, due to the nature of dynamometers, the reference point in the sensors is not constant which leads to gradually shifting measured force values throughout operation. This symptom is called drift and is corrected by subtracting the rolling mean of the signal from the signal. The correction of drift is shown in Fig. 4. Figure 4b also shows that both ends of the signal may contain values that are different compared to the vast majority of the signal. To limit the influence of these varying values, the first and last 1,000 measurements of a signal are pruned. This amount of pruning has been empirically shown to be the optimal value for both models. We believe that this value is the optimal value in the trade-off between removing more deviating values (higher value) and removing less potentially useful measurements (lower value).

4.3   Feature Selection and Hyperparameter Tuning

Gradient Boosting Machine. For the benchmark set, maximum, minimum, median, average, standard deviation, sum of the absolute values and the difference between the current mean and the mean of the previous cut were considered for each of the parameters. For the industrial dataset, the same features except the latter one were taken into account. Feature selection per parameter was performed using a genetic algorithm.

Temporal Convolutional Network. As suggested by the TCN literature, the signals are summarized in multiple time steps by taking the maximum and the average per time step. The number of measurements per time step thus equals the number of measurements in the signal divided by the number of chosen time steps. For the benchmark dataset, 100 time steps were used, whereas for the industrial dataset the number of time steps was limited to 50 due to the more limited number of acceleration measurements per cut. The features were scaled using min-max scaling.

4.4   Results

The results expressed in terms of mean absolute error (MAE) for the benchmark dataset are shown in Table 1. The column names indicate the subset that was used for testing, using the remaining two subsets for training the model. For example, C1 indicates that subset $c1$ was used as test set, while subsets $c4$ and $c6$ were used to train the model. The rows marked with an asterisk indicate that for this model the hyperparameters were optimized using grid search. The results are given in micrometer and are rounded to 1 decimal. The results are the average of 3 predictions with the same parameters.

Table 1. MAE for different models on the benchmark data

Model	C1	C4	C6
GBM	10.9	14.4	13.7
GBM*	10.9	14.4	13.7
TCN	9.3	10.9	16.4
TCN*	9.5	10.9	8.5
CBLSTM [20]	10.8	7.1	9.8
TDConvLSTM [15]	**6.99**	**6.96**	**7.5**

The GBM was also compared with other methods that require manual feature engineering (SVM and Random Forest), showing that the GBM obtained the lowest MAE. Interesting to note is that the hyperparameter optimization did not improve the results for the GBM.

Overall, the TCN with optimized hyperparameters obtains the best results. These results are shown in Fig. 5. The results of the mean MAE are 2.5 μm worse than the state-of-the-art models. The respective papers, however, do not provide any additional statistics regarding the number of runs that were required to obtain these results. The best average MAE for the TCN with optimized hyperparameters is 8.9 μm. Next to the best average MAE, the TCN also achieves the lowest mean standard deviation of errors per predicted set. This means that the size of the errors within a prediction does not differ that much from one another.

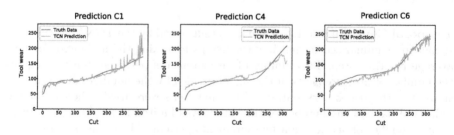

Fig. 5. Prediction of TCN* (Benchmark data)

Similarly, Table 2 shows the obtained results on the industrial dataset. Also for this dataset, the TCN obtains the best results for all calculated statistics. The TCN with optimized hyperparameters not only obtains a good MAE, but also a relatively low standard deviation of 0.09 on the average MAE results. Across the 3 iterations, the average MAE thus remains stable. Figure 6 again visually represents the results for the best performing model.

Table 2. MAE for different models on the industrial data

Model	I1	I2	I3
GBM	17.5	22.3	16.6
GBM*	13.7	21.1	16.6
TCN	14.3	24.1	14.3
TCN*	**13.3**	**20.8**	**12.8**

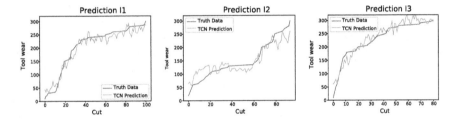

Fig. 6. Prediction of TCN* (Industrial data)

Next to the accuracy, another important criterion for the industrial applicability is the prediction time. On average, training the GBM with optimized hyperparameters took 30 ms on a standard laptop. The prediction takes less than 8 ms (for both datasets). The TCN with optimized hyperparameters needed close to 6 min for training and on average took 0.5 s for predicting on specialized computing infrastructure. For the industrial datasets, this took close to 13 min for training and 1 s for testing. Both models thus warrant a near-real time prediction.

Overall, the results show that the TCN performs similar to the state of the art on the PHM 2010 dataset, which confirms that convolutional networks are a valid alternative for recurrent networks for predictions based on temporal data. Also for the industrial dataset, the TCN was well able to predict the tool wear. Both the TCN and the GBM resulted in the targeted accuracy of less than 20 μm in MAE on the industrial dataset.

The difference in accuracy between both datasets is probably caused by the low sampling frequency of the accelerations in the industrial dataset, due to which the acceleration signals are only taken into account to a limited extent by the models. Also the sampling frequency of the forces and the acoustic emission is about 100 times larger in the benchmark dataset when compared to the industrial dataset, which can have an influence on the resulting model performance. The training time measurements show that the TCN with optimized hyperparameters requires approximately 3 times less time to train and half of the time to predict when compared to the same model without hyperparameter optimization. The hyperparameters thus have a strong influence on the speed of the model. Since the TCN is more accurate than the GBM but requires in comparison much more training and prediction time, the choice between the TCN and GBM for industrial settings needs to be based both on the requirements regarding accuracy as well as on the available hardware to train the models and make the predictions.

5   Conclusion and Future Work

In this paper, we investigated the applicability of two machine learning methods for predicting tool wear in industrial milling processes using sensor data on exerted cutting forces, acoustic emission and acceleration. To this end, the use of Gradient Boosting Machines and Temporal Convolutional Networks was validated on both a benchmark dataset as well as on a real-world industrial dataset. The results show that both methods are able to predict the tool wear within an industrially-relevant error margin of $20\,\mu m$ in an acceptable computation time.

In future work, the generalizability of the approach to additional machine settings (i.e., cutting speed, temperature) will be explored, as this could offer new insights into the factors that determine the wear. Also the influence of the material of the workpiece on the performance of the model will be examined in more detail. Whereas currently only the force, accelerometer and acoustic emission measurements of externally-mounted sensors were taken into account as parameters for the prediction, also the use of parameters that are directly gathered by the machine (i.e., power consumption) will be investigated.

Acknowledgements. This work was partially supported by Flanders Innovation & Entrepreneurship (VLAIO) through the SBO project HYMOP (150033) and by the Brussels-Capital Region - Innoviris through the TeamUp ROADMAP project. Part of the computational resources and services used in this work were provided by the VSC (Flemish Supercomputer Center), funded by the Research Foundation - Flanders (FWO) and the Flemish Government—department EWI.

References

1. Bai, S., Kolter, J.Z., Koltun, V.: An empirical evaluation of generic convolutional and recurrent networks for sequence modeling. CoRR abs/1803.01271 (2018)
2. Barrow, G.: Tool-life equations and machining economics, pp. 481–493. Macmillan Education, London (1972). https://doi.org/10.1007/978-1-349-01397-5_59

3. Bergmeir, C., Hyndman, R.J., Koo, B.: A note on the validity of cross-validation for evaluating autoregressive time series prediction. Comput. Stat. Data Anal. **120**(C), 70–83 (2018). https://doi.org/10.1016/j.csda.2017.11.003
4. Friedman, J.H.: Greedy function approximation: a gradient boosting machine. Ann. Stat. **29**(5), 1189–1232 (2001)
5. Hastie, T., Tibshirani, R., Friedman, J.: Boosting and additive trees. In: The Elements of Statistical Learning. SSS, pp. 337–387. Springer, New York (2009). https://doi.org/10.1007/978-0-387-84858-7_10
6. He, K., Zhang, X., Ren, S., Sun, J.: Deep residual learning for image recognition. In: Proceedings of the IEEE Conference on Computer Vision and Pattern Recognition, pp. 770–778 (2016)
7. Kuljanic, E.: Effect of stiffness of tool wear and new tool life equation. CIRP Ann. **23**, 15–16 (1974). https://doi.org/10.1115/1.3438707
8. Kurada, S., Bradley, C.: A review of machine vision sensors for tool condition monitoring. Comput. Ind. **34**(1), 55–72 (1997). https://doi.org/10.1016/S0166-3615(96)00075-9
9. Dos Santos, A.L.B., Duarte, M., Abrao, A., Machado, A.: An optimisation procedure to determine the coefficients of the extended Taylor's equation in machining. Int. J. Mach. Tools Manuf **39**, 17–31 (1999). https://doi.org/10.1016/S0890-6955(98)00025-X
10. LeCun, Y., et al.: Backpropagation applied to handwritten zip code recognition. Neural Comput. **1**(4), 541–551 (1989)
11. Li, X., et al.: Fuzzy neural network modelling for tool wear estimation in dry milling operation. In: Annual Conference of the Prognostics and Health Management Society, PHM 2009, January 2009
12. Long, J., Shelhamer, E., Darrell, T.: Fully convolutional networks for semantic segmentation. In: Proceedings of the IEEE Conference on Computer Vision and Pattern Recognition, pp. 3431–3440 (2015)
13. Murua, M., Suarez, A., Lacalle, L., Santana, R., Wretland, A.: Feature extraction-based prediction of tool wear of inconel 718 in face turning. Insight Non-Destruct. Testing Condition Monit. **60**, 443–450 (2018). https://doi.org/10.1784/insi.2018.60.8.443
14. PHM Society: 2010 PHM Society Conference Data Challenge. https://www.phmsociety.org/competition/phm/10
15. Qiao, H., Wang, T., Wang, P., Qiao, S., Lan, Z.: A time-distributed spatiotemporal feature learning method for machine health monitoring with multi-sensor time series. Sensors **18**, 2932 (2018). https://doi.org/10.3390/s18092932
16. Taylor, F.W.: On the Art of Cutting Metals. American Society of Mechanical Engineers, New York (1906)
17. Wu, D., Jennings, C., Terpenny, J., Gao, R., Kumara, S.: A comparative study on machine learning algorithms for smart manufacturing: tool wear prediction using random forests. J. Manuf. Sci. Eng. **139** (2017). https://doi.org/10.1115/1.4036350
18. Shi, X., Chen, Z., Wang, H., Yeung, D.-Y., Wong, W., Woo, W.: Convolutional LSTM network: a machine learning approach for precipitation nowcasting. In: Advances in Neural Information Processing Systems, pp. 802–810 (2015)
19. Zhao, R., Yan, R., Chen, Z., Mao, K., Wang, P., Gao, R.X.: Deep learning and its applications to machine health monitoring. Mech. Syst. Signal Process. **115**, 213–237 (2019)
20. Zhao, R., Yan, R., Wang, J., Mao, K.: Learning to monitor machine health with convolutional bi-directional LSTM networks. Sensors **17**, 273 (2017). https://doi.org/10.3390/s17020273

12th International Workshop on Machine Learning and Music (MML 2019)

Cross-version Singing Voice Detection in Opera Recordings: Challenges for Supervised Learning

Stylianos I. Mimilakis[1]([✉]), Christof Weiss[2], Vlora Arifi-Müller[2],
Jakob Abeßer[1], and Meinard Müller[2]

[1] Fraunhofer IDMT, Ilmenau, Germany
`mis@idmt.fraunhofer.de`
[2] International Audio Laboratories Erlangen, Erlangen, Germany

Abstract. In this paper, we approach the problem of detecting segments of singing voice activity in opera recordings. We consider three state-of-the-art methods for singing voice detection based on supervised deep learning. We train and test these models on a novel dataset comprising three annotated performances (versions) of Richard Wagner's opera "Die Walküre." The results of our cross-version experiments indicate that the models do not sufficiently generalize across versions even in the case that another version of the same musical work is available for training. By further analyzing the systems' predictions, we highlight certain correlations between prediction errors and the presence of specific singers, instrument families, and dynamic aspects of the performance. With these findings, our case study provides a first step towards tackling singing voice detection with deep learning in challenging scenarios such as Wagner's operas.

Keywords: Opera · Singing voice detection · Supervised deep learning

1 Introduction

The automatic identification of vocal segments in music recordings—known as singing voice detection (SVD)—is a central problem in music information retrieval (MIR) research [1]. In relevant literature, most SVD approaches are tailored to popular music [6–8,12,13,15,16]. However, Scholz et al. [19] showed that SVD quality considerably depends on the music genre, and that systems do often not generalize well across genres. Partly, this is due to the genre-specific usage of instruments and singing styles. A particular case is Western opera, where singing is often embedded in a rich orchestral accompaniment and instruments often imitate singing techniques such as vibrato [20]. Dittmar et al. [2] studied SVD within an opera scenario involving several versions of Weber's "Der Freischütz." Using carefully selected audio features and random forest classifiers, they showed that bootstrap training [12,22] helps to leverage the genre-dependency problem.

S. I. Mimilakis and C. Weiß—Equally contributing authors

© Springer Nature Switzerland AG 2020
P. Cellier and K. Driessens (Eds.): ECML PKDD 2019 Workshops, CCIS 1168, pp. 429–436, 2020.
https://doi.org/10.1007/978-3-030-43887-6_35

They further demonstrated the benefit of a cross-version scenario by performing late fusion of the individual versions' results. We are not aware of any studies dealing with SVD for Wagner's operas, which constitute a challenging scenario due to their large and complex orchestration and highly expressive singing styles.

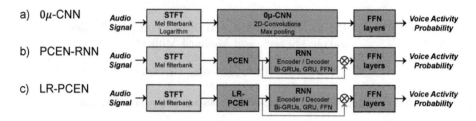

Fig. 1. The three examined DL models. Red modules denote non-trainable, predefined functions. Green modules denote parameterized functions subject to optimization. The cross symbol $\otimes$ denotes the element-wise multiplication introduced in [11]. (Color figure online)

As the system used in [2], early approaches to SVD [14,15] typically consist of two parts—the extraction of audio features and the supervised training of classifiers such as random forests. Recently, SVD based on deep learning (DL) has become popular [7–9,17]. As for the contributions of this paper, we apply several state-of-the-art (SOTA) approaches [11,18,23]—proposed for SVD in popular music—to the opera scenario. We systematically assess their efficacy on a limited dataset comprising three semi-automatically annotated versions of Richard Wagner's opera "Die Walküre" (first act). Our experiments demonstrate that the models do not sufficiently generalize across versions even when the training data contains other versions of the same musical work. Finally, we highlight specific challenges in Wagner's operas, pointing out interesting correlations between errors and the voices' registers as well as the activity of specific instruments.

2   Deep-Learning Methods

In this paper, we examine three SVD approaches based on supervised DL (Fig. 1).[1] Lee et al. [6] give an overview and a quantitative analysis of DL-based SVD systems. Our first model (Fig. 1a) is based on a convolutional neural network (CNN) followed by a classifier module. CNNs have been widely used for SVD [16–18]. To achieve sound-level-invariant SVD, Schlüter et al. [18] introduce zero-mean convolutions—an update rule that constrains the CNN kernels to have zero mean. We use this zero-mean update rule within the specific architecture presented in [18] for our first model (denoted as 0μ-CNN). As an alternative

---

[1] Due to limited space, we only provide an overview of the models. For details, we refer to the relevant literature [7,11,18,23] and our source code: https://github.com/Js-Mim/wagner_vad.

approach to sound-level-invariant SVD, Schlüter et al. [18] suggest per-channel energy normalization (PCEN) [23]. For our second model (Fig. 1b, denoted as PCEN-RNN), we consider this technique as front-end followed by recurrent layers and the classifier, realized by feed-forward network (FFN) layers. Recurrent neural networks (RNNs) have been used for SVD in [7], among others. As our third model, we examine a straightforward extension to PCEN involving a low-rank autoencoder (Fig. 1c, denoted as LR-PCEN). For both RNN-based models (PCEN-RNN and LR-PCEN), we include skip-filtering connections [11], which turns out to be useful for "pin-pointing" relevant parts of spectrograms [10].

For pre-processing, we partition the monaural recording into non-overlapping segments of length 3 s. Inspired by previous approaches [6,7,18], we compute a 250-band mel-spectrogram for each segment. As input to the 0μ-CNN model [17], we use the logarithm of the mel-spectrogram. For the PCEN-RNN model, we use the mel-spectrogram as input to the trainable PCEN front-end [23] followed by a bi-directional encoder with gated recurrent units (GRUs) and residual connections [11]. The decoder predicts a mask (of the original input size) for filtering the output of the PCEN. For the LR-PCEN, we replace the first-order recursion [23, Eq.(2)] with a low-rank (here: rank one) autoencoder that shares weights across mel-bands. The output of the autoencoder is used alongside residual connections with the input mel-spectrogram. We randomly initialize the parameters and jointly optimize these using stochastic gradient descent with binary cross-entropy loss and the Adam [5] solver setting the initial learning rate to 10^{-4} and the exponential decay rates for the first- and second-order moments to 0.9. We optimize over the training data for 100 iterations and adapt the learning rate depending on the validation error. Moreover, we perform early stopping after 10 non-improving iterations.

3   Dataset

We evaluate the systems on a novel dataset comprising three versions of Wagner's opera "Die Walküre" (first act) conducted by Barenboim 1992 (Bar), Haitink 1988 (Hai), and Karajan 1966 (Kar), each comprising 1523 measures and roughly 70 min of music. Starting with the libretto's phrase segments, we manually annotate the phrase boundaries as given by the score (in musical measures/beats). To transfer the singing voice segments to the individual versions, we rely on manually generated measure annotations [24]. Using the measure positions as anchor points, we perform score-to-audio synchronization [3] for generating beat and tatum positions, which we use to transfer the segmentation from the *musical time* of the libretto to the *physical time* of the performances.

Since alignment errors and imprecise singer performance may lead to offsets between the transferred segment boundaries and the actual singing, we manually refined our semi-automatic annotations for the Kar recording, which we use as test version in our experiments. Almost every phrase boundaries was adjusted, thus affecting rouhgly 4% of all frames in total. Due to our annotation strategy, there might be another issue. Since we start from the libretto with its

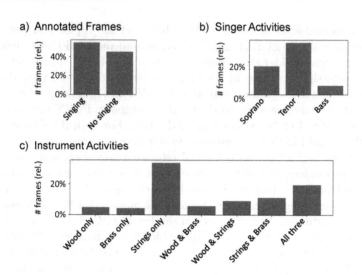

Fig. 2. Percentage of frames (Kar version) with (a) annotated singing voice, (b) activity of individual singers, (c) activity of instrument sections and their combination.

Table 1. Data splits used for the experiments.

Data split	DS-1	DS-2	DS-3
Training	Bar, Hai, Kar	Bar, Hai	Bar
Validation	Kar	Kar	Hai
Test	Kar	Kar	Kar

phrase-level segments, the annotations do not account for smaller musical rests within textual phrases—an issue that is also common for SVD annotations in popular music. To estimate the impact of these gaps within phrases (labeled as "singing"), we compute the overlap between the phrase-level singing regions from the libretto (Kar) and note-level annotation derived from an aligned score. The two annotations match for only 94% of all frames. This suggests that in the opera scenario, phrase-level annotations as well as automatic alignment strategies may not be precise enough for high-quality SVD evaluated on the frame level. We therefore regard an accuracy or F-measure of 94% as a kind of upper bound for our experiments.

In our dataset, singing and non-singing frames are quite balanced (Fig. 2a). Among the three singers performing in the piece, the tenor dominates, followed by soprano and bass (Fig. 2b), while they never sing simultaneously. Regarding instrumentation, the string section alone plays most often, followed by all sections together, and other constellations (Fig. 2c). For systematically testing generalization to unseen versions, we create three data splits (Table 1). In DS-1, the test version (Kar) is available during training and validation. DS-2 only sees the test version at validation. DS-3 is the most realistic and restrictive split.

4   Experiments

For our results, Table 2 reports precision, recall, and F-measure with singing as the relevant class. Let us look at the results of the scenario DS-1 where the `Kar` version is used both for training and testing. All models perform well here and almost reach the upper bound of 94% discussed above. For the more realistic scenario DS-2, where the test version (`Kar`) is only available for validation, the F-measures of all models decrease. Furthermore, the models tend towards more false negatives (precision > recall). Both effects are particularly prominent for 0μ-CNN. In the scenario DS-3, where the `Kar` version is only used for testing, the results further deteriorate. Again, all models show a clear tendency towards false negatives (most prominently 0μ-CNN). This points to detection problems in presence of the orchestra, which become particularly relevant when generalizing to unseen versions with different timbral characteristics and acoustic conditions.

Table 2. SVD results for all models (0μ–CNN, PCEN-RNN, LR-PCEN) and data splits.

Data split	DS-1			DS-2			DS-3		
Models	0μ–CNN	PCEN-RNN	LR-PCEN	0μ–CNN	PCEN-RNN	LR-PCEN	0μ–CNN	PCEN-RNN	LR-PCEN
Precision	**0.95**	0.93	0.94	**0.96**	0.91	0.92	**0.97**	0.87	0.90
Recall	0.91	**0.92**	0.90	0.81	**0.88**	**0.88**	0.69	**0.76**	0.74
F-Measure	**0.93**	**0.93**	0.92	0.88	0.89	**0.90**	0.80	0.81	**0.82**

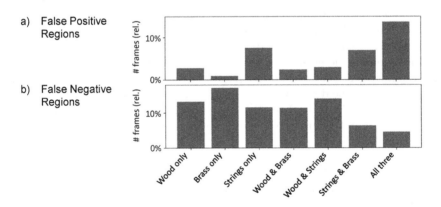

Fig. 3. (a) False positive and (b) false negative frames as detected by the 0μ-CNN model (`Kar` version). We plot the percentage of errors for regions with certain instrument sections or constellations playing, in relation to these regions' total duration.

We want to study such hypotheses in more detail for the realistic split DS-3. Regarding individual singers, the 0μ-CNN model obtains higher recall for the bass (74% of frames detected) than for tenor and soprano (each 68%). Interestingly, both PCEN models behave the opposite way, obtaining low recall (<50%)

for the bass and high recall (almost 80%) for the others. We might conclude that the 0μ-CNN is less affected by the imbalance of singers in the training data. Since segments typically imply a certain length, we conduct a further experiment using median filtering for removing short segments in a post-processing step (not shown in the table). As observed in [2], F-measures improve by 2–4% for all models using a median filter of roughly one second length.

Finally, we want to investigate correlations between errors and specific instrument activities for the LR-PCEN model's results (Fig. 3). For most instrument combinations, we cannot observe any strong preference for producing false positives or negatives, with two interesting exceptions. When only brass instruments are playing without singing, the LR-PCEN practically never produces false positive predictions. In contrast, when brass only occurs together with singing, we observe a strong increase of false negatives. The highest frequency of false positives occurs for tutti passages (all three sections playing). When listening to false-positive regions, we often find expressive strings-only passages. In contrast, false-negative regions often correspond to soft and gentle singing. Examining this in more detail, we observed a slight loudness-dependency for all models. As reported for popular music [18], singing frames are usually louder leading to more "loud" false positives and "soft" false negatives. This indicates that, despite the models' level invariance, confounding factors such as timbre or vibrato might affect SVD quality.

Our experiments and analyses only provide a first step towards understanding the challenges of SVD in complex opera recordings. From the results, we conclude that the systems do not sufficiently generalize across versions due to their different acoustic characteristics—even if the specific musical work is part of the training set. While all models are capable of fitting the data to a reasonable degree (given the reliability and precision of our annotations), generalization becomes problematic as soon as the test version is not seen during training or validation. Even if loudness dependencies are eliminated, our results suggest that more work has to be done to impose further invariances and constraints. A nice example is given in [21] where the generalization performance of the models is optimized. Furthermore, considering techniques such as data augmentation or unsupervised domain adaptation [4] might be useful to achieve robust SVD systems for the opera scenario.

Acknowledgements. This work was supported by the German Research Foundation (AB 675/2-1, MU 2686/11-1, MU 2686/7-2). The International Audio Laboratories Erlangen are a joint institution of the Friedrich-Alexander-Universität Erlangen-Nürnberg (FAU) and Fraunhofer Institut für Integrierte Schaltungen IIS. We thank Cäcilia Marxer and all students who helped preparing the data and annotations.

References

1. Berenzweig, A.L., Ellis, D.P.W.: Locating singing voice segments within music signals. In: Proceedings of the IEEE Workshop on Applications of Signal Processing to Audio and Acoustics (WASPAA), New Paltz, NY, USA, pp. 119–122 (2001)

2. Dittmar, C., Lehner, B., Prätzlich, T., Müller, M., Widmer, G.: Cross-version singing voice detection in classical opera recordings. In: Proceedings of the International Society for Music Information Retrieval Conference (ISMIR), Málaga, Spain, pp. 618–624 (2015)
3. Ewert, S., Müller, M., Grosche, P.: High resolution audio synchronization using chroma onset features. In: Proceedings of IEEE International Conference on Acoustics, Speech, and Signal Processing (ICASSP), Taipei, Taiwan, pp. 1869–1872 (2009)
4. Gharib, S., Drossos, K., Çakir, E., Serdyuk, D., Virtanen, T.: Unsupervised adversarial domain adaptation for acoustic scene classification. Computing Research Repository (CoRR) abs/1808.05777 (2018)
5. Kingma, D.P., Ba, J.: Adam: A method for stochastic optimization. In: Proceedings of the International Conference for Learning Representations (ICLR), San Diego, California, USA (2015)
6. Lee, K., Choi, K., Nam, J.: Revisiting singing voice detection: A quantitative review and the future outlook. In: Proceedings of the International Society for Music Information Retrieval Conference (ISMIR), Paris, France, pp. 506–513 (2018)
7. Leglaive, S., Hennequin, R., Badeau, R.: Singing voice detection with deep recurrent neural networks. In: Proceedings of the IEEE International Conference on Acoustics, Speech, and Signal Processing (ICASSP), Brisbane, Australia, pp. 121–125 (2015)
8. Lehner, B., Schlüter, J., Widmer, G.: Online, loudness-invariant vocal detection in mixed music signals. IEEE/ACM Trans. Audio Speech Lang. Process. **26**(8), 1369–1380 (2018). https://doi.org/10.1109/TASLP.2018.2825108
9. Lehner, B., Widmer, G., Böck, S.: A low-latency, real-time-capable singing voice detection method with LSTM recurrent neural networks. In: Proceedings of the European Signal Processing Conference (EUSIPCO), Nice, France, pp. 21–25 (2015)
10. Mimilakis, S.I., Drossos, K., Cano, E., Schuller, G.: Examining the mapping functions of denoising autoencoders in singing voice separation. IEEE/ACM Trans. Audio Speech Lang. Process. **28**, 266–278 (2020)
11. Mimilakis, S.I., Drossos, K., Virtanen, T., Schuller, G.: A recurrent encoder-decoder approach with skip-filtering connections for monaural singing voice separation. In: Proceedings of the IEEE International Workshop on Machine Learning for Signal Processing (MLSP), Tokyo, Japan, pp. 1–6 (2017)
12. Nwe, T.L., Wang, Y.: Automatic detection of vocal segments in popular songs. In: Proceedings of the International Society for Music Information Retrieval Conference (ISMIR), Barcelona, Spain, pp. 138–144 (2004)
13. Ramona, M., Peeters, G.: Audioprint: An efficient audio fingerprint system based on a novel cost-less synchronization scheme. In: Proceedings of the IEEE International Conference on Acoustics, Speech, and Signal Processing (ICASSP), Vancouver, Canada, May 2013, pp. 818–822. https://doi.org/10.1109/ICASSP.2013.6637762
14. Ramona, M., Richard, G., David, B.: Vocal detection in music with support vector machines. In: Proceedings of the IEEE International Conference on Acoustics, Speech, and Signal Processing (ICASSP), Las Vegas, Nevada, USA, pp. 1885–1888 (2008)
15. Regnier, L., Peeters, G.: Singing voice detection in music tracks using direct voice vibrato detection. In: Proceedings of the IEEE International Conference on Acoustics, Speech, and Signal Processing (ICASSP), Taipei, Taiwan, pp. 1685–1688 (2009)

16. Schlüter, J.: Learning to pinpoint singing voice from weakly labeled examples. In: Proceedings of the International Society for Music Information Retrieval Conference (ISMIR), New York City, USA, pp. 44–50 (2016)
17. Schlüter, J., Grill, T.: Exploring data augmentation for improved singing voice detection with neural networks. In: Proceedings of the International Society for Music Information Retrieval Conference (ISMIR), Málaga, Spain, pp. 121–126 (2015)
18. Schlüter, J., Lehner, B.: Zero-mean convolutions for level-invariant singing voice detection. In: Proceedings of the International Society for Music Information Retrieval Conference (ISMIR), Paris, France, pp. 321–326 (2018)
19. Scholz, F., Vatolkin, I., Rudolph, G.: Singing voice detection across different music genres. In: Proceedings of the AES International Conference on Semantic Audio, Erlangen, Germany, pp. 140–147 (2017)
20. Seashore, C.E.: The natural history of the vibrato. Proc. Nat. Acad. Sci. USA **17**(12), 623–626 (1931)
21. Tolstikhin, I., Bousquet, O., Gelly, S., Schoelkopf, B.: Wasserstein auto-encoders. In: Proceedings of the 6th International Conference on Learning Representations (ICLR), Vancouver, Canada, pp. 1–16 (2018)
22. Tzanetakis, G.: Song-specific bootstrapping of singing voice structure. In: Proceedings of the IEEE International Conference on Multimedia and Expo (ICME), Taipei, Taiwan, vol. 3, pp. 2027–2030 (2004)
23. Wang, Y., Getreuer, P., Hughes, T., Lyon, R.F., Saurous, R.A.: Trainable frontend for robust and far-field keyword spotting. In: Proceedings of the IEEE International Conference on Acoustics, Speech, and Signal Processing (ICASSP), New Orleans, USA, pp. 5670–5674 (2017)
24. Weiß, C., Arifi-Müller, V., Prätzlich, T., Kleinertz, R., Müller, M.: Analyzing measure annotations for Western classical music recordings. In: Proceedings of the International Conference on Music Information Retrieval (ISMIR), New York, USA, pp. 517–523 (2016)

Neural Symbolic Music Genre Transfer Insights

Gino Brunner$^{(\boxtimes)}$, Mazda Moayeri, Oliver Richter, Roger Wattenhofer, and Chi Zhang

ETH Zurich, Zurich, Switzerland
{brunnegi,moayerim,richtero,wattenhofer,czhang}@ethz.ch

Abstract. Transferring a song from one genre to another is most difficult if no instrumentation information is provided and genre is only defined by the timing and pitch of the played notes. Inspired by the CycleGAN music genre transfer presented in [2] we investigate whether recent additions to GAN training like spectral normalization and self-attention can improve transfer. Our preliminary results show that spectral normalization improves audible quality, while self-attention hurts content retention due to its non-locality. We further provide insights into genre attribution, showing that often only few notes are genre-decisive.

1 Introduction

What if you could listen to your favourite Beethoven symphony as a Jazz interpretation at the press of a button? Humans are capable of performing such transcription tasks, but it requires considerable skill, effort and creativity. The goal of music genre transfer is to automate this task by training deep neural networks on large amounts of music data. Unsupervised methods excel at this task by allowing us to find structure in complex data in the absence of explicit ground truth labels. Deep generative models have been particularly successful, exemplified by methods such as Variational Autoencoders [5] and Generative Adversarial Networks [3]. One natural application of deep generative models is domain transfer, in which we learn a mapping function between two domains and thus implicitly parts of the underlying data generating distributions. This has led to many impressive applications such as rendering photographs in the style of different painters [13]. However, most applications have focused on images and only recently approaches for other types of data such as music have been proposed. In this work we focus on the task of transferring pieces of music in the MIDI format between different genres, e.g., from classic to jazz. For that purpose we extend the architecture from [2] with recent advances in GANs, in particular spectral normalization [8] and self-attention layers [12], and present respective transfer performance as measured by an automatic classifier-based metric, as well as inherent problems of using self-attention in domain transfer. With this,

ⓒ Springer Nature Switzerland AG 2020
P. Cellier and K. Driessens (Eds.): ECML PKDD 2019 Workshops, CCIS 1168, pp. 437–445, 2020.
https://doi.org/10.1007/978-3-030-43887-6_36

we introduce a simple content change metric to quantify content retention in transferred pieces. We further give insights in the decisions made by a neural network based genre classifier using a gradient-based attribution method [10] to better understand genre differences.

2   Related Work

Most neural network based domain transfer approaches are build on either VAEs [5] or GANs [3], or a combination of the two. Liu et al. [7] use a pair of GANs to learn the joint distribution of observations. Their method cannot directly perform domain transfer, but it can generate multiple versions of the same image in different domains (e.g., the same face with different hair color). Liu et al. [6] use a VAE architecture with a shared latent space to perform unsupervised image-to-image translation. Zhu et al. [13] introduce an architecture called CycleGAN which consists of a pair of GANs and is trained to perform domain transfer using a cycle consistency loss. While aforementioned methods are generally applicable, they focus their empirical evaluation on images, where best practices are well established.

In contrast, we focus on domain transfer in music. Mor et al. [9] use an autoencoder based architecture with a shared domain-invariant latent space to transfer input sounds to different instruments. While instruments can be indicative of genre, we focus on the task of genre transfer in absence of any instrumentation information. Brunner et al. [1] force one dimension of the latent space of a VAE to encode the genre by attaching a style classifier. Genre transfer can then be achieved by manipulating this latent genre label. They also propose a classifier-based metric to automatically evaluate the genre transfer. In a follow up work, Brunner et al. [2] adapt the original CycleGAN architecture to perform music genre transfer and achieve good results as measured by a slightly improved classifier-based metric. However, GANs are known to be difficult to train and there are many common failure modes, such as mode collapse or the discriminator overpowering the generator. We therefore investigate the effect of two recent advances in GANs that have been shown to improve GAN performance. In particular, we apply spectral normalization [8] to both the generator and discriminator. We further incorporate self-attention, a recent advance in neural network architectures that has been applied successfully for language modeling [11] and music generation [4]. Self-attention has been incorporated into GANs and together with spectral normalization was shown to improve training stability and overall performance [12]. We investigate both self-attention and spectral normalization in our setup and compare with the genre transfer performance of [2] as measured by a classifier-based metric. We further evaluate their individual impact and show that self attention hinders content retention.

3   Methodology

3.1   Dataset

Our dataset is based on polyphonic multi-instrument MIDI (Musical Instrument Digital Interface) files from three genres: jazz, classic and pop. We use the same dataset and preprocessing steps as [2]. That is, we remove the drum track and merge the remaining instrument tracks into a single piano track, resulting in a two dimensional matrix usually referred to as a *piano roll*, where the one dimension represents time steps and the other represents pitches. Each matrix entry indicates whether that note is played at the corresponding time. To acquire a homogeneous dataset, we omit songs whose time signature is not consistently $\frac{4}{4}$. We choose a sampling rate of 16 time steps per bar and combine 4 consecutive bars into one training example. This means that the shortest possible note we consider is the 16th note. While music in MIDI files can have pitch values between 0 and 127, i.e., note pitches ranging from C_{-1} to C_9, a standard piano can only play pitches between 21 to 108, i.e., notes ranging from A_0 to C_8. Since we merge all tracks into a single-instrument piano track we discard pitches beyond that range. Therefore, each input piano roll matrix has dimensions 64×84, corresponding to $16 * 4$ timesteps and 84 possible pitches respectively.

3.2   Architecture

Our neural network architecture is based on Generative Adversarial Networks (GANs [3]), where a generator and a discriminator are optimized by playing a minimax game. Since we want to perform style transfer in two directions, i.e., from domain A to domain B and vice versa, two GANs are arranged in a CycleGAN architecture [13]. In particular we use as baselines the "full" models from [2] with the additional discriminators.[1] We add two self-attention layers each to the discriminator and generator. For the generator, we add them after the second to last and the last residual blocks. For the discriminator, we add the attention layers after both hidden layers. Spectral normalization is applied to all convolution layers of each discriminator and generator. We use a batch size of 16 and the Adam optimizer with a learning rate of 0.0002. The generators and discriminators are both updated at each step. While training the discriminator, we add Gaussian noise with mean 0 and standard deviation σ_D as this was found to improve genre transfer performance in [2].

3.3   Metrics

As discussed in [1,2], human genre transfer evaluation is time consuming and cannot be applied continuously during development. Thus, a classifier based metric was introduced in [1] and slightly adapted in [2]. The classifier is a 5-layer

---

[1] See https://github.com/sumuzhao/CycleGAN-Music-Style-Transfer for more details on the baseline architecture.

CNN that performs binary classification between two genres. To evaluate genre transfer, the classifier is applied before and after transfer. For example, when performing transfer from A to B, the original piece should be classified as A, the transferred piece as B, and the transferred-back piece again as A. We report the transfer strength S^D_{tot}, a measure of average difference in correctly classified samples [2]. Specifically, let $P_A(x)$ be the empirical probability of classifying x as genre A. We then calculate the $A \rightarrow B$ transfer strength as

$$S^D_{A \rightarrow B} = \frac{1}{2}(P_A(x_A) + P_A(\tilde{x}_A) - 2 \cdot P_A(\hat{x}_B))$$

where x_A is a sample from domain A, $\hat{x}_B$ is the same sample transferred to domain B and $\tilde{x}_A$ is the sample transferred back to domain A. S^D_{tot} is then calculated as

$$S^D_{tot} = \frac{1}{2}(S^D_{B \rightarrow A} + S^D_{A \rightarrow B})$$

where $S^D_{B \rightarrow A}$ is defined symmetrically to $S^D_{A \rightarrow B}$. For the sake of brevity we refer to [2] for more details.

Further, as genre classification does not capture content retention, we introduce a new *content change metric*. We quantify content change by counting the number of added/removed notes in the piano roll, divided by the number of non-zero entries in the source piano roll. Specifically, for input sample $x \in \{0, 1\}^{64 \times 84}$ we calculate the content change $c(x)$ as

$$c(x) = \frac{\sum_{t,p} |x_{t,p} - \hat{x}_{t,p}|}{\sum_{t,p} x_{t,p}}$$

where $\hat{x}$ is the transferred sample and t and p are the indices into the time and pitch dimension. For a more fine grained analysis we can additionally look at added/removed notes individually:

$$c_{added}(x) = \frac{\sum_{t,p} \max(\hat{x}_{t,p} - x_{t,p}, 0)}{\sum_{t,p} x_{t,p}} \qquad c_{removed}(x) = \frac{\sum_{t,p} \max(x_{t,p} - \hat{x}_{t,p}, 0)}{\sum_{t,p} x_{t,p}}$$

Note that instead of looking at all notes one could also apply a heuristic for melody extraction, e.g., taking the skyline notes, to quantify melody change (as opposed to overall content change). However, we show that the simple metric based on all nodes already correlates well with human ranking of content retention.

3.4 Genre Attribution

We note that genre is ill defined, but the decisions of deep neural networks could provide insights into its nature. We therefore apply a gradient based input attribution method to the trained genre classifier in order to highlight notes that are most important in deciding genre. For instance, a 1-entry in the piano roll matrix corresponds to a played note, and if the back propagated class activation gradient is high for that note, then removing it would decrease the confidence in the corresponding genre classification, indicating that the presence of the note was significant in determining its genre. We use the saliency map attribution method [10], which multiplies all positive gradients with the original sample element-wise.

4 Experiments and Results

4.1 Genre Transfer

We fix $\sigma_D = 1$ as this worked best in [2] and compare our new models with the corresponding re-trained *full* model from [2], here referred to as *Baseline*. The genre transfer results in Fig. 1 show that self-attention (SA) and spectral normalization (SN) – individually and combined – improve the transfer in two out of the three genre pairings. Moreover, we see that transfer strength mainly depends on the genre pair investigated, as the boundary between some genres is ill defined. Further, the classifier metric does not measure content retention and audible quality, two aspects we are also interested in when performing genre transfer. To preliminarily investigate these aspects we took the classic vs. pop models and asked 12 people of our lab to rank the anonymized and randomly ordered transfers of the 4 models (Baseline, SN, SA, SN + SA) on 8 song snippets (4 transferred from classic to pop and 4 from pop to classic). Each participant thereby ordered for each song the transfers according to (a) content retention and (b) audible quality with respect to the target domain. We aggregated the rankings linearly into a normalized *human ranking score* s_{hr}^M by scoring each model M according to

$$s_{hr}^M = \frac{1}{N} \sum_{r=1}^{K} \#\{\text{rank of } M = \text{r}\} \frac{K - r}{K - 1}$$

where K is the number of models compared (4 in our case) and N is the number of participants. Note that rank one corresponds to the best and rank K to the worst transfer. Figure 1 (right) shows that our content change metric introduced above correlates negatively (Pearson correlation -0.805) with the human content retention ranking, indicating that this is a good heuristic to quantify content retention. Also visible in the figure is that models with self-attention score worse on content retention. This is also reflected in the content change metric over all test samples reported in Table 1, which shows an average content change of 0.92 for the Baseline model, 0.52 for SN, 2.11 for SA and 2.19 for SN + SA.

We therefore suspect that the use of self-attention can actually be harmful, as the generators can encode information in a global manner, as every time step and every pitch level attends to all other time-pitch cells, and hence the generators can alter the content of the source piece more strongly while still being able to achieve cycle-consistency. Explicit regularization techniques to retain parts of the content, e.g., the melody, could be developed in future work. As for audible

Table 1. Results of the content change metric for the different models. Shown is the mean and standard deviation over the test set samples.

	Baseline		SN	
	$C \to P$	$P \to C$	$C \to P$	$P \to C$
Added	0.82 ± 0.45	0.28 ± 0.17	0.57 ± 0.38	0.08 ± 0.09
Removed	0.27 ± 0.17	0.46 ± 0.12	0.91 ± 0.07	0.3 ± 0.11
Total	1.10 ± 0.5	0.75 ± 0.25	0.66 ± 0.38	0.38 ± 0.16
	SA		SN + SA	
Added	1.47 ± 0.41	0.85 ± 0.79	1.78 ± 1.33	0.72 ± 0.66
Removed	0.95 ± 0.04	0.95 ± 0.04	0.95 ± 0.05	0.93 ± 0.05
Total	2.42 ± 0.42	1.79 ± 0.78	2.73 ± 1.33	1.65 ± 0.66

	J vs. P	C vs. P	J vs. C
Baseline	28.49%	**64.62%**	57.64%
SN	32.16%	61.88%	63.98%
SA	**44.85%**	59.35%	63.56%
SN+SA	33.23%	53.07%	**66.76%**

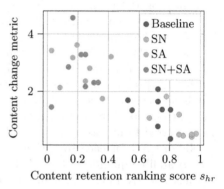

Fig. 1. Top: Genre transfer performance S_{tot}^D. J: Jazz, C: Classic, P: Pop, SN: With spectral normalization, SA: With self-attention. **Right:** Content change metric to human evaluation correlation.

quality, the results of our small user study were less homogeneous. On average, models with spectral normalization where slightly preferred over the others: the Baseline scored 0.47, SN 0.60, SA 0.42 and SN + SA 0.51, where scores are between 1 (always ranked best) and 0 (always ranked worst). Note that the user study only reflects relative audio quality among the studied models, and that there is room for improvement in terms of absolute fidelity. In particular, the genre transfer seems to introduce quite many dissonant notes. However, note that audible quality is already an issue with the original pre-processed pieces, because we reduce music pieces to single-instrument tracks, remove the drums and get rid of some of the dynamics (ignoring velocity, constant tempo). Using a richer input representation as, e.g., done in [1], would already result in more pleasing audio.[2]

---

[2] Additional results and audio samples can be found here: http://bit.ly/31VnTxS.

4.2   Attribution

Figure 2 depicts source piano rolls from the jazz and classical genres, along with the corresponding attributed piano roll. Intensity-thresholded instance normalized saliency maps [10] are presented. Pixels with intensity less than one-fourth of the maximum were removed in order to reduce clutter around more significantly attributed notes. Attribution was conducted on correctly classified samples with high gradient magnitudes to show interesting examples.

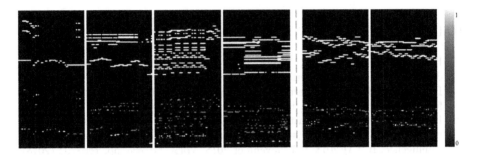

Fig. 2. Piano rolls (top) with corresponding saliency maps (bottom). Jazz samples are to the left of the delimiter, classical to the right.

The saliency maps are dominated by a few hyperintense pixels. Therefore, what distinguishes a sample's genre from the perspective of a deep classifier is truly subtle. We find that in jazz samples, often a sequence of notes in the lower pitch ranges are highlighted. This is somewhat similar to how humans recognize jazz, where genre becomes clear upon hearing a bass play a simple, rhythm-keeping line, over which different melodies are played.

One limitation of gradient-based attribution is that it is only a first order approximation and it is unable to capture complex dependencies across notes. Furthermore, patterns are not always obvious or provable. Nonetheless, the attribution provides a qualitative insight into the decisions of the deep classifier, highlighting certain musical motifs and revealing the nuance of musical genre in its ability to be determined mostly by a small number of notes. Identifying and isolating such motifs would make for fascinating future work in better defining genre and extracting genre specific features.

5   Conclusion

We presented preliminary qualitative insights on automated music genre transfer using MIDI files. We start from the CycleGAN model presented in [2] and show the effect of adding spectral normalization and self-attention on transfer as measured by a classifier-based metric. Further, we find on subsequent inspection that self-attention often makes the transferred songs less recognizable from a human viewpoint, which is emphasized by our simple content change metric which seems to correlate well with human perception. To the best of our understanding this is due to the global attention mechanism scrambling the pitch/time locality of notes. We further show that genre is often a matter of changing a few notes by looking at the attribution of our genre classifier. Our work offers many directions for follow up work, including the development of a better metrics for genre transfer as well as a quantitative analysis of motifs that make up a genre using attribution on classifiers. To stimulate further research in this direction make our code publicly available.[3]

References

1. Brunner, G., Konrad, A., Wang, Y., Wattenhofer, R.: MIDI-VAE: modeling dynamics and instrumentation of music with applications to style transfer. In: Proceedings of the 19th International Society for Music Information Retrieval Conference, ISMIR 2018, Paris, France, 23–27 September 2018, pp. 747–754 (2018)
2. Brunner, G., Wang, Y., Wattenhofer, R., Zhao, S.: Symbolic music genre transfer with cyclegan. In: IEEE 30th International Conference on Tools with Artificial Intelligence, ICTAI 2018, 5–7 November 2018, Volos, Greece, pp. 786–793 (2018)
3. Goodfellow, I., et al.: Generative adversarial nets. Adv. Neural Inf. Process. Syst. **27**, 2672–2680 (2014)
4. Huang, C.A., et al.: An improved relative self-attention mechanism for transformer with application to music generation. CoRR abs/1809.04281 (2018)
5. Kingma, D.P., Welling, M.: Auto-encoding variational bayes. In: 2nd International Conference on Learning Representations, ICLR 2014, Banff, AB, Canada, 14–16 April 2014, Conference Track Proceedings (2014)
6. Liu, M., Breuel, T., Kautz, J.: Unsupervised image-to-image translation networks. In: Advances in Neural Information Processing Systems 30: Annual Conference on Neural Information Processing Systems 2017, 4–9 December 2017, Long Beach, CA, USA, pp. 700–708 (2017)
7. Liu, M., Tuzel, O.: Coupled generative adversarial networks. In: Advances in Neural Information Processing Systems 29: Annual Conference on Neural Information Processing Systems 2016, 5–10 December 2016, Barcelona, Spain, pp. 469–477 (2016)
8. Miyato, T., Kataoka, T., Koyama, M., Yoshida, Y.: Spectral normalization for generative adversarial networks. In: 6th International Conference on Learning Representations, ICLR 2018, Vancouver, BC, Canada, 30 April – 3 May 2018, Conference Track Proceedings (2018)
9. Noam Mor, Lior Wold, A.P., Taigman, Y.: A universal music translation network. In: International Conference on Learning Representations (ICLR) (2019)
10. Simonyan, K., Vedaldi, A., Zisserman, A.: Deep inside convolutional networks: visualising image classification models and saliency maps. In: 2nd International Conference on Learning Representations, ICLR 2014, Banff, AB, Canada, 14–16 April 2014, Workshop Track Proceedings (2014)
11. Vaswani, A., et al.: Attention is all you need. In: Advances in Neural Information Processing Systems 30: Annual Conference on Neural Information Processing Systems 2017, 4–9 December 2017, Long Beach, CA, USA, pp. 6000–6010 (2017)

---

[3] https://github.com/czhang0808/Music-Genre-Transfer-with-Deep-Learning.

12. Zhang, H., Goodfellow, I.J., Metaxas, D.N., Odena, A.: Self-attention generative adversarial networks. In: Proceedings of the 36th International Conference on Machine Learning, ICML 2019, 9–15 June 2019, Long Beach, California, USA, pp. 7354–7363 (2019)
13. Zhu, J., Park, T., Isola, P., Efros, A.A.: Unpaired image-to-image translation using cycle-consistent adversarial networks. In: IEEE International Conference on Computer Vision, ICCV 2017, Venice, Italy, 22–29 October 2017, pp. 2242–2251 (2017)

Familiar Feelings: Listener-Rated Familiarity in Music Emotion Recognition

Lloyd May[(✉)] and Michael Casey

Dartmouth College, Hanover, NH 03755, USA
{Lloyd.May.GR,Michael.A.Casey}@dartmouth.edu

Abstract. The task of music emotion recognition (MER) has previously been explored using a variety of audio, lyrical, and basic meta-data features. As feature extraction and classification algorithms advance, the need for relevant extra-musical features becomes apparent. The efficacy of familiarity as a feature in a MER system was evaluated by a Random Forest feature importance analysis on a novel dataset of 5000 clips with annotated familiarity and valence. Familiarity was correlated to perceived valence ($r = 0.250$) and resulted in a statistically significant increase of 0.011 in the F-score of a baseline MER classifier upon its inclusion.

Keywords: Music emotion recognition · Familiarity · Music perception

1 Introduction

The perceived emotion of a piece of music has long been part of the societal discourse around sound. The algorithmic prediction of this emotion is of increasing interest in the age of streaming. Personalized playlists are generated and customized based on these, and many other, algorithmic predictions about, and features of, music. However, these systems have traditionally used very few extra-musical features in this task other than standard music meta-data, i.e. genre, year of release, artist(s), etc. These extra-musical features are not contained in the actual audio signal and are useful in search and recommendation tasks as the catalogue of recorded music continues to increase. Subjective familiarity has been shown to impact a subject's perception of emotion present in a piece of music [18]. While music emotion recognition (MER) systems have been created using context-based features [5,7] and higher dimension emotional mapping [4], none have included familiarity.

2 Related Work

Labels are inherent in both music making and its consumption, with listeners able to accurately predict some labels in less than a second [8]. Many systems have been developed with the goal of automatically labeling mood or emotion

© Springer Nature Switzerland AG 2020
P. Cellier and K. Driessens (Eds.): ECML PKDD 2019 Workshops, CCIS 1168, pp. 446–453, 2020.
https://doi.org/10.1007/978-3-030-43887-6_37

in music. A written-query corpus study showed that 80% of production music queries contained emotional terms [6], indicating the importance of emotion in semantic music descriptions. Therefore it is in the interest of music recommendation as well as playlist creation services to leverage the power of emotion labels.

The development of these automated MER systems involves a series of decisions regarding the emotion and annotation parameters, leading to difficulties in system comparisons [1]. These decisions require researchers to choose between dynamic or static annotation protocols as well as between perceived or induced emotion annotation. "Perceived emotion" is the emotion a subject thinks the music is communicating, while "induced emotion" is the emotion the subject personally feels while listening to the music. MER systems typically predict perceived emotion as induced emotion tasks require more complex annotation tasks and do not offer the same level of clarity of inter-subject comparisons. For example, a song that communicates a positive perceived emotion may induce a negative emotion in a subject if the music is associated with negative memories. Emotion annotations also vary as they typically include semantic mood classification, "happy" or "angry" etc., as well as dynamic annotation in valence-arousal space [7]. Valence ranges from negative to positive while arousal typically ranges from low to high. Inverse u-shaped trends, originally noted by Berlyne [3], have been widely seen when visualizing emotion annotations in the valence-arousal 2D Cartesian plane [4]. In MER, this trend translates to few songs and semantic concepts existing with neutral valency and high/low arousal [7]. An inverse-U relationship between familiarity and musical preferences was proposed after the trend was widely noted [15].

Familiarity with a piece of music is an important contextual enhancer in the perception of emotion in music and is a function of factors including the number of previous exposures, context of the exposure, and intensity of the elicited emotional reactions. Subjects consistently rank the perceived emotion of a clip as more intense if they are familiar with the song [19]. This phenomenon is also seen in neuroscience as the emotion and reward circuits of the brain have shown increased activity in fMRI studies when subjects are listening to familiar music, regardless of musical preferences [17].

Music that is subjectively familiar to a patient has been effectively used to improve certain music therapy practices [2] and elicits notable physiological responses, such as increases in the speed and consistency of a subject's gait [10] as well as increased affective modulation of subject anxiety states [20,21]. It has also been noted, both by researchers [19] and by the broadcasting community, that people are more likely to rate familiar music favorably when compared to unfamiliar music of a similar style [16]. These different responses to familiarity have been explored in EEG based emotion recognition systems which achieved a classification accuracy of 82% in a 4-class model of emotion [11].

Popular music is often more familiar as listeners are likely to have heard popular songs more often in public spaces or have been biased towards them by recommendation algorithms. While popularity and familiarity are correlated, they are inherently separate concepts that capture different information about the listener's relation to a piece of music. Familiarity is individually subjective while popularity is largely culturally and contextually dependent. As a result of this, familiarity is not a binary variable. There is a gradient of familiarity ranging from no previous exposure to the clip or similar clips, to over-learned familiarity [13]. The subjective familiarity that exists between these binary states is akin to the 'feeling of knowing' (FoK) widely discussed in psychology [13]. FoK-type familiarity allows for a scale of familiarity as well as associations related to the clip. For example, a subject might have heard much of an artist's work, yet is uncertain if they have heard a specific song that is being played. Studies have previously shown a positive correlation between this FoK-type familiarity and differences in the perceived emotion in music [16,17]. Familiarity was noted as a feature of interest in the development of an MER benchmark following the MediaEval 2013–2015 MER tasks [1]. However, the feature could not be studied due to annotators not being familiar with the large public-domain data sets historically used in MER tasks.

State-of-the-art MER systems use low-level audio features and/or meta-data; combined with a ground truth measure of emotion, to train and test classifiers [1]. Modern classifiers typically use a deep-learning framework [17], achieving averaged root-mean square errors below 0.25. Comparable classification accuracies of over 85% have been achieved by support vector machine (SVM) models [14]. Context-based systems, such as mood classification within known genre families [9], have been used to improve mood and emotion classification accuracies in MER systems of 67% combined with a finer grain model of emotion and emotion-state-transitions [5]. Following the conclusion of the 2013–2015 MediaEval MER task, recurrent neural recurrent neural networks with large feature sets were shown to outperform other models [1]. Extra-musical features, such as familiarity, would prove particularly useful as neural network systems become more elaborate as these are features that cannot be extracted or learned from the audio.

The psychological and physiological effects of familiar music have been leveraged in other music information retrieval (MIR) problems. The inference of familiarity in a listener profile has been explored by clustering listeners into different groups based on their individual preferences for novelty and complexity in new music [12]. Other techniques, such as efficient user prompting or correlating number of plays, may also be explored to further leverage psychological phenomena in MER.

2.1 Dataset

50 songs from across 6 genre families and 7 decades of release were collected from Vevo certified YouTube videos with more than 100k streams. Two 20-s clips were taken from each song and root-mean squared (RMS) normalized. One clip contained the hook or chorus of the song and the other included either a verse or bridge section. The use of these two types of clips allowed for reports of perceived emotion to change depending on the section of the song and provided

an additional level of robustness to the self-reported familiarity. The clip length was chosen to exceed the 15-s threshold for perceived emotion reporting [7], but short enough to allow for each annotator to label all of the clips. The Spotify API was used to collect a measure of valence to ensure that no bias relating valence to genre or release year was present in the dataset. This was validated by calculating the Spearman's correlation (-0.032) between year of release and annotated valence by an initial group of 32 annotators.

2.2 Labeling

50 annotators aged of 18–30 with no reported hearing impairments performed individual annotations on all 100 clips. The 50 annotators consisted of 28 males, 21 females, 1 gender non-conforming individual with 52% of annotators identifying as non-musicians, 36% amateur musicians, and 12% professional musicians. Annotators were played a 20-second clip and asked to complete 3 questions at the conclusion of each clip: (1) Rate the subjective familiarity of the clip using a scale of 1–4, (2) If the emotion conveyed in the clip was more positive or negative, and (3) How intensely that emotion was conveyed on a 4 point scale.

A seven-class valence model was developed classifying these annotations into 7 classes, allowing for the reported emotions to be ranked and classified without the need for a semantic explanation. Valence was selected as the annotation parameter as it has previously been shown to be correlated to familiarity [17].

A notable trend in the annotations was that more clips were classified as having a more positive perceived emotion. This phenomenon may be a result of the collection of clips themselves, alluding to the difficulty of constructing a data set for a familiarity task that balances across genre family, year, subjective familiarity, and perceived emotion. Incomplete annotations were removed, resulting in a total dataset of 4722 annotations spread out among the 100 clips. Each clip received on average $47.2 +/- 1.5$ annotations, with a maximum of 50 and minimum of 45 annotations. As subjective familiarity cannot be averaged among annotators, each clip-annotation pair was used an individual data point in the dataset.

2.3 Algorithms and Analysis

The effect of familiarity on annotated valence was assessed via (1) a correlation analysis, (2) a feature importance analysis and (3) the inclusion of familiarity as a feature to a baseline MER system. This analytical framework was selected to assess the effectiveness of familiarity at a variety of applied levels.

A significant Spearman's correlation ($r = 0.250$) was found between familiarity rating and annotated valence class. Other notable and significant correlations were found between familiarity rating and response time (-0.132), familiarity and if the clip contained the hook/chorus section (-0.104), and between year of release and class (-0.094). A suite of 55 audio features were extracted for each clip using the Librosa library. These included the first 13 Mel Frequency Cepstral Coefficents (MFCCs), tempo, zero-crossing rate as well as spectral bandwidth, center, contour and roll-off. Tempo and spectral features were selected to construct a baseline model as they have previously been used to great effect within SVM MER systems. The mean and standard deviation of all features over the clip were both used in the models.

Feature importance was calculated by the change in prediction error when the out-of-bag error of the feature was altered, all else constant, in a Random Forest model. This was repeated for each variable-tree pairing, producing a ranked list of feature importance with percentage inclusion and standard deviation measures. As seen in Fig. 1, self-rated familiarity rating and response time ranked higher than the acoustic features by a statistically significant amount.

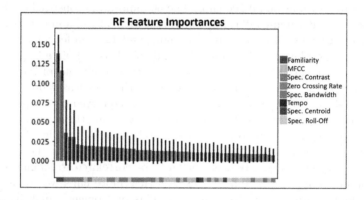

Fig. 1. Feature importance rankings for all 58 features. Familiarity rating and rating time were ranked as significantly more important.

A SVM model was chosen to construct a baseline as it has been implemented successfully in many systems [7,14] and allows for a more meaningful analysis of the effect of including additional features while still non-trivially modeling non-linearities. Neural networks do slightly outperform SVM MER systems [7], however, the lack of feature-effect transparency inherent in neural network systems precluded them from the study. A baseline model was constructed and optimized using all 56 audio features while the proposed model included all 56 audio features as well as 2 additional familiarity features. These features where the familiarity rating as well as the response time to complete the familiarity assessment in seconds. A randomized third of the data was left out of each of the 500 iterations as testing data and the mean of the F-scores across all iterations was calculated. One model was trained on n principal components while another was trained on the same n principal components as well as the reported familiarity. The gamma and penalty parameters were optimized using a grid search and the values remained constant for each model throughout the analysis.

3 Results

Radial basis function (RBF) and second degree polynomial kernels were used to evaluate the effect of the inclusion of reported familiarity's effect on the MER SVM system performance. The improvement in F-score and accuracy is largest when 6 or fewer principal components are used, regardless of the SVM kernel. The model with familiarity included consistently outperformed the base model at each number of included components.

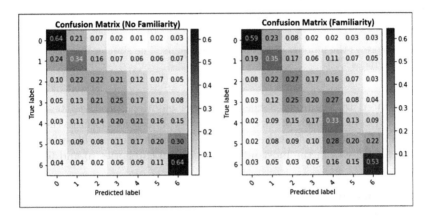

Fig. 2. The normalized mean confusion matrix of the RBF SVM both with and without familiarity. Class labels correspond to 0 - very low valence to 6 - very high valence.

No significant difference between SVM kernels was found past 15 included components. Systems with the familiarity feature included have a statistically significant improvement in F-score of 0.01 and an increase in accuracy of 0.01 once 15 or more principal components are included. This gain in system performance is comparable to those seen in innovations in state-of-the-art MER systems, a decrease in RMSE of 0.02. The systems that use an RBF kernel reached the maximum F-score with fewer features than the systems that use the polynomial kernel. The difference in F-score distribution between the two RBF SVM models yielded a highly significant Wilcoxon T of $1.18e5$ with a p-value of $8.372e - 42$. This increase in performance is hypothesized to be a result of including familiarity rating, a feature which is both personalized on the level of the annotator and has previously been shown to correlate with the perceptions of emotion in music.

Both confusion matrices in Fig. 2 show a strong diagonal trend, indicating that the classifier is not overly biased in classifying clips into only a subset of classes. This strong diagonal trend also indicated that, in the majority of cases when clips were misclassified, they were incorrectly classified into the nearby classes. As the 7 class valence model is ordered and neighboring classes are hierarchically related, the reclassification into neighboring classes was expected.

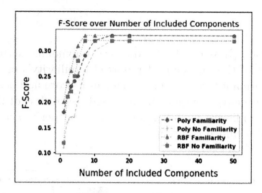

Fig. 3. F-score vs the number of included components. The standard error of all points is less than 10e–4.

When reduced to a simpler 3 class model of emotion, positive, negative and neutral, a system accuracy of 70.42% is achieved. This is far closer to the state-of-the-art SVM accuracies of over 80% [7, 14]. The most accurate classification the model achieved was in the most intense emotional categories (0 and 6) (Fig. 3).

4   Conclusion and Future Work

Through leveraging findings in psychology and neuroscience, the extra-musical feature of subjective familiarity was identified and a dataset constructed, annotated, and analyzed to show the efficacy of the inclusion of subjective familiarity ratings in MER systems. The findings support the notion that subjective familiarity may be a useful feature in the context of MER and possibly other MIR tasks. Given the size of current MER system improvements, the increase in performance generated through the inclusion of familiarity may motivate its inclusion to larger scale MER systems. Future work would include identifying the most efficient means to collect or infer subjective familiarity within an MER system. Possible proxies for familiarity, such as number of plays and other group-filtering features, may also be investigated in the future.

References

1. Aljanaki, A., Yang, Y.H., Soleymani, M.: Developing a benchmark for emotional analysis of music. PloS one **12**(3), e0173392 (2017)
2. Bailey, L.M.: The use of songs in music therapy with cancer patients and their families. Music Ther. J. AAMT **3**(1), 5–17 (1984). https://doi.org/10.1093/mt/4.1.5
3. Berlyne, D.E.: Studies in the New Experimental Aesthetics: Steps Toward an Objective Psychology of Aesthetic Appreciation. Hemisphere, New York (1974)
4. Buccoli, M., et al.: A higher-dimensional expansion of affective norms for English terms for music tagging, pp. 316–322 (2016)

5. Han, B.J., Rho, S., Jun, S., Hwang, E.: Music emotion classification and context-based music recommendation. Multimedia Tools Appl. **47**(3), 433–460 (2010)
6. Inskip, C., Macfarlane, A., Rafferty, P.: Towards the disintermediation of creative music search: analysing queries to determine important facets. Int. J. Digit. Libr. **12**(2–3), 137–147 (2012)
7. Kim, Y.E., et al.: Music emotion recognition: a state of the art review. In: Proceedings of ISMIR, vol. 86, pp. 937–952. Citeseer (2010)
8. Krumhansl, C.L.: Plink: "Thin slices" of music. Music Percept. **27**(5), 337–354 (2010). https://doi.org/10.1525/mp.2010.27.5.337. http://mp.ucpress.edu/cgi/doi/10.1525/mp.2010.27.5.337
9. Laurier, C.F., et al.: Automatic classification of musical mood by content-based analysis. Universitat Pompeu Fabra (2011)
10. Leow, L.A., Rinchon, C., Grahn, J.: Familiarity with music increases walking speed in rhythmic auditory cuing. Ann. N.Y. Acad. Sci. **1337**(1), 53–61 (2015). https://doi.org/10.1111/nyas.12658
11. Lin, Y.P., et al.: EEG-based emotion recognition in music listening. IEEE Trans. Biomed. Eng. **57**(7), 1798–1806 (2010). https://doi.org/10.1109/TBME.2010.2048568
12. Marques, A., Andrade, N., Balby, L.: Exploring the relation between novelty aspects and preferences in music listening. In: Proceedings of the 14th International Society for Music Information Retrieval Conference, pp. 407–412 (2013)
13. Nelson, T.O., Gerler, D., Narens, L.: Accuracy of feeling-of-knowing judgments for predicting perceptual identification and relearning. J. Exp. Psychol. Gen. **113**(2), 282 (1984)
14. Ness, S.R., Theocharis, A., Tzanetakis, G., Martins, L.G.: Improving automatic music tag annotation using stacked generalization of probabilistic SVM outputs. In: Proceedings of the 17th ACM International Conference on Multimedia, pp. 705–708 (2009)
15. North, A., Hargreaves, D.: The Social and Applied Psychology of Music. Oxford University Press, Oxford (2008)
16. North, A.C., Hargreaves, D.J.: Subjective complexity, familiarity, and liking for popular music. Psychomusicology **14**(1–2), 77–93 (1995)
17. Pellegrini, T., Barrière, V., Pellegrini, T., Barrière, V., Estimation, T.C.: Time-continuous Estimation of Emotion in Music with Recurrent Neural Networks To cite this version: HAL Id : hal-01327121 (2016)
18. Pereira, C.S., Teixeira, J., Figueiredo, P., Xavier, J., Castro, S.L., Brattico, E.: Music and emotions in the brain: familiarity matters. PloS one **6**(11), e27241 (2011)
19. Ali, S.O., Peynircioğğlu, Z.F.: Intensity of emotions conveyed and elicited by familiar and unfamiliar music. Music Percept. Interdisc. J. **27**(3), 177–182 (2010)
20. Riegler, S.J., Miller, R.: Degrees of familiar and affective music and their effects on state anxiety. J. Music Ther. **17**(1), 2–15 (1980). https://doi.org/10.1093/jmt/17.1.2
21. Sung, H.C., Lee, W.L., Li, T.L., Watson, R.: A group music intervention using percussion instruments with familiar music to reduce anxiety and agitation of institutionalized older adults with dementia. Int. J. Geriatr. Psychiatry **27**(6), 621–627 (2012). https://doi.org/10.1002/gps.2761

Rhythm, Chord and Melody Generation for Lead Sheets Using Recurrent Neural Networks

Cedric De Boom[(✉)], Stephanie Van Laere, Tim Verbelen, and Bart Dhoedt

IDLab, Department of Information Technology at Ghent University – imec,
Technologiepark-Zwijnaarde 126, 9052 Ghent, Belgium
{cedric.deboom,stephanie.vanlaere,tim.verbelen,bart.dhoedt}@ugent.be

Abstract. Music that is generated by recurrent neural networks often lacks a sense of direction and coherence. We therefore propose a two-stage LSTM-based model for lead sheet generation, in which the harmonic and rhythmic templates of the song are produced first, after which, in a second stage, a sequence of melody notes is generated conditioned on these templates. A subjective listening test shows that our approach outperforms the baselines and increases perceived musical coherence.

Keywords: Music generation · Lead sheets · Neural networks

1 Lead Sheets

Lead sheets are widely used to represent the fundamental musical information about almost any contemporary song: they contain a chord scheme, a melody line, some navigation and repetition markers, and sometimes lyrics. They seldom contain information about the instrumentation or accompaniment, so any band can take a lead sheet as a guideline and make the song their own, sometimes even by improvising over the chord schemes. In this paper we focus on generating chords and melody lines for lead sheets from scratch.

A major difficulty in music generation is that harmony, melody and rhythm all influence each other. For example, a melody note can change whenever the underlying harmony changes, and vice versa. Rhythmic patterns can influence which notes are played, and rhythm and harmony together define the overall groove of the piece. To tackle this issue, we split the generation process into two stages. First, we generate a harmonic progression using chord sequences, while simultaneously picking the most appropriate rhythmic patterns. And in a second step the melody is generated on top of this harmonic and rhythmic template.

We are, however, not the first to tackle the problem of lead sheet generation and, in general, music generation. Briot et al. provide a recent and extensive overview of all deep learning based techniques in this field [1]. Regarding lead sheets specifically, Liu et al. use GAN-based models on piano roll representations, but the melody and chords are still predicted independently by different generators [9]. Roy et al. devise a lead sheet generator with user constraints

P. Cellier and K. Driessens (Eds.): ECML PKDD 2019 Workshops, CCIS 1168, pp. 454–461, 2020.
https://doi.org/10.1007/978-3-030-43887-6_38

i	c_i	r_i	m_i
1	C	quarter	G4
2	C	eighth	E4
3	C	eighth	G4
4	C	half	C5
5					
6	G	quarter	D5
7	G	quarter	B4
8	G	half	G4
9					

Fig. 1. Example of a lead sheet decomposition into chords, rhythms and melodies.

defined by Markov models, and harmonic synchronization between melody and chords through a probabilistic model that encodes which melody notes fit on which chords [11]. There have also been many efforts in the past that learn to generate chords for a given melody [3,8,10] or the other way round [12]. In this paper we want to show that, on the one hand, a two-stage generation process greatly improves the perceived quality of the music. And, on the other hand, we show that melodic coherence improves when the melody generator gets to look ahead at the entire harmonic template of the song.

We formally define a lead sheet $x_{1:n}$ of length n, characterized by a sequence of chords $c_{1:n}$, rhythms $r_{1:n}$ and melody pitches $m_{1:n}$. At each time step i in the piece, each of these quantities take some value:

$$x_{1:n} = \{c_{1:n}, r_{1:n}, m_{1:n}\}, \quad x_i = \{c_i, r_i, m_i\}. \tag{1}$$

In this equation, the compact $x_{j:k}$ notation denotes the sequence $(x_j, x_{j+1} \ldots x_k)$. Figure 1 shows an example of this decomposition. Notice that the chords are repeated until there is a change in harmony, thereby allowing us to model the entire lead sheet using a single shared time scale. We also point out that there is only one melody note per time step, which is not a severe restriction, since most lead sheets only contain monophonic melodies. Finally, we choose to treat the barlines as separate elements in the sequence, which is indicated by the vertical bars in Fig. 1.

2   The Wikifonia Dataset

In this paper we will make use of the Wikifonia dataset, a former public lead sheet repository hosted by wikifonia.org. It contains more than 6,500 lead sheets in MusicXML format, and in all sorts of modern genres. This section goes over the different preprocessing and encoding steps that are executed on the dataset in order to obtain a clean collection of lead sheets.

2.1   Preprocessing

Eliminate Polyphony. Whenever multiple notes sound at the same time, we only retain the note with the highest pitch, as it is often the note that characterizes the melody.

Ignore Ties. Connections between two notes with the same pitch that extend the first note's duration are ignored. The two notes are therefore treated as two separate notes with their original duration.

Delete Anacruses. Incomplete bars that often appear at the start of a piece, are removed from all lead sheets.

Unfold Repetitions. Lead sheets can contain repetition and other navigation markers. If a section should be repeated, we duplicate that particular section, thereby unfolding the piece into a single linear sequence.

Remove Ornaments. Since such ornaments do not contribute much to the overall melody, we leave them out.

2.2 Data Encoding and Features

After preprocessing, we encode the melody, rhythm and chord symbols into feature vectors such that they can be used as input to our generators.

Encoding Rhythms. We retain the 12 most common rhythm types in the dataset, which are given in Appendix B. We remove 184 lead sheets from the dataset that contain other than these 12 types. Together with the representation for a barline, we encode rhythm into a 13-dimensional one-hot vector r_i.

Encoding Chords. A chord is described by both its root and its mode. There are 12 possible roots (C, C♯, D, D♯, ..., B) and we choose to convert all accidentals to either no alteration or one sharp. We count 47 different modes in the dataset, which we map to one of the following four: major, minor, diminished or augmented. This mapping only very slightly reduces musical expressivity and interestingness. The mapping table can be found in Appendix A. The 12 roots and 4 modes give 48 chord options in total, resulting in a 49-dimensional one-hot vector c_i if we include the barline.

Encoding Melody. The MIDI standard defines 128 possible pitches. We assign two additional dimensions for rests and barlines, resulting in a 130-dimensional one-hot encoded melody vector m_i.

3   Recurrent Neural Network Design

As mentioned in Sect. 1, the lead sheet generation process happens in two stages: in stage one the rhythm and chord template of the song is learned, and in stage two the melody notes are learned on top of that template. We will use separate LSTM-based models for both stages [2]; the models are trained independently of each other, but they are combined at inference time to generate an entire lead sheet from scratch. Figure 2 shows the complete architecture.

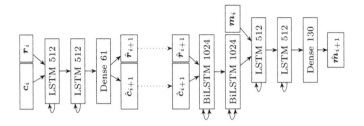

Fig. 2. The RNN architecture for both stage one (left) and stage two (right). The output dimensionality for every layer is written in each of the blocks. Whenever two blocks appear next to each other, the (output) vectors are concatenated.

Stage One. In this stage, the rhythm and chord vectors are first concatenated and are subsequently given as inputs to two LSTM layers followed by a dense layer. All LSTM layers have a output dimensionality of 512 states, as indicated in the figure. The output of the dense layer is cut in two vectors, both on which we apply a softmax nonlinearity with temperature τ, controlling the concentration of the output distribution. This way we are effectively modeling a distribution over the chord and rhythm symbols that come next in the sequence.

Stage Two. The second model will process the generated sequence of predicted chords and rhythms. To this end, each chord and rhythm vector is again concatenated before being processed by two BiLSTM layers. These BiLSTM states allow the pitch generator to look back and also ahead at the harmonic sequence, as inspired by [8]. The dimensionality of the BiLSTM layers is 512 in both directions, adding up to a total of 1024 states. After concatenation with the previous melody vector, the BiLSTM states are fed through another stack of two LSTM layers. The output of the last dense layer is used to predict the next melody note, again using a softmax nonlinearity controlled by a temperature parameter.

3.1 Optimization Details

In this paper we train both stages separately; it is possible to jointly train both models through reparameterization tricks [4], but we leave this as future work. While training is done separately, inference of lead sheets from scratch is easily done by feeding the output of the first model to the input of the second model. We use a standard cross-entropy loss function on all outputs. In stage one we sum the losses on both the chord and rhythm outputs with hyperparameter α:

$$\mathcal{L}_{\text{stage 1}}(\hat{c}, \hat{r}) = \alpha \cdot \mathcal{L}_{\text{CE}}(\hat{c}) + (1 - \alpha) \cdot \mathcal{L}_{\text{CE}}(\hat{r}). \tag{2}$$

In this equation, $\mathcal{L}_{\text{CE}}(\cdot)$ indicates the cross-entropy loss. In stage two the loss function is equal to the cross-entropy loss on the melody output.

$$\mathcal{L}_{\text{stage 2}}(\hat{m}) = \mathcal{L}_{\text{CE}}(\hat{m}). \tag{3}$$

We use Adam with learning rate λ to optimize both models [6]. During optimization, the training data is augmented by shifting the pitches and chords with a random number of semitones between -12 and 12, that is, between plus or minus one octave. This virtually increases the amount of training data by a factor of 25, thereby aiding model generalization towards less frequently appearing keys.

4 Experiments

Hyperparameters. In all experiments we use a batch size of 128 sequences, each of length 100. The learning rate λ is set fixed to 0.001. We empirically found that a value α of 0.5 leads to good results, so we set it fixed to that value. We also set the temperature τ slightly lower than 1 during inference of the melody, which helps to improve the perceived quality of the generated music; we varied it between 0.75 and 1.0 during the experiments. For the rhythm and chord patterns a temperature of 1.0 gives the most pleasing results.

Baselines. We will compare our model against two baselines:

1. An unconditioned LSTM-based model similar to the stage one model in Fig. 2, but now the melody is also concatenated to the input and output. The melody is no longer conditioned on the entire chord and rhythm sequence. We also add an extra LSTM layer, adding up to a total of three.
2. A two-stage model where the BiLSTM layers are replaced by regular LSTM layers, so that the melody cannot look ahead at the harmonic sequence. We keep all other parameters identical to the original model.

Subjective Listening Test. We conducted an online listening test in which we asked 40 participants to score 12 short audio clips, each of approximately one minute long. The following songs were included in the test[1]:

- 3 pieces generated by the two-stage model from scratch,
- 3 pieces generated by the two-stage model, but conditioned on the chord and rhythm scheme of existing songs: *I Have a Dream* (Abba), *Autumn Leaves* (jazz standard) and *Colors of the Wind* (Alan Menken),
- 2 pieces generated by the one-stage baseline model,
- 2 pieces generated by the two-stage baseline model,
- 2 (relatively unknown) human-composed songs: *You Belong to my Heart* (Bing Crosby) and *One Small Photograph* (Kevin Shegog).

As stated in Sect. 1, a lead sheet only encodes the basic template of a song, and it ideally needs to be played by a real musician. We therefore gave all lead sheets to a semi-professional pianist; the pianist stayed true to the sheet music, but was free to create an accompaniment that suited the piece. In our regards, this evaluation method reflects best how a lead sheet, produced by an AI model, would in practice be used and experienced by musicians and listeners.

---

[1] Listen to the audio clips at https://users.ugent.be/~cdboom/music.

Table 1. Results of the subjective listening experiments. We report averaged $\bar{Z}$-scores for each of the questions, along with the standard deviations.

Model	Pleasing $\bar{Z}$	Coherence $\bar{Z}$	Turing $\bar{Z}$
One-stage	-0.23 ± 0.34	-0.24 ± 0.31	-0.22 ± 0.29
Two-stage, without BiLSTM	-0.04 ± 0.35	-0.07 ± 0.33	-0.09 ± 0.33
Two-stage, with BiLSTM	-0.01 ± 0.36	0.01 ± 0.34	-0.02 ± 0.34
Two-stage, with existing chords	$\mathbf{0.15 \pm 0.37}$	0.09 ± 0.36	$\mathbf{0.13 \pm 0.36}$
Human-composed songs	0.03 ± 0.34	$\mathbf{0.13 \pm 0.36}$	$\mathbf{0.13 \pm 0.34}$

The audio clips were presented to each user in randomized order. For each clip we asked the user to rate on a scale of 1 to 5 how much he likes the piece, if the melody is musically coherent, and whether the piece is composed by a computer (1) or a human (5). We also asked the user to indicate if he recognizes the piece. Since each user has his own rating bias and spread [5,7], we converted the ratings for each user to a standardized Z score between -0.5 and 0.5:

$$Z_{c,u} = \frac{R_{c,u} - \mu_u}{\max_{c'} R_{c',u} - \min_{c'} R_{c',u}}. \tag{4}$$

In this formula, $R_{c,u}$ is the rating of user u for clip c, μ_u is the average rating of user u, and $Z_{c,u}$ is the associated standardized score. Table 1 reports the average $\bar{Z}$ score across the audio clips for each of the three questions in the survey, along with the standard deviation. A negative score means that the ratings are below average overall, and a positive score indicates an overall above-average rating.

We observe that the scores are, by far, better for the two-stage models compared to the unconditioned one-stage model. This shows that first sampling a harmonic and rhythmic sequence, and conditioning the melody on top of this sequence, is more beneficial than sampling all quantities simultaneously. Next to this, we also notice that adding the BiLSTM layers improves the score for all three questions. And although by a small margin, we can conclude that the musical quality improves when the melody generator can look ahead in the harmonic sequence. When we condition the melody generator on an existing chord and rhythm scheme, it is remarkable that the human-composed and AI-composed songs perform almost on par. The AI-composed songs are even considered most pleasing. Related to this observation, 4 participants indicated having recognized a piece from the two-stage model, 5 recognized a piece that was generated based on existing chords, and 3 participants recognized a human-composed song.

Finally, we also want to point out that the standard deviations are very substantial, which shows that there is a high level of disagreement between the reviewers. It is however interesting to point out that the standard deviation is slightly higher for better performing models. This might indicate that there is more consensus on what it means for music to sound 'badly', but that the definition of 'good' music is more subjective and person-dependent.

5   Conclusion

We have proposed a two-stage LSTM-based model to generate lead sheets from scratch. In the first stage, a sequence of chords and rhythm patterns is generated, and in the second stage the sequence of melody notes is generated conditioned on the output of the first stage. We conducted a subjective listening test of which the results showed that our approach outperformed the baselines. We can therefore conclude that conditioning helps the quality of the generated music, and that this approach can be explored further in the future.

A   Mode Mapping for Chords

In Table 2 we show how different chord modes are mapped to one of the following four options: major, minor, diminished or augmented.

Table 2. Chord modes are mapped to one of four options.

Original mode	Mapped mode	Original mode	Mapped mode
6	major	major-6-9	major
7	major	major-7	major
9	major	major-9	major
augmented	augmented	major-minor	major
augmented-7	augmented	minor	minor
augmented-9	augmented	minor-11	minor
diminished	diminished	minor-13	minor
diminished-7	diminished	minor-6	minor
dominant	major	minor-7	minor
dominant-11	major	minor-7-b5	diminished
dominant-13	major	minor-9	minor
dominant-7	major	minor-major	minor
dominant-9	major	minor-major-7	minor
half-diminished	diminished	power	major
major	major	sus2	major
major-13	major	sus4	major
major-6	major	sus4-7	major

B   Rhythm types

Table 3 provides an overview of the twelve rhythmic figures that are used.

Table 3. The rhythm types that are considered in this paper.

Textual description	Musical symbol
32nd note	♪
32nd dotted note	♪.
16th note	♪
8th triplet note	♪ 3
8th note	♪
quarter triplet note	♩ 3
8th dotted note	♪.
quarter note	♩
quarter dotted note	♩.
half note	♩
half dotted note	♩.
whole note	o

References

1. Briot, J.P., et al.: Deep Learning Techniques for Music Generation - A Survey. arXiv.org (2017)
2. Hochreiter, S., Schmidhuber, J.: Long short-term memory. Neural Comput. **9**(8), 1735–1780 (1997)
3. Huang, C.Z.A., et al.: Music transformer: generating music with long-term structure. In: ICLR (2019)
4. Jang, E., et al.: Categorical reparameterization with gumbel-softmax. In: ICLR (2017)
5. Jin, R., Si, L.: A study of methods for normalizing user ratings in collaborative filtering. In: SIGIR (2004)
6. Kingma, D., Ba, J.: Adam: a method for stochastic optimization. In: ICLR (2015)
7. Koren, Y., et al.: Matrix factorization techniques for recommender systems. Computer **42**(8), 30–37 (2009)
8. Lim, H., et al.: Chord generation from symbolic melody using BLSTM networks. In: ISMIR (2017)
9. Liu, H.M., Yang, Y.H.: Lead Sheet Generation and Arrangement by Conditional Generative Adversarial Network. arXiv.org (2019)
10. Pachet, F., Roy, P.: Non-conformant harmonization - the real book in the style of take 6. In: ICCC (2014)
11. Roy, P., et al.: Sampling Variations of Lead Sheets. arXiv.org (2017)
12. Yang, L.C., et al.: MidiNet - a convolutional generative adversarial network for symbolic-domain music generation. In: ISMIR (2017)

Bacher than Bach? On Musicologically Informed AI-Based Bach Chorale Harmonization

Alexander Leemhuis[1,2(✉)], Simon Waloschek[1,2], and Aristotelis Hadjakos[1,2]

[1] Center of Music and Film Informatics, Detmold University of Music,
Detmold, Germany
{a.leemhuis,s.waloschek,a.hadjakos}@cemfi.de
[2] OWL University of Applied Sciences and Arts, Lemgo, Germany
http://www.cemfi.de

Abstract. Writing chorales in the style of Bach has been a music theory exercise for generations of music students. As such it is not surprising that automatic Bach chorale harmonization has been a topic in music technology for decades. We suggest several improvements to current neural network solutions based on musicological insights into human choral composition practices. Evaluations with expert listeners show that the generated chorales closely resemble Bach's harmonization style.

Keywords: Bach chorale harmonization · Deep learning · Beam search

1 Introduction

Chorales by J.S. Bach traditionally play an important role in Western music education. Concise voice leading techniques and precepts such as the often quoted *prohibition of parallel fifths* make these chorales interesting as subject in music theory. But they are also interesting for computational music analysis and generation. Especially automatic harmonization of melodies, i.e., producing a four-part chorale given the soprano part, has been a topic for a long time.

In 1986, the first significant attempt was made: The CHORAL system [4] used over 270 hand-engineered rules for harmonization. Later, focus shifted from rule-based systems to neural networks [11,15]. In 2002, the usage of Recurrent Neural Networks (RNN) and Long Short-Term Memory cells (LSTM) [7] by Eck and Schmidhuber [5] specifically addressed the sequential nature of music and produced state-of-the-art results at that time. A decade later, statistical models like Hidden Markov models and Bayesian networks were developed [1,13,16]. Recent solutions such as BachBot [10] and DeepBach [6] again use LSTMs and incorporate metadata such as information on fermatas or metrical positions of notes to enhance the results.

Although various music theory concepts have been applied for evaluation of the resulting chorales, the actual human composition process has not yet been used for modeling neural networks. We therefore propose a Convolutional Neural

P. Cellier and K. Driessens (Eds.): ECML PKDD 2019 Workshops, CCIS 1168, pp. 462–469, 2020.
https://doi.org/10.1007/978-3-030-43887-6_39

Network (CNN) architecture that follows—to some extent—workflows that are documented and commonly recommended in music theory literature and taught in music theory classes for writing four-part chorales.

Expert listening tests with musicologists and music majors indicate that some of our generated harmonizations are more Bach-like than the originals, in the sense that they were believed to be the work of Bach even in direct comparison to the master's original harmonization of the same soprano part.

2 Musicologically Informed Harmonization

Contemporaries of the Baroque epoch as well as modern experts recommend to start four-part harmonization by elaborating a bass part given the soprano part, see for example [3,8]. The bass part is not only considered one of four equitable voices but also an indicator of the tonal skeleton: Once the bass line is determined, the structure of the chorale is mostly set. Only small leeway is left for the middle voices that are formed in a second step and can be very plain, solely blending into the harmonic progression [3, p. 255]. Telemann emphasizes in [14] that the alto part should be written before the tenor part so that the closest possible voicing can be accomplished. The advantages of generating the bass line first in generative systems have already been discussed [16].

Particular attention should be paid to the ends of musical phrases, typically marked by fermatas. Such phrases oftentimes end with rather canonical cadences and thus should be prepared in advance as Daniel suggests [3, p. 159]. Daniel also argues, that in many cases there is only one solution for a valid choice of alto and tenor notes [3, p. 256]. Therefore, sometimes during harmonization the choices of specific notes lead to dead ends in a sense that further voice development breaks common voice leading rules. These problems are commonly solved by simply going back and revising certain notes.

In summary, expert knowledge teaches us to use the following strategies when harmonizing Bach chorales:

- Generate the bass part first given the soprano part
- Support close voicings by choosing tenor notes after the alto
- Give enough context to allow for correct cadences
- Allow changes to previously generated notes

The following sections describe how these insights were integrated in our approach.

2.1 Data Processing and Augmentation

Symbolic score data is retrieved from and processed with the music21 [2] framework for Python. Besides offering various possibilities to process symbol music, it also includes a corpus with numerous chorales composed by Bach. To augment this dataset, all pieces are transposed up and down to different keys. Transpositions are limited in such way that no voice part exceeds the tonal range as used by Bach in order to ensure generation of "singable" results.

The smallest time unit used in Bach chorales is a semiquaver. Therefore we use a semiquaver time resolution to retain all information. For each time step, we compute one-hot vectors per part. The individual vectors can encode one of three slightly different events for each time step:

- **New note** If a new note starts at the given time step, its pitch is encoded.
- **Rest** Rests are handled as if they were notes with a special pitch value.
- **Continuation** In case that a note or rest is tied, i.e., not finished yet, we set a special continuation flag.

Additional score information (hereinafter called metadata) such as the current time position within a measure, the overall key of the choral, its time signature and the position of fermatas are also fed as one-hot vectors into the network.

We are aware, that Bach sometimes used the same melody to compose several different chorales. Therefore, it may happen that a specific melody has been present in the training as well as the test dataset due to random splitting of the dataset. Since the harmonizations in such cases are still different, we follow the practice of similar generative systems [6,10] and do not take this circumstance further into account.

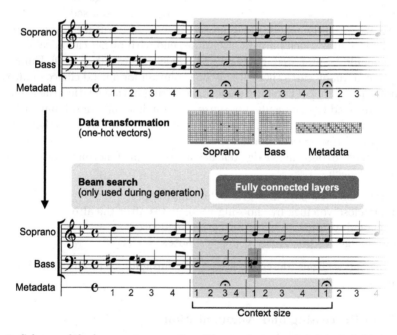

Fig. 1. Scheme of the bass part generation. The one-hot encoded data is fed into several fully connected layers to generate the output for a single time step. Afterwards, the context window is shifted by one step into the future. (Context size shown in blue is deliberately reduced compared to the actual implementation to enhance readability.) (Color figure online)

2.2   Network Architecture

Our proposed architecture features three similar consecutive networks. The first network creates a single bass note. It takes a frame of the soprano part, metadata and the prior bass notes as an input. After the entire bass line is generated, two networks are alternated to generate alto and tenor notes based on the soprano/bass part, metadata and the previously generated middle voices. Each part is generated using only a single hidden layer of size 650. Input and output layer dimensions are defined by the individual pitch range of each part. The output layers use softmax nonlinearities, all other layers use SELUs [9]. The ordering of note generation is as follows:

1. The entire bass line is generated first. A bass event b_i depends on the soprano and metadata in a local context of ± 32 time steps $s_{i-32:i+32}, m_{i-32:i+32}$ and 32 previous bass events $b_{i-32:i-1}$ (see Fig. 1). We use 32 steps (8 quarter notes) as a context as it provides a sufficient look ahead to prepare cadences as suggested by Daniel [3, p. 159]. The probability model for predicting b_i is thus

$$p(b_i | s_{i-32:i+32}, m_{i-32:i+32}, b_{i-32:i-1}).$$

2. After the bass line and thus the harmonic outline is completed, tenor and alto voice are generated from time step to time step. The alto prediction a_i is generated based on soprano and metadata context as above but with current and future bass events, which have been generated in the previous step, as well as previous alto events $a_{i-32:i-1}$ and tenor events $t_{i-32:i-1}$. The underlying probability model is thus

$$p(a_i | s_{i-32:i+32}, m_{i-32:i+32}, b_{i-32:i+32}, a_{i-32:i-1}, t_{i-32:i-1}).$$

3. The tenor is generated similar to the alto, but it also depends on the alto note generated in the current time step i, i.e., it depends on $a_{i-32:i}$:

$$p(t_i | s_{i-32:i+32}, m_{i-32:i+32}, b_{i-32:i+32}, a_{i-32:i}, t_{i-32:i-1}).$$

2.3   Beam Search

For every time step i, our network predicts the probabilities $p(b_i|\cdot)$, $p(a_i|\cdot)$ and $p(t_i|\cdot)$ conditioned on the local context. We want to find the sequence that maximizes the total probability, which is the product of the probabilities for each choice[1]

$$\prod_{i=0}^{N} p(b_i|\cdot) \prod_{i=0}^{N} p(a_i|\cdot) \, p(t_i|\cdot).$$

A greedy approach would select pitches with maximal probability at every prediction step. However, since future predictions depend on previous ones (see

---

[1] The multiplication is split in two parts to emphasize that the entire bass line is created first.

Sect. 2.2), always choosing the highest local probability option can lead to sub-optimal total probability of the sequence.

We therefore use *beam search* [12] to find solutions that help maximizing the total probability of the sequence. Beam search is a best-first search algorithm where only a fixed number of candidate alternatives are maintained to limit run-time and memory requirements. Previous work in Bach chorale harmonization has in fact suggested to use beam search, see [10]. Up to now this has, however, not been implemented and evaluated. Figure 2 provides a graphical example of how we employ beam search for generation of the bass part. The alto and tenor parts are generated in a similar manner.

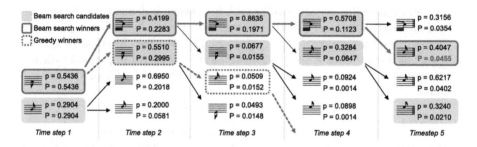

Fig. 2. Example of beam search for bass part with beam width of 2 in comparison to a greedy approach. P denotes the total probability of the branch, p denotes the conditional local probability.

3   Generation Results

At first glance, the chorales produced exhibit similarities to original Bach chorales. Two of the generated chorales were randomly chosen and given to Lydia Steiger, music theory teacher at the Detmold University of Music, for an in-depth musical analysis. She provided the following feedback:

- In several places voice leading rules were violated.
- The algorithm lacked sensitivity for musical tension and therefore sometimes choses a plain solution in places were a more sophisticated composition would have been more appropriate.
- The network uses common musical phrases used by J.S. Bach. In some places, the algorithm split these phrases arbitrarily across voices.

Further development of this approach should aim to address these shortcomings. All generated chorales of the test dataset and other pieces can be reviewed online at the project homepage[2], see Fig. 3 for a generation example.

---

[2] See http://www.cemfi.de/research/bachnet (accessed: 2019-09-03).

Fig. 3. Generated harmonization given the melody from "Ich ruf zu dir, Herr Jesu Christ" (BWV 177.5) by Bach.

4   Evaluation

We also evaluated our network with two online listening tests. The first test *without beam search* was conducted with the help of music majors. Thus we expect a high degree of familiarity with Bach chorale harmonization. The test presented paired samples consisting of (A) the original four-part chorale by Bach and (B) our generated harmonization using the same soprano part. Participants first had to give a self-assessment about their familiarity with Bach chorale harmonization and were then asked to identify the original Bach chorale for each pair. In case participants were unsure, question could be skipped. In 61% of the presented pairs, the participants could correctly identify the Bach work. 39% misjudged our generated pieces to be composed by Bach or skipped questions (see Fig. 4). Interestingly, 5 of 17 generated chorales could not be correctly identified by the majority of the participants.

After implementing beam search, we once more evaluated our solution. Since the results had subjectively improved, we decided to evaluate our network with participants that had an even greater expertise by directly addressing professional musicologists. Apart from the new harmonizations, the same online survey was used. Only 66% could distinguish the Bach pieces from the artificial ones, 34% chose the generated harmonizations or gave no answer. Although some of the musicology experts might be familiar with the *exact* Bach chorale, still 3 of 17 generated chorales were not correctly identified by more than 50% of the participants. One chorale was even preferred over the authentic work.

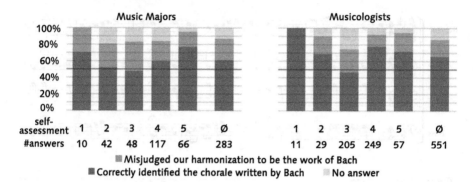

Fig. 4. Summarized results of the online evaluation. The chart shows how both participating groups scored in identifying the original Bach chorale given a generated harmonization as well as the master's work broken down by the self-assessment given. 5 corresponds to a high familiarity with Bach chorale harmonization, 1 corresponds to a low familiarity. 68 music majors and 127 musicologists participated.

To conclude this paper, we encourage future research on harmonization and automatic composition based on neural networks to take the human music creation process into account. The question, why several pieces sound more Bach-like than the original works even to experts could also be an interesting topic: What is it that deceives the listener and makes these chorales sound "bacher than Bach"?

References

1. Allan, M., Williams, C.: Harmonising chorales by probabilistic inference. In: Advances in Neural Information Processing Systems, vol. 17 (2005)
2. Cuthbert, M.S., Ariza, C.: music21: a toolkit for computer-aided musicology and symbolic music data (2010)
3. Daniel, T.: Der Choralsatz bei Bach und seinen Zeitgenossen: Eine historische Satzlehre. Verlag Dohr, Köln-Rheinkassel, Germany (2000)
4. Ebcioglu, K.: An expert system for chorale harmonization. In: Proceedings of the 5th AAAI National Conference on Artificial Intelligence (1986)
5. Eck, D., Schmidhuber, J.: A first look at music composition using LSTM recurrent neural networks. Instituto Dalle Molle di studi sull' intelligenza artificiale (2002)
6. Hadjeres, G., Pachet, F., Nielsen, F.: DeepBach: a steerable model for bach chorales generation. In: Proceedings of the 34th International Conference on Machine Learning (2017)
7. Hochreiter, S., Schmidhuber, J.: Long short-term memory. Neural Comput. **9**, 1735–1780 (1997)
8. Kaiser, U.: Der vierstimmige Satz: Kantionalsatz und Choralsatz, 4th edn. Bärenreiter, Kassel (2015)
9. Klambauer, G., Unterthiner, T., Mayr, A., Hochreiter, S.: Self-normalizing neural networks. In: Advances in Neural Information Processing Systems, vol. 30 (2017)

10. Liang, F.T., Gotham, M., Johnson, M., Shotton, J.: Automatic stylistic composition of bach chorales with deep LSTM. In: Proceedings of the 18th International Society for Music Information Retrieval Conference, Suzhou, China (2017)
11. Mozer, M.C., Soukup, T.: Connectionist music composition based on melodic and stylistic constraints. In: Advances in Neural Information Processing Systems, vol. 3 (1991)
12. Norvig, P.: Paradigms of Artificial Intelligence Programming: Case Studies in Common Lisp. Elsevier Science, Amsterdam (1992)
13. Suzuki, S., Kitahara, T.: Four-part harmonization using Bayesian networks: pros and cons of introducing chord nodes. J. New Music Res. **43**(3), 331–353 (2014)
14. Telemann, G.P.: Fast Allgemeines Evangelisch-Musicalisches Lieder-Buch. Philip Ludwig Stromer, Hamburg (1730)
15. Todd, P.M.: A connectionist approach to algorithmic composition. Comput. Music J. **13**(4), 27–43 (1989)
16. Whorley, R.P., Rhodes, C., Wiggins, G., Pearce, M.T.: Harmonising melodies: why do we add the bass line first? (2013)

Adaptively Learning to Recognize Symbols in Handwritten Early Music

Luisa Micó[✉], Jose Oncina, and José M. Iñesta

Department of Software and Computing Systems,
University of Alicante, Alicante, Spain
{mico,oncina,inesta}@dlsi.ua.es

Abstract. Human supervision is necessary for a correct edition and publication of handwritten early music collections. The output of an optical music recognition system for that kind of documents may contain a significant number of errors, making it tedious to correct for a human expert. An adequate strategy is needed to optimize the human feedback information during the correction stage to adapt the classifier to the specificities of each manuscript. In this paper, we compare the performance of a neural system, difficult and slow to be retrained, and a nearest neighbor strategy, based on the neural codes provided by a neural net, trained offline, used as a feature extractor.

1 Introduction

Optical Music Recognition (OMR) investigates how computers can read music notation in scores. Although research in printed modern notation has achieved good performances [5], the task becomes much harder when dealing with collections of early music handwritten scores. In particular, this work is applied to documents written in the Spanish white mensural notation system from the 16th and 17th centuries (see Fig. 1), for which a perfect recognition cannot be expected as the initial output of the OMR system [4].

The automatic pattern recognition approach has been traditionally focused on accomplishing a fully-automated operation. Nevertheless, in our approach, complete automation is not possible, although a perfect transcription of the original documents is needed for editing and publishing a collection. Therefore, we have to focus on the human-machine interaction tasks and how to optimize the expert user feedback loop [6].

The errors made by the system are usually seen as an issue outside the research process because correcting them is considered as the procedure for converting the system hypothesis into the desired result. However, semi-automatic approaches in which the human operator has the eventual responsibility of verifying and completing the task are the key to an efficient solution [7].

In this paper, we will study how using the user's corrections help the classifier to learn its model incrementally, decreasing the error throughout the task. Deep convolutional neural nets (DCNN) [3] are improving the state of the art

© Springer Nature Switzerland AG 2020
P. Cellier and K. Driessens (Eds.): ECML PKDD 2019 Workshops, CCIS 1168, pp. 470–477, 2020.
https://doi.org/10.1007/978-3-030-43887-6_40

Fig. 1. The kind of documents processed (here a fragment of a page) are single-voice vocal music written in the Spanish variant of the white mensural notation.

in computer image analysis tasks. Although there are works already that permit to modify these recognition models as a continuous learning process as new classes of data arrive [9], the task is still computationally demanding. We explore the possibility of combining the ability of DCNN for extracting good image features, with the simplicity of a nearest-neighbor (1-NN) classifier to adapt its performance to a specific training set along the edition stage.

2   Data Structure

The dataset for this study is a collection of pages $\mathcal{P} = \{P_1, P_2, ..., P_{|\mathcal{P}|}\}$ annotated with their ground-truth categories. This way, each page P_p can be considered as a training subset $\mathcal{X}^{(P_p)} = \{(\mathbf{x}_i, y_i)\}_{i=1}^{|P_p|}$, where the $\mathbf{x}_i$ represent the symbol bounding boxes in page P_p and y_i their corresponding labels.

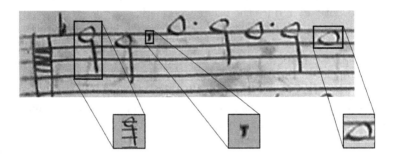

Fig. 2. Examples of bounding boxes for some symbols and how they are adapted to a 30×30 window in different situations. (Left and right:) fit to a 30×30 window, keeping aspect ratio and background padding; (center:) no bounding box stretching is done when it is smaller than the target window.

The symbol bounding boxes were extracted from the image and re-scaled to a 30×30-pixel window (only if the bounding box is bigger than that), keeping the aspect ratio of the box and padding the background with its maximum pixel value (see Fig. 2 for some examples).

When this window is the input for the DCNN, no additional processing is made, since the filters of the network input layer process the window regardless of its size. Nevertheless, the 1-NN will classify every window considering it as a vector $\mathbf{x} \in [0, 255]^{30 \times 30}$. Due to its sensitivity to the dimensionality of the feature space, we have transformed the windows by downsampling, keeping the central pixel of every non-overlapping 3×3 pixel area, assigning to it the mean of the 9 pixels involved. This is equivalent to low-pass filtering of the window, keeping the main features of the image in a smaller space ($[0, 255]^{10 \times 10}$).

3   Incremental Learning

The key point in this work is to study how can we adapt the recognition model to the data and do it incrementally. In real operation, a collection of scores is presented to the user by pages. Each page is processed, and the symbols are classified (see [1] for details). Then, the user makes corrections to the symbols that were incorrectly classified. This happens when the system hypothesis does not match the ground-truth label or when the symbol belongs to previously unseen classes. User corrections will be simulated. These interactions are utilized to improve the model for the classification of the next pages.

The recognition algorithm (the *model $\mathcal{M}$*) is a key issue in any pattern classification system, but in an interactive architecture, the most relevant feature is the ability of the algorithm to adapt to the specificities of the data through the error corrections made by the user. In the interactive paradigm, the efficient exploitation of *human expert knowledge* is the main objective, so the correctness of the system output is no longer the main issue to assess. The challenge now is the development of interactive schemes capable of efficiently exploiting the feedback to eventually reduce the user's workload.

In light of that, we have selected a very simple, but flexible, classification algorithm as the nearest neighbor is. It does not need a parametric analysis of the feature space for operation, and the training set $\mathcal{X}$ can be incrementally built by adding new pairs as they are found in the input in operation time: $\mathcal{X}^{(k+1)} = \mathcal{X}^{(k)} \bigcup \{(\mathbf{x}_i, y_i)\}_{i=1}^{N}$. Only an initial model $\mathcal{M}^{(1)}$, trained offline, is needed to start classifying. This model can be trained with the symbols on the first page $\mathcal{X}^{(P_1)}$, including the labels for the symbols on it, $y_i \in \mathcal{C}^{(P_1)}$, or with an initial subset of pages if the model needs more examples, as explained below.

Also, it is easy to add new classes dynamically by adding new labels, if needed. Besides, editing and condensing methods [8] can be easily applied to the training set if advised by the user corrections. The system must operate in real-time, so the user can interact with it comfortably. This is another feature that advises using simple, adaptive, and fast classification algorithms.

The algorithm outline is shown below (Algorithm 1). As explained, errors in a page P_p can be due to symbols belonging to unseen classes. In such a case, the interaction step includes the addition of the new class to the training set, with the symbols seen on the current page as prototypes. This algorithm will try to minimize the number of errors $\ell^{(p)}$, and so the need of user corrections, as the pages are processed.

Algorithm 1. Outline of the method

Input: A collection of pages $\mathcal{P} = \{P_1, P_2, ..., P_{|\mathcal{P}|}\}$
$\mathcal{X}^{(1)} = \mathcal{X}^{(P_1)}$
Train the model $\mathcal{M}^{(1)}$ with $\mathcal{X}^{(1)}$
for $p = 2$ **to** $|\mathcal{P}|$ **do**
    Apply $\mathcal{M}^{(p-1)}$ to samples in $\mathcal{P}_p$
    $\ell^{(p)} = |\{\mathbf{x}_i \in P_p \mid \hat{y}_i \neq y_i\}|$
    Interaction: the user fixes wrong $\hat{y}_i$ to actual y_i
    $\mathcal{X}^{(p)} = \mathcal{X}^{(p-1)} \bigcup \mathcal{X}^{(p)}$
    Train $\mathcal{M}^{(p)}$ with $\mathcal{X}^{(p)}$
end for

This algorithm does not change independently of the classification model utilized, $\mathcal{M}$. Only the $\mathcal{X}^{(1)}$ considered might be different, as explained below.

We want to explore a trade-off between accuracy in the classification and speed and flexibility in re-training the model. DCNN are state-of-the-art image classification methods, but the usual size of these models make their adaptation difficult and time-consuming. On the other hand, in many classical classification algorithms, like the 1-NN, the adaptation is straightforward, because it needs only updating the training set to adapt to a new situation in real-time (we consider as real-time any situation in which the user does not perceive that he or she has to wait for the system to make a decision).

Taking these considerations into account, we plan to compare three different classification models:

1. DCNN: the model $\mathcal{M}$ is a deep convolutional neural network. It is expected to achieve good performance (low error rates) but long retraining times.
2. 1-NN: $\mathcal{M}$ is a nearest neighbor classifier. Retraining and recognition can be done in real time, but higher error rates are expected.
3. NC+1NN: A DCNN learned on a subset of initial pages of $\mathcal{P}$ is used as a feature extractor (neural codes [2], NC) and the 1-NN is applied to the NC to implement the incremental classification described in the algorithm.

The network utilized is composed of 7 convolutional layers with 100 3×3 filters each. Then, a global max-pooling layer provides a $\mathcal{R}^{100}$ vector that will be the NC features for the nearest neighbor (in the case 3.) or fully connected to a layer with as many neurons as classes that will be classified with a softmax in the case of full DCNN classification (1.). The network architecture is displayed in Fig. 3. For training, Adam optimization has been used with a learning rate of 0.001 during 100 epochs, using minibatches of 64 images. Units have ReLU activations.

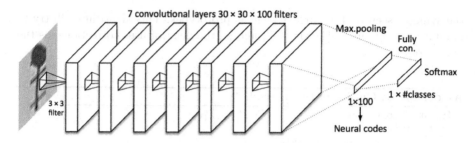

Fig. 3. Architecture of the DCNN used. Neural codes are the activations of the 100-neuron layer after pooling. For that, the last layer is removed and the neuron activations are the input to a nearest neighbor classifier.

4   Data and Results

We have selected a Mass in A minor from a collection of sacred vocal music from the 17th century. In this case, we have $|\mathcal{P}| = 124$ pages. Each page has a maximum of 6 staves of monophonic music, containing between 20 and 30 symbols each on average, for a total of 17,114 samples.

Initial Training Set. According to the instructions in Algorithm 1, the initial training set is $\mathcal{X}^{(1)} = \mathcal{X}^{(P_1)}$, but some considerations about this follow.

The number of prototypes in $\mathcal{X}^{(P_1)}$ is 143 from 23 classes. The C and F clefs received special consideration. Since the initial pages were written in the G clef, 6 prototypes of each of the other clefs were included in $\mathcal{X}^{(1)}$ from other composition. This way, $|\mathcal{X}^{(1)}| = 143 + 12$ prototypes from 25 classes.

When the DCNN is utilized, either as a classifier or for feature extraction, we need more data to train such a large structure properly. For that, an initial subset of the first 16 pages was considered: $\mathcal{X}^{(1)} = \bigcup_{p=1}^{16} \mathcal{X}^{(P_p)}$. This way, $|\mathcal{X}^{(1)}| = 2440 + 12$ prototypes from 44 classes. In this case, the Algorithm 1 runs for $p = 17$ to $|\mathcal{P}|$.

As the algorithm runs, the number of classes will increase. Each time an unseen class appears, it produces errors. As this page is included in the training set for the next step, the unknown class appears for the next iteration. The final number of classes is 53 for this composition.

Results. First, a study of the difference between with and without interaction is shown (see Fig. 4). The graph shows that, when the user corrections are used, a rapid drop in the error rate for the 1-NN is initially observed. The error rises again when another voice of the same composition begins to be processed. When the effect of the new pages is learned by the model, it is able to reduce the error again. Every time a change in the conditions happens the system degrades its performance, but it is able to continue learning later on, improving its performance.

On the contrary, when the training set remains fixed (using the 16 first pages for it), the error not only does not decay but tends to increase, because the system is not able to adapt to the specificities of the new pages.

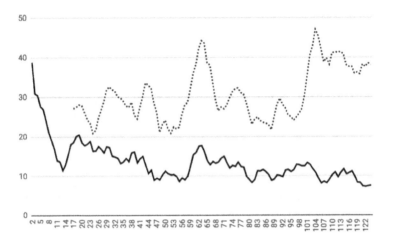

Fig. 4. Smoothed evolutions of the error rate (%) along the 124 pages of the studied composition, using 1-NN for classification. In solid line: error rate when using the interactive scheme. In dotted line: error rate without interaction, using a fixed training set.

One interesting feature is that pages from 60 to 65 have specificities not seen before. In that situation, both curves degrade. But when a similar situation occurs again from page 100 onwards, the incremental method is robust against that particular situation (solid line), while the non-incremental method is not and its performance degrades again.

Figure 5 shows the performance of the three recognition systems comparatively. When only the nearest neighbors are used (dotted line) the results are displayed from page $p = 2$, but when the nets are used they use the first 16 pages for initial training, so the results are displayed from $p = 17$.

The 1-NN adapts the best but performs the worst (an average error of 9.6% for the last 15 pages), but it runs in real time, without the user noticing any delay. On the other hand, the DCNN is able to reach a 1.1% of error at the end, but running the whole Algorithm 1 using that model took 7749 s in a GPU computer (note from Algorithm 1 that the network is retrained with the new training set after each page). The 1-NN applied to the neural codes was able to adapt to the data (less than using the 1-NN alone), reaching a nice 4.4% of error at the end. This approach works in real time, since the network learning is made only once, before starting the classification and adaptation.

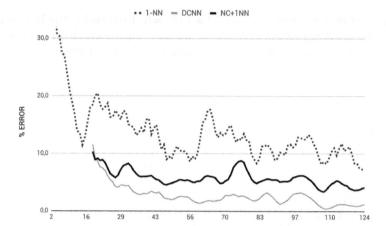

Fig. 5. Evolution of the error rates for the different methods considered. 1-NN (dotted): only nearest neighbors classifying the pixel values; DCNN (thin solid): only deep convolutional nets classifying the bounding boxes; NC + 1NN (thick solid): nearest neighbors classifying the extracted neural codes.

5   Conclusion

The presence of the expert user in the learning loop opens up new possibilities for study adaptive learning algorithms. The presented study shows that a combination of DCNN acting as a feature extractor and a nearest neighbor for classifying the extracted neural codes can provide a good trade off between precision and real-time operation. In any case, there are many ways left to be explored to make the human-in-the-loop approach efficient and effective.

Acknowledgements. Work supported by the Spanish Ministry HISPAMUS project TIN2017-86576-R, partially funded by the EU.

References

1. Calvo-Zaragoza, J., Rizo, D., Iñesta, J.M.: Two (note) heads are better than one: pen-based multimodal interaction with music scores. In: Proceedings of the 17th International Society for Music Information Retrieval Conference, pp. 509–514, August 2016
2. Calvo-Zaragoza, J., Gallego, A., Pertusa, A.: Recognition of handwritten music symbols with convolutional neural codes. In: 14th IAPR International Conference on Document Analysis and Recognition, ICDAR 2017, pp. 691–696 (2017)
3. Krizhevsky, A., Sutskever, I., Hinton, G.E.: ImageNet classification with deep convolutional neural networks. In: Neural Information Processing Systems, vol. 25, pp. 1097–1105 (2012)
4. Pacha, A., Calvo-Zaragoza, J.: Optical music recognition in mensural notation with region-based convolutional neural networks. In: 19th International Society for Music Information Retrieval Conference, pp. 240–247 (2018)

5. Rebelo, A., Fujinaga, I., Paszkiewicz, F., Marçal, A.R.S., Guedes, C., Cardoso, J.S.: Optical music recognition: state-of-the-art and open issues. Int. J. Multimedia Inf. Retrieval **1**(3), 173–190 (2012). https://doi.org/10.1007/s13735-012-0004-6

6. Sober-Mira, J., Calvo-Zaragoza, J., Rizo, D., Iñesta, J.M.: Pen-based music document transcription. In: Proceedings of the 12th IAPR International Workshop on Graphics Recognition, GREC (2017)

7. Toselli, A.H., Vidal, E., Casacuberta, F.: Multimodal Interactive Pattern Recognition and Applications, 1st edn. Springer, London (2011). https://doi.org/10.1007/978-0-85729-479-1

8. Wilson, D.R., Martinez, T.R.: Reduction techniques for instance-based learning algorithms. Mach. Learn. **38**(3), 257–286 (2000). https://doi.org/10.1023/A:1007626913721

9. Xiao, T., Zhang, J., Yang, K., Peng, Y., Zhang, Z.: Error-driven incremental learning in deep convolutional neural network for large-scale image classification. In: Proceedings of the 22nd ACM International Conference on Multimedia, MM 2014, pp. 177–186. ACM (2014)

Feature-Based Classification of Electric Guitar Types

Renato de Castro Rabelo Profeta$^{(\boxtimes)}$ and Gerald Schuller

Ilmenau University of Technology, PF 10 05 65, 98684 Ilmenau, Germany
{renato.profeta,gerald.schuller}@tu-ilmenau.de
https://www.tu-ilmenau.de/en/applied-media-systems-group/

Abstract. The classification of musical instruments of instruments of the same type is a challenging case of study. In this paper we conduct feature-based machine learning experiments to classify electric guitar recordings from different manufacturers and models. The Constant-Q Transform features and the Support Vector Machine algorithm obtained an accuracy of 95% in a binary classification task of guitars from two manufacturers, and 78% in a multiclass problem with four classes, distinguishing specific models from two different manufacturers.

Keywords: Musical instruments classification · Machine learning · Electric guitars · Music information retrieval

1 Introduction

Musical instruments recognition and sound characterization is a common subject of research in many different areas related to audio and music. Up to now several studies explored the identification of musical instruments of different classes (e.g. distinguishing a saxophone from a piano) but to the best of our knowledge, little research has been conducted to identify different types of the same instrument, for example distinguishing a Les Paul guitar from a Stratocaster model.

Johnson and Tzanetakis [1] studied the classification of guitars, mostly acoustic guitars, using hand-crafted features such as Mel-Frequency Cepstral Coefficients (MFCC), spectral moments, zero crossing rate and other features typically used in Musical Information Retrieval. They concluded that guitar models have unique sound characteristics that allows the use of classification algorithms such as k-Nearest Neighbours (kNNs) and Support Vector Machines (SVMs) to identify different guitar types. In their conclusion they state overall results with an accuracy just above 50%.

Setragno et al. [2] used feature-based analysis to investigate timbral characteristics of violins discerning two classes of violins (historical and modern) using machine learning techniques.

The goal of our study presented in this paper is to use feature-based representations and machine learning classification methodologies to classify different models of electrical guitars from two different manufacturers. A guitar model refers to a specific guitar type with certain design and construction characteristics chosen by its manufacturer.

P. Cellier and K. Driessens (Eds.): ECML PKDD 2019 Workshops, CCIS 1168, pp. 478–484, 2020.
https://doi.org/10.1007/978-3-030-43887-6_41

2   Methodology

2.1   Electric Guitars Audio Recordings Dataset

To populate the dataset we recorded single notes of 17 electric guitars: 6 Fender Stratocasters, 3 Fender Telecasters, 3 Epiphone Les Pauls, 1 Epiphone Casino, 1 Epiphone Dot and 3 Epiphone SGs. All recordings were performed by the same musician, using all pickup positions and playing notes in different frets of the neck. The single notes recorded include all open strings, and notes from an A minor scale starting at the fifth fret of the sixth string played using a 3 notes per string pattern.

The audio was acquired using the cleanest signal path possible, without any amplifiers or effects processors and using a sample rate of 44100 Hz and 16-bit depth. Table 1 gives a summary of all recorded models and the number of notes per model.

Table 1. Number of notes recorded from each electric guitar model

Brand	Model	No. of notes
EPIPHONE	LES PAUL STANDARD	202
EPIPHONE	SG	188
FENDER	SQUIER STRATOCASTER MOD. VINTAGE	183
FENDER	SQUIER STRATOCASTER MOD. SM	169
FENDER	AMERICAN STRATOCASTER STD HSS	168
FENDER	AMERICAN STRATOCASTER STD	162
FENDER	STRATOCASTER STD SPECIAL TREMOLO (MEX)	155
FENDER	STRATOCASTER STD (MEX)	152
FENDER	JAP. TELECASTER REI. '62 CUSTOM MOD	99
EPIPHONE	CASINO	96
EPIPHONE	DOT	95
EPIPHONE	LES PAUL STUDIO	94
EPIPHONE	SG PRO	92
FENDER	TELECASTER (MEX)	92
FENDER	SQUIER TELECASTER	92

In this paper we study two machine classification tasks: one classifying each note according to the manufacturer of the guitar model (two classes) and another using 4 classes in order to distinguish Fender Stratocasters, Fender other models, Epiphone Les Pauls and Epiphone other models.

The fact that there are more Fender guitars in the dataset and that the Fender Stratocaster has usually 5 pickup positions whereas the Epiphones have normally 3 pickup positions, results in the dataset to be highly imbalanced. To account for this problem we create two subsets by applying a random under-sampler from the Imbalanced-learn (version 0.4.2) [3] package for Python to under-sample the majority class to achieve an equal number of observations for each class. One subset is used for Binary Classification and another subset is used in the Multiclass Classification problem. We refer to an observation in this paper as a member of the dataset.

The total number of notes is 2040: 1273 notes from Fender guitars and 767 from Epiphone guitars. After random undersampling, the subset used for the binary classification experiment resulted in 767 notes from each guitar manufacturer. For the classification experiment with 4 classes, there are 990 notes from Fender Stratocasters, 296 notes from Epiphone Les Pauls, 471 from other models of Epiphone and 283 notes from Fender Telecasters resulting in a subset after the random under-sampling with 283 notes per class.

We split the subsets into training/testing sets with a 0.25 ratio of audio recordings in the testing set randomly preserving the percentage of observations for each class.

2.2 New Approach: Feature Analysis and Classification Algorithm

To perform the machine learning classification experiments we extracted features typically used in Music Information Retrieval (MIR) such as Mel-Frequency Cepstral Coefficients (MFCC), the Constant-Q Transform (CQT), and spectral features (centroid, bandwidth, contrast, flatness, kurtosis and skewness).

The audio files are loaded into Python 3.5 using Librosa (version 0.5.1) [4] keeping its original sample rate and bit-depth and are processed using overlapping frames with 2048 audio samples and a hop length of 1024, or 50% overlap.

A challenging point in these classification tasks is that we are dealing with audio recordings of different duration. It's a common practice in many cases to trim or zero-pad audio files in order to have a fixed number of audio samples for all audio recordings.

In our experiments we decided to extract the features from the audio recordings using all audio samples resulting in features vectors with a different number of frames. To have features vectors with the same number of features for all observations we average the features from frames that lie inside the same energy band. We refer to an energy band as frames that lie inside two Root-Mean-Square (RMS) energy percentage boundaries (e.g. RMSE percentage between 22% and 11%). The final features vector is a flattened version of the averaged features. Because we are dealing with frame-based calculations, the borders of the energy bands can vary slightly as we are taking the first element bigger than the calculated energy border value.

An example of this averaging procedure is described in Fig. 1. In this example we use an audio recording with a duration of 5.06 s, or 111505 audio samples. We calculate 84 MFCCs, using a frame length of 2048 samples and a hop length equals to 1024. This, results in a feature vector of 84 coefficients $\times$ 218 frames. We then compute 10 RMS energy bands and discard the frames inside the first and the last bands. We average the MFCCs of frames inside the same energy bands resulting in 84 coefficients $\times$ 8 frames. The final features vector used is a normalized flattened version with 672 elements.

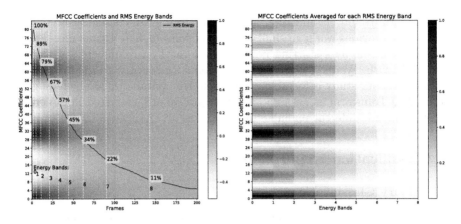

Fig. 1. RMS energy bands averaging procedure example.

We use three different sets of features: Spectral Features, MFCCs Features and the Constant-Q Transform Features. Librosa [4] is used to extract all features with the same frame length of 2048, a hop length of 1024 audio samples and the Hann window. Features are standardized to have zero mean and unit variance.

Spectral Features. The spectral features vector consists of a vector with the spectral centroid, spectral contrast, spectral bandwidth, spectral flatness, spectral kurtosis and spectral skewness calculated for each frame and averaged using the RMS energy bands averaging procedure as explained before. The result is a vector with 48 elements (6 spectral features × 8 energy bands) per observation.

Mel-Frequency Cepstral Coefficients. The MFCC features vector is extracted using 84 coefficients resulting in a features vector with 672 elements per observation.

Constant-Q Transform. The Constant-Q transform is calculated using a total of 84 bins, 12 bins per octave, resulting in a features vector with 672 elements per observation.

Classification Algorithm. The classification algorithm used is the SVM using a Polynomial kernel with degree equals to 3, the Penalty Parameter C equals to 0.01 and the kernel coefficient Gamma of 1.0. The classification experiment is run using Scikit-Learn (version 0.20.0) [5] library for Python.

The features and classification algorithm selected are chosen observing the results from previous works in similar tasks by Johnson and Tzanetakis [1], and Setragno et al. [2] as a starting point and all the hyperparameters are tuned empirically with tryouts from many experiments. In contrast to the works mentioned above, our study is focused solely in electric guitars, we are including the Constant-Q Transform as an audio feature, and we are interested in discerning models from two different manufacturers.

3   Classification Experiments

3.1   Binary Classification

The CQT features outperformed the other features vectors achieving an accuracy of 95%, predicting the correct guitar manufacturer 364 times from the total of 384 guitar recordings present in the test set (Table 2).

Table 2. Binary classification results

	Spectral		MFCC		CQT	
Evaluation	Epiphone	Fender	Epiphone	Fender	Epiphone	Fender
Recall	76.6%	72.4%	75.0%	84.4%	95.3%	94.3%
Precision	73.5%	75.5%	82.8%	77.1%	94.3%	95.3%
F1- Score	75.0%	73.9%	78.7%	80.6%	94.8%	94.8%
Accuracy	74.0% (286/384)		80.0% (306/384)		95.0% (364/384)	

3.2   Multiclass Classification

The Spectral and MFCC features performed poorly in a more complex task with 4 classes achieving less than 60% accuracy. The CQT features obtained an accuracy of 78%. We decided to include a features vector concatenating the CQT features vector with the Spectral features vector multiplied by a factor of 2 (giving it more weight), resulting in a vector with 720 elements (672 CQT features + 48 spectral features) and the accuracy improved slightly to 81%. Table 3 displays the results for the multiclass classification experiments.

Table 3. Multiclass classification results

Evaluation	Spectral Features			
	Epiphone Les Paul	Epiphone Others	Fender Stratocaster	Fender Others
Recall	57.7%	52.1%	47.1%	62.0%
Precision	54.3%	45.1%	63.5%	63.8%
F1- Score	54.3%	48.4%	54.1%	63.0%
Accuracy	55.0% (155/283)			
	MFCC Features			
Evaluation	Epiphone Les Paul	Epiphone Others	Fender Stratocaster	Fender Others
Recall	50.7%	42.3%	32.9%	62.0%
Precision	45.0%	41.1%	40.0%	61.1%
F1- Score	47.7%	41.7%	39.7%	61.1%
Accuracy	47.0% (133/283)			
	CQT Features			
Evaluation	Epiphone Les Paul	Epiphone Others	Fender Stratocaster	Fender Others
Recall	87.3%	74.5%	68.6%	80.3%
Precision	72.9%	79.1%	81.4%	79.2%
F1- Score	79.5%	76.8%	74.4%	79.7%
Accuracy	78.0% (220/283)			
	CQT + (Spectral * 2) Features			
Evaluation	Epiphone Les Paul	Epiphone Others	Fender Stratocaster	Fender Others
Recall	91.5%	71.8%	81.4%	80.3%
Precision	72.2%	83.6%	86.4%	86.4%
F1- Score	80.7%	77.3%	83.8%	83.2%
Accuracy	81.0% (230/283)			

4   Conclusion

In this study we conducted feature-based machine learning classification experiments to distinguish guitars from different manufacturers and models.

We recorded 17 different guitars from two manufacturers: Fender and Epiphone. We used three sets of features (spectral features, MFCCs and CQT) and one classification algorithm (SVM). The CQT features vector achieved an accuracy of 95% in a binary classification problem, correctly predicting the manufacturer of 364 guitar notes out of the total of 384. In the multiclass classification problem, the spectral and MFCC had a very poor performance, and a weighted concatenation of the CQT features with the spectral feature values multiplied by 2 obtained an accuracy of 81% with 230 correct predictions from 283 notes.

Due to the substantial difference between the datasets used, it is not possible to compare the results of our study with the previous studies mentioned. Johnson and Tzanetakis [1] studied mostly acoustic guitars and they addressed, among other things, the influence of the acoustics of the room where the acoustic guitars were recorded in their work. This issue is not present in our work with electric guitars.

It seems to us that common Music Information Retrieval features and standard machine learning classification algorithms have a good performance for binary classification of guitar notes from two quite distinctly sounding manufacturers and models. However, more complex tasks, such as distinguishing similar sounding guitars requires different techniques for features extraction and classification.

The dataset of electric guitars recordings is constantly being expanded with different models and manufacturers. This will allow more complex and reliable experiments. In future work we plan to use Deep Learning techniques and explore generative models for different applications and problems.

References

1. Johnson, D., Tzanetakis, G.: Guitar model recognition from single instrument audio recordings. In: IEEE Pacific Rim Conference on Communications, Computers and Signal Processing (PACRIM), Victoria, BC (2015)
2. Setragno, F., Zanoni, M., Antonacci, F., Sarti, A.: Feature-based characterization of historical and modern violins. In: International Symposium on Musical Acoustics, Montreal (2017)
3. Lemaitre, G., Nogueira, F., Aridas, C.: Imbalanced-learn: a Python toolbox to tackle the curse of imbalanced datasets in machine learning. J. Mach. Learn. Res. **18**, 1–5 (2017)
4. McFee, B., et al.: librosa: audio and music signal analysis in Python. In: Proceedings of the 14th Python in Science Conference, Austin, pp. 18–25 (2015)
5. Pedregosa, F., et al.: Scikit-learn: machine learning in Python. J. Mach. Learn. Res. **12**, 2825–2830 (2011)

RecurSIA-RRT: Recursive Translatable Point-Set Pattern Discovery with Removal of Redundant Translators

David Meredith$^{(\boxtimes)}$ ⓘ

Aalborg University, Aalborg, Denmark
dave@create.aau.dk
http://www.titanmusic.com
http://personprofil.aau.dk/119171

Abstract. We introduce two algorithms, RecurSIA and RRT, designed to increase the compression factor achievable using point-set cover algorithms based on the SIA and SIATEC pattern discovery algorithms. SIA computes the maximal translatable patterns (MTPs) in a point set, while SIATEC computes the translational equivalence class (TEC) of every MTP in a point set, where the TEC of an MTP is the set of translationally invariant occurrences of that MTP in the point set. In its output, SIATEC encodes each MTP TEC as a pair, $\langle P, V \rangle$, where P is the first occurrence of the MTP and V is the set of non-zero vectors that map P onto its other occurrences. RecurSIA recursively applies a TEC cover algorithm to the pattern P, in each TEC, $\langle P, V \rangle$, that it discovers. RRT attempts to remove translators from V in each TEC without reducing the total set of points covered by the TEC. When evaluated with COSIATEC, SIATECCompress and Forth's algorithm on the JKU Patterns Development Database, using RecurSIA with or without RRT increased compression factor and recall but reduced precision. Using RRT alone increased compression factor and reduced recall and precision, but had a smaller effect than RecurSIA.

Keywords: Pattern discovery · Point sets · Music analysis · Data compression · SIATEC · COSIATEC · SIATECCompress · Forth's algorithm · Geometric pattern discovery in music

1 Introduction

The principle of parsimony posits that, when given two models that account equally accurately for a given set of observations (data), then the simpler model is less likely to be an accurate description of the data by chance. That is, the simpler model is more likely to be a faithful representation of the true process that gave rise to the data. This principle, commonly known as "Ockham's razor", has been formalized in various ways in recent times, including Rissanen's minimal description length principle [17] and Kolmogorov's structure function [18]. The

P. Cellier and K. Driessens (Eds.): ECML PKDD 2019 Workshops, CCIS 1168, pp. 485–493, 2020.
https://doi.org/10.1007/978-3-030-43887-6_42

principle has been one of the foundational principles of scientific enquiry since antiquity and recent results in information theory [19] have shown that data compression is almost always the best strategy both for model selection and prediction.

In recent years, we have had some success in using compression-based point-set pattern discovery algorithms, such as COSIATEC [10,13,14,16], SIATE-CCOMPRESS [11,13,14] and Forth's algorithm [4,5], in conjunction with normalized compression distance, to carry out classification tasks such as folk song tune family detection [8,12,13]. Moreover, Louboutin and Meredith [8] found a highly significant correlation between compression factor and performance on the task of automatically discovering fugue subjects and countersubjects [6,7]. This motivates us to search for ways to improve the compression factor achieved by such algorithms in the hope that improving compression factor may also result in improved performance on a variety of musicological tasks. Our research programme is driven by the hypothesis that shorter encodings of data objects represent better ways of understanding those objects. We therefore strive to devise algorithms that compute encodings of musical data objects that are as parsimonious as possible.

Let D be a set of k–dimensional points, such that $D \subset \mathbb{R}^k$ and $|D| = n$. We call D a *dataset*. For any vector, $v \in \mathbb{R}^k$, the *maximal translatable pattern* (MTP) in D is defined as $\mathrm{MTP}(v, D) = D \cap (D - v)$. The SIA algorithm [15] computes all the non-empty MTPs in such a dataset in $\Theta(n^2 \log_2 n)$ time. Two point sets, P_1, P_2, are *translationally equivalent*, denoted by $P_1 \equiv_\mathrm{T} P_2$, if and only if there exists a vector, v, such that $P_1 = P_2 + v$. The translational equivalence relation partitions the powerset of D exhaustively and exclusively into *translational equivalence classes* (TECs), such that the TEC to which a point set, $P \subseteq D$, belongs is defined to be $\mathrm{TEC}(P) = \{Q \mid Q \subseteq D \wedge Q \equiv_\mathrm{T} P\}$. The SIATEC algorithm [15] computes the TEC of every non-empty MTP in a dataset, D, in $\Theta(n^3)$ time. A TEC, $\mathrm{TEC}(P)$, can be encoded in a compressed form as a pair, $\langle P, V \rangle$, where V is the set of non-zero vectors, $\{v \mid P + v \subseteq D\}$. Each TEC in the output of SIATEC is encoded in this form. Given a TEC, $T = \mathrm{TEC}(P) = \langle P, V \rangle$, we define $P(T) = P$ and $V(T) = V$. $P(T)$ is called the TEC's *pattern* and $V(T)$ is called the TEC's *translator set* or *set of translators*. The *covered set* of a TEC, T, is the union of the point sets in the TEC and is given by $C(T) = P \cup \bigcup_{v \in V(T)} (P(T) + v)$. The *compression factor* of a TEC, $T = \mathrm{TEC}(P) = \langle P, V \rangle$ is defined as $\mathrm{CF}(T) = |C(T)| / (|P(T)| + |V(T)|)$. It is the ratio of $|C(T)|$, the number of points whose coordinates need to be explicitly specified if the covered set of the TEC is described *in extenso*, to $|P(T)| + |V(T)|$, the number of points and vectors whose coordinates need to be specified if the TEC is encoded as a pair, $\langle P, V \rangle$, as defined above.

SIATECCOMPRESS and Forth's algorithm use SIATEC to compute the MTP TECs in a dataset, D, and then attempt, using a greedy strategy, to select a subset of these TECs, E, such that $\bigcup_{T \in E} C(T) = D$ and $\sum_{T \in E} (|P(T)| + |V(T)|)$ is minimized. That is, these algorithms attempt to find a minimum-length description of the dataset in terms of a cover constructed from

TEC covered sets. The TEC covered sets in the covers computed by SIATEC-COMPRESS and Forth's algorithm may share points. However, the COSIATEC algorithm typically achieves better compression than these algorithms by partitioning the input dataset exhaustively and exclusively into non-intersecting TEC covered sets. It does this by incrementally constructing an encoding, E, by (1) running SIATEC, (2) adding the TEC with the best compression factor to E, (3) removing the covered set of this TEC from D and then repeating this three-step process on progressively smaller, unencoded subsets of the dataset until all the points in the dataset have been covered.

```
RECURSIA(𝒜, D)
1    E ← 𝒜(D)
2    if |E| = 1 ∧ |E[0][1]| = 1 return E
3    for i ← 0 to |E| − 1
4        e ← RECURSIA(𝒜, E[i][0])
5        if |e| > 1 ∨ |e[0][1]| > 1
6            E[i][0] ← e
7    return E
```

Fig. 1. The RECURSIA algorithm

In this paper, we introduce two novel techniques for improving the compression factor achieved using TEC cover algorithms. First, an algorithm, RECURSIA, is presented, that recursively applies a TEC cover algorithm to the pattern, P, in each TEC in the cover it generates. Second, an approximation algorithm, RRT, is presented, that aims to remove as many translators from each TEC as possible without removing points from its covered set. The two techniques are evaluated separately and in combination on the effect that they have on compression factor, recall and precision, when used with COSIATEC, SIATECCOMPRESS and Forth's algorithm on the JKU Patterns Development Database [2].

2   The RECURSIA Algorithm

Figure 1 gives pseudocode for the RECURSIA algorithm. RECURSIA has two parameters, a TEC cover algorithm, $\mathcal{A}$ (e.g., COSIATEC, SIATECCOMPRESS or Forth's algorithm) and a dataset D. RECURSIA runs $\mathcal{A}$ on D to obtain an *encoding*, $\mathbf{E}$ (line 1 in Fig. 1), which is a list of TECs, $\mathbf{E} = \langle T_1, T_2, \ldots, T_{|\mathbf{E}|} \rangle$. Each TEC, T_i, is encoded as a pair, $\langle P_i, V_i \rangle$, as defined above. If the encoding, $\mathbf{E}$, contains only one TEC and the pattern for this TEC has only one occurrence, then $\mathcal{A}$ failed to find any non-trivial MTPs in D. In this case, $\mathcal{A}$ is not applied to the pattern in this TEC, so RECURSIA returns $\mathbf{E}$ (see line 2 in Fig. 1). If $\mathcal{A}$ finds more than one TEC or at least one TEC whose pattern has more than one occurrence, then RECURSIA is applied recursively to the pattern, $P_i = \mathbf{E}[i][0]$, in each TEC in $\mathbf{E}$ (Fig. 1, lines 3–4). This generates a new encoding, $\mathbf{e}_i$, for each pattern, P_i. If the encoding, $\mathbf{e}_i$, for a pattern, P_i, contains more than one TEC, or a TEC whose pattern occurs more than once, then $\mathbf{e}_i$ is a *compressed* encoding of P_i and $\mathbf{e}_i$ replaces P_i in the TEC, $\mathbf{E}[i]$ (Fig. 1, lines 5–6).

3   The RRT Algorithm

Given a TEC, $T = \text{TEC}(P) = \langle P, V \rangle$, the RRT algorithm attempts to replace V with one of the smallest possible subsets of V—let us call it V'—such that $C(\langle P, V' \rangle) = C(T)$, where $C(T)$ denotes the covered set of T, as defined above. Exhaustively testing every subset of V to determine if the resulting covered set is the same as $C(T)$ would take time exponential in the size of V and would therefore only be practical for relatively small translator sets. RRT therefore uses a greedy approximation strategy with a polynomial time complexity instead of carrying out an exhaustive search.

```
RRT(T)
1    F ← COMPUTEPOINTFREQSET(T)
2    if F[|F| − 1][0] = 1 return T
3    S ← COMPUTESIAMVECTORTABLE(T, F)
4    R ← COMPUTEREMOVABLEVECTORS(T, S)
5    M ← COMPUTEMAXPOINTS(T, R, F)
6    if M = ∅ then T[1] \← R, return T
7    V ← COMPUTEVECTORMAXPOINTSETPAIRS(M)
8    Q ← COMPUTERETAINEDVECTORS(V)
9    return REMOVEREDUNDANTVECTORS(T, Q, R)
```

Fig. 2. The RRT algorithm

Figure 2 provides pseudocode for the RRT algorithm. For convenience, we define the function $V(p, T)$ to be the set of vectors in $V(T)$ that map points in $P(T)$ onto the point p. Formally,

$$V(p, T) = \{p - q \mid p - q \in V(T) \land q \in P(T)\}. \tag{1}$$

The first step in the algorithm is to compute for each $p \in C(T)$ the ordered pair $\langle f(p, T), p \rangle$, where $f(p, T) = |V(p, T)|$. These ordered pairs are placed in a sequence in lexicographical order and stored in the variable, $\mathbf{F}$ (Fig. 2, line 1). We call $f(p, T)$ the *frequency* of p in T. For example, for the TEC,

$$\langle \{\langle 1, 1 \rangle, \langle 2, 2 \rangle, \langle 3, 3 \rangle\}, \{\langle 0, 0 \rangle, \langle 1, 1 \rangle, \langle 2, 2 \rangle, \langle 3, 3 \rangle, \langle 4, 4 \rangle\} \rangle \tag{2}$$

the COMPUTEPOINTFREQSET function would return

$$\langle \langle 1, \langle 1, 1 \rangle \rangle, \langle 1, \langle 7, 7 \rangle \rangle, \langle 2, \langle 2, 2 \rangle \rangle, \langle 2, \langle 6, 6 \rangle \rangle, \langle 3, \langle 3, 3 \rangle \rangle, \langle 3, \langle 4, 4 \rangle \rangle, \langle 3, \langle 5, 5 \rangle \rangle \rangle.$$

If, for some $p \in C(T)$, $f(p, T) > 1$, then we call p a *multipoint*. If $\mathbf{F}$ contains no multipoints, then none of the translators in $V(T)$ can be removed without also removing points from $C(T)$. This will be the case if and only if the frequency of the last entry in $\mathbf{F}$ is one. We therefore check for this in line 2 of Fig. 2 and return the TEC unchanged if it is the case.

The set of translators that can be removed from $V(T)$ is a subset of those vectors that map the whole pattern, $P(T)$, onto multipoints. That is, if a translator, $v \in V(T)$, maps any point in $P(T)$ onto a point in $C(T)$ that is not a multipoint, then we know that v cannot be removed from $V(T)$ without removing points from $C(T)$. We therefore define a *removable vector* to be a translator that maps the TEC's entire pattern, $P(T)$, onto a set of multipoints. In lines 3–4 of Fig. 2 we compute a list, $\mathbf{R}$, of these removable vectors. This is done by using the initial steps of the SIAM algorithm [10,20] to compute the set, $S = \{\langle q - p, p \rangle \mid p \in P(T) \wedge q \in C(T) \wedge f(q,T) > 1\}$. This set S or *vector table* is sorted lexicographically to give the list, $\mathbf{S}$, (line 3 in Fig. 2) from which the maximal matches of the TEC pattern, $P(T)$, to the multipoints in $C(T)$ can be obtained. For example, for the TEC in Eq. 2, COMPUTESIAMVECTORTABLE returns the following sorted SIAM vector table, where each maximal match is printed on its own line:

$$
\begin{aligned}
&\langle\langle\langle -1, -1 \rangle, \langle 3, 3 \rangle\rangle, \\
&\langle\langle 0, 0 \rangle, \langle 2, 2 \rangle\rangle, \langle\langle 0, 0 \rangle, \langle 3, 3 \rangle\rangle, \\
&\langle\langle 1, 1 \rangle, \langle 1, 1 \rangle\rangle, \langle\langle 1, 1 \rangle, \langle 2, 2 \rangle\rangle, \langle\langle 1, 1 \rangle, \langle 3, 3 \rangle\rangle, \\
&\langle\langle 2, 2 \rangle, \langle 1, 1 \rangle\rangle, \langle\langle 2, 2 \rangle, \langle 2, 2 \rangle\rangle, \langle\langle 2, 2 \rangle, \langle 3, 3 \rangle\rangle, \\
&\langle\langle 3, 3 \rangle, \langle 1, 1 \rangle\rangle, \langle\langle 3, 3 \rangle, \langle 2, 2 \rangle\rangle, \langle\langle 3, 3 \rangle, \langle 3, 3 \rangle\rangle, \\
&\langle\langle 4, 4 \rangle, \langle 1, 1 \rangle\rangle, \langle\langle 4, 4 \rangle, \langle 2, 2 \rangle\rangle, \\
&\langle\langle 5, 5 \rangle, \langle 1, 1 \rangle\rangle\rangle
\end{aligned}
\tag{3}
$$

The COMPUTEREMOVABLEVECTORS function (Fig. 2, line 4) scans this sorted SIAM vector table to identify the vectors that map the entire pattern onto multipoints (i.e., the ones for which the maximal matches have the same cardinality as the TEC pattern itself). For the TEC in Eq. 2, the list $\mathbf{R}$ returned by COMPUTEREMOVABLEVECTORS would be $\langle\langle 1, 1 \rangle, \langle 2, 2 \rangle, \langle 3, 3 \rangle\rangle$.

We say that $p \in C(T)$ is a *maxpoint* if and only if all the vectors in $V(p, T)$ (as defined in Eq. 1) are removable vectors, i.e., $V(p, T) \subseteq \mathbf{R}$. If $C(T)$ contains any maxpoints, then it will not be possible to remove all the vectors in $\mathbf{R}$ from $V(T)$ without also removing the maxpoints from the covered set. Indeed, we can remove all the vectors in $\mathbf{R}$ from $V(T)$ if and only if $C(T)$ contains no maxpoints. In line 5 of Fig. 2, the maxpoints are computed and then, in line 6, if there are no maxpoints, all the removable vectors, $\mathbf{R}$, are removed from the TEC's translator set and the modified TEC is returned. The COMPUTEMAXPOINTS function, called in line 5 of the RRT algorithm (line 5 in Fig. 2) actually returns a set of ordered pairs, $M = \{\langle p_1, R_1 \rangle, \langle p_2, R_2 \rangle, \ldots, \langle p_{|M|}, R_{|M|} \rangle\}$, where each $\langle p_i, R_i \rangle$ gives the maxpoint, p_i, and the set of removable vectors, R_i, that map pattern points onto that maxpoint. As an example, the TEC in Eq. 2 has just one maxpoint, so the COMPUTEMAXPOINTS function returns the following: $\{\langle\langle 4, 4 \rangle, \{\langle 1, 1 \rangle, \langle 2, 2 \rangle, \langle 3, 3 \rangle\}\rangle\}$.

If $C(T)$ contains maxpoints, then our goal is to find the smallest subset of $\mathbf{R}$ that contains, for each maxpoint, at least one vector that maps a point in $P(T)$ onto that maxpoint. We first compute a list of $\langle v, P \rangle$ pairs that give, for each removable vector, v, the set of maxpoints, P, onto which v maps points in the TEC pattern, $P(T)$. This is computed by the COMPUTEVECTORMAXPOINTSETPAIRS function in line 7 of the RRT algorithm in Fig. 2. Formally, COMPUTEVECTORMAXPOINTSETPAIRS computes the set, V, defined as follows: $V = \{\langle v, P \rangle \mid v \in \mathbf{R} \wedge P = \{p \mid p \in M \wedge p - v \in P(T)\}\}$. This set is then sorted to give an ordered set, $\mathbf{V}$, so that the $\langle v, P \rangle$ pairs are in decreasing order of maxpoint set size (i.e., pairs in which P is larger appear earlier in the list).

```
COMPUTERETAINEDVECTORS(V)
1    Q ← ∅
2    while V ≠ ⟨⟩
3        Q ← Q ∪ {V[0][0]}
4        for i ← 1 to |V| − 1 do V[i][1] ← V[i][1] \ V[0][1]
5        Y ← ⟨⟩
6        for i ← 1 to |V| − 1
7            if V[i][1] ≠ ∅ then Y ← Y ⊕ ⟨V[i]⟩
8        V ← Y
9    return Q
```

Fig. 3. The COMPUTERETAINEDVECTORS function. ($\mathbf{A} \oplus \mathbf{B}$ concatenates the lists $\mathbf{A}$ and $\mathbf{B}$.)

We then use $\mathbf{V}$ in a greedy strategy to find a small subset of $\mathbf{R}$ that contains, for each maxpoint, at least one vector that maps a point in $P(T)$ onto that maxpoint. This set of *retained vectors* is computed in line 8 of Fig. 2 by the COMPUTERETAINEDVECTORS function (shown in Fig. 3). The first step in this function is to add to the list of retained vectors, $\mathbf{Q}$, the vector associated with the largest set of maxpoints, that is, the first in the list $\mathbf{V}$ (see lines 1–3 of Fig. 3). All the maxpoints mapped to by that vector from points in the TEC pattern can then be removed from the maxpoint sets of the other elements in $\mathbf{V}$ (line 4 in Fig. 3). The effect of lines 5–8 of Fig. 3 is to remove from $\mathbf{V}$ the first element and every other element whose maxpoint set is empty after removing the maxpoint set of the first element. The process is repeated, with the vector of the first pair in the list being selected on each iteration until $\mathbf{V}$ is empty. This results in a list, $\mathbf{Q}$, of retained vectors that constitute a subset of the removable vectors that is sufficient to generate all the maxpoints. Finally, in line 9 of Fig. 2, the REMOVEREDUNDANTVECTORS function removes from the TEC's set of translators all removable vectors that are not retained vectors.

4   Evaluation

Figure 4(a) shows the effect of RECURSIA and RRT on the compression factor achieved using a variety of SIATEC-based TEC cover algorithms, when these algorithms were used to analyse the five pieces in the JKU Patterns Development Database [2]. Three basic algorithms, COSIATEC, SIATECCOMPRESS and Forth's algorithm were run, each with and without compactness trawling [3] (indicated by 'CT') and with or without the SIA algorithm replaced by SIAR [1] (indicated by 'R'). Each of these 12 algorithms was run in its basic form (orange curve), with RECURSIA (blue curve), with RRT (green curve), and with both RECURSIA and RRT (red curve). As expected, using RECURSIA and RRT together nearly always improved compression factor, with particularly large gains being observed on the Beethoven and Mozart sonata movements when Forth's algorithm was used with compactness trawling. Using RRT alone only had a noticeable effect on the Bach fugue and the Beethoven sonata movement. Over all pieces and algorithms, using RECURSIA in combination with RRT improved compression factor by 12.5%, using RECURSIA alone improved it by

9.2% and using RRT alone improved it by 2.1%. Figure 4(b) shows the effect that RecurSIA and RRT had on three-layer precision (TLP) [13], averaged over the pieces in the JKU-PDD and for the same 12 algorithms, each run in "Raw" mode, "BB" mode and "Segment" mode (see [13]). On average, over all pieces, algorithms and modes, using RecurSIA in combination with RRT reduced TLP by 20.3%, using RecurSIA alone reduced it by 21.2% and using RRT alone reduced it by 0.7% (see Fig. 4(b)). On the other hand, on average, over all pieces, algorithms and modes, using RecurSIA and RRT together increased three-layer recall (TLR) [13] by 7.2%, using RecurSIA alone increased it by 10.3%. Using RRT alone reduced TLR by 3.7% (see Fig. 4(c)).

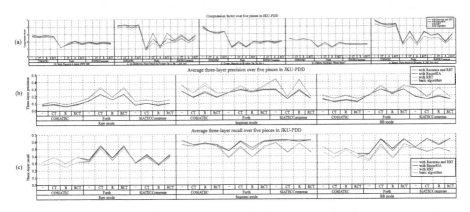

Fig. 4. Effect of RecurSIA and RRT on compression factor (a), three-layer precision (b) and recall (c), over the pieces in the JKU-PDD. (Color figure online)

5 Conclusion

Two algorithms, RecurSIA and RRT, have been presented, designed to increase the compression factor achieved using any TEC cover algorithm. When tested with three basic algorithms and evaluated on the JKU Patterns Development database, using RecurSIA with or without RRT increased compression factor and three-layer recall but reduced three-layer precision. Using RRT alone generally had a smaller effect than using RecurSIA, and, on average, increased compression factor but reduced both recall and precision on the JKU-PDD.

Supplementary Materials
The results reported in this paper were obtained using the implementations of the algorithms in the OMNISIA software [9]. The source code for the version of OMNISIA used here is available on GitHub at https://github.com/chromamorph/omnisia-recursia-rrt-mml-2019. An executable JAR file is also available at http://www.titanmusic.com/software/omnisia/201904151348OMNISIA.zip.

Acknowledgements. The author would like to thank Geraint A. Wiggins for suggesting the idea of applying the COSIATEC algorithm recursively to the patterns in TECs.

References

1. Collins, T.: Improved methods for pattern discovery in music, with applications in automated stylistic composition. Ph.D. thesis, Faculty of Mathematics, Computing and Technology, The Open University, Milton Keynes (2011)
2. Collins, T.: JKU Patterns Development Database (2013). https://dl.dropbox.com/u/11997856/JKU/JKUPDD-Aug2013.zip
3. Collins, T., Thurlow, J., Laney, R., Willis, A., Garthwaite, P.H.: A comparative evaluation of algorithms for discovering translational patterns in baroque keyboard works. In: 11th International Society for Music Information Retrieval Conference (ISMIR 2010), Utrecht, The Netherlands, 9–13 August 2010, pp. 3–8 (2010)
4. Forth, J.C.: Cognitively-motivated geometric methods of pattern discovery and models of similarity in music. Ph.D. thesis, Department of Computing, Goldsmiths, University of London (2012)
5. Forth, J., Wiggins, G.A.: An approach for identifying salient repetition in multidimensional representations of polyphonic music. In: Chan, J., Daykin, J.W., Rahman, M.S. (eds.) London Algorithmics 2008: Theory and Practice, pp. 44–58. College Publications, London (2009)
6. Giraud, M., Groult, R., Levé, F.: Subject and counter-subject detection for analysis of the Well-Tempered Clavier fugues. In: Aramaki, M., Barthet, M., Kronland-Martinet, R., Ystad, S. (eds.) CMMR 2012. LNCS, vol. 7900, pp. 422–438. Springer, Heidelberg (2013). https://doi.org/10.1007/978-3-642-41248-6_24
7. Giraud, M., Groult, R., Levé, F.: Truth file for the analysis of Bach and Shostakovich fugues (2013/12/27 version) (2013). http://www.algomus.fr/truth/fugues.truth.2013.12
8. Louboutin, C., Meredith, D.: Using general-purpose compression algorithms for music analysis. J. New Music. Res. **45**(1), 1–16 (2016)
9. Meredith, D.: Omnisia. http://www.titanmusic.com/software/omnisia/OMNISIA.pptx
10. Meredith, D.: Point-set algorithms for pattern discovery and pattern matching in music. In: Dagstuhl Seminar on Content-based Retrieval (No. 06171, 23–28 April 2006). Schloss Dagstuhl, Germany (2006). http://drops.dagstuhl.de/opus/volltexte/2006/652
11. Meredith, D.: COSIATEC and SIATECCompress: pattern discovery by geometric compression. In: MIREX 2013 (Competition on Discovery of Repeated Themes & Sections) (2013). http://www.titanmusic.com/papers/public/Meredith MIREX2013.pdf
12. Meredith, D.: Using point-set compression to classify folk songs. In: Fourth International Workshop on Folk Music Analysis (FMA 2014), 12–13 June 2014. Bogazici University, Istanbul (2014)
13. Meredith, D.: Music analysis and point-set compression. J. New Music. Res. **44**(3), 245–270 (2015)
14. Meredith, D.: Analysing music with point-set compression algorithms. In: Meredith, D. (ed.) Computational Music Analysis, pp. 335–366. Springer, Cham (2016). https://doi.org/10.1007/978-3-319-25931-4_13

15. Meredith, D., Lemström, K., Wiggins, G.A.: Algorithms for discovering repeated patterns in multidimensional representations of polyphonic music. J. New Music. Res. **31**(4), 321–345 (2002)
16. Meredith, D., Lemström, K., Wiggins, G.A.: Algorithms for discovering repeated patterns in multidimensional representations of polyphonic music. In: Cambridge Music Processing Colloquium (2003). http://www.titanmusic.com/papers/public/cmpc2003.pdf
17. Rissanen, J.: Modeling by shortest data description. Automatica **14**(5), 465–471 (1978)
18. Vereshchagin, N.K., Vitányi, P.M.B.: Kolmogorov's structure functions and model selection. IEEE Trans. Inf. Theory **50**(12), 3265–3290 (2004)
19. Vitányi, P.M.B., Li, M.: Minimum description length induction, Bayesianism and Kolmogorov complexity. IEEE Trans. Inf. Theory **46**(2), 446–464 (2000)
20. Wiggins, G.A., Lemström, K., Meredith, D.: SIA(M)ESE: an algorithm for transposition invariant, polyphonic content-based music retrieval. In: Proceedings of the Third International Conference on Music Information Retrieval (ISMIR 2002), Paris, France, 13–17 October 2002, pp. 283–284 (2002)

Bow Gesture Classification to Identify Three Different Expertise Levels: A Machine Learning Approach

David Dalmazzo[✉] and Rafael Ramírez[✉]

Universitat Pompeu Fabra, 08018 Barcelona, Spain
{david.cabrera,rafael.ramirez}@upf.edu

Abstract. To acquire new skills in a high-level music context, students need many years of conscious dedication and practice. It is understood that precise motor actions have to be incorporated into the musicians' automatic executions, where a repertoire of technical actions must be learned and mastered. In this study, we develop a computer modelled assistant applying machine learning algorithms, for self-practice musicians with the violin as a test case. We recorded synchronized data from the performer's forearms implementing an IMU device with ambient sound recordings. The musicians perform seven standard bow gesture. We tested the model with three different expertise levels to identify relevant dissimilitudes among students and teachers.

Keywords: Machine learning · Music education · Hidden Markov Model

1 Introduction

1.1 Motivation

To become an expert performer in the context of music education is not only needed natural attitudes, as well, many years of conscious practice. It is understood that specific fine-motor actions must become part of the automatic execution (system 1) [10] in other words, a *"learned technique of the body"* [3], known as musical gesture, has to be developed and incorporated through precise practice and repetition. The standard strategy behind new skills development is based on the coupling of sound qualities, expressiveness and motor executions. However, the standard master-apprentice educative model based in imitation by example has some weaknesses, where the students could develop bad habits in self-practising hours. Therefore, in the context of Telmi (Technology Enhanced Learning of Musical Instrument Performance), we are investigating the implications of applying a computer modelled assistant to novice students, particularly at the moment to acquire new skills practising standard classical gestures

Music Technology Group, and Spanish Ministry of Economy and Competitiveness under the Maria de Maeztu Units of Excellence Programme (MDM-2015-0502).

P. Cellier and K. Driessens (Eds.): ECML PKDD 2019 Workshops, CCIS 1168, pp. 494–501, 2020.
https://doi.org/10.1007/978-3-030-43887-6_43

with the test case of violin performers. We intend to stretch the gap of *"good-practice"* feedback, providing immediate information about gestural executions in real-time.

1.2   Gesture Recognition in Musical Context

To address the first stage of recognising specific gestures executions, we implemented Machine Learning (ML) techniques broadly found in the literature such a Hidden Markov Models (HMM) [2].

Bevilacqua et al. [1] presented a study in which an HMM system reports gesture time-progressions and its likelihood windowing. The ML model can be adjusted in states; which estimates Gaussian probabilities inside gesture progressions. Authors are not focused on specific gestural analysis; instead, they presented an optimal "low-cost" algorithm without the need for big datasets. Fiebrink and Cook [6] introduced the open-source multi-platform application called Wekinator, which includes a set of ML algorithms for pattern classifications, as well, dynamic time warping algorithms for time-related events. The tool is broadly used in academics and workshops for prototyping, artistic interactive music applications or as an educative reference of ML applicability in research topics. Fiebrink et al. [7] Executed the Wekinator to analyze bow-stroke articulations in a cello player. Authors embedded an IMU device in the bow-frog called K-Bow. The main goal was to allow the performer to interact in real-time through the gestures with a compositional computer-assistant. Françoise et al. [8,9] First exposed a gestural descriptor applying HMM and introduced the concept of mapping-by-demonstration as a principle of teaching with small amount of data the ML algorithms to then be used in the context of music education or real-time music interaction. In the next publication, authors describe probabilistic models such as Gaussian Mixture Models (GMM), Gaussian Mixture Regression (GMR), Hierarchical HMM (HHMM) and Multimodal Hierarchical HMM (MHMM). Dalmazzo and Ramirez [4] Based on IMU device and EMG data recorded from left-hand violinist players, authors estimated fingering disposition in the violin's neck. Two ML approaches (DT and HMM) were compared to determine accuracy. The main goal is to develop a computer-assisted pedagogical tool for self-regulated learners. Tanaka et al. [14] Based on the mapping-by-demonstration principle, authors describe different ML approaches to interact with generative sound and upper limb gestural patterns, applying techniques such as Static Regression, Temporal Modelling (HMM), Neural Network Regression and Windowed Regression, where the ML was feed using an IMU device including electromyogram (EMG) musician muscle-activity of the forearm signals. Dalmazzo and Ramírez [5] presented an ML approach to describe seven standard bow-stroke articulations (Détaché, Martelé, Spiccato, Ricochet, Sautillé, Staccato and Bariolage). A high-level expert violinist recorded the gestures, and then the system was used as a gestural estimator with an accuracy of 94%. ML model is based on HHMM, which is trained using audio descriptors and inertial motion information from the IMU device called Myo. The primary

Fig. 1. Music score reference for the seven bow-strokes. Gestures 1, 2, 3, 4, and 6 are in G mayor. Gesture 5 in G melodic-minor and gesture 7 in G chromatic scale. All gestures were recorded with a metronome with a fixed tempo of Square-note 80 BPM.

purpose is to develop a computer-assistant for specific real-time feedback provider for self-regulated music students.

2   Methods and Materials

2.1   Music Score

Seven bow-strokes were recorded following a score with a fixed tempo of quarter-note in 80 bpm. Gestures were recorded in the key of G major, except for Tremolo (G minor) and Collegno (Chromatic G scale). In the violin, two octaves starting from G3 covers the whole neck and also the four strings are needed (Fig. 1).

2.2   Recordings and Synchronization

For the study, nine musicians (4 female) were recorded performing all gestures and a final music piece (Kreutzer 4), which include several bow-strokes examples. The data is composed of two expert performers categorized as **L1**, three high-level students categorized with the **L2** with more than nine years of practice, and four middle-level violin students categorized as **L3** with less than eight years of practice (5–7 years of practice). Data from two IMU devices *Myo* placed on both forearms were recorded using a C++ application which receives Bluetooth

HHMM Blocks

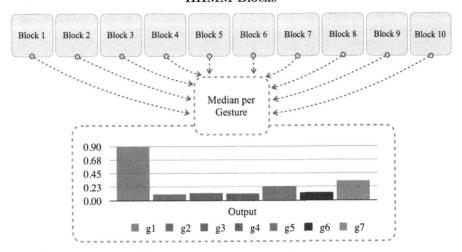

Fig. 2. Each block is an input of an HHMM which then gives as an output seven likelihood progressions and seven classification outputs of the most common number identified by the ten blocks

signals and formats it in a CSV file. Audio samples are synchronized with the *Myo* signals, recording all files with the same length in terms of time-reference. Both files are created and stored in the same time-events triggers. Audio playback has a timing reference in milliseconds, which is directly used to read Myo's data. $-5\,\text{ms}$ offset is needed to synchronize inertial data with audio sampling. A time reference value is stored with the inertial data which is transmitted at a $200\,\text{Hz}$ ratio, that time reference is used from the audio player to sync gestures and sound.

2.3   OpenFrameworks Visualization

An application programmed in C++ using the open-source platform called Openframeworks (OF) [11] is used to visualize the data. From OF the data is send to Max 8 patch (via Open-Sound-Control) which has an HHMM implemented using the MUBU object extension [13] for real-time gesture estimation. For offline analysis, the python library *hmmlearn* is implemented [12].

2.4   Machine Learning Model

In a previous publication, we have implemented an HHMM to recognize gestures based on the *mapping-by-demonstration* principle [5]. In the current model, we intended to design a more generalist probabilistic estimation to be tested by different students. For that we have an architecture based on ten blocks of HHMM sampling ten different dispositions of gestures over the four strings of the violin;

ten sub-blocks are trained with one of the experts L1 and the other ten sub-blocks are trained with the second L1 expert. A median is then extracted as a final output for all likelihood gestures estimations (Fig. 2).

3   Results

Three different performers were selected from the original nine recordings, one for each expertise level, L1, L2, L3, being L1 the expert as a model, L2 high-level students and L3, middle-level student. Confusion Matrix in the Fig. 3 is composed of three different expertise levels: L1 corresponds to a high-level expert. L2 corresponds to an advanced student. L3 corresponds to a beginner-level student. Gestures are distributed as (1) Martelè (2) Staccato (3) Detaché (4) Ricochet (5) Tremolo (6) Collè and (7) Collegno. L1, L2 and L3 identification are at the right part of the matrix.

Weighted probabilities in the Fig. 4 in letters (E), (F), (G) AND (H) plot the output of the average block as a result of the ten HHMM blocks estimations. (E) is Ricochet gesture from L1 and (F) is Ricochet gesture from L2. (G) is the Tremolo gesture from L1, and H) is the Tremolo gesture from L2. Those maps are distributed in a range of 0.0 to 1.0 (normalized), where 1.0 is the highest probability that the current gesture is being recognized.

Confusion Matrix of three different performer's levels

class	1	2	3	4	5	6	7	
1	0.999	0.000	0.000	0.000	0.000	0.001	0.000	
2	0.000	0.650	0.096	0.000	0.008	0.246	0.000	
3	0.000	0.005	0.994	0.000	0.000	0.001	0.000	L1
4	0.024	0.116	0.002	0.695	0.001	0.162	0.000	
5	0.000	0.000	0.002	0.006	0.839	0.001	0.151	
6	0.000	0.029	0.047	0.000	0.005	0.919	0.000	
7	0.000	0.000	0.015	0.000	0.002	0.160	0.823	
1	0.971	0.000	0.000	0.000	0.024	0.004	0.000	
2	0.000	0.261	0.190	0.000	0.220	0.329	0.000	
3	0.002	0.002	0.764	0.005	0.227	0.000	0.000	
4	0.010	0.000	0.000	0.831	0.087	0.072	0.000	L2
5	0.085	0.001	0.009	0.001	0.904	0.001	0.000	
6	0.001	0.011	0.035	0.001	0.264	0.687	0.000	
7	0.000	0.428	0.027	0.011	0.023	0.148	0.363	
1	0.451	0.000	0.000	0.350	0.052	0.147	0.000	
2	0.003	0.204	0.259	0.020	0.337	0.176	0.000	
3	0.017	0.031	0.409	0.309	0.132	0.094	0.008	
4	0.006	0.000	0.011	0.638	0.000	0.344	0.000	L3
5	0.000	0.096	0.073	0.000	0.793	0.032	0.007	
6	0.115	0.077	0.292	0.000	0.439	0.076	0.000	
7	0.000	0.021	0.156	0.281	0.319	0.136	0.089	

Fig. 3. Confusion Matrix figure of the three different levels (L1, L2 and L3) numbers are classes identifications per gesture. The colour code is based on a linear gradient where white is 0.0, and full orange is 1.0 (Color figure online)

Likelihood Comparison and Weighted Maps

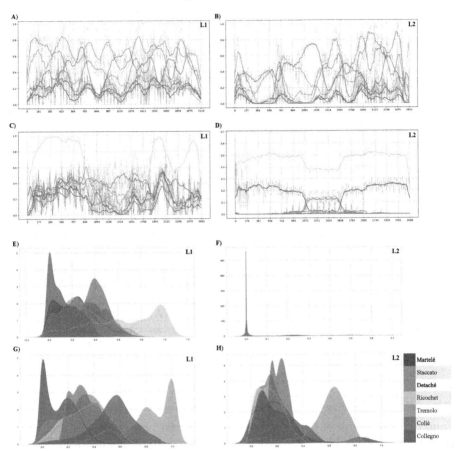

Fig. 4. (A) and (B) corresponds to the second gesture (Staccato) from the L1 and L2 performers; (C) and (D) corresponds to Ricochet from the levels L1 and L2 respectively. (E) and (F) Are weighted-maps (WM) in a range from 0.0 to 1.0 in the X-axis, where 1.0 corresponds to 100% accuracy in gesture estimation. (E) is the WM from gesture 4 (Ricochet) from L1, and (F) is the same WM for gesture 4 in the case of L2. (G) and (H) are WM of the gesture 5 (Tremolo) comparing the levels L1 and L2. Dotted lines in X-axis are markers for each note in the scale where the gesture was performed

4   Discussion and Conclusions

In the case where a small amount of training data is available, HHMM is a robust algorithm for pattern recognition of temporal events. The *mapping-by-demonstration* principles is sufficient for modelling an ML human gestures classifier; as in the case of generative music and gesture interaction [14]. However, for a more generalist model, similar to an MNIST [15], another approach would be needed, perhaps the implementation of Recurrent Neural Networks (RNN), and bigger datasets. The HHMM approach based on blocks reported accurate results in recognizing the seven gestures explained above. Nevertheless, some curious differences among L1 and L2 were observed for the gestures Ricochet (4) and Tremolo (5). The Confusion Matrix in Fig. 3 in the case of L1 reported 69.5% and 83.9% of accuracy in gestures 4 and 5 consecutively, and for the L2 case it was higher 83.1% and 90%, however, in the Fig. 4 different probabilistic weighted-maps (graph (C) and (D), as well, (E) and (F)), are visible, in (C) L1 gesture estimation oscillates between 100% to bellow 20% and L2 in (D) keeps more stable around 50% of certainty. As the HHMM blocks are build using two experts, we consider that both have some dissimilitudes, particularly when the first string of the violin is played. It opens the discussion that strings two, three and four might have a more constrained range of movement as the bow needs to avoid contact with the neighbour's strings, therefor performers permit some execution-freedom in the first string.

In the Fig. 3, the Confusion Matrix give an insight of the variability among the three levels, where L1 is above 82% in gestures Martelé, Detaché, Tremolo, Collè and Collegno, L2 has some variations especially in the gestures Tremolo, Collè and Collegno; and the L3 has a broader variability. Staccato is a gesture commonly confused with Martelé; it is characterized as an isolated distinct sound; it does not have a strong attack; however, it has some similitude with Detaché. In Fig. 4 this similitude can be seen in the (A) and (B) examples, where L1 model mixes Staccato and Detaché; and (B) L2 case Staccato appears at the beginning of some gestures, but the model also detects Detaché, Collè and even Tremolo.

4.1 Future Work

From the perspective of building a general model for bow-stroke gestural detection, it is needed a broader dataset, also to apply data augmentation, as the motion information is based on an imaginary direction in terms of quaternions, it is possible to expand by extrapolating to many other horizontal angles. A new algorithm based on Long-Short Term Memory (RNN) would be tested in a mixture architecture with Hidden Markov Models.

References

1. Bevilacqua, F., Zamborlin, B., Sypniewski, A., Schnell, N., Guédy, F., Rasami-manana, N.: Continuous realtime gesture following and recognition. In: Kopp, S., Wachsmuth, I. (eds.) GW 2009. LNCS (LNAI), vol. 5934, pp. 73–84. Springer, Heidelberg (2010). https://doi.org/10.1007/978-3-642-12553-9_7. http://articles.ircam.fr/textes/Bevilacqua09b/index.pdf
2. Caramiaux, B., Bevilacqua, F., Schnell, N.: Towards a gesture-sound cross-modal analysis. In: Kopp, S., Wachsmuth, I. (eds.) GW 2009. LNCS (LNAI), vol. 5934, pp. 158–170. Springer, Heidelberg (2010). https://doi.org/10.1007/978-3-642-12553-9_14

3. Carrie, N.: Agency and Embodiment: Performing Gestures/Producing Culture. Harvard University Press, Cambridge (2009)
4. Dalmazzo, D., Ramirez, R.: Air violin: a machine learning approach to fingering gesture recognition, November 2017, pp. 63–66 (2017). https://doi.org/10.1145/3139513.3139526
5. Dalmazzo, D., Ramírez, R.: Bowing gestures classification in violin performance: a machine learning approach. Front. Psychol. **10**(MAR), 1–20 (2019). https://doi.org/10.3389/fpsyg.2019.00344
6. Fiebrink, R., Cook, P.R.: The Wekinator: a system for real-time, interactive machine learning in music. In: Proceedings of The Eleventh International Society for Music Information Retrieval Conference (ISMIR 2010), vol. 4, no. 3, p. 2005 (2010). http://ismir2010.ismir.net/proceedings/late-breaking-demo-13.pdf?origin=publicationDetail
7. Fiebrink, R., Cook, P.R., Trueman, D.: Human model evaluation in interactive supervised learning, p. 147 (2011). https://doi.org/10.1145/1978942.1978965
8. Françoise, J., Caramiaux, B., Bevilacqua, F.: A hierarchical approach for the design of gesture-to-sound mappings. In: 9th Sound and Music Computing Conference, pp. 233–240 (2012)
9. Françoise, J., Schnell, N., Borghesi, R., Bevilacqua, F., Stravinsky, P.I.: Probabilistic models for designing motion and sound relationships. In: Proceedings of the International Conference on New Interfaces for Musical Expression, pp. 287–292 (2014)
10. Kahneman, D.: Thinking, Fast and Slow (Kindle Edition) (2011)
11. Openframeworks (2017). https://openframeworks.cc/
12. Rabiner, L.R.: tutorial on hmm and applications.pdf
13. Schnell, N., Röbel, A., Schwarz, D., Peeters, G., Borghesi, R.: MuBu & friends - assembling tools for content based real-time interactive audio processing in MAX/MSP. In: Proceedings of the International Computer Music Conference (ICMC), pp. 423–426 (2009)
14. Tanaka, A., Di Donato, B., Zbyszynski, M.: Designing gestures for continuous sonic interaction, June 2019
15. Zhu, W.: Classification of MNIST handwritten digit database using neural network (2000)

Symbolic Music Classification Based on Multiple Sequential Patterns

Kerstin Neubarth[1] and Darrell Conklin[2,3]([✉])

[1] Europa-Universität Flensburg, Flensburg, Germany
[2] Department of Computer Science and Artificial Intelligence,
University of the Basque Country UPV/EHU, San Sebastián, Spain
`darrell.conklin@ehu.eus`
[3] IKERBASQUE, Basque Foundation for Science, Bilbao, Spain

Abstract. This paper presents a study of associative music classification with sequential patterns. The analysis focuses on class prediction from multiple patterns, considering aggregation of pattern-level measures and evaluation of pattern-set measures. The study complements recent work in music classification which has employed unsupervised pattern discovery followed by instance-based classification.

Keywords: Music classification · Associative classification · Supervised pattern discovery · Sequential pattern · Pattern set

1 Introduction

Research in music classification has explored different symbolic representations, including global features and sequential representations [8,11]. Inter-opus patterns, i.e. patterns recurring across several pieces in a music corpus, offer a hybrid representation: when expressed as boolean predicates (presence vs. absence of a pattern in a music piece), they can be considered global features which capture sequential information about a music piece [3]. Given a data set organised into classes of examples, supervised pattern discovery identifies patterns which describe distinctive properties of classes [18]. Associative classification refers to classification employing patterns associated with a class.

In symbolic music classification, pattern-based approaches have largely focused on instance-based classification combined with unsupervised pattern discovery, i.e. pattern discovery not taking into account class information [1,6,15]. Only few music studies have explored associative classification [2,16]. Those studies follow the classic CBA (Classification Based on Associations) method [14], which learns patterns for each of the classes and predicts the class label of a test example based on the best pattern. Modifications and extensions to CBA evaluate multiple patterns for prediction by integrating their individual contributions [12,13]. Alternatively, multiple patterns can be described by measures at the pattern-set rather than pattern level [7].

© Springer Nature Switzerland AG 2020
P. Cellier and K. Driessens (Eds.): ECML PKDD 2019 Workshops, CCIS 1168, pp. 502–508, 2020.
https://doi.org/10.1007/978-3-030-43887-6_44

The research presented in this paper implements associative classification with sequential musical patterns. It compares different strategies to derive class predictions from multiple patterns and explores pattern-set measures for class-label prediction.

Table 1. Selected viewpoints used in pattern-based classification.

int	melodic interval from previous note (in semitones)
intref	interval from tonic reference note (diatonic interval)
c3(pitch)	3-point pitch contour (ascending, descending, equal)
ioi	inter-onset-interval
c3(dur)	duration contour (longer, equal, shorter)
c3i(level)	metric level contour (stronger, equal, weaker)
phrpos	position in phrase (first, within, last)

2 Pattern-Based Classification

Pattern-based classification methods generally comprise several steps: discovering strong patterns, ranking and selecting discovered patterns (optional), and using the patterns to predict a class label for a given test example. This paper focuses on comparing different prediction strategies.

2.1 Pattern Representation

In the current study, sequences in a symbolic music data set are represented by *viewpoints*, which have been successfully employed in earlier classification studies [4,6]. Each note in a music sequence constitutes an event. A viewpoint maps an event to a more abstract event feature (see Table 1). Two or more viewpoints may be *linked* in order to specify combinations of event features, e.g. int $\otimes$ ioi to represent interval–duration sequences. A *viewpoint pattern* is a sequence of, individual or linked, features over contiguous events, e.g. [int:+2, int:+2, int:-4]. A music sequence can contain patterns based on different viewpoints.

2.2 Pattern Discovery

Patterns identified by supervised descriptive pattern mining can be expressed as *class association rules*, i.e. rules of the form $X \rightarrow C$, where X refers to a pattern and C refers to a class [14]. The *cover* of a pattern is the set of examples in a data corpus covered by the pattern: a music sequence is covered by a sequential pattern if it contains the pattern one or more times. The cover of a class is the set of examples labelled by that class. Let D denote the data set, $cov(C)$ the cover of class C and $cov(X)$ the cover of pattern X. A rule is *frequent* if its support

$$s(X \rightarrow C) = \frac{|cov(X) \cap cov(C)|}{|D|}$$

satisfies a minimum support threshold; it is *strong* if its confidence

$$c(X \rightarrow C) = \frac{|cov(X) \cap cov(C)|}{|cov(X)|}$$

meets a minimum confidence threshold.

To reduce the number of spurious and redundant rules in the discovery output, the current study applies two pruning strategies during pattern discovery: (1) *significance pruning* – retaining only significant frequent and strong rules, whose p-values (here computed by Fisher's exact test) satisfy a given significance level α [12]; (2) *redundancy pruning* – retaining only the most general strong and significant rules while pruning their specialisations, which extend the pattern by another event feature.

2.3   Pattern Ranking and Selection

For pattern-based classification, classifiers are generally built from a subset of discovered strong patterns. Using a sequential-covering strategy to select rules for inclusion in the classifier, rules are first ranked by – in descending priority – decreasing confidence, decreasing support and increasing pattern length, to impose a global order on the set of discovered rules [14]. In a second step, the covering strategy iterates over the ordered list of rules, adding a rule to the classifier if it correctly classifies at least one training example; once an example has been covered by a specified number δ of rules, it is removed from the training set [13]. Finally, a default rule is attached to the end of the list, which assigns the majority class among uncovered training examples – or the majority class of the complete data set if no uncovered examples remain – in order to classify test examples which are not covered by any rule in the list.

2.4   Pattern-Based Prediction

In the prediction phase unlabelled test examples are classified by evaluating those rules in the classifier whose antecedent (pattern) covers the test example.

Prediction Based on Pattern Measures. In most associative classification methods, prediction is based on measures of the individual patterns, either indirectly by exploiting the order of the ranked rules or directly by aggregating individual rule scores. Here we consider the following prediction strategies:

- *Decision list:* Given a test example to classify, the CBA method [14] and its adaptations to music classification [2,16] iterate over the ranked list of rules and assign the class label C predicted by the first, i.e. strongest, rule $X \rightarrow C$ whose antecedent X covers the test example.

- *Majority voting:* The test example is assigned the class label of the class with the highest number of covering rules [17]. Thus majority voting considers the number of covering rules for each class rather than pattern strength; it is nevertheless included here as a frequently used prediction strategy and for comparison with other voting methods.
- *Rank-weighted voting:* Weighted voting strategies assign a weight to each rule and for each class separately sum the weights of the rules covering the test example. The class with the highest sum determines the class label for the test example. Linear Weighted Voting and Inverse Weighted Voting [17] compute the weight w of a rule from the rule's rank in the ordered list of rules as $w = 1 - rank/(|R| + 1)$ and $w = 1/rank$ respectively, where $|R|$ is the total number of rules in the list.
- *Metric-weighted voting:* Instead of using a rule's rank, rule confidence (or another rule metric) can provide the rule weight: $w = c(X \rightarrow C)$. Again the contribution of covering rules is summed separately for the different classes and the test example is assigned to the class with the highest score [12].
- *Probabilistic prediction:* Rule confidence corresponds to the conditional probability of class C given pattern X: $c(X \rightarrow C) = P(C|X)$. Probabilistic prediction evaluates the conditional probabilities for all classes $C_1, ..., C_k$ associated with a pattern X, by attaching the class distribution to each rule: $X \rightarrow C_i \ [P(C_1|X), ..., P(C_k|X)]$. The class with the highest average conditional probability over all rules covering the test example is chosen as the example's class label [9].

Prediction Based on Pattern-Set Measures. In order to define measures at pattern-set level, *cover* is re-defined to accommodate sets of patterns [7]: the cover of a pattern set $\mathcal{X} = \{X_1, ..., X_m\}$ is the union over the single-pattern covers, $cov(\mathcal{X}) = \bigcup_i cov(X_i)$, i.e. the set of examples which each contain one or more patterns in the pattern set. Then common rule measures can be adapted to describe rule sets (for convenience, let $\mathcal{X}_C$ denote a set of association rules $X_i \rightarrow C$ sharing the same class):

- *Pattern-set confidence:* Given the above notations, confidence can be re-defined at pattern-set level as

$$c(\mathcal{X}_C) = \frac{|cov(\mathcal{X}) \cap cov(C)|}{|cov(\mathcal{X})|}$$

- *Pattern-set accuracy:* Accuracy evaluates correctly and incorrectly classified examples [7]. It can be approximated as

$$acc(\mathcal{X}_C) = |cov(\mathcal{X}) \cap cov(C)| - |cov(\mathcal{X}) \cap cov(\neg C)|$$

- *Pattern-set weighted accuracy:* Weighted accuracy takes into account the proportion of covered examples relative to the size of the class C and the background $\neg C$ respectively [7],

$$wacc(\mathcal{X}_C) = \frac{|cov(\mathcal{X}) \cap cov(C)|}{|cov(C)|} - \frac{|cov(\mathcal{X}) \cap cov(\neg C)|}{|cov(\neg C)|}$$

To employ pattern-set measures in the prediction phase of associative classification, the patterns covering a test example are divided according to the different classes, thus deriving rule sets $\mathcal{X}_C$ for each class C, and the test example is assigned to the class with the highest score for the chosen measure.

3   Results

Results are presented for geographic folk song classification on a data set used in an earlier study [2]: 195 folk songs from Austria (102 songs) and Switzerland (93 songs). Figure 1 gives results for four selected viewpoint combinations: in the cases of a single linked viewpoint (Fig. 1, top), all patterns in a music sequence share the same representation; with multiple viewpoints (Fig. 1, bottom), patterns in a sequence can be of different types. The diagrams display the proportion of correctly classified test examples across a range of minimum confidence thresholds; classification performance varying with the choice of confidence threshold is a well-known challenge in associative classification.

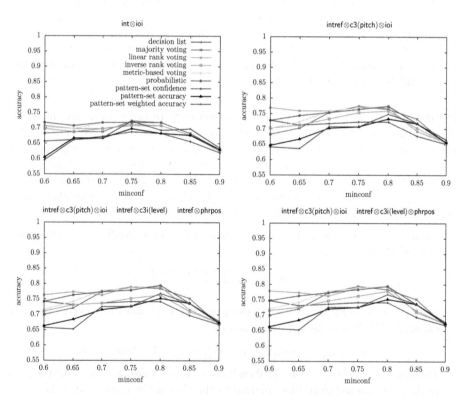

Fig. 1. Proportion of correctly classified test examples at different minimum confidence thresholds ($s_{min} = 0.05$, $\alpha = 0.1$, $\delta = 10$, 10-fold cross-validation).

The results support the following observations: first, considering multiple patterns generally leads to higher classification accuracies than prediction based on a single, strongest pattern. At lower confidence thresholds, however, classification using a decision list outperforms pattern-set based strategies and, here with the exception of the linked viewpoint int $\otimes$ ioi, majority voting. Second, in the current case study strategies integrating individual pattern measures overall give better classification results than those based on pattern-set level measures. Majority voting, which does not consider pattern strength directly, performs well especially at confidence thresholds 0.75 and 0.8. Note that here pattern-set measures are only applied during prediction but not taken into account during pattern discovery or selection. Third, the best classification results reported here (77.4% for homogeneous pattern sets, 79.5% for heterogeneous pattern sets) can be achieved with different prediction strategies.

Previous work on the data set of Austrian and Swiss folk tunes allows to contextualise the results: reference classification accuracies are a baseline of 52% for always predicting the majority class, 66% using logistic regression on global features and 75% using a trigram model on interval-duration sequences [2]. Associative classification with feature-set patterns – using a decision list but without a default rule – led to 77% of classified songs being labelled correctly, albeit at a test set coverage of only 43% [2]. Against these reference results, the classification performance in the current study competes well.

4   Conclusions

To complement existing work on unsupervised pattern discovery and instance-based classification in music, this paper draws attention to associative classification, with a particular focus on class prediction from multiple sequential patterns. Results of an exploratory study on geographical folk song classification compare well against related published work; they also suggest directions for future research, such as integration of viewpoint selection strategies to determine the best single viewpoint or multiple-viewpoint set, parameter-free pattern discovery [5], analysis of the interaction between different phases in associative classification (e.g. pattern evaluation measures, pattern selection criteria and prediction strategies) or optimisation of pattern-set measures during pattern discovery and selection [10].

References

1. Boot, P., Volk, A., de Haas, W.B.: Evaluating the role of repeated patterns in folk song classification and compression. J. New Music Res. **45**(3), 223–238 (2016)
2. Conklin, D.: Melody classification using patterns. In: Proceedings of the 2nd International Workshop on Machine Learning and Music at ECML/PKDD 2009 (MML 2009), Bled, pp. 37–41 (2009)
3. Conklin, D.: Discovery of distinctive patterns in music. Intell. Data Anal. **14**, 547–554 (2010)

4. Conklin, D.: Multiple viewpoint systems for music classification. J. New Music Res. **42**(1), 19–26 (2013)
5. Egho, E., Gay, D., Boullé, M., Voisine, N., Clérot, F.: A user parameter-free approach for mining robust sequential classification rules. Knowl. Inf. Syst. **52**(1), 53–81 (2016). https://doi.org/10.1007/s10115-016-1002-4
6. Goienetxea, I., Neubarth, K., Conklin, D.: Melody classification with pattern covering. In: Proceedings of the 9th International Workshop on Machine Learning and Music at ECML/PKDD 2016 (MML 2016), Riva del Garda, pp. 26–30 (2016)
7. Guns, T., Nijssen, S., De Raedt, L.: k-pattern set mining under constraints. IEEE Trans. Knowl. Data Eng. **25**(2), 402–418 (2013)
8. Hillewaere, R., Manderick, B., Conklin, D.: Global feature versus event models for folk song classification. In: Proceedings of the 10th International Society for Music Information Retrieval Conference (ISMIR 2009), Kobe, Japan, pp. 729–733 (2009)
9. Kavšek, B., Lavrač, N.: APRIORI-SD: adapting association rule learning to subgroup discovery. Appl. Artif. Intell. **20**, 543–583 (2006)
10. Knobbe, A., Valkonet, J.: Building classifiers from pattern teams. In: Proceedings of the International Workshop From Local Patterns to Global Models at ECML/PKDD 2009 (LeGo 09), Bled, pp. 77–93 (2009)
11. van Kranenburg, P., Volk, A., Wiering, F.: A comparison between global and local features for computational classification of folk song melodies. J. New Music Res. **42**(1), 1–18 (2013)
12. Li, J., Zaiane, O.R.: Exploiting statistically significant dependent rules for associative classification. Intell. Data Anal. **21**(5), 1155–1172 (2017)
13. Li, W., Han, J., Pei, J.: CMAR: accurate and efficient classification based on multiple class-association rules. In: Proceedings of the IEEE International Conference on Data Mining (ICDM 01), San Jose, pp. 369–376 (2001)
14. Liu, B., Hsu, W., Ma, Y.: Integrating classification and association rule mining. In: Proceedings of the 4th International Conference on Knowledge Discovery and Data Mining (KDD 98), New York, pp. 80–86 (1998)
15. Louboutin, C., Meredith, D.: Using general-purpose compression algorithms for music analysis. J. New Music Res. **45**(1), 1–16 (2016)
16. Shan, M.K., Kuo, F.F.: Music style mining and classification by melody. In: IEEE International Conference on Multimedia and Expo, Lausanne (2002)
17. Sulzmann, J.N., Fürnkranz, J.: A comparison of techniques for selecting and combining class association rules. In: International Workshop From Local Patterns to Global Models at ECML/PKDD 2008 (LeGo 08), Antwerp (2008)
18. Ventura, S., Luna, J.M.: Supervised Descriptive Pattern Mining. Springer, Cham (2018). https://doi.org/10.1007/978-3-319-98140-6

OPTISIA: An Evolutionary Approach to Parameter Optimisation in a Family of Point-Set Pattern-Discovery Algorithms

Viktor Schmuck[(⊠)] and David Meredith

Aalborg University, Aalborg, Denmark
{vsch,dave}@create.aau.dk

Abstract. We propose a genetic algorithm (GA), OPTISIA, for efficiently finding optimal parameter combinations when running OMNISIA [15], a program that implements a family of analysis and compression algorithms based on the SIA point-set pattern discovery algorithm [20]. The GA, when given a point-set representation of a piece of music as input, runs OMNISIA multiple times, attempting to evolve a combination of parameter values that achieves the highest compression factor on the input piece. When evaluated on two musicological tasks, the system consistently selected well-performing parameters for Forth's algorithm [6] compared to combinations found in published evaluations on the same musicological tasks.

Keywords: Pattern discovery · Genetic algorithm · Parameter optimization · Music analysis · COSIATEC · OMNISIA · Geometric algorithms · Forth's algorithm · Point sets

1 Introduction

Genetic algorithms (GAs) provide a biologically inspired, evolutionary approach to optimisation problems [9]. Previous work suggests that GAs can provide a time-efficient and custom-fit solution when finding optimal parameter combinations in a variety of contexts [7,14]. We propose a decimal-encoding-based GA for efficiently finding optimal parameter combinations when running OMNISIA [15], a program that implements a family of analysis and compression algorithms based on the SIA point-set pattern discovery algorithm [20]. OMNISIA provides implementations of three compression-based pattern mining algorithms, COSIATEC [21], SIATECCompress [18], and Forth's algorithm [6]. Moreover, it allows each of these algorithms to be run with a wide range of options, such as replacing SIA with SIAR [3] or SIACT [5] or using chromatic or morphetic pitch representations [16,17].

In this paper, we present OPTISIA, a GA-based algorithm that runs OMNISIA on a point-set representation of a piece of music multiple times, evolving a combination of parameter values that optimise the achieved compression

P. Cellier and K. Driessens (Eds.): ECML PKDD 2019 Workshops, CCIS 1168, pp. 509–516, 2020.
https://doi.org/10.1007/978-3-030-43887-6_45

factor. The output of this evolutionary process is the analysis of the input piece generated by the particular parameter value combination represented by the simplest chromosome in the final generation that achieves the maximum compression factor on that input piece.

The choice of compression factor as our fitness function is motivated by the widely accepted principle that the shortest (lossless) encodings of a data object represent the best explanations for that object. This parsimony principle (a.k.a. "Ockham's razor") can be traced back to antiquity and has been formalized in more recent times in various ways, including the MDL principle [22] and Kolmogorov's structure function [24]. A number of recent studies in music information retrieval have demonstrated the potential of using the parsimony principle for classification and clustering tasks [2,12,19] and thematic/motivic analysis [18]. We have tested our new approach on two music-analytical tasks: (1) discovering subject and countersubject entries in the fugues of the first book of J. S. Bach's *Das Wohltemperirte Clavier* [8]; and (2) discovering themes and sections in the polyphonic version of the JKU Patterns Development Database [4].

2   Previous Work on Parameter Tuning with Genetic Algorithms

Genetic algorithms present a biologically inspired approach to optimisation problems based on evolution [9]. This approach creates a map of parameters which can be used as possible permutations of genes. To create a population, chromosomes are formed by chaining genes. The population is advanced by computing a fitness score for each parameter combination (chromosome) to perform selection. There are a number of widely used selection methods such as fitness proportionate, roulette-wheel sampling, and elitist selection [10,13]. When the number of chromosomes is reduced to a desired group (parent population), their chromosomes are recombined in pairs (crossover) to produce members for the next generation. Mutation might also be applied to chromosomes, resulting in a potential value change on one or more of the genes. The randomly selected value from the gene types is often allowed to hold its previous value, resulting in no mutation. Genetic algorithms can, with the application of evolutionary principles, evolve optimal or near-optimal parameter combinations over generations [7]. Genetic algorithms are able to reduce the time required for parameter optimisation [7]. Moreover, they allow the encoding of interval values. This is illustrated in [14], where the proposed decimal encoding of nominal and interval parameters results in shorter chromosomes and the accuracy reaches that obtained with binary encoding. Due to shorter chromosomes the search efficiency of the approach is increased relative to other encoding methods.

3   OMNISIA

OMNISIA [15] is a Java program that implements a family of analysis and compression algorithms based on the SIA point-set pattern discovery algorithm [20].

OMNISIA provides implementations of three compression-based pattern discovery algorithms, COSIATEC [21], SIATECCompress [18], and Forth's algorithm [6]. Moreover, it allows each of these algorithms to be run with a wide range of optional parameter settings. The program has been used in a number of previous studies on music analysis and generation [1,11,19].

The descriptions of the various algorithms implemented in OMNISIA and their parameters are given in the original papers describing the algorithms and summarised in [19]. Figure 1 illustrates the effect of some of these switches on the output generated by OMNISIA for the C minor Prelude (BWV 871) from Book 2 of J. S. Bach's *Das Wohltemperirte Clavier*.

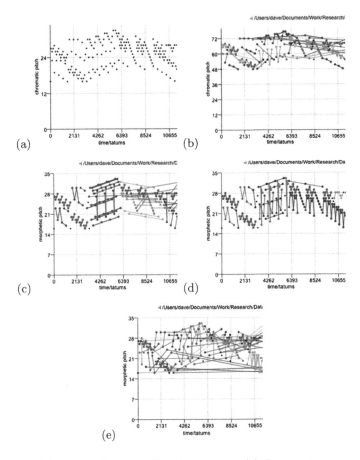

Fig. 1. Example outputs of the OMNISIA program. (a) Point set representation of the prelude from BWV871 given as input. (b) Output generated by COSIATEC using chromatic pitch. (c) Output generated by COSIATEC using morphetic pitch with -d switch selected. (d) Output generated using morphetic pitch and compactness trawler (-ct switch). (e) Output generated when SIA is replaced with SIAR.

4   OPTISIA: An Evolutionary Approach to Parameter Optimisation in OMNISIA

To solve the problem of optimising parameters for the OMNISIA program, the various compression-based pattern mining algorithms and their options were mapped. OMNISIA can use three base algorithms: COSIATEC (COS) [21], SIATECCompress (SCo) [18], and Forth's algorithm (FoA) [6]. The first two require 7 switches, while Forth's algorithm requires 11. The option values of switches were sorted, based on them being nominal (enabled/disabled) or interval (a range of values). The interval values were mapped to ordinal ones to shrink the search space of the optimisation (Table 1).

Table 1. The outline of option switch prefixes, their nominal–interval distinction and presence when using different base algorithms to run OMNISIA.

Switches		Base algorithms	
Switch prefixes	Nominal (N) or Interval (I)	COSIATEC, SIATECCompress	Forth
-d	N	X	X
-ct	N	X	X
-cta	I	X	X
-ctb	I	X	X
-rsd	N	X	X
-r	I	X	X
-rrt	N	X	X
-crlow	I		X
-comlow	I		X
-cmin	I		X
-bbcomp	N		X

To design a gene pool for each element of the chromosomes, the base algorithm options need to be encoded. Following the work of Liu and Wang [14] the values are decimal-encoded. Therefore, gene values range from 0 to 9 instead of multiple genes describing a single parameter that has more than two types. Decimal encoding was chosen to minimise the number of genes the algorithm has to handle during evolution, lowering the required population size and generation count, and consequently the running time of the algorithm.

Some options in OMNISIA are dependent on each other. Therefore, when creating chromosomes, if the nominal values of '-ct' (compactness trawling) [5] or '-rsd' (r superdiagonals) [3] are not set to 'True', the dependent parameters of '-cta' (minimum compactness of trawled patterns), '-ctb' (minimum size of trawled patterns), and '-r' (number of superdiagonals used in SIAR) should not hold values either. This relation between genes was respected during the crossover and mutation operations of the GA. In a chromosome, if the dependent values were not set beforehand, they were initialised at random to complete it.

To create the first population of chromosomes, genes were selected in randomised combinations, discarding repeated ones, therefore no chromosomes were the same at the start. The population size was chosen based on the amount of genes required to encode all chromosomes (7 for COS and SCo, and 11 for FoA). Due to the selection and recombination described below, it is more beneficial to choose population counts divisible by 3. Taking the calculated population sizes based on Gutowski's [9, p. 198] inequality, the previously described divisibility, and the observed population counts [14,23] into account, we tested the execution time and fitness on a single piece from the Fugues database [8] with population sizes of 12, 15 and 18 for COS and SCo, and 18, 21 and 24 for FoA. If the achieved maximum fitness scores were identical in two cases, the lower execution time was used to set the population size, resulting in 12 for COS and SCo, and 21 for FoA.

Fitness scores were acquired by running the parameter combinations and retrieving the resulting compression factors. To create subsequent generations, $1/3$ of the population was kept with elitism-based selection. The genes of these parent-chromosomes were recombined in randomly selected pairs to create 4 offspring chromosomes each, so that the new generation could reach the set population size. An illustration of the selection and recombination can be seen in Fig. 2. Finally, each gene within the offspring chromosomes had its mutation chance set to $100/C_{\text{len}}$ where C_{len} is chromosome length. Genes undergoing mutation were allowed to take their previous values at random.

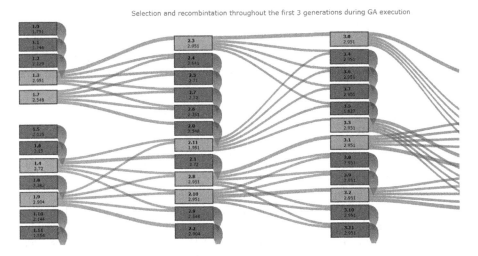

Fig. 2. The figure shows the first 3 generations of chromosomes and their calculated fitness scores. Green cells show parent chromosomes. Blue lines show the recombination. (Color figure online)

The GA optimisation was terminated if the fitness failed to improve (stagnated) for k generations after a minimum of g generations, where $k = 15$ and $g = 30$ for COS and SCo, and $k = 30$ and $g = 40$ for FoA, following the generation

			Fugues (0.0–0.4)	JKU-PDD (0.0–0.5)
SIATECCompress	Raw	GA	0.21	0.22
		M15	0.21 0.22	0.18 0.22
	BB	GA	0.21	0.26
		M15	0.23 0.30	0.27 0.40
	Segment	GA	0.21	0.28
		M15	0.21 0.27	0.28 0.49
Forth	Raw	GA	0.31	0.37
		M15	0.22 0.28	0.23 0.42
	BB	GA	0.30	0.41
		M15	0.23 0.28	0.24 0.43
	Segment	GA	0.25	0.41
		M15	0.23 0.28	0.29 0.42
COSIATEC	Raw	GA	0.13	0.14
		M15	0.12 0.16	0.14 0.20
	BB	GA	0.17	0.25
		M15	0.17 0.18	0.29 0.32
	Segment	GA	0.17	0.38
		M15	0.17 0.18	0.35 0.45

Fig. 3. Results of using OPTISIA to discover subjects and countersubjects in the fugues of the first book of J. S. Bach's *Das Wohltemperirte Clavier* (left-hand side of figure, in black); and discover repeated themes and sections in the JKU Patterns Development Database (right-hand side of figure, in red). Values are for three-layer F_1 score [19, pp. 256–259]. See main text for details. (Color figure online)

count estimation proposed in [9, p. 198]. In most cases, the parameter optimisation of each piece was stopped by the previously described early-termination mechanism.

5   Evaluation

We evaluated our approach on two music-analytical tasks: (1) discovering subject and countersubject entries in the fugues of the first book of J. S. Bach's *Das Wohltemperirte Clavier* [8]; and (2) discovering themes and sections in the polyphonic version of the JKU Patterns Development Database [4]. Figure 3 summarizes the results obtained. In each of the two experiments, the best-compressing chromosome discovered by the GA for each of the three basic algorithms (COS, SCo and FoA) was run in "Raw", "BB" and "Segment" mode (see [19] for an explanation of these terms). The rows headed "GA" in Fig. 3 show the three-layer F_1 (TLF1) scores [19, pp. 256–259] obtained using these nine algorithm–mode combinations on the two experiments. The rows headed "M15" in Fig. 3 show, for each experiment, for each basic algorithm and for each mode, the range of TLF1 scores obtained for that experiment in [19] for parameter combinations using the same algorithm and mode.

For `FoA`, Fig. 3 shows that, for all modes, the chromosome automatically selected by the GA performed well compared with the best of the Forth chromosomes tested in [19] on both the JKU-PDD and the fugues database. However, for `COS` and `SCo`, the GA typically selected a chromosome that performed poorly relative to previously tested parameter combinations. Indeed, in several cases, the GA-selected chromosome performs worse than any of the previously tested parameter combinations for the given algorithm and mode. We speculate that the poorer performance of our GA-based approach on `COS` and `SCo`, is at least partly due to the lower gene count in the chromosomes for these algorithms. Perhaps this problem could be mitigated by increasing the probability of mutation in order to avoid the population converging on a relatively poorly-performing, but locally optimal, parameter combination.

6   Conclusion and Suggestions for Future Work

We used a GA with compression factor as a fitness function to evolve parameter combinations for the SIATEC-based analysis algorithms implemented in the OMNISIA point-set analysis program. When the approach was evaluated on two musicological pattern discovery tasks, it was found that it consistently selected a high performing parameter value combination for Forth's algorithm [6], but relatively poorly-performing parameter combinations for COSIATEC [21] and SIATECCompress [18]. It may be possible to improve the genetic algorithm's efficiency, and possibly increase the achieved compression factor, by mapping the base algorithm options that were mapped to hold ordinal values to interval ones instead. In addition, the effect of using fitness proportionate selection should be investigated. Lastly, it can be hypothesized that this approach, as opposed to elitist selection, would require more computation time, but it would be less prone to stagnation despite the presence of local maxima in the space defined by the fitness function.

References

1. Boot, P., Volk, A., de Haas, W.B.: Evaluating the role of repeated patterns in folk song classification and compression. J. New Music Res. **45**(3), 223–238 (2016)
2. Cilibrasi, R., Vitányi, P.M.B., de Wolf, R.: Algorithmic clustering of music based on string compression. Comput. Music J. **28**(4), 49–67 (2004)
3. Collins, T.: Improved methods for pattern discovery in music, with applications in automated stylistic composition. Ph.D. thesis, Faculty of Mathematics, Computing and Technology, The Open University, Milton Keynes (2011)
4. Collins, T.: JKU Patterns Development Database (2013). https://dl.dropbox.com/u/11997856/JKU/JKUPDD-Aug2013.zip
5. Collins, T., Thurlow, J., Laney, R., Willis, A., Garthwaite, P.H.: A comparative evaluation of algorithms for discovering translational patterns in Baroque keyboard works. In: 11th International Society for Music Information Retrieval Conference (ISMIR 2010), pp. 3–8 (2010)

6. Forth, J.C.: Cognitively-motivated geometric methods of pattern discovery and models of similarity in music. Ph.D. thesis, Department of Computing, Goldsmiths, University of London (2012)

7. Francescomarino, C.D., et al.: Genetic algorithms for hyperparameter optimization in predictive business process monitoring. Inf. Syst. **74**, 67–83 (2018)

8. Giraud, M., Groult, R., Levé, F.: Truth file for the analysis of Bach and Shostakovich fugues (2013/12/27 version) (2013). http://www.algomus.fr/truth/fugues.truth.2013.12

9. Gutowski, M.W.: Biology, physics, small worlds and genetic algorithms. In: Shannon, S. (ed.) Leading Edge Computer Science Research, pp. 165–218. Nova Science Publishers Inc., New York (2005)

10. Hancock, P.J.B.: An empirical comparison of selection methods in evolutionary algorithms. In: Fogarty, T.C. (ed.) AISB EC 1994. LNCS, vol. 865, pp. 80–94. Springer, Heidelberg (1994). https://doi.org/10.1007/3-540-58483-8_7

11. Herremans, D., Chew, E.: MorpheuS: generating structured music with constrained patterns and tension. IEEE Trans. Affect. Comput. (2017). https://doi.org/10.1109/TAFFC.2017.2737984

12. Hillewaere, R., Manderick, B., Conklin, D.: String quartet classification with monophonic models. In: 11th International Society for Music Information Retrieval Conference (ISMIR 2010), Utrecht, The Netherlands, pp. 537–542 (2010)

13. Lipowski, A., Lipowska, D.: Roulette-wheel selection via stochastic acceptance. CoRR abs/1109.3627 (2011). http://arxiv.org/abs/1109.3627

14. Liu, Y., Wang, C.: A modified genetic algorithm based optimisation of milling parameters. Int. J. Adv. Manuf. Technol. **15**(11), 796–799 (1999)

15. Meredith, D.: OMNISIA, software. http://www.titanmusic.com/software/omnisia/OMNISIA20160926.zip

16. Meredith, D.: The ps13 pitch spelling algorithm. J. New Music Res. **35**(2), 121–159 (2006)

17. Meredith, D.: Computing pitch names in tonal music: a comparative analysis of pitch spelling algorithms. Ph.D. thesis, Faculty of Music, University of Oxford (2007)

18. Meredith, D.: COSIATEC and SIATECCompress: pattern discovery by geometric compression. In: MIREX 2013, Competition on Discovery of Repeated Themes & Sections (2013). https://www.music-ir.org/mirex/abstracts/2013/DM10.pdf

19. Meredith, D.: Music analysis and point-set compression. J. New Music Res. **44**(3), 245–270 (2015)

20. Meredith, D., Lemström, K., Wiggins, G.A.: Algorithms for discovering repeated patterns in multidimensional representations of polyphonic music. J. New Music Res. **31**(4), 321–345 (2002)

21. Meredith, D., Lemström, K., Wiggins, G.A.: Algorithms for discovering repeated patterns in multidimensional representations of polyphonic music. In: Cambridge Music Processing Colloquium (2003). http://www.titanmusic.com/papers/public/cmpc2003.pdf

22. Rissanen, J.: Modeling by shortest data description. Automatica **14**(5), 465–471 (1978)

23. Saini, L.M., Aggarwal, S.K., Kumar, A.: Parameter optimisation using genetic algorithm for support vector machine-based price-forecasting model in national electricity market. IET Gener. Transm. Distrib. **4**(1), 36–49 (2010)

24. Vereshchagin, N.K., Vitányi, P.M.B.: Kolmogorov's structure functions and model selection. IEEE Trans. Inf. Theory **50**(12), 3265–3290 (2004)

Predicting Dynamics in Violin Pieces with Features from Melodic Motifs

Fábio Jose Muneratti Ortega$^{(\boxtimes)}$, Alfonso Perez-Carrillo, and Rafael Ramírez

Music Technology Group, Machine Learning and Music Lab, Department of
Communication and Information Technology, Pompeu Fabra University,
Barcelona, Spain
fabiojose.muneratti@upf.edu

Abstract. We present a machine–learning model for predicting the performance dynamics in melodic motifs from classical pieces based on
musically–meaningful features calculated from score–like symbolic representation. This model is designed to be capable of providing expressive
directions to musicians within tools for expressive performance practice,
and for that reason, in contrast with previous research, all modeling is
done on a phrase level rather than note level. Results show the model is
powerful but struggles with the generalization of predictions. The robustness of the chosen summarized representation of dynamics makes its
application possible even in cases of low accuracy.

Keywords: Expressive music performance · Machine learning · Violin

1 Introduction

The development of computer models of music expression has been an active field
of research for over 30 years with a wide range of approaches [9]. Most models
that have been proposed by the research community share the trait of being
designed with the goal of improving computer performance. Our motivation, on
the other hand, is to make use of smart technologies to improve the tools available
for learning to play music, and in particular, to help musicians improve their
expressive performance skills. In our envisioned scenario [12], a computer system
powered by a meaningful expressive performance model could give musicians
expressive directions during performance or visual feedback regarding a recording
based on information extracted from a musical score. As an effort to enable
such scenario, this paper presents a machine learning model for predicting the
dynamics of an ensemble based on high-level features extracted from a score–like
symbolic representation of the musical piece. The proposed method focuses on
modeling long-term dynamics variations, so as to allow a musician to follow the
modeled dynamics suggestions during performance.

1.1 Related Work

Several computer models of expression have been successful in the generation
of convincing performances, particularly of classical piano pieces, as could be

© Springer Nature Switzerland AG 2020
P. Cellier and K. Driessens (Eds.): ECML PKDD 2019 Workshops, CCIS 1168, pp. 517–523, 2020.
https://doi.org/10.1007/978-3-030-43887-6_46

witnessed in the RENCON competition [8]. Most recent are the automatic compositions in the context of project Magenta [7] but the nature of the model makes composition and performance inherently inseparable whereas our learning scenario primarily requires producing performances for already composed and well-established pieces. An approach more related to our own is seen in the YQX system [16], which predicts timing, dynamics and articulation variations in classical piano pieces, and in [6] where a system for predictions of ornamentations in jazz guitar melodies is described. In both cases, melodic lines play an important role, characterized by their Narmour Implication/Realization model classes [11]. In our case, the same type of information is presented to our machine–learning algorithm using a different representation based on pitch curve coefficients. We differ from both, however, by predicting phrase–level instead of note–level expression. Most applicable to our desired scenario are the models reported in [3], which, as in our case, are able to output predictions of expressive parameters based on score information, but a sensible difference is that these models use dynamics markings from the score as a starting point whereas ours seek to generate predictions without using any input indication of expression.

2 Materials and Methods

2.1 Materials

An adequate dataset for the intended model design required a wide variety of melodic themes in both audio and synchronized symbolic representation. Corpi of solo piano pieces such as the MAESTRO [7] were not optimal for the problem since we were interested in mapping the relationship between melody and harmony in the modeled features, and these elements tend to be fully blended in piano parts. The MusicNet dataset [14] provides audio–to–score synchronization as well as the necessary melodic diversity and still allows a clear distinction between main melodic lines and harmonies thanks to the abundance of chamber music pieces with individual instrument parts, and was thus chosen for the task. To distinguish the main melody from harmony, violin parts were treated as melodies and all other instruments, as harmony. Only the subset of pieces which contained a violin were used, resulting in 122 pieces and a total of 874 min of recordings. For estimating the dynamics performed by the ensembles, the momentary loudness in windows of 0.1s according to the EBU R128 standard [4] was computed with the help of the Essentia library [1].

2.2 Methods

The designed model consists of a feed–forward neural network trained to predict the dynamics curve of a musical *motif*, that is, a short phrase of roughly one or two bars. An important aspect of the modeling is that each training instance represents a motif rather than a single note. This design decision is motivated by two beliefs: first, that musicians plan and execute their expressive movements considering a horizon of a few notes rather than momentarily focusing on each one; and second, that in our music learning scenario, performance suggestions based on model outputs can be best visualized and interpreted in that level of granularity. As a consequence of choosing to train the model on motifs, it is necessary to determine musically-relevant motif boundaries in the pieces as well as appropriate features for this representation. The motif boundary detection is done by applying the LBDM [2] algorithm to estimate boundary probabilities, and recursively dividing the piece until one of two conditions is met: either no boundary probability is two standard deviations larger that the rest, or the resulting segment has fewer than 10 notes. Table 1 summarizes the input features used for training. Piece keys and modes were estimated from pitch profiles as detailed in [13]. The output features of the model should represent the dynamics of the motif and its variation on time. We have summarized that information by approximating the performance loudness curve extracted with Essentia by a parabola, fit using the least-squares method. This is consistent with the observation by Todd [10] and other researchers [5,15] that dynamics variations tend to follow a quadratic profile. Given this approximation, the task of the neural network is optimizing the three coefficients that define the dynamics curve.

To facilitate the optimization task, some data conditioning was performed. Loudness measurements of each piece were normalized to zero mean and unit variance to eliminate differences caused by inconsistent recording conditions. Motifs with less than 4 notes and outliers (z-score above 10 in any feature) were discarded, all nominal features were converted to "one-hot" format and all numeric features were standardized. The resulting dataset had around 10.000 instances, which were divided into training and test sets containing 90% and 10% of instances, respectively.

The feed-forward network was programmed in the PyTorch[1] framework and built with two hidden layers of 25 nodes each, using ReLU as an activation function and standard mean-squared error as a loss function. The training was run for 1800 epochs in stochastic gradient descent optimization with batches of 100 instances, learning rate of 0.2 and momentum of 0.1. The learning rate was decreased by a factor of 10 every 600 epochs. All parameters were cross-validated using a subdivision of the training set prior to the final training round.

3   Results and Discussion

3.1   Results

Table 2 shows the obtained correlation coefficients for each of the dynamics coefficients predicted for all instances in the test set. Examples of the loudness curve, ground-truth quadratic approximation and predicted curve for three motifs can be seen in Fig. 1. Table 3 provides some perspective on the accuracy of the modeled dynamics by indicating root-mean-square errors for a deadpan prediction, for the ground-truth approximations, and for the model's output.

---

[1] http://pytorch.org.

Table 1. Input features of the model.

Feature	Data type	Description
Beat in measure	$x \in [0,4]$	The beat where the motif begins
Metric strength	$x \in \{3,2,1,0\}$	How strong the start beat is e.g.: down beat = 3
Number of notes	$x \in \mathbf{N}$	Total of notes in motif
Duration	$x \in [0,\infty)$	Motif duration in beats
Location in piece	$x \in [0,1]$	Where in the piece the motif is played
Pitch curve coefficients	$x_0, x_1, x_2 \in \mathbf{R}$	Quadratic coefficients approximating the MIDI pitches of motif notes
Pitch contour coefficients	$x_0, x_1, x_2 \in \mathbf{R}$	Quadratic coefficients approximating the variation in MIDI pitches of motif notes
Rhythm drops	Boolean	Whether a note with higher duration follows another with shorter duration in the motif
Rhythm rises	Boolean	Whether a note with shorter duration follows another with higher duration in the motif
Rhythm contour coefficients	$x_0, x_1, x_2 \in \mathbf{R}$	Quadratic coefficients approximating the variation in duration of motif notes
Strongest note location	$x \in [0,1]$	Where in the motif is the note with highest metric strength
Piece key	A - G#	Tonality estimation of motif piece
Piece mode	Major/Minor	Mode estimation of motif piece
Chord probabilities	$x_0..x_6 \in [0,1]$	Estimated diatonic chords presence probabilities
Initial chord degree	I - VII	Most likely chord in motif start
Final chord degree	I - VII	Most likely chord in motif end
Has dissonance	Boolean	Whether there are notes from a different tonality
Dissonance location	$x \in [0,1]$	Location of first occurence of dissonant note
Is solo piece	Boolean	Solo or ensemble piece

Table 2. Correlation coefficients for output features.

Output coefficient	Pearson's r (test set)	Pearson's r (training set)
x^2	0.2177	0.5222
x^1	0.2375	0.7054
x^0	0.2383	0.6818

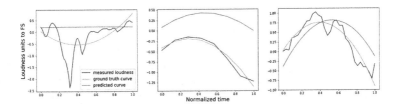

Fig. 1. Comparison of loudness values measured in performance, their ideal (ground-truth) approximation, and model output for three motifs.

3.2   Discussion

The Pearson correlation coefficients obtained for the training set show that the model is sufficiently powerful to predict the complex relationships present in this scenario, but the lower correlation values seen in the test set indicate that some overfitting occurred, and the meaningful correlations detected in the data only partially explain the observed dynamics. The deadpan-level (E_d) and ground-truth-level (E_g) errors in Table 3 can be seen as lower and upper boundaries of accuracy, indicating that this modeling approach offers a potential reduction in prediction error of up to $E_d - E_g = 2.17$ dB. The 3.39 dB value obtained with our predictions implies an error reduction of $E_d - 3.39 = 0.47$ dB compared to the deadpan baseline, which corresponds to $0.47/2.17 = 21.65\%$ of the predicted potential. That is consistent with the correlation coefficient values and shows that the prediction of coefficients translates well into prediction of dynamics levels.

Table 3. RMS error in loudness levels prediction.

Prediction type	Error level
Deadpan performance	3.86 dB
Ground-truth approximation	1.69 dB
Model prediction	3.39 dB

The prediction examples highlighted in Fig. 1 illustrate some relevant conclusions: It can be seen that most of the short–term variation in loudness levels occurs on note boundaries due to note articulation, and in terms of perceived dynamics can be understood as noise. The quadratic approximation (labeled ground-truth) provides a cleaner and more intuitive visualization of the variation of loudness in a phrase, and in most individually–inspected cases represents it quite well. The leftmost example is an exception, as it shows a case in which the phrase boundaries chosen by the algorithm don't seem to match the performer's choice, hence the silence during the phrase and the poor results even in the proposed ground-truth approximation. In many observed cases, as shown in the middle and rightmost graphs, the predicted curve shows robustness, especially with relation to the x^2 and x^1 coefficients, since some variation in their predictions doesn't affect the character of the interpretation. It is reasonable to assume that despite the difference between ground-truth and predicted values in such cases, performances executed according to instructions from the latter could be considered just as pleasing.

Logical improvements to our proposed approach under consideration include adding information that relates to the repetition of motifs, detecting key modulations or modal harmony in pieces, augmenting the training set with multiple different divisions of motifs per piece and experimenting with different treatments of time–series data such as training with long short–term memory networks.

References

1. Bogdanov, D., et al.: Essentia: an audio analysis library for music information retrieval. In: 14th Conference of the International Society for Music Information Retrieval (ISMIR). International Society for Music Information Retrieval (ISMIR) (2013)
2. Cambouropoulos, E.: The local boundary detection model (LBDM) and its application in the study of expressive timing. In: Proceedings of the International Computer Music Conference ICMC01, Havana, Cuba, pp. 232–235 (2001)
3. Cancino-Chacón, C.E.: Computational modeling of expressive music performance with linear and non-linear basis function models. Ph.D. thesis, Johannes Kepler University Linz (2018)
4. EBU Tc Committee: Tech 3341: Loudness metering: 'EBU mode' metering to supplement EBU R 128 loudness normalization. Technical report, EBU, Geneva (2016)
5. Gabrielsson, A., Bengtsson, I., Gabrielsson, B.: Performance of musical rhythm in 3/4 and 6/8 meter. Scand. J. Psychol. **24**(1), 193–213 (1983). https://doi.org/10.1111/j.1467-9450.1983.tb00491.x
6. Giraldo, S.I., Ramirez, R.: A machine learning approach to discover rules for expressive performance actions in jazz guitar music. Front. Psychol. **7**, 1965 (2016). https://doi.org/10.3389/fpsyg.2016.01965
7. Hawthorne, C., et al.: Enabling factorized piano music modeling and generation with the MAESTRO dataset. In: International Conference on Learning Representations (2019)
8. Katayose, H., Hashida, M., De Poli, G., Hirata, K.: On evaluating systems for generating expressive music performance: the Rencon experience. J. New Music Res. **41**(4), 299–310 (2012). https://doi.org/10.1080/09298215.2012.745579
9. Kirke, A., Miranda, E.R.: An overview of computer systems for expressive music performance. In: Kirke, A., Miranda, E. (eds.) Guide to Computing for Expressive Music Performance, pp. 1–47. Springer, London (2013). https://doi.org/10.1007/978-1-4471-4123-5_1
10. McAngus Todd, N.P.: The dynamics of dynamics: a model of musical expression. J. Acoust. Soc. Am. **91**(6), 3540–3550 (1992). https://doi.org/10.1121/1.402843
11. Narmour, E.: The Analysis and Cognition of Melodic Complexity: Theimplication-Realization Model. University of Chicago Press, Chicago (1992)
12. Ramirez, R., Ortega, F.J.M., Giraldo, S.I.: Technology enhanced learning of expressive music performance. In: Proceedings of the 16th Brazilian Symposium on Computer Music (2017)

13. Temperley, D.: What's key for key? The krumhansl-schmuckler key-finding algorithm reconsidered. Music Percept. Interdisc. J. **17**(1), 65–100 (1999). https://doi.org/10.2307/40285812
14. Thickstun, J., Harchaoui, Z., Foster, D.P., Kakade, S.M.: Invariances and data augmentation for supervised music transcription. In: International Conference on Acoustics, Speech, and Signal Processing (ICASSP) (2018)
15. Tobudic, A., Widmer, G.: Relational IBL in music with a new structural similarity measure. In: Proceedings of the 13th International Conference on Inductive Logic Programming, pp. 365–382 (2003)
16. Widmer, G., Flossmann, S., Grachten, M.: YQX plays chopin. AI Mag. **30**(3), 35 (2009). https://doi.org/10.1609/aimag.v30i3.2249

Sequence Generation Using Unwords

Darrell Conklin[1,2(✉)]

[1] Department of Computer Science and Artificial Intelligence, University of the
Basque Country UPV/EHU, San Sebastian, Spain
darrell.conklin@ehu.es
[2] IKERBASQUE, Basque Foundation for Science, Bilbao, Spain

Abstract. Statistical context models for sequence generation provide
probabilities for each event in a sequence, conditioned on the context
or history of the event in the sequence. A fundamentally different type
of generative method inverts this view entirely, stating only what patterns cannot occur, allowing all other possibilities. The two concepts of
statistical models and unexpectedly absent words can be combined for
sequence generation, by sampling from a statistical model while avoiding
generation of any unwords. A desirable feature of this approach is that
an efficient random walk procedure can be used even for long sequences.

Keywords: Music generation · Statistical models · Pattern discovery

1 Introduction

In the context of a project on the reconstruction of Mozarabic chant [6], a statistical generation method has been used to generate pieces that instantiate predefined *templates*. This method allows the specification of several types of features
in templates, including contour relations between successive events, intra-opus
patterns expressing equality relations between distant fragments, and defined
pitches.

Figure 1 shows a fragment from the León antiphoner (E-L 8, Catedral de
León), the most important source of Mozarabic chants, dating from the early
tenth century, containing over 3000 chants preserved in adiastematic neume notation. Although pitch information is absent, the adiastematic neumatic notation
indicates the contours of the melodies, and the lyrics and the adiastematic neume
notation therefore determine a hypothetical contour sequence (o: any; e: equal;
l: lower; h: higher) which can be used as a template for generation [6]. The
opening pitch is fixed to an A3 (midi 57). Two different intra-opus patterns are
highlighted in the figure, each having two instances. The presence of intra-opus
patterns is an important aspect of the generation method as they allow long
range relations between music segments to be specified. For these patterns the
same pitch material must be generated in all instances of the pattern, while
still obeying the contour sequences (and possibly defined pitches) under these
patterns.

© Springer Nature Switzerland AG 2020
P. Cellier and K. Driessens (Eds.): ECML PKDD 2019 Workshops, CCIS 1168, pp. 524–530, 2020.
https://doi.org/10.1007/978-3-030-43887-6_47

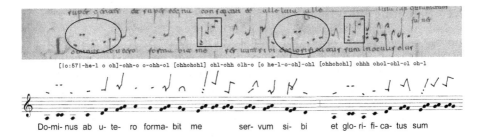

Fig. 1. The first line of the responsory *Dominus ab utero formavit me* for the feast of St. John the Baptist in the early tenth-century León antiphoner (E-L 8, 214r2), with two intra-opus patterns. Below the León image a template of contour letters (also defining the first pitch as 57), with intra-opus patterns in brackets. Bottom: a generated sequence compatible with the template.

The method has successfully been employed to efficiently generate an entire suite of new chants that were performed in a festival (Nederlands Gregoriaans Festival, 's-Hertogenbosch, June 14–16, 2019). Nevertheless, several open points remain to improve the method [10]. One of these is the avoidance of large melodic intervals that may be produced by the left-to-right random walk sampling procedure, in the presence of patterns and defined pitches.

Though samples can be drawn exactly from models with simple unary pitch constraints [8,11], such models are inadequate for chant generation, which requires the modelling of intra-opus patterns. The presence of these patterns implies an intractable sampling problem [4,13,14] because the fraction of times a particular sequence is sampled will not necessarily correspond to its probability. This can mean that low probability sequences, possibly poor and having stylistic violations, can be generated.

More precisely, an *information peak* will arise when the left boundary of a repeated pattern instance, or a defined pitch, restricts the set of possibilities for an event, and only low probability (high information content) continuations are available given previously generated material. These information peaks may be within *unwords* [5]: patterns that should never appear in a generated sequence. This paper presents a method for generating sequences while avoiding *surprising* unwords: those expected to appear in a corpus but in fact never appear. In this method the unavoidable presence of an unword will cause backtracking to the previous position rather than resampling of an entirely new sequence.

2   Methods

This section describes in more detail the problem of information peaks and proposes unwords as a possible solution. The concept of unwords is formally defined, and a method for generating sequences while avoiding unwords is outlined.

2.1   Sampling into Templates

Statistical models for music generation are typically *context models*, defining distributions of next events given a context already generated or provided [2]. They can be guided by a *template* which for chant generation is a specification of patterns, defined pitches, and contours. Sequences are generated by sampling instances of a template from a model: those sequences compatible with all constraints defined in the template. For context models this can be done with the help of constraint satisfaction methods [4], maintaining and adjusting the sets of permissible events during every step of the left-to-right random walk. This can be done without any backtracking, provided that the underlying statistical model is *non-exclusive*: assigning a probability, however tiny, to every possible event at every position.

2.2   Information Peaks

For context models without constraints, the random walk procedure suffices to sample from the correct sequence distribution. However once constraints are posted on the generated sequence, random walk becomes inexact. The basic problem is that when constrained events are encountered, the continuation from the context may be required to "snap" back to the constrained event, with a very high information content.

These *information peaks* are spikes in the information content time series and may be within low probability sampled sequences. Figure 2 shows this phenomenon with a real template and model. It can be seen that a peak of information content occurs exactly at a defined pitch (first magenta vertical bar, pitch E4 note 140). The melody, which has moved earlier to the high end of the range (pitch A6), moves as required down to the area of E4 although with an unnatural cadence A5-B5-E4, and a substantially low probability event at E4 (information content above 10 bits).

Thus information peaks may give rise to *unwords* and a way to define and avoid these unwords can be used productively to avoid information peaks in generated music. It should be noted that some information peaks are acceptable: here only those peaks that cause unwords are avoided.

2.3   Unwords

For pattern discovery one can invert the task and ask: what are the patterns that *do not occur* in a corpus? Patterns that are completely absent from a large dataset may be expected to also be absent in any new pieces in the style or genre under consideration, including generated pieces.

This definition however is too broad to be useful because in any corpus most theoretically possible unwords are absent merely due to statistical sparseness of a corpus, while the interesting unwords are those that were expected to occur but did not. A method for computing the statistical significance of an unword [5] is briefly outlined here.

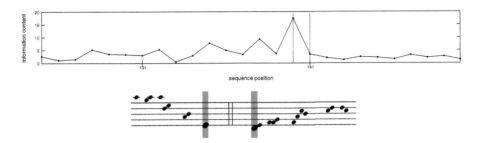

Fig. 2. Information peaks illustrated, from the template *Dum complerentur* (E-L 8, 210r14). Magenta vertical lines refer to the position of defined pitches. Information content: negative log probability. Bottom: focus around the defined pitches at positions 140 and 141 (see top of Fig. 3 for the full sequence information content)

The *piece count* of a pattern Φ in a corpus is the number of pieces in the corpus that instantiate the pattern. An *unword* is any pattern Φ with a piece count of zero, and which is not subsumed by any other such pattern (i.e., unwords are minimal absent patterns). To evaluate the significance of an unword, let X be a random variable modelling the piece count of a pattern, so $P(X = k)$ is the probability that the observed piece count of the pattern is exactly k (≥ 0). The p-value $P(X = 0)$ of an unword can indicate whether the unword is purely statistical (insignificant p-value), versus structural (significant, low p-value): see [5] for details. The p-value is determined by computing the expected number of pieces in the corpus containing the absent pattern: the p-value being lower with higher expectation and therefore indicating a more surprising unword.

The algorithm for unword discovery is a refinement tree search that uses two refinement operators to specialize patterns [3]. For unword discovery, paths are explored until a piece count of zero is reached, at which point it is assessed whether the p-value is lower than a specified threshold α. Paths can be terminated early if they only lead to patterns with a p-value greater than α: see [7] for the p-value bounding conditions.

2.4   Generation Using Unwords

To employ unwords during generation, random walk with backtracking is used. At every position, compatibility with the template is tested and an event is sampled from the compatible distribution. To incorporate unwords, at every sequence position the entire set of unwords is scanned to see if any unword is a suffix of the proposed updated sequence. Such events are not permitted. In the case no events are compatible, backtracking to the previous position occurs.

3   Results

A corpus of 115 Gregorian offertories [9] was used to train a trigram model on absolute pitches occurring in the corpus. Though the data is purely symbolic (spelled pitches) MIDI numbers can be used for convenience as there are no enharmonics. For prediction the PPM (prediction by partial match [1]) algorithm with Method C backoff probabilities was employed.

For unword discovery, the same corpus was used. Unword expectations [5] were determined using a unigram model of absolute pitches (thus, only unwords of length 2 or more are reported). A total of 559 unwords were found at a p-value threshold of $\alpha = 0.001$. Table 1 presents a selection of unwords identified in the corpus.

Table 1. A small selection of significant unwords in Gregorian chant. Boxed: the unword occurring in the fragment shown in Fig. 2

Unword	$-\log(p\text{-value})$
72,62	248.7
72,64	223.9
65,74	219.4
62,71	167.7
72,65,72	160.2
65,72,67	157.4
67,72,65	157.4
65,72,65	144.8
69,71,64	32.3
72,65,65,65	29.9

Sequences were sampled into a long template for the chant *Dum complerentur* (E-L 8, 210r14) which contains several intra-opus patterns and defined pitches. Figure 3 presents these results. The use of unwords is able to reduce the height of information peaks, producing a sequence with no peaks of more than 10 bits.

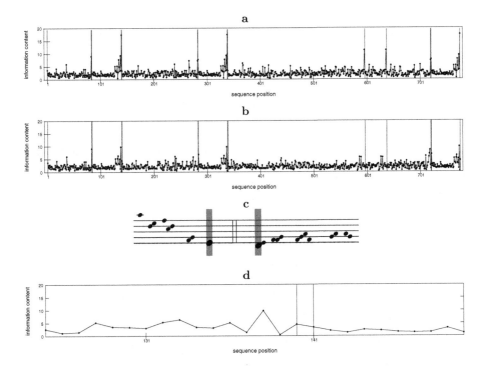

Fig. 3. Sequence generation with unwords for the template *Dum complerentur* (E-L 8, 210r14). (a) basic random walk; (b) with unwords; (c,d) focus around the defined pitches at positions 140 and 141

4   Conclusions

This paper has proposed the incorporation of unwords into statistical music generation, as a way to remove unacceptable information peaks while still retaining the efficiency of random walk sampling. Results are promising and a larger follow-up study will explore a larger catalog of templates, and including *inter-opus* patterns: those occurring across different pieces.

In a related study [12], all patterns of length greater than a specified length are prohibited in generated sequences, with the intention of reducing literal lifting of long material from the corpus. These however are not unwords and provide only weakly applicable constraints on generated sequences.

Other approaches to handle information peaks should be studied, particularly blocked Gibbs sampling (many events sampled concurrently in each step), although such approximate sampling methods may be too inefficient for practical use when inter-opus patterns are considered. On that point, one of the motivations of this work is the generation of an entire suite of Mozarabic chants with global coherence provided by intra- and inter-opus patterns.

Acknowledgments. Thanks to Geert Maessen for providing the chant template and the neumatic notation of Fig. 1, and to Kerstin Neubarth for valuable comments on the manuscript.

References

1. Cleary, J.G., Witten, I.H.: Data compression using adaptive coding and partial string matching. IEEE Trans. Commun. **32**(4), 396–402 (1984)
2. Conklin, D.: Music generation from statistical models. In: Proceedings of the AISB Symposium on Artificial Intelligence and Creativity in the Arts and Sciences, Aberystwyth, Wales, pp. 30–35 (2003)
3. Conklin, D.: Discovery of distinctive patterns in music. Intell. Data Anal. **14**(5), 547–554 (2010)
4. Conklin, D.: Chord sequence generation with semiotic patterns. J. Math. Music **10**(2), 92–106 (2016)
5. Conklin, D.: Music corpus analysis using unwords. In: Montiel, M., Gomez-Martin, F., Agustín-Aquino, O.A. (eds.) MCM 2019. LNCS (LNAI), vol. 11502, pp. 338–343. Springer, Cham (2019). https://doi.org/10.1007/978-3-030-21392-3_28
6. Conklin, D., Maessen, G.: Generation of melodies for the lost chant of the Mozarabic rite. Appl. Sci. **9**(20), 4285 (2019)
7. Conklin, D., Weisser, S.: Pattern and antipattern discovery in Ethiopian bagana songs. In: Meredith, D. (ed.) Computational Music Analysis, pp. 425–443. Springer, Cham (2016). https://doi.org/10.1007/978-3-319-25931-4_16
8. Hadjeres, G., Nielsen, F.: Anticipation-RNN: enforcing unary constraints in sequence generation, with application to interactive music generation. Neural Comput. Appl. **32**(4), 995–1005 (2018). https://doi.org/10.1007/s00521-018-3868-4
9. van Kranenburg, P., Maessen, G.: Comparing offertory melodies of five medieval Christian chant traditions. In: Proceedings of the 18th International Society for Music Information Retrieval Conference (ISMIR 2017), Suzhou, China, pp. 204–210 (2017)
10. Maessen, G.: Aspects of melody generation for the lost chant of the Mozarabic rite. In: Proceedings of the 9th International Workshop on Folk Music Analysis (FMA 2019), Birmingham, UK, pp. 23–24 (2019)
11. Pachet, F., Roy, P., Barbieri, G.: Finite-length Markov processes with constraints. In: Proceedings of the 22nd International Joint Conference on Artificial Intelligence (IJCAI 2011), Barcelona, Spain, pp. 635–642 (2011)
12. Papadopoulos, A., Roy, P., Pachet, F.: Avoiding plagiarism in Markov sequence generation. In: Proceedings of the 28th AAAI Conference on Artificial Intelligence, AAAI 2014, pp. 2731–2737 (2014)
13. Rivaud, S., Pachet, F., Roy, P.: Sampling Markov models under binary equality constraints is hard. In: Journées Francophones sur les Réseaux Bayésiens et les Modéles Graphiques Probabilistes, Clermont-Ferrand, France, June 2016
14. Walder, C., Kim, D.: Computer assisted composition with Recurrent Neural Networks. In: Proceedings of Machine Learning Research, vol. 77, pp. 359–374 (2017)

A Machine Learning Approach to Study Expressive Performance Deviations in Classical Guitar

Sergio Giraldo[1(✉)], Alberto Nasarre[1], Isabelle Heroux[2], and Rafael Ramirez[1]

[1] Pompeu Fabra University, Roc Boronat 138, 08018 Barcelona, Spain
sergio.giraldo@upf.edu
[2] Université du Québec à Montréal (UQAM), C.P. 8888, succ. Centre-ville Montréal (Québec) H3C 3P8, Montreal, Canada
heroux.isabelle@uqam.ca

Abstract. Expression is the added value of a musical performance, in which deviations in timing, energy, and articulation are introduced by musicians. Computational models have been proposed aiming at understanding and modelling the expressive content of music performances, to convey concrete expressive intentions. However, little work has been done to investigate the intrinsic variations that musicians might introduce, i.e. when no specific expressive indications are provided. In this contribution, we present a machine learning approach to study the expressive variations that nine different guitarists introduce when performing the same musical piece, for which no performance indications are provided. We study the correlations on the variations in timing and energy. We extract features from the score to obtain predictive models for each musician to later cross-validate among them. Preliminary results indicate that musicians use similar variations when applying these variations, based on correlation measures. Also, similar correlation indexes are found on the cross-validation exercise.

Keywords: Classical guitar · Expressive performance modelling · Machine learning

1 Introduction

Musicians introduce deviations to the score when performing a musical piece in order to achieve a particular expressive intention. Computational expressive music performance modelling (CEMPM) aims to characterise such deviations using computational techniques (e.g. machine learning techniques). In this context, CEMPM aims to formulate a hypothesis on the expressive devices musicians use when performing (consciously or unconsciously), which can be empirically verified on measured performance data. Empirical models are often obtained from the quantitative analysis of musical performances, based on measurements of timing, dynamics, and articulation (e.g Shaffer et al. 1985; Clarke 1985; Gabrielsson 1987;

© Springer Nature Switzerland AG 2020
P. Cellier and K. Driessens (Eds.): ECML PKDD 2019 Workshops, CCIS 1168, pp. 531–536, 2020.
https://doi.org/10.1007/978-3-030-43887-6_48

Palmer 1996a; Repp 1999; Goebl 2001, to name a few). State of the art reviews are presented in Gabrielsson (2003). Computational models have been implemented as rule-based models (Friberg et al. 2000; the KTH model), mathematical models (Todd 1992), structure-level models (Mazzola 2002).

Machine Learning techniques have been used to predict performance variations in timing, articulation and energy (e.g. Widmer 2002), to model concrete expressive intentions (e.g. mood, musical style, performer etc). Most of the literature focus on classical piano music (e.g. Widmer 2002). The piano keys work as ON/OFF switching devices (e.g. MIDI pianos), which simplifies the process of data acquisition, where performance data has to be converted into machine readable data. Some exceptions can be found in in jazz saxophone music where case-based reasoning (Arcos et al. 1998) and inductive logic programming (Ramirez et al. 2011) have been used. Jazz guitar expressive performance modelling has been studied by Giraldo and Ramirez (2016a, b), in which special emphasis is done in melodic ornamentation.

However, few studies have been done in the context of classical guitar, aiming to study the intrinsic variations performers introduce when no specific expressive intentions are provided. In this study, we present a machine learning approach in which CEMPM techniques are applied to study the expressive variations that nine different guitarists introduce when performing the same musical piece, for which no performance indications are provided. We study the correlations on the variations in timing and energy. We extract features from the score to obtain predictive models for each musician to later cross-validate among them.

2   Materials and Methods

For this study we obtained recordings of nine professional guitarists performing the same musical piece. The piece was written for classical guitar, and was composed specifically for this study. Musicians did not knew the piece before hand, and any particular expressive/performance indications were provided (nor written or verbally). The performers were allowed to freely introduce the expressive variations to their taste/criteria. Musicians were also allowed to practice the piece as long as they wanted, until they were satisfied with their interpretation, before recording. The recordings took place at different studios/institutions and were collected by the Department of Music form the Faculty of Arts of the University of Quebec in Montreal (UQAM) Canada.

2.1   Framework

The general framework of the project is depicted in Fig. 1.

Data Processing. The musical score was created in musicXML format from which we obtained machine readable data (MIDI type) information of each note, i.e. its onset (in seconds), duration (in seconds), pitch, and velocity (which refers to volume). We used the score as the *dead-pan* performance (i.e. robotic or inexpressive performance).

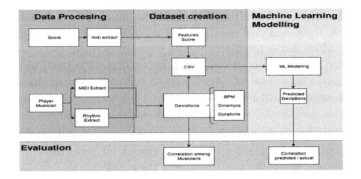

Fig. 1. Framework and data processing flow.

In a second stage we obtained machine readable data of the performance in MIDI type format. This process was performed in a semi-automatic fashion, where we used score-informed Non-negative Matrix Factorisation (NMF). The NMF method decomposes an input spectrogram $X \in^{KxN}$ with K frequency bins and N frames as:

$$X = WH \tag{1}$$

where $W \in^{KxR}$ contains the spectral bases for each of the R pitches and $H \in^{RxN}$ is the pitch activity matrix across time. The number of R pitches and the W and W matrices initial weights were initialised, informed by the score (for an overview see Clarke 1985). Later, manual correction was performed over the spectrum. Finally, energy information (i.e velocity) was obtained from the RMS value, calculated over the audio wave, in between the obtained note boundaries.

Similarly, we performed automatic beat extraction (Zapata et al. 2014), followed by manual correction to obtain the beat information (in seconds) over the audio signal.

Data-set Creation. Feature extraction from the score was performed by extracting *local* information of the notes in the score (e.g pitch, duration, etc.) as well as *contextual* information in which the note occurs (e.g. previous/next interval, metrical strength, harmonic/melodic analysis, etc.). For an overview see Giraldo and Ramirez (2016a, b). A total of 27 descriptors were extracted for each note. Later, deviations in tempo variation, measured in Beats Per Minute (BPM) and Inter Onset Interval (IOI) for each note/performer, were calculated by considering the difference among the theoretical BPM/IOI values in the score and the corresponding values in the performance. Finally, we obtained data-sets for each of the nine performers, as well as for each of the three performance deviations considered (i.e. energy, BPM, and IOI deviations). A total of 27 data-sets were obtained, where each instance is composed by the feature set extracted for each note, and the considered deviations are the value to be predicted.

Machine Learning Modelling. Each of the nine *performer* data-sets were used as both *train* and *test* sets in a all-vs-all cross-validation fashion. This consisted in obtaining a predictive model for each performer (i.e all performer data sets were used as train set) and applying each of them to all the performers (i.e all the performer data sets were used as test set), and finally obtaining a model evaluation for each pair.

Evaluation. A preliminary evaluation consisted in obtaining the correlations among the actual deviations for each note, among all performers. Later, at the Machine Learning stage, the performance of the predictive models was addressed by obtaining the Correlation Coefficient (CC) among the predicted values of the model and the actual values at the test set. The algorithms considered were Support vector regression (SVR, with radial kernel), Regression Trees (RT, with pruning), and Artificial Neural Networks (ANN, fully connected with one hidden layer), for which CC's on preliminary tests are presented in Table 1. Given that ANN out performed at the prediction of the three considered expressive deviations, in this paper we report on the CC obtained with ANN.

Table 1. Mean Correlation Coefficient (CC) comparison among models.

Deviations	SVR (CC)	RT (CC)	ANN (CC)
Energy	0.45	0.42	0.61
IOI	0.58	0.60	0.82
BPM	0.68	0.69	0.87

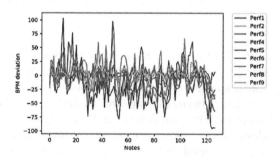

Fig. 2. BPM percentage of deviation among nine performers for each consecutive note

3   Results

Figure 2 show the measured deviations in percentage of BPM of each consecutive note for each performer. It can be noticed the correspondence of peaks and valleys (with different amplitude/deviation degree) among performers. Figure 3 present the scaled graph of the obtained correlation coefficients using ANNs. The numbers on the vertical axis indicate the performer data sets (numbered from 1 to 9) used as train set, whereas the horizontal axis represent the performers data set when used as test set. At each intersection, the colour map represents the correlation coefficient obtained using each pair of train/test data sets. As expected, the diagonal shows higher correlations, representing the performance on the train set (i.e train and test set are the same performer). Higher correlations can be found at the BPM and IOI deviations. Also, a similar pattern can be observed on the CCs obtained among performers. This might indicate that the majority of the performers introduce similar timing variations based on the information provided by the score. This tendency can be observed as well at Fig. 2 (e.g. as seen at the ritardando introduced by most performers at the end of the piece). In contrast, lower level of correlations were obtained on the energy deviation models, which might indicate that the decision on the loudness of a note is less consistent among performers. However, other external factors, such as different recording conditions (e.g. the use of a different guitar, or recordings being done at different studios) might bias this result.

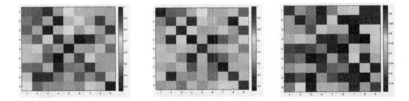

Fig. 3. Scaled graph of the correlations obtained for each performer model, for each of the three expressive deviations considered (from left to right: BPM, IOI and Energy). Vertical axis correspond to performer data used as train set (from 1 to 9), and horizontal axis corresponds to performer data used as test set (from 1 to 9)

4   Conclusion

In this paper we have presented a machine learning approach based on computational modelling of expressive music performance to study the correlations on the intrinsic expressive deviations that musicians introduce when performing a musical piece. We have obtained recordings of the same musical piece by nine professional guitarists, in which any indications of expressiveness is indicated, and performers have freely choose on the expressive actions performed. We have extracted descriptors from the score, and measure the deviations introduced on the performance by each performer in terms of the BPM, IOI and Energy deviations. We have obtained machine learning models using ANNs, and for each performer, and cross-validated the performance among interpreters' models based on CC. Preliminary results indicate, that performer take similar actions in terms of timing deviations, whereas less correlation was obtained in energy deviations.

References

Arcos, J.L., de Mantaras, R.L., Serra, X.: SaxEx: a case-based reasoning system for generating expressive performances. J. New Music Res. **27**, 194–210 (1998)

Clarke, E.F.: Some aspects of rhythm and expression in performances of Erik Saties Gnossienne No. 5. Music perception. Interdisc. J. 2, 299–328 (1985)

Friberg, A., Colombo, V., Fryden, L., Sundberg, J.: Generating musical performances with Director Musices. Comput. Music J. **24**, 23–29 (2000)

Gabrielsson, A.: Once again: the theme from Mozarts piano sonata in a major (K.331). In: A. Gabrielsson (ed.) Action and Perception in Rhythm and Music, vol. 55, pp. 81–103. Publications issued by the Royal Swedish Academy of Music, Stockholm (1987)

Gabrielsson, A.: Music performance research at the millenium. Psychol. Music **31**, 221–272 (2003)

Giraldo, S., Ramirez, R.: A machine learning approach to ornamentation modeling and synthesis in jazz guitar. J. Math. Music **10**(2), 107–126 (2016)

Giraldo, S.I., Ramirez, R.: A machine learning approach to discover rules for expressive performance actions in jazz guitar music. Front. Psychol. **7**, 1965 (2016)

Goebl, W.: Melody lead in piano performance: expressive device or artifact? J. Acoust. Soc. Am. **110**, 563–572 (2001)

Mazzola, G., Goller, S.: Performance and interpretation. J. New Music Res. **31**(221–232), 563–572 (2002)

Palmer, C.: Anatomy of a performance: sources of musical expression. Music Percept. **13**, 433–453 (1996)

Ramirez, R., Maestre, E., Serra, X.: A rule-based evolutionary approach to music performance modeling. IEEE Trans. Evol. Comput. **16**(1), 96–107 (2011)

Repp, B.H.: A microcosm of musical expression: II. Quantitative analysis of pianists dynamics in the initial measures of Chopins Etude in E major. J. Acoust. Soc. Am. 105, 1972–1988 (1999)

Shaffer, L.H., Clarke, E.F., Todd, N.P.M.: Metre and rhythm in piano playing. Cognition **20**, 61–77 (1985)

Todd, N.P.M.: The dynamics of dynamics: a model of musical expression. J. Acoust. Soc. Am. **91**, 3540–3550 (1992)

Widmer, G.: Machine discoveries: a few simple, robust local expression principles. J. New Music Res. **31**, 37–50 (2002)

Zapata, J.R., Davies, M.E., Gmez, E.: Multi-feature beat tracking. IEEE/ACM Trans. Audio Speech Lang. Process. **22**(4), 816–825 (2014)

Enhanced De-Essing via Neural Networks

Simon Hestermann$^{(\boxtimes)}$ ⓘ and Niklas Deffner ⓘ

Baden-Wuerttemberg Cooperative State University (DHBW), Stuttgart, Germany
me@simonhestermann.com

Abstract. De-essing is the process of attenuating vocal sibilance in audio recordings. Especially in audio mastering, conventional de-essers often degrade the clarity of the source signal due to unreliable differentiation between vocal sibilance and other high-pitched sounds. Machine learning poses a promising solution to this problem. In this context, a new de-essing approach based on a convolutional neural network architecture is presented. The introduced prototype de-esser outperforms existing de-esser plugins in terms of erroneous signal attenuation and was rated favorably by audio professionals.

Keywords: De-essing · Audio mastering · CNN

1 Introduction

The combination of machine learning and digital signal processing has led to new automated production tools, such as in the form of intelligent dynamic compression or re-mixing of complete mixes [5,6]. These techniques can be especially beneficial in mastering, where typically only complete mixes are altered [4].

A common time-consuming task in mastering is de-essing. De-essing aims to eliminate sharp vocal sibilance in audio recordings [3]. While conventional de-essers perform this task automatically, they may compromise the audio material beyond the desired sibilant parts, since complete mixes may contain other high-pitched sounds besides vocal sibilance that trigger the de-esser.

Deep convolutional neural networks (CNNs), on the other hand, have been successfully tested for sound classification and may provide new opportunities for the reliable identification of vocal sibilance in complete mixes [9,11]. This may overcome the detection limitations of conventional de-essers and allow for more precise automated de-essing in the mastering process.

In this context, data engineering steps and a CNN architecture for a de-esser prototype named Cytrus are presented. Cytrus surpasses the detection accuracy and precision of three prominent conventional de-esser plugins. In combination with an equalizer, the de-essing performance of Cytrus is favored by a major group of surveyed audio professionals, depending on the music genre.

The remainder of this paper is organized as follows. Conventional de-essing is briefly introduced in Sect. 2. All data related steps are presented in Sect. 3. The prototype architecture is presented in Sect. 4. Evaluation results are presented and discussed in Sect. 5. A summary is given in Sect. 6.

ⓒ Springer Nature Switzerland AG 2020
P. Cellier and K. Driessens (Eds.): ECML PKDD 2019 Workshops, CCIS 1168, pp. 537–542, 2020.
https://doi.org/10.1007/978-3-030-43887-6_49

2   Conventional De-Essing

Conventional de-essers use traditional signal processing techniques to detect vocal sibilance in audio signals [7,8]. For this purpose, frequencies in the upper frequency spectrum of a given audio signal are boosted or isolated and branched off into a side chain.

Whenever a defined dynamic threshold in the side chain is exceeded, sibilance is detected. In order to attenuate detected sibilance, the side chain signal is either processed directly and mixed back into the source signal or the processing of the source signal is controlled according to the side chain analysis. Processing usually consists of dynamic compression or equalization.

Despite advanced de-essing techniques, such as in the form of adaptive filters described in [1], existing approaches offer no reliable distinction between vocal sibilance and other sibilant sounds. Applying a conventional de-esser thus affects any sibilant sound that exceeds the de-essing threshold, such as cymbals or hi-hats, and compromises the clarity of the entire audio material. This is especially relevant in audio mastering, where complete mixes are processed [4]. Hence Cytrus, as presented in the following, meets a demand for de-essers that reliably attenuate vocal sibilance while leaving the remaining audio material unaltered.

3   Data Preparation

Training the CNN of the Cytrus prototype de-esser required a considerable amount of data. The authors annotated 8296 occurrences of vocal sibilance in 51 Pop music mixes from professional mastering projects. The files were sampled at 44.1 kHz and 24 bit resolution. A few files contained solo vocals, e.g. the singer's performance, while the majority were complete mixes with a fully instrumented playback and multiple vocal layers.

Binary classification was used to label vocal sibilance. Sibilance labelled as 1 was considered to require processing by the de-esser, while all the remaining audio material was automatically labelled 0. The annotations were linked to their exact sample positions in the respective source file, i.e. the labels refer to sample regions, not FFT windows. The authors discussed edge cases to ensure consistent labelling. The annotated sibilance regions ranged from 210 to 41584 samples in length, roughly corresponding to a right-skewed Gaussian distribution.

The audio data was converted to the frequency domain via fast Fourier transform. A FFT window size of 1024 samples was chosen to especially take short sibilance peaks into account. The windows overlapped by one quarter of their length and were modified by the von Hann window function. Zero-padding was applied to each window to double the resolution of the calculated auto power spectrum. The sample positions from the annotation process were converted to window labels accordingly.

The unfrequent occurrence of vocal sibilance compared to non-sibilant parts in the training data set required data augmentation. A duplicate of each positively labelled window was shifted by five frequency bins, as expressed in Eq. 1 with the binary shift matrix C:

$$[a_1 \, a_2 \, \ldots \, a_{1024}] \cdot C = [a_{1020} \, \ldots \, a_{1024} \, a_1 \, \ldots \, a_{1019}]. \tag{1}$$

Another set of duplicates of all positively labelled windows was created by shifting all frequency bins between the 200th and 750th index one index backwards. All other bins were replaced by uniformly distributed random noise. This followed the idea of confronting the model with the same sibilant frequencies in a different context of other surrounding frequencies. These two data augmentation steps improved the validation accuracy by about ten percentage points, but may become obsolete through more labelled sibilance.

The training data was optimized for the *sigmoid* activation function, as shown in Eq. 2. The auto power spectrum windows were shifted and compressed to center the spectrum at 0 and reduce the range of the most relevant sibilance peaks in the identified interval $[190, 350]$ to the interval $[-2.22, 4.88]$.

$$[190.00, 350.00] \xrightarrow[\text{shift}]{-240} [-50.00, 110.00] \xrightarrow[\text{compression}]{\cdot\frac{2}{45}} [-2.22, 4.88] \tag{2}$$

$$\downarrow sigmoid(x) \qquad sigmoid(x) \downarrow \qquad sigmoid(x) \downarrow$$

$$[1.00, 1.00] \qquad\qquad [0.00, 1.00] \qquad\qquad [0.10, 0.99]$$

The optimization for the *sigmoid* activation function led to a better model performance than using more linear activation functions, such as *relu*. The final interval of $[-2.22, 4.88]$ also yielded better results than smaller intervals, such as $[-1.00, 1.00]$, within the more linear slope of the *sigmoid* function.

In a last step, the 1024×1 auto power spectrum vectors were reshaped to 32×32 matrices for the two-dimensional convolutional input layer of the network. The reshaping filled the 32×32 matrix with the consecutive entries of the 1024×1 vector in row-major order. This improved the validation accuracy by three more percentage points.

The good performance of the network with these two-dimensional input matrices is surprising, given a horizontal distance of 21.5 Hz, but a vertical distance of approximately 688 Hz between frequency bins in the 32×32 matrices. Comparable one-dimensional networks with 1024×1 input vectors and a constant frequency bin distance of about 21.5 Hz, however, only yielded significantly lower validation and test scores. This may be explained by vocal sibilance characteristic spanning across a wide frequency range which may be easier to detect with this frequency bin arrangement.

4 Prototype Architecture

Cytrus consists of a CNN for sibilance detection and a high shelf filter for sibilance attenuation. A great variety of neural network architectures was implemented and evaluated in Keras. The two-dimensional CNN schematically depicted in Eq. 3 yielded the best validation accuracy of about 95% after optimization on the training data.

$$\text{Input (32,32,1)} \xrightarrow{} \text{Batch Norm.} \xrightarrow{\text{sigmoid}} \text{Conv2D 64, 5x5} \xrightarrow{} \text{Batch Norm.} \xrightarrow{\text{relu}} \text{Conv2D 64, 5x5} \xrightarrow{} \text{Batch Norm.} \xrightarrow{\text{relu}} \text{Conv2D 64, 5x5} \xrightarrow{} \text{Batch Norm.} \xrightarrow{\text{relu}} \text{Max Pooling 5x5} \xrightarrow{} \text{Dropout 50\%} \xrightarrow{} \text{Batch Norm.} \xrightarrow{\text{relu}} \text{Dense 256} \xrightarrow{} \text{Batch Norm.} \xrightarrow{\text{softmax}} \text{Output (2,)} \qquad (3)$$

Three two-dimensional convolutional layers with sigmoid and relu activations are followed by one max pooling layer, a dropout layer and one dense layer before the output layer with softmax activation. All convolutional layers have 64 filters and a kernel size of 5. The max pooling layer uses a stride of 1.

Batch normalization is applied before each layer except the dropout layer. Dropout and batch normalization were identified to prevent early overfitting during training [2,10]. This also led to better predictions for music genres that the network had not been trained on.

A high shelf filter was connected to the output of the CNN for sibilance attenuation. The windows that were classified as sibilant by the CNN were converted to sample ranges. The frequency response of these ranges was then attenuated by the high shelf filter by 6 dB from 7000 Hz upwards. As presented in Sect. 5, this simple approach already produced promising listening results. However, more sophisticated attenuation techniques can be expected to yield a more pleasing frequency attenuation in future iterations of Cytrus.

5   Evaluation

The detection accuracy of Cytrus was compared to the FabFilter Pro-DS, the DMG Audio Essence and the Logic Pro X de-esser plugins. In addition, a group of audio professionals was asked to rank the performance of all four de-essers.

For the quantitative evaluation, a selection of short music snippets of different music genres was assembled and annotated. About half of these music snippets notably differ from the Pop music genre the CNN was trained on.

The three conventional de-esser plugins provide a monitoring mode which only outputs the attenuated audio. This output was used for comparison to the true annotated labels. The obtained comparison metrics are denoted in Table 1.

Cytrus detects true positives and true negatives more reliably than the other de-essers with an accuracy of approximately 93%. From the standpoint of this paper, however, precision is more relevant, since false positives degrade the quality of faultless audio material. In this regard, Cytrus surpasses the conventional de-essers with a precision of about 61%. From these presented metrics, the set goal of reducing the degradation of non-sibilant audio material can be considered achieved. The recall score of Cytrus may be improved through a greater variety of training material in the future.

Table 1. Detection metrics in percentage. Cytrus achieves better accuracy, precision and F-score, but a slightly lower recall score than the conventional de-essers.

	Essence	Pro-DS	Logic	Cytrus
Accuracy	75.56	75.60	84.22	**92.84**
Precision	22.59	24.67	27.64	**60.54**
Recall	67.76	**79.67**	43.55	65.00
F-score	33.88	37.67	33.81	**62.67**

For the subjective evaluation of Cytrus, a Hip Hop, Acoustic Folk, Funk and Pop music snippet between twelve and 20 s were processed by the three conventional de-essers and Cytrus. The processed snippets were randomly renamed and sent to ten audio professionals who are familiar with the audio mastering process, as well as the mailing list members of the Music DSP mailing list by the Columbia University Computer Music Center.

An online questionnaire asked survey participants which processed version of each snippet they favor. The participants were introduced to the two goals of reliable vocal sibilance detection and no degradation of the remaining audio material. The 16 responses to the questionnaire are denoted in Table 2.

Table 2. Survey results. The percentages show how often each reference file of the respective de-esser was favored by survey participants, alongside the percentage of participants who commented that they had no preference.

	Essence	Pro-DS	Logic	Cytrus	Undecided
Hip Hop	18.75	18.75	6.25	**37.50**	18.75
Folk	25.00	18.75	0.00	**43.75**	12.50
Funk	0.00	**31.25**	25.00	12.50	31.25
Pop	**37.50**	6.25	12.50	18.75	25.00
Average	20.31	18.75	10.94	**28.13**	21.88

Despite its simple high shelf filter design, the Hip Hop and Folk snippet from the medley processed by Cytrus were favored by about 38% and 44% of the participants, respectively. Regarding the Funk and Pop snippet, Cytrus was outperformed by the FabFilter Pro-DS and DMG Audio Essence plugin. In these cases, however, the number of participants who had no preference were equal or slightly below the favored option. This may indicate that the Funk and Pop snippet alternatives were difficult to distinguish from one another.

Overall, the snippets processed by Cytrus were favored most often by about 28% of the participants. On average, about 22% of the participants had no preference among each set of snippet alternatives, which indicates that very close attention or personal experience might have been crucial for confident decisions and might not have applied to all participants. If undecided answers are not considered, however, Cytrus reaches an average preference of about 36%.

6 Summary

A prototype architecture and data engineering approach for the neural network driven detection of vocal sibilance was presented. The CNN architecture surpasses the accuracy and precision of three conventional de-essers in the performed test. The combination of the CNN and a simple filter for frequency attenuation convinced most surveyed audio professionals on average and was favored by a substantial margin in two of four cases. The presented results suggest that the use of neural networks is a promising approach to the reliable detection of vocal sibilance in complete mixes. If trained with a larger variety of data and combined with more advanced signal processing, future iterations of Cytrus are likely to produce even more convincing results. In future versions, the sequential nature of the audio signal may also be taken into account through the use of other neural network architectures, such as long short-term memory.

References

1. Flaks, J.S.: Apparatus and method for De-esser using adaptive filtering algorithms, US Patent 6,373,953, 16 Apr 2002
2. Ioffe, S., Szegedy, C.: Batch normalization: accelerating deep network training by reducing internal covariate shift. arXiv preprint arXiv:1502.03167 (2015)
3. Izhaki, R.: Mixing Audio. Taylor & Francis Group, London (2017)
4. Katz, B., Katz, R.A.: Mastering Audio: The Art and The Science. Butterworth-Heinemann, Oxford (2003)
5. Mimilakis, S.I., Cano, E., Abeßer, J., Schuller, G.: New sonorities for jazz recordings: separation and mixing using deep neural networks. In: Audio Engineering Society 2nd Workshop on Intelligent Music Production (2016)
6. Mimilakis, S.I., Drossos, K., Virtanen, T., Schuller, G.: Deep neural networks for dynamic range compression in mastering applications. In: Audio Engineering Society Convention 140. Audio Engineering Society (2016)
7. Oliveira, A.J.: A feedforward side-chain limiter/compressor/de-esser with improved flexibility. J. Audio Eng. Soc. **37**(4), 226–240 (1989). http://www.aes.org/e-lib/browse.cfm?elib=6092
8. Orban, R.A.: Combined De-esser and high-frequency enhancer using single pair of level detectors, US Patent 5,574,791, 12 Nov 1996
9. Salamon, J., Bello, J.P.: Deep convolutional neural networks and data augmentation for environmental sound classification. IEEE Signal Process. Lett. **24**(3), 279–283 (2017)
10. Srivastava, N., Hinton, G., Krizhevsky, A., Sutskever, I., Salakhutdinov, R.: Dropout: a simple way to prevent neural networks from overfitting. J. Mach. Learn. Res. **15**(1), 1929–1958 (2014)
11. Zhang, H., McLoughlin, I., Song, Y.: Robust sound event recognition using convolutional neural networks. In: 2015 IEEE International Conference on Acoustics, Speech and Signal Processing (ICASSP), pp. 559–563. IEEE (2015)

Representation, Exploration
and Recommendation of Playlists

Piyush Papreja[✉], Hemanth Venkateswara, and Sethuraman Panchanathan

Arizona State University, Tempe, AZ 85281, USA
{ppapreja,hemanthv,panch}@asu.edu

Abstract. Playlists have become a significant part of our listening experience because of digital cloud-based services such as Spotify, Pandora, Apple Music, making playlist recommendation crucial to music services today. With an aim towards playlist discovery and recommendation, we leverage sequence-to-sequence modeling to learn a fixed-length representation of playlists in an unsupervised manner. We evaluate our work using a recommendation task, along with embedding-evaluation tasks, to study the extent to which semantic characteristics such as genre, song-order, etc. are captured by the playlist embeddings and how they can be leveraged for music recommendation.

Keywords: Playlists · Sequence-to-sequence · Recommendation

1 Introduction

In this age of cloud-based music streaming services such as Spotify, Pandora, Apple music among others, users have grown accustomed to extended music listening experiences typically provided by playlists. As a result, playlist recommendation has been getting a lot of attention over the past couple of years. However, the playlist recommendation task has so far been analogous to playlist prediction [1] and continuation [2] rather than discovery. With billions of playlists already out there, and thousands being added every day, playlist discovery forms a significant part of the overall playlist recommendation pipeline. This work focuses on finding and recommending these existing playlists. We take inspiration from research in the domain of natural language processing to model playlist embeddings the way sentences are embedded by leveraging the relationship playlist:songs :: sentences:words, and model playlists using the sequence-to-sequence [3] learning technique.

In this work, we learn playlist embeddings in an unsupervised manner. We consider two main kinds of embedding models for this work: (a) Seq2seq models and (b) Bag of Words (BoW) models. We evaluate the models using recommendation and embedding-evaluation tasks, with the goal of analyzing the extent of information encoded by different models, and assessing the suitability of our approach for the purpose of recommendation. To the best of our knowledge, our

© Springer Nature Switzerland AG 2020
P. Cellier and K. Driessens (Eds.): ECML PKDD 2019 Workshops, CCIS 1168, pp. 543–550, 2020.
https://doi.org/10.1007/978-3-030-43887-6_50

work is the first attempt at modeling and extensively analyzing compact playlist representations for playlist recommendation. The demo, dataset, and slides for our work can be accessed online at http://www.playlist2vec.com/.

2   Seq2Seq Learning

Here we briefly describe the RNN Encoder-Decoder framework, proposed first in [4] and later improved in [3], upon which our model is based. Given a sequence of input vectors $x = \{x_1, x_2, x_3...x_T\}$, the encoder reads this sequence and outputs a vector c called the context vector. The context vector represents a compressed version of the input sequence which is then fed to the decoder which predicts tokens from the target sequence. One of the significant limitations of this approach was that the model was not able to capture long term dependencies for relatively longer sequences [5]. This problem was partially mitigated in [3] by using LSTM units instead of vanilla RNN units and feeding the input sequence in the reversed order to solve for lack of long-term dependency capture.

Bahdanau et al. [6] introduced the attention mechanism to solve this problem which involved focussing on a specific portion of the input sequence when predicting the output at a particular time step. The attention mechanism ensures the encoder doesn't have to encode all the information into a single context vector. In this setting, the context vector c is calculated using weighted sum of hidden states h_j:

$$c_i = \sum_{j=1}^{T_x} \alpha_{ij} h_j \tag{1}$$

where α_{ij} is calculated as follows:

$$\alpha_{ij} = \frac{\exp(e_{ij})}{\sum_{k=1}^{T_x} \exp(e_{ik})} \tag{2}$$

where $e_{ij} = a(s_{i-1}, h_j)$ and s_{i-1} is the decoder state at time step $i-1$ and h_j is the encoder state at time step j. $a(.)$ is the alignment model which scores how well the output at time step i aligns with the input at time step j. The alignment model a is a shallow feed forward neural network which is trained along with the rest of the network.

3   Embedding Models

In this section, we present the embedding models that we consider for this work:

1. **Bag-of-words Model (BoW):** For baseline comparison we apply a variant [7] of BoW, which uses a weighted averaging scheme to get the sentence embedding vectors followed by their modification using singular-value decomposition (SVD). This method of generating sentence embeddings proves to be a stronger baseline compared to traditional averaging.

2. **Base Seq2seq Encoder (base-seq2seq):** We use a deep, unidirectional RNN-based model with global attention for our base seq2seq model.
3. **Bidirectional Seq2seq Encoder (bi-seq2seq):** For this model, the encoder generated hidden states h_t, where $t \in \{1, \ldots, n\}$ are the concatenation of a forward RNN and a backward RNN that read the sentences in two opposite directions. Global attention is used for this model as well.

4   Experimental Setup

4.1   Data: Source and Filtering

We created the corpus by downloading 1 million publicly available playlists from Spotify using the Spotify developer API. As part of cleaning up the data before training, we follow [8] in discarding the less frequent, and less relevant items from our dataset. First, we remove the tracks occurring in less than 3 playlists, thereby removing rare songs from the corpus. This is a common preprocessing step in NLP-based works, which is equivalent of denoising the data by making the association weight between the more popular words stronger through the removal of their associations with less frequent words, as mentioned in [9]. All duplicate tracks from playlists are also removed. Finally, playlists with lengths in the range $\{10 \ldots 5000\}$ are retained and the rest are discarded. This resulted in a total of 745,543 unique playlists, 2,470,756 unique tracks, and 2680 unique genres, which we consider as training data.

4.2   Data Labeling: Genre Assignment

The songs in our dataset do not have genre labels, however artists do. Despite there being a 1:1 mapping between an artist and their song, we do not use the artist genre for the song because (1) an artist can have songs of different genres and (2) since genres are subjective in their nature (*rock* vs. *soft-rock* vs. *classic rock*), having a large number of genres for songs would result in an ambiguity between the genres with respect to empirical evaluation (classification) and add to the complexity of the problem. Hence, we aim to bring down the number of genres such that they are relatively mutually disjoint.

To achieve this we train a word-2-vec model [10][1] on our corpus to get song embeddings which capture the semantic characteristics (such as genre) of the songs by virtue of their co-occurrence in the playlists. Separate models are trained for embedding sizes $k = \{500, 750, 1000\}$. For each of the embedding sizes, the resulting song embeddings are then clustered into 200 clusters[2]. For each cluster, the artist genre is applied to the corresponding song and a genre-frequency (count) dictionary is created. From this dictionary, the genre having

---

[1] Word2vec details: algorithm: Skipgram, playlists length range: $\{30 \ldots 3000\}$, min. frequency threshold of the songs: 5, negative sampling: 5, window size: 5.
[2] This number was chosen to get maximum feasible clusters while keeping the number within limit which makes it feasible for annotating the data.

a clear majority[3] is assigned as the genre for all the songs in that cluster. All the songs in a cluster with no clear genre majority are discarded from the corpus. Based on the observed genre-distribution in the data, and as a result of clustering sub-genres (such as *soft-rock*) into parent genres (such as *rock*), the genres finally chosen for annotating the clusters are: *Rock, Metal, Blues, Country, Classical, Electronic, Hip Hop, Reggae and Latin*. To validate our approach, we train a classifier on our dataset consisting of annotated song embeddings. With training and test set kept separate at the time of training, we achieve a 94% test[4] accuracy.

For **playlist-genre annotation**, only the playlists having annotations for all the songs are considered, which leaves us with 339,998 playlists in total. This is done to perform a confident evaluation of the playlist embeddings by not making any assumptions about the genre information of songs that are not annotated. Further, since we use hard-labels [11] for the annotation process to make the evaluation task simpler, only those playlists are assigned genres for which more than 70% of the songs have the same genre. These playlists are used for the GDPred and the Recommendation evaluation tasks described in Sect. 5.

4.3 Training

We now outline our approach for estimating playlist embeddings using the following models:

1. **BoW Model:** We experiment with a weighted BoW model where the weight assigned to each song w is $a/(a + p(w))$. Here, a is the control parameter between $[e^{-3}, e^{-5}]$, and $p(w)$ is the (estimated) song frequency.
2. **Seq2seq-based Models:** We train our seq2seq models as autoencoders (where the target sequence is the same as the source sequence, a playlist) where the encoders and decoders are 3-layer networks with hidden state $k \in \{500, 750, 1000\}$. We experiment with both LSTM and GRU units, using Adam and SGD optimizers. We also set the maximum gradient norm to 1 to prevent exploding gradients.

5   Evaluation Tasks

In this section, we outline the criteria to evaluate our playlist embeddings for information content and playlist recommendation. As per the definition of a playlist [12] and the characteristics which make up for a good playlist [13], a good playlist embedding should encode information about the genre of the songs it contains, the order of songs, length of playlist (which directly shapes and impacts the listening experience of the user), and songs themselves, among many other traits. Based on that, we propose the following experiments for embedding playlist evaluation:

---

[3] This was a subjective decision. For example, a dictionary having {rock: 5, indie-rock: 3, blues: 2, soft-rock: 7} is assigned the genre *rock*.

[4] Result achieved for embedding size 750. Comparable results achieved for other sizes.

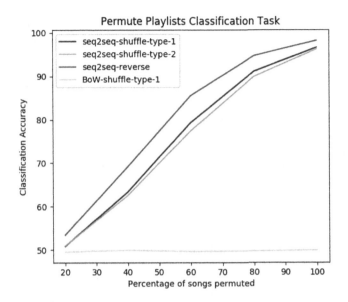

Fig. 1. Permute Classification Task Results: Seq2seq models outperform the BoW model in capturing the order of songs in a playlist. Also, the performance of the seq2seq models improve with the increasing proportion of permuted songs.

- **Genre Diversity Prediction Task (GDPred-Task):** This task measures the extent to which the playlist embedding captures the homogeneity and diversity of the songs (with regards to their genre) constituting it. Given a playlist embedding, the goal of the classifier is to predict the number of genres spanned by the songs in that playlist. The task is formulated as multi-class classification, with 3 output classes being low diversity (0–3 genres), medium diversity (3–6 genres) and high diversity (6–9 genres).
- **Song-content Task (SC-Task):** This closely follows the Word Content (WC) task [14] which evaluates whether it is possible to recover information about the original words in the sentence from the sentence embedding. We pick 750 mid-frequency songs (the middle 750 songs in our corpus of songs sorted by their occurrence count), and sample equal numbers of playlists that contain one and only one of these songs. We formulate it as a 750-way classification problem where the aim of the classifier is to predict which of the 750 songs does a playlist contain, given the playlist embedding.
- **Permute Classification Task:** Through this task we aim to answer the question: *Can the proposed embedding models capture song order, and if they can, to what extent?* We split this task into two sub tasks: **(i)** **Shuffle Task**, and **(ii)** **Reversal** task. In the **Shuffle** task, for each playlist in our task-specific dataset[5], we randomly select a fraction of all the songs in that playlist and shuffle them to create a permuted playlist. We then train a binary

---

[5] A list of 38168 playlists with lengths in the range $\{50, \ldots, 100\}$.

classifier to distinguish between the original and the permuted playlist embedding. The **Reversal** task is similar to the Shuffle task except that the randomly selected sub-sequence of songs is reversed.

- **Recommendation Task:** Recommendation being inherently subjective in nature is best evaluated by having user-labeled data. However, in the absence of such annotated datasets, we evaluate our proposed approach by measuring the extent to which the playlist space created by the embedding models is relevant, in terms of the similarity of genre and length information of closely-lying playlists. We quantify the *relevance of the embedding space* by calculating the precision recall scores in terms of genre and length labels, for a set of query playlists selected from the embedding space.

We use the Approximate Nearest Neighbors Algorithm using Spotify ANNOY library [15] to populate the tree structure with the genre-annotated playlist embeddings mentioned in Sect. 4.2. A query playlist is randomly selected and the search results are compared with the queried playlist in terms of genre and length information. There are nine possible genre labels. For comparing length, ten output classes (spanning the range $\{30 \ldots 250\}$) corresponding to bins of size 20 are created. The final precision value is calculated by taking an average of precision values for 100 queries for each recall value.

6   Results

The results for the GDPred-Task and the SC-Task are outlined in Table 1. In the **GDPred-Task**, the BoW model performs better than the seq2seq models achieving 80% accuracy while seq2seq models achieve an accuracy of 76%.

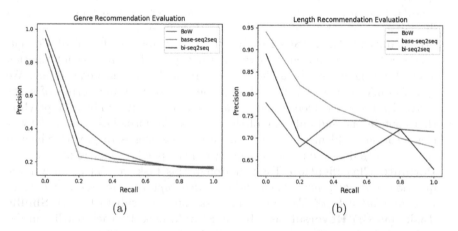

(a)   (b)

Fig. 2. Recommendation Tasks Results. (a) Genre Recommendation (b) Length Recommendation. BoW model captures genre information better, whereas seq2seq models capture length information better.

For the **SC-Task**, the seq2seq models perform poorly compared to the BoW model. However, our results for the seq2seq models closely match the results for the same task in [14], where the authors cite the inability of the seq2seq models to capture the content-based information due to the complexity of the way the information is encoded.

As seen in Fig. 1, for the **Permute Classification Task**, seq2seq model is able to distinguish correctly the permuted playlists from the original playlists as the proportion of the permutation is increased[6].

BoW model, on the other hand, fails the task as it is not able to capture the order information, thus making the seq2seq models better for capturing the song-order in the playlist.

The Recommendation task, as shown in Fig. 2a and b, captures some interesting insights about the effectiveness of different models for capturing different characteristics. Firstly, high precision values demonstrate the relevance of the playlist embedding space which is the first and foremost expectation from a recommendation system. Also, BoW models capture genre information[7] better than seq2seq models (Fig. 2a), while length information is better captured by the seq2seq models (Fig. 2b), demonstrating the suitability of different models for different tasks.

Table 1. Evaluation task accuracies for the embedding models for size 750.

	GDPred-Task	SC-Task
BoW	80.5	44.3
seq2seq	75.8	15.3
bi-seq2seq	76.2	21.7

7 Conclusions

We have presented a sequence-to-sequence based approach for learning playlist embeddings, which can be used for tasks such as playlist comparison and recommendation. First we define the problem of learning a playlist-embedding and describe how we formulate it as a seq2seq-based problem. We compare our proposed model with the weighted BoW model on embedding evaluation tasks as well as on a recommendation task. We show that our proposed approach is effective in capturing the semantic properties of playlists, and suitable for recommendation purposes.

Acknowledgement. The authors thank ASU, Adidas, and the National Science Foundation for their funding support. This material is partially based upon work supported by Adidas and by the National Science Foundation under Grant No. 1828010.

---

[6] Results for bi-seq2seq model follow a similar trend.

[7] Since BoW created playlist embeddings lie in the song space (as calculated using arithmetic mean of song embeddings) where genre annotation happens, they perform better.

References

1. Andric, A., Haus, G.: Automatic playlist generation based on tracking user's listening habits. Multimed. Tools Appl. **29**(2), 127–151 (2006)
2. Volkovs, M., Rai, H., Cheng, Z., Wu, G., Lu, Y., Sanner, S.: Two-stage model for automatic playlist continuation at scale. In: Proceedings of the ACM Recommender Systems Challenge 2018, p. 9. ACM (2018)
3. Sutskever, I., Vinyals, O., Le, Q.V.: Sequence to sequence learning with neural networks. In: Advances in NIPS (2014)
4. Cho, K., Van Merriënboer, B., Gulcehre, C., Bahdanau, D., Bougares, F., Schwenk, H., Bengio, Y.: Learning phrase representations using RNN encoder-decoder for statistical machine translation. arXiv preprint arXiv:1406.1078 (2014)
5. Bengio, Y., Simard, P., Frasconi, P., et al.: Learning long-term dependencies with gradient descent is difficult. IEEE Trans. Neural Netw. **5**(2), 157–166 (1994)
6. Bahdanau, D., Cho, K., Bengio, Y.: Neural machine translation by jointly learning to align and translate. arXiv preprint arXiv:1409.0473 (2014)
7. Arora, S., Liang, Y., Ma, T.: A simple but tough-to-beat baseline for sentence embeddings (2016)
8. De Boom, C., Agrawal, R., Hansen, S., Kumar, E., Yon, R., Chen, C.-W., Demeester, T., Dhoedt, B.: Large-scale user modeling with recurrent neural networks for music discovery on multiple time scales. Multimed. Tools Appl. **77**(12), 15385–15407 (2018)
9. Levy, O., Goldberg, Y., Dagan, I.: Improving distributional similarity with lessons learned from word embeddings. Trans. Assoc. Comput. Linguist. **3**, 211–225 (2015)
10. Mikolov, T., Chen, K., Corrado, G., Dean, J.: Efficient estimation of word representations in vector space. arXiv preprint arXiv:1301.3781 (2013)
11. Galstyan, A., Cohen, P.R.: Empirical comparison of "hard" and "soft" label propagation for relational classification. In: Blockeel, H., Ramon, J., Shavlik, J., Tadepalli, P. (eds.) ILP 2007. LNCS (LNAI), vol. 4894, pp. 98–111. Springer, Heidelberg (2008). https://doi.org/10.1007/978-3-540-78469-2_13
12. Fields, B., Lamere, P.: Finding a path through the juke box: the playlist tutorial. In: 11th International Society for Music Information Retrieval Conference (ISMIR). Citeseer (2010)
13. De Mooij, A.M., Verhaegh, W.F.J.: Learning preferences for music playlists. Artif. Intell. **97**(1–2), 245–271 (1997)
14. Conneau, A., Kruszewski, G., Lample, G., Barrault, L., Baroni, M.: What you can cram into a single vector: probing sentence embeddings for linguistic properties. arXiv preprint arXiv:1805.01070 (2018)
15. Bernhardsson, E.: Annoy: approximate nearest neighbors in C++/Python optimized for memory usage and loading/saving to disk (2013). https://github.com/spotify/annoy

Large-Scale Biomedical Semantic Indexing and Question Answering (BioASQ)

Results of the Seventh Edition
of the BioASQ Challenge

Anastasios Nentidis[1,2]($\boxtimes$), Konstantinos Bougiatiotis[1,3], Anastasia Krithara[1],
and Georgios Paliouras[1,4]

[1] National Center for Scientific Research "Demokritos", Athens, Greece
`{tasosnent,bogas.ko,akrithara,paliourg}@iit.demokritos.gr`
[2] Aristotle University of Thessaloniki, Thessaloniki, Greece
[3] National and Kapodistrian University of Athens, Athens, Greece
[4] University of Houston, Houston, TX, USA

Abstract. The results of the seventh edition of the BioASQ challenge are presented in this paper. The aim of the BioASQ challenge is the promotion of systems and methodologies through the organization of a challenge on the tasks of large-scale biomedical semantic indexing and question answering. In total, 30 teams with more than 100 systems participated in the challenge this year. As in previous years, the best systems were able to outperform the strong baselines. This suggests that state-of-the-art systems are continuously improving, pushing the frontier of research.

Keywords: Semantic indexing · Question answering · Biomedical knowledge

1 Introduction

The aim of this paper is twofold. First, we aim to give an overview of the data issued during the BioASQ challenge in 2019. In addition, we aim to present the systems that participated in the challenge and evaluate their performance. To achieve these goals, we begin by giving a brief overview of the tasks, which took place from February to May 2019, and the challenge's data. Thereafter, we provide an overview of the systems that participated in the challenge. Detailed descriptions of some of the systems are given in the workshop proceedings. The evaluation of the systems, which was carried out using state-of-the-art measures or manual assessment, is the last focal point of this paper, with remarks regarding the results of each task. The conclusions sum up this year's challenge.

2 Overview of the Tasks

The challenge comprised two tasks: (1) a large-scale biomedical semantic indexing task (Task 7a) and (2) a biomedical question answering task (Task 7b). In this section a brief description of the tasks is provided focusing on differences from previous years and updated statistics about the corresponding datasets. A complete overview of the tasks and the challenge is presented in [58].

P. Cellier and K. Driessens (Eds.): ECML PKDD 2019 Workshops, CCIS 1168, pp. 553–568, 2020.
https://doi.org/10.1007/978-3-030-43887-6_51

Table 1. Statistics on test datasets for Task 7a.

Batch	Articles	Annotated articles	Labels per article
1	7,358	7,194	11.67
	7,166	7,021	12.95
	11,019	10,831	13.04
	5,566	5,482	12.32
	6,729	6,353	12.96
Total	37,838	36,881	12.31
2	6,380	6,098	12.51
	6,785	6,621	12.75
	6,207	5,927	12.75
	7,382	7,079	13.00
	7,240	6,756	12.65
Total	33,994	32,481	12.27
3	6,266	5,835	12.58
	11,455	10,386	12.86
	4,750	3,947	12.67
	7,338	5,021	12.70
	6,920	4,554	12.63
Total	36,729	29,743	12.14

2.1 Large-Scale Semantic Indexing - 7a

In Task 7a the goal is to classify documents from the PubMed digital library into concepts of the MeSH hierarchy. Here, new PubMed articles that are not yet annotated by MEDLINE indexers are collected and used as test sets for the evaluation of the participating systems. Similarly to task 5a and 6a, articles from all journals were included in the test data sets of task 7a. As soon as the annotations are available from the MEDLINE indexers, the performance of each system is calculated using standard flat information retrieval measures, as well as, hierarchical ones. As in previous years, an on-line and large-scale scenario was provided, dividing the task into three independent batches of 5 weekly test sets each. Participants had 21 h to provide their answers for each test set. Table 1 shows the number of articles in each test set of each batch of the challenge. 14,200,259 articles with 12.69 labels per article, on average, were provided as training data to the participants.

2.2 Biomedical Semantic QA - 7b

The goal of Task 7b was to provide a large-scale question answering challenge where the systems had to cope with all stages of a question answering task for four types of biomedical questions: "yes/no", "factoid", "list" and "summary" questions [5]. As in previous years, the task comprised two phases: In phase A, BioASQ released 100 questions and participants were asked to respond with relevant elements from specific resources, including relevant MEDLINE articles, relevant snippets extracted from the articles, relevant concepts and relevant RDF triples. In phase B, the released questions were enhanced with relevant articles and snippets selected manually and the participants had to respond with *exact answers*, as well as with summaries in natural language (dubbed *ideal answers*). The task was split into five independent batches and the two phases for each batch were run with a time gap of 24 h. In each phase, the participants received 100 questions and had 24 h to submit their answers. Table 2 presents the statistics of the training and test data provided to the participants. The evaluation included five test batches.

Table 2. Statistics on the training and test datasets of Task 7b. All the numbers for the documents and snippets refer to averages.

Batch	Size	Documents	Snippets
Train	2,747	11.14	13.91
Test 1	100	3.07	3.93
Test 2	100	2.64	3.22
Test 3	100	3.08	4.05
Test 4	100	2.78	3.71
Test 5	100	2.39	2.62
Total	**3,247**	9.85	12.31

3   Overview of Participants

3.1   Task 7a

For this task, 12 teams participated and results from 30 different systems were submitted. In the following paragraphs we describe those systems for which a description was available, stressing their key characteristics. An overview of the systems and their approaches can be seen in Table 3.

The National Library of Medicine (NLM) team, in its *"ceb"* systems [48], adopts an end-to-end deep learning architecture with Convolutional Neural Networks (CNN) [27] to improve the results of the Medical Text Indexer (MTI) [35]. In particular, they combine text embeddings with journal information. They also consider information about the years of publication and indexing, to capture concept drift and variations in the MeSH vocabulary respectively. They also experiment with an ensemble of independently trained DL models.

The Fudan University team builds upon their previous *"DeepMeSH"* systems, which are based on document to vector (*d2v*) and tf-idf feature embeddings [43], the MESHLabeler system [28] and learning to rank (LTR). This year,

Table 3. Systems and approaches for Task 7a. Systems for which no description was available at the time of writing are omitted.

System	Approach
ceb	CNN, embeddings, ensembles
DeepMesh	d2v, tf-idf, MESHlabeler, attention scheme, PLT
Iria	bigrams, Luchene Index, k-NN, ensembles, UIMA ConceptMapper
MeSHProbeNet-P	Bidirectional RNN (GRU), attention scheme, encoder-decoder architecture
Semantic NoSQL KE	UIMA ConceptMapper, par2vec, DeepLearning4j[a]

[a]https://deeplearning4j.org/ Accessed June 2019

they incorporate AttentionXML [66], a deep-learning-based extreme multi-label text classification model, in the *"DeepMeSH"* framework. In particular, AttentionXML combines a multi-label attention mechanism, to capture label-specific information, with a shallow and wide probabilistic label tree (PLT) [18], for improved efficiency.

The *"Iria"* systems [52] are based on the same techniques used by their systems for the previous version of the challenge which are summarized in Table 3 and described in the corresponding challenge overview [38].

The *"MeSHProbeNet-P"* systems are upgraded versions of MeSH-ProbeNet [61], which participated in BioASQ6 with the name *"xgx"*. Their approach is based on an end-to-end deep learning model with an encoder-decoder architecture. The encoder consists of a recurrent neural network with multiple attentive MeSH probes to extract different aspects of biomedical knowledge from each input article. In *"MeSHProbeNet-P"* the attentive MeSH probes are also personalized for each biomedical article, based on the domain of each article as expressed by the journal where it has been published.

Finally, the *"Semantic NoSQL KE"* system variants [37] were developed extending previous year's *"SNOKE"* systems. The systems are based on the ZB MED Knowledge Environment [36], utilizing the Snowball Stemmer [1] and the UIMA [56] ConceptMapper to find matches between MeSH terms and words in the title and abstract of each target document, adopting different matching strategies. Paragraph Vectors [24] trained on the BioASQ corpus are used to rank and filter all the MeSH headings suggested by the UIMA-based framework for each document.

Similarly to the previous year, two systems developed by NLM to assist the indexers in the annotation of MEDLINE articles, served as baselines for the semantic indexing task of the challenge. MTI [35] with some enchantments introduced in [67] and an extension of it, incorporating features of the winning system of the first BioASQ challenge [59].

3.2   Task 7b

The question answering task was tackled by 73 different systems, developed by 18 teams. In the first phase, which concerns the retrieval of information required to answer a question, 6 teams with 23 systems participated. In the second phase, where teams are requested to submit exact and ideal answers, 13 teams with 52 different systems participated. An overview of the technologies employed by each team can be seen in Table 4.

Table 4. Systems and approaches for Task7b. Systems for which no information was available at the time of writing are omitted.

Systems	Phase	Approach
AUTH	A, B	MetaMap, BeCAS, Lucene Index, ElasticSearch, Wordnet, ELMo, SentiWordnet, w2vec, BiLSTM
AUEB	A	BM25, w2vec, BERT, DL (BCNN, PACRR, PDRMM)
MindLab	A	ElasticSearch, BM25, QuickUMLS, w2vec, WMD, DL (CNN)
_sys	A	Word and Sentence embeddings, Pseudo Relevance Feedback, BM25, LSI
BJUTNLP	B	SQUAD, GloVe, BiLSTM, Pointer Network
BIOASQ_VK	B	ELMo, DMN attention mechanisms, NLTK-VADER
DMIS	B	BioBERT, SQUAD, transfer learning
google	B	BERT, CoQA, Natural Questions
L2PS	B	SQUAD, Quasar-T, DRQA (RNN, LSTM), PSPR (LSTM), BioBERT
LabZhu	B	PubTator, Stanford POS tool, SPARQL
MQU	B	w2vec, tf-idf, DL (LSTM), Reinforcement Learning
UNCC	B	BioBERT, SQUAD, Stanford POS tool, AllenNLP entailment
unipi-quokka-QA	B	ELMo, ELMo-PubMed, BERT, BioBERT, SciSpacy

The "*AUTH*" team participated in both phases of Task 7B, with focus on phase B. For the document retrieval task, they experimented with approaches based on the BioASQ search services and ElasticSearch, querying with the conjunction of words in each question for the top 10 documents. In Phase B, for factoid and list questions they used updated versions of their BioASQ6 system [11], based on word embeddings, MetaMap [3], BeCAS [40] and WordNet. For yes/no questions they experiment with different deep learning methods, based on ELMo embeddings [46], SentiWordnet [12] and similarity matrices to represent the question/answer pairs and use them as input for different BiLSTM architectures [11].

The "*AUEB*" team participated in Phase A on document and snippet retrieval tasks yielding great results. They built upon their BioASQ6 document retrieval systems [6,29], which they modify to yield a relevance score for each sentence and experiment with BERT and PACRR [30] for this task. For snippet retrieval, they utilize a BCNN [64] model and a model based on POSIT-DRMM (PDRMM) [30]. They also introduce JPDRMM, a novel deep learning approach for joint document and snippet ranking, based on PDRMM [42].

Another approach based on deep learning methodologies for Phase A, focusing again on document and snippet retrieval, was proposed by the "*MindLaB*" team from the National University of Colombia [47]. For the document retrieval they use the BM25 model [53] and ElasticSearch [15] for efficiency, along with a Word Mover's Distance [22] based re-ranking scheme. For snippet retrieval, as in the previous approach, they utilized a very large collection of PubMed articles to train a CNN with similarity matrices of question-answer pairs. More specifically, they employ the BioNLPLab[1] w2vec embeddings that take into account the Part of Speech of each word. Also, they deploy the QuickUMLS [55] tool to create a cui2vec embedding for each snippet.

The "*_sys*" systems also participated in Phase A of Task 7B. These systems filter the queries, using stop-word lists and regular expressions, and expand them using word embeddings and pseudo-relevance feedback. Relevant documents are retrieved, utilizing Query Likelihood with bigrams and BM25, and reranked, based on Latent Semantic Indexing (LSI) and document vectors. In particular, document vectors based on averaging sentence embeddings are adopted. Finally, different lists of documents are merged to form the final result, considering the position of the documents in each list.

In phase B, most systems focused on using embeddings and deep learning methodologies to tackle the tasks. For example the "*BJUTNLP*" system utilizes the SQUAD Dataset for pre-training. The system uses both GloVe embeddings [45] (fine tuned during training) and character-level word embeddings (through a 1-dimensional CNN) as input to a BiLSTM model and for each question a Pointer Network [54] is finally responsible for pinpointing the exact start and end position of the answer in the relevant snippets.

The "*BIOASQ_VK*" systems were based on BioBERT [25], but with novel modifications to allow the model to cope with yes/no, factoid and list questions [41]. They pre-trained the model on the SQUAD dataset (for factoid and list questions) and SQUAD2 (for yes/no questions) to leverage the small size of the BioASQ dataset and by exploiting different pre-/post-processing techniques they obtained great results on all subtasks.

The "*DMIS*" systems focused on the importance of the information (words, phrases and sentences) for a given question [65]. To this end, sentence level embeddings based on ELMo embeddings [46] and attention mechanisms facilitated by Dynamic Memory Networks (DMN) [21] are deployed. Moreover, sentiment analysis is performed on yes/no questions to guide the classification (positive corresponds to yes) using the NLTK-VADER [17] tool.

---

[1] http://bio.nlplab.org Accessed June 2019.

The *"google"* systems [16], focus on factoid questions and are based on BERT based models [9], specifically the one in [2] trained on the Natural Questions [23] dataset, while also utilizing the CoQA [50] and the BioASQ datasets. They experiment with different input to the models, including the abstracts of relevant articles, the provided gold snippets and predicted relevant snippets. In particular, they focus on error propagation in end-to-end information retrieval and question answering systems, reaching the interesting conclusion that the information retrieval part is a bottleneck for such end-to-end QA systems.

Interesting results come from the *"L2PS"* team where they quantify the importance of pre-training and fine-tuning models for question answering and view the task under different regimes, namely Reading Comprehension (RC) and Open QA [19]. For the RC regime they use DRQA's document reader [7] while for the Open QA they utilize the PSPR model [26]. They experiment with different datasets (SQUAD [49] for RC and Quasar-T [10] for Open QA) for fine-tuning the models, as well as BioBert [25] embeddings to gain insights on the effect of the context length in this task.

The *"LabZhu"* [44] systems improved upon their systems from BioASQ6, with focus on exact answer generation. In particular, for factoid and list questions they developed two distinct approaches. One based on traditional information retrieval approaches, involving candidate answer generation and ranking, and one Knowledge-Graph based approach. In the latter approach, the answer type and the topic entity of the question are predicted and a SPARQL query is generated based on them and used to retrieve some results from the Knowledge Graph. Finally, the results of the two approaches are combined for the final answer of the question.

The Macquarie University (*"MQU"*) team focused on ideal answers and approached the task under a classification approach for snippet relevance [33]. Extending their previous work [31,32] the snippets are marked as summary relevant or not, utilizing w2vec embeddings and tf-idf vectors of the question-sentence pairs, showcasing that a classification scheme is more appropriate than a regression one. Also, based on their previous work [34], they conduct experiments using reinforcement learning towards the ROUGE score of the ideal answers and a correlation analysis between various ROUGE metrics and the BioASQ human evaluation scores, observing poor correlation of the ROUGE-Recall score with human evaluation.

The *"UNCC"* team focused on factoid, list and yes/no questions [57]. Their work is based on the BioBERT [25] embeddings fine-tuned on previous years of BioASQ. They also utilize the SQUAD dataset for factoid answers and incorporated the Lexical Answer Type (LAT) [13] and POS-tags along with hand made rules to address specific errors of the system. Furthermore, they incorporated the entailment of the candidate sentences in yes/no questions using the AllenNLP library [14].

Finally, the *"unipi-quokka-QA"* system tackled all the different question types in phase B [51]. Their work focused on experimenting with different Transformer models and embeddings, namely: ELMo, ELMo-Pumbed, BERT and BioBERT.

They used different strategies depending on the question type, such as ensembles on yes/no questions, biomedical named entity extraction (using SciSpacy [39]) on list questions and different pre-/post-processing procedures.

In this challenge too, the open source OAQA system proposed by [63] served as baseline for phase B. The system which achieved among the highest performances in previous versions of the challenge remains a strong baseline for the exact answer generation task. The system is developed based on the UIMA framework. ClearNLP is employed for question and snippet parsing. MetaMap, TmTool [60], C-Value and LingPipe [4] are used for concept identification and UMLS Terminology Services (UTS) for concept retrieval. The final steps include identification of concept, document and snippet relevance, based on classifier components and scoring, ranking and reranking techniques.

4   Results

4.1   Task 7a

Each of the three batches of Task 7a were evaluated independently. The classification performance of the systems were measured using flat and hierarchical evaluation measures [5]. The micro F-measure (MiF) and the Lowest Common Ancestor F-measure (LCA-F) were used to choose the winners for each batch [20].

According to [8] the appropriate way to compare multiple classification systems over multiple datasets is based on their average rank across all the datasets. On each dataset the system with the best performance gets rank 1.0, the second best rank 2.0 and so on. In case two or more systems tie, they all receive the average rank. Table 5 presents the average rank (according to MiF and LCA-F) of each system over all the test sets for the corresponding batches. Note, that the average ranks are calculated for the 4 best results of each system in the batch according to the rules of the challenge.

The results in Task 7a show that in all test batches and for both flat and hierarchical measures, some systems outperform the strong baselines. In particular, The *"MeSHProbeNet-P"* systems achieve the best performance in the first batch, outperformed by the *"DeepMeSH"* systems in the last two batches. More detailed results can be found in the online results page[2]. Comparison of these results with corresponding system results from previous years reveals the improvement of both the baseline and the top performing systems through the years of the competition as shown in Fig. 1.

4.2   Task 7b

Phase A: For phase A and for each of the four types of annotations: documents, concepts, snippets and RDF triples, we rank the systems according to the Mean Average Precision (MAP) measure. The final ranking for each batch is calculated

---

Table 5. Average system ranks across the batches of the Task 7a. A hyphenation symbol (-) is used whenever the system participated in fewer than 4 tests in the batch. Systems with fewer than 4 participations in all batches are omitted.

System	Batch 1		Batch 2		Batch 3	
	MiF	LCA-F	MiF	LCA-F	MiF	LCA-F
DeepMeSH5	-	-	1,00	1,00	1	1
DeepMeSH4	-	-	9,50	9,50	2,25	1,75
DeepMeSH3	8,25	8,50	3,50	5,00	2,5	2,75
DeepMeSH1	5,00	6,25	2,00	2,63	3,75	4,13
DeepMeSH2	7,25	7,25	3,50	4,50	4,75	4,38
MeSHProbeNet-P2	2,63	2,63	4,63	5,88	6,5	8,25
MeSHProbeNet-P1	3,25	2,13	6,38	4,25	6,88	6,5
MeSHProbeNet-P3	5,00	4,63	8,38	7,25	7,5	7,38
MeSHProbeNet-P	2,38	3,25	7,00	4,38	8,13	7,75
MeSHProbeNet-P0	1,50	1,25	6,25	5,63	8,75	7,88
ceb 1 ensemble	-	-	-	-	11	11
Default MTI	9,75	8,75	12,00	11,75	12,25	12,25
ceb1	8,75	9,25	11,00	11,25	12,25	13,5
MTI First Line Index	11,50	11,25	13,00	12,50	13,25	12
iria-mix	-	-	14,00	14,00	14,5	14,75
Semantic NoSQL KE 2	-	-	-	-	16	16
Semantic NoSQL KE 1	-	-	-	-	17	17,75

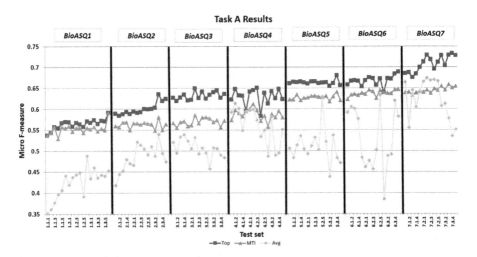

Fig. 1. The micro f-measure achieved by systems across different years of the BioASQ challenge. For each test set the micro F-measure is presented for the best performing system (Top) and the MTI, as well as the average micro f-measure of all the participating systems (Avg).

Table 6. Results for snippet retrieval in batch 4 of phase A of Task 7b.

System	Mean precision	Mean recall	Mean F-measure	MAP	GMAP
aueb-nlp-2	0.2060	0.4039	0.2365	**0.2114**	0.0075
aueb-nlp-1	0.2124	0.4083	0.2440	0.2086	0.0065
aueb-nlp-5	**0.2157**	**0.4235**	**0.2467**	0.1821	**0.0098**
MindLab QA Reloaded	0.1587	0.2760	0.1723	0.1527	0.0013
Deep ML methods for	0.1331	0.2692	0.1589	0.1234	0.0009
MindLab Red Lions++	0.1371	0.2538	0.1535	0.1187	0.0014
aueb-nlp-3	0.1488	0.3427	0.1779	0.1149	0.0053
MindLab QA System ++	0.1288	0.2049	0.1364	0.1136	0.0010
aueb-nlp-4	0.1520	0.3237	0.1791	0.1116	0.0056
MindLab QA System	0.1297	0.2536	0.1478	0.1094	0.0016
lh_sys1	0.0399	0.0810	0.0478	0.0178	0.0001
lh_sys3	0.0233	0.0437	0.0266	0.0151	0.0001
lh_sys5	0.0233	0.0437	0.0266	0.0151	0.0001
lh_sys4	0.0233	0.0437	0.0266	0.0148	0.0001
lh_sys2	0.0182	0.0281	0.0193	0.0051	0.0001

Table 7. Results for document retrieval in batch 3 of phase A of Task 7b. Only the top-10 systems are presented.

System	Mean precision	Mean recall	Mean F-measure	MAP	GMAP
aueb-nlp-4	0.1750	**0.6266**	0.2471	**0.1199**	0.0151
aueb-nlp-2	0.1740	0.6139	0.2449	0.1121	0.0156
aueb-nlp-5	**0.3599**	0.6128	**0.4034**	0.1102	**0.0164**
aueb-nlp-1	0.1700	0.5912	0.2380	0.1041	0.0118
auth-qa-1	0.2675	0.3896	0.2894	0.1033	0.0018
aueb-nlp-3	0.1600	0.5806	0.2266	0.0986	0.0104
lh_sys4	0.1420	0.5490	0.2081	0.0920	0.0069
Ir_sys1	0.1410	0.5365	0.2059	0.0907	0.0059
lh_sys1	0.1420	0.5449	0.2076	0.0881	0.0063
MindLab QA Reloaded	0.1330	0.5288	0.1950	0.0863	0.0062

as the average of the individual rankings in the different categories. In Tables 6 and 7 some indicative results from batches 3 and 4 are presented. Full results are available in the online results page of Task 7b, phase A[3]. These results are

---

[3] http://participants-area.bioasq.org/results/7b/phaseA/.

Table 8. Results for batch 5 for exact answers in phase B of Task 7b. Only the top-10 systems are presented along with the BioASQ baseline.

System	Yes/No		Factoid			List		
	Acc.	F1	Str. Acc.	Len. Acc.	MRR	Prec.	Rec.	F1
BioBERT-DMIS-3	**0.8286**	**0.8250**	**0.2857**	0.4286	0.3452	**0.5653**	0.4131	**0.4619**
BioBERT-DMIS	0.8000	0.7822	0.2571	0.4571	0.3224	0.5236	0.3714	0.4202
unipi-quokka-QA-5	0.8000	0.7939	0.0857	0.1714	0.1152	0.1713	**0.5873**	0.2537
BioBERT-DMIS-2	0.7429	0.7200	0.2571	0.4571	0.3271	0.5486	0.3992	0.4468
BioBERT-DMIS-4	0.7429	0.7351	0.2286	0.4571	0.3238	0.5069	0.3575	0.4051
google-gold-input-ab	0.7143	0.6941	0.2286	0.2857	0.2571	0.1774	0.4175	0.2415
unipi-quokka-QA-4	0.7143	0.6941	0.0857	0.1714	0.1152	0.1713	**0.5873**	0.2537
unipi-quokka-QA-3	0.6857	0.6578	0.0857	0.1714	0.1152	0.1713	**0.5873**	0.2537
google-gold-input	0.6571	0.6023	**0.2857**	0.3714	0.3167	0.2159	0.4452	0.2824
DMIS	0.6571	0.6023	**0.2857**	**0.5143**	**0.3638**	0.5050	0.3714	0.4124
BioASQ_Baseline	0.4857	0.4643	0.0571	0.1429	0.0867	0.2127	0.3619	0.2573

preliminary. The final results for Task 7b, phase A will be available after the manual assessment of the system responses.

Phase B: In phase B of Task 7b the systems were asked to produce exact and ideal answers. For ideal answers, the systems will eventually be ranked according to manual evaluation by the BioASQ experts [5]. Regarding exact answers[4], the systems were ranked according to accuracy, F1 score on prediction of yes answer, F1 on prediction of no and macro-averaged F1 score for the yes/no questions, mean reciprocal rank (MRR) for the factoids and mean F-measure for the list questions. Table 8 shows the results for exact answers for the last batch of Task 7b. These results are preliminary. The full results of phase B of Task 7b are available online[5]. The final results for Task 7b, phase B will be available after the manual assessment of the system responses.

The results presented in Fig. 2 show that this year the performance of systems in the yes/no questions, has clearly improved. In batch 5 for example, presented in Table 8, some systems outperformed the strong baseline based on previous versions of the OAQA system, with the top system achieving almost double the score of the baseline. Some improvement is also observed in the performance of the top systems for factoid and list questions in the preliminary results. However, there is even more room for improvement in these types of question as can be seen in Fig. 2.

---

[4] For summary questions, no exact answers are required.
[5] http://participants-area.bioasq.org/results/7b/phaseB/.

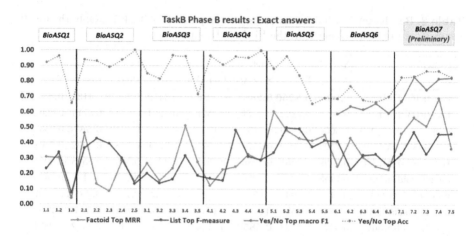

Fig. 2. The performance achieved by systems in exact answer generation part of Task B, Phase B, across different years of the BioASQ challenge. For each test set the performance of the best performing system (Top) is presented based on the official evaluation measures. Since BioASQ6 the macro-averaged F1 score (macro F1) is the official measure for Yes/No questions, but accuracy (Acc), the former official measure, is also presented. The results for BioASQ7 are preliminary. The final results for Task 7b, phase B will be available after the manual assessment of the system responses.

5   Conclusions

In this paper, an overview of the seventh BioASQ challenge is presented. The challenge consisted of two tasks: semantic indexing and question answering. Overall, as in previous years, the best systems were able to outperform the strong baselines provided by the organizers. This suggests that advances over the state of the art were achieved through the BioASQ challenge but also that the benchmark in itself is challenging. Moreover, the shift towards systems that incorporate ideas based on deep learning models observed in the previous year, is even more clear. Novel ideas have been tested and state-of-the-art deep learning methodologies have been adapted to biomedical question answering with great results. Specifically, the breakthroughs in different NLP tasks using clever techniques with the advent of new language-models, such as BERT and gpt-2, gave birth to new approaches that significantly boost the performance of the systems. In the future, we expect novel methodologies, such as the newly proposed XLNet [62], to further cultivate research in the biomedical information systems field. Consequently, we believe that the challenge is successfully pushing the research frontier of this domain. In future editions of the challenge, we aim to provide even more benchmark data derived from a community-driven acquisition process.

Acknowledgments. Google was a proud sponsor of the BioASQ Challenge in 2018. The seventh edition of BioASQ is also sponsored by the Atypon Systems inc. BioASQ is grateful to NLM for providing baselines for task 7a and to the CMU team for providing the baselines for task 7b. Finally, we would also like to thank all teams for their participation.

References

1. Agichtein, E., Gravano, L.: Snowball: extracting relations from large plain-text collections. In: Proceedings of the Fifth ACM Conference on Digital Libraries, DL 2000, pp. 85–94. ACM, New York (2000). https://doi.org/10.1145/336597.336644
2. Alberti, C., Lee, K., Collins, M.: A BERT baseline for the natural questions. arXiv preprint arXiv:1901.08634 (2019)
3. Aronson, A.R., Lang, F.M.: An overview of MetaMap: historical perspective and recent advances. J. Am. Med. Inform. Assoc. **17**, 229–236 (2010)
4. Baldwin, B., Carpenter, B.: Lingpipe (2003). Available from World Wide Web. http://alias-i.com/lingpipe
5. Balikas, G., et al.: Evaluation framework specifications. Project deliverable D4.1, UPMC, May 2013 (2013)
6. Brokos, G.I., Liosis, P., McDonald, R., Pappas, D., Androutsopoulos, I.: AUEB at BioASQ 6: Document and Snippet Retrieval, September 2018. http://arxiv.org/abs/1809.06366
7. Chen, D., Fisch, A., Weston, J., Bordes, A.: Reading Wikipedia to answer open-domain questions. arXiv preprint arXiv:1704.00051 (2017)
8. Demsar, J.: Statistical comparisons of classifiers over multiple data sets. J. Mach. Learn. Res. **7**, 1–30 (2006)
9. Devlin, J., Chang, M.W., Lee, K., Toutanova, K.: BERT: pre-training of deep bidirectional transformers for language understanding. arXiv preprint arXiv:1810.04805 (2018)
10. Dhingra, B., Mazaitis, K., Cohen, W.W.: Quasar: datasets for question answering by search and reading. arXiv preprint arXiv:1707.03904 (2017)
11. Dimitriadis, D., Tsoumakas, G.: Word embeddings and external resources for answer processing in biomedical factoid question answering. J. Biomed. Inform. **92**, 103118 (2019). https://doi.org/10.1016/j.jbi.2019.103118
12. Esuli, A., Sebastiani, F.: SENTIWORDNET: a publicly available lexical resource for opinion mining. In: Proceedings of the 5th Conference on Language Resources and Evaluation, LREC 2006, pp. 417–422 (2006)
13. Ferrucci, D., et al.: Building Watson: an overview of the DeepQA project. AI Mag. **31**(3), 59–79 (2010)
14. Gardner, M., et al.: AllenNLP: a deep semantic natural language processing platform. arXiv preprint arXiv:1803.07640 (2017)
15. Gormley, C., Tong, Z.: Elasticsearch: The Definitive Guide: A Distributed Real-Time Search and Analytics Engine. O'Reilly Media, Inc., Sebastopol (2015)
16. Hosein, S., Andor, D., Mcdonald, R.: Measuring domain portability and error propagation in biomedical QA. In: Seventh BioASQ Workshop: A Challenge on Large-Scale Biomedical Semantic Indexing and Question Answering (2019)
17. Hutto, C.J., Gilbert, E.: VADER: a parsimonious rule-based model for sentiment analysis of social media text. In: Eighth International AAAI Conference on Weblogs and Social Media (2014)
18. Jain, H., Prabhu, Y., Varma, M.: Extreme multi-label loss functions for recommendation, tagging, ranking & other missing label applications. In: Proceedings of the 22nd ACM SIGKDD International Conference on Knowledge Discovery and Data Mining - KDD 2016, pp. 935–944. ACM Press, New York (2016). https://doi.org/10.1145/2939672.2939756

19. Kamath, S., Grau, B., Ma, Y.: How to pre-train your model? Comparison of different pre-training models for biomedical question answering. In: Seventh BioASQ Workshop: A Challenge on Large-Scale Biomedical Semantic Indexing and Question Answering (2019)

20. Kosmopoulos, A., Partalas, I., Gaussier, E., Paliouras, G., Androutsopoulos, I.: Evaluation measures for hierarchical classification: a unified view and novel approaches. Data Min. Knowl. Disc. **29**(3), 820–865 (2015)

21. Kumar, A., et al.: Ask me anything: dynamic memory networks for natural language processing. In: International Conference on Machine Learning, pp. 1378–1387 (2016)

22. Kusner, M., Sun, Y., Kolkin, N., Weinberger, K.: From word embeddings to document distances. In: International Conference on Machine Learning, pp. 957–966 (2015)

23. Kwiatkowski, T., et al.: Natural questions: a benchmark for question answering research. Trans. Assoc. Comput. Linguist. **7**, 453–466 (2019). https://www.mitpressjournals.org/doi/full/10.1162/tacl_a_00276

24. Le, Q.V., Mikolov, T.: Distributed representations of sentences and documents, May 2014. http://arxiv.org/abs/1405.4053

25. Lee, J., et al.: BioBERT: pre-trained biomedical language representation model for biomedical text mining. arXiv preprint arXiv:1901.08746 (2019)

26. Lin, Y., Ji, H., Liu, Z., Sun, M.: Denoising distantly supervised open-domain question answering. In: Proceedings of the 56th Annual Meeting of the Association for Computational Linguistics (Volume 1: Long Papers), pp. 1736–1745 (2018)

27. Liu, J., Chang, W.C., Wu, Y., Yang, Y.: Deep learning for extreme multi-label text classification. In: Proceedings of the 40th International ACM SIGIR Conference on Research and Development in Information Retrieval, pp. 115–124. ACM (2017)

28. Liu, K., Peng, S., Wu, J., Zhai, C., Mamitsuka, H., Zhu, S.: MeSHLabeler: improving the accuracy of large-scale mesh indexing by integrating diverse evidence. Bioinformatics **31**(12), i339–i347 (2015)

29. McDonald, R., Brokos, G.I., Androutsopoulos, I.: Deep relevance ranking using enhanced document-query interactions, September 2018. http://arxiv.org/abs/1809.01682

30. McDonald, R., Brokos, G.I., Androutsopoulos, I.: Deep relevance ranking using enhanced document-query interactions. arXiv preprint arXiv:1809.01682 (2018)

31. Molla, D.: Macquarie University at BioASQ 5B query-based summarisation techniques for selecting the ideal answers. In: Proceedings BioNLP 2017 (2017)

32. Molla, D.: Macquarie University at BioASQ 6B: deep learning and deep reinforcement learning for query-based summarisation. In: Proceedings of the 6th BioASQ Workshop. A Challenge on Large-Scale Biomedical Semantic Indexing and Question Answering, pp. 22–29 (2018)

33. Molla, D., Jones, C.: Classification betters regression in query-based multi-document summarisation techniques for question answering. In: Seventh BioASQ Workshop: A Challenge on Large-Scale Biomedical Semantic Indexing and Question Answering (2019)

34. Mollá-Aliod, D.: Towards the use of deep reinforcement learning with global policy for query-based extractive summarisation. In: Proceedings of the Australasian Language Technology Association Workshop 2017, pp. 103–107 (2017)

35. Mork, J.G., Demner-Fushman, D., Schmidt, S.C., Aronson, A.R.: Recent enhancements to the NLM medical text indexer. In: Proceedings of Question Answering Lab at CLEF (2014)

36. Müller, B., Poley, C., Pössel, J., Hagelstein, A., Gübitz, T.: LIVIVO – the vertical search engine for life sciences. Datenbank-Spektrum **17**(1), 29–34 (2017). https://doi.org/10.1007/s13222-016-0245-2
37. Mller, B., Rebholz-Schuhmann, D.: Selected approaches ranking contextual term for the BioASQ multi-label classification (Task6a and 7a). In: Seventh BioASQ Workshop: A Challenge on Large-Scale Biomedical Semantic Indexing and Question Answering (2019)
38. Nentidis, A., Bougiatiotis, K., Krithara, A., Paliouras, G., Kakadiaris, I.: Results of the fifth edition of the BioASQ challenge. In: BioNLP 2017, pp. 48–57 (2017)
39. Neumann, M., King, D., Beltagy, I., Ammar, W.: ScispaCy: fast and robust models for biomedical natural language processing. In: Proceedings of the 18th BioNLP Workshop and Shared Task, pp. 319–327. Association for Computational Linguistics, Stroudsburg (2019). https://doi.org/10.18653/v1/W19-5034, https://www.aclweb.org/anthology/W19-5034
40. Nunes, T., Campos, D., Matos, S., Oliveira, J.L.: BeCAS: biomedical concept recognition services and visualization. Bioinformatics **29**(15), 1915–1916 (2013). https://doi.org/10.1093/bioinformatics/btt317
41. Oita, M., Vani, K., Oezdemir-Zaech, F.: Semantically corroborating neural attention for biomedical question answering. In: Seventh BioASQ Workshop: A Challenge on Large-Scale Biomedical Semantic Indexing and Question Answering (2019)
42. Pappas, D., McDonald, R., Brokos, G.I., Androutsopoulos, I.: AUEB at BioASQ 7: document and snippet retrieval. In: Seventh BioASQ Workshop: A Challenge on Large-Scale Biomedical Semantic Indexing and Question Answering (2019)
43. Peng, S., You, R., Wang, H., Zhai, C., Mamitsuka, H., Zhu, S.: DeepMeSH: deep semantic representation for improving large-scale mesh indexing. Bioinformatics **32**(12), i70–i79 (2016)
44. Peng, S., You, R., Xie, Z., Zhang, Y., Zhu, S.: The Fudan participation in the 2015 BioASQ challenge: large-scale biomedical semantic indexing and question answering. In: CEUR Workshop Proceedings, vol. 1391. CEUR Workshop Proceedings (2015)
45. Pennington, J., Socher, R., Manning, C.: Glove: global vectors for word representation. In: Proceedings of the 2014 Conference on Empirical Methods in Natural Language Processing (EMNLP), pp. 1532–1543 (2014)
46. Peters, M.E., et al.: Deep contextualized word representations, February 2018. http://arxiv.org/abs/1802.05365
47. Pineda-Vargas, M., Rosso-Mateus, A., Gonzlez, F., Montes-Y-Gmez, M.: A mixed information source approach for biomedical question answering: MindLab at BioASQ 7B. In: Seventh BioASQ Workshop: A Challenge on Large-Scale Biomedical Semantic Indexing and Question Answering (2019)
48. Rae, A., Mork, J., Demner-Fushman, D.: Convolutional neural network for automatic MeSH indexing. In: Seventh BioASQ Workshop: A Challenge on Large-Scale Biomedical Semantic Indexing and Question Answering (2019)
49. Rajpurkar, P., Zhang, J., Lopyrev, K., Liang, P.: Squad: 100,000+ questions for machine comprehension of text. arXiv preprint arXiv:1606.05250 (2016)
50. Reddy, S., Chen, D., Manning, C.D.: CoQA: a conversational question answering challenge. Trans. Assoc. Comput. Linguist. **7**, 249–266 (2019)
51. Resta, M., Arioli, D., Fagnani, A., Attardi, G.: Transformer models for question answering at BioASQ 2019. In: Seventh BioASQ Workshop: A Challenge on Large-Scale Biomedical Semantic Indexing and Question Answering (2019)

52. Ribadas-Pena, F.J., de Campos, L.M., Bilbao, V.M.D., Romero, A.E.: Cole and UTAI at BioASQ 2015: experiments with similarity based descriptor assignment. In: Working Notes of CLEF 2015 - Conference and Labs of the Evaluation Forum, Toulouse, France, 8–11 September 2015 (2015). http://ceur-ws.org/Vol-1391/84-CR.pdf

53. Robertson, S.E., Jones, K.S.: Relevance weighting of search terms. J. Am. Soc. Inf. Sci. **27**(3), 129–146 (1976)

54. See, A., Liu, P.J., Manning, C.D.: Get to the point: summarization with pointer-generator networks. arXiv preprint arXiv:1704.04368 (2017)

55. Soldaini, L., Goharian, N.: QuickUMLS: a fast, unsupervised approach for medical concept extraction. In: MedIR Workshop, SIGIR (2016)

56. Tanenblatt, M.A., Coden, A., Sominsky, I.L.: The conceptmapper approach to named entity recognition. In: LREC (2010)

57. Telukuntla, S.K., Kapri, A., Zadrozny, W.: UNCC biomedical semantic question answering systems. BioASQ: Task-7B, Phase-B. In: Seventh BioASQ Workshop: A Challenge on Large-Scale Biomedical Semantic Indexing and Question Answering (2019)

58. Tsatsaronis, G., et al.: An overview of the BIOASQ large-scale biomedical semantic indexing and question answering competition. BMC Bioinformatics **16**, 138 (2015). https://doi.org/10.1186/s12859-015-0564-6

59. Tsoumakas, G., Laliotis, M., Markontanatos, N., Vlahavas, I.: Large-scale semantic indexing of biomedical publications. In: 1st BioASQ Workshop: A Challenge on Large-Scale Biomedical Semantic Indexing and Question Answering (2013)

60. Wei, C.H., Leaman, R., Lu, Z.: Beyond accuracy: creating interoperable and scalable text-mining web services. Bioinformatics (Oxford, England) **32**(12), 1907–1910 (2016). https://doi.org/10.1093/bioinformatics/btv760

61. Xun, G., Jha, K., Yuan, Y., Wang, Y., Zhang, A.: MeSHProbeNet: a self-attentive probe net for MeSH indexing. Bioinformatics, 1–8 (2019). https://doi.org/10.1093/bioinformatics/btz142

62. Yang, Z., Dai, Z., Yang, Y., Carbonell, J.G., Salakhutdinov, R., Le, Q.V.: XLNet: generalized autoregressive pretraining for language understanding. CoRR abs/1906.08237 (2019). http://arxiv.org/abs/1906.08237

63. Yang, Z., Zhou, Y., Eric, N.: Learning to answer biomedical questions: OAQA at BioASQ 4B. In: ACL 2016, p. 23 (2016)

64. Yin, W., Schütze, H., Xiang, B., Zhou, B.: ABCNN: attention-based convolutional neural network for modeling sentence pairs. Trans. Assoc. Comput. Linguist. **4**, 259–272 (2016)

65. Yoon, W., Lee, J., Kim, D., Jeong, M., Kang, J.: Pre-trained language model for biomedical question answering. In: Seventh BioASQ Workshop: A Challenge on Large-Scale Biomedical Semantic Indexing and Question Answering (2019)

66. You, R., Dai, S., Zhang, Z., Mamitsuka, H., Zhu, S.: AttentionXML: extreme multi-label text classification with multi-label attention based recurrent neural networks, pp. 1–16, November 2018. http://arxiv.org/abs/1811.01727

67. Zavorin, I., Mork, J.G., Demner-Fushman, D.: Using learning-to-rank to enhance NLM medical text indexer results. In: ACL 2016, p. 8 (2016)

Selected Approaches Ranking Contextual Term for the BioASQ Multi-label Classification (Task6a and 7a)

Bernd Müller$^{(\boxtimes)}$ (iD) and Dietrich Rebholz-Schuhmann (iD)

ZB MED - Information Centre for Life Sciences,
Gleueler Str. 60, 50931 Cologne, Germany
{muellerb,rebholz}@zbmed.de
https://www.zbmed.de

Abstract. MeSH annotations are attached to the Medline abstracts to improve retrieval and this service is provided from the curators at the National Library of Medicine (NLM). Efforts to automatically assign such headings to Medline abstracts have proven difficult, on the other side, such approaches would increase throughput and efficiency. Trained solutions, i.e. machine learning solutions, achieve promising results, however these advancements do not fully explain, which features from the text would suit best the identification of MeSH Headings from the abstracts. This manuscript describes new approaches for the identification of contextual features for automatic MeSH annotations, which is a Multi-Label Classification (BioASQ Task6a): more specifically, different approaches for the identification of compound terms have been tested and evaluated. The described system has then been extended to better rank selected labels and has been tested in the BioASQ Task7a challenge. The tests show that our recall measures (see Task6a) have improved and in the second challenge, both the performance for precision and recall were boosted. Our work improves our understanding how contextual features from the text help reduce the performance gap given between purely trained solutions and feature-based solutions (possibly including trained solutions). In addition, we have to point out that the lexical features given from the MeSH thesaurus come with a significant and high discrepancy towards the actual annotations of MeSH Headings attributed by human curators, which also hinders improvements to the automatic annotation of Medline abstracts with MeSH Headings.

Keywords: Paragraph Vectors · Named Entity Recognition · Semantic Retrieval · UIMA · DeepLearning4j · BioASQ

1 Introduction

The scientific biomedical literature is being collected and archived by the National Library of Medicine (NLM) over the past 150 years. Documents have

© Springer Nature Switzerland AG 2020
P. Cellier and K. Driessens (Eds.): ECML PKDD 2019 Workshops, CCIS 1168, pp. 569–580, 2020.
https://doi.org/10.1007/978-3-030-43887-6_52

manually been annotated with Medical Subject Headings[1] in order to search and access the documents efficiently. The process of manually assigning indexing terms is very time consuming and thus tedious work. Furthermore, the biomedical literature in PubMed has grown from 12 Million citations in 2004 [4] to 29 Million citations in 2019[2] having a growth rate of 4% per year [23] leading to high pressure in delivering the MeSH annotations.

The growth in published biomedical literature as well as the difficulties in manually assigning indexing terms shows the need for routines that automatically annotate and index the scientific articles in order to use metadata terms for information retrieval purposes. At best such supporting automatic solutions should also contribute clues to the curators about the selection of most relevant and best supported terms throughout all the stages of their work. Such clues could be difficult to derive, e.g., from the scientific text, since the MeSH Headings cover mostly compound terms, which – at best – have complex representations in the text.

The Medical Text Indexer (MTI) has been developed by the NLM to provide an automated indexing system for the Medical Subject Headings to the curators. From 2000 onward, the NLM indexing initiative has been initiated, in particular due to the availability of the electronic versions of the scientific articles since the mid 90s [2]. However, the newly introduced automated indexing systems had to be evaluated to compare and improve the performance against benchmarks. The ongoing developments of the MTI then introduced machine learning components that have been tested across different document types, e.g., clinical health records that require different indexing approaches than merely assignment of MeSH Headings. The performance of the MTI on clinical health records has been evaluated in 2007 for the assignment of ICD-9 codes with promising results [3].

Solutions for the automated assignment of MeSH Headings have barely ever been evaluated, neither for their performance nor for their reproducibility. Conceptually, the evaluation of six different MeSH taggers showed that the k-nearest neighbour (k-NN) approach outperforms all other solutions [33]. Apart from the NLM's critical response with regards to reproducibility, NLM still emphasizes "that current challenges in MeSH indexing include an increase of the scope of the task" [26].

The demands for such evaluation has motivated the NLM improving MTI as well as organizing large-scale evaluation challenges. Now MTI incorporates k-NN clustering showing a boost in the performance of the system [15], and in 2012, the BioASQ challenge was initiated (funding horizon of 5 years) leading to the evaluation of systems for large-scale biomedical indexing and question answering [34].

---

[1] https://www.ncbi.nlm.nih.gov/mesh. Accessed May 2019.

[2] https://www.ncbi.nlm.nih.gov/pubmed/. Accessed May 2019.

2   Related Work

In 2013, the first BioASQ challenge was comprising two tasks, one on large-scale semantic indexing for the automated assignment of MeSH Headings to unlabeled Medline citations, the other one on question answering for scientific research questions in the biomedical domain [27]. In the first BioASQ challenge, 11 teams participated in Task A with 40 systems. In Task B, three different teams participated with 11 systems.

In Task A, there were two baselines of Task A for large-scale semantic indexing, the first one was an unsupervised machine learning approach, the second one was based on NLM's MTI. The evaluation was conducted using the metrics Micro F-measure (MiF) and Lowest Common Ancestor F-measure (LCA-F). The best-performing system, even outperforming the MTI baseline, called AUTH [36], is based on a binary Support Vector Machine (SVM) predicting N top labels for each article with a certain confidence score to rank the predicted labels.

In Task B, two baselines were created as the top 50 and top 100 predictions of an ensemble system that combines predictions of factoid and list questions, yes/no questions, and summary questions. The evaluation metric was the Mean Average Precision (MAP). The Wishart [10] system was able to outperform the two baselines. It uses the PolySearch[3] tool for query expansion and the retrieval of candidate documents from which either entities or sentences are extracted as answers for the respective questions.

The BioASQ challenge was then executed every year until today bringing about a variety of approaches in both tasks A and B [5,8,19,25]. In Task A, MeSHLabeler performed best the challenges 2014, 2015 and 2016 [21] using an ensemble approach of k-NN, the MTI itself as well as further MeSH classification solutions.

In recent years, term vector space representations have been introduced exceeding classical bag-of-words approaches, since they are able to capture the context of words in the text and to prioritize words in the vector representation according to given similarity scores [24,30]. In addition, the word vectorization allows for better use of sentence and paragraph representations [20] in the machine learning approaches, e.g., deep learning. The organizers of the BioASQ challenge also published a word2vec representation of PubMed articles [28] for participants to improve their systems.

In 2017, the first deep learning based approach called DeepMeSH participated in the challenge of Task A and performed best [29]. In 2018, DeepMeSH outperformed others in 2 out of 3 batches while the third batch was won by a set of systems called "xgx" that is potentially associated to the AttentionMeSH system [16]. This system uses end-to-end DeepLearning incorporating an attention layer to emphasize predictions towards commonly used MeSH labels.

Further systems participated in the Task A employing named entity recognition with lexical features such as a dictionary, and using Paragraph Vectors also [22]. It has been shown that machine learning-based approaches based on

---

[3] http://wishart.biology.ualberta.ca/polysearch/.

k-NN and Paragraph Vectors can be used to boost the performance in the BioASQ challenge [17]. This paper describes the participation of a system for Multi-Label Classification based on named entity recognition with lexical features that incorporate Label Ranking derived from Paragraph Vectors in order to achieve a conjoint system for Multi-Label Ranking.

3   Methodology

Task A of the BioASQ challenge is a Multi-Label Classification task, which in addition can be subdivided into the two sub-tasks of Multi-Label Classification and Label Ranking [35]. The resulting classification of multiple labels with an assigned confidence score for each label is a Multi-Label Ranking [9].

Initially, a subset of MeSH Headings is attributed to each document in the test set of Medline citations. Then, each MeSH Heading combined with its confidence score representing the probability for the MeSH Heading being correctly assigned to the respective Medline citation. The resulting set of MeSH Headings is filtered according to the minimum confidence score.

The two sub-tasks, i.e. Multi-Label Classification and Label Ranking for a Multi-Label Ranking, are given in the system architecture of this paper. The first component creates an initial set of MeSH Headings for each document in the test set for the Medline citations. Then, all MeSH Headings receive a confidence score to generate the scored MeSH Headings.

The first component for the task of Multi-Label Classification is described in Sect. 3.1 and the second component for the Label Ranking is described in Sect. 3.2. The combined system for the Multi-Label Ranking is described in Sect. 3.3.

3.1   Multi-label Classification

The Multi-Label Classification task is based on lexical features for the named entity recognition solution that has been developed within the Unstructured Information Management Architecture (UIMA)[4] [11–14,31]. In the framework, a reader for the BioASQ JSON format processes the document stream through the Common Analysis System (CAS) of the pipeline. Tokenization is conducted using an Offset Tokenizer that splits tokens at their whitespaces and punctuations. Stemming of the tokens is conducted using the Snowball Stemmer [1]. The stemmed tokens are analyzed using the analysis engine ConceptMapper [32] that uses a dictionary to annotate matching synonyms in the text with offset information onto concept identifiers. In the last part of the UIMA-pipeline, the documents with their annotated MeSH Headings are written with a CAS-Writer into the BioASQ submission format. The implemented workflow is shown in Fig. 1.

The lexical features for the ConceptMapper are provided as a dictionary that is created from the current MeSH (version 2019). In the dictionary, concepts are

---

[4] https://uima.apache.org/. Accessed May 2019.

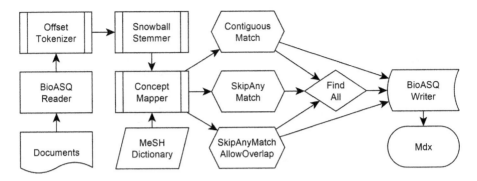

Fig. 1. The UIMA-based workflow with different combinations of configurations for the ConceptMapper to produce the result set of MeSH Headings for Medline citations.

created for each MeSH Heading and synonyms are added from the MeSH Entry Terms for the MeSH Heading. Further synonyms are created with the Snowball Stemmer by stemming the concept name as well as each of the synonyms. The resulting dictionary for the ConceptMapper contains 29,351 different concepts with 251,463 synonyms.

The analysis engine ConceptMapper [32] provides various dictionary look-up solutions that can match against different sequences of tokens. Before applying a matching strategy, stop words and punctuation are removed. Then, one of the three lookup strategies are applied with a flag for allowing partial matches or allowing only complete matches. The different look-up solutions with the flag for finding also partial matches of synonyms will result in 5 different pattern matching configurations. For the BioASQ Task6a in 2018, each of the different dictionary look-up approaches are listed as separate system enumerated as SNOKE1 to SNOKE5. For the BioASQ Task7a in 2019, all the results from the five different systems have been merged together into a union set.

3.2 Label Ranking

Paragraph Vectors allow for capturing contextual information of words in text. The contextual information is trained by calculating the probability of certain words preceding or succeeding the contextual word. The resulting Paragraph Vector model enables the calculation of similarities of different texts according to their probability of occurring close to each other.

The task of Label Ranking is conducted by creating such a Paragraph Vector model to score all MeSH Headings for each document in the test set. Each MeSH Headings gets an assignment of a confidence score ranging from -1 to 1. In order to provide such a system for assigning confidence scores, an unsupervised machine learning model is trained according to the algorithm described in [20] (Fig. 2).

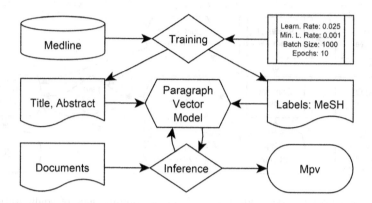

Fig. 2. The Paragraph Vector model is trained on the BioASQ corpus with $225,127$ documents published from 2018 until January 2019 using the given MeSH Headings from the documents as training labels. Documents during the challenge were inferred using the trained model that resulted in a set M_{pv} of scored MeSH Headings for each document.

The Paragraph Vectors were trained using the BioASQ corpus with $225,127$ citations that have been published either in 2018 or 2019 in conjunction with their corresponding MeSH Headings of $25,363$ different labels in total. The Paragraph Vector model is PV-DBOW based on Skip-Grams (length 4) and have been trained using a configuration with 10 epochs, a learning rate of 0.025, a minimum learning rate of 0.001, and a batch size of 1000. The model is available online[5]. The training as well as the predictions during the BioASQ challenge were implemented using the DeepLearning4j framework[6].

3.3   Multi-label Ranking

The task of Multi-Label Ranking is achieved by combining both systems for Multi-Label Classification and for Label Ranking. The first system uses five different vectors of MeSH Headings according to the pattern matching algorithm. For each document, a single set of MeSH Headings is created by taking the union set from the five different vectors.

Similarly, the Paragraph Vector model is used for assigning the confidence scores to each of the MeSH Headings. This results in a set of $25,363$ Headings for each document with a confidence score of -1 to 1 assigned to each Heading. Then, both the union set as well as the confidence scores for the MeSH Headings are joined by filtering the union set for only the top-k scored terms. K was chosen for 500 and for $1,0000$ resulting in two different BioASQ Task7a submissions. The algorithm for creating the sets M_{top500} and $M_{top1000}$ for each document is shown in Algorithm 1.

---

[5] https://gitlab.zbmed.de/mueller/dl4j-models/blob/master/15000000. Accessed May 2019.

[6] https://deeplearning4j.org/. Accessed May 2019.

Data: M_{dx}; M_{pv}; $m \leftarrow length(D)$;
Result: M_{top500}; $M_{top1000}$;
$M_{top500} \leftarrow \{\}$;
$M_{top1000} \leftarrow \{\}$;
for $(i \leftarrow 0; \; i < m; \; i \leftarrow i+1)$ **do**
    $counter \leftarrow 0$;
    for p in $M_{pv}[i]$ **do**
        if $M_{dx}[i].contains(p)$ *AND counter* < 500 **then**
           | $M_{top500}[i] \leftarrow M_{top500}[i].add(p)$;
        end
        if $M_{dx}[i].contains(p)$ *AND counter* < 1000 **then**
           | $M_{top1000}[i] \leftarrow M_{top1000}[i].add(p)$;
        end
        $counter \leftarrow counter + 1$;
    end
end

Algorithm 1: Algorithm for harmonizing the results by taking only MesH Headings that are either scored in the top500 or the top1000 by the predictions with the Paragraph Vector model.

4 Results

In BioASQ TaskA, the systems have been challenged to outperform the MTI for the annotation of Medline citations with MeSH Headings. In the challenge, there have been three test batches leading to 5 runs for each batch. For each run, a test set of Medline citations has been published that have not yet been annotated with MeSH Headings by human curators. The evaluation of the participating systems for each of the runs in every batch is an automated process implemented within the BioASQ infrastructure [6,7].

The evaluation infrastructure computes the results with two different classes of measurements, flat and hierarchical. The comparison of the performance of the participating systems are assessed with one flat and one hierarchical measure: the Lowest Common Ancestor F1-measure `LCA.F` [18] and the Label-Based Micro F1-measure `MiF`. Besides the two main evaluation F1-measures, there is also the Example-Based F1-Measure, Accuracy, Label-Based Macro F1-Measure, and Hierarchical F1-Measure. For each F1-Measure, the respective precision and recall measures are calculated.

The system for the Multi-Label Classification participated in the Task6a in 2018 and is explained in Sect. 4.1. The conjoint system that uses the initial Multi-Label Classification for Label-Ranking in order to produce a Multi-Label Ranking participated in the Task7a in 2019 and is explained in Sect. 4.2.

4.1 System for Multi-label Classification in Task6a

In the 2018 BioASQ Task6a, the maximum MiF score of 0.6880 was achieved by the system xgx and the maximum LCA-F score of 0.5596 was achieved by

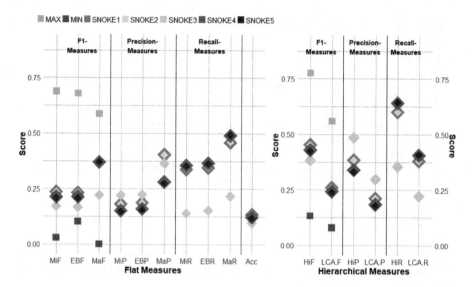

Fig. 3. The 10 flat and the 6 hierarchical measures are sorted according to the F1 measure, Precision, and Recall for each of the different configuration setups in Task6a. The maximum score achieved by the best participating system is given as green squares, the minimum score achieved by the worst participating system as red squares. (Color figure online)

the system xgx0. The maximum label-based micro-recall (MiR) was 0.6751 and the maximum label-based micro-precision (MiP) was 0.8110. The highest lowest common ancestor recall (LCA-R) was 0.5563 and the highest lowest common ancestor precision (LCA-P) was 0.6212. For all participating systems, the tendency was towards having a higher precision than having a higher recall.

Contrastingly, the submissions for the Multi-Label Classification system with SNOKE1 to SNOKE5 reached higher recall measures than precision measures. The maximum MiF score of 0.236 was achieved by both the submissions for SNOKE1 and SNOKE2. The maximum LCA-F score of 0.261 was achieved by the submission for SNOKE1. The highest MiR was 0.356 while the MiP was 0.221. A similar picture was shown for the highest LCA-R having a 0.408 while the highest LCA-P was 0.298. In Fig. 3, the 10 flat and the 6 hierarchical measurements for SNOKE1 to SNOKE5 are visualized.

4.2   System for Multi-label Ranking in Task7a

In the 2019 BioASQ Task7a, both the maximum MiF score of 0.733 and the maximum LCA-F score of 0.612 was achieved by the system DeepMeSH5. The maximum (MiR) was 0.707 and the maximum MiP was 0.791. The highest LCA-R was 0.6 and the highest LCA-P was 0.663. Similar to the Task6a in 2018, the tendency was again more towards having a higher precision than having a higher recall.

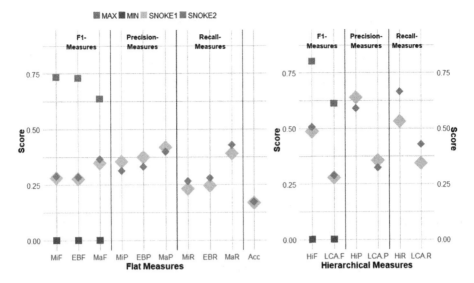

Fig. 4. The 10 flat and 6 hierarchical measures arranged according to their F1-Measure, Precision, and Recall for each configuration setup in Task7a. The maximum score achieved by the best participating system as green squares, the minimum score achieved by the worst participating system as red squares. (Color figure online)

In comparison to the 2018 participation, the submissions for the Multi-Label Ranking system SNOKE1 and SNOKE2 reached higher precision measures than for the recall measures for the label-based micro-measures. The maximum MiF score was 0.288 with a maximum MiP of 0.354 and a maximum MiR of 0.267. For the measures of the lowest common ancestor, the recall was higher than the precision. The maximum $LCA - F$ score was 0.288 with a maximum LCA-P of 0.356 and a maximum LCA-R with 0.428. In Fig. 4, the 10 flat and the 6 hierarchical measurements for the submissions of SNOKE1 and SNOKE2 are visualized.

5   Conclusion

This paper describes the participation of two different systems and their combined solution in the Task6a and the Task7a of the BioASQ challenge. The first system that participated in Task6a is a Multi-Label Classification system that incorporates lexical features from MeSH. The system was extended for the participation in the Task7a for the functionality of introducing Label Ranking for the assignment of confidence scores to MeSH Headings resulting in a conjoint system for Multi-Label Ranking.

The first system for Multi-Label Classification participated in the Task6a of the BioASQ challenge. The results indicate that the recall for the system are higher than the precision although the general tendency of the other participating systems is the opposite. Nevertheless, the label-based macro F1-measure shows better performance than the label-based micro F1-measure.

The second system that incorporates both Multi-Label Classification and Label-Ranking for Multi-Label Ranking participated in the Task7a of the BioASQ challenge. All performance measures were improved in comparison to the first system that participated in the Task6a of the BioASQ challenge. Generally, the precision has been boosted in comparison to the earlier participation.

6   Discussion

In the BioASQ challenge, systems are supposed to outperform the baseline of the MTI for the assignment of MeSH Headings to Medline citations. Two different participations in the BioASQ challenge are described in this paper, one for Task6a and one for Task7a. The initially developed system that incorporates lexical features from the MeSH thesaurus is extended to a ranking of MeSH Headings according to the confidence values.

The participation of the first system in Task6a generally shows higher recall than precision performance. The extended system that exploits the ranking of MeSH Headings was able to increase both, precision and recall, resulting in better F1-measures overall. However, the initially assigned MeSH Headings based on lexical features still have a low overlap in comparison to the assigned MeSH Headings by the human curators.

References

1. Agichtein, E., Gravano, L.: Snowball: extracting relations from large plain-text collections. In: Proceedings of the Fifth ACM Conference on Digital Libraries, pp. 85–94. ACM, New York (2000). https://doi.org/10.1145/336597.336644
2. Aronson, A., et al.: The NLM indexing initiative. In: AMIA 2000, American Medical Informatics Association Annual Symposium, Los Angeles, CA, USA, 4–8 November 2000 (2000)
3. Aronson, A., et al.: From indexing the biomedical literature to coding clinical text: experience with MTI and machine learning approaches. In: Biological, Translational and Clinical Language Processing, BioNLP@ACL Prague, Czech Republic, pp. 105–112 (2007)
4. Aronson, A., Mork, J., Gay, C., Humphrey, S., Rogers, W.: The NLM indexing initiative's medical text indexer. Stud. Health Technol. Inform. **107**, 268–272 (2004)
5. Balikas, G., Kosmopoulos, A., Krithara, A., Paliouras, G., Kakadiaris, I.: Results of the BioASQ tasks of the question answering lab at CLEF. In: Conference and Labs of the Evaluation forum, Toulouse, France (2015). http://ceur-ws.org/Vol-1391/inv-pap7-CR.pdf
6. Balikas, G., Partalas, I., Baskiotis, N., Artieres, T., Gausier, E., Gallinari, P.: Evaluation infrastructure software for the challenges 2nd version. Technical report D4.7 (2014)
7. Balikas, G., Partalas, I., Baskiotis, N., Artieres, T., Gaussier, E., Gallinari, P.: Evaluation infrastructure. Technical report (2013)

8. Balikas, G., Partalas, I., Ngonga-Ngomo, A., Krithara, A., Paliouras, G.: Results of the BioASQ track of the question answering lab at CLEF. In: Working Notes for CLEF 2014 Conference, Sheffield, UK, pp. 1181–1193 (2014). http://ceur-ws.org/Vol-1180/CLEF2014wn-QA-BalikasEt2014.pdf

9. Brinker, K., Fürnkranz, J., Hüllermeier, E.: A unified model for multilabel classification and ranking. In: ECAI 2006, 17th European Conference on Artificial Intelligence, Riva del Garda, Italy, Proceedings, pp. 489–493 (2006). https://dblp.org/rec/bib/conf/ecai/BrinkerFH06

10. Cheng, D., Knox, C., Young, N., Stothard, P., Damaraju, S., Wishart, D.: PolySearch: a web-based text mining system for extracting relationships between human diseases, genes, mutations, drugs and metabolites. Nucleic Acids Res. **36**(Web Server issue), 399–405 (2008). https://doi.org/10.1093/nar/gkn296

11. Ferrucci, D., Lally, A.: Accelerating corporate research in the development, application and deployment of human language technologies. In: Proceedings of the HLT-NAACL 2003 Workshop on Software Engineering and Architecture of Language Technology Systems (SEALTS), Morristown, NJ, USA, pp. 67–74 (2003). https://doi.org/10.3115/1119226.1119236

12. Ferrucci, D., Lally, A.: UIMA: an architectural approach to unstructured information processing in the corporate research environment. Nat. Lang. Eng. **10**(3–4), 327–348 (2004). https://doi.org/10.1017/S1351324904003523

13. Ferrucci, D., Lally, A., Verspoor, K., Nyberg, E.: Unstructured information management architecture (UIMA) version 1.0. OASIS Standard (2009). https://docs.oasis-open.org/uima/v1.0/uima-v1.0.html

14. Götz, T., Suhre, O.: Design and implementation of the UIMA common analysis system. IBM Syst. J. **43**(3), 476–489 (2004). https://doi.org/10.1147/sj.433.0476

15. Jimeno-Yepes, A., Mork, J., Wilkowski, B., Demner-Fushman, D., Aronson, A.: MEDLINE mesh indexing: lessons learned from machine learning and future directions. In: ACM International Health Informatics Symposium, IHI 2012, Miami, FL, USA, pp. 737–742 (2012). https://doi.org/10.1145/2110363.2110450

16. Jin, Q., Dhingra, B., Cohen, W., Lu, X.: AttentionMeSH: simple, effective and interpretable automatic MeSH indexer. In: Proceedings of the 6th BioASQ Workshop. A Challenge on Large-Scale Biomedical Semantic Indexing and Question Answering, pp. 47–56 (2018). https://www.aclweb.org/anthology/W18-5306

17. Kosmopoulos, A., Androutsopoulos, I., Paliouras, G.: Biomedical semantic indexing using dense word vectors in BioASQ (2015). http://nlp.cs.aueb.gr/pubs/jbms_dense_vectors.pdf

18. Kosmopoulos, A., Partalas, I., Gaussier, É., Paliouras, G., Androutsopoulos, I.: Evaluation measures for hierarchical classification: a unified view and novel approaches. CoRR abs/1306.6802 (2013). https://dblp.org/rec/bib/journals/corr/KosmopoulosPGPA13

19. Krithara, A., Nentidis, A., Paliouras, G., Kakadiaris, I.: Results of the 4th edition of BioASQ challenge. In: Proceedings of the Fourth BioASQ workshop, Berlin, Germany, pp. 1–7 (2016). https://doi.org/10.18653/v1/W16-3101

20. Le, Q., Mikolov, T.: Distributed representations of sentences and documents. arXiv preprint arXiv:1405.4053 (2014). https://arxiv.org/abs/1405.4053v2

21. Liu, K., Peng, S., Wu, J., Zhai, C., Mamitsuka, H., Zhu, S.: MeSHLabeler: improving the accuracy of large-scale MeSH indexing by integrating diverse evidence. Bioinformatics **31**(12), 339–347 (2015). https://doi.org/10.1093/bioinformatics/btv237

22. Longwell, S.: Distributed representations for automating mesh indexing (2016). https://cs224d.stanford.edu/reports/Longwell.pdf

23. Lu, Z.: PubMed and beyond: a survey of web tools for searching biomedical literature. Database (Oxford), p. baq036 (2011). https://doi.org/10.1093/database/baq036
24. Mikolov, T., Sutskever, I., Chen, K., Corrad, G., Dean, J.: Distributed representations of words and phrases and their compositionality. CoRR abs/1310.4546 (2013). https://dblp.org/rec/bib/journals/corr/MikolovSCCD13
25. Nentidis, A., Bougiatiotis, K., Krithara, A., Paliouras, G., Kakadiaris, I.: Results of the fifth edition of the BioASQ challenge. In: BioNLP 2017, Vancouver, Canada, pp. 48–57 (2017). https://doi.org/10.18653/v1/W17-2306
26. Neveol, A., Mork, J., Aronson, A.: Comment on 'MeSH-up: effective MeSH text classification for improved document retrieval'. Bioinformatics **25**(20), 2770–2771 (2009). https://doi.org/10.1093/bioinformatics/btp483
27. Partalas, I., Gaussier, É., Ngonga-Ngomo, A.: Results of the first BioASQ workshop. In: Proceedings of the First Workshop on Bio-Medical Semantic Indexing and Question at CLEF, Valencia, Spain (2013). http://ceur-ws.org/Vol-1094/bioasq2013_overview.pdf
28. Pavlopoulos, I., Kosmopoulos, A., Androutsopoulos, I.: Continuous space word vectors obtained by applying word2vec to abstracts of biomedical articles (2014). http://bioasq.lip6.fr/info/BioASQword2vec/
29. Peng, S., You, R., Wang, H., Zhai, C., Mamitsuka, H., Zhu, S.: DeepMeSH: deep semantic representation for improving large-scale MeSH indexing. Bioinformatics **32**(12), 70–79 (2016). https://doi.org/10.1093/bioinformatics/btw294
30. Pennington, J., Socher, R., Manning, C.: Glove: global vectors for word representation. In: Proceedings of the 2014 Conference on Empirical Methods in Natural Language Processing, EMNLP, Doha, Qatar, pp. 1532–1543 (2014). https://dblp.org/rec/bib/conf/emnlp/PenningtonSM14
31. Schor, M.: An effective, Java-friendly interface for the unstructured management architecture (UIMA) common analysis system. Technical report IBM RC23176, IBM T. J. Watson Research Center (2004)
32. Tanenblatt, M., Coden, A., Sominsky, I.: The ConceptMapper approach to named entity recognition. In: Proceedings of the International Conference on Language Resources and Evaluation, LREC, Valletta, Malta (2010). http://www.lrec-conf.org/proceedings/lrec2010/summaries/448.html
33. Trieschnigg, D., Pezik, P., Lee, V., de Jong, F., Kraaij, W., Rebholz-Schuhmann, D.: MeSH Up: effective MeSH text classification for improved document retrieval. Bioinformatics **25**(11), 1412–1418 (2009). https://doi.org/10.1093/bioinformatics/btp249
34. Tsatsaronis, G., et al.: BioASQ: a challenge on large-scale biomedical semantic indexing and question answering. In: Information Retrieval and Knowledge Discovery in Biomedical Text, Papers from the 2012 AAAI Fall Symposium, Arlington, Virginia, USA (2012). http://www.aaai.org/ocs/index.php/FSS/FSS12/paper/view/5600
35. Tsoumakas, G., Katakis, I., Vlahavas, I.: Mining multi-label data. In: Data Mining and Knowledge Discovery Handbook, 2nd edn., pp. 667–685 (2010). https://doi.org/10.1007/978-0-387-09823-4_34
36. Tsoumakas, G., Laliotis, M., Markantonatos, N., Vlahavas, I.: Large-scale semantic indexing of biomedical publications. In: Proceedings of the First Workshop on Bio-Medical Semantic Indexing and Question Answering at CLEF, Valencia, Spain (2013). http://ceur-ws.org/Vol-1094/bioasq2013_submission_6.pdf

Convolutional Neural Network
for Automatic MeSH Indexing

Alastair R. Rae[✉], James G. Mork, and Dina Demner-Fushman

National Library of Medicine, 8600 Rockville Pike, Bethesda, MD 20894, USA
alastair.rae@nih.gov

Abstract. MEDLINE is the indexed subset of the National Library of Medicine's (NLM) journal citation database. It currently contains over 25 million biomedical citations, each indexed with a controlled vocabulary called MeSH. Since 1990, there has been a sizable increase in the number of articles indexed each year for MEDLINE, and since 2002, the NLM has been using automatic MeSH indexing systems to assist indexers with their increasing workload. This paper explores a deep learning approach to the automatic MeSH indexing problem. We present a Convolutional Neural Network (CNN) for automatic MeSH indexing and evaluate its performance by participating in the BioASQ 2019 task on large-scale online biomedical semantic indexing. The CNN model demonstrates competitive performance and outperforms the NLM's Medical Text Indexer (MTI) by about 3%. The paper presents a preliminary analysis comparing the results of the CNN model to MTI and also outlines the advantages of end-to-end deep learning approaches to automatic MeSH indexing.

Keywords: Automatic MeSH indexing · Medical text indexing · Convolutional neural network · Deep learning

1 Introduction

MEDLINE®/PubMed® is the National Library of Medicine's (NLM) premier bibliographic database, and MEDLINE is the indexed subset of the database. MEDLINE currently covers more than 5,200 international journals and contains over 25 million indexed biomedical citations. The database is freely available online and can be searched via the PubMed[1] web interface. From an information retrieval perspective, the unique value of MEDLINE is that citations are manually indexed with a hierarchical controlled vocabulary called Medical Subject Headings (MeSH®).[2] The assigned MeSH descriptors can be used in PubMed to define advanced search queries.

Indexing of MEDLINE articles is a time-consuming and highly specialized activity. NLM indexers review the full text of an article and then assign MeSH

---

[1] https://www.ncbi.nlm.nih.gov/pubmed/.
[2] https://www.nlm.nih.gov/mesh/.

This is a U.S. government work and not under copyright protection in the U.S.;
foreign copyright protection may apply 2020
P. Cellier and K. Driessens (Eds.): ECML PKDD 2019 Workshops, CCIS 1168, pp. 581–594, 2020.
https://doi.org/10.1007/978-3-030-43887-6_53

descriptors that represent the central concepts as well as every other topic that is discussed to a significant extent. Indexers are required to have a working knowledge of the large MeSH vocabulary and also scientific expertise in the subject indexed.

Since 1990, there has been a steady and sizable increase in the number of articles indexed each year for MEDLINE. Between 1990 and 2018 the number of articles indexed per year has increased from about 400,000 to over 900,000, and the NLM expects to index over one million articles annually within a few years. To help indexers cope with their increasing workload, the NLM has developed an automated indexing system called the Medical Text Indexer (MTI) [11]. MTI is a machine learning and rule-based system that takes the article title and abstract as its input and returns predicted MeSH terms as its output. The system improves productivity by providing a pick list of recommended MeSH terms that can be quickly selected by indexers.

Automatic MeSH indexing is a difficult machine learning problem. It is usually treated as a multi-label text classification problem, and the main challenges are the large number of MeSH descriptors and their highly imbalanced frequency distribution. There are over 29,000 MeSH descriptors in the 2019 MeSH vocabulary. At the end of 2018, the most frequent descriptor 'Humans' had been indexed more than 17 million times, whereas the 20,000th most frequent descriptor 'Ananas' had only been indexed 454 times.

Despite these challenges, effective systems for automatic MeSH indexing have been developed at the NLM and elsewhere. Since 2013, much of the progress in the field has been driven by the large-scale online biomedical semantic indexing task of the BioASQ challenge [14]. For this task, test sets of soon-to-be indexed articles are provided to participants and MeSH descriptor predictions must be submitted within 24 h (i.e. before the articles have been indexed).

Deep learning is a type of machine learning algorithm, based on artificial neural networks, that uses multiple processing layers to learn representations of data with multiple levels of abstraction [7]. In the last few years, deep learning approaches have demonstrated state-of-the-art (SOTA) performance in a wide variety of natural language processing (NLP) tasks, including text classification. Deep learning technologies have also been used to improve automatic MeSH indexing performance. Initially, these technologies were used to enhance existing systems [12], but more recently, end-to-end deep learning models (e.g. [15]) have demonstrated SOTA performance.

This paper presents an end-to-end deep learning model for automatic MeSH indexing that uses a Convolutional Neural Network (CNN) architecture. The model is evaluated by participating in task 7a of the BioASQ 2019 challenge and is shown to have competitive performance - outperforming the current NLM indexing system (MTI) in terms of micro F1 score. The presented CNN architecture has a number of customizations for the MeSH indexing task and these are shown to improve performance in an ablation study. We perform a preliminary analysis comparing the BioASQ challenge results of the CNN and MTI systems and also highlight the advantages of end-to-end deep learning approaches to automatic MeSH indexing.

2   Related Work

Automatic MeSH indexing is a well-studied multi-label classification problem, and since 2013, the BioASQ challenge has provided a useful benchmark for automatic MeSH indexing research. In recent challenges, two high-performing approaches have emerged: learning to rank based approaches (e.g. MTI [11], DeepMeSH [12]) and end-to-end deep learning approaches (e.g. MeSHProbeNet [15], AttentionMeSH [4]).

Learning to rank [10] is a supervised machine learning technique that is used to solve ranking problems. For automatic MeSH indexing the algorithm is used to rank candidate MeSH descriptors by integrating multiple sources of evidence. MTI uses learning to rank to boost its prediction performance [17], and candidate MeSH descriptors are obtained using MetaMap [1], PubMed Related Citations (PRC) [8], and machine learning algorithms. MetaMap maps biomedical text to UMLS® Metathesaurus[3] concepts. These concepts are then mapped to MeSH descriptors using the Restrict to MeSH [2] algorithm. PRC is a nearest neighbor algorithm that identifies similar articles based on their title and abstracts. MeSH descriptors from similar articles are considered as candidates. There are some special MeSH descriptors called Check Tags and these cover concepts that are mentioned in almost every article (e.g. Human, Animal, Male, Female, Child, etc.). 12 of the 40 Check Tags are identified by individually trained binary classifiers. The final list of MeSH descriptors is obtained after applying indexing rules to the ranked candidate descriptors.

The DeepMeSH system has demonstrated consistently high performance in recent BioASQ challenges. It combines learning to rank with a separate model to predict the number of MeSH descriptors. Like MTI, DeepMeSH uses nearest neighbors and binary classifier algorithms to identify candidate descriptors. A novel aspect of DeepMeSH is that it represents the title and abstract as the concatenation of term frequency inverse document frequency (TFIDF) and document to vector [6] (Doc2Vec) features.

Deep neural networks have been shown to be very effective for many NLP problems. Universal language models (e.g. [3]) are currently the SOTA for many tasks, but convolutional neural networks and recurrent neural networks (RNN) still provide excellent performance, often with lower computational cost (e.g. [5,16]). MeSHProbeNet was the best performing system in the BioASQ 2018 challenge, and it uses an end-to-end deep learning model with an RNN architecture. Specifically, it uses two bidirectional gated recurrent unit (GRU) layers followed by an attention layer to obtain a fixed length embedding of the concatenated title and abstract. The attention layer is novel because it uses multiple independent query vectors (MeSH probes) to generate different embedded representations (views) of the input text. The network output layer has one node for each MeSH descriptor and uses a sigmoid activation function to generate confidence scores between zero and one.

---

[3] https://www.nlm.nih.gov/research/umls/.

A CNN is a type of neural network that uses convolution operations to extract salient features. They are most commonly applied to image processing problems, but they have also been shown to be effective for many NLP tasks, including text classification. An effective CNN architecture for text classification is presented by Kim et al. [5] in their paper on sentence classification. This architecture uses a convolution layer followed by a max pooling layer to generate a fixed length text representation. A major advantage of CNNs is that they are quick to train because the CNN architecture is highly parallelizable.

Rios et al. [13] were the first to use a CNN model for MeSH indexing, and they trained independent binary classifiers for 12 Check Tags and another 17 hard-to-classify MeSH descriptors. The paper shows an improvement in performance compared to previous work, but it would not be practical to scale their binary relevance approach to the full MeSH vocabulary. Liu et al. [9] propose a CNN for extreme multi-label text classification, and they use a compression layer to allow the simultaneous prediction of up to 670,000 labels. The presented model demonstrated very competitive performance when evaluated on 6 extreme multi-label text classification benchmarks.

3   Methods

The CNN architecture (Fig. 1) is based on the architecture proposed by Liu et al. [9] for extreme multi-label classification. We customize their architecture with separate text inputs for the title and abstract and by adding journal, publication year, and year indexed inputs to the hidden layer. The outputs of the model are confidence scores for each MeSH descriptor. The following sections describe the different aspects of the model architecture in detail.

3.1   Title and Abstract Embeddings

The article title and abstract serve very different purposes, and we therefore choose to process them as separate inputs. The idea is to make it easy for the model to learn different rules for the title and abstract, if necessary. For example, one might expect the model to assign higher importance to features detected in the title compared to the abstract.

A CNN component is used to process the title and abstract and generate the fixed length embeddings required by the hidden layer. Let $x_i \in \mathbb{R}^k$ be the k-dimentional word embedding corresponding to the i-th word in the input text. Input text of length m is represented as the concatenation of word embeddings $x_{1:m} = [x_1, x_2, ..., x_m] \in \mathbb{R}^{mk}$. Let $x_{i:i+j}$ refer to the concatenation of words $x_i, x_{i+1}, ..., x_{i+j}$. A convolution operation applies a filter $w \in \mathbb{R}^{hk}$ to a window of h words to produce a new feature. Feature c_i is generated from a window of words $x_{i:i+h-1}$ by

$$c_i = f(w * x_{i:i+h-1} + b), \tag{1}$$

where f is the ReLU activation function, $*$ is the convolution operation, and $b \in \mathbb{R}$ is the bias term. A feature vector is obtained by applying the filter to each possible window position:

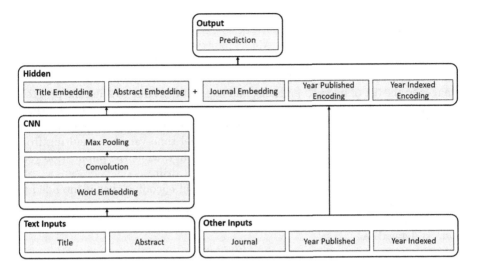

Fig. 1. The model architecture

$$c = [c_1, c_2, ..., c_{m-h+1}] \in \mathbb{R}^{m-h+1}. \tag{2}$$

Next, a dynamic max pooling operation [9] is used to select r features for this particular filter (assuming m is dividable by r):

$$\hat{c} = [max\{c_{1:\frac{m}{r}}\}, ..., max\{c_{m-\frac{m}{r}+1:m}\}] \in \mathbb{R}^r. \tag{3}$$

The advantage of dynamic max pooling over standard max pooling is that some position information is retained.

This section has described the process by which r features are extracted by one filter. The model uses multiple filters with different window sizes. The title and abstract text are processed using the same word embeddings and convolution layer weights to produce embeddings e_{title} and $e_{abstract}$ respectively. Standard max pooling is used for the title due to its short length.

3.2 Journal Embedding

Most MEDLINE journals have a narrow topic and are indexed with a small subset of the MeSH vocabulary. Providing the article journal as a model input was found to improve performance, and we expect this is because the model can learn the MeSH descriptor distribution for each journal.

In the presented architecture, the article journal is treated as a categorical input. As such, it could be represented using a sparse one-hot vector, but instead we represent each journal with a fixed size embedding $e_{journal} \in \mathbb{R}^d$, where d is the embedding size. We hope that an embedded representation will allow for better generalization between journals. The journal embeddings are randomly initialized and learned during training.

3.3   Year Encoding

One of the challenges of automatic MeSH indexing is that the MEDLINE dataset has significant time-variance. There are many factors that cause time-variance, and these include changes to the MeSH vocabulary, changes to indexing policy, changes to the list of indexed journals, and concept drift due to scientific progress and trends. In order to allow the neural network to effectively model the time-variance of the dataset, both the publication and indexing year are provided as inputs. The indexing year is required to model time-variance resulting from changes to the MeSH vocabulary and indexing policy, while the publication year is required to model time-variance due to concept drift.

The year inputs are represented using a special encoding that is intended to capture the sequential nature of time and to facilitate generalization between years. The encoding for a year $e_{year} \in \{0,1\}^s$ from a consecutive range of s years is defined as

$$e_{year}^i = \begin{cases} 0 \ i > \Delta \\ 1 \ i \leq \Delta \end{cases}, \tag{4}$$

where Δ is the difference between the year and the minimum year that needs to be encoded. Figure 2 is an illustration of this encoding for years between 2014 and 2018.

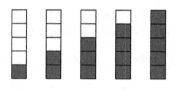

year = { 2014, 2015, 2016, 2017, 2018 }

Fig. 2. Illustration of the special encoding used for year inputs. The example shows how years between 2014 and 2018 would be encoded.

3.4   Hidden and Classification Layers

The embedded inputs are concatenated to form the input e to the hidden layer:

$$e = [e_{title}, e_{abstract}, e_{journal}, e_{pub\_year}, e_{year\_indexed}]. \tag{5}$$

Next, the hidden layer activations a_h are computed as

$$a_h = f(W_h e + b_h), \tag{6}$$

where f is the ReLU activation function, and W_h, b_h are the hidden layer weights. The final confidence scores $\hat{p} \in [0,1]^L$ for each of the L MeSH descriptors are computed as

$$\hat{p} = \sigma(W_c a_h + b_c), \tag{7}$$

where σ is the sigmoid activation function, and $\boldsymbol{W}_c$, $\boldsymbol{b}_c$ are the classification layer weights.

Dropout regularization is implemented to reduce overfitting to the training data, and it is applied to the title and abstract embeddings ($\boldsymbol{e}_{title}, \boldsymbol{e}_{abstract}$) and the hidden layer activations ($\boldsymbol{a}_h$).

3.5 Optimization

A binary-cross entropy objective function is formulated as

$$\min_{\Theta} -\frac{1}{N} \sum_{n=1}^{N} \sum_{l=1}^{L} [y_{nl} log(\hat{p}_{nl}) + (1 - y_{nl}) log(1 - \hat{p}_{nl})], \tag{8}$$

where Θ represents the model parameters, N is the number of training examples, and y are the indexer annotations. This objective was minimized using mini-batch gradient descent and the Adam optimizer. Batch normalization was found to improve the model performance and was implemented for the hidden and convolution layers.

4   Experiments

4.1   Dataset

The dataset is comprised of citation data for MEDLINE articles published from 2004 onward. Only articles with both a title and abstract were included, and fully or semi-automatically indexed articles were excluded. Semi-automatic indexing is when MTI has been used as the "first line indexer," and the results have later been reviewed (and potentially modified) by human indexers.[4] Semi-automatically indexed articles were excluded (in addition to fully automatically indexed articles) because we believe that the indexing may be biased towards MTI's predictions. The indexing method of an article is provided as an attribute in the latest PubMed XML format.[5] The final dataset contains about 8.5 million articles: 20,000 2018 articles were randomly selected for the validation set, and 40,000 2018 articles were randomly selected for the ablation study test set. Citation data was downloaded from the MEDLINE/PubMed 2019 annual baseline[6] because the BioASQ training data does not include the indexing year or indexing method.

The model performance was evaluated by participating in the large-scale online biomedical semantic indexing task of the BioASQ 2019 challenge. The challenge required participants to make predictions for 15 test sets of approximately 10,000 articles. The tests sets were released weekly, and the overall challenge was divided into 3 batches of 5 consecutive tests sets. Citation data for

---

[4] https://ii.nlm.nih.gov/MTI/MTIFL.shtml.
[5] https://www.nlm.nih.gov/pubs/techbull/ja18/ja18_indexing_method.html.
[6] https://www.nlm.nih.gov/databases/download/pubmed_medline.html.

the test set articles was downloaded from the MEDLINE/PubMed daily update files.[7] It would have been possible to use the BioASQ test set citation data, but it was simpler for us to process the daily update files as our system is designed to import data in PubMed XML format.

4.2   Evaluation Metric

Binary MeSH descriptor predictions $\hat{y} \in \{0,1\}^L$ are obtained by applying a single decision threshold to all model outputs. The evaluation metric is the micro F1 score (MiF) and this is defined as the harmonic mean of the micro precision (MiP) and the micro recall (MiR):

$$MiF = \frac{2 \cdot MiP \cdot MiR}{MiP + MiR}, \tag{9}$$

where

$$MiP = \frac{\sum_{n=1}^{N} \sum_{l=1}^{L} y_{nl} \cdot \hat{y}_{nl}}{\sum_{n=1}^{N} \sum_{l=1}^{L} \hat{y}_{nl}}, \tag{10}$$

$$MiR = \frac{\sum_{n=1}^{N} \sum_{l=1}^{L} y_{nl} \cdot \hat{y}_{nl}}{\sum_{n=1}^{N} \sum_{l=1}^{L} y_{nl}}. \tag{11}$$

There is an optimum decision threshold that results in the highest F1 score, and this threshold was determined by a linear search on the validation set.

4.3   Configuration

The model was implemented in Keras (v2.1.6) with a Tensorflow (v1.12.0) backend, and its hyperparamters are listed in Table 1. The word embeddings were randomly initialized and trained with the model. The learning rate was reduced by a factor of 3 if the validation set micro F1 score did not improve by more than 0.01 between epochs, and training was stopped early if the F1 score did not improve by more than 0.01 over two epochs. Training the model takes about 1 day on a single NVIDIA Tesla V100 (16 GB) GPU. Making predictions for a test set of 40,000 articles takes about 30 s.

4.4   Evaluation Results

We participated in task 7a of the BioASQ 2019 challenge with two systems. The first system ('CNN') is the model described in this paper, and the second system ('CNN Ensemble') is the same model with ensembling. Ensembling was implemented by training 9 separate 'CNN' models and then taking the average of their predictions. Predictions from the 'CNN' model were submitted for all 15 test sets of the challenge, while predictions from the 'CNN Ensemble' model were only submitted for the last 4 test sets.

---

[7] ftp://ftp.ncbi.nlm.nih.gov/pubmed/updatefiles/.

Table 1. CNN hyperparameters

Hyperparameter	Value
Vocabulary size	400,000
Word embedding size (k)	300
Title max words	64
Abstract max words	448
Number of convolution filters	350
Convolution filter sizes (h)	2, 5, 8
Dynamic max pooling number of regions (r)	5
Activation function for classification layer	Sigmoid
Activation function for all other layers	Relu
Hidden layer size	3365
Journal embedding size (d)	50
Dropout rate	0.15
Batch size	128
Learning rate	0.001

Table 2 shows the evaluation results for the top performing systems in the challenge. The average and sample standard deviation of the test set micro F1 score is shown for each system and batch. For teams participating with multiple versions of the same system, this analysis considers only the best performing configuration in each test set. The 'CNN' system predictions for batch 3 week 1 were erroneous and are therefore ignored in the analysis. The full results of the 2019 challenge are available on the BioASQ website,[8] and on the website our systems are named 'ceb 1' and 'ceb 1 ensemble'. The challenge results show that, as expected, the CNN model with ensembling outperforms the same model without ensembling. The 'CNN Ensemble' system is also found to outperform the current MTI implementations ('Default MTI' and 'MTI First Line Index') by about 3%. Compared to the other systems in the challenge, the CNN model demonstrated competitive performance, and it was typically the 3rd best performing system across all evaluations.

We were interested in evaluating the performance of a pure deep learning approach to automatic MeSH indexing, and for this reason our challenge systems did not make use of MTI's predictions. This may have put our systems at a disadvantage because approximately 20% of the articles in the challenge test sets were from semi-automatically indexed journals, and we suspect that the NLM's semi-automatic indexing methods are biased towards MTI's predictions. To get an idea of the performance improvement that can be achieved by making use of MTI's predictions, the performance of a hybrid system was evaluated on the articles in the last four test sets of the challenge. The hybrid system uses 'MTI

---

[8] http://bioasq.org/.

Table 2. Average and sample standard deviation of test set micro F1 scores for the top performing systems in the BioASQ 2019 challenge.

System	Batch		
	1	2	3
CNN	0.648 ± 0.006	0.654 ± 0.003	0.656 ± 0.005
CNN Ensemble	N/A	N/A	0.672 ± 0.005
DeepMeSH	0.667 ± 0.022	**0.714 ± 0.012**	**0.724 ± 0.012**
Default MTI	0.639 ± 0.004	0.643 ± 0.004	0.653 ± 0.005
MeSHProbeNet	**0.683 ± 0.005**	0.689 ± 0.003	0.693 ± 0.005
MTI First Line Index	0.624 ± 0.004	0.633 ± 0.007	0.649 ± 0.007

First Line Index' predictions for articles that were semi-automatically indexed ('Curated' indexing method in the PubMed XML files) and 'CNN Ensemble' predictions for all other articles. The hybrid system is found to achieve a micro F1 score of 0.687, an improvement of approximately 2% over the micro F1 score of the 'CNN Ensemble' model.

5   Ablation Study

This section presents the results of an ablation study (Table 3) that explores how different aspects of the model architecture contribute to its overall performance. Five different models with ablations were trained and their performance was evaluated on the ablation test set (see Sect. 4.1). The study considers three ablations concerning task specific architectural features (removing separate title and abstract inputs, removing journal input, removing year inputs) and two ablations concerning more general architectural features (removing batch normalization, removing dynamic max pooling). The study finds that the task specific architectural features increase the model performance by 1.2–1.5%, and batch normalization also increases model performance by 1.1%. Dynamic max pooling offers the smallest performance improvement of 0.4%.

Table 3. Model performance with ablations

Model description	MiF	Difference (%)
CNN	0.6548	
CNN with single text input (concat. title & abstract)	0.6449	−1.5
CNN without journal input	0.6466	−1.3
CNN without year inputs	0.6471	−1.2
CNN without batch normalization	0.6475	−1.1
CNN without dynamic max pooling	0.6523	−0.4

Even if the model architecture is unchanged, there will be some variation in performance due to the random initialization of parameters and stochastic gradient descent. To understand the scale of this variation, the micro F1 score performance of 10 independently trained 'CNN' models was measured on the ablation test set. The sample standard deviation of the micro F1 score was found to be 0.0004, and this gives us confidence that the ablation study results are significant.

6 Discussion

We were able to perform a preliminary analysis comparing the results of the 'CNN' and 'Default MTI' systems using the BioASQ results and the final indexed MEDLINE citations. We reviewed the results from 78,574 fully human indexed citations containing 979,014 NLM indexed MeSH descriptors. From this set we can see that the 'Default MTI' system had slightly higher recall with 615,347 correct MeSH descriptors versus 603,973 for the 'CNN' system. But, the 'CNN' system was much more precise with only 288,146 incorrect MeSH descriptors compared to 'Default MTI' providing 371,658 incorrect MeSH descriptors. Note that the CNN decision threshold was selected for best micro F1 score, but this can be adjusted to increase recall at the cost of precision. Looking closer at the MeSH descriptors that both systems used, we see that the 'Default MTI' system used 21,158 distinct MeSH descriptors, while the 'CNN' system only used 19,745 distinct MeSH descriptors. Of these, 18,710 MeSH descriptors were common to both systems, 'Default MTI' had 2,448 unique MeSH descriptors which were correct 43.45% of the time, and the 'CNN' system had 1,035 unique MeSH descriptors which were correct 34.65% of the time.

Looking at the final indexed MeSH descriptors that both systems missed, we see the usual suspects including age related Check Tags (e.g. Young Adult, Adolescent) and sex related Check Tags (e.g. Female, Male). We looked at a small sample of 10 cases where both systems missed the most MeSH descriptors and found that, in 9 out of the 10 cases, the missing information was only available in the full text of the article. This confirmed our suspicions that the information was just not available to the two systems, since the NLM indexers index the article from the full text, while the automated systems only have access to the title and abstract. Our overall impression is that the 'CNN' system tends to predict more general MeSH descriptors, whereas the 'Default MTI' system does well on certain important MeSH descriptors that we have specifically focused on detecting in the past.

This paper has shown that the CNN model outperforms the current MTI implementation, and from a software engineering perspective, the deep learning approach also has a number of other advantages. A key advantage is that it is an end-to-end system without any dependencies. This will make it easier to deploy and maintain than a complex multi-component system, like MTI. For example, when the MeSH vocabulary needs to be updated at the end of the year, it will take approximately one day to retrain the CNN model. In comparison, it usually takes about a week to update MTI with a new MeSH vocabulary. Another advantage of the CNN model is that it will be possible to support the NLM indexers with a single model instance running on a GPU enabled server. For comparison, MTI uses parallel processing with 70 clients on 6 servers to achieve the level of required throughput. The close to real-time predictions of the CNN model will also make it possible to develop more interactive indexer tools in the future.

A pure deep learning solution does have some disadvantages. A major disadvantage is that a learning based system is unable to make predictions for new MeSH descriptors or follow new indexing rules. MTI also suffers from this problem because it has machine learning components, however, its MetaMap component is able to detect new descriptors based on the information in the UMLS. MTI also uses a lookup list of new terms and their synonyms to ensure they are recommended, and current indexing policy is enforced using manually coded rules.

Another disadvantage of the CNN model is that it is a black box - it does not provide any explanation for why a particular set of MeSH descriptors were predicted. Having more interpretable predictions would be of great benefit to indexers and is something that we plan to look into in the future. For example, it should be possible to determine which n-grams most influence the CNN model's predictions, and these n-grams could then be highlighted in the text.

7   Conclusion

This paper has presented a CNN model for automatic MeSH indexing. The model demonstrated competitive performance in the BioASQ 2019 task on large-scale online biomedical semantic indexing, outperforming the NLM's current automatic indexing system, MTI, by about 3%. The paper has also presented an ablation study highlighting how task specific customizations to the model architecture result in improved performance.

In the future, we will explore the possibility of replacing MTI with a deep learning based system. We plan to complete our analysis comparing the strengths and weaknesses of the CNN and MTI systems, and our goal is to achieve higher performance by combining the best aspects of the two systems. Providing the reasons why a particular MeSH descriptor was recommended would be very useful for indexing staff, and we therefore plan to research deep learning architectures that offer more interpretable predictions.

Acknowledgments. This research was supported by the Intramural Research Program of the National Institutes of Health (NIH), National Library of Medicine (NLM), and Lister Hill National Center for Biomedical Communications (LHNCBC).

The author would like to thank Daniel Le for his support and guidance during the initial phase of this work and also Xiaoli Zhang and Chan Moon for useful discussions and suggestions.

References

1. Aronson, A.R., Lang, F.M.: An overview of MetaMap: historical perspective and recent advances. J. Am. Med. Inform. Assoc. **17**(3), 229–236 (2010)
2. Bodenreider, O., Nelson, S.J., Hole, W.T., Chang, H.F.: Beyond synonymy: exploiting the UMLS semantics in mapping vocabularies. In: AMIA Annual Symposium, Lake Buena Vista, Florida, 7–11 November 1998. Proceedings of the AMIA Symposium, p. 815. Hanley & Belfus, Philadelphia (1998)
3. Devlin, J., Chang, M., Lee, K., Toutanova, K.: BERT: pre-training of deep bidirectional transformers for language understanding (2018). https://arxiv.org/abs/1810.04805v1, arXiv:1810.04805v1 [Preprint]
4. Jin, Q., Dhingra, B., Cohen, W., Lu, X.: AttentionMeSH: simple, effective and interpretable automatic MeSH indexer. In: 6th BioASQ Workshop, Brussels, Belgium, 1 November 2018. Proceedings of the 6th BioASQ Workshop, pp. 47–56. ACL (2018)
5. Kim, Y.: Convolutional neural networks for sentence classification. In: EMNLP 2014, Doha, Qatar, 25–29 October 2014. Proceedings of the 2014 Conference on Empirical Methods in Natural Language Processing, pp. 1746–1751. ACL (2014)
6. Le, Q., Mikolov, T.: Distributed representations of sentences and documents. In: 31st International Conference on Machine Learning, Beijing, China, 22–24 June 2014. Proceedings of Machine Learning Research, vol. 32, pp. 1188–1196. PMLR, Cambridge (2014)
7. LeCun, Y., Bengio, Y., Hinton, G.E.: Deep learning. Nature **521**(7553), 436–444 (2015)
8. Lin, J., Wilbur, W.J.: PubMed related articles: a probabilistic topic-based model for content similarity. BMC Bioinform. **8**(1), 423 (2007)
9. Liu, J., Chang, W.C., Wu, Y., Yang, Y.: Deep learning for extreme multi-label text classification. In: SIGIR 2017, Tokyo, Japan, 7–11 August 2017. Proceedings of the 40th International ACM SIGIR Conference on Research and Development in Information Retrieval, pp. 115–124. ACM, New York (2017)
10. Liu, T.Y.: Learning to rank for information retrieval. Found. Trends Inf. Retr. **3**(3), 225–331 (2009)
11. Mork, J., Aronson, A., Demner-Fushman, D.: 12 years on - is the NLM medical text indexer still useful and relevant? J. Biomed. Semant. **8**(1), 8 (2017)
12. Peng, S., You, R., Wang, H., Zhai, C., Mamitsuka, H., Zhu, S.: DeepMeSH: deep semantic representation for improving large-scale MeSH indexing. Bioinformatics **32**(12), i70–i79 (2016)
13. Rios, A., Kavuluru, R.: Convolutional neural networks for biomedical text classification: application in indexing biomedical articles. In: BCB 2015, Atlanta, Georgia, 9–12 September 2015. Proceedings of the 6th ACM Conference on Bioinformatics, Computational Biology and Health Informatics, pp. 258–267. ACM, New York (2015)
14. Tsatsaronis, G., et al.: An overview of the BioASQ large-scale biomedical semantic indexing and question answering competition. BMC Bioinform. **16**(1), 138 (2015)
15. Xun, G., Jha, K., Yuan, Y., Wang, Y., Zhang, A.: MeSHProbeNet: a self-attentive probe net for MeSH indexing. Bioinformatics **35**, 3794–3802 (2019)

16. Yang, Z., Yang, D., Dyer, C., He, X., Smola, A., Hovy, E.: Hierarchical attention networks for document classification. In: NAACL HLT 2016, San Diego, California, 12–17 June 2016. ACL Proceedings of the 2016 Conference of the North American Chapter of the Association for Computational Linguistics: Human Language Technologies, pp. 1480–1489. ACL (2016)
17. Zavorin, I., Mork, J., Demner-Fushman, D.: Using learning-to-rank to enhance NLM medical text indexer results. In: 4th BioASQ Workshop, Berlin, Germany, 12–13 August 2016. Proceedings of the Fourth BioASQ Workshop, pp. 8–15. ACL (2016)

A Mixed Information Source Approach for Biomedical Question Answering: MindLab at BioASQ 7B

Mónica Pineda-Vargas[1](✉), Andrés Rosso-Mateus[1](✉), Fabio A. González[1](✉), and Manuel Montes-y-Gómez[2](✉)

[1] Universidad Nacional de Colombia, Bogotá, Colombia
{mppinedav,aerossom,fagonzalezo}@unal.edu.co
[2] Laboratorio de Tecnologías del Lenguaje, INAOE, Puebla, Mexico
mmontesg@inaoep.mx

Abstract. This paper describes the participation of the MindLab research group in the BioASQ 2019 Challenge for task 7b, document retrieval and snippet retrieval. For document retrieval, Elastic Search was used for the initial document retrieval step with BM25 as a scoring function. In the second stage, the top 100 retrieved documents were re-ranked with several strategies to exploit embedding semantic similarity. For the snippets retrieval subtask, the proposed approach was based on textual and conceptual information similarity patterns that were combined into a feature matrix that was subsequently processed by a convolutional neural network architecture. Our approach reached the third and second positions for the document retrieval and snippet retrieval task respectively.

Keywords: BioASQ · Snippet retrieval · Biomedical document retrieval

1 Introduction

In the biomedical domain, experts constantly search in previous works to support their research hypothesis, investigate causes, diseases symptoms, etc. The number of published documents is growing continuously, more than 3000 articles are indexed every day in biomedical journals [17], making it harder to find and access valuable information.

Question Answering (QA) systems can help to retrieve concise information naturally, given the precise answer and supporting passages for any information need. The interest in QA systems in the biomedical domain has been growing [1,17] and is playing an important role in the closed domain information access and is considered to be the next step in information retrieval systems [19].

BioASQ is a closed domain information retrieval challenge over biomedical articles [17], this challenge has helped to advance the research in the biomedical

© Springer Nature Switzerland AG 2020
P. Cellier and K. Driessens (Eds.): ECML PKDD 2019 Workshops, CCIS 1168, pp. 595–606, 2020.
https://doi.org/10.1007/978-3-030-43887-6_54

information retrieval field. Mindlab team has participated in the last two editions. Here we will describe our second participation for the seventh edition in task B.

The goal of the target task is: given a question the system must return relevant concepts, relevant documents (from 2018 PubMed articles baseline [11]), relevant snippets (extracted from articles), and relevant Resource Description Framework (RDF) triples from designated ontologies [17]. In this year, our focus was document and snippet retrieval. Our method was based on a convolutional neural network model that takes as input a question-snippet similarity matrix. It combines different embeddings of words and medical concepts with the purpose of building a more meaningful representation.

The structure of this paper is as follows: Sect. 2 describes the system architecture, the strategies used for document retrieval are presented in Sect. 3 using Okapi-BM25 and Elastic Search as the first filter for efficiency and then representing documents and questions as word embeddings [10] for Doc Centroid Rerank and Word Mover's Distance as re-ranking functions. In Sect. 3.1, we present the passage retrieval module with the proposed method. Some performance analysis experiments were performed with the BioASQ6 data to evaluate and compare the proposed strategies for document and snippet retrieval, these results are shown in Sect. 4. The results of the current challenge are presented in Sect. 4.1 and finally, conclusions and future works in Sect. 5.

2   The MindLab System at a Glance

Our approach consists of two main components: the document retrieval and the snippet retrieval modules, as shown in Fig. 1.

The first module has the goal of producing a set of documents where the answer for a posed questions can reside. The Elastic Search (ES) information retrieval platform [6] was configured to index and query the PubMed Baseline Repository (MBR) document set [11]. This year the BioASQ challenge used the 2018 MBR. Those documents are bio-medical papers with title and abstract sections, also they contain meta-information such as MESH Terms, year of publication and keywords.

Based on a posed question, a query string is submitted to the ES search engine, the engine returns a set of 100 documents that are relevant to the query. The next step is a fine-grained document filtering using the query and document terms, the goal is to reduce the number of documents to 10. Our approach for document filtering is based on Word Mover's Distance (WMD) and Document Centroid semantic match.

The selected relevant documents are analyzed in depth. Snippets of the 10 most relevant documents are extracted and ranked with our Convolutional Neural Network (CNN) model that exploits semantic similarity patterns.

In the end, the top 10 scored documents and snippets are submitted in descending order to the BioASQ server.

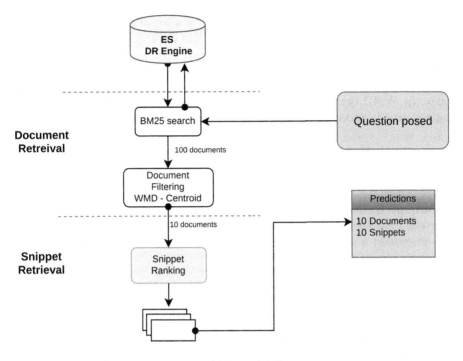

Fig. 1. BioASQ model diagram

2.1 Document Re-Ranking

The results produced by ES suffer from lack of precision. To alleviate this, we perform a re-ranking of the top-n documents using a more precise, but costly, semantic matching method based on semantic similarity. In our experiments, we evaluated two embedding similarity measures: Word Mover's Distance and Document Centroid.

Word Mover's Distance: The first was Word Mover's Distance (WMD) [8], a particular case of the Earth Mover's Distance [13]. The query and each document are represented as a weighted point cloud of embedded words as shown in Fig. 2. The distance between them is the minimum cumulative distance that words from the query need to travel to match exactly the point cloud of the document.

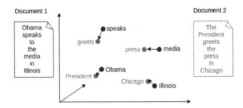

Fig. 2. Word Mover's Distance between two documents.

Let q be the query user, $d \in D$ where D is a set of n relevant documents, and $|q|$, $|d|$ the number of distinct tokens in q and d respectively. Let $\boldsymbol{T}$ be a flow matrix where $\boldsymbol{T}_{ww'}$ denotes how much the word w in q travels to word w' in d and C is the transportation cost with $C_{w,w'} := dist(\boldsymbol{v}_{q_w}, \boldsymbol{v}_{d'_w})$ normally provided by their Euclidean distance in the word2vec embedding space. Finally, we can define the WMD between the query and document as the minimum cumulative cost required to move all words from q to d.

$$\min_{\boldsymbol{T} \geq 0} \sum_{w,w'}^{n} \boldsymbol{T}_{ww'} C(w, w') \tag{1}$$

Document Centroid Similarity: For a given query q and each document d we compute the centroid of the corresponding word vectors. The similarity of a query and a document corresponds to the cosine similarity between their corresponding centroids.

2.2   Snippet Retrieval

Traditional passage retrieval methods use only textual information to identify the semantic match between the question and the answer. However, in domains such as biomedicine, there is a good number of structured knowledge resources in the form of ontologies, thesaurus, and taxonomies. The use of structured sources could provide advantages such as unambiguous knowledge representation, the possibility of applying automated reasoning methods [18], and the facility of linking different information facts, among others. The proposed approach for passage retrieval takes advantage of the huge amount of textual data in the biomedical domain in conjunction with structured-knowledge data sources.

Our passage retrieval model is based on two main hypothesis: first, that question and answer passages are semantically correlated term by term and concept by concept; second, that structured and unstructured information are complementary modalities that can jointly represent, in a better way, the semantic content of questions and passages.

The proposed method has two stages as Fig. 3 shows. The first one (training phase) has the objective to learn the similarity patterns for question-answer pairs. In the second stage (testing), the trained similarity model is used to obtain the ranking scores of a set of candidate answers (snippets) for a particular question. The method uses two representations schemes, textual and conceptual, for both answers and questions. Both representations are used as input independently. The representations are combined using a different fusion strategy.

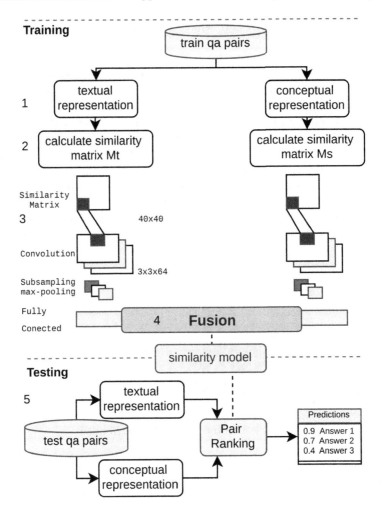

Fig. 3. Passage retrieval process

The details of the steps depicted in Fig. 3 are discussed next.

- **Step 1 - Extract the representation**: The question and answer pairs are transformed to feed the neural network, the process is different for each modality.
    - **Textual Representation**: First the text is cleaned and tokenized, a grammatical tagging is carried out with NLTK POS-tagger to extract syntactical information that will be used in salience weighting; each term is transformed later in a vector embedding using a pre-trained word2vec

model provided by NLPLab, which is trained on Wikipedia and PubMed documents.[1]

- **Conceptual representation**: To identify medical concepts we use QuickUMLS [15] which is an unsupervised biomedical concept extraction tool. Those identified concepts are then transformed into a continuous vector representation using a cui2vec embedding. This embedding maps medical concepts instead of words. Concepts are referred by their concept unique identifier (CUI) from the Unified Medical Language System (UMLS) thesaurus [2]. In contrast with textual representation, there are fewer words, 4 concepts in average per question, in the text fragments that can be embedded in the conceptual representation as it is shown in Fig. 4. This has to do with the reduced size of the cui2vec vocabulary. To overcome this restriction we applied expansion to question CUI embeddings following the centroid method proposed by Kuzi et al. [9].

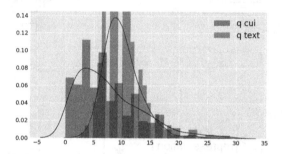

Fig. 4. Question terms and cuis distribution

- **Step 2 - Calculate the similarity matrices**: Each i, j-entry of the similarity matrices M_t and M_s, represents the semantic relatedness of the i-th question term (or concept) and the j-th answer term (or concept) according to the embedding (nlplab or cui2vec).
  - **Textual similarity Matrix M_t:** In the case of textual representation the cosine similarity between terms is weighted based on the grammatical function of the term pair, this grammatical weighting is called a salience score $sal(qt_i, at_j)$. The similarity between element i-th and j-th is calculated as Eq. 2 shows.

$$M_{i,j} = scos(qt_i, at_j) * sal(qt_i, at_j) \tag{2}$$

$$scos(qt_i, at_j) = 0.5 + \frac{qt_i \cdot at_j}{2 \left\| qt_i \right\|_2 \left\| at_j \right\|_2} \tag{3}$$

---

[1] BioNLP word vector representation, trained with biomedical and general-domain texts http://bio.nlplab.org.

$$sal(qt_i, at_j) = \begin{cases} 1 & if\ imp(qt_i) + imp(at_j) = 2 \\ 0.6\ if\ imp(qt_i) + imp(at_j) = 1 \\ 0.3\ if\ imp(qt_i) + imp(at_j) = 0 \end{cases} \quad (4)$$

Where $imp(qt_i)$ and $imp(at_j)$ are the importance weighting for every question and answer term. The related function returns 1 if the term is a verb, noun or adjective, otherwise, returns 0.

- **Concept similarity matrix M_s:** In the case of conceptual information we calculate just the cosine similarity between cui2vec concept vectors.
- **Step 3. Convolutional model:** The architecture of the convolutional model is shown in Fig. 3 step 3. A convolutional layer is fed with both similarity matrices M_t and M_s, CNN layer will identify element-similarity patterns to rank the relevance of a question-answer pair using both knowledge representations. Patterns identified by each CNN filter are sub-sampled by a pooling layer. The pooling layer for all the filters is merged with two fully connected layers, first one with 64 units and the second one with 16 units, a regularized dropout of 10% is used between them. The total number of parameters for the fusion model is 6,173.
- **Step 4. Multimodal fusion:** The dense outputs of the modalities are merged in a unique dense layer, which feeds another dense layer. Finally, the output score of the model is generated by a sigmoid unit on top of the last dense layer.
- **Step 5. Pair Ranking:** Candidate answers $(a_1, a_2, ..., a_k)$ are ranked against the query q using the trained similarity model. The model produces the final similarity score taking into account information from both modalities.

Information Fusion Approaches: As we have used information that comes from textual representation and conceptual representation the combination of those modalities is also a model parameter to explore. In that way we have evaluated four different configurations to measure the performance involving different information representation approaches:

- Approach 1: Only textual representation. Questions and candidate answers are represented using only the textual embedding.
- Approach 2: Mixed data representation intermediate method –MIF. In this model, the fusion of textual and conceptual representations is carried out in an intermediate dense layer after the textual and conceptual patterns are identified by CNN's layers Fig. 3. The merged layer is then connected to the sigmoidal output unit with dropout as regularization strategy.
- Approach 3: Mixed Data Representation Late Fusion –MLF. In this approach each model (textual and conceptual) independently calculates a score for each question-answer pair, $score\_t$ for textual representation, and $score\_s$ for conceptual representation. Lastly, a linear combination produces the final score f_score, as shown in Eq. 5. The *alpha* value was found using cross-validation with the validation partition; it was set to 0.73.

$$f\_score(q, a_k) = (1 - \alpha) * score_t(q, a_t) + \alpha * score_s(q, a_s) \quad (5)$$

3   Model Performance Tuning

To evaluate the performance of the proposed methods, we used the BioASQ 6 dataset. The experimentation process is divided into two phases, the first one focused on the document retrieval process and the second one for snippets retrieval.

3.1   Document Retrieval

We indexed the full data of the 2018 MBR in ES version 6.2.2 with the default configuration, this is our baseline. For b and $k1$ BM25 parameters we evaluated the default values and the values proposed by CLEF eHealth evaluation lab 2016 [4].

The number of processed files was 928 and the total number of medical articles was 26,759,399. For each article, we extracted the title, MESH concepts and abstract to be indexed. The indexing time was around 18 h in an Intel Xeon processor Intel(R) at 2.60 GHz with 82 GB RAM and GeForce GTX TITAN X.

We evaluated four experiments: retrieve the 10 most relevant documents with *BM25 default configuration* (**BM25_v1**), the second one is to retrieve documents with *BM25 clef tuned parameters* (**BM25_v2**), the third is using *BM25 clef tuned parameters* and re-rank using Word Mover's Distance (**BM25_v2_WMD**). The last one is to retrieve documents with *BM25 clef tuned parameters* and re-rank using Doc Centroid Rerank (**BM25_v2_centroid**).

The averaged results over the five 6b batches are presented in Table 1. Adjusting BM25 parameters (b) for document length scaling and ($k1$) for document term scaling have a positive impact on the overall ES document retrieval performance. The centroid re-ranking strategy was not successful, but the use of Word Mover's Distance has a slightly better performance over the BM25 initial document set.

Table 1. Document retrieval results for BioASQ 6 (summarized)

Model	Mean precision	Recall	F-Measure	MAP	GMAP
BM25_v1	0.2074	0.4724	0.2174	0.1319	0.0257
BM25_v2	0.2147	0.4810	0.2241	0.1354	0.0279
BM25_v2_WMD	0.2256	0.4910	0.2384	0.1483	0.0286
BM25_v2_centroid	0.2084	0.4706	0.2186	0.1332	0.0228

3.2   Snippet Retrieval

The training was done with the question and answer pairs from 2016, 2017 and 2018 BioASQ Task B training datasets. The total number of used question-answer pairs were 124,144. The obtained dataset was very unbalanced, only 18%

of the total number of pairs were labeled as a relevant answer. To balance the dataset, the sample extraction in the training phase was done with the same number of positive and negative samples, this strategy is also applied in the validation phase.

The model training was done using RMSprop optimization algorithm with 32 samples per mini-batch and the loss function was binary cross entropy. The number of maximum epochs was set to 50. In each epoch, we evaluated MAP and MRR, and after 5 epochs without any improvement in MAP metric, we applied early stopping to avoid over-fitting.

We have conducted our experimentation with the released batches for BioASQ 6, as it was done for the document retrieval stage. The scores obtained from BioASQ results submission page [17] are presented in Table 2. We have included the results from the winning team at BioASQ 6 challenge snippet retrieval sub-task [3] for performance comparison.

Table 2. Snippet retrieval results averaged at BioASQ 6b

Method	Mean precision	Mean recall	F-measure	MAP	GMAP
aueb-nlp-5	0.3807	0.3655	0.3452	0.3320	0.0536
mindlab	0.2074	0.2437	0.2021	0.2102	0.0076
Only Textual	0.2074	0.2437	0.2021	0.2102	0.0161
MIF	**0.2181**	**0.2517**	**0.2161**	**0.2201**	**0.0098**
MLF	0.2014	0.2217	0.2100	0.2095	0.0086

The experimental evaluation shows that the incorporation of conceptual information can improve the performance in passage retrieval. Moreover, multimodal intermediate fusion outperforms the use of each modality individually and late fusion approaches.

Model Parameters. The model hyper-parameters were tuned using hyper-parameter exploration. The parameters chosen are listed next.

- **Convolution parameters:** The number of convolutional filters used are 64, width 3 and length 3, the stride used is 1 without padding.
- **Convolution activation function:** After a convolutional layer, it is useful to apply a nonlinear layer [5]. We tested different activation functions and RELU gave us the best performance.
- **Pooling layers:** For the pooling layer, we used max pooling.
- **Dropout layer:** We add a dropout layer as a regularization strategy [16], setting the parameter in 10%.

4   Results at BioASQ 2019 and Discussion

4.1   Document Retrieval

The results are shown in Table 3. It shows that our document retrieval implementation could still improve. In most batches, the top competitor gets approximately the double of our score.

The best result was obtained in batch 3 (Recall = 0.5213 and GMAP = 0.0070), with the tuned parameters for BM25, *Index-v2*. This improves our results obtained in the past competition [12], but the team leader in this batch reached 0.6128 in Recall and 0.0164 in GMAP, an important difference.

Document retrieval is relevant for snippet retrieval task, because it is the first information filter. The snippet retrieval method works on the top 10 documents retrieved in this phase, if it has low precision, the performance of the snippet retrieval method is affected.

Table 3. BioASQ 7 document retrieval results

Batch	Model	Mean precision	Recall	F-Measure	MAP	GMAP
7b1	Our model	0.1120	0.5087	0.1660	0.0742	0.0039
	Top competitor	0.1190	0.5216	0.1746	0.0809	0.0047
7b2	Our model	0.0950	0.4733	0.1444	0.0579	0.0021
	Top competitor	0.1260	0.5967	0.1905	0.0771	0.0075
7b3	Our model	0.1280	0.5213	0.1887	0.0803	0.0070
	Top competitor	0.3599	0.6128	0.4034	0.1102	0.0164
7b4	Our model	0.1040	0.5103	0.1573	0.0726	0.0033
	Top competitor	0.3332	0.6141	0.3783	0.1015	0.0116
7b5	Our model	0.0550	0.3476	0.0888	0.0326	0.0005
	Top competitor	0.0710	0.3937	0.1120	0.0425	0.0010

4.2   Snippet Retrieval

As Table 4 shows, the snippet retrieval approach obtained a good performance despite the low precision and recall in document retrieval. The results were enough to reach the second position in the last 4 batches and the first position in the first one.

It is important to highlight that the number of parameters to learn in our model is not large (5,192) compared to other QA Deep Learning approaches which are in order of millions and hundreds of thousands [7,14].

Part of the error in the snippet retrieval phase is propagated from document retrieval. If we improve the results in document retrieval, we could have better results in this phase because we only retrieve the snippets present in the top 10 documents from the document retrieval phase. Even though, the incorporation of conceptual information can improve the performance in passage retrieval because the medical terms are referred by their concept unique identifier, which is unique for different words associated to the same concept.

Table 4. BioASQ 7 snippet retrieval results

Batch	Model	Mean precision	Recall	F-Measure	MAP	GMAP
7b1	Our model	0.0951	0.2447	0.1253	0.0808	0.0008
	Top competitor	-	-	-	-	-
7b2	Our model	0.0900	0.2243	0.1212	0.0893	0.0004
	Top competitor	0.1447	0.3722	0.1855	0.1438	0.0019
7b3	Our model	0.1371	0.2519	0.1617	0.1404	0.0009
	Top competitor	0.2159	0.3634	0.2472	0.2206	0.0081
7b4	Our model	0.1587	0.2760	0.1723	0.1527	0.0013
	Top competitor	0.2060	0.4039	0.2365	0.2114	0.0075
7b5	Our model	0.0440	0.1823	0.0656	0.0499	0.0001
	Top competitor	0.0542	0.2411	0.0818	0.0631	0.0003

5 Conclusion

In this paper we have presented the approaches and results obtained in our second participation for the seventh BioASQ challenge version. The proposed method for document retrieval was based on the Elastic Search platform with BM25 as scoring function, then a second document filtering was carried out using Word Mover's Distance (WMD) and Document Centroid semantic match.

For the snippet retrieval sub-task, the selected approach was based on a convolutional neural network that extracts similarity patterns over mixed data input representation. We have tested different fusion approaches to combine information coming from conceptual and textual sources.

The results obtained are promising for snippets retrieval, where we reach the second position despite the not very good performance of our approach for document retrieval. This motivates our future work which will focus on improving the document retrieval phase.

References

1. Bauer, M.A., Berleant, D.: Usability survey of biomedical question answering systems. Hum. Genomics **6**(1), 17 (2012)
2. Beam, A.L., et al.: Clinical concept embeddings learned from massive sources of medical data. arXiv preprint arXiv:1804.01486 (2018)
3. Brokos, G.-I., Liosis, P., McDonald, R., Pappas, D., Androutsopoulos, I.: AUEB at BioASQ 6: document and snippet retrieval. arXiv preprint arXiv:1809.06366 (2018)

4. Goeuriot, L., et al.: Overview of the CLEF eHealth evaluation lab 2015. In: Mothe, J., et al. (eds.) CLEF 2015. LNCS, vol. 9283, pp. 429–443. Springer, Cham (2015). https://doi.org/10.1007/978-3-319-24027-5_44

5. Goodfellow, I., Bengio, Y., Courville, A.: Deep Learning. MIT Press, Cambridge (2016)

6. Gormley, C., Tong, Z.: Elasticsearch: The Definitive Guide: A Distributed Real-Time Search and Analytics Engine. O'Reilly Media, Inc., Sebastopol (2015)

7. He, H., Lin, J.J.: Pairwise word interaction modeling with deep neural networks for semantic similarity measurement. In: HLT-NAACL, vol. 1, pp. 937–948 (2016)

8. Kusner, M., Sun, Y., Kolkin, N., Weinberger, K.: From word embeddings to document distances. In: International Conference on Machine Learning, pp. 957–966 (2015)

9. Kuzi, S., Shtok, A., Kurland, O.: Query expansion using word embeddings. In: Proceedings of the 25th ACM International on Conference on Information and Knowledge Management, pp. 1929–1932. ACM (2016)

10. Mikolov, T., Sutskever, I., Chen, K., Corrado, G.S., Dean, J.: Distributed representations of words and phrases and their compositionality. In: Burges, C.J.C., Bottou, L., Welling, M., Ghahramani, Z., Weinberger, K.Q. (eds.) Advances in Neural Information Processing Systems 26, pp. 3111–3119. Curran Associates, Inc. (2013). http://papers.nips.cc/paper/5021-distributed-representations-of-words-and-phrases-and-their-compositionality.pdf

11. National Institutes of Health. Pubmed baseline repository

12. Rosso-Mateus, A., González, F.A., Montes-y Gómez, M.: Mindlab neural network approach at bioasq 6b. In: Proceedings of the 6th BioASQ Workshop A Challenge on Large-Scale Biomedical Semantic Indexing and Question Answering, pp. 40–46 (2018)

13. Rubner, Y., Tomasi, C., Guibas, L.J.: The earth mover's distance as a metric for image retrieval. Int. J. Comput. Vision 40(2), 99–121 (2000)

14. Severyn, A., Moschitti, A.: Learning to rank short text pairs with convolutional deep neural networks. In: 38th ACM SIGIR (2015)

15. Soldaini, L., Goharian, N.: QuickUMLS: a fast, unsupervised approach for medical concept extraction. In: MedIR Workshop, SIGIR (2016)

16. Srivastava, N., Hinton, G.E., Krizhevsky, A., Sutskever, I., Salakhutdinov, R.: Dropout: a simple way to prevent neural networks from overfitting. J. Mach. Learn. Res. 15(1), 1929–1958 (2014)

17. Tsatsaronis, G., et al.: An overview of the BIOASQ large-scale biomedical semantic indexing and question answering competition. BMC Bioinform. 16(1), 138 (2015)

18. Tunstall-Pedoe, W.: True knowledge: open-domain question answering using structured knowledge and inference. AI Mag. 31(3), 80–92 (2010)

19. Zadeh, L.A.: From search engines to question answering systems-the problems of world knowledge, relevance, deduction and precisiation. Capturing Intell. 1, 163–210 (2006)

AUEB at BioASQ 7: Document and Snippet Retrieval

Dimitris Pappas[1,2(✉)], Ryan McDonald[1,3], Georgios-Ioannis Brokos[1], and Ion Androutsopoulos[1]

[1] Department of Informatics, Athens University of Economics and Business, Athens, Greece
{pappasd,ion}@aueb.gr, g.brokos@gmail.com
[2] Institute for Language and Speech Processing, Research Center 'Athena', Marousi, Greece
dpappas@ilsp.gr
[3] Google Research, New York, USA
ryanmcd@google.com

Abstract. We present the submissions of AUEB to the BIOASQ 7 document and snippet retrieval tasks (parts of Task 7b, Phase A). Our systems build upon the methods we used in BIOASQ 6. This year we also experimented with models that jointly learn to retrieve documents and snippets, as opposed to using separate pipelined models for document and snippet retrieval. We also experimented with models based on BERT [5]. Our systems obtained the best document and snippet retrieval results for all batches of the challenge that we participated in.

Keywords: Information retrieval · Document retrieval · Document reranking · Snippet retrieval · Snippet extraction · Sentence selection · Biomedical question answering · Machine learning · Deep learning

1 Introduction

BIOASQ [26] is a biomedical document classification, retrieval, and question answering competition. It provides, among other information, tuples containing questions, gold relevant documents, and gold relevant snippets of relevant documents. All questions are expressed in natural language by human experts of the biomedical field. For the document and snippet retrieval tasks, the competitors receive a set of questions and must retrieve relevant documents and then extract relevant snippets from the retrieved documents. The available documents are abstracts from a collection of approx. 28 million MEDLINE/PUBMED biomedical articles. In this paper, we provide an overview of the submissions of AUEB to the document and snippet retrieval tasks (parts of Task 7b, Phase A) of BIOASQ 7.[1]

---

[1] See http://bioasq.org/participate/challenges.

P. Cellier and K. Driessens (Eds.): ECML PKDD 2019 Workshops, CCIS 1168, pp. 607–623, 2020.
https://doi.org/10.1007/978-3-030-43887-6_55

Most related research for biomedical document retrieval and snippet retrieval (or 'snippet extraction' or 'sentence selection') focuses mainly on one of the two tasks. When tackling both tasks, researchers usually follow a pipelined architecture where two models are trained separately and then run in sequence: document retrieval followed by snippet extraction from the retrieved documents. A major novel research direction of our participation this year is a new deep learning model which is jointly trained for both document and snippet retrieval. We build upon our BIOASQ 6 models for document retrieval [3,17] and modify them to also yield a relevance score for each *sentence* of the documents; we treat each sentence as a snippet, hence we use the two terms as synonyms. Since neural document retrieval methods are computationally intensive, we rely on conventional information retrieval (IR) methods to pre-fetch a list of possibly relevant documents, and then rerank the top retrieved documents and their sentences using the neural models. To our knowledge this is the first work on deep learning for joint document and snippet reranking. We also experimented with pipelined and joint models for document and snippet retrieval that employ BERT [5].

Our systems scored at the top for all batches of the challenge we participated in. Although a plain BERT model outperformed the other methods we considered for document retrieval, our joint document and snippet retrieval model obtained substantially better snippet retrieval results, even without using BERT and even though it uses much fewer parameters than the corresponding pipelined models. We make publicly available the database, code, and trained models.[2]

2   Document Retrieval Models

BIOASQ requires competitors to return a list of 10 relevant documents and 10 relevant snippets (from the 10 documents) per query. As already noted, we use conventional (BM25-based) IR methods to pre-fetch possibly relevant documents, which we then rerank using neural models. We experimented with two neural models for document reranking, one based on PACRR [17] and one based on BERT [5]. PACRR was one of the best document retrieval methods in BIOASQ 6; and BERT has led to state of the art results in several tasks [5].

2.1   Term-PACRR

The first model we use for document retrieval is TERM-PACRR [3,17], a modification of PACRR [9].[3] To train TERM-PACRR, we use mini-batches containing randomly selected relevant and irrelevant documents (in equal numbers) from the top N documents that the IR engine retrieves per training query, and we minimize binary cross-entropy. As in [3], we use a final linear layer that combines the TERM-PACRR score with traditional IR features like BM25, unigram and bigram overlap, and IDF-weighted unigram overlap. Consult [3] for details.

---

[2] See https://github.com/nlpaueb/aueb-bioasq7.

[3] TERM-PACRR is called PACRR-DRMM in [17].

2.2 BERT Based Document Retrieval

and right contexts, and (b) the next sentence. In the second document retrieval model, we employ BERT [5], which has recently led to state of the art results in several tasks, including document reranking on other datasets [20,23,30]. We add a task-specific logistic regression classifier on top of BERT, similar to other deep learning architectures [17,19,25]. We pre-trained our own BERT model on the PUBMED corpus using the titles and abstracts of the articles, and identical parameter settings to the uncased BERT LARGE model [5].[4] This is similar to BIOBERT [13], however unlike that work, we do not initialize the model with the public pre-trained BERT BASE instance, and we use a custom wordpiece model [29] also trained on PUBMED.

When fine-tuning BERT on BIOASQ data or when using it a test time, we feed it with the concatenation of a question and a (relevant or irrelevant) document. As standard, a special [CLS] token is added to the start of the concatenation, while a [SEP] token separates the question from the document (concatenated title and abstract), as illustrated in Fig. 1. The output vector of BERT for the [CLS] token is passed through a logistic regression layer (linear layer with sigmoid) to obtain a BERT-based score for the document. This score is then concatenated to extra features of the document (BM25 score and string overlap features), which are the same as in TERM-PACRR. Finally, another logistic regression layer is applied to the concatenated vector to get the final score of the document.

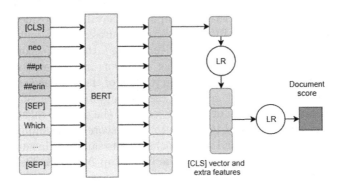

Fig. 1. BERT document ranking, with extra features added in a final layer.

During fine-tuning, negative samples (irrelevant documents to be concatenated with a training query) are drawn randomly from the non-relevant (according to the expert annotators) documents in the list of top N documents that the conventional IR system (the same as in the other methods) returned for the particular query. Critically, we found that incurring two losses per training instance helped accuracy. The first loss is standard, the binary cross-entropy of the final

---

[4] Max length was set to 512. Instances exceeding this threshold where truncated.

document score. The second loss is also binary cross-entropy, but computed on the BERT-based score, before concatenating it with the extra features. The two losses are summed. We found that this forced the model to use the BERT layers more effectively, otherwise the system tended to rely almost exclusively on the additional features during training.

Similarly to [3], we also experiment with a *high-confidence* version of BERT.[5] In this model, only documents with scores (probabilities of being relevant) greater than 0.01 were returned as relevant, hence fewer than 10 documents (the maximum allowed in BIOASQ) might be returned. This helped improve snippet retrieval as it focused that component only on the most relevant documents.

3   Snippet Retrieval Models

For snippet retrieval, we used two deep learning methods. The first one is the 'basic CNN' of [32], BCNN for short. It had the best snippet retrieval results in BIOASQ 6 [3]. The second method, POSIT-DRMM, PDRMM for short, had the best *document* retrieval results in the experiments of [17], but here we use it to score *snippets*, as a first step towards a joint model for document and snippet retrieval.

3.1   BCNN

BCNN [32] is a CNN-based snippet scoring method, which is fed with pairs consisting of a query and a snippet (relevant or irrelevant). For each pair, it returns an estimate of the probability that the snippet is relevant to the query. Following [3], we concatenate the score of BCNN to extra features of the snippet (sentence): the length of the sentence and the query (in tokens), the BM25 score of the sentence compared to the query, the number of tokens in the sentence excluding stopwords, the unigram and bigram token overlap of the sentence and the query, and finally the sum of the IDF scores of the overlapping tokens of the sentence and query divided by the sum of the IDF scores of the query's tokens. A logistic regression layer is then applied to obtain the final score of the snippet. As in [3], BCNN is trained on (relevant and irrelevant) snippets sampled (in equal numbers) from the relevant (gold) documents in the list of top N documents returned by the IR engine. Consult [3] for further details.

3.2   PDRMM

The second model we investigate is a modification of POSIT-DRMM [17], henceforth PDRMM. PDRMM was proposed for document scoring (reranking of documents retrieved by a conventional IR system), but here we use it for snippet scoring (reranking the sentences of retrieved documents). We first describe the original PDRMM, and then how we modified it to score snippets.

Given a query $q = \langle q_1, \ldots, q_n \rangle$ of n query terms (*q-terms*) and a document $d = \langle d_1, \ldots, d_m \rangle$ of m terms (*d-terms*), PDRMM computes context-sensitive term

---

[5] In [3], the high-confidence models were for another model, ABEL-DRMM.

embeddings $c(q_i)$ and $c(d_i)$ from the static (e.g., WORD2VEC) embeddings $e(q_i)$ and $e(d_i)$ by applying two stacked convolutional layers with trigram filters, residuals [8], and zero padding to q and d, respectively.[6]

PDRMM then computes three similarity matrices S_1, S_2, S_3, each of dimensions $n \times m$ (Fig. 2). Each element $s_{i,j}$ of S_1 is the cosine similarity between $c(q_i)$ and $c(d_j)$. S_2 is similar, but uses the static word embeddings $e(q_i), e(q_j)$. S_3 uses one-hot vectors for q_i, d_j, signaling exact matches. To each matrix (S_1, S_2, or S_3) we apply three row-wise pooling operators to extract 9 features for each q-term: max-pooling (to obtain the similarity of the best match between the q-term of the row and any of the d-terms), average pooling (to obtain the average match of each q-term to all d-terms), and average of k-max (to obtain the average similarity of the k best matches per q-term).[7] We concatenate the three features extracted from each row of the three similarity matrices (9 features in total) and concatenate them to obtain a new matrix S' of dimensions $n \times 9$ (Fig. 2, right). Each row of S' indicates the similarity of the corresponding q-term to any of the d-terms, through three different views of the terms (one-hot, static, context-aware embeddings). Each row of S' is then passed to a Multi-Layer Perceptron (MLP) to obtain a single match score per q-term.[8]

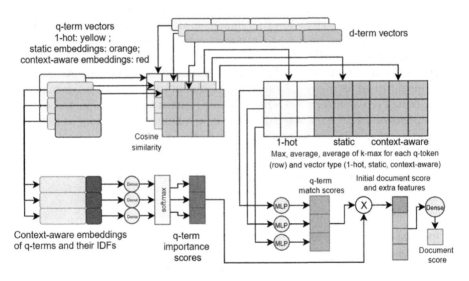

Fig. 2. PDRMM scoring documents with respect to a query. The same model can be used to score individual sentences with respect to a query, with different extra features.

---

[6] In [17], a BILSTM is used instead of convolutions, but the latter are faster and do not degrade performance.

[7] In our experiments, $k = 5$. We added the average pooling to PDRMM to balance the other two pooling operators that favor long documents.

[8] This MLP consists of one dense layer with 8 neurons and leaky RELU activation function, followed by a second dense layer with 1 output and no activation function.

Each context-aware q-term embedding is also concatenated with the corresponding IDF score (Fig. 2, bottom left) and passed to another linear layer that computes a score for each q-term. A SoftMax activation function is then applied across all the q-term scores to compute the importance of each q-term (e.g., words with low IDFs may not be helpful to answer the question).[9] Let v be the vector containing the n match scores of the q-terms, and u the vector of the corresponding n importance scores (Fig. 2, bottom). We extract an initial relevance score for the document as $\hat{r}(q, d) = v^T u$, which is then concatenated with four extra features: z-score normalized BM25 [24]; percentage of q-terms with exact match in d (regular and IDF weighted); percentage of q-term bigrams matched in d. An MLP computes the final relevance $r(q, d)$ from the five features.[10]

In its original form, PDRMM is trained on triples $\langle q, d, d' \rangle$, where d is a relevant document from the top N that the IR engine (the same as in the other methods) returned for query q, and d' a randomly sampled irrelevant document among the top N. Hinge loss is used, requiring $r(q, d)$ to exceed $r(q, d')$ by a margin.

In this work, we also use PDRMM to score snippets (sentences) by feeding it with query-snippet pairs, instead of query-document pairs, i.e., the d-terms (top of Fig. 2) are now the tokens of a particular snippet s from a retrieved document (relevant or irrelevant), and the output (bottom right of Fig. 2) is now the relevance score $r(q, s)$ of s. We also use different extra features when scoring snippets with PDRMM, which are the same as in BCNN (Sect. 3.1), instead of the extra features that are used when PDRMM scores documents. PDRMM is again trained on triples $\langle q, d, d' \rangle$, where q is a query, while d and d' are relevant and irrelevant documents, respectively, sampled from the top N documents that the IR engine returned for query q. Unlike the original PDRMM that scores documents, we use binary cross-entropy loss when training PDRMM to score snippets, treating the snippets of d that were selected by BIOASQ's human annotators as relevant, and all the other snippets from d and d' as irrelevant.

4   Joint Document and Snippet Retrieval Models

4.1   JPDRMM

As PDRMM can be used for both tasks, we create a joint PDRMM-based model, called JPDRMM, which given a query and a document, outputs relevance scores for each sentence (snippet) of the document, along with a relevance score for the entire document. JPDRMM applies the same process described in Sect. 3.2 to compute a score for each sentence in the document (Fig. 2, now operating on sentences). Then the maximum score of all the sentences is selected and concatenated to the extra features of the document (left part of Fig. 3), which

---

[9] The importance scores of the q-terms can also be viewed as self-attention scores.

[10] This MLP also consists of one dense layer with 8 neurons and leaky RELU activation function, followed by a second dense layer with no activation function.

are the same as when PDRMM scores documents.[11] The score of the document is computed by applying an MLP to the concatenated features.[12] The scores of the sentences are then revised to take into account the score of the entire document; the intuition is that snippets from relevant documents are more likely to be relevant. To do so, we concatenate the score of each sentence to the document score (Fig. 3, right part), and pass each pair of sentence-document scores through a logistic regression layer to obtain the final sentence score.

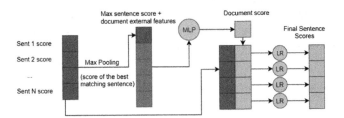

Fig. 3. The final layers of JPDRMM. The scores of the sentences (left) are generated by PDRMM (Fig. 2) operating on sentences. The maximum sentence score is concatenated with the external features of the document. An MLP produces the document score. A logistic regression layer then revises the score of each sentence, taking into account the original score of the sentence and the score of the document.

Like the original PDRMM, JPDRMM is trained on triples $\langle q, d, d' \rangle$, where q is a query, and d, d' are relevant and irrelevant documents, respectively, sampled from the top N documents returned by the IR engine for q. In this case, however, we apply a sentence splitter to d and d', and use JPDRMM to obtain relevance scores for d, d', and each one of their sentences. We compute a document hinge loss from the scores of d and d' as when PDRMM scores documents, and a binary cross-entropy loss for each sentence (relevant or irrelevant) of d and d' as when PDRMM scores sentences. The document hinge loss is added to the average sentence cross-entropy loss (averaged over all the sentences of d and d'), and their sum is used to train the entire model via backpropagation.

We create two versions of JPDRMM: one using pre-trained WORD2VEC embeddings, and one using pre-trained embeddings obtained from the top layer of the publicly available BERT BASE instance [5].[13] We call W2V-JPDRMM and BERT-JPDRMM the two versions, respectively. BERT's tokenizer splits words into subword units (wordpieces) [29]. In BERT-JPDRMM, in order to use IDF scores of entire words and compute exact matches across entire words, as in W2V-JPDRMM, we reconstruct the words from the subword units before feeding them to the rest

---

[11] We also experimented with other pooling operators to obtain the document score from the sentence scores, including combinations of max-pooling, average pooling, average of top k pooling, but they did not improve performance.

[12] This MLP consists of one dense layer with 8 neurons and leaky RELU activation function, followed by a second dense layer with no activation function.

[13] We also experimented with BIOBERT [13], but there was no notable improvement.

of the model. Also, we use BERT's top-level embedding for the first wordpiece of each reconstructed word as the pretrained embedding of that word.

5   Overall System Architecture

Figure 4 presents the architecture of our pipelined systems. The first step is retrieving N documents using a conventional BM25-based IR engine given a user question; see Sect. 6.1 below for details. Then a neural document retrieval model reranks the N documents and selects the top K_d. The K_d documents are reranked by a neural snippet retrieval model, which returns the top K_s snippets. BIOASQ requires $K_d = K_s = 10$, and we set $N = 100$.[14]

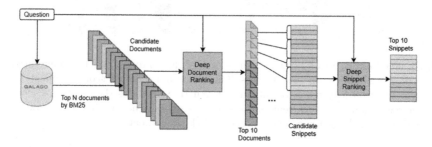

Fig. 4. Architecture of our pipelined document and snippet retrieval systems. The IR engine retrieves candidate relevant documents (left). A neural document retrieval model ranks the retrieved documents and returns the top 10. Then a neural snippet retrieval model ranks the snippets from the 10 documents and returns the top 10 snippets.

Figure 5 illustrates the architecture of our joint document and snippet retrieval models. The same IR engine is used to retrieve N documents. Then a joint model assigns relevance scores to the N documents and their snippets. We return the K_d documents with the highest relevance scores, and the K_s snippets with the highest relevance scores among all the snippets of the K_d documents. We use the same N, K_d, K_s values as in the pipelined models.

6   Experiments

6.1   Data and Experimental Setup

The document collection consists of approx. 29M 'articles' (titles and abstracts) from the 'MEDLINE/PUBMED baseline 2019' collection.[15] We discarded approx. 10M 'articles' that contained only titles, since very few of them had been judged as relevant by the expert annotators for any question. We created an index of

---

[14] Setting N to larger values had no impact on the final results.
[15] See https://www.nlm.nih.gov/databases/download/pubmed_medline.html.

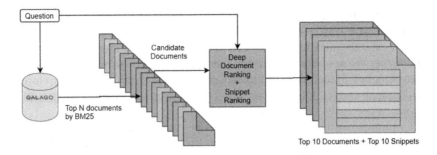

Fig. 5. Architecture of our joint document and snippet retrieval systems. The IR engine retrieves candidate relevant documents. The joint neural model assigns scores to the sentences of the retrieved documents and their snippets. We return the 10 documents with the highest scores, and the 10 snippets of those documents with the highest scores.

the remaining approx. 19M articles using Galago.[16] For indexing purposes, we removed stopwords and applied Krovetz's stemmer [12]. To train the neural models (or to fine-tune the BERT-based ones), we used years 1–6 of the BIOASQ data (2,647 questions), using batch 5 of year 6 as development set (100 questions).

In TERM-PACRR, BCNN, PDRMM (as sentence ranker), and W2V-JPDRMM, we use the biomedical WORD2VEC embeddings and the pre-computed IDF scores of [17]. We do not update the word embeddings when training these models. Similarly, when training BERT-JPDRMM, we use the same IDF scores and we do not update the BERT instance that provides the wordpiece embeddings; recall that we reconstruct the tokens from the wordpieces and use the first wordpiece embedding for each reconstructed token in BERT-JPDRMM. For tokenization, in methods that do not use wordpieces, we rely on the 'bioclean' tool provided by BIOASQ. Even for models using wordpieces, the features for the final logistic regression layer are derived using 'bioclean' in order to have tokens consistent with the same IDF table used in the other models. A slightly modified version of 'bioclean' is also used when constructing the index of the IR engine. The same 'bioclean' version (plus stopword removal) is also applied to the question when passing it as a query to the IR engine. In snippet retrieval, we use NLTK's English sentence splitter.[17]

TERM-PACRR was trained using default settings from the public release.[18] The BERT based document ranker (Sect. 2.2) was also trained using default settings from the public release.[19] The only exception was that we used a learning rate of $5e-6$ for fine-tuning, based on development set performance.

---

[16] We used Galago version 3.10. Consult http://www.lemurproject.org/galago.php.

[17] We used NLTK v3.2.4. See https://www.nltk.org/api/nltk.tokenize.html.

[18] See https://github.com/nlpaueb/aueb-bioasq6.

[19] See https://github.com/google-research/bert.

For BCNN, we used the publicly available code.[20] BCNN was trained using the settings that won last year's snippet extraction task [3], using Adagrad [6], with learning rate 0.08, and batch size 200. The model was trained for a maximum of 50 epochs. We keep for testing the parameters of BCNN from the epoch with the best snippet Mean Average Precision (MAP) score [16] on the development set. Document and snippet MAP are the official scores for document and snippet retrieval, respectively, in BIOASQ.

We re-implemented PDRMM (as a sentence ranker) in PYTORCH [21] replicating the code of [3]; we also implemented JPDRMM in PYTORCH.[21] We trained PDRMM and W2V-JPDRMM for a maximum of 20 epochs, and BERT-JPDRMM for a maximum of 4 epochs, selecting the parameters from the epoch with the best development snippet MAP for PDRMM and document MAP the two JPDRMM versions. We also applied early stopping and stopped training when development performance (snippet or document MAP) stopped improving for 4 consecutive epochs. For the two JPDRMM versions, one could also monitor *snippet* MAP on development data, instead of document MAP, or a combination of the two. We plan to examine how this affects the performance of JPDRMM in future work. W2V-JPDRMM, BERT-JPDRMM, and PDRMM were trained using Adam [11] with a learning rate of 0.01, $\beta_1/\beta_2 = 0.9/0.999$, and a batch size of 32.

6.2   Official Submissions

We submitted five different systems to BIOASQ 7 (Task 7b, Phase A), all of which consist of components described above.

AUEBNLP-1: W2V-JPDRMM for both document retrieval and snippet extraction.
AUEBNLP-2: BERT-JPDRMM for both document retrieval and snippet extraction.
AUEBNLP-3: pipeline consisting of TERM-PACRR for document retrieval, followed by BCNN for snippet retrieval in batches 2 and 3, or PDRMM in batches 4 and 5.
AUEBNLP-4: pipeline consisting of BERT for document retrieval, followed by BCNN for snippet retrieval in batches 2 and 3, or PDRMM in batches 4 and 5.
AUEBNLP-5: pipeline of BERT high confidence (Sect. 2.2, last paragraph) for document retrieval, followed by BCNN for snippet retrieval in batches 2 and 3, or PDRMM in batches 4 and 5.

In all five systems, after obtaining the top 10 documents and top 10 snippets, we reranked the top 10 snippets by the scores of the documents they came from. The goal was to promote snippets coming from highly relevant documents. In the first two systems, which use JPDRMM versions, this final reranking of the snippets made almost no difference, since JPDRMM internally revises the scores of the snippets taking into account the scores of the documents they come from.

---

[20] BCNN's code is also available from https://github.com/nlpaueb/aueb-bioasq6.
[21] The original code of PDRMM is also available from https://github.com/nlpaueb/aueb-bioasq6. All the additional code of this paper will also be made available.

6.3   Results

Table 1 reports our test results (F1, MAP) for batches 2–5 of BIOASQ 7. We did not participate in batch 1. We observe that the BERT document ranker (used in AUEB-NLP-4) has the best document MAP scores in all batches; recall that document MAP is the official document retrieval measure of BIOASQ and also the measure we monitored on the development data to select the best training epoch. The BERT high confidence document ranker (used in AUEB-NLP-5) has the second best document MAP overall, but with greatly improved F1.

Table 1. Performance on BIOASQ Task 7b, Phase A (batches 2–5) for document and snippet retrieval. Top Comp. is the top scoring submission from other teams.

Document retrieval					Snippet retrieval				
System	Rank	F-M.	MAP	GMAP	System	Rank	F-M.	MAP	GMAP
Batch 2					Batch 2				
AUEB-NLP-1	9	17.84	7.41	0.66	AUEBNLP-1	1	18.55	**14.38**	0.19
AUEB-NLP-2	10	18.23	7.41	0.62	AUEB-NLP-2	3	17.64	12.90	**0.28**
AUEB-NLP-3	5	19.05	7.71	**0.75**	AUEB-NLP-3	6	11.11	6.25	0.13
AUEBNLP-4	1	19.11	**8.49**	0.67	AUEB-NLP-4	5	10.96	6.43	0.14
AUEB-NLP-5	2	**34.43**	8.30	0.49	AUEB-NLP-5	2	**19.01**	13.62	0.23
Top Comp.	3	18.77	7.91	0.48	Top Comp.	4	12.12	8.93	0.04
Batch 3					Batch 3				
AUEB-NLP-1	4	23.80	10.41	1.18	**AUEBNLP-1**	1	24.72	**22.06**	0.81
AUEB-NLP-2	2	24.49	11.21	1.56	AUEB-NLP-2	2	**25.63**	21.97	**0.89**
AUEB-NLP-3	6	22.66	9.86	1.04	AUEB-NLP-3	7	14.43	9.90	0.28
AUEBNLP-4	1	24.71	**11.99**	1.51	AUEB-NLP-4	5	15.44	11.26	0.37
AUEB-NLP-5	3	**40.34**	11.02	**1.64**	AUEB-NLP-5	3	24.56	19.21	0.85
Top Comp.	5	28.94	10.33	0.18	Top Comp.	4	16.17	14.04	0.09
Batch 4					Batch 4				
AUEB-NLP-1	4	20.56	9.51	1.01	AUEB-NLP-1	2	24.40	20.86	0.65
AUEB-NLP-2	3	20.51	9.68	0.83	**AUEBNLP-2**	1	23.65	**21.14**	0.75
AUEB-NLP-3	5	19.42	9.09	0.83	AUEB-NLP-3	7	17.79	11.49	0.53
AUEBNLP-4	1	21.48	**10.34**	1.12	AUEB-NLP-4	9	17.91	11.16	0.56
AUEB-NLP-5	2	**37.83**	10.15	**1.16**	AUEB-NLP-5	3	**24.67**	18.21	**0.98**
Top Comp.	6	18.53	8.35	0.51	Top Comp.	4	17.23	15.27	0.13
Batch 5					Batch 5				
AUEB-NLP-1	3	9.90	3.68	0.06	AUEB-NLP-1	3	8.04	5.81	0.02
AUEB-NLP-2	6	9.00	3.55	0.06	**AUEBNLP-2**	1	8.18	**6.31**	**0.03**
AUEB-NLP-3	5	9.91	3.66	0.07	AUEB-NLP-3	8	5.81	3.87	0.02
AUEBNLP-4	1	11.20	**4.25**	**0.10**	AUEB-NLP-4	6	6.53	4.16	0.02
AUEB-NLP-5	2	**20.12**	3.99	0.08	AUEB-NLP-5	2	**9.89**	6.17	**0.03**
Top Comp.	4	9.27	3.68	0.05	Top Comp.	4	6.56	4.99	0.01

Interestingly, the joint model (used in AUEB-NLP-1/2) outperformed comparable pipelined systems (AUEB-NLP-1 vs. AUEB-NLP-3, AUEB-NLP-2 vs. AUEB-NLP-4) by a wide margin in snippet MAP. It obtained very competitive results in snippet MAP even without using BERT embeddings (AUEB-NLP-1) and against pipelines that used BERT for document retrieval (AUEB-NLP-4) and additional reranking heuristics (AUEB-NLP-5). Recall, also, that in the joint model we selected the best training epoch by monitoring the *document* MAP on development data, whereas for the snippet retrieval components of the pipelined models (AUEB-NLP-3/4/5) *snippet* MAP was monitored; hence, the snippet MAP scores of the joint model might improve further by monitoring *snippet* MAP. We also note that the joint models use much fewer trainable parameters than the pipeline models (Table 1); and they outperform AUEB-NLP-3, which was one of the best systems of BIOASQ 6. It is also interesting that in both document and snippet retrieval, there is no clear difference between AUEB-NLP-1, which does not rely on BERT at all, and AUEB-NLP-2, which uses BERT to obtain word embeddings (Table 2).

Table 2. Number of trainable parameters for systems submitted.

Model	Number of parameters
AUEB-NLP-1	5,793
AUEB-NLP-2	3,541,551
AUEB-NLP-3	16,519
AUEB-NLP-4/5 (with BCNN for snippets)	109,499,902
AUEB-NLP-4/5 (with PDRMM for snippets)	109,489,455

It is particularly interesting is that the joint model (AUEB-NLP-1/2) outperforms the BERT based high-confidence model (AUEB-NLP-5). Similarly to [3], we observed that passing only high-confidence retrieved documents to the snippet ranking component in pipeline systems improved snippet retrieval greatly (compare the snippet scores of AUEB-NLP-4 vs. AUEB-NLP-5), because it allowed the snippet retrieval component to operate only on documents that were likely to be relevant. However, JPDRMM did not require such heuristics. Instead, since it models the fact that good snippets come from good documents and vice-versa, it naturally selected snippets mostly from high confidence documents. Thus the empirical results validate the hypothesis that joint modeling is beneficial. An open question is why the joint models do worse on document ranking compared to the pipelined models (AUEB-NLP-4/5). This is likely due to BERT (the document scorer of AUEB-NLP-4/5) being such a powerful model. A future line of investigation is to build joint models that integrate BERT to a larger extent, instead of just providing word embeddings to JPDRMM as in BERT-JPDRMM.

7   Related Work

7.1   Document Retrieval

Neural document ranking models [7,9,10,17,18] have only recently managed to improve upon the rankings of traditional IR systems (e.g., rankings based on BM25). See also [14] for caveats.

PACRR [9] uses a matrix containing the cosine similarities between each query term embedding and each document term embedding; the multiple similarity matrices of PDRMM [17] (Sect. 3.2, Fig. 3.2) are an extension of PACRR's similarity matrix. PACRR applies convolutions with multiple filters of kernel size 2 and 3 to its similarity matrix to capture bigram and trigram matches, respectively; PDRMM skips these convolutions, since one of its similarity matrices already contains similarities between context-aware embeddings. PACRR then employs max-pooling (over the outputs of kernels of the same size) followed by row-wise k-max pooling to obtain the k-best unigram, bigram, and trigram matches between each query term and the entire document, producing $3k$ document-aware features per query term; these pooling operations are again very similar to the ones of PDRMM. The IDF score of each query term is then appended to its $3k$ features, and the features of all the query terms are then concatenated into a single vector, which is passed to an MLP that produces the relevance score of the document. The only difference between PACRR and TERM-PACRR [3,17] (Sect. 2.1) is that the latter passes the features of each query term separately to the MLP, obtaining a separate relevance score per query term, and then uses a linear layer to combine the relevance scores.[22] By contrast, PDRMM computes a weighted sum of the feature vectors of the query terms (weighted by their importance scores, bottom right of Fig. 1) and passes the weighted sum to the MLP that produces the document's relevance score.

TERM-PACRR's final layers, which apply an MLP separately to document-aware features of each query term and then combine the resulting relevance scores of the query terms using a linear layer, are very similar to the corresponding layers of DRMM [7]. In DRMM, however, the document-aware features of each query term represent a histogram of (frequencies of buckets of) the cosine similarities between the embedding of the query term and all the terms of the document. These histogram representations are non-differentiable, hindering the end-to-end training of the model via backpropagation. By contrast, all the models used in our work are fully differentiable, following [17].

The document retrieval model of Zhu et al. [33] was designed to handle medical questions. It uses a BIGRU with self-attention [2,4] to produce a single query embedding from the query's word embeddings; and a hierarchical BIGRU [31] to produce a single document embedding. The word-level BIGRU of the hierarchical BIGRU reads the word embeddings of a single sentence of the document at a time, turning each sentence into a sentence embedding; it also employs a cross-attention mechanism between the word embeddings of the query and those of

---

[22] We note again that TERM-PACRR is called PACRR-DRMM in [17].

the sentence. The sentence-level BIGRU reads the sentence embeddings and produces the document embedding using a self-attention mechanism. The relevance score of the document is then computed by taking the element-wise product of the query and document embeddings and feeding it to an MLP. Although we hope to compare to the model of Zhu et al. in future work, we note that their experiments were conducted on a dataset much smaller than BIOASQ's, containing only 7.5k documents and 7.5k queries with only one relevant document per query. Furthermore, the documents of Zhu et al.'s dataset were article sections from a healthcare portal and the queries were produced by annotators looking at a particular article. The annotators were not biomedical experts, hence their queries and terminology were much simpler compared to BIOASQ's, where the annotators are biomedical experts and queries reflect real needs.

BERT based models have recently been explored for document ranking. Most approaches train shallow task-specific layers on top of BERT [20,30], much as in our BERT based document retrieval model (Sect. 2.2, Fig. 1). MacAvaney et al. [15] explored ways to combine ELMO [22] and BERT [5] with complex neural IR models such as DRMM [7] and PACRR [9]. It would be interesting to explore similar ways to improve BERT-JPDRMM (Sect. 4.1), e.g., by using cosine similarity matrices (Fig. 2) computed on wordpiece embeddings coming from different layers of BERT, or by concatenating the embedding of BERT's [CLS] token (Fig. 1) with the extra document features in the final layers of JPDRMM (Fig. 3).

7.2 Snippet Extraction

BCNN (Sect. 3.1) is one of the several CNN-based models explored by Yin et al. [32]. We used BCNN in our pipeline systems (Sect. 6.2), because it had the best snippet retrieval results in BIOASQ 6 [3]. As we demonstrated with PDRMM, however, neural document retrieval models can also be used to rank snippets, and in our experiments PDRMM performed better than BCNN for snippet retrieval, which is why it replaced BCNN in our pipeline systems in batches 4 and 5.

Amiri et al. [1] use context-sensitive autoencoders to create question and sentence vectors. They compute the cosine similarity between the question and sentence vectors and rank the sentences in the dataset. They experiment on three datasets, including TREQ QA [27], which includes biomedical data. Their method is unsupervised and performed competitively compared to former state-of-the-art supervised models. However, it does not take into account the relevance of the documents when ranking sentences.

Other neural models have also been proposed for snippet extraction in biomedical question answering. Wang et al. [28] use a stacked BILSTM that reads the concatenation of the question and a candidate sentence. In each timestep, the model produces a relevance score for the sentence, taking into account the tokens read so far. Then a mean pooling operation extracts the final relevance score of the sentence. Wang et al. combine the relevance score of the neural model with a keyword matching score, in order to distinguish tokens with similar embeddings and to favor exact token matches. In PDRMM and JPDRMM, this effect is achieved by external overlap features in the final linear layer, but also by including in the neural model a similarity matrix (view) with one-hot token embeddings (Fig. 2). As in the model of Amiri et al. discussed above, the model of Wang et al. does not take into account the relevance of the documents when ranking sentences.

8   Discussion and Future Work

We presented the models, experiments, and results of the submissions of AUEB for the document and snippet retrieval tasks of BIOASQ 7. Our systems obtained the best document and snippet retrieval results in the four batches we participated in.

We introduced a new jointly trained model for document and snippet retrieval. The joint model outperformed comparable pipelined architectures by a wide margin in snippet retrieval. It obtained very competitive results in snippet retrieval even without using BERT at all, and against pipelines that used BERT for document retrieval and additional reranking heuristics. On the other hand, a BERT based document ranker performed better at the document retrieval level than the joint model. We aim to investigate if tuning the weights of the document and snippet losses of the joint model could help it perform better in document retrieval too. We also aim to integrate more tightly BERT into our joint model, e.g., by using similarity matrices based on embeddings coming from different levels of BERT, instead of using only the top-level BERT embeddings (as in one version of our joint model), and by adding the embedding of BERT's [CLS] token to the extra features of the joint model. Finally, we aim to extend the joint model to also perform exact answer extraction (part of BIOASQ Task 7b, Phase B).

References

1. Amiri, H., Resnik, P., Boyd-Graber, J., Daumé III, H.: Learning text pair similarity with context-sensitive autoencoders. In: Proceedings of the 54th Annual Meeting of the Association for Computational Linguistics (vol. 1: Long Papers), Berlin, Germany, pp. 1882–1892 (2016)
2. Bahdanau, D., Cho, K., Bengio, Y.: Neural machine translation by jointly learning to align and translate. In: 3rd International Conference on Learning Representations, San Diego, California (2015)
3. Brokos, G., Liosis, P., McDonald, R., Pappas, D., Androutsopoulos, I.: AUEB at BioASQ 6: document and snippet retrieval. In: Proceedings of the 6th BioASQ Workshop A Challenge on Large-scale Biomedical Semantic Indexing and Question Answering, Brussels, Belgium (2018)
4. Cho, K., et al.: Learning phrase representations using RNN encoder-decoder for statistical machine translation. In: Proceedings of the Conference on Empirical Methods in Natural Language Processing, Doha, Qatar, pp. 1724–1734 (2014)
5. Devlin, J., Chang, M.W., Lee, K., Toutanova, K.: BERT: pre-training of deep bidirectional transformers for language understanding. arXiv:1810.04805 (2018)
6. Duchi, J., Hazan, E., Singer, Y.: Adaptive subgradient methods for online learning and stochastic optimization. J. Mach. Learn. Res. **12**, 2121–2159 (2011)

7. Guo, J., Fan, Y., Ai, Q., Croft, W.B.: A deep relevance matching model for ad-hoc retrieval. In: Proceedings of the 25th ACM International on Conference on Information and Knowledge Management, Indianapolis, Indiana, USA, pp. 55–64 (2016)

8. He, K., Zhang, X., Ren, S., Sun, J.: Deep residual learning for image recognition. In: IEEE Conference on Computer Vision and Pattern Recognition, pp. 770–778 (2016)

9. Hui, K., Yates, A., Berberich, K., de Melo, G.: PACRR: a position-aware neural IR model for relevance matching. In: Proceedings of the Conference on Empirical Methods in Natural Language Processing, Copenhagen, Denmark, pp. 1049–1058 (2017)

10. Hui, K., Yates, A., Berberich, K., de Melo, G.: Co-PACRR: a context-aware neural IR model for ad-hoc retrieval. In: Proceedings of the 11th ACM International Conference on Web Search and Data Mining, Marina Del Rey, CA, pp. 279–287 (2018)

11. Kingma, D.P., Ba, J.: Adam: A method for stochastic optimization. CoRR abs/1412.6980 (2015)

12. Krovetz, R.: Viewing morphology as an inference process. In: Proceedings of the 16th Annual International ACM SIGIR Conference on Research and Development in Information Retrieval, Pittsburgh, PA, pp. 191–202 (1993)

13. Lee, J., et al,.: Biobert: a pre-trained biomedical language representation model for biomedical text mining. arXiv preprint arXiv:1901.08746 (2019)

14. Lin, J.: The neural hype and comparisons against weak baselines. SIGIR Forum **52**(2), 40–51 (2019)

15. MacAvaney, S., Yates, A., Cohan, A., Goharian, N.: CEDR: contextualized embeddings for document ranking. CoRR abs/1904.07094 (2019)

16. Manning, C.D., Raghavan, P., Schütze, H.: Introduction to Information Retrieval. Cambridge University Press, New York (2008)

17. McDonald, R., Brokos, G.I., Androutsopoulos, I.: Deep relevance ranking using enhanced document-query interactions. In: Proceedings of the Conference on Empirical Methods in Natural Language Processing, Brussels, Belgium (2018)

18. Mitra, B., Craswell, N.: An Introduction to Neural Information Retrieval. Now Publishers, Boston (2018)

19. Mohan, S., Fiorini, N., Kim, S., Lu, Z.: Deep learning for biomedical IR: learning textual relevance from click logs. BioNLP **2017**, 222–231 (2017)

20. Nogueira, R., Cho, K.: Passage re-ranking with BERT. CoRR abs/1901.04085 (2019)

21. Paszke, A., et al.: Automatic differentiation in PyTorch. In: NIPS-W (2017)

22. Peters, M., et al.: Deep contextualized word representations. In: Proceedings of the 2018 Conference of the North American Chapter of the Association for Computational Linguistics: Human Language Technologies, New Orleans, Louisiana, vol. 1. pp. 2227–2237 (2018)

23. Qiao, Y., Xiong, C., Liu, Z.H., Liu, Z.: Understanding the behaviors of BERT in ranking. CoRR abs/1904.07531 (2019)

24. Robertson, S., Zaragoza, H.: The probabilistic relevance framework: BM25 and beyond. Found. Trends Inf. Retrieval **3**(4), 333–389 (2009)

25. Severyn, A., Moschitti, A.: Learning to rank short text pairs with convolutional deep neural networks. In: Proceedings of the 38th International ACM SIGIR Conference on Research and Development in Information Retrieval, pp. 373–382. ACM (2015)

26. Tsatsaronis, G., et al.: An overview of the BIOASQ large-scale biomedical semantic indexing and question answering competition. BMC Bioinform. **16**, 138 (2015)

27. Voorhees, E.M.: Question answering in TREC. In: Proceedings of the Tenth International Conference on Information and Knowledge Management, New York, NY, USA, pp. 535–537 (2001)

28. Wang, D., Nyberg, E.: A long short-term memory model for answer sentence selection in question answering. In: Proceedings of the 53rd Annual Meeting of the Association for Computational Linguistics and the 7th International Joint Conference on Natural Language Processing (Volume 2: Short Papers), Beijing, China, pp. 707–712 (2015)

29. Wu, Y., et al.: Google's Neural Machine Translation System: Bridging the Gap between Human and Machine Translation. CoRR abs/1609.08144 (2016)

30. Yang, W., Zhang, H., Lin, J.: Simple applications of BERT for ad hoc document retrieval. CoRR abs/1903.10972 (2019)

31. Yang, Z., Yang, D., Dyer, C., He, X., Smola, A., Hovy, E.: Hierarchical attention networks for document classification. In: Proceedings of the 2016 Conference of the NA Chapter of the Association for Computational Linguistics: Human Language Technologies, pp. 1480–1489 (2016)

32. Yin, W., Schütze, H., Xiang, B., Zhou, B.: ABCNN: Attention-based convolutional neural network for modeling sentence pairs. Trans. Assoc. Comput. Linguist. **4** (2016)

33. Zhu, M., Ahuja, A., Wei, W., Reddy, C.K.: A hierarchical attention retrieval model for healthcare question answering. In: The World Wide Web Conference, San Francisco, CA, USA, pp. 2472–2482 (2019)

Classification Betters Regression in Query-Based Multi-document Summarisation Techniques for Question Answering
Macquarie University at BioASQ7b

Diego Mollá[(✉)] and Christopher Jones

Macquarie University, Sydney, NSW 2109, Australia
Diego.Molla-Aliod@mq.edu.au, Christopher.Jones4@students.mq.edu.au

Abstract. Task B Phase B of the 2019 BioASQ challenge focuses on biomedical question answering. Macquarie University's participation applies query-based multi-document extractive summarisation techniques to generate a multi-sentence answer given the question and the set of relevant snippets. In past participation we explored the use of regression approaches using deep learning architectures and a simple policy gradient architecture. For the 2019 challenge we experiment with the use of classification approaches with and without reinforcement learning. In addition, we conduct a correlation analysis between various ROUGE metrics and the BioASQ human evaluation scores.

Keywords: Deep learning · Reinforcement learning · Evaluation · Query-based summarisation

1 Introduction

The BioASQ Challenge[1] includes a question answering task (Phase B, part B) where the aim is to find the "ideal answer"—that is, an answer that would normally be given by a person [12]. This is in contrast with most other question answering challenges where the aim is normally to give an exact answer, usually a fact-based answer or a list. Given that the answer is based on an input that consists of a biomedical question and several relevant PubMed abstracts[2], the task can be seen as an instance of query-based multi-document summarisation.

As in past participation [6,7], we wanted to test the use of deep learning and reinforcement learning approaches for extractive summarisation. In contrast with

---

[1] http://www.bioasq.org.
[2] https://www.ncbi.nlm.nih.gov/pubmed/.

Code associated to this paper is available as a Docker container in https://hub.docker.com/r/dmollaaliod/bioasq7b.

© Springer Nature Switzerland AG 2020
P. Cellier and K. Driessens (Eds.): ECML PKDD 2019 Workshops, CCIS 1168, pp. 624–635, 2020.
https://doi.org/10.1007/978-3-030-43887-6_56

past years where the training procedure was based on a regression set up, this year we experiment with various classification set ups. The main contributions of this paper are:

1. We compare classification and regression approaches and show that classification produces better results than regression but the quality of the results depends on the approach followed to annotate the data labels.
2. We conduct correlation analysis between various ROUGE evaluation metrics and the human evaluations conducted at BioASQ and show that Precision and F1 correlate better than Recall.

Section 2 briefly introduces some related work for context. Section 3 describes our classification and regression experiments. Section 4 details our experiments using deep learning architectures. Section 5 explains the reinforcement learning approaches. Section 6 shows the results of our correlation analysis between ROUGE scores and human annotations. Section 7 lists the specific runs submitted at BioASQ 7b. Finally, Sect. 8 concludes the paper.

2 Related Work

The BioASQ challenge has organised annual challenges on biomedical semantic indexing and question answering since 2013 [12]. Every year there has been a task about semantic indexing (task a) and another about question answering (task b), and occasionally there have been additional tasks. The tasks defined for 2019 are:

BioASQ Task 7a: Large Scale Online Biomedical Semantic Indexing.
BioASQ Task 7b: Biomedical Semantic QA involving Information Retrieval (IR), Question Answering (QA), and Summarisation.
BioASQ MESINESP Task: Medical Semantic Indexing in Spanish.

BioASQ Task 7b consists of two phases. Phase A provides a biomedical question as an input, and participants are expected to find relevant concepts from designated terminologies and ontologies, relevant articles from PubMed, relevant snippets from the relevant articles, and relevant RDF triples from designated ontologies. Phase B provides a biomedical question and a list of relevant articles and snippets, and participant systems are expected to return the exact answers and the ideal answers. The training data is composed of the test data from all previous years, and amounts to 2,747 samples.

There has been considerable research on the use of machine learning approaches for tasks related to text summarisation, especially on single-document summarisation. Abstractive approaches normally use an encoder-decoder architecture and variants of this architecture incorporate attention [10] and pointer-generator [11]. Recent approaches leveraged the use of pre-trained models [2]. Recent extractive approaches to summarisation incorporate recurrent neural networks that model sequences of sentence extractions [8] and may incorporate an abstractive component and reinforcement learning during the training

Table 1. Summarisation techniques used in BioASQ 6b for the generation of ideal answers. The evaluation result is the human evaluation of the best run.

System	Abstractive approaches	Extractive approaches
[7]	(none)	Regression & Reinforcement learning
[4]	Fusion	Maximum marginal relevance
[1]	(none)	Lexical chains
[9]	Fine-tuned pointer generator coverage	Learning to rank

stage [13]. But relatively few approaches have been proposed for query-based multi-document summarisation. Table 1 summarises the approaches presented in the proceedings of the 2018 BioASQ challenge.

3   Classification *vs.* Regression Experiments

Our past participation in BioASQ [6,7] and this paper focus on extractive approaches to summarisation. Our decision to focus on extractive approaches is based on the observation that a relatively large number of sentences from the input snippets has very high ROUGE scores, thus suggesting that human annotators had a general tendency to copy text from the input to generate the target summaries [6]. Our past participating systems used regression approaches using the following framework:

1. Train the regressor to predict the ROUGE-SU4 F1 score of the input sentence.
2. Produce a summary by selecting the top n input sentences.

A novelty in the current participation is the introduction of classification approaches using the following framework.

1. Train the classifier to predict the target label ("summary" or "not summary") of the input sentence.
2. Produce a summary by selecting all sentences predicted as "summary".
3. If the total number of sentences selected is less than n, select n sentences with higher probability of label "summary".

Introducing a classifier makes labelling the training data not trivial, since the target summaries are human-generated and they do not have a perfect mapping to the input sentences. In addition, some samples have multiple reference summaries. [3] showed that different data labelling approaches influence the quality of the final summary, and some labelling approaches may lead to better results than using regression. In this paper we experiment with the following labelling approaches:

threshold t**:** Label as "summary" all sentences from the input text that have a ROUGE score higher than a threshold t.
top m**:** Label as "summary" the m input text sentences with highest ROUGE score.

As in [3], The ROUGE score of an input sentence was the ROUGE-SU4 F1 score of the sentence against the set of reference summaries.

We conducted cross-validation experiments using various values of t and m. Table 3 shows the results for the best values of t and m obtained. The regressor and classifier used Support Vector Regression (SVR) and Support Vector Classification (SVC) respectively. To enable a fair comparison we used the same input features in all systems. These input features combine information from the question and the input sentence and are shown in Fig. 1. The features are based on [5], and are the same as in [6], plus the addition of the position of the input snippet. The best SVC and SVR parameters were determined by grid search.

- $tf.idf$ vector of the candidate sentence.
- Cosine similarity between the $tf.idf$ vector of the question and the $tf.idf$ vector of the candidate sentence.
- The largest cosine similarity between the $tf.idf$ vector of candidate sentence and the $tf.idf$ vector of each of the snippets related to the question.
- Cosine similarity between the sum of word2vec embeddings of the words in the question and the word2vec embeddings of the words in the candidate sentence. We used vectors of dimension 200 pretrained using PubMed documents provided by the organisers of BioASQ.
- Pairwise cosine similarities between the words of the question and the words of the candidate sentence. We used word2vec to compute the word vectors. We then computed the pairwise cosine similarities and selected the following features:
  - The mean, median, maximum, and minimum of all pairwise cosine similarities.
  - The mean of the 2 highest, mean of the 3 highest, mean of the 2 lowest, and mean of the 3 lowest.
- Weighted pairwise cosine similarities where the weight was the $tf.idf$ of the word.

Fig. 1. Features used in the SVC and SVR experiments.

Preliminary experiments showed a relatively high number of cases where the classifier did not classify any of the input sentences as "summary". To solve this problem, and as mentioned above, the summariser used in Table 3 introduces a backoff step that extracts the n sentences with highest predicted values when the summary has less than n sentences. The value of n is as reported in our prior work and shown in Table 2.

Table 2. Number of sentences returned by the regression-based summarisers and the backoff step of the classification-based summarisers, for each question type

	Summary	Factoid	Yesno	List
n	6	2	2	3

The results confirm [3]'s finding that classification outperforms regression. However, the actual choice of optimal labelling scheme was different: whereas in [3] the optimal labelling was based on a labelling threshold of 0.1, our experiments show a better result when using the top 5 sentences as the target summary. The reason for this difference might be the fact that [3] used all sentences from the abstracts of the relevant PubMed articles, whereas we use only the snippets as the input to our summariser. Consequently, the number of input sentences is now much smaller. We therefore report the results of using the labelling schema of top 5 snippets in all subsequent classifier-based experiments of this paper.

Table 3. Regression vs. classification approaches measured using ROUGE SU4 F-score under 10-fold cross-validation. The table shows the mean and standard deviation across the folds. "firstn" is a baseline that selects the first n sentences. SVR and SVC are described in Sect. 3. NNR and NNC are described in Sect. 4.

Method	Labelling	ROUGE-SU4 F1 Mean ± 1 stdev	
firstn		0.252 ± 0.015	
SVR	SU4 F1	0.239 ± 0.009	
SVC	threshold 0.2	0.240 ± 0.012	
SVC	top 5	0.253 ± 0.013	
NNR	SU4 F1	0.254 ± 0.013	
NNC	SU4 F1	0.257 ± 0.012	
NNC	top 5	0.262 ± 0.012	

0.22   0.23   0.24   0.25   0.26   0.27

4   Deep Learning Models

Based on the findings of Sect. 3, we apply minimal changes to the deep learning regression models of [7] to convert them to classification models. In particular, we add a sigmoid activation to the final layer, and use cross-entropy as the loss function.[3] The complete architecture is shown in Fig. 2.

The bottom section of Table 3 shows the results of several variants of the neural architecture. The table includes a neural regressor (NNR) and a neural classifier (NNC). The neural classifier is trained in two set ups: "NNC top 5" uses classification labels as described in Sect. 3, and "NNC SU4 F1" uses the regression labels, that is, the ROUGE-SU4 F1 scores of each sentence. Of interest

---

[3] We also changed the platform from TensorFlow to the Keras API provided by TensorFlow.

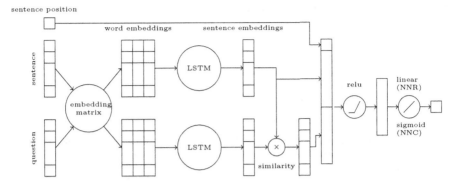

sentence position

Fig. 2. Architecture of the neural classification and regression systems. A matrix of pre-trained word embeddings (same pre-trained vectors as in Fig. 1) is used to find the embeddings of the words of the input sentence and the question. Then, LSTM chains are used to generate sentence embeddings—the weights of the LSTM chains of input sentence and question are not shared. Then, the sentence position is concatenated to the sentence embedding and the similarity of sentence and question embeddings, implemented as a product. A final layer predicts the label of the sentence.

is the fact that "NNC SU4 F1" outperforms the neural regressor. We have not explored this further and we presume that the relatively good results are due to the fact that ROUGE values range between 0 and 1, which matches the full range of probability values that can be returned by the sigmoid activation of the classifier final layer.

Table 3 also shows the standard deviation across the cross-validation folds. Whereas this standard deviation is fairly large compared with the differences in results, in general the results are compatible with the top part of the table and prior work suggesting that classification-based approaches improve over regression-based approaches.

5   Reinforcement Learning

We also experiment with the use of reinforcement learning techniques. Again these experiments are based on [7], who uses REINFORCE to train a global policy. The policy predictor uses a simple feedforward network with a hidden layer.

The results reported by [7] used ROUGE Recall and indicated no improvement with respect to deep learning architectures. Human evaluation results are preferable over ROUGE but these were made available after the publication of the paper. When comparing the ROUGE and human evaluation results (Table 4), we observe an inversion of the results. In particular, the reinforcement learning approaches (RL) of [7] receive good human evaluation results, and as a matter of fact they are the best of our runs in two of the batches. In contrast, the regression systems (NNR) fare relatively poorly. Section 6 expands on the comparison between the ROUGE and human evaluation scores.

Table 4. Results of ROUGE-SU4 Recall (R) and human (H) evaluations on BioASQ 6b runs, batch 5. The human evaluation shows the average of all human evaluation metrics.

Run	System	Batch 1		Batch 2		Batch 3		Batch 4		Batch 5	
		R	H	R	H	R	H	R	H	R	H
MQ-1	First n	0.46	3.91	0.50	**4.01**	0.45	**4.06**	0.51	4.16	0.59	4.05
MQ-2	Cosine	0.52	**3.96**	0.50	3.97	0.45	3.97	0.53	4.15	0.59	4.06
MQ-3	SVR	0.49	3.87	0.51	3.96	0.49	**4.06**	0.52	4.17	0.62	3.98
MQ-4	NNR	**0.55**	3.85	**0.54**	3.93	**0.51**	4.05	**0.56**	**4.19**	**0.64**	4.02
MQ-5	RL	0.38	3.92	0.43	**4.01**	0.38	4.04	0.46	4.18	0.52	**4.14**

Encouraged by the results of Table 4, we decided to continue with our experiments with reinforcement learning. We use the same features as in [7], namely the length (in number of sentences) of the summary generated so far, plus the $tf.idf$ vectors of the following:

1. Candidate sentence;
2. Entire input to summarise;
3. Summary generated so far;
4. Candidate sentences that are yet to be processed; and
5. Question.

The reward used by REINFORCE is the ROUGE value of the summary generated by the system. Since [7] observed a difference between the ROUGE values of the Python implementation of ROUGE and the original Perl version (partly because the Python implementation does not include ROUGE-SU4), we compare the performance of our system when trained with each of them. Table 5 summarises some of our experiments. We ran the version trained on Python ROUGE once, and the version trained on Perl twice. The two Perl runs have different results, and one of them clearly outperforms the Python run. However, given the differences of results between the two Perl runs we advice to re-run the experiments multiple times and obtain the mean and standard deviation of the runs before concluding whether there is any statistical difference between the results. But it seems that there may be an improvement of the final evaluation results when training on the Perl ROUGE values, presumably because the final evaluation results are measured using the Perl implementation of ROUGE.

We have also tested the use of word embeddings instead of $tf.idf$ as input features to the policy model, while keeping the same neural architecture for the policy (one hidden layer using the same number of hidden nodes). In particular, we use the mean of word embeddings using 100 and 200 dimensions. These word embeddings were pre-trained using word2vec on PubMed documents provided by the organisers of BioASQ, as we did for the architectures described in previous sections. The results, not shown in the paper, indicated no major improvement,

Table 5. Experiments using Perl and Python versions of ROUGE. The Python version used the average of ROUGE-2 and ROUGE-L, whereas the Perl version used ROUGE-SU4.

Training on	Python ROUGE	Perl ROUGE
Python implementation	0.316	0.259
Perl implementation 1	0.287	0.238
Perl implementation 2	0.321	0.274

and re-runs of the experiments showed different results on different runs. Consequently, our submission to BioASQ included the original system using *tf.idf* as input features in all batches but batch 2, as described in Sect. 7.

6 Evaluation Correlation Analysis

As mentioned in Sect. 5, there appears to be a large discrepancy between ROUGE Recall and the human evaluations. This section describes a correlation analysis between human and ROUGE evaluations using the runs of all participants to all previous BioASQ challenges that included human evaluations (Phase B, ideal answers). The human evaluation results were scraped from the BioASQ Results page, and the ROUGE results were kindly provided by the organisers. We compute the correlation of each of the ROUGE metrics (recall, precision, F1 for ROUGE-2 and ROUGE-SU4) against the average of the human scores. The correlation metrics are Pearson, Kendall, and a revised Kendall correlation explained below.

The Pearson correlation between two variables is computed as the covariance of the two variables divided by the product of their standard deviations. This correlation is a good indication of a linear relation between the two variables, but may not be very effective when there is non-linear correlation.

The Spearman rank correlation and the Kendall rank correlation are two of the most popular among metrics that aim to detect non-linear correlations. The Spearman rank correlation between two variables can be computed as the Pearson correlation between the rank values of the two variables, whereas the Kendall rank correlation measures the ordinal association between the two variables using Eq. 1.

$$\tau = \frac{(\text{number of concordant pairs}) - (\text{number of discordant pairs})}{n(n-1)/2} \tag{1}$$

It is useful to account for the fact that the results are from 28 independent sets (3 batches in BioASQ 1 and 5 batches each year between BioASQ 2 and BioASQ 6). We therefore also compute a revised Kendall rank correlation measure that only considers pairs of variable values within the same set. The revised metric is computed using Eq. 2, where S is the list of different sets.

$$\tau' = \frac{\sum_{i \in S} \left[(\text{number of concordant pairs})_i - (\text{number of discordant pairs})_i \right]}{\sum_{i \in S} \left[n_i(n_i - 1)/2 \right]}$$

$$(2)$$

Table 6 shows the results of all correlation metrics. Overall, ROUGE-2 and ROUGE-SU4 give similar correlation values but ROUGE-SU4 is marginally better. Among precision, recall and F1, both precision and F1 are similar, but precision gives a better correlation. Recall shows poor correlation, and virtually no correlation when using the revised Kendall measure. For reporting the evaluation of results, it will be therefore more useful to use precision or F1. However, given the small difference between precision and F1, and given that precision may favour short summaries when used as a function to optimise in a machine learning setting (e.g. using reinforcement learning), it may be best to use F1 as the metric to optimise.

Table 6. Correlation analysis of evaluation results

Metric	Pearson	Spearman	Kendall	Revised Kendall
ROUGE-2 precision	0.61	0.78	0.58	0.73
ROUGE-2 recall	0.41	0.24	0.16	−0.01
ROUGE-2 F1	0.62	0.68	0.49	0.42
ROUGE-SU4 precision	0.61	0.79	0.59	0.74
ROUGE-SU4 recall	0.40	0.20	0.13	−0.02
ROUGE-SU4 F1	0.63	0.69	0.50	0.43

Figure 3 shows the scatterplots of ROUGE-SU4 recall, precision and F1 with respect to the average human evaluation[4]. We observe that the relation between ROUGE and the human evaluations is not linear, and that Precision and F1 have a clear correlation.

---

[4] The scatterplots of ROUGE-2 are very similar to those of ROUGE-SU4.

7 Submitted Runs

Table 7 shows the results and details of the runs submitted to BioASQ. The table uses ROUGE-SU4 Recall since this is the metric available at the time of writing this paper. However, note that, as explained in Sect. 6, these results might differ from the final human evaluation results. Therefore we do not comment on the results, other than observing that the "first n" baseline produces the same results as the neural regressor. As mentioned in Sect. 3, the labels used for the classification experiments are the 5 sentences with highest ROUGE-SU4 F1 score.

Table 7. Runs submitted to BioASQ 7b

Batch	Run	Description	ROUGE-SU4 R
1	MQ1	First n	0.4741
	MQ2	SVC	0.5156
	MQ3	NNR batchsize = 4096	0.4741
	MQ4	NNC batchsize = 4096	0.5214
	MQ5	RL tf.idf & Python ROUGE	0.4616
2	MQ1	First n	0.5113
	MQ2	SVC	0.5206
	MQ3	NNR batchsize = 4096	0.5113
	MQ4	NNC batchsize = 4096	0.5337
	MQ5	RL embeddings 200 & Python ROUGE	0.4787
3	MQ1	First n	0.4263
	MQ2	SVC	0.4512
	MQ3	NNR batchsize = 4096	0.4263
	MQ4	NNC batchsize = 4096	0.4782
	MQ5	RL tf.idf & Python ROUGE	0.4189
4	MQ1	First n	0.4617
	MQ2	SVC	0.4812
	MQ3	NNR batchsize = 1024	0.4617
	MQ4	NNC batchsize = 1024	0.5246
	MQ5	RL tf.idf & Python ROUGE	0.3940
5	MQ1	First n	0.4952
	MQ2	SVC	0.5024
	MQ3	NNR batchsize = 1024	0.4952
	MQ4	NNC batchsize = 1024	0.5070
	MQ5	RL tf.idf & Perl ROUGE	0.4520

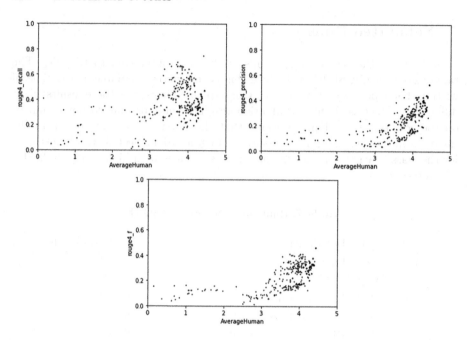

Fig. 3. Scatterplots of ROUGE SU4 evaluation metrics against the average human evaluations.

8   Conclusions

Macquarie University's participation in BioASQ 7 focused on the task of generating the ideal answers. The runs use query-based extractive techniques and we experiment with classification, regression, and reinforcement learning approaches. At the time of writing there were no human evaluation results, and based on ROUGE-F1 scores under cross-validation on the training data we observed that classification approaches outperform regression approaches. We experimented with several approaches to label the individual sentences for the classifier and observed that the optimal labelling policy for this task differed from prior work.

We also observed poor correlation between ROUGE-Recall and human evaluation metrics and suggest to use alternative automatic evaluation metrics with better correlation, such as ROUGE-Precision or ROUGE-F1. Given the nature of precision-based metrics which could bias the system towards returning short summaries, ROUGE-F1 is probably more appropriate when using at development time, for example for the reward function used by a reinforcement learning system.

Reinforcement learning gives promising results, especially in human evaluations made on the runs submitted to BioASQ 6b. This year we introduced very small changes to the runs using reinforcement learning, and will aim to explore more complex reinforcement learning strategies and more complex neural models in the policy and value estimators.

Acknowledgements. This research is partly funded by CSIRO Data61.

References

1. Bhandwaldar, A., Zadrozny, W.: UNCC QA: a biomedical question answering system. In: Proceedings BioASQ Workshop at EMNLP 2018, pp. 66–71 (2018)
2. Hoang, A., Bosselut, A., Celikyilmaz, A., Choi, Y.: Efficient Adaptation of Pretrained Transformers for Abstractive Summarization. Arxiv pre-print 1906.00138, May 2019
3. Kaur, M., Mollá, D.: Supervised machine learning for extractive query based summarisation of biomedical data. In: Proceedings of the Louhi 2018 (2018)
4. Li, Y., Gekakis, N., Chandu, K.R., Nyberg, E.: Extraction meets abstraction: ideal answer generation for biomedical questions. In: Proceedings BioASQ Workshop at EMNLP 2018, pp. 57–65 (2018)
5. Malakasiotis, P., Archontakis, E., Androutsopoulos, I.: Biomedical question-focused multi-document summarization: ILSP and AUEB at BioASQ3. In: CLEF 2015 Working Notes (2015)
6. Mollá, D.: Macquarie University at BioASQ 5b – query-based summarisation techniques for selecting the ideal answers. In: Proceedings of the BioNLP2017 (2017)
7. Mollá, D.: Macquarie University at BioASQ 6b: Deep learning and deep reinforcement learning for query-based multi-document summarisation. In: Proceedings BioASQ Workshop at EMNLP 2018 (2018)
8. Nallapati, R., Zhai, F., Zhou, B.: SummaRuNNer: a recurrent neural network based sequence model for extractive summarization of documents. In: AAAI 2017, November 2017
9. Naresh Kumar, A., et al.: Ontology-based retrieval & neural approaches for BioASQ ideal answer generation. In: Proceedings BioASQ Workshop at EMNLP 2018, pp. 79–89 (2018)
10. Rush, A.M., Chopra, S., Weston, J.: A neural attention model for abstractive sentence summarization. In: Proceedings of the Conference on Empirical Methods in Natural Language Processing (EMNLP), pp. 379–389, September 2015
11. See, A., Liu, P.J., Manning, C.D.: Get to the point: summarization with pointer-generator networks. In: ACL 2017 (2017)
12. Tsatsaronis, G., et al.: An overview of the BIOASQ large-scale biomedical semantic indexing and question answering competition. BMC Bioinform. **16**(1), 138 (2015). https://doi.org/10.1186/s12859-015-0564-6
13. Zhang, X., Lapata, M., Wei, F., Zhou, M.: Neural latent extractive document summarization. In: EMNLP 2018. August 2018

Structured Summarization of Academic Publications

Alexios Gidiotis[1,2]($\boxtimes$) and Grigorios Tsoumakas[1]($\boxtimes$)

[1] School of Informatics, Aristotle University of Thessaloniki, Thessaloniki, Greece
{gidiotis,greg}@csd.auth.gr
[2] Atypon Hellas, Vasilissis Olgas 212, Thessaloniki, Greece

Abstract. We propose SUSIE, a novel summarization method that can work with state-of-the-art summarization models in order to produce structured scientific summaries for academic articles. We also created PMC-SA, a new dataset of academic publications, suitable for the task of structured summarization with neural networks. We apply SUSIE combined with three different summarization models on the new PMC-SA dataset and we show that the proposed method improves the performance of all models by as much as 4 ROUGE points.

Keywords: Text summarization · Natural language processing · Deep learning

1 Introduction

Having informative summaries of scientific articles is crucial for dealing with the avalanche of academic publications in our times. Such summaries would allow researchers to quickly and accurately screen retrieved articles for relevance to their interests. More importantly, such summaries would lead to high quality indexing of the articles by (academic) search engines, leading to more relevant academic search results.

Currently, the role of such summaries is played by the abstracts produced by the authors of the articles. However, authors usually include in the abstract only the contributions and information of the paper that they consider important and ignore others that might be equally important to the scientific community [6].

A solution to the above problem would be to employ state-of-the-art abstractive summarization approaches [13,15], in order to automatically create short informative summaries of the articles to replace and/or accompany author abstracts for machine indexing and human inspection. However, these approaches have focused on the summarization of newswire articles, while academic articles exhibit several differences and pose major challenges compared to news articles.

First of all, news articles are much shorter than scientific articles and the news headlines that serve as summaries are much shorter than scientific abstracts. Secondly, scientific articles usually include several different key points that are

P. Cellier and K. Driessens (Eds.): ECML PKDD 2019 Workshops, CCIS 1168, pp. 636–645, 2020.
https://doi.org/10.1007/978-3-030-43887-6_57

scattered throughout the paper and need to be accurately included in a summary. These problems make it difficult to use summarization models that achieve state-of-the-art performance on newswire datasets for the summarization of academic articles.

We propose SUSIE (StrUctured SummarIzEr), a novel training method that allows us to effectively train existing summarization models on academic articles that have structured abstracts. Our method uses the XML structure of the articles and abstracts in order to split each article into multiple training examples and train summarization models that learn to summarize each section separately. We call such a task *structured summarization*. We further contribute a novel dataset consisting of open access PubMed Central articles along with their structured abstracts. SUSIE can easily be combined with different summarization models in order to address the problem of long articles and has been found to improve the performance of state-of-the-art summarization models by 4 ROUGE points.

We also created *PMC-SA* (PMC Structured Abstracts), a novel dataset that consists of academic articles from the biomedical domain. The articles for this dataset were collected from the *PubMed Central Open Access* (PMC-OA) repository and follow the *IMRD* (Introduction, Methods, Results, Discussion) structure. The abstracts in this dataset are also structured in a similar manner and each section of the full text can be paired with the corresponding section of the abstract.

2   Related Work

2.1   Summarization Methods

Automatic text summarization methods fall into two categories. Extractive methods [4, 10] select the most informative sentences from the source text and use them to construct a summary. On the other hand, abstractive methods [2, 13, 15] compose a coherent summary by generating new text and paraphrasing. In this work our main focus will be on the latter, because it is similar to the way that humans summarize text.

Advances in recurrent neural networks (RNNs) have demonstrated impressive capabilities of generating fluent language [1, 16]. State-of-the-art summarization methods use RNNs with the encoder-decoder architecture (or sequence-to-sequence architecture). These methods usually treat the whole source text as an input sequence, encode it into their hidden state and generate a complete summary from that hidden state.

Strong results have been achieved by such models when combined with an *attention* mechanism [3, 11, 14]. Adding a *pointer-generator* mechanism has been shown to further improve results [15]. The pointer-generator mechanism gives the model the ability to copy important words from the source text in addition to generating words from a predefined vocabulary. Adding a *coverage* mechanism has been shown to lead to even better results. [15]. The coverage mechanism prevents the model from repeating itself, which is a common problem with sequence-to-sequence models. The LSTM cells in the model of [15], were replaced in [5] with a new type of RNN unit, called rotational unit of memory, in order to overcome the fundamental limitation of LSTM cells in dealing with long sequences. Recent work utilizes reinforcement learning and policy gradient to further improve the performance of baseline models [2,13].

2.2 Summarization Datasets

Most of the summarization datasets that are found in the literature such as *Newsroom* [7], *Gigaword* [12] and *CNN/Daily Mail* [8] are focused on newswire articles. The average article lengths are relatively small and range from 50 words (Gigaword) to a few hundred words (CNN/Daily Mail, Newsroom). The average summary lengths are also rather small and range from a single sentence (Gigaword, Newsroom) to a few sentences (CNN/Daily Mail).

TAC 2014 (Text Analysis Conference 2014) is a well known dataset that focuses on the summarization of (biomedical) academic articles. The articles have an average of 9,759 words and the summaries an average of 235 words. However, as it consists of just 20 articles, it is not useful for training complex neural network summarization models. Another dataset of academic articles is CSPubSum [4] which exploits ScienceDirect[1] and uses the highlight statements submitted by authors as target summaries for each article. CSPubSum consists of approximately 10,000 articles and thus was mainly used for extractive summarization.

Finally, the BioASQ challenge [17] includes a sub-task where participants are given a question and a set of snippets, taken from academic biomedical publications, containing the correct answer and are asked to produce paragraph-sized summaries of theses snippets as *ideal* answers. BioASQ 2019 released a training set of 2747 pairs of snippets with ideal answers. This could be considered as a related dataset concerning query-focused summarization of academic papers. Again this is too small to be helpful for training state-of-the-art abstractive summarization methods.

3   Summarizing Academic Papers

3.1   Flat Abstract Summarization

A simple approach to summarizing academic papers would be to train sequence-to-sequence models using the full text of the article as source input and the abstract as reference summary. However, sequence-to-sequence models face multiple difficulties when given long input texts. A very long input sequence requires the encoder RNN to run for a lot of time steps. This greatly increases the computational complexity of the forward pass. To make things worse, the training of the encoder on very long input sequences becomes increasingly difficult due to the computational complexity of the backward pass. The training becomes

---

[1] https://www.sciencedirect.com/.

Table 1. The different sections that we annotate and the keywords associated with them.

Section	Keywords
introduction	introduction, case
literature	background, literature, related
methods	methods, method, techniques, methodology
results	result, results, experimental, experiments, experiment
discussion	discussion, limitations
conclusion	conclusion, conclusions, concluding

increasingly slower and in many cases the vanishing gradients prevent the model from learning useful information.

A solution to this problem would be to truncate very long sequences (more than 600 words), but this can result in serious information loss which would severely affect the quality of the produced summaries.

Even harder is the training of a decoder with very long output sequences. In this case, the computational complexity and memory requirements of the decoder make it pointless to try and train a model with very long reference summaries.

Another problem of this straightforward approach, is that the different parts of an academic paper are not equally important for the task of summarization. Sections like the *introduction* include core information for the summary, while others like the *experiments* are noisy and usually include little useful information.

3.2   SUSIE

SUSIE (StrUctured SummarIzEr) is a novel summarization method that exploits structured abstracts in order to address the aforementioned problems.

Many academic articles, especially in the life sciences domain follow the typical *IMRD* structure with sections like *introduction, background, methods, results* and *conclusion*. When the abstract of the article is structured it usually includes similar sections too. We employed a very simple method that looks for specific keywords in the header of each section in order to annotate both the article and abstract sections. For example, sections that include keywords like *methods, method, techniques* and *methodology* in their header are annotated as *methods*. Table 1 presents the different section types and the keywords associated with them.

Once the article and abstract sections are annotated, we pair each section of the full text with the corresponding section of the abstract and create one training example per section. We can then use one of the existing summarization methods and train a model for the summarization of single sections. Summarizing a single section of an article is a much easier task since the input and output sequences are a lot shorter and the information is more compact and focused on

Table 2. Per section type number of words for the articles in the PMC-SA dataset.

Section type	Source length		Abstract length	
	mean	std	mean	std
introduction	570.26	381.40	58.25	41.00
methods	1,133.32	638.90	80.26	38.98
conclusion	152.08	178.14	49.92	23.83

specific aspects of the article. In addition, section annotation allows us to filter out particular sections that are not useful for summarization.

At test time we extract the specified sections of the article and run the summarization model for each of them in order to produce section summaries. Then we combine those summaries in order to get the full summary of the article.

4 PMC Structured Abstracts

PubMed Central (PMC) is a free digital repository that archives publicly accessible full-text scholarly articles that have been published within the biomedical and life sciences journal literature. The PMC-SA (PMC Structured Abstracts) dataset was created from the open access subset of PMC, comprising approximately 2 million articles. We used the XML format downloaded from the PMC FTP server to create the dataset. Only the articles that have abstracts structured in sections were selected and included in the dataset. PMC-SA has a total of 712,911 full text articles along with their abstracts. The full texts of the articles have an average length of 2,514 words and are used as source texts for the summarization, while the abstracts have an average length of 260 words and are used as reference summaries. Code and instructions for the creation of the PMC-SA dataset will be made available online.[2] When compared with the existing datasets discussed in Sect. 2.2 PMC-SA is clearly different in multiple ways. The articles and summaries are significantly longer compared to the different newswire datasets and this makes it a much harder task. Also, the new dataset is a lot larger than both the TAC 2014 dataset and CSPubSum [4] that focus on academic publications. This makes it suitable for the training of state-of-the-art summarization models.

We can easily apply SUSIE on PMC-SA since the XML format allows us to effectively split the full text and abstract into annotated sections. In Table 2 we show detailed statistics about the source and abstract length for each section type.

5 Experiments

As we mentioned, SUSIE can be combined with a number of different summarization models. In order to evaluate the effectiveness of SUSIE the three different

---

[2] https://github.com/AlexGidiotis/PMC-StructuredAbstracts-Dataset.

Table 3. Experimental results. Best result per evaluation measure is highlighted in bold typeface.

Model	ROUGE-1 F1		ROUGE-2 F1		ROUGE-L F1	
	Flat	SUSIE	Flat	SUSIE	Flat	SUSIE
attention sequence-to-sequence	0.2833	**0.3341**	0.1043	**0.1261**	0.2619	**0.3026**
pointer-generator	0.3020	**0.3591**	0.1020	**0.1416**	0.2726	**0.3179**
pointer-generator + coverage	0.3300	**0.3716**	0.1142	**0.1466**	0.2893	**0.3296**

Table 4. Statistics about the training sets for the two experiments. In the flat abstract experiment each training example is an article and the whole abstract is used as reference summary. With SUSIE we create an average of 2 examples per article. The source inputs are article sections and the corresponding abstract sections are the reference summaries.

	Flat	SUSIE
# training articles	641,994	641,994
# training examples	641,994	1,211,826
avg. source length (words)	1,451	677
avg. summary length (words)	260	130

summarization models that were described in Sect. 2.1 are trained and evaluated on PMC-SA using both the flat abstract method from Sect. 3.1 and SUSIE.

The training set has 641,994 articles, the validation set has 35,309 articles and the test set 10,111 articles. In all experiments we included for summarization only the *introduction*, *methods* and *conclusion* sections because we have found that these particular section selection gives us the best performing models. For the flat abstract method, the selected sections are concatenated and used as source input paired with the concatenation of the corresponding abstract sections as reference summary. For SUSIE one example is created for each of the selected sections with the corresponding abstract section as reference summary. In Tables 4 we provide detailed statistics about the training data used in the two different methods.

5.1 Experimental Setup

We used the implementation of the three models provided by [15][3]. The hyperparameter setup used for the models is similar to that of [15].

In order to speed up the training process we start off with highly truncated input and output sequences. In more detail, we begin with input and output sequences truncated to 50 and 10 words respectively and train until convergence. Then we gradually increase the input and output sequences up to 500 and 100 words respectively.

---

[3] https://github.com/abisee/pointer-generator.

When using the flat abstract method, we truncate each section to $\frac{L}{n}$ words before concatenating them to get the input and output sequences, where L is the required article length and n is the number of extracted sections from this article.

The truncation of an academic article to a total of 500 words is definitely going to result in some severe information loss but we deemed it necessary due to the difficulties described in Sect. 3.1. To get the coverage model we simply add the coverage mechanism to the converged pointer-generator model and continue training.

At test time, for the flat abstract method, we truncate each input section to $\frac{L}{n}$ with $L = 500$ words and concatenate them to get an input sequence of 500 words. Then we run beam search for 120 decoding steps in order to generate a summary. For SUSIE each of the selected sections is truncated to 500 words before we run beam search for 120 decoding steps to get a summary for each one of them. Then we concatenate the individual summaries to get the summary of the full article.

5.2   Results

We evaluate the performance of all models with the ROUGE family of metrics [9] using the *pyrouge* package[4]. In specific, we report F1 scores for ROUGE-1, ROUGE-2 and ROUGE-L. ROUGE-1 and ROUGE-2 measure the overlap, in unigrams and bigrams respectively, between the generated and the reference summary. ROUGE-L measures the longest common subsequence overlap.

Table 3 presents the results of our experiments. We can see that the pointer-generator model achieves higher scores than the simple attention sequence-to-sequence and adding the coverage mechanism further improves those scores which is in line with the experiments of [15].

We also notice that SUSIE improves the scores of the flat summarization approach for all three models by as much as 4 ROUGE points. The performance of the best model, pointer-generator with coverage, is improved by approximately 13%, 28% and 14% in terms of ROUGE-1, ROUGE-2 and ROUGE-L F1 score respectively. It is clear that the flat approach suffers from information loss due to the truncation of the source input. In the appendix we illustrate the difference in the quality of the summaries produced by the two different methods by presenting generated examples for a real article.

6   Conclusion

This work focused on the summarization of academic publications. We have shown that summarization models that perform well on smaller articles have difficulties when applied on longer articles with a lot of diverse information like academic articles. We proposed SUSIE, a novel approach that allowed us to successfully adapt

---

[4] https://pypi.org/project/pyrouge/0.1.3.

existing summarization models to the task of structured summarization of academic articles. Also, we created PMC-SA, a new dataset of academic articles that is suitable for the training of summarization models using SUSIE. We found that training with SUSIE on the PMC-SA greatly improves the performance of summarization models and the quality of the generated summaries.

Acknowledgements. We would like to thank the anonymous reviewers for their comments that helped us significantly improve this work.

A   Appendix

Here we will provide an example of summaries generated by the best performing model, namely pointer-generator with coverage, for a sample article from the test set. We provide two summaries, one generated from a model trained with the flat method and another generated from a model trained with SUSIE. We also provide the original abstract of the article for reference. One can find the original article with *PMCID PMC5051331* at the PMC website[5].

Comparing the two generated summaries, we can see that the one generated with SUSIE is superior to the flat one in terms of structure, readability and factual correctness. When compared with the original abstract, we can see that both summaries are not perfect but the one generated with SUSIE is in many cases acceptable.

A.1   Reference Summary

Objective. To examine the efficacy of psychological and psychosocial interventions for reductions in repeated self-harm.

Design. We conducted a systematic review, meta-analysis and meta-regression to examine the efficacy of psychological and psychosocial interventions to reduce repeat self-harm in adults. We included a sensitivity analysis of studies with a low risk of bias for the meta-analysis. For the meta-regression, we examined whether the type, intensity (primary analyses) and other components of intervention or methodology (secondary analyses) modified the overall intervention effect.

Data Sources. A comprehensive search of medline, psycinfo and embase (from 1999 to june 2016) was performed.

Eligibility Criteria for Selecting Studies. Randomised controlled trials of psychological and psychosocial interventions for adult self-harm patients.

Conclusions. Consideration of a psychological or psychosocial intervention over and above treatment as usual is worthwhile; with the public health benefits of ensuring that this practice is widely adopted potentially worth the investment. However, the specific type and nature of the intervention that should be delivered is not yet clear. Cognitive–behavioural therapy or interventions with an interpersonal focus and targeted on the precipitants to self-harm may be the best candidates on the current evidence. Further research is required.

---

[5] https://www.ncbi.nlm.nih.gov/pmc/articles/PMC5051331.

A.2   Flat P-Gen + Coverage

Clinically treated non-fatal self-harm (we use the term self-harm henceforth) is common in terms of adverse outcomes such as repetition of self-harm, suicide and all-cause mortality; mental health morbidity, quality of life and physical, psychological and social functioning, mental health morbidity, quality of life and physical, psychological and social functioning. It is also costly in terms of immediate and ongoing treatment of self-harm of 16.3% after 1 year and 7% after 9 years. A more recent review of 177 studies from western and non-western countries indicated little change in these estimates, with reported repetition of self-harm of 16%, while the suicide rate was 1.6.

A.3   SUSIE P-Gen + Coverage

Objective. To evaluate the efficacy of psychological and pharmacological interventions for reducing repetition of self-harm in unselected populations, but again highlight the poor quality of the evidence base in pooling data. In addition to these cochrane reviews, a number of trials have been published and several systematic reviews produced that aim to highlight what interventions are most often efficacious interventions (cbt) (and not problem-solving therapy).

Methods. We searched medline, embase, and the cochrane central register of controlled trials (rct) published up to february 2016 to identify randomized controlled trials evaluating the efficacy of psychological and psychosocial interventions to reduce repeat self-harm, (primary outcome) and to reduce suicidal ideation, depression and hopelessness (secondary outcomes) using meta-analysis; and (2) examine whether the type, intensity or other specific components of the interventions, or study methodology, modify the pooled intervention effect using meta-regression analysis.

Conclusions. our study is consistent with the updated cochrane review, which in contrast to the original version showed no support for problem-focused (predominantly problem-solving therapy) interventions but a significant effect of interventions. Our study has shown that psychological or psychosocial interventions are effective overall, with cbt and psychodynamic interpersonal therapy currently the most promising for implementation.

References

1. Bahdanau, D., Cho, K., Bengio, Y.: Neural machine translation by jointly learning to align and translate. arXiv preprint arXiv:1409.0473 (2014)
2. Celikyilmaz, A., Bosselut, A., He, X., Choi, Y.: Deep communicating agents for abstractive summarization. In: Proceedings of the 2018 Conference of the North American Chapter of the Association for Computational Linguistics: Human Language Technologies, Volume 1 (Long Papers), New Orleans, Louisiana, June 2018, pp. 1662–1675. Association for Computational Linguistics (2018). https://doi.org/10.18653/v1/N18-1150

3. Chopra, S., Auli, M., Rush, A.M.: Abstractive sentence summarization with attentive recurrent neural networks. In: Proceedings of the 2016 Conference of the North American Chapter of the Association for Computational Linguistics: Human Language Technologies, pp. 93–98 (2016)
4. Collins, E., Augenstein, I., Riedel, S.: A supervised approach to extractive summarisation of scientific papers. In: Proceedings of the 21st Conference on Computational Natural Language Learning (CoNLL 2017), pp. 195–205 (2017). https://doi.org/10.18653/v1/k17-1021
5. Dangovski, R., Jing, L., Nakov, P., Tatalović, M., Soljačić, M.: Rotational unit of memory: a novel representation unit for RNNs with scalable applications. Trans. Assoc. Comput. Linguist. (2019). https://doi.org/10.1162/tacl_a_00258
6. Elkiss, A., Shen, S., Fader, A., Erkan, G., States, D., Radev, D.: Blind men and elephants: what do citation summaries tell us about a research article? J. Am. Soc. Inf. Sci. Technol. (2008). https://doi.org/10.1002/asi.20707
7. Grusky, M., Naaman, M., Artzi, Y.: Newsroom: a dataset of 1.3 million summaries with diverse extractive strategies. In: Proceedings of the 2018 Conference of the North American Chapter of the Association for Computational Linguistics: Human Language Technologies, Volume 1 (Long Papers), pp. 708–719 (2018)
8. Hermann, K.M., et al.: Teaching machines to read and comprehend. In: Advances in Neural Information Processing Systems, pp. 1693–1701 (2015)
9. Lin, C.Y.: ROUGE: a package for automatic evaluation of summaries. Text Summarization Branches Out (2004)
10. Nallapati, R., Zhai, F., Zhou, B.: SummaRuNNer: a recurrent neural network based sequence model for extractive summarization of documents. In: AAAI, pp. 3075–3081 (2017)
11. Nallapati, R., Zhou, B., dos Santos, C., Gulcehre, C., Xiang, B.: Abstractive text summarization using sequence-to-sequence RNNs and beyond. In: Proceedings of The 20th SIGNLL Conference on Computational Natural Language Learning, pp. 280–290 (2016)
12. Napoles, C., Gormley, M., Van Durme, B.: Annotated gigaword. In: Proceedings of the Joint Workshop on Automatic Knowledge Base Construction and Web-scale Knowledge Extraction, pp. 95–100 (2012)
13. Paulus, R., Xiong, C., Socher, R.: A deep reinforced model for abstractive summarization. arXiv preprint arXiv:1705.04304 (2017)
14. Rush, A.M., Chopra, S., Weston, J.: A neural attention model for abstractive sentence summarization. arXiv preprint arXiv:1509.00685 (2015)
15. See, A., Liu, P.J., Manning, C.D.: Get to the point: summarization with pointer-generator networks. In: Proceedings of the 55th Annual Meeting of the Association for Computational Linguistics (Volume 1: Long Papers), Stroudsburg, PA, USA, pp. 1073–1083. Association for Computational Linguistics (2017). https://doi.org/10.18653/v1/P17-1099
16. Sutskever, I., Vinyals, O., Le, Q.V.: Sequence to sequence learning with neural networks. In: Advances in Neural Information Processing Systems, pp. 3104–3112 (2014)
17. Tsatsaronis, G., et al.: An overview of the BioASQ large-scale biomedical semantic indexing and question answering competition. BMC Bioinform. (2015). https://doi.org/10.1186/s12859-015-0564-6

How to Pre-train Your Model? Comparison of Different Pre-training Models for Biomedical Question Answering

Sanjay Kamath[1,3]([✉]), Brigitte Grau[1,2]([✉]), and Yue Ma[3]([✉])

[1] LIMSI, CNRS, Université Paris-Saclay, Orsay, France
bg@limsi.fr
[2] ENSIIE, Université Paris-Saclay, Évry, France
[3] LRI, Univ. Paris-Sud, CNRS, Université Paris-Saclay, Orsay, France
{sanjay,ma}@lri.fr

Abstract. Using deep learning models on small scale datasets would result in overfitting. To overcome this problem, the process of pre-training a model and fine-tuning it to the small scale dataset has been used extensively in domains such as image processing. Similarly for question answering, pre-training and fine-tuning can be done in several ways. Commonly reading comprehension models are used for pre-training, but we show that other types of pre-training can work better. We compare two pre-training models based on reading comprehension and open domain question answering models and determine the performance when fine-tuned and tested over BIOASQ question answering dataset. We find open domain question answering model to be a better fit for this task rather than reading comprehension model.

Keywords: Deep learning · Reading comprehension · Open domain question answering

1 Introduction

Deep learning models have been widely used in several NLP tasks since the emergence of large scale labelled datasets. In Question Answering (QA) specifically on open domain, several neural network models have been introduced, such as Convolutional Neural Networks (CNN), Recurrent Neural Networks (RNN) using GRUs or LSTMs and attention mechanisms, Self-attention networks (Transformers), and Pretrained language models like ELMO, BERT which can be fine-tuned to Question Answering task. Several kinds of Question Answering (QA) related tasks are widely studied such as *Answer Sentence Selection, Reading Comprehension* and *Open QA*.

Reading Comprehension (RC) is a QA task where a question and a relevant paragraph are given and the goal is to extract the answer string present in the

© Springer Nature Switzerland AG 2020
P. Cellier and K. Driessens (Eds.): ECML PKDD 2019 Workshops, CCIS 1168, pp. 646–660, 2020.
https://doi.org/10.1007/978-3-030-43887-6_58

paragraph. The main assumption of this task is that the answer is present in the paragraph, like in SQUAD v1.0 [12]. Variants of this task include unanswerable questions such as in [11]. Answers are usually short phrases or entities. There is a leaderboard on SQUAD dataset[1] which showcases lot of models built for this task [1,2,14].

Open QA is a QA task where a question is given and the goal is to retrieve an answer. An answer has to be retrieved from a set of documents or passages of textual sources as Wikipedia articles or news. Answers are also usually short phrases or entities. In NN approaches for Open QA, generally answers are extracted using a reading comprehension model on the subset of the retrieved documents or passages considered as relevant [3,4].

One of the main differences between *Reading Comprehension (RC)* and *Open QA* tasks is that the answer must be present in the paragraphs (or documents) for Reading Comprehension, but for Open QA this condition might not hold true because the retrieved documents considered to be relevant to the question might not contain the answer. Another characteristic is that in the Open QA task, several paragraphs or documents contain the answer.

The BIOASQ Phase B task provides dataset for biomedical question answering which is a small scale labelled dataset for factoid questions (779 question in BIOASQ 7). Each question is associated with multiple relevant paragraphs, some irrelevant ones, and one or several answers. The work of [15] transforms the BIOASQ Phase B dataset into the format of a Reading Comprehension task where each question has an answer text along with the offset in a paragraph which contains the answer. If a paragraph does not contain an answer, it is discarded. This modification of the BIOASQ dataset enables to use a RC model off-the-shelf.

By using such a model on BIOASQ dataset which is a small scale labelled dataset, it will not result in similar performance as on the large scale open domain datasets due to overfitting. One way of overcoming this problem as reported by [6,15] is by pre-training a deep learning model on a large scale dataset and fine-tuning the same model to the target small scale dataset. The intuition is that the model learns better representations when learnt on a large scale dataset than having a randomly initialized model trained only on the small scale dataset.

However the BIOASQ task resembles more towards an *Open QA* task than a *RC* task because of the existence of paragraphs without answers even though they are considered relevant. Thus we propose a new way to tackle the BIOASQ task by using an Open QA model that takes into account this particularity.

We present a comparison of using different pre-training models (reading comprehension and open QA models) for BIOASQ question answering task and also report the performance of a single model without pre-training and without fine-tuning to show the importance of this process.

We report the performance of our model on different datasets and show that in some cases it outperforms the state-of-the-art systems of BIOASQ [5,7,15] in average.

---

[1] https://rajpurkar.github.io/SQuAD-explorer/.

2   State of the Art

Since BIOASQ 5, deep learning methods were introduced by [15] by automatically adapting BIOASQ QA task as *Reading Comprehension* task and pretraining the model with SQUAD v1.0 dataset. Similar approach of pre-training and fine-tuning are used by [5] who pre-train their models using DRQA [1] and BioBert [7] using Bert [2].

The models discussed in this article for the BIOASQ task use automatic annotations as done by [16] who transform the BIOASQ dataset into reading comprehension dataset by using the gold standard answer strings by searching them in the snippets for exact match and are treated as answers if only they are found in the snippets, i.e., the answer string must be a substring of the snippet.

As the NN models participating to BIOASQ are based on domain adaptation, we first present general approaches used for that purpose before presenting how it is done by the BIOASQ NN models.

2.1   Domain Adaptation

Pre-training is a training process started from randomly initialized model weights. Fine-tuning is also a training process but started from the model weights of pre-trained model and not randomly initialized model weights. Both pre-training and fine-tuning together can be termed as *Domain Adaptation* when the domain of data used for pre-training and fine-tuning are different. For example, open domain and biomedical domain.

Pre-training and fine-tuning or domain adaptation can be done in several ways. The general approaches are listed below.

Type 1 - The task remains the same for pre-training and fine-tuning. Pre-training should be done on a large scale dataset from random initialization of parameters. Fine-tuning should be done on a small scale dataset by loading the model parameters from pre-trained model rather than random initialization. This approach is used when a target dataset is small scaled and using it to train a deep neural network would result in overfitting. This type of pre-training is common in computer vision field where models are pre-trained on Imagenet [13] and fine-tuned on target image classification datasets.

Type 2 - The tasks are different for pre-training and fine-tuning. Pre-training should be done on a large scale dataset from random initialization of parameters. Fine-tuning should be done on a different model which uses certain parameters from the pre-trained model which are frozen (non-trainable) and learns some parameters which are randomly initialized on a different task. These approaches in NLP were initially proposed for sequence labelling tasks by [9] which were later evolved into ELMO (Embedding Language Models) by [10] which significantly improved the state of the art across a broad range of challenging NLP tasks such as question answering, textual entailment and sentiment analysis. This type of method uses special contextual text embeddings obtained from the pre-trained models that are added as features into downstream models built for another task.

Type 3 - The tasks are different for pre-training and fine-tuning. Pre-training should be done on a large scale dataset from random initialization of parameters. Fine-tuning should be done on the pre-trained model by modifying certain layers to fit to the new task. Newly added layers can be randomly initialised and pre-trained model layers together with newly added ones are trained on the new task. This approach is similar to *Type 2* approach with a difference that the reference model can be slightly modified for target task rather than building a different model. This type of approach proposed by [2] is being widely used in NLP tasks such as question answering, textual entailment, sentiment analysis, named entity recognition, relation extraction etc. which are easily done by modifying a final output layer of the original model and fine-tuned. Fine-tuning can be done either by learning the whole model parameters or learning only a part of the model by freezing the rest.

Our work uses *Type 1* domain adaptation and compares the results with *Type 3* domain adaptation results reported by [7].

2.2 Deep Learning and Domain Adaptation in BIOASQ

The work by [16] comes under *Type 1* domain adaptation using SQUAD v1.0 pre-training. Since the introduction of pre-trained language models by [9], works by [2,10] have been used in several NLP tasks and have been proven to outperform many prior state of the art models. BioBert by [7] have been shown to be useful in biomedical domain tasks such as named entity recognition, relation extraction and BIOASQ question answering. This work belongs to *Type 3* domain adaptation methods where the authors use Bert model by [2] and re-train it on the same task but on biomedical domain texts. Later this model is modified for different biomedical tasks and tested. This method, as reported in the paper [7], fetches state of the art scores on BIOASQ QA task which is listed in Table 3 under *BioBert* column.

3 Question Answering Tasks and Models

In this section, we describe the two kinds of question answering tasks and the related models we used for domain adaptation towards biomedical domain.

3.1 Tasks

Question Answering (QA) is a field of research which lies in the intersection of Natural Language Processing and Information Retrieval disciplines. Several types of tasks exists which are commonly referred as Question Answering.

In this article, we focus on Reading Comprehension (RC) task a.k.a Machine Reading task, and Open Domain Question Answering a.k.a Open QA or Open Question Answering.

Reading Comprehension task contains questions, a relevant paragraph and answers from the paragraph. RC is also called as Answer Extraction because the answer is known to be present in the paragraph.

Open QA task contains questions and their short answers without any paragraphs. Open QA can be formulated as a parent task which involves two child tasks, (1) Retrieving the relevant paragraphs for a question and (2) Extracting a short answer from the paragraphs. In Open QA, the first task is generally referred as paragraph selection or answer sentence selection and the second task is often modelled as Reading Comprehension although there exists several relevant and irrelevant paragraphs. Open QA models should distinguish if the paragraph is relevant and then extract the answer unlike the RC models.

BIOASQ phase B task is a question answering task with questions in biomedical domain. For a question, there are relevant documents, paragraphs, answers given. Below is an example from the dataset.

Q: Which calcium channels does ethosuximide target?
A: T-type calcium channels
P1: ..neuropathic pain is blocked by ethosuximide, known to block T-type calcium channels...
P2: Theta rhythms remained disrupted during a subsequent week of withdrawal but were restored with the T-type channel blocker ethosuximide.

The goal of BIOASQ question answering task is to extract the correct answer from supporting data. As shown in the example, one paragraph (P1) has the gold standard answer and the other (P2) does not (i.e. it does not contain the exact match of the answer string). Therefore this resembles more like an *Open QA* task than a *Reading Comprehension* task.

The evolution of deep learning methods led to the emergence of large scale datasets for these two types of tasks in open domain Question Answering. We use two models for *RC* and *Open QA* which are shown in the Fig. 1 and are described below.

3.2   Reading Comprehension - DRQA Model

DRQA's document reader developed by [1] is a simple LSTM model for *Reading Comprehension* task which takes as input a question and a paragraph and aims at extracting an answer from the paragraph. As per the assumption of the *RC task*, the answer is always present inside the paragraph as a substring. An overview of the model can be seen on the left figure of Fig. 1. Both the question and the paragraph are tokenized and their word embeddings are used for the model i.e. question words $Q = \{q_1, ..., q_m\}$ and paragraph words $S = \{s_1, ..., s_n\}$ are sequences which are encoded using an embedding layer of dimension D.

$$E(Q) = \{E(q_1), .., E(q_m)\} \tag{1}$$

$$E(S) = \{E(s_1), .., E(s_n)\} \tag{2}$$

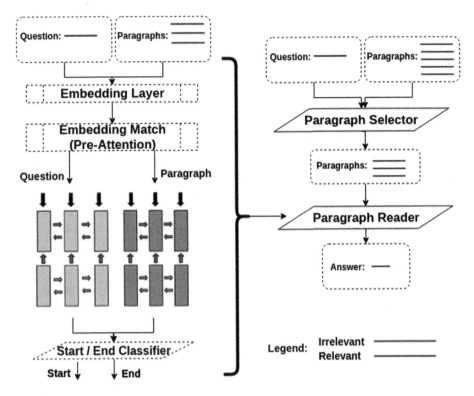

Fig. 1. Left: DRQA - Paragraph Reader (*RC task*). Right: PSPR - Paragraph Selector and Paragraph Reader model (*Open QA task*)

A pre-attention mechanism captures the similarity between paragraph words and question words in the same layer. For this purpose, a feature $\mathcal{F}align$ shown in Eq. 3 is added as a feature to the LSTM layer.

$$\mathcal{F}align(p_i) = \Sigma_j a_{i,j} E(q_j) \tag{3}$$

Where $a_{i,j}$ is,

$$a_{i,j} = \frac{exp\left(\alpha(E(s_i)) \cdot \alpha(E(q_j))\right)}{\Sigma_{j'} \ exp(\alpha(E(s_i)) \cdot \alpha(E(q_{j'})))} \tag{4}$$

which computes the dot product between nonlinear mappings of word embeddings of question and paragraph.

They are followed by a 3-layer Bidirectional LSTM layers for both question and sentence encodings.

$$\{E(q_1), .., E(q_n)\} = \text{Bi-LSTM}(\{\tilde{E}(q_1), .., \tilde{E}(q_n)\}) \tag{5}$$

$$\{E(s_1), .., E(s_n)\} = \text{Bi-LSTM}(\{\tilde{E}(s_1), .., \tilde{E}(s_n)\}) \tag{6}$$

These LSTM states are connected to two independent classifiers that use a bilinear term to capture the similarity between paragraph words and question words and compute the probabilities of each token being start or end of the answer span.

$$P_{\text{start}}(i) \propto \exp\left(\mathbf{p}_i \mathbf{W}_s \mathbf{q}\right) \tag{7}$$

$$P_{\text{end}}(i) \propto \exp\left(\mathbf{p}_i \mathbf{W}_e \mathbf{q}\right) \tag{8}$$

During prediction, we choose the best span from token i to token i' such that $i \leq i' \leq i + 15$ and $P_{start}(i) \times P_{end}(i')$ is maximized.

To make the scores compatible across paragraphs in one or several retrieved documents, unnormalized exponential is used and an argmax is taken over all considered paragraph spans for the final predictions which are offset to start and end of an answer span in the paragraph.

Answer probability for the answer span can be computed among several other answer spans as shown in Sect. 3.3 in Eq. 9. This probability score can be used to return top 5 answers for BIOASQ task which is explained in Sect. 4.2.

3.3 Open QA - PSPR Model

PSPR model is an Open QA model by [8] whose overview is presented in the right figure of Fig. 1. This model has two parts, namely Paragraph Selector (PS) and Paragraph Reader (PR) in a cascade fashion. Although *PSPR* model contains a *Reading Comprehension* task submodule for answer extraction, the main difference comes from learning the answer extraction module using the paragraph probabilities computed by the paragraph selector. Paragraphs for the questions are retrieved using an information retrieval technique. Then the Paragraph Selector model predicts a probability distribution $Pr\left(p_i|q, P\right)$ over all the retrieved paragraphs where P is the set of paragraphs for the question.

The Paragraph Reader model extracts answer spans as shown in the *DRQA* model and predicts a probability $Pr\left(a|q, p_i\right)$ for each answer span where p_i is i^{th} paragraph in Paragraph set P.

The reader model gives two probabilities (one for start and one for end token given by two classifiers) as described in Eqs. 7 and 8. The answer probability $Pr(a|q, P)$ is computed as shown below:

$$Pr\left(a|q, p_i\right) = \sum_j Pr\left(a_s^j\right) Pr\left(a_e^j\right) \tag{9}$$

The answer with highest probability is returned as the final prediction.

The Paragraph Selector uses tokenized question words $Q = \{q_1, \ldots, q_m\}$ and tokenized paragraph words $P = \{p_1, \ldots, p_n\}$ which are encoded using an embedding layer of dimension D.

$$E(Q) = \{E(q_1), \ldots, E(q_m)\} \tag{10}$$

$$E(P) = \{E(p_1), \ldots, E(p_n)\} \tag{11}$$

A RNN layer encodes the contextual information of the sequence.

$$\{E(q_1), \ldots, E(q_m)\} = \text{RNN}(\{\tilde{E}(q_1), \ldots, \tilde{E}(q_m\}) \tag{12}$$

$$\{E(p_1), .., E(p_n)\} = \mathrm{RNN}(\{\tilde{E}(p_1), .., \tilde{E}(p_n\}) \tag{13}$$

Using this hidden representation, a self attention operation is applied to get the question representation q:

$$\hat{\mathbf{q}} = \sum_j \alpha^j \hat{\mathbf{q}}^j \tag{14}$$

where α^j encodes the importance of each question word against the other question words which is calculated as:

$$\alpha_i = \frac{\exp\left(\mathbf{w}_b \mathbf{q}_i\right)}{\sum_j \exp\left(\mathbf{w}b\mathbf{q}_j\right)} \tag{15}$$

Where w is the learnt weight vector. Finally, the probability of each paragraph is calculated via a max-pooling and a softmax layer as shown below:

$$Pr\left(p_i|q, P\right) = softmax\left(\max_j \left(\hat{\mathbf{p}}_i^j \mathbf{W} \mathbf{q}\right)\right) \tag{16}$$

where $\mathbf{W}$ is a learnt weight matrix.

Since not all the paragraphs contain an answer in the Open QA setting, the probability scores from Eq. 16 should indicate if there exists an answer or not.

While training, the paragraphs containing the answer are highlighted as 1 and the rest as 0. And while testing, the paragraph with highest probability is chosen to extract the answer. Combining the two probabilities, the overall answer is chosen by choosing the highest probable answer from $Pr(a|q, P)$ for a question q which is calculated as :

$$Pr(a|q, P) = \sum_{p_i \in P} Pr\left(a|q, p_i\right) Pr\left(p_i|q, P\right) \tag{17}$$

For Paragraph Reader, the $DRQA$ model can be used directly as shown in the Fig. 1 with small differences. During training the reader model extracts the answer only when there is an answer in it. i.e. when the paragraph probability score (Eq. 16) of that paragraph is 1. During testing, the reader model extracts the answer from the paragraph which has the highest probability score.

Only difference while adapting this to BIOASQ is that number of answers to be extracted for BIOASQ is top 5 and not 1. Because of this, instead of choosing from only the top most probable paragraph, we select top 5 answers from combined probability scores in Eq. 17, which might consider 1 or several paragraphs to extract answers from.

3.4   Domain Adaptation for BIOASQ Task

Pre-training and fine-tuning or domain adaptation can be done in several ways, the following is a general abstraction of the two processes:

- **Type 1**
  - Build a reference model (Model R) for task A.
  - Train the Model R with a sufficiently large scale dataset on task A.
  - Save the Model R weights and model.
  - DO NOT randomly initialize Model R weights, instead use the saved weights of model R from the previous step.
  - Train the Model R again on a small scale target dataset on task A.
  - Predict and Evaluate.
- **Type 2**
  - Build a reference model (Model R) for task A.
  - Train the Model R with a sufficiently large scale dataset on task A.
  - Save the Model R weights and model.
  - Build a model from scratch (Model S) for task B.
  - Model S is built to use some features from model R.
  - Initialize the model S weights randomly (apart from the Model R features which are used right away).
  - Train the Model S on a large or small scale target dataset on task B.
  - Predict and Evaluate.
- **Type 3**
  - Build a reference model (Model R) for task A.
  - Train the Model R with a sufficiently large scale dataset on task A.
  - Save the Model R weights and model.
  - Build a Model S, which is built upon Model R (by Adding or Modifying the input or output layers of Model R to work on task B).
  - Initialize the model S weights randomly (All of Model R features can be used right away, or some partially used and partially randomized for training).
  - Train the Model S on a large or small scale target dataset on task B.
  - Predict and Evaluate.

In this work, we apply *Type 1* domain adaptation.

The Data for Pre-training. Two datasets correspond to each of the two tasks: SQUAD V1.0 dataset for *RC* task and QUASAR-T dataset for *Open QA* task and we show below their differences.

- QUASAR-T which is based on Trivia questions is generated synthetically, and SQUAD is annotated manually by humans on a crowd-sourcing platform.
- Each question in QUASAR-T is associated to 100 sentence-level passages retrieved from ClueWeb09 dataset based on Lucene, whereas SQUAD 1.0 has 1 relevant paragraph.
- Some paragraphs in QUASAR-T do not have an answer.[2]

---

[2] SQUAD 2.0 is a variant of SQUAD dataset which contains questions without answers. We do not use this because the reference models also do not use v2.0 to pre-train.

Comparing the above differences with BIOASQ dataset, the QUASAR-T dataset resembles more closely to BIOASQ than that of SQUAD v1.0 due to the following reasons.

- BIOASQ data has more than 1 relevant paragraphs per question.
- Some paragraphs do not have an answer.

4   Experiments and Results

For fine-tuning the models we use BIOASQ datasets. The statistics of different sets are reported in the Table 1.

Table 1. Datasets used in the experiments along with their splits. The numbers represent number of questions.

Datasets	Train	Dev	Test
BIOASQ 4b	427	59	161
BIOASQ 5b	544	75	150
BIOASQ 6b	685	94	161
SQUAD v1.0	87,599	10,570	9,533
QUASAR-T	37,012	3,000	3,000

Table 2. Results reporting the importance of Pre-Training and Fine-Tuning a model. DRQA model by [1] is used for these experiments. S.Acc is Strict Accuracy, L.Acc is Lenient Accuracy (the correct answer is in the top 5) and MRR is Mean Reciprocal Rank.

Datasets	Metrics	No-Pre	No-Fine	Pre+Fine
4b	S.Acc	08.98	23.96	**24.00**
	L.Acc	16.56	35.26	**39.21**
	MRR	11.36	28.40	**29.34**
5b	S.Acc	25.91	32.17	**32.43**
	L.Acc	34.86	45.58	**47.73**
	MRR	29.27	36.75	**38.37**
6b	S.Acc	13.40	26.24	**26.72**
	L.Acc	27.08	40.60	**43.72**
	MRR	19.20	32.57	**33.80**
Average	S.Acc	16.09	27.45	**27.71**
	L.Acc	26.16	40.48	**43.55**
	MRR	19.94	32.57	**33.83**

4.1   Importance of *Pre-training* and *Fine-Tuning*

To show the importance of *Pre-Training* and *Fine-Tuning* for domain adaptation to biomedical domain, we experimented three approaches on a single model *DRQA* without altering any hyperparameters. (Default parameters as used by [1][3])

(1) *No-Pre* model is the *DRQA* model trained on BIOASQ dataset only.

(2) *No-Fine* model is the *DRQA* model trained on SQUAD v1.0 dataset only.

(3) *Pre+Fine* is the *DRQA* model trained on SQUAD v1.0 dataset and fine-tuned on BIOASQ dataset.

Results are shown in the Table 2 for different BIOASQ test sets. When a model is only trained on a small dataset like BIOASQ, the results are very low as shown in the column *No-Pre*. When a model is trained on a large dataset like SQUAD v1.0, the model can be straight away used to predict results on the biomedical dataset. *No-Fine* shows an improvement doing so, against *No-Pre*. Lastly, *Pre+Fine* pre-training on a large scale dataset and fine-tuning on the biomedical dataset shows an improvement over the other approaches. This set of experiments show that the best approach is to do pre-training and fine-tuning on smaller domain specific datasets.

4.2   Experiments with the Two QA Modellings: DRQA and PSPR

In this section, we experiment mainly with the two models for a *Reading Comprehension* task and an *Open QA* task.

For studying the modelling of the BIOASQ QA task as a *Reading Comprehension* task, we use SQUAD v1.0 dataset for pre-training and experiment with the *DRQA* model. For studying its modelling as an *Open QA* task, we use QUASAR-T dataset for pre-training and experiment with *PSPR* model.

We also experiment using the paragraph probabilities for reranking the DRQA answers and choosing the top 5. As the *PSPR* model is a cascaded model with paragraph selector and paragraph reader, we use the paragraph probabilities predicted by the paragraph selector and multiply them with the answer probabilities obtained using *DRQA* model to select the top 5 answers which have combined higher probabilities. Note that this approach is not the same as cascaded *PSPR* model because the *PSPR* model's reader model uses the paragraph probabilities to learn the extraction of answers which might have a different impact.

For obtaining top 5 answers with *DRQA* model, if there are more than 5 paragraphs for the question, we take 1 answer from each paragraph and choose top 5 based on the answer probabilities. We do this to make sure each paragraph's top answer is contributed towards top 5 answers. If there are less than 5 paragraphs, we take top 5 answers based on answer probabilities to keep it simple.

Results are shown in Table 3 for different BIOASQ test sets. We compare different model results with BioBert scores reported in [7]. The scores from

---

[3] https://github.com/facebookresearch/DrQA.

Table 3. *DRQA* is the Reading Comprehension model by [1], *PSPR* is the Open QA model by [8], *DRQA+PS* is answers chosen with scores by multiplying answer probabilities of DRQA with Paragraph Selector probabilities of PSPR. SOTA scores are reported by [7] who average the best scores from each batch (possibly from multiple different models). Results from BIOASQ 4b, 5b and 6b test sets. 7b test set cannot be evaluated yet due to lack of gold standard answers. S.Acc is Strict Accuracy, L.Acc is Lenient Accuracy and MRR is Mean Reciprocal Rank. Experiments are done with the original BIOASQ data.

Datasets	Metrics	BioBert by [7]	DRQA	DRQA+PS	PSPR
4b	S.Acc	**36.48**	24.00	26.22	30.28
	L.Acc	**48.89**	39.21	32.33	40.34
	MRR	**41.05**	29.34	26.54	34.19
5b	S.Acc	41.56	32.43	30.62	**46.59**
	L.Acc	**54.00**	47.73	47.86	53.76
	MRR	46.32	38.37	36.96	**49.55**
6b	S.Acc	35.58	26.72	26.50	**43.91**
	L.Acc	**51.39**	43.72	42.16	51.34
	MRR	42.51	33.80	32.07	**45.70**
Average	S.Acc	37.87	27.71	27.78	**40.26**
	L.Acc	**51.43**	43.55	40.78	48.48
	MRR	**43.29**	33.83	31.85	43.14

PSPR model shows that the performance is better than BioBert on Strict and Lenient accuracy on 5b and 6b test sets. By taking paragraph probability into account *PSPR* allows to better rank top 1 correct answer than BioBert which extracts answers only from the longer pre-processed paragraphs which have correct answers.

Although *PSPR* has a reader model similar to *DRQA*, considering the paragraph probability seems to improve the answer extraction in *PSPR* model.

4.3 Experiments with Longer Contexts (Modified BIOASQ Data)

For the BIOASQ task we noted that the method used by [7] with BioBert modifies the original paragraphs. For computing the BioBert model, the authors have retrained the original Bert model by [2], using Pubmed and PMC articles. For applying it on the BIOASQ task, the authors use longer documents (instead of the actual snippets) from Pubmed corresponding to the data given by BIOASQ in the "documents" field to access the Pubmed documents for each question. Therefore the modification of the dataset leads to different results for BioBert compared to the performance on the regular BIOASQ dataset. The exact pre-processing of BIOASQ dataset in order to do this is not very clear from the paper, however the authors release the modified dataset in their repository[4].

---

[4] https://github.com/dmis-lab/biobert.

658   S. Kamath et al.

Table 4. Experiments with data containing longer contexts (Document level) by [7]. *DRQA* is a Reading Comprehension model by [1]. *BioBert-Unaltered* is the original BIOASQ dataset with questions and paragraphs which contain answers. *Biobert by* [7] is the modified BIOASQ dataset where the paragraphs are longer paragraphs (documents from respective articles), where all the models are pre-trained on SQUAD v1.0 dataset and finetuned on BIOASQ dataset. SOTA scores are reported by [7] who average the best scores from each batch (possibly from multiple different models). Results from BIOASQ 4b, 5b and 6b test sets. 7b test set cannot be evaluated yet due to lack of gold standard answers. S.Acc is Strict Accuracy, L.Acc is Lenient Accuracy and MRR is Mean Reciprocal Rank.

Datasets	Metrics	SOTA	DRQA	BioBert-Unaltered	BioBert by [7]
4b	S.Acc	20.59	18.49	13.08	**36.48**
	L.Acc	29.24	32.51	18.54	**48.89**
	MRR	24.04	23.88	15.48	**41.05**
5b	S.Acc	**41.82**	28.92	22.84	41.56
	L.Acc	57.43	46.54	32.46	**54.00**
	MRR	**47.73**	35.88	25.94	46.32
6b	S.Acc	25.12	21.70	16.35	**35.58**
	L.Acc	40.20	41.51	22.61	**51.39**
	MRR	29.28	28.60	18.72	**42.51**
Average	S.Acc	29.18	23.03	17.42	**37.87**
	L.Acc	42.29	40.18	24.53	**51.43**
	MRR	33.68	29.45	20.04	**43.29**

In order to evaluate the importance of this data modification, we did three experiments: (1) DRQA with longer contexts (2) BioBert with unaltered data from BIOASQ (3) BioBert results with modified paragraphs and as reported by [7] in their paper.

The results are shown in Table 4. The results of *BioBert* is as presented in [7] where the authors have fine-tuned the models first using SQUAD v1.0 dataset and adapted it to BIOASQ data. We use the modified dataset to experiment it with the *DRQA* model to determine if it would improve the performance of the pre+fine *DRQA* model as reported in Table 2. We got lower performances to that of the *DRQA* model trained on the original BIOASQ data.

For comparison, we try the BioBert model on the original BIOASQ data i.e. paragraphs given by BIOASQ data and not pre-processed. The results in Table 4, under the column *BioBert-Unaltered* represents these results. It is evident that the modification performed on the BIOASQ data fetches better results using BioBert model.

5   Conclusion

In this work we have shown the importance of pre-training and fine-tuning process a.k.a domain adaptation for biomedical domain question answering. We have also compared two QA models based on i.e. (1) Reading Comprehension task (2) Open QA task, and found that the performance is better when using an Open QA model than a Reading Comprehension model.

Based on a different pre-processing done by [7] on the biomedical dataset by using longer contexts from documents than shorter contexts, we found that the Reading Comprehension model performs worse on the pre-processed longer contexts compared to the shorter contexts originally given by BIOASQ data.

On the other end, a large pre-trained language model such as BERT performs much better on the pre-processed longer contexts than shorter contexts. Future work shall focus on training BERT model on Open QA task which will better suit the BIOASQ dataset.

Acknowledgements. This work is funded by the ANR project GoAsQ (ANR-15-CE23-0022).

References

1. Chen, D., Fisch, A., Weston, J., Bordes, A.: Reading Wikipedia to answer open-domain questions. In: Proceedings of the 55th Annual Meeting of the Association for Computational Linguistics (Volume 1: Long Papers) (2017)
2. Devlin, J., Chang, M.W., Lee, K., Toutanova, K.: BERT: pre-training of deep bidirectional transformers for language understanding. In: Proceedings of the 2019 Conference of the North American Chapter of the Association for Computational Linguistics: Human Language Technologies, Volume 1 (Long and Short Papers), pp. 4171–4186. Association for Computational Linguistics, Minneapolis, June 2019. https://www.aclweb.org/anthology/N19-1423
3. Dhingra, B., Mazaitis, K., Cohen, W.W.: Quasar: datasets for question answering by search and reading. CoRR abs/1707.03904 (2017). http://arxiv.org/abs/1707.03904
4. Joshi, M., Choi, E., Weld, D., Zettlemoyer, L.: TriviaQA: a large scale distantly supervised challenge dataset for reading comprehension. In: Proceedings of the 55th Annual Meeting of the Association for Computational Linguistics (Volume 1: Long Papers), pp. 1601–1611 (2017)
5. Kamath, S., Grau, B., Ma, Y.: An adaption of BIOASQ question answering dataset for machine reading systems by manual annotations of answer spans. In: Proceedings of the 6th BioASQ Workshop A challenge on Large-Scale Biomedical Semantic Indexing and Question Answering, pp. 72–78. Association for Computational Linguistics, Brussels, November 2018. https://www.aclweb.org/anthology/W18-5309
6. Kamath, S., Grau, B., Ma, Y.: Predicting and integrating expected answer types into a simple recurrent neural network model for answer sentence selection. In: 20th International Conference on Computational Linguistics and Intelligent Text Processing, La Rochelle, France, April 2019. https://hal.archives-ouvertes.fr/hal-02104488

7. Lee, J., et al.: BioBERT: a pre-trained biomedical language representation model for biomedical text mining. CoRR abs/1901.08746 (2019). http://arxiv.org/abs/1901.08746

8. Lin, Y., Ji, H., Liu, Z., Sun, M.: Denoising distantly supervised open-domain question answering. In: Proceedings of the 56th Annual Meeting of the Association for Computational Linguistics (Volume 1: Long Papers), pp. 1736–1745 (2018)

9. Peters, M., Ammar, W., Bhagavatula, C., Power, R.: Semi-supervised sequence tagging with bidirectional language models. In: Proceedings of the 55th Annual Meeting of the Association for Computational Linguistics (Volume 1: Long Papers), pp. 1756–1765 (2017)

10. Peters, M.E., et al.: Deep contextualized word representations. arXiv preprint arXiv:1802.05365 (2018)

11. Rajpurkar, P., Jia, R., Liang, P.: Know what you don't know: unanswerable questions for SQUAD. arXiv preprint arXiv:1806.03822 (2018)

12. Rajpurkar, P., Zhang, J., Lopyrev, K., Liang, P.: SQuAD: 100,000+ questions for machine comprehension of text. arXiv preprint arXiv:1606.05250 (2016)

13. Russakovsky, O., et al.: ImageNet large scale visual recognition challenge. Int. J. Comput. Vis. **115**(3), 211–252 (2015)

14. Seo, M., Kembhavi, A., Farhadi, A., Hajishirzi, H.: Bidirectional attention flow for machine comprehension. arXiv preprint arXiv:1611.01603 (2016)

15. Wiese, G., Weissenborn, D., Neves, M.: Neural domain adaptation for biomedical question answering. In: Proceedings of the 21st Conference on Computational Natural Language Learning (CoNLL 2017), pp. 281–289. Association for Computational Linguistics, Vancouver, August 2017. https://doi.org/10.18653/v1/K17-1029. https://www.aclweb.org/anthology/K17-1029

16. Wiese, G., Weissenborn, D., Neves, M.: Neural question answering at BioASQ 5B. In: BioNLP, pp. 76–79 (2017)

Yes/No Question Answering in BioASQ 2019

Dimitris Dimitriadis$^{(\boxtimes)}$ ⓘ and Grigorios Tsoumakas ⓘ

School of Informatics, Aristotle University of Thessaloniki, Thessaloniki, Greece
{dndimitri,greg}@csd.auth.gr

Abstract. The field of question answering has gained greater attention with the rise of deep neural networks. More and more approaches adopt paradigms which are based primarily on the powerful language representations models and transfer learning techniques to build efficient learning models which are able to outperform current state of the art systems. Endorsing this current trend, in this paper, we strive to take a step towards the goal of answering yes/no questions in the field of biomedicine. Specifically, the task is to give a short answer (yes or no) for a question written in natural language, finding clues including in a set of snippets that are related with this question. We propose three different deep neural network models, which are free of assumptions about predefined specific feature functions, while the key elements of these are the ELMo embeddings, the similarity matrices and/or sentiment information. The results have shown that incorporating the sentiment, we can improve the performance of a yes/no question answering system while the proposed learning models significantly outperform the BioASQ baseline.

Keywords: Yes/no question answering · BioASQ challenge · ELMo embeddings · Deep neural networks

1   Introduction

The recent rise of deep neural networks is having a significant impact on the field of question answering. Especially after the introduction of the SQuAD benchmark [8], more and more approaches adopt deep learning techniques, while a lot of effort has been put into building powerful and general language representation models, such as BERT [1] and ELMo [7]. Furthermore, using transfer learning, models built on a specific classification task can be reused on another task with improved results compared to building a model from scratch trained on the latter task [15].

This interesting view of solving tasks has influenced biomedical question answering too. In the BioASQ [10] challenge, which provides a benchmark for the evaluation of biomedical question answering systems, more and more approaches adopt the above paradigm (i.e. language representations and transfer learning) to build efficient models that overcome the previous state of the art. For example,

© Springer Nature Switzerland AG 2020
P. Cellier and K. Driessens (Eds.): ECML PKDD 2019 Workshops, CCIS 1168, pp. 661–669, 2020.
https://doi.org/10.1007/978-3-030-43887-6_59

BioBERT [5], a fined-tuned version of BERT in biomedical text, has achieved state of the art results in biomedical question answering, while previously, pre-trained word embeddings were being used in the task of biomedical question answering [2,12].

In this paper, endorsing this current perspective in question answering, we deal with this task focusing on yes/no question type. Especially, using the definition provided by BioASQ, our aim is to give an answer (yes or no) given a set of snippets that are related with a question. This task is quite similar with the reading comprehension (RC) task but it differs in some points:

1. In RC, only one snippet is related with a question and it is important that the answer is included in this. In contrast, we must cope with several snippets written by different authors and no one guarantees that the answer is part of the snippets or inferred from these.
2. The sub language of biomedical domain is complex and there are plenty of biomedical terms making the task of building a representative language representation model a difficult task. On the other hand, the RC is based on general English which means that a large amount of resources around the web can be used to build a useful language representation model.

An additional issue, we must address, is the nature of the problem of yes/no question answering. Particularly, most of the current approaches (excluding those in yes/no question answering) are focused on finding part of text in the given textual sources (i.e. snippets), whereas in our case, the answer is inferred by the given textual sources.

The main contribution of this paper dealing with the above challenges is the introduction of three different deep learning architectures. The first one is based on ELMo embeddings. The second one extends the first one by enriching the feature space with sentiment information. The last one exploits the similarity between the words of a question and the snippets to build a similarity matrix that is given as input to a deep neural network. Furthermore, we show that sentiment has impact to yes/no question answering. To the best of our knowledge, these architectures have not been used in yes/no question answering.

The rest of this paper is organized as follows. Section 2 describes our methods. Section 3 presents experimental results in BioASQ 2019 along with results on the dataset provided by the BioASQ. Section 4 makes an overview of the existing approaches in yes/no question answering focusing on systems participated in the BioASQ challenge. Finally, Sect. 5 presents the conclusions of this work.

2   Methods

We present three methods for yes/no question answering. We use ELMo embeddings in two of our methods to represent the textual sources, one of which incorporates sentiment information by leveraging SentiWordnet [3]. Our last approach uses a similarity matrix, where each cell is the cosine similarity between a word from the question and a word from the snippets, which is passed as input to a neural network.

2.1 ELMo Embeddings

In the first step, the question and the related snippets are passed through the ELMo layers (one layer that gets the question as input and the other that gets the snippets). These layers are responsible for converting the question and each snippet to multi-dimensional vectors. Let us denote the question vector as q and each snippet vector as p_i where $1 \leq i \leq m$ and m is the number of snippets. Next, we concatenate all vectors $(X = [q; p_1; p_2; ...; p_m])$ to build a joint representation of question and snippets. The produced vector is then passed through a bidirectional LSTM that is fully connected with a two-layered neural network:

$$H = BILSTM(X)$$
$$dense_1 = ReLU(W_1 * H + b_1)$$
$$dense_2 = Sigmoid(W_2 * dense_1 + b_2)$$

where W_1, W_2 are the weights of each layer and b_1, b_2 the corresponding offsets (biases). Because yes/no question answering can be considered as a binary classification problem, the last layer $(dense_2)$ consists of one unit which corresponds to the target of the learning model (no $= 0$, yes $= 1$).

We used two ELMo layers instead of one, without sharing the weights across the network because the training parameters must be updated independently. Particularly, the ELMo layer getting the question as input, should pay more attention to words such as "do", "does", "is", "are" etc. and to the syntax of the question which is different from the syntax of a snippet.

2.2 ELMo Embeddings and Sentiment

As previously, we converted the given question and snippets to multi-dimensional vectors. However, we also used SentiWordnet to get the sentiment scores for each word included in the question and snippets. SentiWordnet maps each word to a triple of sentiment scores (positive, negative, neutral score).

To build the question and snippets sentiment vectors we considered Algorithm 1. Let us denote the sentiment question vector as $qs = (qs_1, qs_2, ..., qs_n)$ where qs_i is the sentiment score of the i-th word contained in the question. For snippets, we denote the snippets sentiment vectors as $ps_i = (ps_{i1}, ps_{i2}..., ps_{im})$, where ps_{ij} is the sentiment score of the j-th word contained in the i-th snippet. The sentiment vectors update the question vector as follows:

$$a = ReLU(W_1 * qs + b_1)$$
$$b = tanh(W_2 * a + b_2)$$
$$probs = Softmax(W_3 * b + b_3)$$
$$mult = q \circ probs$$

where $\circ$ denotes the element-wise multiplication between question vector and question sentiment vector. With a similar way, we update the snippets with the sentiment scores. However, on top of the last equation we apply bidirectional LSTM. Defining the last function as H we concatenated the outputs as follows: $X = [mult; H]$, which is fully connected with a two-layered neural network as in the first method.

Result: Sentiment Vector
$vectorS = []$;
for $word \in text$ **do**
    pos,neg,neut[a] = SentiWordnetWrapper(word)[b];
    if $pos > neg$ **then**
        if $pos > neut$ **then**
        | vectorS.append(pos);
        else
        | vectorS.append(neut);
        end
    else
        if $neg > neut$ **then**
        | vectorS.append(-neg);
        else
        | vectorS.append(neut);
        end
    end
end

Algorithm 1. Text to Sentiment Vector

---

[a] The neutral score in SentiWordnet is referred as objective score
[b] SentiWordnet returns the sentiment scores of a specific synset but a word could correspond to many of these synsets, thus, we built a wrapper function that finds the most common synset corresponds to the given word and returns the sentiment scores of this synset.

Sentiment is an important information for yes/no question answering because it helps us to recognize agreements/contradictions between the given question and the related passages. Considering the question "Is the protein Papilin secreted?", the passage "the protocadherin cdh-3, and two genes encoding secreted extracellular matrix proteins, mig-6/papilin and him-4/hemicentin." agrees with the question because there aren't negative words to transform the passage to a negative statement.

2.3 Similarity Matrix

Instead of passing the question and snippets as input to a neural network, we built a similarity matrix. We first use pre-trained word vectors to represent the words of both the question and the snippets. Then, we estimate the cosine similarity for each pair of question and snippets words. Thus, each row in the similarity matrix corresponds to the similarities of a question word with all the words contained in the snippets. This similarity matrix $(Smatrix)$ passes through the following equations:

$$a = BILSTM(Smatrix)$$
$$dense_1 = tanh(W_1 * a + b_1)$$
$$dense_2 = Sigmoid(W_2 * dense_1 + b_2)$$

The inspiration of this work was from [13] which proposes a QA Matrix where each cell is the semantic similarity between a term of a question and a term of an answer. However, our similarity matrix encodes the similarity between words of the question and words from snippets. Furthermore, our bidirectional LSTM captures the dependencies between the words in snippets where each word is a vector and each dimension of this vector corresponds to the similarity of this word with a word of the question.

Although, recurrent neural networks aim to process sequences, the similarity matrix fits as input to these networks, considering as timesteps the rows of the similarity matrix and as dimensionality of the input, the columns of the matrix.

3   Experimental Setup and Results

To build our models, we used the BioASQ benchmark[1], which contains 745 yes/no questions along with their related snippets. 67% of these pairs of questions and snippets was used as training set and the rest 33% as validation set. We used the ELMo embeddings available at TensorFlow Hub[2] and the pre-trained word2vec embeddings provided by BioASQ[3]. Our architectures were built with the Keras framework[4]. We set the batch size to 2^4, because a larger number would lead to fewer updates of the model weights slowing down convergence. We used the Adam optimizer [4] because it works well in practice using the default learning rate (0.001), while larger learning rates cause divergence of the training criterion. Binary cross entropy was used as loss function for training the supervised neural network via the back-propagation algorithm. We used the SentiWordnet from the nltk[5].

Figure 1 shows the training and validation loss of our methods. Typically, the validation loss should be similar to, but slightly higher than, the training loss. However, in our cases, this doesn't happen. The reason is the class weight that is used during training while, in the validation step, it is not defined. Thus, during training the "no" class gets more attention than during the validation. Furthermore, the convergence of the first and the last methods happens earlier than in the second method. We believe that this happens because the second method incorporates additional information (i.e. sentiment scores) to the model. Thus, the model must put more effort to incorporate this information in its

---

[1] http://participants-area.bioasq.org/Tasks/7b/trainingDataset/.
[2] https://www.tensorflow.org/hub.
[3] http://participants-area.bioasq.org/tools/BioASQword2vec/.
[4] https://keras.io/.
[5] http://www.nltk.org/.

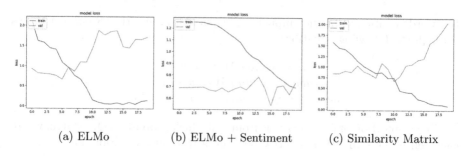

(a) ELMo   (b) ELMo + Sentiment   (c) Similarity Matrix

Fig. 1. Training and validation loss of our methods

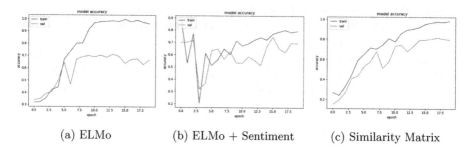

(a) ELMo   (b) ELMo + Sentiment   (c) Similarity Matrix

Fig. 2. Training and validation accuracy for our methods

feature space. In Fig. 2, we present the accuracy of each method both for training and validation. We observe that with ELMo and sentiment scores, we achieve the best accuracy before the model overfits on the training set. One phenomenon that we also observe is the increase of the validation accuracy despite the overfitting. This happens because the dataset is imbalanced, consequently, answering some questions randomly as yes, the accuracy is being increased. To participate in BioASQ 2019, we selected those models with the best accuracy before the model overfits on the training set.

Table 1 summarizes the results of our participated methods against the BioASQ baseline. As we observe, in 3/5 batches the architecture of ELMo embeddings fits better to the test sets rather than that architecture with the similarity matrix. Furthermore, sentiment seems to improve the MaF1 score in test batch 5. Finally, all methods overcome the BioASQ baseline excluding test batch 2 where our approach on Similarity Matrix is slightly worse than the baseline.

Based on the BioASQ leaderboard[6], our team (auth-qa-*[7]) is at the 2nd place in the first three batches, 5th in the fourth batch and 4th in the fifth and final batch. Furthermore, we observed that in some cases the performance of our systems is worse than the performance of other participated systems (e.g. BioBERT-DMIS, google-*-input) for a test batch, while there are some batches in which our systems overcome them. This means that there aren't clear evidences about a state of the art system in the challenge.

---

[6] http://participants-area.bioasq.org/results/7b/phaseB/.

[7] This is the prefix of our systems' names in the BioASQ Leaderboard.

Table 1. MaF1 score for each approach on each test batch - BioASQ 2019. Bold indicates the best score in a particular batch.

Systems	Batches - MaF1				
	1	2	3	4	5
ELMo embeddings	**.5397**	**.6296**	.4866	**.5490**	.5658
ELMo embeddings + Sentiment	–	–	–	–	**.6274**
Similarity matrix	–	.4223	**.5165**	.5461	.4697
BioASQ baseline	.4727	.4258	.1481	.4348	.4643

4   Related Work

Our work shares the high-level goal of answering yes/no questions with many works before us. Due to the fact that we cannot do full justice of related works given space constraints, we focus on two works participating in the BioASQ challenge.

Yes/No question answering can be considered as a binary classification problem where a supervised model learns to predict the truthiness of a question. In this direction, the OAQA system [14] uses a set of hand-written features that were extracted from the given question and snippets to build a binary classifier. Our work shares the main idea with the OAQA system. Particularly, we also consider the yes/no question answering as a binary classification task as well as we incorporate sentiment in one of our methods which helped to improve the accuracy. However, we enforce non-linear functions with millions of parameters to better map the input textual sources to the answer. Furthermore, we use a language representation model to capture the syntax and semantics of the raw input sources (i.e. questions and snippets) letting the model to learn from these representations to predict the answer rather than from a predefined set of features provided by an expert. Finally, instead of incorporating sentiment as a single feature in our model, we firstly find the sentiment of each word of question and snippets and next we input question and snippets sentiment vectors in the model where each dimension corresponds to the sentiment of a specific word either in question or in a snippet from the set of snippets.

A score mechanism was enforced by [9] to answer yes/no questions. Particularly, they used SentiWordnet to get the sentiment score for each word of each snippet. Then, they calculated the sentiment score for each snippet while the decision for the answer either as "yes" or "no" is based on the number of positive and negative snippets. We also use SentiWordnet to get the sentiment scores for each word of a question and snippets, however our aim is to use these sentiment scores as additional information in the feature space of our learning model rather than making these scores the central part of our methods.

Although, answering yes/no questions is very challenging in biomedicine, a few works have been proposed to solve this task in BioASQ challenge, either because the dataset provided by BioASQ was extremely imbalanced and those participants who answered yes to the questions got very good results (e.g. [6]), or because the dataset was quite small and one cannot build efficient learning models. However, the rise of transfer learning and fine-tuned language representation models as well as the introduction of MaF1 to BioASQ challenge as additional measure to evaluate yes/no question answering systems, motivated the participants to deal with the task this year.

5   Conclusions

In this work, we present three methods for solving the yes/no question answering task. The incorporation of sentiment improved the final results w.r.t the MaF1 score. We expect that if we used language representation models fine-tuned on biomedical texts (e.g. BioBERT), the results would be better. Grid search, random search or even hyper-parameter optimization could be considered for tuning our models. The presented methods overcome the BioASQ baseline while we observed that despite the imbalanced dataset and without exhausted tuning, the models can capture some negative cases presented in the test sets.

References

1. Devlin, J., Chang, M.W., Lee, K., Toutanova, K.: BERT: pre-training of deep bidirectional transformers for language understanding. arXiv preprint arXiv:1810.04805 (2018)
2. Dimitriadis, D., Tsoumakas, G.: Word embeddings and external resources for answer processing in biomedical factoid question answering. J. Biomed. Inform. **92**, 103118 (2019)
3. Esuli, A., Sebastiani, F.: SentiWordNet: a publicly available lexical resource for opinion mining. In: LREC, vol. 6, pp. 417–422. Citeseer (2006)
4. Kingma, D.P., Ba, J.: Adam: a method for stochastic optimization. arXiv preprint arXiv:1412.6980 (2014)
5. Lee, J., et al.: BioBERT: pre-trained biomedical language representation model for biomedical text mining. arXiv preprint arXiv:1901.08746 (2019)
6. Mao, Y., Wei, C.H., Lu, Z.: NCBI at the 2014 BioASQ challenge task: large-scale biomedical semantic indexing and question answering. In: CLEF (Working Notes), pp. 1319–1327 (2014)
7. Peters, M.E., et al.: Deep contextualized word representations. arXiv preprint arXiv:1802.05365 (2018)
8. Rajpurkar, P., Zhang, J., Lopyrev, K., Liang, P.: SQuAD: 100,000+ questions for machine comprehension of text. arXiv preprint arXiv:1606.05250 (2016)
9. Sarrouti, M., El Alaoui, S.O.: A yes/no answer generator based on sentiment-word scores in biomedical question answering. Int. J. Healthc. Inf. Syst. Inform. (IJHISI) **12**(3), 62–74 (2017)
10. Tsatsaronis, G., et al.: An overview of the BioASQ large-scale biomedical semantic indexing and question answering competition. BMC Bioinformatics **16**(1), 138 (2015)
11. Weissenborn, D., Wiese, G., Seiffe, L.: Making neural QA as simple as possible but not simpler. arXiv preprint arXiv:1703.04816 (2017)

12. Wiese, G., Weissenborn, D., Neves, M.: Neural question answering at BioASQ 5B. arXiv preprint arXiv:1706.08568 (2017)
13. Yang, L., Ai, Q., Guo, J., Croft, W.B.: aNMM: ranking short answer texts with attention-based neural matching model. In: Proceedings of the 25th ACM International on Conference on Information and Knowledge Management, pp. 287–296. ACM (2016)
14. Yang, Z., Zhou, Y., Nyberg, E.: Learning to answer biomedical questions: OAQA at BioASQ 4B. In: Proceedings of the Fourth BioASQ Workshop, pp. 23–37 (2016)
15. Yosinski, J., Clune, J., Bengio, Y., Lipson, H.: How transferable are features in deep neural networks? In: Advances in Neural Information Processing Systems, pp. 3320–3328 (2014)

Semantically Corroborating Neural Attention for Biomedical Question Answering

Marilena Oita[1](✉), K. Vani[2](✉), and Fatma Oezdemir-Zaech[1](✉)

[1] Novartis Institutes for Biomedical Research (NIBR), Novartis, Basel, Switzerland
{marilena.oita,fatma.oezdemir-zaech}@novartis.com
[2] Dalle Molle Institute for Artificial Intelligence Research (IDSIA),
Lugano, Switzerland
vanik@idsia.ch

Abstract. Biomedical question answering is a great challenge in NLP due to complex scientific vocabulary and lack of massive annotated corpora, but, at the same time, is full of potential in optimizing in critical ways the biomedical practices. This paper describes the work carried out as a part of the BioASQ challenge (Task-7B Phase-B), and targets an integral step in the question answering process: extractive answer selection. This deals with the identification of the exact answer (words, phrases or sentences) from given article snippets that are related to the question at hand. We address this problem in the context of factoid and summarization question types, using a variety of deep learning and semantic methods, including various architectures (e.g., Dynamic Memory Networks and Bidirectional Attention Flow), transfer learning, biomedical named entity recognition and corroboration of semantic evidence. On the top of candidate answer selection module, answer prediction to yes/no question types is also addressed by incorporating a sentiment analysis approach. The evaluation with respect to Rouge, MRR and F1 scores, in relation to the type of question answering task being considered, exhibits the potential of this hybrid method in extracting the correct answer to a question. In addition, the proposed corroborating semantics module can be added on top of the typical QA pipeline to gain a measured 5% improvement in identifying the exact answer with respect to the gold standard.

Keywords: Question Answering · Text comprehension · Attention · Dynamic Memory Networks · Biomedical embeddings · Semantic analysis · Entity corroboration · Transfer learning

1 Introduction

Question Answering (QA) is one of the most challenging Natural Language Processing (NLP) tasks, which requires advanced capabilities such as causal reasoning

---

M. Oita, K. Vani—Contributed equally to this work.

P. Cellier and K. Driessens (Eds.): ECML PKDD 2019 Workshops, CCIS 1168, pp. 670–685, 2020.
https://doi.org/10.1007/978-3-030-43887-6_60

and semantic inferences, generally referred to as Natural Language Understanding (NLU).

In the life science domains, such as biomedical, there exists a huge potential for both improving the existing practices and developing optimized methods for automatic discovery. Neural-based extraction techniques are gaining significant importance with the fast advancements of deep learning methodology and architectures. With the Stanford's freely available machine comprehension generic dataset, SQUAD[1], the interest in machine comprehension increased at an exponential rate [21]. But the availability of training data becomes a restriction for specific domains, one being the biomedical domain.

Transfer learning[2], weakly supervised automated annotation[3], and synthetic data generation [23] are proposed for solving such issues, but the extent of the improvement is still to be defined. In the biomedical domain for instance, a major challenge is to deal with a highly complex vocabulary, specific abstractions and polysemy.

The availability of specific domain information contributes however heavily to domain language understanding. BioASQ [24] is one such initiatives in the biomedical domain which provides datasets for the evaluation of the main steps involved in the question answering process. The BioASQ challenge[4] includes multiple tasks, providing an end-to-end pipeline to improve the way we discover relevant information from the scientific literature (in this case, Pubmed articles), such as semantic indexing, information retrieval (IR), question answering (QA) and summarization.

This paper presents results for the BioASQ Task B-Phase B, where the focus is to predict an answer to a given question, given the contexts (i.e., relevant snippets of support information from Pubmed) with respect to the question made available. Based on the question type, the goal is to predict the exact answer. The challenge includes four types of questions: yes/no, factoid, list, and summary types. In the summary type, the ideal answer is assumed to be the precise span or rephrase, while in the factoid type of question, the answer is considered to be one of the top-5 candidates representing the text spans extracted from the relevant snippets. Among the various modules in a QA system, answer extraction represents the module which decides the candidates, that is the contexts' sub-spans from which the best answer can be derived. Existing approaches towards this decision often contain a ranking mechanism of candidates using either typical NLP processing methods like feature-based cues [17], or more recently, neural methods and embedding spaces. The goal of this work is to integrate the best deep learning approaches for question answering, while outlining and mitigating their limits in the biomedical domain. We explore deep learning machine comprehension modules, which outputs the candidate answers ranked in the order of their probability to be most relevant to the question at hand. The evaluation

---

[1] https://stanford-qa.com.

[2] http://ruder.io/transfer-learning/.

[3] https://hazyresearch.github.io/snorkel/.

[4] http://bioasq.org/.

is done separately for each question type, and, in this work, we propose adapted techniques to each type of question in the BioASQ challenge, excluding list, for which deeper semantic analysis is deemed necessary.

One problem that we identify in general deep QA systems is that the answer contexts are targeted, partially by most candidates, but this may not be always done with the best precision: additional residual contextual text can remain as a part of the candidate answer.

The efficacy of deep learning methods made it possible to obtain very good results even in specific domains, even in the lack of sufficient data (since transfer learning we can mitigate some of these issues). But, at the same time, the learned representation denotes still a lack of understanding of the semantics behind complex biomedical expressions, compound abbreviations, and how to solve ambiguous cases. In the quest of targeting the exact answers, our aim is to improve the precision of the deep learning systems by semantically re-ranking predictions (in the factoid QA task), and extracting the best sub-spans (in the summarization QA task).

In the next section, the relation between the approaches developed in this work and the existing state-of-the-art is presented, with a focus on the biomedical domain.

2   Literature Review

Machine Comprehension has been addressed using a large variety of approaches e.g., using information retrieval (IR)/search-based methods, information extraction (IE)/semantic methods, neural or even hybrid [2,27]. In an IR-based QA system, the information indexed from large collections are used to derive the answers. While in a knowledge-based system, semantic representations of user queries are formulated, and then used to query semantic graphs.

In the recent years, deep learning approaches have heavily contributed to the fast advancements towards obtaining usable results for such a complex task. Attention mechanisms [5] are heavily contributing to the success of recent neural architectures like bidirectional attention flow (bidaf) [10,22], and transformer architectures [7,29] fine-tuned for question answering. Vector representations (aka. embeddings) constructed at character, word or sentence level are found to automatically capture to a certain extent the syntactic and semantic properties of the text, leaving behind the need for tedious feature engineering step of more traditional approaches. These embeddings range from Word2Vec [16], Glove [19], to domain-specific transfer embeddings for the biomedical task [26], or BioBERT [13] etc. Further, the embeddings at sentence-level (i.e., for multi-token expressions) like bioSent2Vec [4] have also been transferred to the biomedical domain.

In open domain QA, i.e, machine comprehension in the generic domain, accuracy results are around 90% on datasets such as SQUAD[5]. The same successful

---

[5] https://github.com/google-research/bert.

algorithms, but applied in specific domains, where the data availability is an issue (in addition to dealing with a vocabulary that is different in structure and expression), construct models which are less performant. Transfer learning through fine-tuning to questions answering, or to the biomedical domain, has proven a reliable method [26] to bridge the specificity gap. The benefit is most visible for models that are meant to perform in specific domains, because they can benefit from the generically-learned patterns, especially when the available training dataset is quite small compared to the generic domain.

Adapted solutions to the biomedical consider specific training data sources (e.g., Pubmed), datasets such as BioASQ, and semantic resources like biomedical ontologies and knowledge bases which are rigorously curated and maintained. Although semantic technologies can support the information seeking, they cannot solve by themselves machine comprehension, in the same flexible and efficient way deep learning approaches manage to.

Having built a deep learning biomedical model for QA, further incorporating best practices in the area of NLP such as Named Entity Recognitions (NER), syntactic parsing, or predicate labelling [8] appears to be a promising way to bridge the semantic gap [12,14,25] and improve the results. This paper explores a variety of techniques including deep learning in collaboration with semantic analysis for answering to the factoid, yes/no, and summary type of questions in the BioASQ Task-B Phase-B challenge.

In the next section, we describe the proposed approach for each type of studied question. Section 4 presents the experimental settings, analysis and discussion of the results. Finally, conclusions and future work insights are briefed in Sect. 5.

3   Biomedical QA Methodology

This section is structured around the three types of questions existing in the BioASQ challenge for question answering: factoid, summarization and yes/no. For more information about the types of questions and the standard evaluation associated, we direct the reader to the reference description of the challenge[6].

Factoid QA has been addressed using bidirectional attention flow mechanism integrated with semantic corroboration. For the summary-based QA, we have approached the problem from an extractive summarization point-of-view, and applied dynamic memory networks (DMN) in conjunction with ELMo embeddings. For yes/no types, the problem is viewed as a sentiment analysis task, where positivity towards the question represents a yes, and, the contrary case, a no.

We further detail these methods.

3.1   Factoid QA

For Factoid QA, a deep learning approach is chosen due to its astonishing success, as demonstrated by recent [13,25,26], even in the biomedical domain which is specific and complex.

---

[6] http://bioasq.org/participate/challenges.

In recent years on BioASQ, the accuracy of state-of-the-art models has increased from an MRR of 28% [26] to almost 50% claimed by BioBert [13] within two years, just by considering the neural architecture innovations.

Recently Bert [7]-based architectures achieve on SQUAD an F1 metric of 90%. With a number of less than 1000 annotated factoid questions in the BioASQ dataset, we are still very far from the more than 100,000 annotations of SQUAD. The biomedical domain suffers from a lack of annotated data which limits the ability to obtain the same range of accuracies with the same type of performant algorithms. Additionally, in the biomedical domain the decisions derived from the QA systems can have critical consequences, therefore boosting the precision and robustness of models is critical.

Having the neural architectures evolving towards deriving more distilled models, we focus our attention of how can we semantically analyze the predicted answer candidates. to mitigate the problem that in the biomedical domain the QA model predictions are not so precise as in the generic domain.

Specifically, for factoid QA, we adapt Jack the Reader [25] machine comprehension framework (initially presented and tested for generic domain QA) to the biomedical domain using transfer learning, and applying a bidirectional attention flow process similar to [22]. The proposed approach for factoid QA is depicted in Fig. 1.

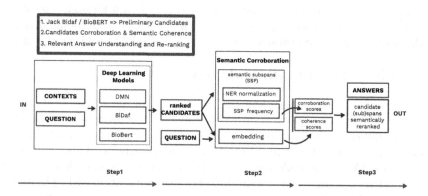

Fig. 1. Factoid question answering approach

The system integrates various domain-adaptation methods: first, the use of transfer learning to infer appropriate embeddings for the biomedical [26]; second, the use of an appropriate semantic analysis and parsing/tokenization using biomedical lexicons and specific biomedical expressions extractor; third, we add a post-processing module which leverages the different forms in which the answer is presented, through normalization of recognized entities, and additional candidate-to-question coherency study, using bio-phrase embeddings.

Although deep neural networks bring an important improvement with respect to the identification of answer precise location across multiple contexts, even state-of-the-art models have difficulties in precisely identifying the best answer. Statistical analysis of candidates with respect to the gold standard shows that the 'expected answer' is present in the candidates set in 83% of cases. This is an important finding, denoting that the neural model gets the answer, but either not in the best order (with respect to the MRR metric), or not in the ideal/compatible format (the same concept is expressed differently), and possibly in a less concise form. However, the typical measures for evaluating factoid QA systems give a high score to a test result if all the candidates are in the right order of relevance, and when the candidate text is strictly equal to the gold standard. These conditions are very hard to achieve, and with respect to the neural model predictions, these measures are not capturing in the best way what we expect from a biomedical QA application, from a usability point of view.

Measuring token overlap is not always appropriate, due to the complex vocabulary that needs to be first normalized across all the data sources.

Studying empirically the root cause, we observe that, most of the time, the context of the answer is correctly targeted by the model, but the indications of the attention pointers are not perfectly precise. As a result, the candidates might be ranked in the wrong order. MRR expects the first candidate to be precisely the gold standard answer, and the subsequent good answers are evaluated decrementally. In addition, the typical evaluation does not take into account synonyms and variation of expression: first, expression of the same concept or entity in different forms, and second, lack of normalization across texts (e.g., of biomedical abbreviated expressions); finally, compositionality is not taken into account by the typical tokenizers used. To resume, the lack of uniformity in expression, and mis-parsing complex expressions takes away the opportunity for the neural model to develop clear intuitions around strong patterns denoting complex concepts.

A human can however leverage this high conceptual overlap of the neural candidates, when targeting the right answer contexts. To mitigate these problems, considering that in the majority of cases the answer is 'hidden' somewhere in the predicted candidates, then rather than considering the output of the deep model as the final step, an originality of this work is to consider the resulting candidate answer spans to a question in a buffer space in which corroboration of information can be applied.

The first step towards this is semantically normalizing the candidates (e.g., 'alpha-synuclein' to be equivalent to 'α-synuclein', '(mmp)-9' to be equivalent to 'matrix metalloproteinase-9', etc.), which in turn needs entity identification, aggregation of the semantic evidence, and correlation to the intent of the question. The semantic analysis is expensive in general, but considering applying it to quite short candidate answer spans (in our model, that would be an average length of 25 characters, but in general this is dependent on the neural model parameters), the processing is much more affordable and ultimately, scalable. In order to mitigate the above mentioned issues, we introduce two main original steps: corroboration of entities within candidates, and candidate-to-question coherence measurement.

Corroborating Semantics. This module has been introduced to diminish the importance of out-of-biodomain candidate answers, with respect to the question intent, and to improve the robustness of results. Corroboration is a typical notion for interpreting evidence applied in a possible unreliable setting (e.g., decentralized systems) when trying to assess the "truth" of a hypothesis, currently in question. In our case, we can understand it as a voting mechanism that occurs between the agents involved in the 'decision', that is, the candidates outputted by the neural model for a question.

The corroboration of entities makes sure the answers that we get are being "agreed" as *relevant for the biomedical domain* by all candidates. To exemplify, Fig. 2 presents a typical output of candidates from a neural QA model, given a test question, and how corroborating entities extracted from all (typically, top 5) candidates, directly by aggregating concept occurrence, can help in re-ranking the candidates towards the rightful order expected by the gold standard. The corroboration score is normalized to have values between 0 and 1.

Note that here we work with what the neural models give as results, therefore this corroboration logic works only because the neural model reaches already a reasonable precision in identifying the answer location, and conceptual overlap, even though it might be introducing surrounding text, and the same concepts might be expressed differently in different candidates. This is the reason for applying Named Entity Recognition (NER) first, so that we target the core of an answer, and make abstraction of differences in candidate answer length. Since we focus on the semantic relevance scores of candidates with respect to the target biomedical domain, we open the possibility of linking more evidence and involving more elaborate semantic analysis (e.g., link prediction and graph mining), by augmenting at the same time the interpretability of the results. The first step towards this corroboration is to apply a biomedical NER to assess whether there are entities recognized in the candidates text. This step also normalizes the candidates, in order for them to express in the same way the formulated concepts.

Candidate-to-Question Coherence. Cui2Vec [1] is a recent multimodal biomedical concept embedding scheme, which associates an embedding vector to a CUI as defined by UMLS thesaurus[7].

Additionally, recognizing that the coverage will never be perfect since biodata are enriched and evolve constantly, we conclude that Cui2Vec embeddings are useful in the coherency computation of a candidate text with respect to the question, but they need to be used as a boost, and unfortunately not as a reliable resource: in the case of missing embedding for a CUI, we need to have a back-up strategy.

If no entities are identified at all in the candidates, the approach turns towards considering n-grams in the corroboration process. This ensures that the corroboration and coherence study can be applied between the neural candidates

---

[7] https://www.nlm.nih.gov/research/umls/.

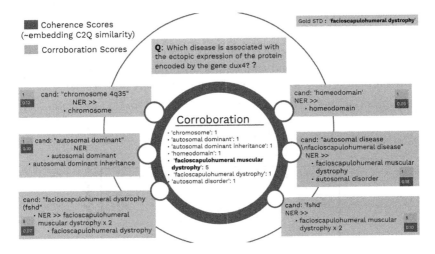

Fig. 2. Factoid questions approach

and the question even in the absence of knowledge from available lexicons, or CUIs embeddings.

The intent coherence step has been introduced to diminish the importance of out-of-scope candidate answers. This verification step consists in measuring the neural candidate span coherence with the question, computed as the cosine similarity between their embeddings. This enforces the condition of relevancy to the question intent, since the neural model has a strong bias towards the contexts from which it has learned from.

The embedding scheme is chosen depending on whether or not there are biomedical concepts mapped to Cui2Vec; in the positive case, we retrieve the conceptual embeddings for the text, and we sum them up. We do the same for the question, and next, their similarity in the embedding space is measured. In the case in which 1. The candidates and question do not contain biomedical concepts at all (rare, but existing cases), 2. The concepts haven't been recognized, or there exists no Cui2Vec embedding associated, the default case applies: we backup to overall n grams corroboration and candidate n-grams coherence study. For that, the alternative embedding model must be able to operate at the level of multi-token expressions/sentence, therefore BioSent2Vec [4] is chosen as the most appropriate option.

Finally, the re-rank score is the multiplication of the corroboration and coherence scores, and each candidate is ranked in accordance with these new probabilities.

3.2 Extractive Answer Selection Module

The focus of this section is to produce extractive summaries using dynamic memory networks adapted for biomedical data. For the generation of a proper ideal answer, a combination of extractive and abstractive summarization needs to be applied. While for question types such as yes/no, list and factoid, an extractive summarization alone may suffice. But if we view the generation of ideal answer as a stand-alone task, the end goal will be to produce a comprehensive human readable summary. Here we focus on the summary QA and yes/no QA.

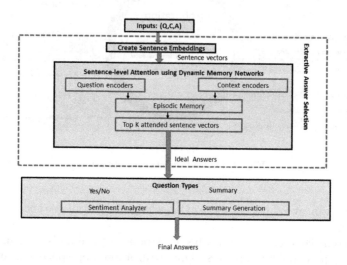

Fig. 3. Workflow for answer selections using neural extractive summarization

The general workflow for this task is depicted in Fig. 3. The work focuses on the extractive answer selection module, which selects the sentences that are possible elements of the ideal answers. The task here is synonymous to finding an extractive summary from the contexts given, with respect to the question at hand. Hence the aim is to train a model that attends to the sentences that can possibly be the candidates of the ideal answer space. In the proposed work, Dynamic Memory Networks (DMN) are utilized, which are the attention models specifically trained for QA problems. DMNs refine the attention mechanism, so that questions trigger an iterative attention process which allows the model to condition its attention on the inputs and the result of previous iterations [11,28]. The sub-modules are briefly described in the subsequent sections.

Sentence Embedding Module. The input to this module is a triplet (Question, Context, Ideal Answer), referred to as (Q, C, A). All these elements are subjected to sentence tokenization. For simplicity, the ideal answer is just referred to as 'Answer'. The proposed approach uses sentence-level attentions for which sentence vectors for (Q, C, A) are computed using ELMo: Deep contextualized word representations [20]. ELMo embeddings captures syntax and semantic characteristics and the variations across linguistic contexts. A deep bidirectional language model (biLM) pre-trained on a large text corpus is used to learn these vectors. In the current work, contextualized ELMo embeddings provided by Tensorflow hub is used to construct the sentence vectors[8]. The inputs represented with ELMo embeddings are fed to attention module.

Sentence-level Attention using Dynamic Memory Networks. DMN comprises an input module, question module, episodic memory module and an answer module. The sentence vectors from the embedding module is encoded to form the question and context encoders using gated recurrent units (GRU) [6]. In general, given a training set of input sequences (here context or snippets)and questions, DMN forms episodic memories, and use them to generate relevant answers. This module consists of an attention mechanism and a recurrent network that updates its memory based on given facts. The attention in this model is computed by constructing similarity measures between each fact, the current memory, and the original question. So during each iteration, the attention mechanism produces and episode based on the facts or contexts, question and the previous memory. The memory is updated by doing a weighted pass with a GRU over the input facts. In order to avoid adding incorrect information into memory when the context is shorter then the full length of the matrix, an attention mask is created. The final state of GRU is the input to the answer module. The resulting sentence vector is the sentence in the context that is closest in distance to the result in our final output. This sentence is selected by assigning a score for each sentence, which is the final result's distance from the sentence. Consequently, the top k (here k = 2) sentences are extracted, which form the candidate sentences for the ideal answer[9].

3.3 Yes/No Answering Model

In the proposed work, a preliminary model for yes/no QA is also tried out. In this model, the output of the candidate answer selection module, i.e., the ideal answers are used as the input. If the exact word 'Yes' or 'No' is already in the selected sentences, then it is a definitive 'Yes' or 'No' respectively. Otherwise, a sentence based sentiment analysis is performed, to check whether the sentence conveyed a positive or negative sentiment. This is done using the NLTK-VADER sentiment analyzer [9]. A senti-score is computed, and if it is positive, an 'Yes' is assigned, or a 'No' in the contrary case. In the neutral cases, an 'Yes' is assigned (this is a crude approach, observing that most of the BioASQ answers relate to 'Yes'). The next section presents the experimental settings, analysis and discussions of the results obtained in Task-B Phase-B.

4   Experimental Analysis

We further detail in the experiments section the method choices and experiments, for the studied question types.

---

[8] https://tfhub.dev/google/elmo.

[9] https://www.oreilly.com/ideas/question-answering-with-tensorflow.

4.1 Factoid QA

For factoid QA, we apply deep learning by training on the BioASQ dataset 7b for question answering, which contains around 800 proper factoid questions annotated with relevant snippets, and gold standard answers.

In the experiments for the factoid QA we first train the neural QA model. The best obtained bidirectional attention flow model has been trained with the following parameters: repr_dim of 25, max_span_size of 25, Adam value 0.001, dimension of vectors 500, all trained over 30 epochs, with a biomedical word2vec (transfer embeddings) vocabulary size of 1,943,494. The results exhibit in terms of MRR surpass the 26% of [26] with 2%. Knowing that the best model of the 2018 competition on factoid achieved a best of 30% [3], these results are statistically significant. Better results are obtained with the recent Biobert [13].

In our experiments, we have employed an adapted parser for the biochemical space[10], and included a couple of biomedical NER tools, in-house or generic e.g., BioBert for NER[11]).

The UMLS concept identification technique[12] has been used as alternative to typical entity recognition. UMLS concepts identified, we can retrieve their CUI (Concept Unique Identifier) mapping, needed further for retrieval of Cui2Vec [1] embeddings. With more than 100,000 concepts covered, Cui2Vec is considered the largest collection of embeddings for the biomedical domain. However, in our experiments, we observed that the Cui2Vec collection is still incomplete with respect to all possible biomedical concepts present in text.

The intuition behind the post-processing of candidates for factoid QA relies on the availability of accurate embedding schemes for the biomedical domain, and the completeness of knowledge bases which can assist the corroboration.

The experiments show an improvement of 5% on the exact answer identification, which is statistically significant. This approach has the merit of being potentially applicable to various machine comprehension models, since it starts operating once we have the candidates answers given by the neural model, thus boosting the results of any neural QA model in a post-prediction phase.

Further experiments include exploring more candidates re-ranking schemes and different bio-embeddings types, which has a direct influence on the accuracy of results, and beyond, on the different QA system usecases.

4.2 Summary and Yes/No QA

For Summary and Yes/No QA, the DMN neural model is trained with the following parameters for extractive answer selections: a learning rate of 0.005, dropout probability 0.5, 4 number of passes in episodic memory, and a recurrent layer size of 128, 1024 dimension of vectors, and a hidden layer size of 256, considering 30,000 training-iterations, and number of attended sentences, k equal to 2.

---

[10] http://chemdataextractor.org/.
[11] https://github.com/dmis-lab/biobert.
[12] https://github.com/Georgetown-IR-Lab/QuickUMLS.

A demo of the candidate sentence selection task done using the above app-roach is shown in Figs. 4 and 5.

QUESTION: which gene controls the expression of gata-1 isoforms?
EXPECTED: pu.1
mutations in exon 2 interfere with the synthesis of the full-length isoform of gata-1 and lead to the production of a shortened isoform, gata-1s. in this study, we report a transcriptional network in which pu.1 positively regulates gata-1 expression in mast cell development. this isoform contains an alternatively spliced first exon (ib) that is distinct from the first exon (ie) incorporated in the major erythroid mrna transcript. reintroduction of pu.1 restores variant ib isoform and upregulates total gata-1 protein expression, which is concurrent with mast cell differentiation. novel combinatorial interactions of gata-1, pu.1, and c/ebpepsilon isoforms regulate transcription of the gene encoding eosinophil granule major basic protein. furthermore, we observe that in pu.1(-/-) fetal liver cells, low levels of the ie gata-1 isoform is expressed, but the variant ib isoform is absent. our findings identify novel combinatorial protein-protein interactions for gata-1, pu.1, and c/ebpepsilon isoforms in eosinophil gene transcription that include gata-1/pu.1 synergy and repressor activity for c/ebpepsilon(27).

Fig. 4. Attention mechanism: case 1 (Color figure online)

QUESTION: which tool is used for the identification of recurrent variants in noncoding regions?
EXPECTED: larva
larva: an integrative framework for large-scale analysis of recurrent variants in noncoding annotations. in cancer research, background models for mutation rates have been extensively calibrated in coding regions, leading to the identification of many driver genes, recurrently mutated more than expected. noncoding regions are also associated with disease; however, background models for them have not been investigated in as much detail. this is partially due to limited noncoding functional annotation. also, great mutation heterogeneity and potential correlations between neighboring sites give rise to substantial overdispersion in mutation count, resulting in problematic background rate estimation. here, we address these issues with a new computational framework called larva. it integrates variants with a comprehensive set of noncoding functional elements, modeling the mutation counts of the elements with a β-binomial distribution to handle overdispersion. larva, moreover, uses regional genomic features such as replication timing to better estimate local mutation rates and mutational hotspots. we demonstrate larva's effectiveness on 760 whole-genome tumor sequences, showing that it identifies well-known noncoding drivers, such as mutations in the tert promoter. furthermore, larva highlights several novel highly mutated regulatory sites that could potentially be noncoding drivers. we make larva available as a software tool and release our highly mutated annotations as an online resource (larva.gersteinlab.org).

Fig. 5. Attention mechanism: case 2 (Color figure online)

The highlighted portion (red color) in each of these figures, represent the sentences attended by the attention mechanism used in proposed approach. In both cases, it can be observed that the attended sentences potentially represent the answer space.

In Fig. 4, it can be observed that some of the sentences which appear to be more important are less attended. For instance, here the first sentence needs to be given a high priority based on human understanding. In future, these problems will be focused, and the intepretability also will be investigated.

In BioASQ challenge, for Summary QA, Rouge score is used to automatically evaluate the ideal answers. Rouge measures the overlap between the ideal answer generated to the reference answer constructed by human experts. The widely used versions of Rouge, R2 and RSU4 are used in the challenge. R2 computes the n-gram overlap, where n=2. RSU4 is the version of ROUGESU with the maximum distance between words of any skip bigram is limited to 4 [15].

In the BioASQ Task-7B Phase-B challenge, the proposed approach for yes/no and summary QA using extractive answer selections was tested in Batch 4 and 5.

682   M. Oita et al.

For the Yes/No questions, accuracy is computed by comparing against the exact answers provided by the experts. Here the accuracy is defined as the ratio of correctly answered yes/no questions (c) to the total number of yes/no questions (n).

Analyzing the experimental results of different BioASQ challenges from 1–7, it was noted that in different batches, the performances varied subjective to the test data set. For Yes/No questions, the top performances varied from 0.87 to 1.00 while for summary the top performance range was from 0.3 to 0.75. Taking into account these factors of variability, the comparison is done only with the top score systems in challenge-7 (batch 4 and 5), where the proposed approach was tested for yes/no and summary type QA. In Table 1, the comparison of proposed approach on the SUmmary and Yes/No QA types is compared with the top-performing systems of BioASQ challenge-7 (batch 4 and 5).

Table 1. Comparison of proposed approaches with BioASQ challenge top ranked systems.

Batch	System	Yes/No		Summary	
		Accuracy	Macro F1	Rouge-2	Rouge-SU4
Batch 4	BioBERT-DMIS-4	0.8696	0.7928	*	*
	MQ-4	0.7391	0.425	0.5177	0.5246
	Proposed	0.6087	0.5801	0.368	0.3813
Batch 5	BioBERT-DMIS-4	0.8286	0.825	*	*
	MQ-4	0.5429	0.3519	0.5035	0.507
	Proposed	0.5143	0.386	0.4818	0.4896

In Batch 4 and 5, BioBERT-DMIS-4 is top ranked in Yes/No QA while MQ-4 in Summary QA. For the Yes/No QA, the proposed approach used a simple model using the NLP based sentiment analysis over the results obtained after extractive answer selections, which gives a macro-F1 score of 0.58 in Batch 5 and 0.35 in Batch 5. The result indicates that the approach is not sufficient enough to derive to exact Yes/No answer from the attended sentences. Analyzing the performance of MQ-4 on the same, it is found that performance is almost comparable.

In Summary QA, the proposed approach exhibited a reasonable performance on both batches, and was second ranked according to the task evaluation after the MQ-4 systems[13]. An improvement in system performance from Batch 4 to 5 is also noted, which can be attributed to training with larger amount of data. Still the complete data was not used, which will be addressed in future. We believe that the performance can be improved by using bio-embeddings and integration of semantic approaches.

---

[13] *BioBERT-DMIS-4 results on Summary was blank, assuming they didn't participate for the same.

5   Conclusion and Future Insights

The main focus of this work is to explore the different deep learning and knowledge based approaches for biomedical QA. This paper analyzes the potential of using various deep learning models and embeddings, in collaboration with a novel method of semantic post-processing of results to enhance the accuracy of results. For extractive answer selection, a sentence-level attention mechanism with contextualized vector representations is studied. The automatic evaluation with Rouge scores for the summarization part exhibits good performance.

A novel approach is presented for the semantic re-ranking of candidates using the concepts of corroboration and coherence. Experiments show this post-processing step of neural candidates can improve with 5% the identification of exact answer, on top of the neural model performance.

More experiments are ongoing for the assessment of the degree in which semantics can help, what is the relative importance of the tools we use, especially related to the embeddings, and other sources of knowledge considered in the process.

Semantic information can be extracted from multiple biomedical knowledge bases and ontologies, such as PubMed Mesh[14], EBI[15] etc. Aggregating these semantic sources and computing graph embeddings can be beneficial towards seemingly incorporating semantics into neural networks and enhancing the representation learning. Computing better embeddings for the biomedical domain is a very promising direction, and aggregating more knowledge is also expected to further improve the results.

BioASQ dataset is still very small in comparison with state-of-the-art machine comprehension datasets such as SQUAD. A fair pursuit would be to explore the potential of weakly supervised techniques for biomedical QA, and other automatic and semi-automatic annotation methods.

We believe that the deep neural models and knowledge-based systems complement the biomedical domain QA system needs, especially in terms of interpetability: therefore our vision goes towards a tighter collaboration between the two, at architecture level [18], for more precise and reliable biomedical text comprehension.

References

1. Beam, A.L., et al.: Clinical concept embeddings learned from massive sources of medical data. CoRR abs/1804.01486 (2018). http://arxiv.org/abs/1804.01486
2. Bouziane, A., Bouchiha, D., Doumi, N., Malki, M.: Question answering systems: survey and trends. Procedia Comput. Sci. **73**, 366–375 (2015)
3. Chandu, K., Naik, A., Chandrasekar, A., Yang, Z., Gupta, N., Nyberg, E.: Tackling biomedical text summarization: OAQA at BioaSQ 5B. In: BioNLP 2017, pp. 58–66 (2017)

---

[14] https://www.ncbi.nlm.nih.gov/pubmed/.
[15] https://www.ebi.ac.uk/ols/ontologies/efo.

4. Chen, Q., Peng, Y., Lu, Z.: BioSentVec: creating sentence embeddings for biomedical texts. CoRR abs/1810.09302 (2018). http://arxiv.org/abs/1810.09302

5. Choi, E., Bahadori, M.T., Sun, J., Kulas, J., Schuetz, A., Stewart, W.: Retain: an interpretable predictive model for healthcare using reverse time attention mechanism. In: Lee, D.D., Sugiyama, M., Luxburg, U.V., Guyon, I., Garnett, R. (eds.) Advances in Neural Information Processing Systems, vol. 29, pp. 3504–3512. Curran Associates, Inc. (2016). http://papers.nips.cc/paper/6321-retain-an-interpretable-predictive-model-for-healthcare-using-reverse-time-attention-mechanism.pdf

6. Chung, J., Gulcehre, C., Cho, K., Bengio, Y.: Empirical evaluation of gated recurrent neural networks on sequence modeling. arXiv preprint arXiv:1412.3555 (2014)

7. Devlin, J., Chang, M., Lee, K., Toutanova, K.: BERT: pre-training of deep bidirectional transformers for language understanding. CoRR abs/1810.04805 (2018). http://arxiv.org/abs/1810.04805

8. Eckert, F., Neves, M.: Semantic role labeling tools for biomedical question answering: a study of selected tools on the BioASQ datasets. In: Proceedings of the 6th BioASQ Workshop A Challenge on Large-scale Biomedical Semantic Indexing and Question Answering, pp. 11–21 (2018)

9. Hutto, C.J., Gilbert, E.: VADER: a parsimonious rule-based model for sentiment analysis of social media text. In: Eighth international AAAI Conference on Weblogs and Social Media (2014)

10. Ke, J., Wang, Y., Xia, F.: Question answering system with bi-directional attention flow. CS224N Report (2017)

11. Kumar, A., et al.: Ask me anything: dynamic memory networks for natural language processing. In: International Conference on Machine Learning, pp. 1378–1387 (2016)

12. Kumar, A.N., et al.: Ontology-based retrieval & neural approaches for BioASQ ideal answer generation. In: Proceedings of the 6th BioASQ Workshop A Challenge on Large-scale Biomedical Semantic Indexing and Question Answering, pp. 79–89 (2018)

13. Lee, J., et al.: BioBERT: pre-trained biomedical language representation model for biomedical text mining. arXiv preprint arXiv:1901.08746 (2019)

14. Li, Y., Gekakis, N., Wu, Q., Li, B., Chandu, K., Nyberg, E.: Extraction meets abstraction: ideal answer generation for biomedical questions. In: Proceedings of the 6th BioASQ Workshop A Challenge on Large-Scale Biomedical Semantic Indexing and Question Answering, pp. 57–65 (2018)

15. Lin, C.Y.: ROUGE: a package for automatic evaluation of summaries. Text Summarization Branches Out (2004)

16. Mikolov, T., Sutskever, I., Chen, K., Corrado, G., Dean, J.: Distributed representations of words and phrases and their compositionality. In: Proceedings of the 26th International Conference on Neural Information Processing Systems, NIPS 2013, vol. 2, pp. 3111–3119. Curran Associates Inc., USA (2013). http://dl.acm.org/citation.cfm?id=2999792.2999959

17. Mishra, A., Jain, S.K.: A survey on question answering systems with classification. J. King Saud Univ. Comput. Inf. Sci. **28**(3), 345–361 (2016)

18. Oita, M.: Reverse engineering creativity into interpretable neural networks. In: Arai, K., Bhatia, R. (eds.) FICC 2019. LNNS, vol. 70, pp. 235–247. Springer, Cham (2020). https://doi.org/10.1007/978-3-030-12385-7_19

19. Pennington, J., Socher, R., Manning, C.: GloVe: global vectors for word representation. In: Proceedings of the 2014 Conference on Empirical Methods in Natural Language Processing (EMNLP), pp. 1532–1543 (2014)

20. Peters, M.E., et al.: Deep contextualized word representations. arXiv preprint arXiv:1802.05365 (2018)
21. Rajpurkar, P., Zhang, J., Lopyrev, K., Liang, P.: SQuAD: 100,000+ questions for machine comprehension of text. arXiv preprint arXiv:1606.05250 (2016)
22. Seo, M.J., Kembhavi, A., Farhadi, A., Hajishirzi, H.: Bidirectional attention flow for machine comprehension. CoRR abs/1611.01603 (2016). http://arxiv.org/abs/1611.01603
23. Tremblay, J., et al.: Training deep networks with synthetic data: bridging the reality gap by domain randomization. In: Proceedings of the IEEE Conference on Computer Vision and Pattern Recognition Workshops, pp. 969–977 (2018)
24. Tsatsaronis, G., et al.: An overview of the BioASQ large-scale biomedical semantic indexing and question answering competition. BMC Bioinformatics **16**(1), 138 (2015)
25. Weissenborn, D., et al.: Jack the reader - a machine reading framework. CoRR abs/1806.08727 (2018). http://arxiv.org/abs/1806.08727
26. Wiese, G., Weissenborn, D., Neves, M.L.: Neural domain adaptation for biomedical question answering. CoRR abs/1706.03610 (2017). http://arxiv.org/abs/1706.03610
27. Wimalasuriya, D.C., Dou, D.: Ontology-based information extraction: an introduction and a survey of current approaches. J. Inf. Sci. **36**(3), 306–323 (2010). https://doi.org/10.1177/0165551509360123
28. Xiong, C., Merity, S., Socher, R.: Dynamic memory networks for visual and textual question answering. In: International Conference on Machine Learning, pp. 2397–2406 (2016)
29. Yang, Z., Dai, Z., Yang, Y., Carbonell, J., Salakhutdinov, R., Le, Q.V.: XLNet: generalized autoregressive pretraining for language understanding (2019). http://arxiv.org/abs/1906.08237

Measuring Domain Portability and Error Propagation in Biomedical QA

Stefan Hosein$^{(\boxtimes)}$, Daniel Andor$^{(\boxtimes)}$, and Ryan McDonald$^{(\boxtimes)}$

Google, Mountain View, USA
{smhosein,andor,ryanmcd}@google.com

Abstract. In this work we present Google's submission to the BioASQ 7 biomedical question answering (QA) task (specifically Task 7b, Phase B). The core of our systems are based on BERT QA models, specifically the model of [1]. In this report, and via our submissions, we aimed to investigate two research questions. We start by studying how domain portable are QA systems that have been pre-trained and fine-tuned on general texts, e.g., Wikipedia. We measure this via two submissions. The first is a non-adapted model that uses a public pre-trained BERT model and is fine-tuned on the Natural Questions data set [4]. The second system takes this non-adapted model and fine-tunes it with the BioASQ training data. Next, we study the impact of error propagation in end-to-end retrieval and QA systems. Again we test this via two submissions. The first uses human annotated relevant documents and snippets as input to the model and the second predicted documents and snippets. Our main findings are that domain specific fine-tuning can benefit Biomedical QA. However, the biggest quality bottleneck is at the retrieval stage, where we see large drops in metrics – over 10pts absolute – when using non gold inputs to the QA model.

Keywords: Biomedical · Question answering · BERT

1 Introduction

BioASQ [11] is a large-scale online biomedical research competition. There are many tasks within the competition: question answering (QA), information retrieval and semantic indexing. Our submissions focus on Task 7b, Phase B which requires participating systems to generate ideal or exact answers to biomedical questions using mainly PubMed articles. We focus on exact answers which can include factoid, list, and yes/no question types.

The systems we used for QA were all BERT-based [2] models using the public available *large* pre-trained models and fine-tuned on the Natural Questions corpus [1,4] and Conversational Question Answering dataset [10]. Additionally, three of the four systems we submitted were further fine-tuned on the BioASQ training data. The difference between the biomedical specific models is the input into the models: using only snippets, using snippets from the previous information retrieval phase (Task 7b, Phase A) and a mixture of snippets and abstracts.

© Springer Nature Switzerland AG 2020
P. Cellier and K. Driessens (Eds.): ECML PKDD 2019 Workshops, CCIS 1168, pp. 686–694, 2020.
https://doi.org/10.1007/978-3-030-43887-6_61

This work-flow has no pre-processing of the data necessary and uses very little in-domain knowledge to achieve successful results.

Our systems focused mainly on factoid questions and their results. The evaluation metrics for factoid were strict accuracy, lenient accuracy, and Mean Reciprocal Rank (MRR) [11]. The results of the competition show that all our models are always in the top half of systems for factoid questions which indicate that neural QA models based on large pre-trained language models are very robust across domains. In addition, since our system used snippets from the previous information retrieval phase and had a lower but still competitive accuracy indicated that the limiting factor of this neural model is the document and snippet retrieval architecture and not the QA model itself.

In this paper we start with a literature review which explains our reasoning for using BERT-based models and the architectures of previous entrants for the BioASQ challenge, then we go in-depth into explaining the differences between our 4 systems that were submitted, lastly we discuss the performance of our systems and how error propagates between retrieval and QA systems.

2   Related Work

The use of BERT-based models [2] is becoming ubiquitous in the field of question answering (QA). At the time of this writing, out of the top 5 systems in SQuAD 2.0 [9], 4 are BERT models. For the CoQA [10] challenge, all of the top 5 systems are BERT models. With the success of BERT models, many papers are tuning these models to their specific domain. One such paper is BioBERT [5], where the authors created a domain specific language representation biomedical BERT model for a few biomedical tasks, one being question answering. They evaluated their models on BioASQ test sets for BioASQ 4, 5 and 6. They saw a an absolute improvement of 9.61% with the models.

The BioASQ [11] competition has been very popular amongst researchers. Some of the early systems in BioASQ were not neural architectures. For the 2nd BioASQ challenge, [7] developed a system that tries to extract the lexical answer type of the question. Then, they selected the relevant snippets for each question and provided these as inputs to MetaMap[1] which extracted candidate answers for each factoid question. For the 3rd iteration of the challenge [14] used a three layer architecture for factoid and list questions. The architecture is based on the framework [13] and including many components like MetaMap and ClearNLP[2]. In BioASQ 4 both [7] and [13] improved their models using more biomedical information into their systems. Neural architecture systems started to appear more frequently from BioASQ 5, with the *DeepQA* systems using the

---

[1] https://metamap.nlm.nih.gov/.
[2] https://github.com/clir/clearnlp/.

then state-of-the-art QA model, FastQA [12]. The FastQA was extended by using biomedical word embeddings and pre-training on QA datasets (SQuAD) then fine-tuning on the BioASQ training set. In the last BioASQ challenge (BioASQ 6), there were numerous systems that used neural architectures like LSTMs [3,6].

3   BERT Model

Recent work on learning word representations have focused on learning context dependent representations. An example, the word bank, it could mean the land alongside the river/lake or a financial establishment. Previous methods would have a single representation of the word bank unlike more modern methods which will have two representations for the word based on its context in the sentence. BERT [2] is one such method to produce contextualized word embeddings. The most common instantiation of BERT is pre-trained using bidirectional trans-formers to predict randomly masked words in a sequence, thus removing the limitation that previous bidirectional language models had: the fact that future words should not be seen. In addition, BERT predicts the next sentence given a previous sentence and these two tasks allow BERT to obtain state-of-the-art performance on many NLP tasks.

Our QA model follows the Natural Questions (NQ) baseline model [1], an extractive QA model based on BERT [2]. In the context of the BioASQ data: given a pair of question (the body) Q and context/body (the snippets or some augmentation of the snippets) S, the model predicts the answer by scoring all the sub-spans (candidate answers taken from S) and then ranking all these sub-spans by their score. For more in-depth details, see [1].

4   Systems Overview

There were four systems that we submitted for evaluation in BioASQ Task 7b, Phase B. Below is a brief overview of each system, we give more details in further sub-sections.

- **google-gold-input**: fine-tuned on BioASQ training data, used the provided gold snippets as input to the QA model (see Fig. 1)
- **google-gold-input-ab**: fine-tuned on BioASQ training data, used the provided gold snippets and the abstract of the top ranked document as input to the QA model
- **google-gold-input-nq**: no in-domain training, used the provided gold snippets as input to the QA model
- **google-pred-input**: fine-tuned on BioASQ training data, used snippets from the top-ranked submission from Task 7b, Phase A as input to the QA model

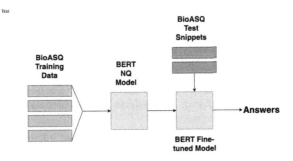

Fig. 1. A figure showing a model which was fine-tuned on BioASQ training data using provided BioASQ test (gold) snippets from test batches to generate the answers for questions (google-gold-input).

4.1 No In-Domain Training

To give our baseline system, google-gold-input-nq, exposure to a broad set of domains, we trained on both the NQ [4] and CoQA [10] datasets. Both NQ and CoQA contain Wikipedia data, while CoQA adds four additional domains, covering news and fiction.

After training on NQ as in [1], we further fine-tuned on CoQA with a learning rate of 5e−5, batch size of 32, for 2 epochs.

4.2 BioASQ Fine-Tuning

Two of our models – accounting for three of our systems – were fine-tuned using the BioASQ training data. The difference between these two models is that one uses a concatenation of relevant snippets as model context (google-gold-input) while the other uses the abstract of the most relevant document concatenated with any remaining snippets (google-gold-input-ab), see Table 1 for an example. We used only one abstract as using abstracts from lower ranked documents would dramatically increase the noise-to-signal ratio.

Starting with the model trained in Sect. 4.1, we fine-tuned on the BioASQ training set using a learning rate of 1e−7, batch size of 32, for 10 epochs. The large number of epochs was necessary due to the very small training dataset size of ∼2700 questions.

4.3 Snippet Retrieval

The model, google-gold-input, and the model used for snippet retrieval, google-pred-input, is the same, however, the difference between them is at test time. Instead of using the gold-standard test snippets provided by BioASQ, google-pred-input used snippets from the top ranking submission to Task 7b, Phase A [8]. This allows us to analyze the effect of information retrieval on the QA system since the only difference between google-pred-input and google-gold-input is the

Table 1. Table showing the differences between the input at test time for each model.

System	Context
google-gold-input	Vaspin expression is increased in white adipose tissue \n Visceral adipose tissue-derived serine protease inhibitor (Vaspin) is an adipocytokine that has been shown to exert anti-inflammatory effects and inhibits apoptosis under diabetic conditions
google-gold-input-ab	Vaspin suppresses cytokine-induced inflammation in 3T3-L1 adipocytes via inhibition of NFu\03baB pathway.\n Vaspin expression is increased in white adipose tissue (WAT) of diet-induced obese mice and rats and is supposed to compensate HFD-induced inflammatory processes and insulin resistance in adipose tissue by ... \n Visceral adipose tissue-derived serine protease inhibitor (Vaspin) is an adipocytokine that has been shown to exert anti-inflammatory effects and inhibits apoptosis under diabetic conditions

context given to the QA system. One interesting property is that the predicted set of snippets is often much larger than the gold set. This is partly due to the nature of the data, where the annotators were tasked with finding enough relevant snippets to support the correct answer – not all the relevant snippets.

4.4 Yes/No and List Question Types

Even though our systems participated in some yes/no and list batches, these were heuristic based and not a core part of our model. For yes/no questions, if *yes* or *no* was present in the candidate answers, then we selected the one with the higher log probability. If we could not find *yes* or *no* in the candidate set, we selected yes by default. For list type questions, we selected the top 5 candidates and split the results into single words or phrases by punctuation and then selected the top 5 results from those. Since these were heuristic based, we do not discuss these results in the paper.

5   Results

We took part in the last three batches of Task 7b, Phase B. More specifically: the answers of google-gold-input and google-pred-input were evaluated on batches 3, 4 and 5 and google-gold-input-nq and google-gold-input-ab were evaluated on batches 4 and 5. For batch 3 our google-gold-input was always in the top two system scores for all factoid evaluations, while google-pred-input had the lowest place of 6th for factoid evaluations. For batches 4 and 5 our scores were generally in the top ten for factoids.

For a comparison of the best system's score and our models see Table 2. The table alludes to a number of interesting results some we discuss in later

Table 2. Performance on BioASQ Task 7b, Phase B for batches 3 to 5. The underlined system is the best scoring system from Google's entries and bold indicates the system is the top from all official entries in that batch. Best Score is the top scoring entry that is not among Google's submissions.

	Batch 3			Batch 4			Batch 5		
	Strict	Lenient	MRR	Strict	Lenient	MRR	Strict	Lenient	MRR
Best Score	0.4483	0.6552	0.5115	0.5882	0.8235	0.6912	0.2857	0.5143	0.3638
gold-input	0.4138	**0.6552**	0.5023	0.4706	0.7059	0.5495	**0.2857**	0.3714	0.3167
pred-input	0.3448	0.5517	0.4322	0.3529	0.5882	0.4338	0.1429	0.2857	0.2057
gold-input-ab	–	–	–	0.4706	0.6471	0.5255	0.2286	0.2857	0.2571
gold-input-nq	–	–	–	0.4706	0.5882	0.5132	**0.2857**	0.3714	0.3057

Table 3. Performance on BioASQ Task 4 and 5b, Phase B averaged over all batches. The bold system is the top scoring. *Best Participant* and *BioBERT* results are from [5].

	BioASQ 4			BioASQ 5		
	Strict	Lenient	MRR	Strict	Lenient	MRR
Best Participant	0.206	0.294	0.240	0.418	0.574	0.477
BioBERT [5]	**0.365**	0.489	**0.411**	0.416	0.540	0.463
google-gold-input	0.311	**0.540**	0.400	**0.458**	**0.615**	**0.520**

subsections. One of those results is that adding abstracts was not significantly helpful and indicates that there is a noise-to-signal issue where the system might get diminishing or negative gains after a certain amount of data is used for the context.

It should be noted that these results are preliminary. Humans have yet to judge the outputs off all participating systems. As a precursor to participating in BioASQ7, we investigated the performance of our model on prior year's data. The advantage of doing this is that the test annotations are much more complete, since they also include all the correct answers from the systems that participated that year. We compare to two baselines. The first is the best system that participated in that specific year's challenge. The second is a recent state-of-the-art model BioBERT [5][3]. This model is similar in nature to our model, with some differences. First, it is pre-trained on biomedical data. Second, it is only fine-tuned on the BioASQ training data and does not use any additional fine-tuning data, i.e., natural questions. Note that all models are comparable: (1) they are trained with the specific training data for the year being tested; and (2) they use provide gold snippets as input.

Table 3 shows the results. We can see here that our model is very competitive with previous models on this data, including other BERT-based models. The main take-away here is that adding domain general fine-tuning data (i.e., the Natural Questions data) can lead to gains in performance.

---

[3] The authors of this system also participated in BioASQ7 and preliminary have the highest scoring submission.

Table 4. Domain portability for factoid biomedical QA

	No-Biomedical fine-tuning google-gold-input-nq			Biomedical fine-tuning google-gold-input-ab			Δ		
	Strict	Lenient	MRR	Strict	Lenient	MRR	Strict	Lenient	MRR
Batch 4	0.4706	0.6471	0.5255	0.4706	0.5882	0.5132	–	0.0589	0.0123
Batch 5	0.2286	0.2857	0.2571	0.2857	0.3714	0.3057	−0.0571	−0.0857	−0.0486

Table 5. Performance on BioASQ Task 4 and 5b, Phase B averaged over all batches to measure domain portability. The bold system is the top scoring. *This model is slightly different from the submitted system as it uses only gold snippets as input.

	BioASQ 4			BioASQ 5		
	Strict	Lenient	MRR	Strict	Lenient	MRR
BioBERT [5]	**0.365**	0.489	**0.411**	0.416	0.540	0.463
google-gold-input	0.311	**0.540**	0.400	**0.458**	**0.615**	**0.520**
google-gold-input-nq*	0.302	0.488	0.376	0.451	0.603	0.509

5.1   Domain Portability

To measure domain portability we investigate the model fine-tuned only on the NQ dataset (google-gold-input-nq) and the model that was further fine-tuned on BioASQ training data (google-gold-input-ab). For this experiment, these models use the top-ranked abstract concatenated with snippets from other documents as input. Results for factoid QA are shown in Table 4. We can see that as of the preliminary results, there is no clear pattern to determine which system is best. This suggests that the QA model, while trained on non-biomedical data, has learned at least as well as a domain-specific model to generalize matching questions to spans of text using the context of the match. Also, when looking at the accuracy of the models against the field of submissions, the non-ported NQ QA model is fairly strong - easily in the top third of submitted systems. This suggest that even general domain QA models can do a reasonable job on new domains, including hyper-specialized ones like biomedical literature.

Again, these results are preliminary, we can again look at previous BioASQ batches with more compete test annotations. Table 5 has the results. From here we can see that the biomedical specific model (google-gold-input) outperforms the domain general model (google-gold-input-nq) consistently, but not by a large margin. Furthermore, the domain general model is competitive with the previous state-of-the-art BioBERT models. These results present stronger empirical evidence that large-scale domain general models do port well to new domains.

It should be noted that we did not measure the effect of in-domain pre-training. BioBERT [5] tested this and did find that for BioASQ 4–6 significant increases in factoid QA metrics could be achieved when using in-domain pre-training. This could suggest that pre-training and not fine-tuning are the keys to improving domain portability of BERT-based QA models.

Table 6. Error propagation for factoid biomedical QA

| | Gold inputs | | | Noisy inputs | | | Δ | | |
| | google-gold-input | | | google-gold-pred | | | | | |
	Strict	Lenient	MRR	Strict	Lenient	MRR	Strict	Lenient	MRR
Batch 3	0.4138	0.6552	0.5023	0.3448	0.5517	0.4322	0.0690	0.1035	0.0701
Batch 4	0.4706	0.7059	0.5495	0.3529	0.5882	0.4338	0.1177	0.1177	0.1157
Batch 5	0.2857	0.3714	0.3167	0.1429	0.2857	0.2057	0.1428	0.0857	0.1110

5.2 Error Propagation

To test error propagation we used our main model: snippets as input; pre-trained BERT; fine-tuned on NQ; and further fine-tuned on BioASQ training data. We then tested two scenarios,

- Gold inputs (google-gold-input): we used gold standard snippets generated by humans as input to the QA model. This is the standard setting for almost all participants in the track, as these were provided by the organizers.
- Noisy inputs (google-gold-pred): We used predicted snippets as input to the QA model. This was provided by [8], a team that participated in 7b Phase A and whose document and snippet retrieval were the highest scoring submissions. Specifically, we used there BERT-based high-confidence document reranker plus snippet extractor.

Table 6 contains the results. We measure error propagation only for factoid QA for batches 3–5, which were the batches that we participated in. We can see from these results that feeding the QA model non-gold inputs leads to a dramatic drop in all metrics: from 7pts up to 14pts absolute. In one case (batch 5, strict accuracy), the metric is halved.

These results strongly suggest that when considering the QA system holistically – retrieval followed by QA – the largest bottleneck is the quality of the retrieval system, and not necessarily the QA model. For batch 3, our model was at the top or near the top for all metrics. However, for batches 4 and 5, our model was significantly lower than the top reporting system and we can see that error propagation is amplified for these batches. It would be useful to measure error propagation against the best reporting BioASQ models for these batches.

6   Conclusion

In this paper, we set out to investigate the domain portability of neural QA systems [1] and to determine what is the impact of error propagation in end-to-end retrieval and QA systems. We found that even though our base QA model was trained on non-biomedical data, it was able to generalize matching questions to spans of text and gave very good results compared to systems that were trained with biomedical data. In addition, our results suggest that when using end-to-end QA systems the bottleneck is the quality of the retrieval system and not necessarily the QA model itself.

References

1. Alberti, C., Lee, K., Collins, M.: A bert baseline for the natural questions. arXiv preprint arXiv:1901.08634 (2019)
2. Devlin, J., Chang, M.W., Lee, K., Toutanova, K.: BERT: pre-training of deep bidirectional transformers for language understanding. arXiv preprint arXiv:1810.04805 (2018)
3. Hochreiter, S., Schmidhuber, J.: Long short-term memory. Neural Comput. **9**, 1735–1780 (1997)
4. Kwiatkowski, T., et al.: Natural questions: a benchmark for question answering research. Trans. Assoc. Comput. Linguist. **7**, 453–466 (2019)
5. Lee, J., et al.: BioBERT: pre-trained biomedical language representation model for biomedical text mining. arXiv preprint arXiv:1901.08746 (2019)
6. Nentidis, A., Krithara, A., Bougiatiotis, K., Paliouras, G., Kakadiaris, I.: Results of the sixth edition of the BioASQ challenge. In: Association for Computational Linguistics, pp. 1–10 (2018)
7. Papanikolaou, Y., Dimitriadis, D., Tsoumakas, G., Laliotis, M., Markantonatos, N., Vlahavas, I.P.: Ensemble approaches for large-scale multi-label classification and question answering in biomedicine. In: CLEF (2014)
8. Pappas, D., McDonald, R., Androutsopoulos, I.: AUEB at BioASQ 7: document and snippet retrieval (2019, in submission)
9. Rajpurkar, P., Zhang, J., Lopyrev, K., Liang, P.S.: SQuAD: 100, 000+ questions for machine comprehension of text. In: EMNLP (2016)
10. Reddy, S., Chen, D., Manning, C.D.: CoQA: a conversational question answering challenge. Trans. Assoc. Comput. Linguist. **7**, 249–266 (2018)
11. Tsatsaronis, G., et al.: An overview of the BioASQ large-scale biomedical semantic indexing and question answering competition. BMC Bioinform. **16**, 138 (2015)
12. Weissenborn, D., Wiese, G., Seiffe, L.: FastQA: a simple and efficient neural architecture for question answering. CoRR abs/1703.04816 (2017)
13. Yang, Z., Garduño, E., Fang, Y., Maiberg, A., McCormack, C., Nyberg, E.: Building optimal information systems automatically: configuration space exploration for biomedical information systems. In: CIKM (2013)
14. Yang, Z., Gupta, N., Sun, X., Xu, D., Zhang, C., Nyberg, E.: Learning to answer biomedical factoid & list questions: OAQA at BioASQ 3B. In: CLEF (2015)

UNCC Biomedical Semantic Question Answering Systems. BioASQ: Task-7B, Phase-B

Sai Krishna Telukuntla$^{(\boxtimes)}$, Aditya Kapri, and Wlodek Zadrozny

College of Computing and Informatics (CCI),
UNC Charlotte, Charlotte, NC 28223, USA
{stelukun,akapri,wzadrozn}@uncc.edu

Abstract. In this paper, we detail our submission to the 7th year BioASQ competition. We present our approach for Task-7b, Phase B, Exact Answering Task. These Question Answering (QA) tasks include Factoid, Yes/No, List Type Question answering. Our system is based on a contextual word embedding model. We have used a Bidirectional Encoder Representations from Transformers (BERT) based system, fined tuned for biomedical question answering task using BioBERT. In the third test batch set, our system achieved the highest 'MRR' score for Factoid Question Answering task. Also, for List type question answering task our system achieved the highest recall score in the fourth test batch set. Along with our detailed approach, we present the results for our submissions, and also highlight identified downsides for our current approach and ways to improve them in our future experiments.

Keywords: BioASQ · Question answering · Factoid · List-type · UNCC

1 Introduction

BioASQ[1] is a biomedical document classification, document retrieval, and question answering competition, currently in its seventh year. We provide an overview of our submissions to semantic question answering task (7b, Phase B) of BioASQ 7 (except for 'ideal answer' test, in which we did not participate this year). In this task systems are provided with biomedical questions and are required to submit ideal and exact answers to those questions. We have used BioBERT [9] based system, see also Bidirectional Encoder Representations from Transformers (BERT) [4], and we fine tuned it for the biomedical question answering task. Our system scored near the top for factoid questions for all the batches of the challenge. More specifically, in the third test batch set, our system achieved highest 'MRR' score for Factoid Question Answering task. Also, for List-type question answering task our system achieved highest recall score in the fourth test batch

---

[1] http://BioASQ.org/participate/challenges.

© Springer Nature Switzerland AG 2020
P. Cellier and K. Driessens (Eds.): ECML PKDD 2019 Workshops, CCIS 1168, pp. 695–710, 2020.
https://doi.org/10.1007/978-3-030-43887-6_62

set. Along with our detailed approach, we present the results for our submissions and also highlight identified downsides for our current approach and ways to improve them in our future experiments.

The QA task is organized in two phases. Phase A deals with retrieval of the relevant document, snippets, concepts, and RDF triples, and phase B deals with exact and ideal answer generations. Exact answer generation is required for factoid, list, and yes/no type question.

BioASQ competition provides the training and testing datasets. The training data consists of questions, golden standard documents, snippets, concepts, and ideal answers (which we did not use in this paper, but we used last year [2]). The test data is split between phase A and phase B. The phase A dataset consists of the questions, unique ids, question types. The phase B dataset consists of the questions, golden standard documents, snippets, unique ids and question types. Exact answers for factoid type questions are evaluated using strict accuracy (consider the top answer), lenient accuracy (consider the top 5 answers), and MRR (Mean Reciprocal Rank) which takes into account the ranks of returned answers. Answers for the list type question are evaluated based on precision, recall, and F-measure.

2 Related Work

2.1 BioASQ

Sharma et al. [16] describe a system with two stage process for factoid and list type question answering. Their system extracts relevant entities and then runs supervised classifier to rank the entities. Wiese et al. [18] propose neural network based model for Factoid and List-type question answering task. The model is based on Fast QA and predicts the answer span in the passage for a given question. The model is trained on SQuAD data set and fine tuned on the BioASQ data. Dimitriadis et al. [5] proposed two stage process for Factoid question answering task. Their system uses general purpose tools such as Metamap, BeCas to identify candidate sentences. These candidate sentences are represented in the form of features, and are then ranked by the binary classifier. Classifier is trained on candidate sentences extracted from relevant questions, snippets and correct answers from BioASQ challenge. For factoid question answering task highest 'MRR' achieved in the 6th edition of BioASQ competition is '0.4325'. Our system is a neural network model based on contextual word embeddings [4] and achieved a 'MRR' score '0.6103' in one of the test batches for Factoid Question Answering task.

2.2 A Minimum Background on BERT

BERT stands for "Bidirectional Encoder Representations from Transformers" [4] is a contextual word embedding model. Given a sentence as an input, contextual embedding for the words are returned. The BERT model was designed so it can be fine tuned for 11 different tasks [4], including question answering tasks. For a question answering task, question and paragraph (context) are given as an input. A BERT standard is that question text and paragraph text are separated by a separator [Sep]. BERT question-answering fine tuning involves adding softmax layer. Softmax layer takes contextual word embeddings from BERT as input and learns to identity answer span present in the paragraph (context). This process is represented in Fig. 1. For detailed understanding of BERT Architecture, please refer to the original BERT paper [4].

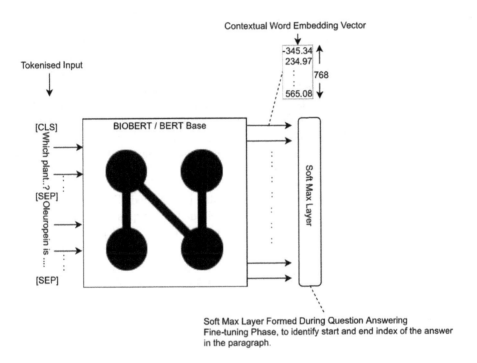

Fig. 1. BioBERT fine tuned for question answering task

Comparison of Word Embeddings and Contextual Word Embeddings.
A 'word embedding' is a learned representation. It is represented in the form of vector where words that have the same meaning have a similar vector representation. Consider a word embedding model 'word2vec' [12] trained on a corpus. Word embeddings generated from the model are context independent that is, word embeddings are returned regardless of where the words appear in a sentence and regardless of e.g. the sentiment of the sentence. However, contextual word embedding models like BERT also takes context of the word into consideration.

2.3   Comparison of BERT and Bio-BERT

'BERT' and BioBERT are very similar in terms of architecture. Difference is that 'BERT' is pretrained on Wikipedia articles, whereas BioBERT version used in our experiments is pretrained on Wikipedia, PMC and PubMed articles. Therefore BioBERT model is expected to perform well with biomedical text, in terms of generating contextual word embeddings.

BioBERT model used in our experiments is based on BERT-Base Architecture; BERT-Base has 12 transformer Layers where as BERT-Large has 24 transformer layers. Moreover contextual word embedding vector size is 768 for BERT-Base and more for BERT-large. According to [4] Bert-Large, fine-tuned on SQuAD 1.1 question answering data [13] can achieve F1 Score of 90.9 for Question Answering task where as BERT-Base Fine-tuned on the same SQuAD question answering [13] data could achieve F1 score of 88.5. One downside of the current version BioBERT is that word-piece vocabulary[2] is the same as that of original BERT Model, as a result word-piece vocabulary does not include biomedical jargon. Lee et al. [9] created BioBERT, using the same pre-trained BERT released by Google, and hence in the word-piece vocabulary (vocab.txt), as a result biomedical jargon is not included in word-piece vocabulary. Modifying word-piece vocabulary (vocab.txt) at this stage would loose original compatibility with 'BERT', hence it is left unmodified.

In our future work we would like to build pre-trained 'BERT' model from scratch. We would pretrain the model with biomedical corpus (PubMed, 'PMC') and Wikipedia. Doing so would give us scope to create word piece vocabulary to include biomedical jargon and there are chances of model performing better with biomedical jargon being included in the word piece vocabulary. We will consider this scenario in the future, or wait for the next version of BioBERT.

3   Experiments: Factoid Question Answering Task

For Factoid Question Answering task, we fine tuned BioBERT [9] with question answering data and added new features. Figure 1 shows the architecture of BioBERT fine tuned for question answering tasks: Input to BioBERT is word tokenized embeddings for question and the paragraph (Context). As per the 'BERT' [4] standards, tokens '[CLS]' and '[SEP]' are appended to the tokenized input as illustrated in the figure. The resulting model has a softmax layer formed for predicting answer span indices in the given paragraph (Context). On test data, the fine tuned model generates n-best predictions for each question. For a question, n-best corresponds that n answers are returned as possible answers in the decreasing order of confidence. Variable n is configurable. In our paper, any further mentions of 'answer returned by the model' correspond to the top answer returned by the model.

---

[2] vocab.txt and all other software and data is available in our GitHub Repo https://github.com/telukuntla/BioMedicalQuestionAnswering_UNCC.

3.1 Setup

BioASQ provides the training data. This data is based on previous BioASQ competitions. Train data we have considered is aggregate of all train data sets till the 5th version of BioASQ competition. We cleaned the data, that is, question-answering data without answers are removed and left with a total count of 530 question answers. The data is split into train and test data in the ratio of 94 to 6; that is, count of 495 for training and 35 for testing.

The original data format is converted to the BERT/BioBERT format, where BioBERT expects 'start_index' of the actual answer. The 'start_index' corresponds to the index of the answer text present in the paragraph/Context. For finding 'start_index' we used built-in python function find(). The function returns the lowest index of the actual answer present in the context(paragraph). If the answer is not found '−1' is returned as the index. The efficient way of finding start_index is, if the paragraph (Context) has multiple instances of answer text, then 'start_index' of the answer should be that instance of answer text whose context actually matches with what's been asked in the question.

Example (Question, Answer and Paragraph from [17]):
Question: Which drug should be used as an antidote in benzodiazepine overdose?
Answer: 'Flumazenil'
Paragraph(context):

"Flumazenil use in benzodiazepine overdose in the UK: a retrospective survey of NPIS data. OBJECTIVE: Benzodiazepine (BZD) overdose (OD) continues to cause significant morbidity and mortality in the UK. **Flumazenil** is an effective antidote but there is a risk of seizures, particularly in those who have co-ingested tricyclic antidepressants. A study was undertaken to examine the frequency of use, safety and efficacy of flumazenil in the management of BZD OD in the UK. METHODS: A 2-year retrospective cohort study was performed of all enquiries to the UK National Poisons Information Service involving BZD OD. RESULTS: Flumazenil was administered to 80 patients in 4504 BZD-related enquiries, 68 of whom did not have ventilatory failure or had recognised contraindications to flumazenil. Factors associated with flumazenil use were increased age, severe poisoning and ventilatory failure. Co-ingestion of tricyclic antidepressants and chronic obstructive pulmonary disease did not influence flumazenil administration. Seizure frequency in patients not treated with flumazenil was 0.3%".

Actual answer is 'Flumazenil', but there are multiple instances of word 'Flumazenil'. Efficient way to identify the start-index for 'Flumazenil' (answer) is to find that particular instance of the word 'Flumazenil' which matches the context for the question. In the above example 'Flumazenil' highlighted in bold is the actual instance that matches question's context. Unfortunately, we could not identify readily available tools that can achieve this goal. In our future work, we look forward to handling these scenarios effectively.

Note: The creators of 'SQuAD' [13] have handled the task of identifying answer's start_index effectively. But 'SQuAD' data set is much more general and does not include biomedical question answering data.

3.2   Training and Error Analysis

During our training with the BioASQ data, learning rate is set to 3e-5, as mentioned in the BioBERT paper [9]. We started training the model with 495 available train data and 35 test data by setting number of epochs to 50. After training with these hyper-parameters training accuracy (exact match) was 99.3% (overfitting) and testing accuracy is only 4%. In the next iteration we reduced the number of epochs to 25 then training accuracy is reduced to 98.5% and test accuracy moved to 5%. We further reduced number of epochs to 15, and the resulting training accuracy was 70% and test accuracy 15%. In the next iteration set number of epochs to 12 and achieved train accuracy of 57.7% and test accuracy 23.3%. Repeated the experiment with 11 epochs and found training accuracy to be the same 57.7% and the test accuracy 22%. In the next iteration we set number of epochs to 9 and found training accuracy of 48% and test accuracy of 15%. Hence optimum number of epochs is taken as 12 epochs.

During our error analysis we found that on test data, model tends to return text in the beginning of the context (paragraph) as the answer. On analysing train data, we found that there are 120 (out of 495) question answering data instances having start_index:0, meaning 120 (about 25%) question answering data has first word(s) in the context(paragraph) as the answer. We removed 70% of those instances in order to make train data more balanced. In the new train data set we are left with 411 question answering data instances. This time we got the highest test accuracy of 26% at 11 epochs. We have submitted our results for BioASQ test batch-2, got strict accuracy of 32% and our system stood in 2nd place. Initially, hyper-parameter- 'batch size' is set to 400. Later it is tuned to 32. Although accuracy (exact answer match) remained at 26%, model generated concise and better answers at batch size 32, that is wrong answers are close to the expected answer in good number of cases.

Example.(from [17])
Question: Which mutated gene causes Chediak Higashi Syndrome?
Exact Answer: 'lysosomal trafficking regulator gene'.

The answer returned by a model trained at 400 batch size is *'Autosomal-recessive complicated spastic paraplegia with a novel lysosomal trafficking regulator'*, and from the one trained at 32 batch size is *'lysosomal trafficking regulator'*.

In further experiments, we have fine tuned the BioBERT model with both 'SQuAD' dataset (version 2.0) and BioASQ train data. For training on 'SQuAD', hyper parameters- Learning rate and number of epochs are set to '3e-3' and 3 respectively as mentioned in the paper [4]. Test accuracy of the model boosted to 44%. In one more experiment we trained model only on 'SQuAD' dataset, this time test accuracy of the model moved to 47%. The reason model did not perform up to the mark when trained with 'SQuAD' alongside BioASQ data could be that in formatted BioASQ data, start_index for the answer is not accurate, and affected the overall accuracy.

4   Our Systems and Their Performance on Factoid Questions

We have experimented with several systems and their variations, e.g. created by training with specific additional features (see next subsection). Here is their list and short descriptions. Unfortunately we did not pay attention to naming, and the systems evolved between test batches, so the overall picture can only be understood by looking at the details.

When we started the experiments our objective was to see whether BioBERT and entailment-based techniques can provide value for in the context of biomedical question answering. The answer to both questions was a yes, qualified by many examples clearly showing the limitations of both methods. Therefore we tried to address some of these limitations using feature engineering with mixed results: some clear errors got corrected and new errors got introduced, without overall improvement but convincing us that in future experiments it might be worth trying feature engineering again especially if more training data were available.

Overall we experimented with several approaches with the following aspects of the systems changing between batches, that is being absent or present:

* training on BioASQ data vs. training on SQuAD
* using the BioASQ snippets for context vs. using the documents from the provided URLs for context
* adding or not the LAT, i.e. lexical answer type, feature (see [3,8] and an explanation in the subsection just below).

For Yes/No questions (only) we experimented with the entailment methods.

We will discuss the performance of these models below and in Sect. 6. But before we do that, let us discuss a feature engineering experiment which eventually produced mixed results, but where we feel it is potentially useful in future experiments.

4.1   LAT Feature Considered and Its Impact (Slightly Negative)

During error analysis we found that for some cases, answer being returned by the model is far away from what it is being asked in the Question.

Example: (from [17])
Question: Hy's law measures failure of which organ?
Actual Answer: 'Liver'.

The answer returned by one of our models was *'alanine aminotransferase'*, which is an enzyme. The model returns an enzyme, when the question asked for the organ name. To address this type of errors, we decided to try the concepts of 'Lexical Answer Type' (LAT) and Focus Word, which was used in IBM Watson, see [6] for overview; [3] for technical details, and [8] for details on question analysis. In an example given in the last source we read:

POETS & POETRY: He was a bank clerk in the Yukon before he published "Songs of a Sourdough" in 1907.
The focus is the part of the question that is a reference to the answer. In the example above, the focus is "he". LATs are terms in the question that indicate what type of entity is being asked for.
(...) In the example, LATs are "he", "clerk", and "poet".

For example in the question *"Which plant does oleuropein originate from?"* ([17]). The LAT here is 'plant'. For the BioASQ task we did not need to explicitly distinguish between the focus and the LAT concepts. In this example, the expectation is that answer returned by the model is a plant. Thus it is conceivable that the cosine distance between contextual embedding of word 'plant' in the question and contextual embedding for the answer present in the paragraph(context) is comparatively low. As a result model learns to adjust its weights during training phase and returns answers with low cosine distance with the LAT.

We used Stanford CoreNLP [11] library to write rules for extracting lexical answer type present in the question, both 'parts of speech'(POS) and dependency parsing functionality was used. We incorporated the Lexical Answer Type into one of our systems, UNCC_QA1 in Batch 4. This system underperformed our system FACTOIDS by about 3% in the MRR measure, but corrected errors such as in the example above.

Assumptions and Rules for Deriving Lexical Answer Type. There are different question types: 'Which', 'What', 'When', 'How' etc. Each type of question is being handled differently and there are commonalities among the rules written for different question types. Question words are identified through parts of speech tags: 'WDT', 'WRB', 'WP'. We assumed that LAT is a 'Noun' and follows the question word. Often it was also a subject (NSUBJ). This process is illustrated in Fig. 2.

LAT computation was governed by a few simple rules, e.g. when a question has multiple words that are 'Subjects' (and 'Noun'), a word that is in proximity to the question word is considered as 'LAT'. These rules are different for each "Wh" word. Perhaps because of using only very simple rules, the accuracy for 'LAT' derivation is 75% that is, in the remaining 25% of the cases LAT word is being identified wrong. And similarly the overall performance the system that used LATs was slightly inferior to the system without LATs, but the types of errors changed. We need to improve our 'LAT' derivation logic, and then perhaps with the neural network techniques they will yield better results.

Overall, the impact of training BioBERT with the LAT feature (as part of the input string) has been slightly negative. However, it works mostly as expected. The errors it introduces usually involve finding the wrong element of the correct type e.g. wrong enzyme when two similar enzymes are described in the text, or 'neuron' when asked about a type of cell with a certain function, when the answer calls for a different cell category, adipocytes, and both are mentioned in the text. We feel with more data and additional tuning or perhaps using an

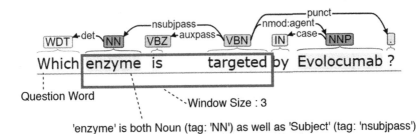

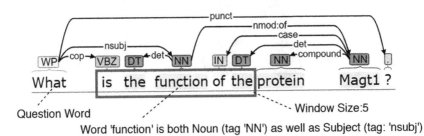

Fig. 2. A simple way of finding the lexical answer types, LATs, of factoid questions: using POS tags to find the question word (e.g. 'which'), and a dependency parse to find the LAT within the window of 3 words. If a noun is not found near the "Wh-" word, we iterate looking for it, as in the second panel.

ensemble model, we might be able to keep the correct answers, and improve the results on the confusing examples like the one mentioned above.

4.2 Impact of Training Using BioASQ Data (Slightly Negative)

Training on BioASQ data in our entry in Batch 1 and Batch 2 under the name QA1 showed it might lead to overfitting. This happened both with (Batch 2) and without (Batch 1) hyperparameters tuning: abysmal 18% MRR in Batch 1, and slightly better one, 40% in Batch 2 (although in Batch 2 it was overall the second best result in MRR but 16% lower than the highest score).

In Batch 3 (only), our UNCC_QA3 system was fine tuned on BioASQ and SQuAD 2.0 [13], and for data preprocessing Context paragraph is generated from relevant snippets provided in the test data. This system underperformed, by about 2% in MRR, our other entry UNCC_QA1, which was also an overall category winner for this batch. The latter was also trained on SQuAD, but not on BioASQ. We suspect that the reason could be the simplistic nature of the *find()* function described in Sect. 3.1. So, this could be an area where a better algorithm for finding the best occurrence of an entity could improve performance.

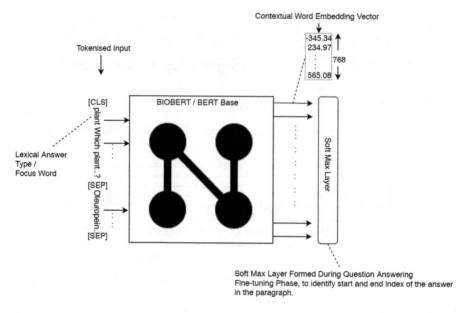

Fig. 3. An example of a using BioBERT with additional features: Contextual word embedding for Lexical Answer Type (LAT) given as feature along with the actual contextual embeddings for the words in question and the paragraph. This change produced mixed results and no overall improvement.

4.3 Impact of Using Context from URLs (Negative)

In some experiments, for context *in testing*, we used documents for which URL pointers are provided in BioASQ. However, our system UNCC_QA3 underperformed our other system tested only on the provided snippets.

In Batch 5 the underperformance was about 6% of MRR, compared to our best system UNCC_QA1, and by 9% to the top performer (Fig. 3).

5 Performance on Yes/No and List Questions

Our work focused on Factoid questions. But we also have done experiments on List-type and Yes/No questions.

5.1 Entailment Improves Yes/No Accuracy

We started by answering always YES (in batch 2 and 3) to get the baseline performance. For batch 4 we used entailment. Our algorithm was very simple: Given a question we iterate through the candidate sentences and try to find any candidate sentence is contradicting the question (with confidence over 50%), if so 'No' is returned as answer, else 'Yes' is returned. In batch 4 this strategy produced better than the BioASQ baseline performance, and compared to our other systems, the use of entailment increased the performance by about 13% (macro F1 score). We used 'AllenNlp' [7] entailment library to find entailment of the candidate sentences with question.

5.2   For List-Type the URLs Have Negative Impact

Overall, we followed the similar strategy that's been followed for Factoid Question Answering task. We started our experiment with batch 2, where we submitted 20 best answers (with context from snippets). Starting with batch 3, we performed post processing: once models generate answer predictions (n-best predictions), we do post-processing on the predicted answers. In test batch 4, our system (called FACTOIDS) achieved highest recall score of '0.7033' but low precision of 0.1119, leaving open the question of how could we have better balanced the two measures.

In the post-processing phase, we take the top 20 (batch 3) and top 5 (batch 4 and 5), predicted answers, tokenize them using common separators: 'comma', 'and', 'also', 'as well as'. Tokens with characters count more than 100 are eliminated and rest of the tokens are added to the list of possible answers. BioASQ evaluation mechanism does not consider snippets with more than 100 characters as a valid answer. Considering lengthy snippets in to the list of answers would reduce the mean precision score. As a final step, duplicate snippets in the answer pool are removed. For example, consider these top 3 answers predicted by the system (before post-processing):

```
{   "text": "dendritic cells",
    "probability": 0.7554540733426441,
    "start_logit": 8.466046333312988,
    "end_logit": 9.536355018615723   },
{   "text": "neutrophils, macrophages and
             distinct subtypes of dendritic cells",
    "probability": 0.13806867348304214,
    "start_logit": 6.766478538513184,
    "end_logit": 9.536355018615723   },
{   "text": "macrophages and distinct subtypes of dendritic",
    "probability": 0.013973475271178242,
    "start_logit": 6.766478538513184,
    "end_logit": 7.24576473236084   },
```

After execution of post-processing heuristics, the list of answers returned is as follows:

```
["dendritic cells"],["neutrophils"],
["macrophages"],["distinct subtypes of dendritic cells"]
```

6 Summary of Our Results

The tables below summarize all our results. They show that the performance of our systems was mixed. The simple architectures and algorithm we used worked very well only in Batch 3. However, we feel we can built a better system based on this experience. In particular we observed both the value of contextual embeddings and of feature engineering (LAT), however we failed to combine them properly (Table 1).

Table 1. Factoid Questions. In Batch 3 we obtained the highest score. Also the relative distance between our best system and the top performing system shrunk between Batch 4 and 5.

System	Strict accuracy	Lenient accuracy	MRR
Batch 1			
QA1	0.1538	0.2308	0.1761
Top Competitor	0.4103	0.5385	0.4637
Batch 2			
QA1	0.36	0.48	0.4033
Top Competitor	0.52	0.64	0.5667
Batch 3			
UNCC_QA1	0.4483	0.5862	**0.5115**
UNCC QA2	0.4138	0.5862	0.4856
UNCC_QA3	0.4138	0.5862	0.4943
Top Competitor	0.36	0.48	0.5023
Batch 4			
FACTOIDS	0.5294	0.7353	0.6103
UNCC QA1	0.4706	0.7353	0.5833
Top Competitor	0.5882	0.8235	0.6912
Batch 5			
UNCC_QA1	0.2857	0.4286	0.3305
UNCC_QA3	0.2286	0.3143	0.2643
QA1	0.2286	0.3714	0.2938
Top Competitor	0.2857	0.5143	0.3638

6.1 Factoid Questions

Systems Used in Batch 5 Experiments

System description for 'UNCC_QA1': The system was finetuned on the SQuAD 2.0. For data preprocessing Context/paragraph was generated from relevant snippets provided in the test data.

System description for 'QA1': 'LAT' feature was added and finetuned with SQuAD 2.0. For data preprocessing Context/paragraph was generated from relevant snippets provided in the test data.

System Description for 'UNCC_QA3': Fine tuning process is same as it is done for the system 'UNCC_QA1' in test batch-5. Difference is during data preprocessing, Context/paragraph is generated from the relevant documents for which URLS are included in the test data.

6.2   List Questions

For List-type questions, although post processing helped in the later batches, we never managed to obtain competitive precision, although our recall was good (Table 2).

Table 2. List questions

System	Mean precision	Recall	F-measure
Batch 2			
QA1	0.0471	0.2898	0.0786
Top Competitor	0.5826	0.4839	0.4732
Batch 3			
UNCC_QA1	0.0780	0.4711	0.1297
Top Competitor	0.4267	0.3058	0.3298
Batch 4			
FACTOIDS	0.1119	0.7033	0.1893
UNCC QA1	0.1087	0.6968	0.1846
UNCC_QA3	0.1087	0.6968	0.1846
Top Competitor	0.4841	0.5051	0.4604
Batch 5			
UNCC_QA1	0.2051	0.5127	0.2862
Top Competitor	0.5653	0.4131	0.4619

6.3   Yes/No Questions

The only thing worth remembering from our performance is that using entailment can have a measurable impact (at least with respect to a weak baseline). The results (weak) are in Table 3.

Table 3. Yes/No questions

System	Accuracy	F1 Yes	F1 No	Macro F1
Batch 1				
QA1	0.7931	0.8846	–	0.4423
Top Competitor	0.8276	0.8980	0.4444	0.6712
Batch 2				
QA1	0.5667	0.7234	–	0.3617
Top Competitor	0.8333	0.8387	0.8276	0.8331
Batch 3				
QA1	0.7826	0.8780	–	0.4390
UNCC_QA3	0.7826	0.8780	–	0.4390
Top Competitor	0.8696	0.9231	0.5714	0.7473
Batch 4				
UNCC_QA1	0.6087	0.7097	0.4000	0.5548
FACTOIDS	0.7391	0.8500	–	0.4250
UNCC_QA3	0.7391	0.8500	–	0.4250
Top Competitor	0.8696	0.9143	0.7273	0.8208
Batch 5				
UNCC QA2	0.5429	0.7037	–	0.3519
Top Competitor	0.8286	0.8500	0.8000	0.8250

7   Discussion, Future Experiments, and Conclusions

Summary. In contrast to 2018, when we submitted [2] to BioASQ a system based on extractive summarization (and scored very high in the ideal answer category), this year we mainly targeted factoid question answering task and focused on experimenting with BioBERT. After these experiments we see the promise of BioBERT in QA tasks, but we also see its limitations. The latter we tried to address with mixed results using feature engineering. Overall these experiments allowed us to secure a best and a second best score in different test batches. Along with Factoid-type question, we also tried 'Yes/No' and 'List'-type questions, and did reasonably well with our very simple approach.

For Yes/No the moral worth remembering is that reasoning has a potential to influence results, as evidenced by our adding the AllenNLP entailment [7] system increased its performance.

All our data and software is available at Github, in the previously referenced URL (end of Sect. 2).

Future Experiments. In the current model, we have a shallow neural network with a softmax layer for predicting answer span. Shallow networks however are not good at generalizations. In our future experiments we would like to create dense question answering neural network with a softmax layer for predicting answer span. The main idea is to get contextual word embedding for the words present in the question and paragraph (Context) and feed the contextual word embeddings retrieved from the last layer of BioBERT to the dense question answering network. The mentioned dense layered question answering neural network need to be tuned for finding right hyper parameters.

In one more experiment, we would like to add a better version of 'LAT' contextual word embedding as a feature, along with the actual contextual word embeddings for question text, and Context and feed them as input to the dense question answering neural network. By this experiment, we would like to find if 'LAT' feature is improving overall answer prediction accuracy. Adding 'LAT' feature this way instead of feeding this word piece embedding directly to the BioBERT (as we did in our above experiments) would not downgrade the quality of contextual word embeddings generated form 'BioBERT'. Quality contextual word embeddings would lead to efficient transfer learning and chances are that it would improve the model's answer prediction accuracy.

We also see potential for incorporating domain specific inference into the task e.g. using the MedNLI dataset [15]. For all types of experiments it might be worth exploring clinical BERT embeddings [1], explicitly incorporating domain knowledge (e.g. [10]) and possibly deeper discourse representations (e.g. [14]).

References

1. Alsentzer, E., et al.: Publicly available clinical BERT embeddings. arXiv preprint arXiv:1904.03323 (2019)
2. Bhandwaldar, A., Zadrozny, W.: UNCC QA: biomedical question answering system. In: Proceedings of the 6th BioASQ Workshop A Challenge on Large-Scale Biomedical Semantic Indexing and Question Answering, pp. 66–71 (2018)
3. Brown, E.W., Ferrucci, D., Lally, A., Zadrozny, W.W.: System and method for providing answers to questions, US Patent 8,275,803, 25 September 2012
4. Devlin, J., Chang, M.W., Lee, K., Toutanova, K.: BERT: pre-training of deep bidirectional transformers for language understanding. In: NAACL-HLT (2018)
5. Dimitriadis, D., Tsoumakas, G.: Word embeddings and external resources for answer processing in biomedical factoid question answering. J. Biomed. Inform. **92**, 103118 (2019)
6. Ferrucci, D.A., et al.: Building Watson: an overview of the DeepQA project. AI Mag. **31**, 59–79 (2010)
7. Gardner, M., et al.: AllenNLP: a deep semantic natural language processing platform. CoRR abs/1803.07640 (2018)
8. Lally, A., et al.: Question analysis: how Watson reads a clue. IBM J. Res. Dev. **56**(3.4), 2:1 (2012)
9. Lee, J., et al.: BioBERT: a pre-trained biomedical language representation model for biomedical text mining. CoRR abs/1901.08746 (2019)
10. Lu, M., Fang, Y., Yan, F., Li, M.: Incorporating domain knowledge into natural language inference on clinical texts. IEEE Access **7**, 57623–57632 (2019)
11. Manning, C.D., Surdeanu, M., Bauer, J., Finkel, J.R., Bethard, S., McClosky, D.: The Stanford CoreNLP natural language processing toolkit. In: ACL (2014)

12. Mikolov, T., Chen, K., Corrado, G., Dean, J.: Efficient estimation of word representations in vector space. arXiv preprint arXiv:1301.3781 (2013)
13. Rajpurkar, P., Zhang, J., Lopyrev, K., Liang, P.S.: SQuAD: 100, 000+ questions for machine comprehension of text. In: EMNLP (2016)
14. Rao, S., Marcu, D., Knight, K., Daumé, H.: Biomedical event extraction using abstract meaning representation. In: BioNLP 2017, pp. 126–135 (2017)
15. Romanov, A., Shivade, C.: Lessons from natural language inference in the clinical domain. arXiv preprint arXiv:1808.06752 (2018)
16. Sharma, V., Kulkarni, N., Pranavi, S., Bayomi, G., Nyberg, E., Mitamura, T.: BioAMA: towards an end to end BioMedical question answering system. In: BioNLP (2018)
17. Tsatsaronis, G., et al.: An overview of the BIOASQ large-scale biomedical semantic indexing and question answering competition. BMC Bioinform. **16**, 138 (2015). https://doi.org/10.1186/s12859-015-0564-6. http://www.biomedcentral.com/content/pdf/s12859-015-0564-6.pdf
18. Wiese, G., Weissenborn, D., Neves, M.L.: Neural question answering at BioASQ 5B. In: Cohen, K.B., Demner-Fushman, D., Ananiadou, S., Tsujii, J. (eds.) BioNLP 2017, Vancouver, Canada, 4 August 2017, pp. 76–79. Association for Computational Linguistics (2017). https://doi.org/10.18653/v1/W17-2309

Transformer Models for Question Answering at BioASQ 2019

Michele Resta$^{(\boxtimes)}$, Daniele Arioli, Alessandro Fagnani, and Giuseppe Attardi

Dipartimento di Informatica, Università di Pisa, Pisa, Italy
{m.resta5,d.arioli,a.fagnani}@studenti.unipi.it, attardi@di.unipi.it

Abstract. We describe our experiments in building a system to tackle task B of the BioASQ 2019 challenge on semantic question answering. We built separate systems to handle the five different types of questions in the dataset. We explored using transformer-based models using both ELMo, BERT and BioBERT. For the *yesno* questions, the results of our submissions using BERT ranked first in batches 3 and 4, while second best in batch 5.

Keywords: Question-answering · Transformer · ELMo

1 Introduction

Along the years the BioASQ challenge has been growing in popularity as well in the difficulty of the tasks to perform. The three tasks of the BioASQ 2019 challenge [7] concern biomedical semantic indexing and question answering.

Task *B* on Biomedical Semantic Question Answering requires creating an automated system capable of responding to a set of biomedical questions with relevant concepts, articles, snippets, and RDF triples, from designated resources, as well as exact and 'ideal' answers. Questions are divided into various types according to the expected answer (yes/no, a single fact, a list of entities, or a summary). Systems often exploit different strategies to address each type of questions.

2 Dataset

The BioASQ training dataset for Task 7b consists of 2747 questions in the biology and medical domain. All questions were constructed by biomedical experts from around Europe. Each dataset item consists of several fields, the most relevant of which are:

- *type:* type of the question;
- *exact_answer:* exact answer (absent in summary questions);
- *ideal_answer:* an answer summarizing the most relevant information;

© Springer Nature Switzerland AG 2020
P. Cellier and K. Driessens (Eds.): ECML PKDD 2019 Workshops, CCIS 1168, pp. 711–726, 2020.
https://doi.org/10.1007/978-3-030-43887-6_63

- *documents:* PubMed articles relevant to the question. The supplied answer snippets are extracted from these documents.

Questions are classified into the following types:

- *yes-no*: systems must provide a "yes" or "no" answer;
- *list*: systems must provide a list of entity names (e.g. a list of gene names);
- *factoid*: similar to list answers, these require as an answer a single entity name (e.g. a disease, drug, or gene), a number, or a similar short expression. Differently from list questions though there are lesser entities in the answer: often, only a single entity or up to 3–4 in a few cases;
- *summary*: these questions must be answered by producing a short text summarizing the most relevant information.

Besides answers of the types described above, submissions may provide an *ideal answer*: a single text paragraph that best summarizes the most relevant information answering the question.

For each question, a list of *snippets* is provided. Snippets are short texts that are expected to contain the information needed to answer the corresponding question. However, some of them may not be useful for extracting the answer.

Table 1 shows in detail the composition of the training dataset. Most of the questions have 1 to 15 associated snippets while only a few of them have a larger number of snippets (>40).

Table 1. Dataset composition

Question type	Number	Dataset %
List	556	20,2
Summary	667	24,2
Yesno	745	27,1
Factoid	779	28,3
Total questions	2747	100

Yesno Questions. There is a great imbalance in the answers to these questions: 82% of the answers are *"yes"* and only 18% have a *"no"* answer.

List Questions. List answers contain between 1 and 38 entities, but over 80% have less than 10 elements.

Factoid Questions. The answer to a factoid question might be phrased differently, therefore for some questions, there may be multiple answers with different wording. This is important since it may affect the scoring, which is based on MRR (Mean Reciprocal Rank).

Summary Questions. For these types of questions no *exact answer* is requested; but the system is expected to provide just a kind of *ideal answer*.

3   Models

Since the challenge expects different types of answers according to the type of questions, we developed specialized systems for each of them. We explored several solutions, some of which had to be discarded because they did not produce good results. Simpler approaches were tried first, such as Sentiment Analysis or information retrieval based on a bag-of-words model, and more complex ones were attempted as soon as the simpler ones showed unsatisfactory results.

In the following sections, a detailed description of the models employed for the competition is presented. The discussion is organized into sections corresponding to the type of questions to handle.

3.1   Models for YesNo Questions

In order to answer this type of questions, we exploited different versions of word embeddings. We explored embeddings created from the BioASQ dataset itself using *word2vec* [13]. Given the small size of the dataset however this type of embeddings showed poor results.

Embeddings of ELMo, BERT, and ELMo-Pubmed (described below) have been extracted using Flair [3] since it offers the same pre-trained weights and simple programming interfaces.

ELMo. Elmo [15] is a deep contextualized word representation that models both complex characteristics of word use (e.g., syntax and semantics), and how these uses vary across linguistic contexts (i.e., to model polysemy). These word vectors are learned functions of the internal states of a deep bidirectional language model (biLM), which is pre-trained on a large text corpus.

This model consists of two layers of 4096 LSTM units. Of these units, half handle the forward language modeling, and the remaining the backward language modeling.

We exploited a pre-trained model on the "One billion word Benchmark" [5]. Figure 1 shows the network architecture.

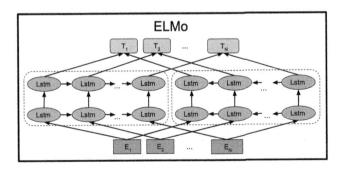

Fig. 1. Elmo architecture

ELMo-Pubmed. This model is identical to the one described in Sect. 3.1, except that it is trained on PubMed abstracts.

BERT. BERT [6] stands for Bidirectional Encoder Representations from Transformers. The *BERT Large* model consists of a stack of 24 Transformer encoders. Each encoder incorporates an attention mechanism to help the model focus on the most relevant parts of the input.

In this model, the attention mechanism is a multi-headed one with 16 attention heads.

The model was trained on the entire Wikipedia and BookCorpus [18] for a total of 1 million update steps.

BERT architecture is showed in Fig. 2.

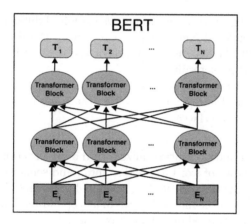

Fig. 2. Bert architecture

BioBERT. The BioBERT model [11] has the same architecture as Sect. 3.1 but it was trained on PubMed abstracts.

After evaluating the performance of this model on the validation set we decided not to use it for the yes-no questions.

Data Pre-processing. Before feeding the models with data, the following preprocessing steps were performed on all questions and snippets:

- removal of spurious newlines,
- tokenization.

Each of the pre-trained models handles tokenization differently. All models are capable of handling case sensitive text, so there was no need to lowercase inputs. The inputs were normalized by each model in its own way, and the resulting embeddings are of fixed size.

To extract the embeddings from the BioBERT model we prepared the input concatenating the question, a separator token, and its snippets. However, since the max token length for BioBERT is 512, we built multiple inputs by pairing a question with each sentence from the snippets. Scispacy [14] was used to perform sentence splitting on the snippets.

Classifier Inputs. Given a dataset consisting of questions $\{Q_n, n < N\}$, their corresponding answer snippets $\{S_{n,k}, k < K\}$ and outputs $\{y_n, n < N\}$, a training set is created consisting of inputs $\{x_i = \langle Q_n, S_{n,k} \rangle, n < N, k < K\}$ and outputs $\{y_i\}$, where i is the index n of the Q_n corresponding to x_i. In other words we create pairs of each question with all its related snippets and assume that they would all have the same answer. This is realistic, since all snippets are assumed to have been chosen in the previous stage of question answering as candidates for containing the answer.

Questions and snippets are transformed into a vector representation by using a language model as follows.

For all models except the one based on BioBERT, for each input pair $x_i = < Q_i, S_i >$ we pass separately through the language model the sequence of *tokens* from the question Q_i and those from the snippet S_i.

By means of functions from the Flair library, we extract the embeddings from all layers of the language model for each token of the input sequence. The vectors for the question tokens are added together and similarly those from the snippet, obtaining fixed length vectors irrespective of the length of the sentences. A *mean*-pooling is applied to the question vectors and snippet vectors so produced for each layer, to obtain the final vectors representing the i-th question, $Emb(Q_i)$, and the i-th snippet, $Emb(S_i)$. These two vectors are concatenated to obtain vector

$$[Emb(Q_i); Emb(S_j)]$$

to be used as input to the classifier described below. We tested also using different types of pooling (min and max), but they led to poorer results with respect to $mean$-pooling. Embeddings vectors computed through BioBERT did not provide improvements with respect to the previous models, therefore we did not employ them in our submissions for the yes-no questions. Even though BioBERT is trained on the biomedical domain, the fact that has fewer parameters than $BERT_{LARGE}$ is probably the cause of performance improvements lack.

Classifiers. Since the rules of the BioASQ challenge allow participants to send up to 5 submissions, we trained 5 different classifiers with different embeddings and different architectures.

Before training, the dataset yes-no ratio was re-balanced as shown in the column *Yes-No* in Table 3 to avoid bias towards "yes" answers.

The classifier consists of fully connected feed-forward networks with the architecture described in Fig. 3.

The classifier was implemented using Keras [4] on a TensorFlow [2] backend.

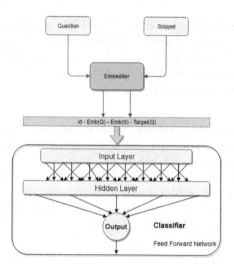

Fig. 3. Yes-No model with feed-forward network. The activation functions are *tanh* for the input and hidden layers and the *sigmoid* for the output layer. The vector dimensions are shown in Table 2.

The hyperparameters for each model were tuned by performing grid searches, in order to find the best combination of:

- number of hidden layers → [1 . . . 5]
- size of vectors in the first layer → [50 . . . 120]
- size of vectors in the hidden layers → [50 . . . 150]
- type of optimizer → [$Sgd, RmsProp, Adam$]
- activation functions for the hidden layers → [$tanh, ReLU$]
- activation function for the input layer → [$tanh, ReLU$]

We used 80% of the training set as development set and the remaining 20% as the validation set. We also performed a *4-fold cross-validation* on the development set in order to select the best models for each task.

We then trained each classifier on the development dataset and tested it on the validation set, to estimate the ability of the model to generalize to unseen data.

The results of the grid searches are summarized in Table 2, while Table 3 shows the scores on yes-no questions on the validation and test datasets.

Note. By inspecting the answers of the models and the training curves as well, we noticed an increase in the validation score respect to training score. Since in the 4-fold CV all the models had less data, more epochs were needed to fit. During training we noticed a classifier tendency to overfit. To limit this problem, we implemented an early stopping technique with patience equal to 4.

Table 2. Grid search results.

System	QA-1	QA-2	QA-3	QA-4	QA-5
Embedding	Elmo-pubmed	BERT_$LARGE$			
Pooling	Mean				
Num h. layers	1				
Neurons h. layer	120	120	120	120	80
Neurons 1-st layer	90	90	90	90	50
Act. hidden	tanh				
Act. input	tanh				
Optimizer	RMSProp				
Loss	Binary-crossentropy				

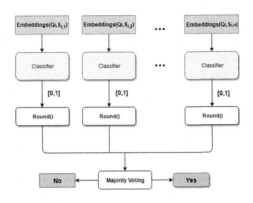

Fig. 4. Ensemble of classifiers for the Yes/No Model. The result is obtained by majority voting

Ensemble Classifier. A ensemble of K classifiers is used to classify each pair of question/snippet $< Q_i, S_i >$ for the same Q_i and the answer is obtained through a majority vote among these classifiers. More precisely, the ensemble outputs:

– *model QA-1*: "yes" if $(votes \geq floor(K/2))$
– *other models*: "yes" if $(votes > floor(K/2))$

The last system, QA-5, was trained on an augmented dataset constructed by manually annotating the yes-no datasets of the previous editions of the BioASQ challenge. Since previous models appeared biased towards giving positive answers, we chose to add to the training set just the questions with negative answers.

3.2 Models for List Questions

To answer this type of questions we explored two approaches: one based on frequency and *tf-idf* and one based on the analysis of the dependency trees of questions and snippets.

Table 3. Scores before retraining.

System	Yes-No%	Epochs	4-fold VL score	Held-out TS score
QA-1	50-50	40	0.9861	0.9612
QA-2	50-50	40	0.9861	0.9564
QA-3	50-50	50	0.9840	0.9668
QA-4	50-50	50	0.8801	0.9429
QA-5	60-40	50	0.8201	0.8455

The processing pipeline to produce the list of entities for the answer consists in the following steps:

– *data pre-processing*: stop words such as "list", punctuation and parentheses were removed from the input texts;
– *entity extraction*: entities from both questions and snippets were extracted using scispacy [14] and then converted to lowercase;
– *ranking*: the extracted entities were scored with the metrics discussed below;
– *filtering*: unrelated entities were dropped;
– *rank-boosting:* entities that were more likely to be an answer received an increase in their score;
– *list-trimming*: entities with a score below a given threshold are discarded.

Entity Extraction. We explored two approaches for entity extraction. The first one is the simplest: the scispaCy mention detector was used to extract entities from questions and snippets. We used the en_core_sci_md model available from the scispaCy website.[1]

This model was trained on a collection of biomedical data, recognizes a wide variety of entity types, has a vocabulary of 101,678 tokens and includes 98,131 word vectors.

Entities that are already present in the question were dropped, assuming that a question will typically look for answers not already known and hence terms appearing in the question are unlikely to be part of an answer.

The second approach is based on the analysis of parse trees. First we determine the *main entity* in the question, that is the entity required as answer to the question. For example, consider the following question: "*Which miRNAs could be used as potential biomarkers for epithelial ovarian cancer?*". The main entity is *miRNAs*, since a list of entities of this type is the required answer.

The next step splits the snippets into sentences and then it parses them with scispaCy. From the parse trees, we extract the entities that:

– are child nodes of an entity similar to the main entity;
– are child nodes of a verb that is the same of the main one;

---

[1] https://allenai.github.io/scispacy/.

- are child nodes of an entity similar to the father of the main entity in the question's parse tree;
- are child nodes of an entity similar to the verb of the main entity's father in the question's parse tree.

The similarity between two entities is based on string comparisons. An approach with distances between vector representation has been taken into account but it was not possible to finish it before the deadline of the last batch.

All the resulting entities are merged with those obtained with the first approach described at the beginning of this section. The whole list is sent to the ranking phase.

Ranking. In this phase all elements of the entity list are ranked based on the frequency of occurrence in the snippet list. So e.g. the entity *genes* appear three times in question's snippets, then $Score(genes) = 3$.

We investigated also a different type of scoring scheme based on *tf-idf*. In this case the collection of documents was composed of all snippets in the training set, and the terms to be scored are the elements of the list created in the previous steps.

The entities list is then sorted according to the scores.

Filtering. Filtering of entities is based mainly on Part of Speech tags from scispacy. If the considered entity is composed of a single word, and this word is a verb or an adjective we delete this element from the list. The same is done if the entity is composed of 2 words and POS are verbs and adposition. In this way we discard entity like *"associated with"*.

Score Boosting. In this phase the score of each entity in the resulting list is increased according to the context of the question.

First, there is an ontology checking, where the entities are boosted if they appear in a local database of the biomedical domain (the choice of the database depends on the main entity of the question). Available databases are: *bacteria* [1], *viruses* [1], *genes* [9,10], *drugs* [8], *proteins, human symptoms, human organs.*

We check if the main entity concern one of the domains above. If it happens, all the entities of the list are searched in the databases and, if an entity is found, its score is increased. This process increases the probability of finding the entities required for the answer with greater rank in respect to those that are unrelated.

In the presented code the score of an entity is increased by the mean of all the ranks $>= 1$.[2]

After the ontology checking, **tf-idf** boosting is performed. Thanks to the tfidf() function we can extract the value of the entities based on an analysis of all the snippets in the dataset. With some analysis and test, we found that

---

[2] This is only an empirical choice due to the short time of development, but it could easily changed with more in-depth work.

entities which have a value between 3 and 4 have more chances of being a good answer to the question.

Based on these considerations, we boosted another time the scores of these entities. In the presented code this function doubles the entity's rank. This choice is empirical and could be refined through a more accurate and in-depth analysis of the problem.

Normalization. Before the trimming phase, scores of all the entities are normalized in the [0,1] range with the formula:

$$score' = \frac{score - min\_score}{max\_score - min\_score}$$

Where $score$ is the actual score of the entity and max and min score are the maximum and minimum score of entities in the list.

List Trimming. The last phase of the process is based on a threshold. All entities that have a score below this threshold are removed. Anyway, if the list exceeds the 100 elements then only the first 100 are returned as the answer.

The threshold has been determined experimentally by running the entire pipeline with different thresholds and evaluating the final Mean F1 Score, since it is the official metrics for these types of questions.

In the current implementation, the threshold is fixed at 0.15.

3.3   Model for Factoid Questions

The approach used for this kind of answer was to exploit the very well known BERT [6] and how it is used for the challenge SQuAD [17]. As a matter of fact, SQuAD is, to some extent, a factoid task.

We employed the pre-trained BioBERT [11] model and the technique of *fine-tuning*, in order to better fit the results to our dataset.

Pre-processing and Augmentation. We needed to build the training data in the form of a SQuAD task. To do so, it was necessary to create a unique context, formed by the concatenation of the snippets. The exact answer had to be inside of this context. Indeed, SQuAD models need also the index of the answer inside the context, found by *lower-casing* both the answer and the context. At the end of the process, a JSON file is created, where each question has this format, e.g.:

```
{"qas": [{
    "question": "Which is the target protein
                of the drug nivolumab?",
    "id": "56af9f130a360a5e45000015",
    "answers": [{"text": "plasma membrane",
                "answer_start": 827}],
    "is_impossible": "false"}],
"context": "..."
}
```

The field *is_impossible* is used to define a test/validation question (when is "false", means that there is no exact answer inside the context).

As already mentioned in the description of the dataset, some factoid questions have more than one answer, to avoid misinterpretation. Each of them has been cloned a number of times equal to the provided exact answers, but with a different id.

For the final training, the dataset was augmented with the second and the fifth version of the BioASQ challenge data, available thanks to [11] and the BioASQ team itself. Unfortunately, the two datasets were the only others available, so there was no possibility to integrate more training sets of past editions.

Post-processing and Evaluation. Up to 5 possible answers are allowed in a submission and the evaluation metrics for the factoid task is the MRR (Mean Reciprocal Rank). This means that the system should aim to produce exact answers with a confident estimate since an answer at a lower ranking position is highly penalized.

By analyzing the answers returned by the model on the validation set, we decided to introduce two post-processing steps:

- *Parentheses' fix*: correcting or removing a sentence that has unbalanced open or closed parenthesis.
- *Dashes' fix*: if there is an answer with a dashed word in the first position (e.g. "S-adenosyl-L-methionine"), a variant without dashes is added in the last ranking position (in order to minimize the misinterpretation of the answer given by the system).

Due to the complexity of the model, the evaluation of the fine-tuning was done only by varying the number of epochs (and only with a very small batch size). A validation set of 127 answers was created from the given training set, in order to validate the model.

3.4   Model for Summary and Ideal Questions

Since there are no exact answers in the summary questions, they have been treated with the ideal answers of the entire dataset. Similarly to the factoid system, BERT, BioBert and SQuAD approach have been used also for the generation of the ideal answers.

Nevertheless, the answer does not have to appear (always) inside the context, but it has to be a completely "original" summary of the paragraph, related to the question. OpenAI's GPT architecture [16] is promising, but the released model has fewer parameters than BERT and its later more sophisticated version, GPT-2, has not been released by its authors.

Our idea was to keep exploiting BERT model, in order to find the most attentive part of the context.

Pre-processing and Augmentation. In order to select as answer the most relevant part of the text and at the same time follow the structure of the SQuAD approach, we exploit the same evaluation metric used in the BioASQ ideal answer's task, ROUGE [12]. In fact, the `answer_start` index is found by calculating the ROUGE score between the returned answer and all the sentences of the snippets. The same idea was also used in the testing/validating phase. In addition, the dataset of the second and the fifth version of the BioASQ challenge were also added to the training, as we did for the factoid question model.

Post-processing and Evaluation. The post-processing phase aims mainly at finding the proper sentence within the text of the full answer returned by the model, and ROUGE metric has been used between all the sentences. Our model achieved a ROUGE-2 score of 0.4137 on the validation set. We tried increasing the number of epochs, but the score did not improve.

We tried also to build the answer starting from the factoid model but the results were worse (ROUGE-2 score of 0.0345).

4   Results

The competition benchmarks were split into 5 batches. We participated in batch 3, 4 and 5 shown in Tables 4, 5 and 6 respectively. Below results are reported for each of the submissions and the ranks obtained against other participating systems.

Table 4. Batch 3 results. Best result indicates the best scoring system across all participant systems. The score is the average of each question type. Highest scores are in bold.

Type	Metric	System					Best result
		QA-1	QA-2	QA-3	QA-4	QA-5	BioBERT-DMIS
Yes/No	Accuracy	**0.8261**	**0.8696**	–	–	–	0.6087
	F1 Yes	**0.9000**	**0.9231**	–	–	–	0.7429
	F1 No	**0.3333**	**0.5714**	–	–	–	0.1818
	Macro-F1	**0.6167**	**0.7473**	–	–	–	0.4623
List	Mean Prec.	–	–	–	–	–	0.4267
	Recall	–	–	–	–	–	0.3058
	F-Measure	–	–	–	–	–	0.3298
Factoid	Strict acc.	–	–	–	–	–	0.6207
	Lenient acc.	–	–	–	–	–	0.4724
	MRR	–	–	–	–	–	0.4267
Average		0.2055	0.2491	–	–	–	**0.4215**

Table 5. Batch 4 results. Best result indicates the best scoring system across all participant systems. The score is the average of each question type. Highest scores are in bold.

Type	Metric	System					Best Result
		QA-1	QA-2	QA-3	QA-4	QA-5	BioBERT-DMIS
Yes/No	Accuracy	0.7391	0.7391	**0.8261**	**0.8696**	-	0.7391
	F1 Yes	**0.8421**	**0.8421**	**0.8889**	**0.9143**	–	0.8125
	F1 No	0.2500	0.2500	**0.6000**	**0.7273**	–	0.5714
	Macro-F1	0.5461	0.5461	**0.7444**	**0.8208**	–	0.6920
List	Mean Prec.	0.1442	0.1442	0.1442	0.1442	–	**0.4841**
	Recall	**0.6268**	**0.6268**	**0.6268**	**0.6268**	–	0.5051
	F-Measure	0.2163	0.2163	0.2163	0.2163	–	**0.4604**
Factoid	Strict acc.	0.2059	0.2059	0.2059	0.2059	–	**0.882**
	Lenient acc.	0.3824	0.3824	0.3824	0.3824	–	**0.8235**
	MRR	0.2730	0.2730	0.2730	0.2730	–	**0.6912**
Average		0.3451	0.3451	0.4112	0.4367	–	**0.6145**

4.1 Ideal Answers

Table 7 summarizes the results obtained by our systems in the ideal answer type of questions. The Manual Scores were not yet published at the time of this writing.

Table 6. Batch 5 results. Best result indicates the best scoring system across all participant systems. The score is the average of each question type. Highest scores are in bold.

Type	Metric	System					Best result
		QA-1	QA-2	QA-3	QA-4	QA-5	BioBERT-DMIS-3
Yes/No	Accuracy	0.5429	0.5429	0.6857	0.7143	0.8000	**0.8286**
	F1 Yes	0.6800	0.6800	0.7556	0.7727	0.8293	**0.8500**
	F1 No	0.2000	0.2000	0.5600	0.6154	0.7586	**0.8000**
	Macro-F1	0.4400	0.4400	0.6578	0.6941	0.7939	**0.8250**
List	Mean Prec.	0.1713	0.1713	0.1713	0.1713	0.1713	**0.5653**
	Recall	**0.5873**	**0.5873**	**0.5873**	**0.5873**	**0.5873**	0.4131
	F-Measure	0.2537	0.2537	0.2537	0.2537	0.2537	**0.4619**
Factoid	Strict acc.	0.0857	0.0857	0.0857	0.0857	0.0857	**0.2857**
	Lenient acc.	0.1714	0.1714	0.1714	0.1714	0.1714	**0.4286**
	MRR	0.1152	0.1152	0.1152	0.1152	0.1152	**0.3452**
Average		0.2696	0.2696	0.3422	0.3543	0.3876	**0.5440**

Table 7. Rouge scores on Ideal Answers. Best result indicates the best scoring system.

Batches	Automatic Scores		Manual Scores				Best Result (MQ-4)	
	Rouge-2	Rouge-SU4	Readability	Recall	Precision	Repetition	Rouge-2	Rouge-SU4
Batch 4	0.3511	0.3638	–	–	–	–	**0.5177**	**0.5246**
Batch 5	0.4265	0.4275	–	–	–	–	**0.5035**	**0.5070**

5   Conclusions

Our participation in task B of the BioASQ 2019 competition focused mainly on answering the *yesno* questions.

We started developing our system just 5 days before the scheduled release of the first test set. Despite the time constraints, our yes-no answering systems achieved top scores ones in Batch 3 and Batch 4 and ranked second in Batch 5. We exploited an ensemble of classifiers, whose input was obtained by processing questions and snippets though transformer-based language models. Our approach was incremental, we tried to exploit simpler word embeddings first, and as soon as they showed limitations due to task complexity, we moved to more powerful embeddings. Contextual embeddings obtained from deep models like ELMo and transformer-based language model (BERT) provided better results thanks to the built in attention mechanisms and to their ability to capture long term dependencies. These deep models are the core of our submitted systems, BERT specifically is employed for three out of four question types. Questions and snippets were processed separately. Concatenating them before processing might allow exploiting better the attention mechanism.

Submissions for factoid and list achieved lower scores. The latter type of question were more challenging given the presence of duplicate entities, lexical variation among entities, and the number of entities to be returned as answer. We noticed by inspecting a number of question that sci-spacy was able to extract almost all the required answer items, among *noisy* entities. Based on this empirical observation, we decided to try the described approach instead of a more computationally demanding fine-tuning of BERT. We had several ideas under development for improving these answers that we could not complete in time for the submission deadline: using tf-idf as main ranking metrics, removing duplicated entities using a clustering algorithm on their vector representation, improved entity extraction from parse trees.

The overall ranking of our submissions on all question types was fourth in Batch 4 and third in Batch 5.

The approach to semantic question answering using classifiers on top of transformer-based language models has proved quite effective and there are still margins for improvements.

Acknowledgments. The experiments were run on a server equipped with 4 Nvidia P100 GPUs, partly funded by the University of Pisa under grant "Grandi Attrezzature 2016".

References

1. GBIF—The Global Biodiversity Information Facility (2019). https://www.gbif.org/
2. Abadi, M., et al.: TensorFlow: large-scale machine learning on heterogeneous systems (2015). Software available from http://tensorflow.org/
3. Akbik, A., Blythe, D., Vollgraf, R.: Contextual string embeddings for sequence labeling. In: 27th International Conference on Computational Linguistics, COLING 2018, pp. 1638–1649 (2018)
4. Chollet, F., et al.: Keras (2015). https://github.com/fchollet/keras
5. Chelba, C., et al.: One billion word benchmark for measuring progress in statistical language modeling (2014)
6. Devlin, J., Chang, M., Lee, K., Toutanova, K.: BERT: pre-training of deep bidirectional transformers for language understanding. CoRR abs/1810.04805 (2018). http://arxiv.org/abs/1810.04805
7. Balikas, G., Krithara, A., Partalas, I., Paliouras, G.: BioASQ: a challenge on large-scale biomedical semantic indexing and question answering (2015)
8. Wishart Research Group. https://www.drugbank.ca/
9. HGNC: HUGO Gene Nomenclature Committee at the European Bioinformatics Institute. http://www.genenames.org/
10. MacArthur Lab: Lists of gene lists. https://github.com/macarthur-lab/gene_lists
11. Lee, J., et al.: BioBERT: a pre-trained biomedical language representation model for biomedical text mining. arXiv preprint arXiv:1901.08746 (2019)
12. Lin, C.Y.: ROUGE: a package for automatic evaluation of summaries. In: Text Summarization Branches Out (2004)

13. Mikolov, T., Sutskever, I., Chen, K., Corrado, G., Dean, J.: Distributed representations of words and phrases and their compositionality. CoRR abs/1310.4546 (2013)
14. Neumann, M., King, D., Beltagy, I., Ammar, W.: ScispaCy: fast and robust models for biomedical natural language processing. CoRR abs/1902.07669 (2019). http://arxiv.org/abs/1902.07669
15. Peters, M.E., et al.: Deep contextualized word representations. In: Proceedings of NAACL (2018)
16. Radford, A., Wu, J., Child, R., Luan, D., Amodei, D., Sutskever, I.: Language models are unsupervised multitask learners. OpenAI Blog **1**, 8 (2019)
17. Rajpurkar, P., Jia, R., Liang, P.: Know what you don't know: unanswerable questions for squad. CoRR abs/1806.03822 (2018). http://arxiv.org/abs/1806.03822
18. Zhu, Y., et al.: Aligning books and movies: towards story-like visual explanations by watching movies and reading books. arXiv preprint arXiv:1506.06724 (2015)

Pre-trained Language Model
for Biomedical Question Answering

Wonjin Yoon⬤, Jinhyuk Lee⬤, Donghyeon Kim⬤, Minbyul Jeong⬤,
and Jaewoo Kang$^{(\boxtimes)}$⬤

Korea University, Seoul, Korea
{wjyoon,jinhyuk_lee,donghyeon,minbyuljeong,kangj}@korea.ac.kr

Abstract. The recent success of question answering systems is largely
attributed to pre-trained language models. However, as language models
are mostly pre-trained on general domain corpora such as Wikipedia,
they often have difficulty in understanding biomedical questions. In
this paper, we investigate the performance of BioBERT, a pre-trained
biomedical language model, in answering biomedical questions including factoid, list, and yes/no type questions. BioBERT uses almost the
same structure across various question types and achieved the best performance in the 7th BioASQ Challenge (Task 7b, Phase B). BioBERT
pre-trained on SQuAD or SQuAD 2.0 easily outperformed previous state-
of-the-art models. BioBERT obtains the best performance when it uses
the appropriate pre-/post-processing strategies for questions, passages,
and answers.

Keywords: Biomedical question answering · Pre-trained language
model · Transfer learning

1 Introduction

Language models pre-trained on large-scale text corpora achieve state-of-the-art
performance in various natural language processing (NLP) tasks when fine-tuned
on a given task [4,13,15]. Language models have been shown to be highly effective
in question answering (QA), and many current state-of-the-art QA models often
rely on pre-trained language models [20]. However, as language models are mostly
pre-trained on general domain corpora, they cannot be generalized to biomedical
corpora [1,2,8,29]. Hence, similar to using Word2Vec for the biomedical domain
[14], a language model pre-trained on biomedical corpora is needed for building
effective biomedical QA models.

Recently, Lee et al. [8] have proposed BioBERT which is a pre-trained language model trained on PubMed articles. In three representative biomedical
NLP (bioNLP) tasks including biomedical named entity recognition, relation
extraction, and question answering, BioBERT outperforms most of the previous state-of-the-art models. In previous works, models were used for a specific
bioNLP task [9,18,24,28]. However, the structure of BioBERT allows a single

P. Cellier and K. Driessens (Eds.): ECML PKDD 2019 Workshops, CCIS 1168, pp. 727–740, 2020.
https://doi.org/10.1007/978-3-030-43887-6_64

model to be trained on different datasets and used for various tasks with slight modifications in the last layer.

In this paper, we investigate the effectiveness of BioBERT in biomedical question answering and report our results from the 7th BioASQ Challenge [7,10, 11,21]. Biomedical question answering has its own unique challenges. First, the size of datasets is often very small (e.g., few thousands of samples in BioASQ) as the creation of biomedical question answering datasets is very expensive. Second, there are various types of questions including factoid, list, and yes/no questions, which increase the complexity of the problem.

We leverage BioBERT to address these issues. To mitigate the small size of datasets, we first fine-tune BioBERT on other large-scale extractive question answering datasets, and then fine-tune it on BioASQ datasets. More specifically, we train BioBERT on SQuAD [17] and SQuAD 2.0 [16] for transfer learning. Also, we modify the last layer of BioBERT so that it can be trained/tested on three different types of BioASQ questions. This significantly reduces the cost of using biomedical question answering systems as the structure of BioBERT does not need to be modified based on the type of question.

The contributions of our paper are three fold: (1) We show that BioBERT pre-trained on general domain question answering corpora such as SQuAD largely improves the performance of biomedical question answering models. Wiese et al. [25] showed that pre-training on SQuAD helps improve performance. We test the performance of BioBERT pre-trained on both SQuAD and SQuAD 2.0. (2) With only simple modifications, BioBERT can be used for various biomedical question types including factoid, list, and yes/no questions. BioBERT achieves the overall best performance on all five test batches of BioASQ 7b Phase B[1], and achieves state-of-the-art performance in BioASQ 6b Phase B. (3) We further analyze the role of pre- and post-processing in our system and show that different strategies often lead to different results.

The rest of our paper is organized as follows. First, we introduce our system based on BioBERT. We describe task-specific layers of our system and various pre- and post-processing strategies. We present the results of BioBERT on BioASQ 7b (Phase B), which were obtained using two different transfer learning strategies, and we further test BioBERT on BioASQ 6b on which our system was trained.

2   Methods

In this section, we will briefly discuss BioBERT[2] [8] and our modifications[3] for the BioASQ Challenge (Fig. 1).

---

[1] http://participants-area.bioasq.org/results/7b/phaseB/.

[2] The source code for BioBERT is available at https://github.com/dmis-lab/biobert.

[3] The source code and pre-processed datasets are available at https://github.com/dmis-lab/bioasq-biobert.

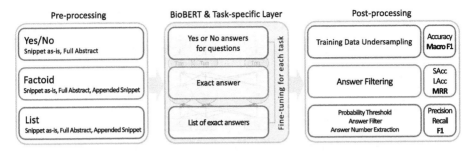

Fig. 1. Overview of our system

2.1   BioBERT

Word embeddings are crucial for various text mining systems since they represent semantic and syntactic features of words [14,22]. While traditional models use context-independent word embeddings, recently proposed models use contextualized word representations [4,13,15]. Among them, BERT [4], which is built upon multi-layer bidirectional Transformers [23], achieved new state-of-the-art results on various NLP tasks including question answering. BioBERT [8] is the first domain-specific BERT based model pre-trained on PubMed abstracts and full texts. BioBERT outperforms BERT and other state-of-the-art models in bioNLP tasks such as biomedical named entity recognition, relation extraction, and question answering [6,19].

An input representation of BioBERT for a given token is composed of the corresponding token, segment, and position embeddings. BioBERT utilizes Word-Piece embeddings [26] which use sub-word units to address the out-of-vocabulary (OOV) problem. Broken sub-word units are denoted by ## (e.g. organoid = organ + ##iod). Positional embeddings are learned during training and segment embeddings are used to mark the location of question and passage tokens in the input sequence. Following the design of BERT, a special token embedding for [CLS] was added to the beginning of every sequence to process yes/no type questions.

2.2   Task-Specific Layer

The BioBERT model for QA is illustrated in Fig. 2. Following the approach of BioBERT [8], a question and its corresponding passage are concatenated to form a single sequence which is marked by different segment embeddings. The task-specific layer for factoid type questions and the layer for list type questions both utilize the output of the passage whereas the layer for yes/no type questions uses the output of the first [CLS] token.

Factoid and List Questions. In (Bio)BERT, the only additional trainable parameters needed for factoid and list type questions are the softmax layer for a

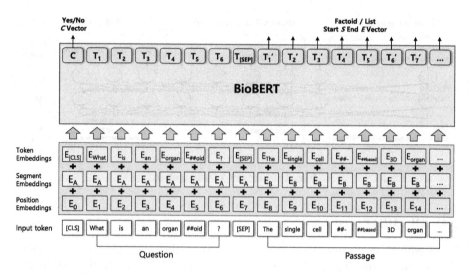

Fig. 2. Example of a single sequence (Question-Passage pair) processed by the BioBERT.

linear transformation of hidden vectors from BioBERT. Following the notation used in the BERT study, we denote the trainable start vector as $S \in \mathbb{R}^H$ and the trainable end vector as $E \in \mathbb{R}^H$ where H denotes the hidden size of BioBERT. The probabilities of the i-th token being the start of the answer token and the j-th token being the end of the answer token can be calculated by the following equations:

$$P_i^{start} = \frac{e^{S \cdot T_i}}{\sum_k e^{S \cdot T_k}}, \quad P_j^{end} = \frac{e^{E \cdot T_j}}{\sum_k e^{E \cdot T_k}}$$

where $T_l \in \mathbb{R}^H$ denotes l-th token representation from BioBERT and $\cdot$ denotes the dot product between two vectors.

Yes/No Questions. We use the first [CLS] for the classification of yes/no questions. Here, we denote the representation of the [CLS] token from BioBERT as $C \in \mathbb{R}^H$. The parameter learned during training is a sigmoid layer consisting of $W \in \mathbb{R}^H$ which is used for binary classification. The probability for the sequence to be "yes" is calculated using the following equation.

$$P_{yes} = \frac{1}{1 + e^{-CW}}$$

Loss. For the factoid/list question layer, we minimize $Loss$ during training, which is defined below. $Loss$ is the arithmetic mean of the $Loss_{start}$ and $Loss_{end}$, which correspond to the negative log-likelihood for the correct start and end positions, respectively. The ground truth start/end positions are denoted as y_s

for the start token, and y_e for the end token. The losses are defined as follows:

$$Loss_{start} = -\frac{1}{N}\sum_{k=1}^{N} \log P_{y_s}^{start,k}, \quad Loss_{end} = -\frac{1}{N}\sum_{k=1}^{N} \log P_{y_e}^{end,k}$$

$$Loss = (Loss_{Start} + Loss_{End})/2$$

where k iterates for a mini-batch of size N.

For yes/no questions, the binary cross entropy between probability P_{yes} and the corresponding ground truth was used as the training loss.

$$Loss = -(y_{yes}\log P_{yes} + (1 - y_{yes})\log(1 - P_{yes}))$$

2.3   Pre-processing

To solve the BioASQ 7b Phase B dataset as extractive question answering, the challenge datasets containing factoid and list type questions were converted into the format of the SQuAD datasets [16,17]. For yes/no type questions, we used 0/1 labels for each question-passage pair.

The dataset in the SQuAD format consists of *passages* and their respective question-answer sets. A passage is an article which contains answers or clues for answers and is denoted as the *context* in the dataset. The length of a passage varies from a sentence to a paragraph. An exact answer may or may not exist in the passage, depending on the task. According to the rules of the BioASQ Challenge, all the factoid and list type questions should be answerable with the given passages [21]. An exact answer and its starting position are provided in the *answers* field. We used various sources including snippets and PubMed abstracts, as passages. Multiple passages attached to a single question were divided to form question-passage pairs, which increased the number of question-passage pairs. The predicted answers of the question-passage pairs which share the same question are later combined in the post-processing layer.

Yes/no type questions are in the same format as the questions in the SQuAD dataset. However, binary answers are given to yes/no type questions, rather than answers selected based on their location in passages. Instead of providing an exact answer and its starting position in the *answers* field, we marked yes/no type questions using the strings "yes" or "no" and the Boolean values "false" and "true" in the *is_impossible* field. Since the distribution of yes/no answers in the training set is usually skewed, we undersampled the training data to balance the number of "yes" and "no" answers.

We used the following strategies for developing the datasets: *Snippet as-is* Strategy, *Full Abstract* Strategy, and *Appended Snippet* Strategy.

- *Snippet as-is Strategy* Using snippets in their original form is a basic method for filling passages. The starting positions of exact answers indicate the positional offsets of exact matching words. If a single snippet has more than one exact matching answer word, we form multiple question-passage pairs for the snippet.

- *Full Abstract Strategy* In the Full Abstract Strategy, we use an entire abstract, including the title of an article, as a passage. Full abstracts are retrieved from PubMed using their provided PMIDs. The *snippets* field of the original dataset is used to find the location of the correct answer. First, we look for the given snippet (e.g., a sentence in a typical case) from the retrieved abstract. Then, we search for the offset of the first exact matching words in the snippet, and add it to the offset of the snippet in the paragraph. In this way, we can find a plausible location of the answer within the paragraph.

- *Appended Snippet Strategy* The Appended Snippet Strategy is a compromise between using snippets as-is and full abstracts. We first search a given snippet from an abstract and concatenate $N \in \mathbb{N}$ sentences before and after the given snippet, forming $2N + k$ sentences into a passage (k denotes the number of sentences in a snippet, which is usually 1).

2.4   Post-processing

Since our pre-processing step involves dividing multiple passages with a same single question into multiple question-passage pairs, a single question can have multiple predicted answers. The probabilities of predicted answers for question-passage pairs sharing the same question, were merged to form a single list of predicted answers and their probabilities for a question. The answer candidate with the highest probability is considered as the final answer for a given factoid type question. For list type questions, probability thresholding was the default method for providing answers. Answer candidates with a probability higher than the threshold were included in the answer list. However, a considerable number (28.6% of BioASQ 6b list type questions) of list type questions contain the number of required answers. From the training example "Please list 6 symptoms of Scarlet fever," we can extract the number 6 from the given question. We extracted the number provided in the question and used it to limit the length of the answer list for the question. For questions that contain the number of answers, the extracted number of answers were yielded.

   For factoid and list type questions, we also filtered incomplete answers. Answers with non-paired parenthesis were removed from the list of possible answers. Pairs of round brackets and commas at the beginning and end of answers were removed.

3   Experimental Setup

3.1   Dataset

For factoid and list type questions, exact answers are included in the given snippets, which is consistent with the extractive QA setting of the SQuAD [17] dataset. Only binary answers are provided for yes/no questions. For each question, regardless of the question type, multiple snippets or documents are provided as corresponding passages.

The statistics of the BioASQ datasets are listed in Table 1. A list type question can have one or more than one answer; question-context pairs are made for every answer of a list type question. In our pre-processing step, 3,722 question-context pairs were made from 779 factoid questions in the BioASQ 7b training set. For yes/no questions, we undersampled the training data to balance the number of "yes" and "no" answers.

About 28.2% of factoid type questions and 5.6% of list type questions in the BioASQ 7b training set do not have an answer in their corresponding snippets. We excluded unanswerable questions, following the approach of Wiese et al. [24].

Table 1. Statistics of the BioASQ training set.

Question type	BioASQ version	# of questions in original datasets	# of pre-processed question-passage pairs
Factoid	6b	618	3,121
	7b	779	3,722
List	6b	485	6,896
	7b	556	7,716
Yes/No	6b	612	5,921
	7b	745	6,676

3.2 Training

Our system is composed of BioBERT, task-specific layers, and a post-processing layer. The parameters of BioBERT and a task-specific layer are trainable. Our training procedure starts with pre-training the system on the SQuAD dataset. The trainable parameters for factoid and list type questions were pre-trained on the SQuAD 1.1 dataset, and the parameters for yes/no type questions were pre-trained on the SQuAD 2.0 dataset. The pre-trained system is then fine-tuned on each task.

We tuned the hyperparameters on the BioASQ 4/5/6b training and test sets. We used a probability threshold of 0.42 as one of the hyperparameters for list type questions. The probability threshold was decided using the tuning procedure.

4 Results and Discussion

In this section, we first report our results for the BioASQ 7b (Phase B) Challenge, which are shown in Table 2. Please note that the results and ranks were obtained from the leaderboard of BioASQ 7b [3]. Then we evaluate our system and other competing systems on the validation set (BioASQ 6b). The results are presented in Table 3. Finally, we investigate the performance gain due to the sub-structures of the system (Tables 5 and 6). Mean reciprocal rank (MRR) and mean average F-measure (F_1) were used as official evaluation metrics to measure the performance on factoid and list type questions from BioASQ, respectively. We reported strict accuracy (SAcc), lenient accuracy (LAcc) and MRR for factoid questions and mean average precision, mean average recall, and mean average F1 score for list questions [4]. Since the label distribution was skewed, macro average F1 score was used as an evaluation metric for yes/no questions.

4.1   Results on BioASQ 7b

Our results on Task 7b (Phase B) of the BioASQ Challenge are reported in Table 2. Each participant can submit up to 5 systems per batch. We submitted 1 to 5 systems which use different combinations of pre- and post-processing strategies. We report the rankings and scores of our best performing system and those of other competing systems for each task in Table 2. Competing systems are the best and second best systems, other than our system, from distinct participants. Manually corrected gold-standard answers are not yet available at the time of writing; therefore, we report the scores based on the online leaderboard[5].

Table 2. Batch results of the BioASQ 7b Challenge. We report the rank of the systems in parentheses.

Batch	Yes/no		Factoid		List		# of systems
	Participating system	Mac F1	Participating system	MRR	Participating system	F1	
1	(1) Ours	**67.12**	(1) Ours	**46.37**	(3) Ours	30.51	17
	(2) auth-qa-1	53.97	(2) BJUTNLPGroup	34.83	(1) Lab Zhu, Fudan Univer	**32.76**	
	(3) BioASQ_Baseline	47.27	(3) auth-qa-1	27.78	(4) auth-qa-1	25.94	
2	(1) Ours	**83.31**	(1) Ours	**56.67**	(1) Ours	**47.32**	21
	(2) auth-qa-1	62.96	(3) QA1	40.33	(3) LabZhu, FDU	25.79	
	(4) BioASQ_Baseline	42.58	(4) transfer-learning	32.67	(5) auth-qa-1	23.21	
3	(5) Ours	46.23	(6) Ours	47.24	(1) Ours	**32.98**	24
	(1) unipi-quokka-QA-2	**74.73**	(1) QA1/UNCC_QA_1	**51.15**	(2) auth-qa-1	25.13	
	(3) auth-qa-2	51.65	(3) google-gold-input	50.23	(4) BioASQ_Baseline	22.75	
4	(2) Ours	79.28	(1) Ours	**69.12**	(1) Ours	**46.04**	36
	(1) unipi-quokka-QA-1	**82.08**	(4) FACTOIDS/UNCC	61.03	(2) google-gold-input-nq	43.64	
	(8) bioasq_experiments	58.01	(9) google-gold-input	54.95	(9) LabZhu, FDU	32.14	
5	(1) Ours	**82.50**	(1) Ours	**36.38**	(1) Ours	**46.19**	40
	(2) unipi-quokka-QA-5	79.39	(3) BJUTNLPGroup	33.81	(6) google-gold-input-nq	28.89	
	(6) google-gold-input-ab	69.41	(4) UNCC_QA_1	33.05	(7) UNCC_*	28.62	

---

[4] For more details, please visit http://participants-area.bioasq.org/Tasks/b/eval_meas_2018/.

[5] The official results of the competition will be provided at http://bioasq.org.

4.2   Validating on the BioASQ 6b Dataset

We compared the performance of existing systems and our system on the BioASQ 6b dataset from the last year (2018), which is shown in Table 3. We micro averaged the scores from five experiments and reported the scores in Table 3. Similarly, the leaderboard scores of the best performing system for each batch were micro averaged and reported as the *Best System* scores [5,12,27]. Our system obtained much higher scores on the BioASQ 6b dataset than the top systems from leaderboard of BioASQ 6b Challenge.

Table 3. Performance comparison between existing systems and our system on the BioASQ 6b dataset (from last year). Note that our system obtained a 20% to 60% performance improvement over the best systems.

System	Factoid (MRR)	List (F1)	Yes/no (Macro F1)
Best System	27.84%	27.21%	62.05%
Ours	**48.41%**	**43.16%**	**75.87%**

Pre-training. In Table 4, we compare the performance of the pre-trained models. BioBERT fine-tuned on the BioASQ 6b dataset outperformed $BERT_{BASE}$ fine-tuned on BioASQ in both factoid and list type questions. BioBERT first pre-trained on SQuAD and then fine-tuned on BioASQ 6b obtained the best performance over other two experiments, demonstrating the effectiveness of pre-training BioBERT on SQuAD, a comprehensive and large-scale question answering corpus.

Table 4. Performance comparison between pre-trained models.

Pre-trained models	Factoid			List		
	SAcc	LAcc	MRR	Prec	Recall	F1
$BERT_{BASE}$+BioASQ Finetune	24.84%	36.03%	28.76%	42.41%	35.88%	35.37%
BioBERT+BioASQ Finetune	34.16%	47.83%	39.64%	44.62%	39.49%	38.45%
BioBERT+SQuAD+BioASQ Finetune	**42.86%**	**57.14%**	**48.41%**	**51.58%**	**43.24%**	**43.16%**

Pre-/Post-processing. The performance of our system is largely affected by how the data is pre-processed (Table 5). However, the effectiveness of the pre-processing strategy varies depending on the type of question. For example, the Appended Snippet strategy and Full Abstract strategy obtained good performance on factoid questions, while the Snippet As-is strategy achieved the highest performance on list and yes/no type questions. Table 6 shows the effect of post-processing on the performance of a system evaluated on list type questions. In our study, both extracting the number of answers from questions and filtering predicted answers were effective.

Table 5. Performance comparison between pre-processing methods. Scores on the BioASQ 6b dataset.

Strategy	Factoid			List			Yes/no
	SAcc	LAcc	MRR	Prec	Recall	F1	MacroF1
Snippet	40.99	55.90	47.38	**51.58**	**43.24**	**43.16**	75.10
Full Abstract	**42.86**	57.14	**48.41**	42.66	32.58	33.52	66.76
Appended Snippet	39.75	**58.39**	48.00	44.04	41.26	39.36	–

Table 6. Ablation study on the post-processing methods. Scores for list type questions in the BioASQ 6b dataset.

Strategy	Precision	Recall	F1
Baseline (Snippet)	**51.58**	43.24	**43.16**
Baseline without filter	50.79	43.24	42.64
Baseline without answer # extraction	50.01	**44.32**	42.58

Ensemble. Starting from test batch 4 of BioASQ 7b, we submitted model ensemble results as one of our systems. The performance gain of the model ensemble on our evaluation set was relatively small; the performance ranged from 0.2% to 2% depending on the task. The model ensemble improved the performance on factoid questions the most (2% gain), but applying the model ensemble to list questions did not obtain higher performance than the single model. Although the model ensemble obtained high scores in the BioASQ 7b Challenge, it could only obtain the highest score on factoid type questions in batch 5.

Qualitative Analysis. In Table 7, we show three predictions generated by our system on the BioASQ 6b factoid dataset. Due to the space limitation, we show only small parts of a passage, which contain the answers (predicted answers might be contained in other parts of the passage). We show the top five predictions generated by our system which can also be used for list type questions. In the first example, our system successfully finds the answer and other plausible answers. The second example shows that most of the predicted answers are correct and have only minor differences. In the last example, we observe that the ground truth answer does not exist in the passage. Also, the predicted answers are indeed correct despite the incorrect annotation.

The prediction result of list question from the BioASQ 6b is presented in Table 8. We found that our system is more likely to produce incorrect predictions on list questions than on factoid questions. Our system internally outputs a list of predictions and the list is likely to include prediction with erroneous span. Even though incorrect prediction ("JBP") with erroneous span has a lower

Table 7. Predictions by our BioBERT based QA system on the BioASQ 6b factoid dataset

No	Type	Description
1	Question	What causes "puffy hand syndrome?"
	Passage	Puffy hand syndrome is a complication of intravenous drug abuse, which has no current available treatment
	Ground Truth	"intravenous drug abuse"
	Predicted Answer	"intravenous drug abuse", "drug addiction", "Intravenous drug addiction", "staphylococcal skin infection", "major depression"
2	Question	In which syndrome is the RPS19 gene most frequently mutated?
	Passage	A transgenic mouse model demonstrates a dominant negative effect of a point mutation in the RPS19 gene associated with Diamond-Blackfan anemia
	Ground Truth	"Diamond-Blackfan Anemia", "DBA"
	Predicted Answer	"Diamond-Blackfan anemia", "Diamond-Blackfan anemia (DBA)", "DBA", "Diamond Blackfan anemia", "Diamond-Blackfan anemia. Diamond-Blackfan anemia"
3	Question	What protein is the most common cause of hereditary renal amyloidosis?
	Passage	We suspected amyloidosis with fibrinogen A alpha chain deposits, which is the most frequent cause of hereditary amyloidosis in Europe, with a glomerular preferential affectation
	Ground Truth	"Fibrinogen A Alpha protein"
	Predicted Answer	"fibrinogen", "fibrinogen alpha-chain. Variants of circulating fibrinogen", "fibrinogen A alpha chain (FGA)", "Fibrinogen A Alpha Chain Protein. Introduction: Fibrinogen", "apolipoprotein AI"

probability than the true prediction ("JBP1" and "JBP2"), it can have considerable absolute probabilities. On factoid questions, selecting a top one answer is required. Hence we can ignore incorrect prediction on factoid questions. On the

738   W. Yoon et al.

contrary, on list questions, prediction with erroneous span gets higher probability through merging predictions in post-processing step. Since our model utilizes fixed threshold value, prediction with erroneous span is imperfect but achieved a higher possibility than the threshold.

Table 8. Prediction by our BioBERT based QA system on the BioASQ 6b list dataset

No	Type	Description
1	Question	Which enzymes are responsible for base J creation in Trypanosoma brucei?
	Passage	JBP1 and JBP2 are two distinct thymidine hydroxylases involved in J biosynthesis in genomic DNA of African trypanosomes
		Here we discuss the regulation of hmU and base J formation in the trypanosome genome by JGT and base J-binding protein
	Ground Truth	"JBP1", "JBP2", "JGT"
	Predicted Answer	"JBP1", "JBP", "thymidine hydroxylase", "JGT", "hmU", "JBP2"

5   Conclusion

In this paper, we proposed BioBERT based QA system for the BioASQ biomedical question answering challenge. As the size of the biomedical question answering dataset is very small, we leveraged pre-trained language models for biomedical domain which effectively exploit the knowledge from large biomedical corpora. Also, while existing systems for the BioASQ challenge require different structures for different question types, our system uses almost the same structure for various question types. By exploring various pre-/post-processing strategies, our BioBERT based system obtained the best performance in the 7th BioASQ Challenge, achieving state-of-the-art results on factoid, list, and yes/no type questions. In future work, we plan to further systematically analyze the incorrect predictions of our systems, and develop biomedical QA systems that can eventually outperform humans.

Acknowledgements. We appreciate Susan Kim for editing the manuscript. This work was funded by the National Research Foundation of Korea (NRF-2017R1A2A1A17069645, NRF-2016M3A9A7916996) and the National IT Industry Promotion Agency grant funded by the Ministry of Science and ICT and Ministry of Health and Welfare (NO. C1202-18-1001, Development Project of The Precision Medicine Hospital Information System (P-HIS)).

References

1. Alsentzer, E., et al.: Publicly available clinical BERT embeddings. arXiv preprint arXiv:1904.03323 (2019)
2. Beltagy, I., Cohan, A., Lo, K.: SciBERT: pretrained contextualized embeddings for scientific text. arXiv preprint arXiv:1903.10676 (2019)
3. BioASQ Participants Area BioASQ, May 2019. http://participants-area.bioasq. org/results/7b/phaseB/
4. Devlin, J., Chang, M.W., Lee, K., Toutanova, K.: BERT: pre-training of deep bidirectional transformers for language understanding. arXiv preprint arXiv:1810.04805 (2018)
5. Dimitriadis, D., Tsoumakas, G.: Word embeddings and external resources for answer processing in biomedical factoid question answering. J. Biomed. Inform. **92**, 103118 (2019)
6. Kim, D., et al.: A neural named entity recognition and multi-type normalization tool for biomedical text mining. IEEE Access **7**, 73729–73740 (2019)
7. Krithara, A., Nentidis, A., Paliouras, G., Kakadiaris, I.: Results of the 4th edition of BioASQ challenge. In: Proceedings of the Fourth BioASQ Workshop, Berlin, Germany, August 2016, pp. 1–7. Association for Computational Linguistics (2016). https://doi.org/10.18653/v1/W16-3101, https://www.aclweb.org/ anthology/W16-3101
8. Lee, J., et al.: BioBERT: a pre-trained biomedical language representation model for biomedical text mining. Bioinformatics (2019). https://doi.org/10.1093/ bioinformatics/btz682
9. Lim, S., Kang, J.: Chemical–gene relation extraction using recursive neural network. Database **2018** (2018)
10. Nentidis, A., Bougiatiotis, K., Krithara, A., Paliouras, G., Kakadiaris, I.: Results of the fifth edition of the BioASQ challenge. In: BioNLP 2017, pp. 48–57 (2017)
11. Nentidis, A., Krithara, A., Bougiatiotis, K., Paliouras, G., Kakadiaris, I.: Results of the sixth edition of the BioASQ challenge. In: Proceedings of the 6th BioASQ Workshop A Challenge on Large-Scale Biomedical Semantic Indexing and Question Answering, Brussels, Belgium, November 2018, pp. 1–10. Association for Computational Linguistics (2018). https://www.aclweb.org/anthology/W18-5301
12. Peng, S., Zhang, Y., You, R., Xie, Z., Wang, B., Zhu, S.: The fudan participation in the 2015 BioASQ challenge: large-scale biomedical semantic indexing and question answering. In: CEUR Workshop Proceedings, vol. 1391. CEUR Workshop Proceedings (2015)
13. Peters, M., et al.: Deep contextualized word representations. In: Proceedings of the 2018 Conference of the North American Chapter of the Association for Computational Linguistics: Human Language Technologies, Volume 1 (Long Papers), pp. 2227–2237 (2018)

14. Pyysalo, S., Ginter, F., Moen, H., Salakoski, T., Ananiadou, S.: Distributional semantics resources for biomedical text processing. In: Proceedings of LBM, pp. 39–44 (2013)
15. Radford, A., Narasimhan, K., Salimans, T., Sutskever, I.: Improving language understanding with unsupervised learning. Technical report, OpenAI (2018)
16. Rajpurkar, P., Jia, R., Liang, P.: Know what you don't know: unanswerable questions for squad. arXiv preprint arXiv:1806.03822 (2018)
17. Rajpurkar, P., Zhang, J., Lopyrev, K., Liang, P.: SQuAD: 100,000+ questions for machine comprehension of text. arXiv preprint arXiv:1606.05250 (2016)
18. Rosso-Mateus, A., González, F.A., Montes-y Gómez, M.: MindLab neural network approach at BioASQ 6B. In: Proceedings of the 6th BioASQ Workshop A Challenge on Large-Scale Biomedical Semantic Indexing and Question Answering, pp. 40–46 (2018)
19. Sousa, D., Lamurias, A., Couto, F.M.: Using neural networks for relation extraction from biomedical literature. arXiv preprint arXiv:1905.11391 (2019)
20. Talmor, A., Berant, J.: MultiQA: an empirical investigation of generalization and transfer in reading comprehension. arXiv preprint arXiv:1905.13453 (2019)
21. Tsatsaronis, G., et al.: An overview of the BioASQ large-scale biomedical semantic indexing and question answering competition. BMC Bioinform. 16(1), 138 (2015)
22. Turian, J., Ratinov, L., Bengio, Y.: Word representations: a simple and general method for semi-supervised learning. In: Proceedings of the 48th Annual Meeting of the Association for Computational Linguistics, pp. 384–394. Association for Computational Linguistics (2010)
23. Vaswani, A., et al.: Attention is all you need. In: Advances in Neural Information Processing Systems, pp. 5998–6008 (2017)
24. Wiese, G., Weissenborn, D., Neves, M.: Neural domain adaptation for biomedical question answering. arXiv preprint arXiv:1706.03610 (2017)
25. Wiese, G., Weissenborn, D., Neves, M.: Neural question answering at BioASQ 5B. arXiv preprint arXiv:1706.08568 (2017)
26. Wu, Y., et al.: Google's neural machine translation system: bridging the gap between human and machine translation. arXiv preprint arXiv:1609.08144 (2016)
27. Yang, Z., Zhou, Y., Nyberg, E.: Learning to answer biomedical questions: OAQA at BioASQ 4B. In: Proceedings of the Fourth BioASQ Workshop, pp. 23–37 (2016)
28. Yoon, W., So, C.H., Lee, J., Kang, J.: CollaboNet: collaboration of deep neural networks for biomedical named entity recognition. BMC Bioinform. 20(10), 249 (2019)
29. Zhu, H., Paschalidis, I.C., Tahmasebi, A.: Clinical concept extraction with contextual word embedding. arXiv preprint arXiv:1810.10566 (2018)

Author Index

Printed in the United States
By Bookmasters